HANDBOOK OF THE HYPOTHALAMUS

VOLUME 3 — Part A

Behavioral Studies of the Hypothalamus

HANDBOOK OF THE HYPOTHALAMUS

Editors

Peter J. Morgane

Laboratory of Neurophysiology and Neuropsychopharmacology
Worcester Foundation for Experimental Biology
Shrewsbury, Massachusetts

Jaak Panksepp

Department of Psychology
Bowling Green State University
Bowling Green, Ohio

Volume 1 Anatomy of the Hypothalamus
Volume 2 Physiology of the Hypothalamus
Volume 3 Behavioral Studies of the Hypothalamus (in two parts)

HANDBOOK OF THE HYPOTHALAMUS
(in three volumes)

VOLUME 3 — Part A

Behavioral Studies of the Hypothalamus

edited by

Peter J. Morgane
Laboratory of Neurophysiology and Neuropsychopharmacology
Worcester Foundation for Experimental Biology
Shrewsbury, Massachusetts 01545

Jaak Panksepp
Department of Psychology
Bowling Green State University
Bowling Green, Ohio 43403

MARCEL DEKKER, INC. New York and Basel

Library of Congress Cataloging in Publication Data

Main entry under title:

Behavioral studies of the hypothalamus.

    (Handbook of the hypothalamus ; v. 3, pt. A)
    Includes indexes.
    1. Hypothalamus. 2. Body temperature–Regulation.
I. Morgane, P. J. II. Panksepp, Jaak, [date]
III. Series. [DNLM: 1. Hypothalamus–Physiology.
2. Behavior–Physiology. 3. Body temperature
regulation. W1 HA513E v. 3]
QP383.7.H36 vol. 3, pt. A [QP383.7] 612'.8262s
ISBN 0-8247-6904-X   [599.01'88]   80-22563

COPYRIGHT © 1980 by MARCEL DEKKER, INC. ALL RIGHTS RESERVED

Neither this book nor any part may be reproduced or transmitted in any form or by any means, electronic or mechanical, including photocopying, microfilming, and recording, or by any information storage and retrieval system, without permission in writing from the publisher.

MARCEL DEKKER, INC.
270 Madison Avenue, New York, New York 10016

Current printing (last digit):
10 9 8 7 6 5 4 3 2 1

PRINTED IN THE UNITED STATES OF AMERICA

Introduction to the Series

Zuerst also Anatomie und dann Physiologie, wenn aber erst Physiologie, dann nicht ohne Anatomie.
von Gudden

Modern approaches to understanding brain function are increasingly embracing the view that physiological analyses of brain circuits which interrelate levels of functional specialization are more important than attempts to derive and dissect so-called "local" functions of anatomical regions, subareas, levels, or "centers." Accordingly, works such as this Handbook, primarily dealing with a single, though complexly organized region of the brain, i.e., the hypothalamus, may on the surface seem somewhat atavistic. They certainly could be were not full respects paid in such works as this to the various relationships of individual hypothalamic components to the brain as a whole. Contributors to this Handbook have, accordingly, provided alternative explanations to those views which plot activity in terms of functional loci and "centers."

The hypothalamus is a nodal link embedded in major loops of circuitry forming components of the limbic forebrain-limbic midbrain system and the limbic-hypothalamic-reticular axis. As emphasized long ago by Nauta, the functional state of the hypothalamus, as manifested by visceral endocrine and motor phenomena, reflects in large part the activation or inhibition of neural mechanisms represented in the limbic forebrain-midbrain circuit and the extensions of the lower brainstem reticular fields, via the hypothalamus, into the forebrain. Hypothalamic functions should be considered in the context of operations of these and other reciprocating circuitries that inextricably relate it to the limbic forebrain and lower brainstem. Obviously, the hypothalamus does not simply sit astride these multitudinous circuitries and serve merely as a conduit. Rather, a variety of evidence shows that it participates in many activities represented in these rostrally and caudally directed circuits. Further, there is considerable cross talk across its entire core from the ventricle laterally through the internal capsule to the striatum, etc. Thus, the functional state of the hypothalamus, as manifested by visceral, endocrine, and behavioral changes, in large part reflects the excitation or inhibition of neural mechanisms represented in the limbic forebrain-limbic midbrain and reticulolimbic circuits. Hence, hypothalamic function must be interpreted on the basis of intersegmental laws and principles, and these are of quite a different order than operations at one brain "level" or segment. Nevertheless, as conceptions of neural organization evolve beyond the search for the functions of anatomically circumscribed parcels of neural tissue or ensembles of "centers," a focus on such a structure as the hypothalamus remains appropriate as long as its interrelations with other brain levels form an integral part of the hypothalamic orientation. In other words, studies should not become arrested at a purely hypothalamic level of behavioral-functional mapping exercises. Simple spatial parcellations of functional domains or behavioral geography of the hypothalamus has implied a staking out of territories related to specific functions. Throughout this Handbook it is emphasized that we cannot conceive a neurobiological basis of behavior as divisible into moieties which are somehow amalgamated into directed or purposeful behavior. Further, there has been a long history assigning to the hypothalamus disproportionate importance in the governance of the organisms's adaptive capacities in the entire regulatory homeostatic sphere. In this regard, the hypothalamus appears to contain a high density of interoreceptors registering the status of the internal environment of the brain and the body and it has, accordingly, been viewed too narrowly over the years somewhat in isolation as merely some sort of headwater for the inner or "visceral" life. Nevertheless, since so many basic regulatory mechanisms are apparently represented within circuits emanating in and transversing this small area of the brain, a study of the hypothalamus, keeping its multiple links with the limbic system and reticular formations in mind, provides a useful starting point for analyzing the overall neural organizations representing a variety of fundamental vital processes. Implied in these considerations of circuits are re-representation of functions at several brain levels, of which the hypothalamus is but one. In a large number of chapters in this Handbook the area of anatomical neuroendocrinology is also given considerable prominence and the concept of the "brain as a gland" is explored from many points of view. Finally, at the behavioral end of the spectrum, various chapters are developed around the theme of how, by various alchemical transformations, purely psychological phenomena such as urges, drives, motivations, and the like might be transformed into "neurologies" of behavior.

The overall aim of this Handbook is to bring together in one series of volumes, with as much organization and unity as is possible in a multiauthored work, the most recent knowledge about the hypothalamus—its anatomical, biochemical, physiological, and functional organization. The Handbook has provided a mechanism for its contributors, who are active researchers in the field, to develop fully integrated essays on major aspects of hypothalamic organization and function from a wide variety of points of view. In large part, the Handbook constitutes a forum mostly for ideas rather than a comprehensive summarization of unassembled empirical findings about the hypothalamus. Contributors were encouraged to aim for broad area reviews and overviews that would not simply be catalogue-like compilations of existing facts and references. Our goal was not to develop a compendium or "dictionary" of the hypothalamus nor to exhaustively review the huge literature on all facets of this subject. These reviews encompass organizational principles which make the best sense of diverse and often fragmentary empirical findings. Critical perspectives have been sought in order to discuss underlying assumptions of various research approaches, and to highlight the strengths and weaknesses of ideas that have been proposed in the literature. These chapters also point toward future directions for research on the hypothalamus and they develop new ideas, not yet fully supported by data, concerning the nature of hypothalamic organization.

A guiding principle of this Handbook series has been to represent a large number of disciplines among the contributors. There is no doubt that understanding of the nervous system, including some of its major components such as the hypothalamus, requires effort at several levels of analysis. Not only does the inquiry require the traditional analytical approaches of studying simpler and simpler units of the organism, which is the accepted hallmark of all the traditional scientific disciplines, but also a deliberate movement in the other direction—toward a study of overall integrative processes at the organismic level. Both approaches are represented in this Handbook. Besides the topics covered, there were other areas of research which we had originally hoped to include but eventually decided to de-emphasize because of extensive coverage in other recent sources. For instance, many aspects of neuroendocrinology that have been recently reviewed in the book *The Hypothalamus* (edited by Reichlin, Baldessarini, and Martin, Raven, New York, 1978) are given somewhat less emphasis here, though other endocrinological areas are covered in considerable detail, particularly anatomical neuroendocrinology. The absence of a specific chapter on electron microscopic analysis of the hypothalamus represents a gap we hope to close in possible future volumes in this series. Even so, the present work rapidly expanded to three volumes dealing with anatomy, endocrinology and physiology, and function—labels of convenience that often do injustice to the interdisciplinary flavor of many of the chapters. In general, we endeavored to organize this Handbook in terms of breadth and depth of coverage so it would continue in historical progression the three main treatises on the hypothalamus which have appeared in this century, namely, *The Hypothalamus: Morphological, Functional, Clinical and Surgical Aspects* (edited by LeGros, Clark, Beattie, Riddoch, and Dott), Oliver and Boyd, Edinburgh, 1938; *The Hypothalamus and Central Levels of Autonomic Function* (edited by Fulton), Association for Research in Nervous and Mental Diseases, Volume 20, Williams and Wilkins, Baltimore, 1940; *The Hypothalamus* (edited by Haymaker, Anderson, and Nauta), Thomas, Springfield, Ill., 1969.

We trust these volumes will be of value to workers in the field for quite a few years to come. Of course, with the rate of new work appearing in this area of neurobiology, it will not be at all surprising if the time rapidly approaches when another comprehensive analysis of the hypothalamus and its neural relations will be in order. No field in neurobiology, no matter from what perspective it is written, will remain "updated" for long. As the feverish pace of new work continues, it will be necessary from time to time to gather it all together in as integrative a manner as possible. This we have attempted to do and, to the extent of its success, we remain indebted to the participants who have devoted their valuable time and considerable creativity to putting this work together.

Peter J. Morgane
Jaak Panksepp

Contents of Volume 3—Part A

	Introduction to the Series	iii
	Contents of Other Volumes	ix
	Contributors to Volume 3—Part A	xi
	Preface to Volume 3—Part A	xiii

1 Hypothalamic Control of Thermoregulation
Neurophysiological Basis   1
Jack A. Boulant

I.	Introduction	1
II.	Hypothalamic and Extrahypothalamic Control of Responses	11
III.	The Role of Hypothalamic Neurons in Thermoregulation	46
	References	70

2 Hypothalamic Control of Thermoregulation
Neurochemical Mechanisms   83
R. D. Myers

I.	Introduction	83
II.	Serotonergic Pathways	88
III.	Catecholaminergic Pathways	100
IV.	Cholinergic Pathways Mediating Thermogenesis	118
V.	Role of Cyclic AMP, Peptides, and Other Humoral Factors	130
VI.	Hypothalamic Ions and the Concept of a Temperature Set Point	135
VII.	Fever and Pyrogens	147
VIII.	Hypothalamic Effect of Drugs That Alter Body Temperature	165
IX.	Conclusion	175
	References	181

3 The Role of the Hypothalamus in Energy Homeostasis   211
*Terry L. Powley, Charles A. Opsahl, James E. Cox,
and Harvey P. Weingarten*

I.	Introduction	211
II.	Disturbances of Energy Homeostasis Produced by Ventromedial Hypothalamic Manipulations	214

	III. Disturbances of Energy Homeostasis Produced by Lateral Hypothalamic Manipulations	256
	IV. Overview	274
	References	277

4 Neurochemical Systems of the Hypothalamus
Control of Feeding and Drinking Behavior
and Water-Electrolyte Excretion — 299
Sarah Fryer Leibowitz

I.	Introduction	299
II.	Noradrenergic and Adrenergic Systems	300
III.	Dopaminergic Systems	353
IV.	Serotonergic Systems	368
V.	Cholinergic Systems	380
VI.	Histaminergic Systems	388
VII.	GABAergic Systems	395
VIII.	Peptidergic Systems	399
IX.	Summary and Conclusions	403
	Abbreviations	405
	References	407

5 Activity of Hypothalamic and Related Neurons in the Alert Animal — 439
Edmund T. Rolls

I.	Introduction	439
II.	The Activity of Neurons in the Hypothalamus and Substantia Innominata of the Alert Animal	442
III.	Comparison of the Activity of Neurons in the Hypothalamus with the Activity of Neurons in Sensory Input Pathways	456
IV.	Comparison of the Activity of Neurons in the Hypothalamus with the Activity of Neurons in Motor Structures	457
V.	Synthesis: Nature of the Activity of Neurons Recorded in the Hypothalamus and Substantia Innominata of the Alert Animal	459
	References	461

Author Index — 467
Subject Index — 493

Contents of Other Volumes

Volume 1 Anatomy of the Hypothalamus

1. Historical and Modern Concepts of Hypothalamic Organization and Function
   Peter J. Morgane

2. Development of the Hypothalamus in Mammals:
   An Investigation into Its Morphological Position during Ontogenesis
   A. Keyser

3. A Cytoarchitectonic Atlas of the Hypothalamus and Hypothalamic Third Ventricle of the Rat
   Ruth Bleier, Perry Cohn, and Inge R. Siggelkow

4. A Golgi Anatomy of the Rodent Hypothalamus
   O. Eugene Millhouse

5. The Blood Supply of the Hypothalamus in the Rat
   György Ambach and Miklós Palkovits

6. Neural Connections of the Hypothalamus
   Miklós Palkovits and László Záborszky

7. Anatomical Organization of Monoamine- and Acetylcholinesterase-Containing Neuronal Systems in the Vertebrate Hypothalamus
   André Parent

8. Limbic and Brainstem Connections of the Hypothalamus
   Anatomical and Electrophysiological Studies
   Jerome Sutin and Russell L. McBride

9. Neurophysiology and Neuropharmacology of Medial Hypothalamic Neurons and Their Extrahypothalamic Connections
   Leo P. Renaud

Author Index
Subject Index

Volume 2 Physiology of the Hypothalamus

1. Biochemical Aspects of Hypothalamic Function
   Jeffrey F. McKelvy, Jay A. Glasel, and Mark Foreman

2. Recent Advances in Structure and Function of the Endocrine Hypothalamus
   Karl M. Knigge, Gloria E. Hoffman, Shirley A. Joseph, David E. Scott, Celia D. Sladek, and John R. Sladek, Jr.

3. Physiology and Pharmacology of Hypothalamic Regulatory Peptides
   Wylie Vale, Catherine Rivier, and Marvin Brown

4. Role of Neurotransmitters in the Control of Adenohypophyseal Secretion
   Claude Kordon, Alain Enjalbert, Micheline Hery, Patricia I. Joseph-Bravo, William Rotsztejn, and Merle Ruberg

5. Hypothalamic-Pituitary Oligopeptides and Behavior
   Albert Witter and David de Wied

6. The Central State of the Hypothalamus in Health and Disease: Old and New Concepts
   Novera Herbert Spector

7. Hypothalamic Control of Metabolism
   Lawrence A. Frohman

8. Input-Output Organization in the Hypothalamus Relating to Food Intake Behavior
   Yutaka Oomura

Author Index
Subject Index

Volume 3 Behavioral Studies of the Hypothalamus

Part B

1. Hypothalamic Function in the Behavioral and Physiological Control of Body Fluids
   Glenn I. Hatton and William E. Armstrong

2. The Lateral Hypothalamus and Adjunctive Behavior
   Matthew J. Wayner, Frank C. Barone, and Costas C. Loullis

3. Hypothalamic Control of Autonomic Functions
   Giuseppe Mancia and Alberto Zanchetti

4. Neural Control of Aggression and Rage Behavior
   Allan Siegel and Henry M. Edinger

5. Theoretical Issues Regarding Hypothalamic Control of Reproductive Behavior
   Donald W. Pfaff

6. Role of Transhypothalamic Pathways in Social Communication
   Paul D. MacLean

7. Hypothalamic Integration of Behavior: Rewards, Punishments, and Related Psychological Processes
   Jaak Panksepp

Author Index
Subject Index

Contributors to Volume 3—Part A

Jack A. Boulant, Ph.D. Associate Professor of Physiology, Department of Physiology, College of Medicine, Ohio State University, Columbus, Ohio 43210

James E. Cox,* Ph.D. Department of Psychology, Yale University, New Haven, Connecticut 06520

Sarah Fryer Leibowitz, Ph.D. Associate Professor, The Rockefeller University, New York, New York 10021

R. D. Myers, Ph.D. Professor, Departments of Psychiatry and Pharmacology, University of North Carolina School of Medicine, Chapel Hill, North Carolina 27514

Charles A. Opsahl, Ph.D. Research Associate, Department of Psychiatry, Yale University Medical School, New Haven, Connecticut 06508

Terry L. Powley,† Ph.D. Department of Psychology, Yale University, New Haven, Connecticut 06520

Edmund T. Rolls, Ph.D. University Lecturer, Department of Experimental Psychology, University of Oxford, Oxford, England OX1 3UD

Harvey P. Weingarten, ‡ Ph.D. Department of Psychology, Yale University, New Haven, Connecticut 06520

**Present affiliation:* Bourne Behavioral Research Laboratory, New York Hospital-Cornell Medical Center, Westchester Division, White Plains, New York 10605
†*Present affiliation:* Professor, Department of Psychological Sciences, Purdue University, West Lafayette, Indiana 47907
‡ *Present affiliation:* Assistant Professor, Department of Psychology, McMaster University, Hamilton, Ontario, Canada L 8S 4K1

Preface to Volume 3—Part A

This volume of the Handbook deals primarily with the behavioral functions of the hypothalamus. Hypothalamic functions must be understood in the context of its relationships with the rest of the brain and in terms of the behavioral capacities that are facilitated and inhibited. It has been generally believed that one of the major overall functions of the hypothalamus is to provide primitive catering capacities to the body—to generate behavioral tendencies to help sustain those homeostatic balances that can be achieved only by periodic interactions with the outside world. Accordingly, as summarized in these chapters, a great deal of work has been devoted to analyzing the role of the hypothalamus in regulating body temperature and water and energy balance as well as in the control of sexual and emotional behaviors. Of course, our knowledge in each of these areas is commensurate with the complexity of the problem. More fundamental data is available on hypothalamic governance of thermoregulation than of thirst. More basic information has appeared on thirst than on hunger. Certainly more is known about sexual and aggressive behaviors than about such ephemeral emotions as fear, panic, and joy.

Physiological psychology is the discipline most involved in these questions but most of the useful organizing principles have been handed down from neurophysiological and neurochemical studies of the brain (see this Handbook, Volumes 1 and 2). While earlier views of hypothalamic function tended to emphasize the behavioral (and what has been termed "motivational") changes resulting from hypothalamic manipulations, more recently there has been a greater tendency to view the behavioral changes arising from hypothalamic manipulations as secondary to basic physiological changes in the body. Thus, it is not presently clear to what extent the hypothalamus directly modifies behavioral propensities as opposed to physiological parameters. Of course, it probably organizes both concurrently. The behavioral changes that result from hypothalamic activity are probably best manifested in the context of changes in the activity of various peripheral organs. Hunger yields a bias toward muscular ("exploratory") activity, while ingestion of food prepares the body for processing nutrients (gastric and pancreatic secretions), and satiety shifts the body to quiescence and perhaps general "parasympathetic" or metabolic dominance. A similar pattern holds for other behaviors controlled by the hypothalamus. Indeed, as emphasized in many of these chapters, perhaps one of the major overall functions of the hypothalamus is to bring behavioral propensities into line with existing physiological conditions and, conversely, to bring physiological conditions into line with environmental demands.

In any case, in order to correlate anatomical substrates (see Volume 1) precisely with the overall functions of a system, an accurate knowledge of the underlying circuitry and cellular and subcellular dynamics of the system must be understood (see Volume 2). Thus, a study of cellular functions has provided one major starting point for the analysis of hypothalamic functions. However, though careful physiological studies set the pace for translating the fruits of neuroanatomy into an understanding of functional organization of brain circuitry, such approaches are often too molecular to make sense of complex circuit characteristics. Indeed, there are presently few tools with which we may efficiently and convincingly probe functional interactions among dynamically interacting neural circuits. Most neurochemical and neurophysiological techniques generally do not tap the ongoing functions of neural circuits; however, as indicated in this volume, their use with in vivo procedures, such as push-pull brain perfusion techniques in the functioning organism, makes neurochemistry a truly incisive tool for functional analysis. Perhaps recent approaches that employ brain energy metabolism to highlight functional activity (e.g., the 2-deoxy-D-glucose autoradiography technique) will also prove to be major new tools for systems analysis. This remains to be demonstrated for psychoneurological processes of any subtlety. Until such tools are further refined and additional ones developed, the primitive techniques of ablation and gross electrical stimulation, especially when they are correlated with identified neurochemical systems and combined with biochemical studies, are assured a continuing role in the mainstream of brain research.

Although brain science has not yet developed to a point where a coherent understanding of the functional organization of brain systems has been achieved, it is truly remarkable how far we have come in the last few decades. The accumulation of basic information and knowledge that the present chapters summarize sets the stage for the understanding of the functional organization of the hypothalamus in the years ahead.

Peter J. Morgane
Jaak Panksepp

HANDBOOK OF THE HYPOTHALAMUS

VOLUME 3 — Part A

Behavioral Studies of the Hypothalamus

1
Hypothalamic Control of Thermoregulation
Neurophysiological Basis

Jack A. Boulant / Ohio State University, Columbus, Ohio

I. Introduction

Much of the nervous system is composed of integrative neuronal networks which receive afferent information through synaptic inputs, integrate this information, and produce efferent signals that either control body functions or relay further information to other neuronal networks. This statement adequately describes the role of a hypothalamic network which serves in autonomic and behavioral homeostatic systems controlling body temperature. Certain hypothalamic neurons, especially those in the preoptic region and anterior hypothalamus (PO/AH) are temperature sensitive and change their firing rates when hypothalamic temperature is altered. In addition, many of these same PO/AH neurons receive afferent synaptic inputs from skin and other deep-body thermoreceptors. The result of this integration of central and peripheral thermal information is the initiation of appropriate thermoregulatory responses necessary for the maintenance of normal body temperature. The hypothalamus is one of the primary integrative and controlling structures in thermoregulation. It exerts thermoregulatory control over a diversity of body functions, including the cardiovascular system, respiration, muscle tone (shivering thermogenesis), metabolic endocrines, sweating, salivation, piloerection, and thermoregulatory behavior. In addition to thermosensitivity, neurons in this same area of the brain are sensitive to a variety of other factors, including glucose, endocrines, and osmotic pressure. These neurons participate in other homeostatic systems, most of which are ultimately related to temperature regulation. In reality, a vast neuronal network extending throughout the hypothalamus and related structures regulates the body's entire internal environment, both autonomically and behaviorally. The separation of the thermoregulatory system from all other regulatory systems is a separation necessarily contrived for simplification.

It is the limited goal of this chapter to describe the hypothalamic control of body temperature. The first half of this chapter will consider the receptor and effector structures involved in each individual thermoregulatory response. While the hypothalamus is considered of primary importance, it is recognized that other peripheral, spinal, and brain stem structures are vital both in thermoreception and in the production of specific thermoregulatory responses. Therefore, whenever it is appropriate, the contribution of these extrahypothalamic structures will be considered in discussing the hypothalamic role in each of the thermoregulatory responses. The second half of this chapter considers the integrative neuronal network specifically within the hypothalamus. It concentrates on hypothalamic single-unit studies and attempts to explain whole-body thermoregulatory responses in terms of the behavior of hypothalamic neurons.

A. Hypothalamic Control of Set-Point Temperature

Recently, numerous authors have formulated neuronal models to explain the control of thermoregulatory responses by hypothalamic and brain stem neurons. The basis of these models stems from a model proposed by Hammel in 1965, which showed how recently discovered, hypothalamic temperature-sensitive neurons could account for the regulation of an apparent hypothalamic set-point temperature. This model, shown in Figure 1, has had a major impact on the field of thermoregulation for over a decade. First of all, it represents the basis of nearly all subsequent neuronal models, which are really only more complex variations of this same concept proposed by Hammel. Second, this model, when proposed, offered a very good explanation of the proportional changes in thermoregulatory responses and the apparent shifts in set-point temperature regulation during fever, exercise, and changes in peripheral temperature. The importance of this model should not be underestimated. It not only explained current observations but also predicted numerous possibilities for future thermoregulatory and single-unit experiments. In this regard, it suggested which experiments should be done, as well as how the results of these experiments could be interpreted. Hammel's model is presented here in detail to orient the reader to the concept of set point and the rationale behind subsequent studies. It should be noted that, at the end of this chapter, other neuronal models are presented, which are based on more recent studies of thermoregulatory responses and single-unit activity.

The model in Figure 1 was primarily offered to explain earlier studies of shivering, vasoconstriction, and panting in dogs that were implanted with water-perfused, hypothalamic thermodes (Hammel et al., 1960, 1963; Fusco et al., 1961). Using these thermodes, the hypothalamus alone could be warmed or cooled to any temperature, and the relative amount of shivering (Sh), cutaneous vasoconstriction (V-c), and panting (Pa) could be observed. As an example, at a neutral skin temperature as in Figure 1(c), warming the hypothalamus past 39°C elicits panting, a heat-loss response. This threshold temperature of 39°C is called the hypothalamic set-point temperature for panting, or $T_{set_{Pa}}$. The amount of panting is proportional to the level of hypothalamic temperature, T_H, above $T_{set_{Pa}}$. If the hypothalamus is cooled past 37°C, shivering will occur, which will increase metabolic heat production and oxygen consumption. The 37°C would then be the hypothalamic set-point temperature for shivering, $T_{set_{Sh}}$; the amount of shivering will be proportional to the level of T_H below $T_{set_{Sh}}$. Thus, each thermoregulatory response has its own hypothalamic set-point temperature, T_{set}; the difference between the level of each thermoregulatory response (R) and some basal level (R_o) will be proportional to the difference between T_{set} and T_H (Hammel et al., 1963). That is: $R - R_o = \alpha_R (T_{set} - T_H)$, where α_R is simply the slope of the line representing the thermoregulatory response as a function of hypothalamic temperature. The α_R is a proportionality constant or simply the hypothalamic thermosensitivity for that response, i.e., it is the amount of change in that response for a given change in hypothalamic temperature once T_{set} has been reached.

Hammel (1965) suggested that each type of thermoregulatory response is controlled by different pools of effector neurons, as shown in Figure 1(a). These are also referred to as motor neurons or set-point neurons. They may be located in the hypothalamus or elsewhere in the brain stem or spinal cord. The firing rate of these effector neurons determines the level of the thermoregulatory response. Thus, an increased firing rate in the Pa effector neurons causes a proportional increase in panting. An increased firing rate in the V-c effector neurons causes a proportional increase in cutaneous vasoconstriction, while an increased firing rate in the Sh effector neurons causes a proportional increase in shivering intensity. The same concept would apply for the effector neurons controlling other thermoregulatory responses.

In the model in Figure 1(a), the firing rate of the effector neurons is determined only by the synaptic inputs from two different pools of thermosensitive neurons located in the PO/AH. One pool is composed of high-Q_{10}* neurons, which are quite thermosensitive over a wide range of hypothalamic temperatures. These neurons are commonly called warm-sensitive neurons. They increase their firing rates when the hypothalamus is warmed and decrease their firing rates when the hypothalamus is cooled. The other pool of "sensory" neurons are low-Q_{10} neurons. These are also called temperature-insensitive neurons because they show very little change in firing rate when the hypothalamus is warmed or cooled. Prior to Hammel's model, single-unit studies in cats had actually identified these two types of sensory neurons (Nakayama et al., 1963). Out of more than 1000 neurons recorded in the anterior and posterior hypothalamus, about 20% were considered to be warm sensitive and the other 80% were considered to be temperature insensitive.

Hammel's model, Figure 1(a), predicts that these two types of sensory neurons send mutually antagonistic synaptic inputs to each of the effector neurons. This antagonistic input then determines the firing rate in each effector neuron. Whether these synaptic inputs travel to the effector neurons directly or through other interneurons is immaterial. In the model, these mutually antagonistic inputs determine the T_{set} for each effector neuron, as well as the T_H range in which each neuron will be facilitated or inhibited. The panting or heat-loss effector neuron, for example, receives excitatory inputs from the high-Q_{10} or warm-sensitive neurons and inhibitory inputs from the low-Q_{10} or temperature-insensitive neurons. As shown in Figure 1(c), at some T_H the relative amount of excitation from the high-Q_{10} neurons will be equal to the amount of effective inhibition from the low-Q_{10} neurons. At this T_H there will be no output firing rate from the Pa neuron, based solely on the inputs from these two types of sensory neurons. This T_H is the T_{set} for the Pa neuron. At all temperatures below this T_{set} there is, again, no output firing rate from this effector neuron, since the input firing rate from the inhibitory, low-Q_{10} neurons is always effectively greater than the excitatory input from the high-Q_{10} neurons. This is because, with a decrease in T_H, the firing rate substantially decreases in the high-Q_{10} neurons but the firing rate of the low-Q_{10} neurons remains about the same. When T_H increases above this T_{set}, however, there is a proportional increase in the output firing rate in the

---

*Q_{10} is a term to describe the effect of temperature on chemical reactions or biological activity. In its simplest form, it is the ratio of a reaction before and after a 10°C change in temperature. If the reaction doubles during a 10°C increase, the Q_{10} is 2.0; if it triples, the Q_{10} is 3.0, and so on. A Q_{10} of 2-3 is generally considered to be normal for most biological activity.

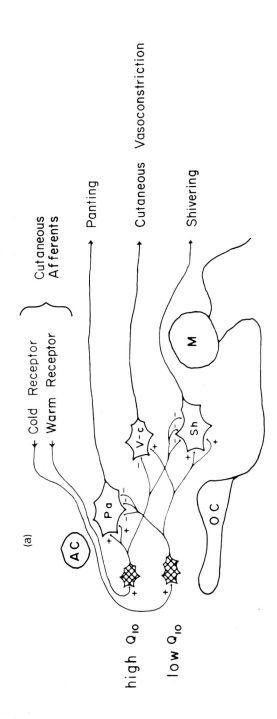

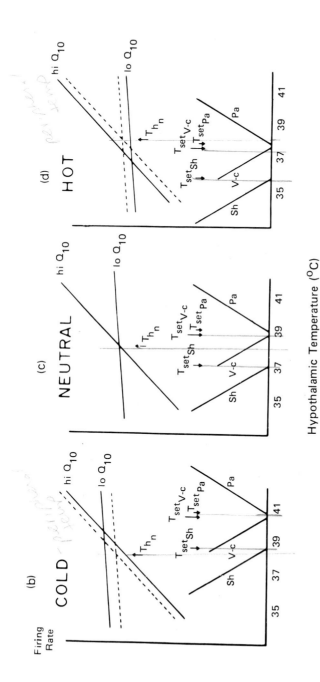

Figure 1 A model explaining the neuronal basis of set-point temperature (T_{set}) and the effect of peripheral temperature on set-point temperature regulation. AC, anterior commissure; OC, optic chiasm; M, mamillary body; Pa, effector neuron for panting; Sh, effector neuron for shivering; V-c, effector neuron for cutaneous vasoconstriction; cross-hatched cell bodies, low-Q_{10} and high-Q_{10} sensory neurons; T_{h_n}, neutral hypothalamic temperature. (Adapted from Hammel, 1965.)

effector neuron controlling panting. This is because an increase in T_H causes the firing rate of the excitatory, high-Q_{10} input to become greater and greater compared with the relatively unchanging input from the inhibitory, low-Q_{10} neuron. Therefore, for a Pa effector neuron such as this, the firing rate would be minimal at all temperatures below T_{set}; above T_{set}, the firing rate would be proportional to the increase in T_H. Neurons having this same type of thermoresponse characteristic (as predicted for the Pa effector neurons) have been recorded in the PO/AH and at other locations (Nakayama et al., 1963; Eisenman and Jackson, 1967; Hellon, 1967; Edinger and Eisenman, 1970). Such neurons are often called warm-sensitive interneurons. As in the model, the apparent T_{set} is often attributed to the mutually antagonistic synaptic inputs from sensory neurons having different hypothalamic thermosensitivities.

The model also predicts that other types of effector neurons must become active when the hypothalamus is cooled. These neurons would control the heat-retention responses like cutaneous vasoconstriction and the heat-production responses like shivering. Figure 1(a) shows how these neurons might be affected. In the Sh effector neurons, for example, the firing rate is determined by excitatory inputs from low-Q_{10} neurons and inhibitory inputs from high-Q_{10} neurons. Again at some T_H, the amount of effective excitatory input will equal the amount of effective inhibitory input, and the output firing rate of the Sh effector neuron will be minimal. This T_H is the T_{set} for the Sh effector neuron. At all hypothalamic temperatures above T_{set}, the output firing rate will remain minimal, because the firing rate of the inhibitory, high-Q_{10} neuron becomes greater with an increase in T_H. Therefore, inhibition predominates and the firing rate of the Sh effector neuron remains at a minimal level. When T_H drops below T_{set}, however, the amount of inhibitory input diminishes compared with the amount of excitation, because the firing rate of the inhibitory high Q_{10} neuron decreases, but the firing rate of the excitatory, low-Q_{10} neuron remains about the same. Therefore, the output firing rate of the Sh effector neuron becomes greater and greater as T_H drops below T_{set}. This is the same type of response as whole-body shivering during hypothalamic cooling. Again, neurons having the same type of thermoresponse characteristic (as predicted for the Sh effector neurons) have been recorded in the PO/AH and at other locations. While Nakayama et al., (1963) did not observe such neurons, subsequent studies have recorded neurons which increase their firing rates when the T_H is cooled, often below an apparent T_{set} (Hardy et al., 1964; Cunningham et al., 1967; Eisenman and Jackson, 1967; Hellon, 1967). These neurons are called cold-sensitive neurons or cold-sensitive interneurons. Over the years, the reported proportions of PO/AH warm-sensitive, cold-sensitive, and temperature-insensitive neurons has varied among the different studies. The cold-sensitive neurons have always remained the smallest group, but the reported proportions of warm-sensitive neurons appears to be higher in more recent studies. One study of only those PO/AH units that could be observed under stable conditions for long periods of time reported that 53% of the units were warm sensitive, 17% were cold sensitive, and 31% were temperature insensitive (Boulant and Bignall, 1973a).

B. Effect of Skin Temperature on Hypothalamic Control

So far, Hammel's model has explained the neuronal basis of hypothalamic set-point temperature regulation and the proportional change in each thermoregulatory response when T_H deviates above or below each T_{set}. The model will also explain the apparent shifts in hypothalamic T_{set} when the skin temperature is either increased or decreased, as reported by Hammel et al. (1963). As shown in Figure 1(b), when the skin temperature is cooled, the hypothalamic T_{set} increases for each thermoregulatory response. Even though the hypothalamus remains at a thermoneutral temperature (T_{h_n}), vasoconstriction and shivering are initiated because of this shift in T_{set}. In a cold environment, the T_H would have to be warmed to an even higher level (for example, almost 41°C) before panting will occur. Thus, with a cold skin, all set-point temperatures are elevated. On the other hand, with a warm skin in a hot environment, as in Figure 1(d), all hypothalamic set-point temperatures appear to be decreased (Hammel et al., 1963). Consequently, panting will occur even though the hypothalamus is at T_{h_n}, but panting can be suppressed by hypothalamic cooling. Also, in a hot environment, the hypothalamus would have to be cooled to a much lower level to elicit vasoconstriction and shivering (for example, the $T_{set_{Sh}}$ = 36°C).

Hammel's model, Figure 1(a), suggests that these shifts in T_{set} can occur simply by excitatory synaptic inputs to the high-Q_{10} and the low-Q_{10} PO/AH neurons from thermoreceptors in the skin. Two general types of thermoreceptors exist in the skin; both may have bell-shaped thermoresponse curves. Warm receptors increase their firing rates with skin warming, often over a range from about 35 to about 44°C. Cold receptors increase their firing rates during skin cooling, often over a range from about 38 to about 28°C. These ranges vary depending on receptor and skin location, but generally the thermal ranges of both receptors tend to overlap at neural skin temperatures. Recent surveys of peripheral thermosensitivity have been presented by Hensel (1973) and Bligh (1973). Skin thermoreceptor fibers synapse in the dorsal horn of the spinal cord (Hellon and Misra, 1973a). Spinal fibers then ascend in the lateral spinothalamic tract to synapse again at various levels of the brain stem reticular formation (Crosby et al., 1962) and also in the posterior ventral thalamus (Hellon and Misra, 1973b). The thalamic neurons project to the somatosensory cortex (Hellon et al., 1973). It is likely that the hypothalamus receives its afferent cutaneous and deep-body thermoreceptive information through ascending fibers from the reticular formation (Nakayama and Hardy, 1969) and possibly the midbrain raphe nuclei (Weiss and Aghajanian, 1971; Dickenson, 1976; Cronin and Baker, 1976).

Hammel's model suggests that the warm-sensitive or high-Q_{10} PO/AH neurons receive excitatory afferent input from the cutaneous warm receptors. Accordingly, the firing rate of the high-Q_{10} neurons would increase during skin warming and decrease during skin cooling, because this is what happens to the activity of the cutaneous warm receptors during skin warming and cooling. The model also indicates that the temperature-insensitive or low-Q_{10} neurons are excited by cutaneous cold receptors. Therefore, the firing rate of the low-Q_{10} neurons would increase during skin cooling and decrease during skin warming.

Given these conditions, Figure 1(b) shows the effect of skin cooling on the set-point temperatures of the three different thermoregulatory effector neurons. Skin cooling will increase the level of firing rate in the low-Q_{10} neurons and decrease the level of firing rate in the high-Q_{10} neurons. Either or both of these effects will result in a high T_{set} for each of the different types of effector neurons. This would be one explanation for the apparent elevation of hypothalamic set-point temperatures observed during peripheral cooling. Figure 1(d) shows that skin warming will decrease the level of firing rate in the low-Q_{10} neurons and increase the level of firing rate in the high-Q_{10} neurons. Either or both of these effects will decrease the T_{set} for each of the thermoregulatory effector neurons. Again, this would be an explanation for the apparent decrease in hypothalamic set-point temperatures observed during peripheral warming. Thus, warming either the hypothalamus or the skin can elicit panting. Warming both areas can produce a greater level of panting, and warming one area while cooling the other may reduce or inhibit panting. This same concept has been used to explain similar apparent T_{set} changes during fever, exercise, sleep, hibernation, etc. In each case, a change in T_{set} can be explained by having some synaptic input or some endogenous substance either increase or decrease the firing rate of PO/AH high-Q_{10} or low-Q_{10} neurons. Finally, it should be noted that recent single-unit and thermoregulatory response studies indicate that PO/AH neuronal integration is more complicated than is presented here in Hammel's model. These points are discussed throughout this chapter and especially in Section III.

Hammel's model is presented here because it ties together the activity of hypothalamic neurons with actual thermoregulatory responses. As mentioned, it has influenced subsequent single-unit and thermoregulatory studies, particularly in terms of their interpretation. In addition, several recent neuronal models are based on the same concepts of effector set points being determined by antagonistic inputs from neurons having different thermosensitivities. Hammel's model also introduces important concepts that will be presented throughout this chapter, including the proportional control of thermoregulatory responses, T_{set} regulation, and the hypothalamic integration of central and peripheral thermal information. Regarding this latter point, Hammel's model proposes that the effect of peripheral temperature is "additive" to the effect of hypothalamic temperature. The firing rate of a peripheral thermoreceptive input is simply added to (or subtracted from) the firing rate of a hypothalamic temperature-sensitive neuron. The effect of this addition of firing rates is that, for each thermoregulatory response, the T_{set} is shifted, but the hypothalamic thermosensitivity (α_R) remains constant. While there are studies to support this concept, it presently remains a matter of controversy. Several other thermoregulatory and single-unit studies indicate that peripheral temperatures (and other factors affecting thermoregulation) not only change T_{set} but also may increase or decrease the hypothalamic thermosensitivity (α_R) for various responses. This implies that the afferent input to PO/AH neurons alters not only their level of firing rate but also their sensitivity to changes in their own hypothalamic temperature. For example, a change in skin temperature may cause a PO/AH thermosensitive neuron to be more sensitive or less sensitive to a change in hypothalamic temperature. Thus, some studies indicate that the PO/AH integration of hypothalamic and extrahypothalamic temperature is simply additive (i.e., shifting T_{set}), while other studies indicate that the α_R is also altered. Specific examples to support both of these points are present throughout this chapter.

It is senseless to consider the hypothalamic control of body temperature without also considering the other neural structures necessary for the wide variety of thermoregulatory responses. The first part of this chapter will consider the role of the hypothalamus and some of these other neural structures in each of the major heat-production, heat-retention, and heat-loss responses. The remainder of this chapter concentrates on single-unit studies of the hypothalamus and, as with Hammel's model, will attempt to explain the neuronal basis of thermoregulatory responses.

C. Critical Review of Basic Experimental Procedures

Three basic experimental procedures have been used to study the neural control of thermoregulation: lesions, stimulations (including electrical, chemical, and thermal stimulations), and single-unit recordings. Each procedure has its own advantages and disadvantages, and any one type of study should not always be taken as proof that a particular neural area does or does not have a role in thermoregulation. Before continuing, some words of caution should be exercised regarding the validity of each type of study.

If an area is lesioned, for example, and body temperature changes, this is not proof that the neurons in this area are important in thermoregulation. It is possible that a lesion eliminates important efferent or afferent pathways passing through this area. This has been a criticism of early lesion studies in the posterior parts of the hypothalamus, which led to the concept of separate hypothalamic "heat-loss" and "heat-production centers." If a lesion does not impair thermoregulation, this does not indicate that the lesioned area is not important in thermoregulation. Lesions, even in the PO/AH, sometimes do not impair thermoregulation. This is primarily because, in the hierarchy of neural control, when one area is lesioned, other areas can often assume more control over thermoregulatory responses.

In electrical stimulation studies it should be remembered that such stimulation in no way duplicates the normal electrical activity in neural tissue. Such variables as tissue adaptation, current spread, pulse durations, and the frequency of single pulses and trains of pulses can account for some of the wide discrepancies between different experiments. Studies of preoptic and septal neurons have shown that afferent stimulation can cause a particular neuron to be initially inhibited, followed by a prolonged period of excitation. Other neurons may be initially excited, followed by a prolonged inhibitory period (Poletti et al., 1973; McLennan and Miller, 1974; Boulant and Demieville, 1977). Thus, for the same stimulation site, the firing rate of such neurons may be increased, decreased, or unaffected, depending on the frequency of stimulation. If electrical stimulation at a given site does affect thermoregulation, it may be because of its effect on afferent and efferent fibers passing through that site and not because of its effect on neurons in that site. Conversely, if electrical or chemical stimulation does not affect thermoregulation, this does not mean that that area is unimportant in thermoregulation. As shown in Figure 1(a), for example, if a stimulating electrode or an excitatory substance is placed in the PO/AH, it might excite both the high-Q_{10} and low-Q_{10} neurons simultaneously and, therefore, would not elicit any thermoregulatory response.

The same arguments given regarding electrical stimulation also apply to local injections of drugs, neurotransmitters, and cations (which are often given in very high concentrations). The posterior hypothalamus, for example, appears to be an important effector area for heat production. Electrical stimulation in the posterior hypothalamus can elicit shivering in cats (Stuart et al., 1961). If the posterior hypothalamus is perfused with a solution containing either no calcium ions or an excess sodium ion concentration, cats will shiver, vasoconstrict, and show a rise in body temperature. Similar perfusions in the anterior hypothalamus produce variable changes or no changes in body temperature (Myers and Veale, 1970, 1971). A reduction in extracellular calcium can increase sodium influx in neurons and at the same time block cholinergic and adrenergic synaptic transmission. This could temporarily excite a large group of possible heat-production effector neurons and at the same time block any synaptic input to these neurons. The resulting change in effector activity could lead to an increased body temperature. This illustrates that electrical and chemical stimulations, while often unphysiological, can demonstrate the importance of some neural areas. They must not be taken to imply, however, that other unresponsive areas are not equally important in thermoregulation. As mentioned, for example, nonspecific ionic stimulation of both high- and low-Q_{10} neurons in the PO/AH would not be expected to drastically alter T_{set} or thermoregulation.

Central thermal stimulation is by no means a physiological stimulation either. Under most conditions, the natural variations in core temperature are not nearly as great as the changes of several degrees imposed in most thermode studies. Even when localized heating or cooling does produce a physiological response, this does not mean that a particular neural area has a normal role in thermoregulation. Some studies for example, have shown that brain-stem heating and cooling near cardiovascular and respiratory "areas" can produce moderate changes in heart rate, blood pressure, breathing frequency, etc. In these cases, it is probable that temperature is simply a means of exciting or inhibiting a physiological response that is either nonthermoregulatory or inappropriate for thermoregulation.

It is true that most stimulation studies can be regarded as unphysiological. This should not detract from the importance of these studies, however. Stimulations should be viewed as a tool for unraveling the function of a neural area. An intense thermal stimulation, for example, may produce a response that can be measured, while the response to a more "physiological" stimulation might go undetected. This is the benefit of stimulation studies. We can recognize that a stimulation is unphysiological and can make inferences about the responses to physiological stimulations.

Single-unit studies also do not prove the thermoregulatory importance of a given area. It is well recognized now that temperature will affect the activity of neurons throughout the brain, even in areas that elicit no appropriate thermoregulatory responses when local temperature is altered. Other areas, like ventrobasal thalamus and somatosensory cortex receive cutaneous thermosensory afferents but probably have no role, at least in autonomic thermoregulatory responses.

The main shortcoming of single-unit studies is that, too often, they have been used simply to confirm the predictions of thermoregulatory response studies. Thermoregulatory studies indicate which neural areas are important and the type of integrative response that is produced. The unit studies simply show that some neurons in these areas behave in a predictable manner. Rarely have unit studies been used to predict the results of physiological studies. The problem stems from the great variability displayed by individual neurons. Each study accumulates a "collection" of single units and groups these units into categories based on one or two of a multitude of possible characteristics. The examples and the results that are reported are the ones that can be easily characterized and explained—that is, explained in terms of previous neuronal models. Therefore, to a degree, the models and the unit studies tend to be mutually reinforcing.

We have mentioned only a few of the criticisms applied to each type of study. Despite these criticisms, there must be some criteria to judge the probable thermoregulatory role of a neural structure. If a combination of different types of studies implicates a given area, it would seem reasonable to consider that area thermally important. Certainly if lesions impair responses to central and peripheral temperatures; if a change in local temperature or the local injection of a substance (like a pyrogen) produce appropriate thermoregulatory responses; and if single units can be recorded which are both temperature sensitive and receive cutaneous and deep-body thermal afferents, then it would seem reasonable to assign a thermoregulatory role to that area. In this regard, the PO/AH definitely qualifies as such an area. It is also reasonable to make predictions regarding the probable functions of neurons in such an area, based on their local and peripheral thermosensitive characteristics. In the sections that follow, the evidence to suggest that certain hypothalamic and extrahypothalamic structures have specific roles in thermoregulation is considered. The previously mentioned criticisms should aid the reader in evaluating the significance of these suggestions.

II. Hypothalamic and Extrahypothalamic Control of Responses

A. General

When compared with other homeostatic systems, thermoregulation is unique in its diversity of responses. A constant body temperature is maintained by a wide variety of physiological and behavioral responses. The hypothalamus has at least some control over each one of these responses. In the sections that follow, each of these responses is considered separately to determine the interaction between hypothalamic and extrahypothalamic structures involved in their control.

1. Physical Principles of Thermoregulation

The effectiveness of each thermoregulatory response has its basis in physical principles. If an animal's body is not storing heat and central temperature remains constant, then the body is in thermal equilibrium with its environment. Under this condition, the amount of heat produced by cellular metabolism and muscular work is equal to the amount of heat lost by radiant, convective, conductive, and evaporative heat loss. Evaporative heat loss (EHL) occurs when water from the respiratory tract or on the skin evaporates into the air. This removes heat energy from the body. Panting and sweating are the most common physiological responses to promote EHL. By regulating blood flow in the skin, the body regulates the amount of radiant, conductive, and convective heat transferred between the body surface and the environment. The level of blood flow in the skin determines the amount of central body heat transported to the skin. The temperature difference between the skin and the surrounding environment determines the net amount of heat the body loses to or gains from the environment. Thus, to conserve central body heat in a cold environment, cutaneous vasoconstriction reduces the difference between skin and environmental temperature. This effectively reduces heat loss from the body surface, especially net radiant heat loss. On the other hand, during exercise, much of the increased body-heat load can be lost by sweating and by increasing blood flow in the skin.

2. Hierarchy of Neural Integration and Control

As stated, the hypothalamus has at least some control over each of the thermoregulatory responses. In recent years, other thermoregulatory areas have been identified throughout the brain stem and in the spinal cord. It is now apparent that a hierarchy of neural integration and control probably exists for each thermoregulatory response. In discussing the neural control of shivering, Chambers et al., (1974) have described this hierarchical organization as follows:

> ... it had been believed that control and effector mechanisms reside only in the hypothalamus, and that other areas only supply information to it. However, the thermoregulatory system may be organized such that the hypothalamus directs the activity of miniature temperature regulation areas, each modulated by the hypothalamus but capable of independent thermosensitive responses. In such a system the hypothalamus would be dominant, for it would amalgamate, adjust, and regulate lower centers; in its absence only a fraction of the capability of the system could be utilized. However, lower centers in the medulla and spinal cord would have some degree of thermosensitivity and a capacity to transduce temperature changes into effector output. Inhibitory regions associated with these centers might form part of the means by which the hypothalamus could exert dominance over them.

To some degree, this aptly describes the hypothalamic control over most of the thermoregulatory responses.

Physiological studies as well as single-unit studies indicate that thermosensitive and thermoregulatory structures exist throughout the brain stem—in the PO/AH and septum, posterior hypothalamus, midbrain, pons, medulla, and even in the spinal cord. The importance of the PO/AH may lie in the fact that, in many instances, it is somewhat more thermosensitive than the other structures. Therefore, a change in central temperature is first "sensed" in the PO/AH, and appropriate thermoregulatory responses can be elicited to correct this change in temperature. In this way, a thermal imbalance may be corrected by the PO/AH before other thermosensitive structures reach their thermal set points, which are necessary to elicit responses. Extrahypothalamic thermosensitivity may be somewhat redundant in an intact animal as long as hypothalamic thermosensitivity can account for the maintenance of normal body temperature. As we will see, however, this is not always the case; under some circumstances, extrahypothalamic thermosensitivity is necessary for normal thermoregulation. It is also quite possible that the PO/AH exerts some tonic influence over the local thermosensitivity of these extrahypothalamic structures such that, when the PO/AH is lesioned, some of these structures may become more thermosensitive while others might become less thermosensitive.

In addition to local thermosensitivity, single-unit studies indicate that, at each brainstem level, certain neurons can integrate local thermal information with afferent information from peripheral thermoreceptors. The PO/AH again appears to be quite important in this integrative role. This is obvious from lesion studies, particularly those that progress from rostral to caudal levels. In intact animals, appropriate thermoregulatory responses can usually be evoked simply by changing peripheral temperature; but, as more and more of the rostral brain stem is removed, peripheral temperature becomes less important, and greater and greater changes in central temperature must occur before any thermoregulatory response is apparent.

B. Metabolic Heat Production

A decrease in either peripheral temperature or central temperature will produce an increase in metabolic heat production, reflected by an increased oxygen consumption. Figure 1 shows that, when hypothalamic temperature, T_H, is cooled past a particular T_{set}, metabolic heat production due to shivering will increase in proportion to the decrease in T_H. Hammel's model suggested a summation of the afferent input from peripheral thermoreceptors with hypothalamic thermosensitivity. Thus, the hypothalamic T_{set} for heat production appeared to increase with a cool skin and decrease with a warm skin, but the α_{sh} or the slope of the T_H response curve remained constant. According to the model, peripheral temperature affected just T_{set} and not the hypothalamic thermosensitivity for this heat production response. This concept was later substantiated in a thermode study in which oxygen consumption was measured in dogs during changes in T_H at various ambient temperatures (Hellstrøm and Hammel, 1967). In 1963, Benzinger et al., reported that, in humans, skin warming can apparently decrease the internal T_{set} for eliciting metabolic heat production, but in this study, the primary effect, at least of skin cooling, was to increase the central thermosensitivity for heat production. Later, Jacobson and Squires (1970) showed, in cats, that peripheral cooling will increase the preoptic thermosensitivity for heat production, while peripheral warming will virtually eliminate the preoptic control of this response. They have also shown that the data from the Hellstrøm and Hammel (1967) study could suggest that peripheral cooling increases and peripheral warming decreases the hypothalamic thermosensitivity (α) for heat production (Jacobson and Squires, 1970). Several recent thermode studies in rabbits have shown that peripheral temperature produces similar changes in the apparent hypothalamic thermosensitivity for metabolic heat production (Gonzalez et al., 1974; Stitt et al., 1974; Boulant and Gonzalez, 1977). As shown in Figure 2, for example, peripheral cooling in a conscious rabbit will not only increase the level of heat production (along with T_{set}) but also will tend to increase the α or hypothalamic thermosensitivity (watts per kilogram per degree centigrade). Peripheral warming, on the other hand, will markedly decrease both heat production and the hypothalamic thermosensitivity for this response.

There are two general types of heat production responses: shivering and nonshivering thermogenesis. During cold exposure, increased muscle activity (including shivering) is the primary source of heat production in most adult species, including humans. Nonshivering thermogenesis (NST) includes every other means of heat production besides muscular contractions and often relies heavily on the sympathetic adrenergic innervation of brown adipose tissue. The relative contribution of shivering versus nonshivering thermogenesis can be determined experimentally by giving curare-like drugs to block shivering (Heming-

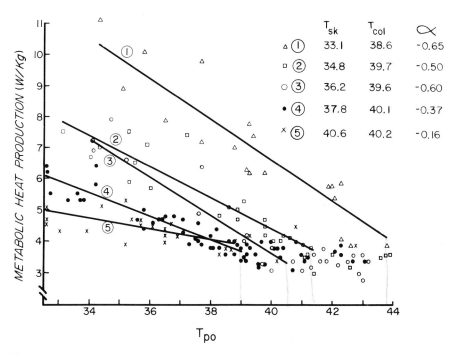

Figure 2 Metabolic heat production in a rabbit as a function of preoptic temperature, T_{po}. Each numbered line and symbol refers to a different level of trunk skin (T_{sk}) and colonic temperature (T_{col}). Thermosensitivity (α) refers to the slope of each line, in watts per kilogram per degree centigrade. (From Boulant and Gonzalez, 1977.)

way and Stuart, 1963) or adrenergic β-receptor blocking agents to block NST (Brück et al., 1969). Such study of shivering versus NST deserves much more attention, not only in a variety of species but under different central and peripheral thermal conditions. Nonshivering thermogenesis is totally lacking in some species, such as the miniature pig. In other species, such as the guinea pig and rat, NST is especially important in neonates and cold-adapted animals. However, in most adult animals—even in cold-adapted animals— shivering appears to contribute more than NST to at least initial, cold-induced heat production (Brück et al., 1969). As mentioned, however, further study of these differences is certainly warranted.

1. Shivering Thermogenesis

Shivering is a cold-induced tremor in the skeletal muscles. It can be blocked by curare and is often suppressed during voluntary movements of the muscles. Heat produced by shivering is derived from rhythmic contractions of antagonistic muscles. The oscillations of these contractions (6-30 hz; Kosaka and Simon, 1968) apparently involve reverberating circuits of neurons at the spinal level. The frequency of this tremor is probably determined solely by these spinal neurons, since high-frequency (25-300 hz) electrical stimulation at sites throughout the brain stem can elicit shivering but will not influence the tremor frequency (Birzis and Hemingway, 1957a). Before discussing the hypothalamic control, the evidence for this spinal organization of thermal shivering will be considered first.

a. Spinal cord

Localized spinal cooling (i.e., 1-3°C) especially in upper spinal cord, will elicit shivering in dogs, rabbits, and cats. This spinal cooling will elicit shivering in intact animals as well as animals having bilateral dorsal root sections or animals having high spinal transections (Meurer et al., 1967; Thauer, 1970; Chambers et al., 1974). During spinal cooling, the frequency range of the shivering tremor is the same for both intact animals and those having spinal sections, further indicating that the rhythm of the oscillating tremor is determined at the spinal level (Kosaka and Simon, 1968; Thauer, 1970). Thus, in some animals, the neuronal network within the isolated spinal cord is capable of sensing decreases in core temperature and eliciting shivering, independent of supraspinal structures.

According to Chambers et al., (1974), however, the isolated spinal cord is not as thermosensitive as the intact cord, since the isolated cord must be cooled to a lower temperature before shivering is evoked. This implies that in intact animals, ascending afferents from spinal thermoreceptors or tonic facilitation from supraspinal levels can enhance spinal-induced shivering. It should be noted that shivering in animals that have had spinal sections is primarily in response to decreases in spinal temperature, not to decreases in skin temperature. Consequently, when these animals are exposed to cold, the core temperature must drop low enough to stimulate the spinal thermoreceptors directly.

Dorsal horn spinal neurons do receive afferent input from cutaneous warm and cold receptors (Hellon and Misra, 1973a; Burton, 1975). Presumably fibers from these neurons ascend in the lateral spinothalamic tract. Studies have identified spinal warm-sensitive and cold-sensitive neurons within this ascending anterolateral tract (Simon and Iriki, 1971; Necker, 1975). Simon (1972) also showed that a high proportion of these spinal thermosensitive neurons were influenced by cutaneous thermoreceptors, indicating that integration of spinal and skin thermal information probably occurs in this ascending anterolateral system. The question becomes: do these integrating spinal neurons have a role in local spinal shivering, or do they simply relay spinal and cutaneous thermal information to higher neuronal networks such as in the pons (Wünnenberg and Brück, 1968) and PO/AH (Guieu and Hardy, 1970b; Boulant and Hardy, 1974)? Since the anterolateral system is an ascending sensory system and skin cooling does not elicit shivering in spinal animals, the latter alternative seems more probable. If this is true, then which neurons in the isolated spinal cord sense temperature and elicit shivering? Recently Pierau et al., (1976) have suggested that local cooling directly increases the excitability of spinal motor neurons innervating skeletal muscle; therefore, it is likely that cold-induced increases in motor neuron activity account for shivering during spinal cooling. Earlier, Klussmann (1969) and Klussmann and Henatsch (1969) showed that spinal cooling will increase the firing rate of, first, the smaller motor neurons, followed by the larger motor neurons. With spinal cooling, the γ motor neurons become activated first, followed by the "tonic" α motor neurons, and then the "phasic" α motor neurons. The γ motor neurons probably do not control shivering because they are inactive at the lower spinal temperatures when shivering occurs. In addition, sectioning the dorsal roots does not eliminate spinal-induced shivering; it only makes the tremor more irregular (Meurer et al., 1967; Thauer, 1970). For these reasons, it is probable that the cold-induced activation of phasic α motor neurons accounts for the shivering tremor during spinal cooling. With intact dorsal roots, the alternating contraction of antagonistic muscles may account for the "regularity" of the tremor. However, this is not necessary for shivering. Without the dorsal roots, the tremor simply becomes more irregular, possibly indicative of the lack of afferent signals from antagonistic muscles.

b. Septum and PO/AH

Despite this apparent independence of the spinal cord, there is much evidence that the septum, the hypothalamus, and other brain-stem structures have a role in the shivering response. This evidence will be considered for each of these neural areas, beginning with the septum and proceeding caudally.

The septum has many synaptic connections with the preoptic region and lies immediately dorsal to this area. The median forebrain bundle provides vast ascending and descending synaptic connections between the septum, anterior and posterior hypothalamus, and midbrain. Whether reported or not, in all previous PO/AH thermode studies, the changes in septal temperature are equivalent to the changes in PO/AH temperature; therefore, the thermoregulatory significance of temperature-sensitive septal neurons should be considered along with that of the PO/AH neurons.

In goats, electrical stimulation of the septum will produce shivering, which continues until core temperature rises about $0.5°C$. At this point, shivering is suppressed, apparently by hypothalamic or other deep-body thermoreceptors (Andersson, 1957). Stuart et al., (1961) showed that shivering could be evoked in cats by electrically stimulating the ventrolateral region of the septum; however, this shivering was not as intense as that due to stimulation in the dorsomedial part of the posterior hypothalamus. Stimulation of the ventromedial septum could suppress shivering, although this suppression was, again, not as great as that produced by stimulation in the PO/AH or ventrolateral posterior hypothalamus. It is quite probable that the septum does have a role in shivering, possibly by some modifying input to the PO/AH and posterior hypothalamus. The true importance of the septum in shivering is unknown, however, particularly in view of the fact that heat production and shivering are not impaired by lesions in the septum (Stuart et al., 1962a) and forebrain structures rostral to the hypothalamus (Bazett et al., 1933; Pinkston et al., 1934; Keller and McClaskey, 1964).

Thermosensitive neurons in the preoptic region and anterior hypothalamus have a strong influence on shivering. As noted earlier, in Figure 2, PO/AH cooling will rapidly increase metabolic heat production. In addition, peripheral cooling will increase heat production and the hypothalamic thermosensitivity for this response, while peripheral warming will have the opposite effect. Much of this initial metabolic heat production is due to increased activity in skeletal muscle, which would include shivering thermogenesis. Other thermode studies have noted that observable shivering can be evoked by PO/AH cooling at neutral or cool ambient temperatures (Hammel et al., 1960; Satinoff, 1964; Jessen et al., 1968; Murgatroyd and Hardy, 1970; Jessen and Mayer, 1971). In addition, localized PO/AH warming will suppress shivering already elicited by skin cooling (Hemingway et al., 1940; Fusco et al., 1961; Brück and Wünnenberg, 1970), spinal cooling (Jessen and Simon, 1971), or whole-body cooling (Sundsten, 1967). Evidently, shivering may be elicited in more caudal areas of the central nervous system (CNS), but the PO/AH temperature can either accentuate (Jessen and Ludwig, 1971) or inhibit this response.

Electrical stimulation at hypothalamic and brain-stem sites can either elicit or inhibit shivering, depending on the location, frequency, and intensity of stimulation. Studies indicate that electrical stimulation of the PO/AH produces the same effect as localized warming of this area, that is, an inhibition of shivering induced by peripheral cooling (Hemingway et al., 1954; Nutik, 1973a). In fact, Andersson and Persson (1957) have shown that preoptic stimulation in goats will suppress cold-induced shivering, even after rectal temperature has dropped as much as 10°C. In cats, electrical stimulation in the PO/AH will strongly suppress shivering. Stimulation in the median forebrain bundle of the lateral hypothalamus will also suppress shivering. Conversely, shivering can be induced by stimulation of the dorsomedial parts of the posterior hypothalamus and midbrain (Hemingway et al., 1954; Stuart et al., 1961).

PO/AH lesions often impair heat loss responses but can also impair shivering thermogenesis and thermoregulation in the cold. Much of this impairment may depend on the species of animal and the degree to which the animal is thermally stressed. Certainly, transient temperature variations following surgical lesions must be viewed skeptically, since infection, rigidity, immobility, lack of nutrients, local hemorrhage, and postsurgical trauma will affect thermoregulation. These experimental factors should be considered when judging the significance of early lesion studies. In earlier studies, rostral hypothalamic lesions near the PO/AH often produced transient hyperthermia and impaired heat-loss responses but did not seem to impair shivering and heat production. In 1940, for example, Ranson summarized the observations of anterior hypothalamic lesions in cats, monkeys, and humans, noting that the probable reason for this transient hyperthermia was the destruction of a heat-loss center near the anterior hypothalamus and a temporary irritation of heat-production areas located more caudally in the hypothalamus. Bilateral lesions in the posterior hypothalamus impaired thermoregulation in both the heat and cold. Thus, many early lesion experiments tended to support a concept of a heat-loss area in the rostral hypothalamus and a separate heat-production area in the caudal hypothalamus and midbrain (Meyer, 1913; Keller, 1933, 1935; Ranson, 1940). In fairness, however, it should be noted that some of these early studies did report that rostral hypothalamic lesions could impair thermoregulation in the cold, but this was not as consistent as the impairment in a warm environment. In 1964, Keller and McClaskey showed that, in dogs, the more rostral parts of the hypothalamus may also be important in shivering and heat production. Some dogs, having transections below the preoptic area and through the rostral thalamus (low-thalamic preparation), could not effectively maintain body temperatures in either 38 or 3°C environments. The impairment in the heat, however, was more consistent in all of the dogs studied. On the other hand, when all tissue rostral and dorsal to the hypothalamus was removed, including the preoptic area (low-hypothalamic preparation), there was very little impairment in the 3°C environment. In these dogs, shivering was initiated once the core temperature reached 38°C. Thus, in some cases, lesions between the preoptic area and the anterior hypothalamus may impair shivering, but in other cases they will not.

More recently, Squires and Jacobson (1968) showed that medial preoptic lesions reduce oxygen consumption and core temperature in undisturbed cats. When exposed to cold, these lesioned cats shivered and maintained their core temperatures. This heat-production response, however, was not as consistent or as effective as that of normal cold-exposed cats. When exposed to heat, such lesioned cats showed above-normal elevations in core temperature. Lipton et al. (1974) have shown in rats that complete PO/AH lesions greatly impair thermoregulation in both warm and cold environments. In addition, discrete cuts between the preoptic region and anterior hypothalamus decrease thermoregulation in either warm or cold environments but not in both. This suggests that there is a fine balance within the PO/AH. A cut through the PO/AH may possibly upset this balance in favor of either heat production or heat loss. These investigators offer this as an explanation of past studies in which some PO/AH lesions produced hyperthermia while others produced hypothermia.

The role of the PO/AH in shivering remains controversial. Electrical and thermal stimulations indicate that the PO/AH can thermally regulate shivering and can certainly suppress this response. Lesion studies often suggest that the posterior hypothalamus is the important structure for eliciting shivering, and, in the absence of the PO/AH, more caudal thermosensitive structures control this response. Most PO/AH lesion studies, however, have not adequately tested the integrative ability of the remaining neural structures, particularly their ability to suppress ongoing shivering. This suppressive role of the PO/AH may be quite important.

c. Lateral and posterior hypothalamus

In heat production, a descending pathway from the PO/AH runs laterally through the median forebrain bundle in the lateral hypothalamus. This median forebrain bundle contains both ascending and descending fibers, quite likely carrying both afferent and efferent information to and from PO/AH thermosensitive neurons. Lipton et al. (1974) have shown in rats that lateral parasagittal cuts, between the PO/AH and the median forebrain bundle, do not disturb thermoregulation in the heat. These cuts, however, do impair regulation in the cold, such that body temperature can drop as much as 2.9°C during cold exposure (i.e., 5°C). On the other hand, medial cuts in the hypothalamus and midbrain (including the dorsolateral fasciculus) did not impair thermoregulation in rats. These investigators suggest that the descending PO/AH pathway for heat production is much more discrete than the more diffuse and redundant pathways controlling heat-loss responses. Using electrical stimulation at various sites, Hemingway et al. (1954) have indicated that a pathway that suppresses shivering descends from the PO/AH through the lateral parts of the hypothalamus and midbrain. In addition, a study in which acetylcholine was microinjected into various hypothalamic areas, favors the existence of a discrete heat-production pathway (but not necessarily a heat-loss pathway) in the lateral hypothalamus and median forebrain bundle (Myers and Yaksh, 1969). As mentioned in Section I.C, these lesion and stimulation studies affect both ascending and descending pathways and are, therefore, not absolute proof of an efferent pathway. A study incorporating a rostral lesion and a caudal stimulation, or a rostral thermode and a caudal lesion, might better define a specific descending, efferent pathway.

It is known that spinal warming will suppress shivering during cold exposure. In guinea pigs, discrete lesions in the dorsolateral part of the middle hypothalamus will block this spinal suppression of shivering. Lesions immediately rostral or caudal to this area apparently do not affect shivering in these animals, nor do they affect the suppression of shivering by spinal warming (Wünnenberg and Brück, 1970). This suggests that this lateral hypothalamic area is necessary for the ascending spinal signals which suppress shivering. This area might simply represent the location of the greatest concentration of appropriate ascending (e.g., spinal) and descending (e.g., PO/AH) fibers in the median forebrain bundle.

Hardy (1973) has recently reviewed the thermoregulatory significance of the posterior hypothalamus, which is an important neural area in the control of shivering. Electrical stimulation of the posterior hypothalamus, particularly the dorsomedial region, readily produces shivering in cats (Stuart et al., 1961). Ranson (1940) has reported, however, that lesions in this medial part of the posterior hypothalamus do not impair thermoregulation, but, in cats and monkeys, lateral lesions in the posterior hypothalamus do impair regulation in both cold and hot environments. On the other hand, Stuart et al. (1962) have shown that, in cats, discrete lesions in the dorsomedial posterior hypothalamus either abolish or markedly impair the shivering response. Such medial lesions, though, do not impair other cold-induced responses, such as cutaneous vasoconstriction, piloerection, and behavioral responses like huddling. Lesions in the lateral and dorsolateral posterior hypothalamus, however, do impair cold-induced cutaneous vasoconstriction but leave shivering and other cold-induced responses intact. These investigators have, therefore, offered an explanation for the hypothermia observed in previous studies following lateral posterior hypothalamic lesions. Such lateral lesions could increase heat loss due to impaired cutaneous vasoconstriction and result in hypothermia, regardless of the shivering intensity. A lesion in the dorsomedial posterior hypothalamus, however, might simply impair shivering but not nonshivering thermogenesis and heat-retention responses. This might be one reason why, in the study by Lipton and co-workers (1974), medial cuts in the posterior hypothalamus and midbrain did not impair thermoregulation in the rat, an animal that uses both shivering and nonshivering thermogenesis.

While the posterior hypothalamus does contain temperature-sensitive neurons (Edinger and Eisenman, 1970; Wünnenberg and Hardy, 1972), this area generally has not been considered to be a thermosensitive structure like the PO/AH. As early as 1940, Hemingway et al. showed that, while PO/AH heating markedly suppressed shivering, posterior hypothalamic heating caused only a slight reduction in cold-induced shivering (and this might have been due to conductive heating of the PO/AH). In fact, even in anesthetized cats, alterations in anterior hypothalamic temperature elicit appropriate responses and changes in rectal temperature; however, similar alterations in posterior hypothalamic temperature often elicit inappropriate thermoregulatory responses and inappropriate changes in rectal temperature. That is, local posterior hypothalamic warming often produces slight increases in rectal temperature, while local cooling often produces slight decreases in rectal temperature (Freeman and Davis, 1959).

It is probable that the posterior hypothalamus is not normally a thermosensitive structure. Rather, it receives input signals concerning PO/AH and extrahypothalamic temperature to control certain thermoregulatory responses like shivering. Nutik (1973a) has shown, for example, that about 9% of the posterior hypothalamic single units are affected by changes in PO/AH temperature (T_{po}); most of these are inhibited by PO/AH warming and excited by PO/AH cooling. In addition, 45% of these T_{po}-sensitive neurons (in the posterior hypothalamus) respond to shivering-suppressive electrical stimulation in the PO/AH. Again, the predominant response is inhibition. Other studies have shown that individual neurons in the posterior hypothalamus are affected by both preoptic and skin temperatures (Nutik, 1973b), as well as preoptic, spinal cord and local posterior hypothalamic temperatures (Wünnenberg and Hardy, 1972).

It is not known how much of this skin and spinal thermoreceptive information projects directly to the posterior hypothalamus or how much is relayed there through the PO/AH. Probably both routes exist. In an experiment by Myers and co-workers, PO/AH and peripheral cooling both increased the efflux of Ca^{2+} in the posterior hypothalamus. However, if the PO/AH neurons were locally anesthetized with procaine, the Ca^{2+} efflux in the posterior hypothalamus was prevented during peripheral cooling (Myers et al., 1976). This would support the conclusion that a substantial amount of peripheral thermal signals ascend to the PO/AH before an integrated signal projects to the posterior hypothalamus. Certainly continued studies are necessary. In fact, by using a combination of lesions and stimulations, it should be possible for future single-unit and thermoresponse studies to demonstrate whether or not both routes exist.

d. Lower brain stem

A diffuse shivering pathway descends from the lateral part of the posterior hypothalamus into the midbrain reticular formation. This pathway continues in a lateral position through the reticular portions of the pons, medulla, and spinal cord. At every level, acute lesions along this pathway abolish shivering, but extensive medial lesions and lesions of the pyramidal and periventricular systems have no effect on shivering (Birzis and Hemingway, 1956). Midbrain and pontine electrical stimulation of this pathway will elicit shivering in cats (Birzis and Hemingway, 1957a). In addition, unit recordings from within this pathway reveal descending fibers (firing at 6-26 impulses per second) whose firing rate is correlated with shivering intensity. These fibers are excited by peripheral cooling, inhibited by peripheral and central warming, and represent a descending pathway, since caudal hemisections and midsagittal cuts do not affect their activity (Birzis and Hemingway, 1957b).

In 1949, Lindsley et al. described a complex interaction among extrapyramidal neural areas controlling the excitability of the stretch reflex. Certain areas in the hypothalamus and much of the rostral reticular formation facilitate these spinal reflexes. Other areas, including the caudal reticular formation, suppress these spinal reflexes. This extrapyramidal motor control of muscle movement and tone involves a great deal of complex brain-stem control over "established" spinal networks. It is quite possible that the neural control of shivering involves a similar balance between facilitory and suppressor areas of the brain stem, which exert control over established spinal networks for shivering.

As mentioned previously, the isolated spinal cord, when cooled, is capable of eliciting shivering (Thauer, 1970). Lipton (1973) has also shown in rats that local cooling in the medulla oblongata elicits shivering and a rise in rectal temperature. This response is unaffected by PO/AH lesions. Evidently, caudal thermosensitive structures have some control over shivering even without hypothalamic thermosensitive structures. On the other hand, we have also seen that discrete dorsomedial lesions in the posterior hypothalamus completely abolish shivering (Stuart et al., 1961). Therefore, it is likely that a brain-stem reticular area exists below the hypothalamus that tonically inhibits the spinal or medullary control of shivering. Otherwise, spinal shivering, for example, would still be evident in animals with posterior hypothalamic lesions. Chambers et al. (1974) have localized this inhibitory network in the midbrain and upper pontine tegmentum. They found that, in cats, spinal cooling elicited shivering, piloerection, cutaneous vasoconstriction, and a rise in rectal temperature. A decerebration anywhere from the caudal hypothalamus to the upper pons abolished all of these responses to spinal cooling. However, by lowering the transection to the lower pons or medulla, all of these responses to spinal cooling returned. These investigators suggest that this explains the conflicting reports of studies describing the effect of decerebrations on thermoregulation. High transections (which leave the midbrain and upper pons intact) tend to abolish thermoregulation in cats (Chambers et al., 1949; Bard and Macht, 1958; Stuart et al., 1962b; Bard et al., 1970), dogs (Keller, 1935; Keller and McClasky, 1964), rats (Woods, 1964), and monkeys (Denny-Brown, 1966). Animals with lower transections at the pons, however, still retain certain thermoregulatory responses, like shivering (Dworkin, 1930; Keller, 1933; Chambers et al., 1949), and some of these animals even became hyperthermic (Keller, 1933; Conner and Crawford, 1969). An exception to the loss of thermoregulation in high decerebrates may occur in newborn animals. Bignall and Schramm (1974), for example, have shown that a kitten having a high transection below the hypothalamus still displays normal shivering, piloerection, for behavioral thermoregulation in a cold environment. This suggests that neural integration and control may progress rostrally during development. The reliance on the hypothalamus for these functions may become more permanent only with maturation.

In addition to a midbrain inhibitory area, Chambers and co-workers (1974) have also indicated that a lower pontine-medullary area exists, which is facilitory to the spinal control of shivering, vasoconstriction, and piloerection. Although unable to maintain core temperature, low-decerebrate cats (having their lower pons and medulla intact) still show these thermoregulatory responses during spinal cooling. Transection of the upper spinal cord, however, abolishes piloerection and reduces the intensity of shivering during spinal cooling. This implies that a pontine-medullary area serves to facilitate the spinal control of shivering.

2. Nonshivering Thermogenesis

Nonshivering thermogenesis (NST) encompasses every means of metabolic heat production except the heat produced by the contraction of skeletal muscles. Brown adipose tissue and the metabolic endocrines are the principal determinants of NST. In recent years, the importance of the PO/AH in shivering has been questioned, particularly in those animals that rely heavily on nonshivering thermogenesis, such as guinea pigs and rats. It has been suggested that, in these animals, the PO/AH tends to control NST, while the spinal cord tends to control shivering. In guinea pigs, for example, moderate cooling of the cervical spinal cord will produce shivering; however, moderate cooling of the skin or

PO/AH will produce NST but not shivering (Brück and Wünnenberg, 1970). Similarly, in rats, moderate PO/AH cooling will produce NST, not shivering, but more severe PO/AH cooling will produce shivering (Banet and Hensel, 1976). These investigators have pointed out that, even in animals that rely on NST, it is probably simply a question of differences in the hypothalamic thresholds needed to elicit shivering or nonshivering thermogenesis. Accordingly, the hypothalamic T_{set} for shivering may be lower than the T_{set} for NST. In this way, moderate PO/AH cooling would elicit NST, but substantial PO/AH cooling would be necessary to elicit shivering.

The following sections consider the hypothalamic integration of central and peripheral temperatures in controlling different components of nonshivering thermogenesis.

a. Brown adipose tissue

Brück and Wünnenberg (1970) have suggested that the PO/AH and cutaneous control of NST can actually suppress the spinal control of shivering in rats, guinea pigs, certain cold-adapted animals, and newborns. In these animals, the cervical spinal cord is enclosed by interscapular and cervical brown adipose tissue (BAT). With moderate peripheral or internal cooling, the PO/AH will initiate NST, and much of the heat will be generated by this brown adipose tissue. This, in effect, preferentially warms the cervical spinal cord and insures that spinal thermoreceptors do not evoke shivering. On the other hand, intense peripheral or central cooling will excite both the spinal thermoreceptors and those thermosensitive PO/AH neurons having the lower set-point temperatures. This evokes shivering. The loss of this cervical brown adipose tissue with maturation might explain the corresponding shift from nonshivering to shivering thermogenesis. Conversely, the increase in cervical BAT with cold adaptation might explain the accompanying shift from shivering to nonshivering thermogenesis.

These anatomical and threshold (i.e., T_{set}) differences between shivering and nonshivering thermogenesis have been further illustrated in a recent study by Fuller et al. (1975). In this case, rats were exposed to cold, producing some shivering and nonshivering thermogenesis. When the PO/AH of these cold-exposed rats was warmed 2-3°C, not only did total metabolic heat production and rectal temperature decrease but shivering actually increased. This paradoxical situation in which warming the PO/AH increases shivering can be explained by the fact that, in some animals, the PO/AH has primary control of NST and the cervical spinal cord has primary control of shivering. Therefore, PO/AH warming could suppress NST, but the resulting drop in spinal temperature would enhance shivering.

In young, cold-adapted animals, PO/AH and cutaneous thermoreceptors seem to be equivalent in their influence on nonshivering thermogenesis (Brück and Wünnenberg, 1970). The PO/AH appears to have a suppressive control over nonshivering thermogenesis, possibly in much the same manner as it suppressed shivering thermogenesis. Not nearly as much is known about the more caudal neural structures which promote NST. PO/AH lesions in guinea pigs can evoke a strong increase in NST resulting in a marked hyperthermia, even in a neutral environment (Brück and Wünnenberg, 1970). This hyperthermia can be inhibited by adrenergic β-receptor or ganglionic blocking agents, indicating that it is due to sympathetically induced NST. Moreover, this hyperthermic response to a PO/AH lesion is only very slight in normal adult guinea pigs but is very strong in cold-adapted and newborn guinea pigs, having large amounts of BAT. In addition, this NST hyperthermic response is reduced by skin warming and enhanced by skin cooling. This indicates that afferents from cutaneous thermoreceptors can directly affect more caudal neural structures and can modify NST even without an intact PO/AH. Apparently, under normal circumstances these same neural structures are tonically inhibited by the PO/AH.

b. Metabolic endocrines

Also important in NST is the increase in cellular metabolism throughout the entire body in response to increased circulating levels of metabolic endocrines, such as thyroxine, adrenal catecholamines, and glucocorticoids. In response to cold exposure, an increased circulating thyroxine and triiodothyroxine will increase whole-body heat production by increasing oxygen consumption in most metabolic tissues. An intact hypothalamohypophyseal pathway is necessary for the increased TSH and thyroid activity during cold exposure (Von Euler and Holmgren, 1956; Knigge and Bierman, 1958). Cold exposure will also increase circulating levels of norepinephrine and epinephrine, which have a general calorigenic action in most tissues and act synergistically with thyroxine to increase whole-body metabolism. (Lutherer et al., 1969; Tanche and Therminarias, 1969; Ganong, 1975). These catecholamines also increase blood glucose due to liver and muscle glycogenolysis and mobilize free fatty acids. In addition, a cold-induced increase in glucocorticoids not only maintains levels of blood metabolites but also enhances the catecholamine mobilization of FFA and lipolysis in adipose tissue (Ganong, 1975).

It appears that the hypothalamus is important in integrating peripheral and central thermal information in controlling the levels of these metabolic endocrines. Changes in PO/AH temperature can produce the same changes in metabolic endocrines as the changes in peripheral temperature. PO/AH cooling will increase thyroid activity (plasma protein-bound iodine) (Andersson et al., 1962, 1963a; Evans and Ingram, 1974); it will increase the urinary excretion of catecholamines (Andersson et al., 1963b); and it will increase the levels of plasma glucocorticoids as well as plasma glucose (Chowers et al., 1964; Gale et al., 1970). Conversely, PO/AH warming will block any increase in these endocrine levels during peripheral cooling. Even in a neutral environment, PO/AH warming can decrease the circulating levels of epinephrine, norepinephrine, plasma glucose, and glucocorticoids (Proppe and Gale, 1970). Generally, temperature changes in other parts of the hypothalamus do not produce these responses. Like PO/AH warming, electrical stimulation of the preoptic area inhibits thyroid activity; however, electrical stimulation in the medial parts of the hypothalamus, from the median eminence to the supraoptic region, tends to increase TSH and thyroid activity (Martin and Reichlin, 1970; Thomas and Anand, 1970). Again, here is a possible example of preoptic control over heat production by inhibiting the output of a more caudal neural area.

In summary, cutaneous thermoreceptive information is integrated with PO/AH thermoreceptive information. This integration probably occurs within the PO/AH and possibly in more caudal neural structures. Much of the output from the PO/AH seems to suppress other neural areas which control metabolic heat production. This would include the dorsomedial posterior hypothalamus and the reticular core of the brain stem which influence the spinal control of shivering. For nonshivering thermogenesis, PO/AH fibers undoubtedly influence structures which send sympathetic efferents to the adrenal medulla and brown adipose tissue. Other PO/AH neurons probably innervate the median eminence to control endocrine release in the anterior pituitary. Thus, a variety of mechanisms exist to increase heat production. In large part, the locations and set-point temperatures of the various thermoreceptors influence which mechanisms will be employed in a given circumstance.

Finally, in metabolic heat production, the integration of hypothalamic and cutaneous temperature is not simply additive, as suggested in Figure 1. Rather, this integration is generally multiplicative—i.e., cooling the skin not only increases heat production but also increases the hypothalamic thermosensitivity (α) for this response. This multiplicative effect of skin temperature is illustrated in Figure 2 and has been shown in several thermoregulatory studies. In these studies, most of the metabolic heat produced has been attributed to shivering thermogenesis. This same type of multiplicative effect has also been demonstrated strictly for nonshivering thermogenesis in 2- to 7-day-old, cold-adapted guinea pigs (Brück and Schwennicke, 1971). In these animals also, skin cooling increases both the level of NST plus the PO/AH thermosensitivity or α for this response. Conversely, skin warming not only decreases the level of NST but also decreases the PO/AH thermosensitivity for this response. Thus, regardless of the mechanism for heat production, the central neuronal integration of hypothalamic and peripheral thermal information appears to be the same.

C. Skin Blood Flow

Skin blood flow (SBF) is the body's principle physiological means of regulating core temperature, especially in an environment that is near thermoneutrality. Along with thermoregulatory behavior, cutaneous vasomotor tone is the first response to change when either ambient or core temperature deviates from neutral. When either hypothalamic or skin temperature begins to decrease, cutaneous vasoconstriction will occur before shivering or nonshivering thermogenesis. With an increase in either hypothalamic or skin temperature, cutaneous vasodilation will occur before sweating and panting. As long as body temperature is greater than the temperature of the surrounding environment, an increase in SBF means that more heat is transported from the body core to the skin surface, where there is a net loss of heat by radiation, conduction, and convection. By warming the skin, increased SBF also increases the vaporization of sweat on the skin surface and thus contributes to evaporative heat loss. In contrast, by decreasing SBF down to minimal or nutrient levels, much of the metabolic heat is retained within the core and is not lost to the environment.

1. Receptor and Effector Areas

The hypothalamus, septum, and reticular formation constitute an extensive neural complex which regulates blood pressure and other cardiovascular parameters (Gebber and Snyder, 1970; Hilton and Spyer, 1971; Calaresu and Mogenson, 1972). Electrical stimulation in various hypothalamic and brain-stem locations can frequently increase blood flow in skeletal muscles while decreasing blood flow in the skin and most abdominal organs (Abrahams et al., 1960; Clarke and Rushmer, 1967; Forsyth, 1970; Coote et al., 1973). Accompanying the hypothalamically induced blood flow changes are changes in the electrical activity of the sympathetic fibers innervating these various organs; this includes an increase in the activity of the postganglionic cutaneous vasoconstrictor fibers (Folkow et al., 1959; Ninomiya et al., 1970; Horeyseck et al., 1976). Most of the complexities of this cardiovascular, hypothalamic-brain-stem network are beyond the scope of this chapter. Some stimulation studies, however, pertain directly to the regulation of body temperature.

In goats, for example, preoptic electrical stimulation can elicit cutaneous vasodilation that will persist even in a cold environment and after a drop of several degrees in rectal temperature (Andersson and Persson, 1957). On the other hand, stimulation in the medial septum can cause cutaneous vasoconstriction and a reduction in skin thermal conductance even in a 32°C environment (Andersson, 1957). Thus, it is not surprising that the hypothalamus and septum have a great deal of control over thermoregulatory adjustments in skin blood flow. Neuronal networks in the hypothalamus and brain stem can integrate central and peripheral thermoreceptive signals. As a result of this integration, appropriate cutaneous vasomotor changes are initiated to control heat loss and maintain the constant core temperature.

a. PO/AH

PO/AH neurons sense changes in local hypothalamic temperature and can elicit appropriate changes in skin blood flow. Several thermode studies have shown that PO/AH cooling produces cutaneous vasoconstriction with an accompanying drop in skin temperature. Localized PO/AH warming produces cutaneous vasodilation and a rise in skin temperature (Hemingway et al., 1940; Andersson and Persson, 1957; Freeman and Davis, 1959; Hammel et al., 1960; Hellstrøm and Hammel, 1967; Sundsten, 1967; Jacobson and Squires, 1970). During these changes in skin blood flow, arterial pressure usually remains fairly constant. This is partly due to antagonistic blood flow changes in other structures, such as the visceral organs. For example, Schönung et al. (1971b) have shown in anesthetized dogs that PO/AH warming increases both skin blood flow and paw-skin temperature, but it also decreases intestinal blood flow; blood pressure decreases only slightly. When the PO/AH is cooled, SBF and paw-skin temperature decrease, but intestinal blood flow increases and blood pressure remains constant. Thus, body temperature and arterial pressure are regulated together.

b. Lateral and posterior hypothalamus

In the control of skin blood flow, as in the control of heat-production responses, the role of the PO/AH must be considered as part of an entire receptor-effector complex that includes other areas of the hypothalamus, reticular formation, and spinal cord. Extrahypothalamic structures are capable of sensing local temperatures and initiating cardiovascular responses. Apparently, the temperature-sensitive neurons of the PO/AH communicate with those hypothalamic and brain-stem areas that regulate blood pressure and the regional distribution of blood flow. PO/AH lesions in rats impair thermoregulation in warm and cold environments, while bilateral cuts between the PO/AH and the lateral hypothalamus impair thermoregulation only in the cold (Lipton et al., 1974). This might suggest that a pathway controlling cold-induced vasoconstriction descends from the PO/AH in the lateral hypothalamus, possibly in the median forebrain bundle. Supporting this is a lesion study in cats, which suggests that lateral and dorsolateral lesions of the posterior hypothalamus selectively impair cutaneous vasoconstriction, while dorsomedial lesions of the posterior hypothalamus abolish shivering (Stuart et al., 1962).

c. Lower brain stem and spinal cord

Extrahypothalamic thermoreceptors also affect skin blood flow. In rats, for example, localized warming at a variety of sites throughout the medial hypothalamus and midbrain as well as the lateral and dorsal pons elicits vasodilation in the tail (Roberts and Mooney, 1974). Ingram and Legge (1972) have shown in pigs that heating the scrotum markedly increases tail blood flow. In addition, this response can be inhibited by cooling either the hypothalamus or the spinal cord. Chai and Lin (1972, 1973) have shown that, in both intact and midcollicular decerebrate animals, cooling either the medulla oblongata or the spinal cord produces moderate peripheral vasoconstriction along with increases in heart rate and blood pressure. These vasomotor responses, however, are not as strong as those produced by altering skin or hypothalamic temperature. In these same animals, medullary or spinal warming produces moderate peripheral vasodilation along with decreases in heart rate and blood pressure.

Other investigators have shown that arterial blood pressure can be maintained fairly well even when skin blood flow is altered by changes in spinal temperature. In the anesthetized dog, spinal warming will increase SBF and decrease intestinal blood flow, with only moderate or no increases in blood pressure. Spinal cooling will decrease SBF, increase intestinal blood flow, and have no effect on blood pressure (Kullmann et al., 1970). Thus, skin blood flow can be influenced by hypothalamic, spinal, and skin temperatures. In each case, arterial pressure can be maintained, primarily by antagonistic changes in the activity of sympathetic vasoconstrictor inputs to the cutaneous and visceral circulatory beds. As an example, Walther et al. (1970) have shown that spinal warming decreases the electrical activity of the sympathetic nerve fibers innervating the ear but increases the activity of sympathetic fibers innervating the intestine. Spinal cooling has the opposite effect on the activity of these two types of sympathetic fibers. These antagonistic changes in cutaneous and intestinal sympathetic activity can be evoked not only by spinal and hypothalamic temperatures but also by warming and cooling the skin (Simon, 1971). Thus, the central integration of both peripheral and deep-body thermoreceptors can produce appropriate thermoregulatory changes in skin blood flow and can maintain a constant arterial pressure by altering blood flow in other vascular beds.

2. Hypothalamic Role in the Integration of Central and Peripheral Temperature

To what extent does this integration of central and peripheral thermal signals occur within the hypothalamus? Iriki and Kozawa (1976) suggest that, while the hypothalamus is necessary for normal cardiovascular responses to hypoxia, it may not be necessary for appropriate responses to thermal stimulation in the spinal cord. Even in rabbits with decerebrate mid- or infracollicular transections, spinal warming decreases ear sympathetic activity but increases splanchnic and cardiac sympathetic activity. Conversely, spinal cooling has the opposite effect on the electrical activity of the sympathetic fibers in these decerebrate rabbits. Chai and Lin (1973) have reported that, in rabbits, the cutaneous vasomotor responses to warming and cooling the skin are reduced after midcollicular decerebration, while the moderate responses to changes in medullary and spinal temperature are reportedly not affected by decerebration. This is based simply on the changes in ear temperature, however. The thermal conductance of the ear more accurately reflects the vasomotor changes that enable heat to be transported from the core to the skin, and lost from the skin surface to the environment. Kluger et al. (1973) have shown that a more precise indication of SBF in the rabbit ear may be determined by the ratio of the difference between ear and ambient temperature relative to the difference between core and ear temperature, that is, $T_{ear} - T_{air}/T_{core} - T_{ear}$. By calculating this ratio for the rabbits in the study of Chai and Lin (1973), one finds that, in intact rabbits, the effect of spinal or medullary temperature is very slight when compared with the vascular effect produced by warming and cooling the skin. Decerebration virtually eliminates and may even reverse any appropriate change in SBF during changes in skin temperature. In addition, decerebration can reduce by about one-half most of the minor vasomotor changes produced by changes in spinal and medullary temperature. These calculations from Chai and Lin's (1973) study suggest that, in the control of SBF, spinal and medullary temperatures are not nearly as effective as skin and probably hypothalamic temperatures. These calculations also reveal that a high decerebration (caudal to the hypothalamus) drastically impairs appropriate SBF responses to both peripheral and deep-body temperatures. It should be noted, however, that some of the values used for these calculations may not be completely accurate. For example, in a 35°C environment, the reported ear temperatures were 0.4°C greater than core temperatures—an impossible situation.

The important study by Chambers et al. (1974) shows that a neural organization for the control of SBF is similar to that described for shivering. Removal of the hypothalamus by a high decerebratate transection can abolish cutaneous vasoconstriction, shivering, and piloerection in response to spinal cooling. Lowering the decerebration to the lower pons and medulla restores these responses during spinal cooling. Therefore, there appears to be an area in the midbrain and upper pons that tonically inhibits thermoregulatory responses, at least when the hypothalamus is removed. Undoubtedly hypothalamic and reticular networks integrate afferent information and tonically facilitate or inhibit more caudal structures. The extent to which this decerebration effect is due to either or both of these functions is not known. This study by Chambers and co-workers also suggests that an area in the lower pons and medulla actually facilitates the skin and spinal thermal control of SBF, shivering, and piloerection. In a spinal animal without these lower brain-stem areas, the vasoconstriction and shivering responses to spinal cooling are reduced and the piloerection response is abolished. Although not complete, this study by Chambers and co-workers is quite important to thermal physiology, because it clearly suggests the course of future studies. An entire series of animals should receive discrete lesions and transections from the hypothalamus to the spinal cord. Hypothalamic, brain-stem, spinal cord, and skin temperatures should be individually increased and decreased, while careful measurements of SBF, metabolism, shivering, and evaporative heat loss are made in each preparation. In this way, the neural areas necessary for the central and peripheral control of each thermoregulatory response can be determined in the same preparation. One of the greatest difficulties in interpreting the large number of previous studies is the experimental variations between studies. These variations include differences in the species, the location of thermodes and lesions, the means of measuring physiological responses, and whether a response is to warming or to cooling a particular site. It is important that future studies be complete for each stimulation and lesion preparation; that a given site be tested for both warming and cooling; and that a measurement be made of as many physiological responses as possible.

As noted earlier, in the unlesioned intact animal, both hypothalamic and extrahypothalamic thermosensitive structures affect the regulation of SBF and arterial pressure. In this complex cardiovascular response, the relative importance of each neural structure remains unclear. In conscious dogs, changes in spinal cord or hypothalamic temperature produce appropriate changes in cutaneous temperature (i.e., SBF) (Jessen and Mayer, 1971).

Skin blood flow responses to hypothalamic thermal stimulation, however, may be somewhat predominant over spinal thermal stimulation. For example, compared with spinal cooling, hypothalamic cooling causes the greatest skin-temperature decreases. In addition, a 1°C increase in hypothalamic temperature can increase skin temperature despite a simultaneous spinal cooling of 5°C (Jessen et al., 1968). On the other hand, Schönung et al. (1971) have shown that, compared with hypothalamic thermal stimulation, spinal thermal stimulation produced equivalent or even greater changes in recorded skin blood flow. Again, these SBF changes were opposite to changes in intestinal blood flow, and arterial pressure remained fairly constant. Without question, PO/AH thermosensitive neurons have a substantial influence on skin blood flow. Based on these previous studies, it is not certain whether hypothalamic thermoreceptors are more important than extrahypothalamic thermoreceptors in controlling skin blood flow in the intact animal.

3. Differences in the Effects of Skin and Hypothalamic Temperature

Skin temperature also determines the amount of skin blood flow and may, in fact, have a direct effect on the reactivity of innervated cutaneous blood vessels. Local cooling of innervated cutaneous veins, for example, increases venoconstrictor tone and favors a decrease in blood flow through superficial veins. This might not only affect SBF but also favor a venous return through the venae comitantes and thereby increase the heat-conserving countercurrent heat exchange with the associated arteries (Webb-Peploe and Shepherd, 1968; Abdel-Sayed et al., 1970; Vanhoutte and Shepherd, 1970). An increased constriction in arterioles and precapillary sphincters also would be necessary to substantially reduce cutaneous capillary flow. There is the possibility that skin thermoreceptors alter SBF by spinal reflexes such that peripheral temperature affects vasomotor tone by way of this spinal reflex. Randall et al. (1966) have shown that, in humans, skin warming will decrease and skin cooling will increase the vasoconstrictor tone in the skin innervated below the level of a spinal transection.

The neural control of the cutaneous circulation is further complicated by the differences between capillary blood flow and blood flow that is shunted through the arteriovenous anastomoses (AVA). In the dog leg, for example, β-adrenergic drugs affect only capillary flow, while α-adrenergic drugs affect both capillary and AVA circulation (Spence et al., 1972). The use of injected radiolabeled microspheres has recently provided an important index of relative capillary and AVA blood flow through a variety of body tissues. Using this technique, Hales and Iriki (1975) and Hales et al. (1975) have shown differences between the effects of central and peripheral thermoreceptors on blood flow in the skin and in other tissues. Central warming of the hypothalamus or spinal cord causes an increased SBF primarily by increasing skin AVA blood flow but not by increasing skin capillary blood flow. In addition, spinal warming increases blood flow through the nasal mucosa and tongue, which are important tissues in heat loss during panting. Spinal warming also increases blood flow to the respiratory muscles used in panting. At the same time, cardiac output remains fairly constant, primarily due to a decreased blood flow to the abdominal organs. Hypothalamic and spinal cooling tend to decrease skin AVA blood flow but, again, have little effect on skin capillary blood flow. On the other hand, warming only the skin increases skin capillary blood flow but does not appreciably affect skin AVA blood flow. Apparently, this cutaneous effect is more involved than just the direct effect of temperature on superficial blood vessels, since the blood flow to a variety of deep tissues is also altered. Skin warming, for example, will increase blood flow to nasal mucosa and respiratory muscles but will decrease blood flow to abdominal organs as well as to the kidneys and thyroid gland (Hales and Iriki, 1975).

Similar findings were noted in a previous study by Zanick and Delaney (1973). Radiolabeled microspheres were injected into the femoral artery of dogs subjected to either whole-body or local skin (hind-limb) warming and cooling. The microspheres that accumulated in the lungs determined the shunted hind-limb AVA blood flow. It was found that heating the core (and trunk skin) greatly increased AVA flow and had little effect on paw capillary flow. Cooling the core greatly reduced both AVA and paw capillary flow. Peripheral hind-limb warming had little or no effect on both AVA flow and paw capillary flow. Peripheral cooling, however, greatly reduced paw capillary flow but had little or no effect on hind-limb AVA flow. Of course, the recorded change in flow through a particular blood vessel will depend on the initial flow through that vessel. This may account for some of the differences between this study and the studies of Hales and co-workers (1975).

To summarize, it appears that both central and skin thermoreceptors initiate a complex series of cardiovascular responses that affect the peripheral circulation. The type of response depends on whether it is initiated by changes in central or peripheral temperatures. Central thermoreceptors in the hypothalamus and spinal cord primarily affect cutaneous AVA blood flow but may also somewhat inhibit capillary flow. It is probable that the peripheral thermoreceptors act through neural control over precapillary sphincters and arterioles. Peripheral thermoreceptors primarily affect cutaneous capillary blood flow but may also affect AVA flow. Bligh (1973), in fact, has suggested how these two controls might function together. In a fairly thermoneutral environment, the main determinant of SBF would be skin temperature affecting capillary blood flow. An increase in capillary flow would allow more blood to lose heat and cool down as it passes through the capillary beds. With the AVAs closed, countercurrent heat exchange between the deep veins and the central artery would rewarm the venous blood as it returns to the core. During heat stress, however, an increase in core temperature would increase AVA blood flow. This would not affect the already increased capillary blood flow, but it could maintain an elevated skin temperature and thus reduce the amount of cooling of capillary and venous blood. The countercurrent heat exchange would be reduced because of the increased rate of blood flow through arteries and veins. Thus, the temperature of arterial and venous blood would remain elevated. Bligh suggests that during skin cooling, arteriolar constriction would greatly reduce capillary blood flow and heat loss. If core temperature remains warm, however, AVA flow might remain high and insure that at least some heat is maintained in the deeper parts of the extremities to prevent frostbite. In fact, Hales (1973) has suggested that such an increase in AVA blood flow may account for the occurrence of cold-induced vasodilation.

4. Effect of Skin Temperature on Hypothalamic Control

Skin blood flow, then, is influenced by many factors, among them the temperatures of the skin and the hypothalamus. In the hypothalamic control of skin blood flow, the question becomes: what is the effect of skin temperature on the hypothalamic control of this thermoregulatory response? This is difficult to answer, since skin temperature can directly or at least reflexively affect skin blood flow. Cutaneous thermal afferents do reach hypothalamic and brainstem areas which control SBF. It is likely, therefore, that these cutaneous afferents can modify the hypothalamic control of SBF. Presently, the relative contribution of skin temperature to this central regulation of SBF is a matter of controversy. In humans, when central or core temperature rises above 37°C, there is a proportional increase in SBF, for example in the fingers and forearm. Some studies indicate that ambient or skin temperature substantially affects this central response by altering the hypothalamic T_{set} to elicit an elevated SBF. Apparently, a reduction in mean skin temperature to about 30°C increases the central T_{set} by about 0.5°C, such that SBF does not begin to increase until core temperature reaches 37.5°C. Increasing the mean skin temperature to about 35°C effectively decreases the central T_{set} and increases the level of SBF even when central temperature is 37°C (Wenger et al., 1975a,b). Other investigators indicate that the central temperature is the primary determinant of SBF and is, in fact, 20 times more effective in controlling SBF than skin temperature (Wyss et al., 1974). It should be noted that, in this second study, skin temperatures were increased from 34 to 40°C. Therefore, it is possible that cutaneous warm receptors, sensing increases in skin temperature, have little effect on the hypothalamic control of SBF. It is also possible that at skin temperatures below 33-34°C, cutaneous cold receptors become active and inhibit central neurons responsible for increasing SBF. This would effectively increase the hypothalamic T_{set} for this response. A concept quite similar to this has been suggested by Benzinger et al.(1963) for the integration of central and peripheral thermal signals in the control of sweating.

In addition to affecting T_{set}, skin temperatures may also alter the local hypothalamic thermosensitivity for the SBF response. Figure 3 shows the effect of PO/AH temperature on the blood flow in a rabbit ear maintained in a neutral environment. As mentioned previously, the ratio $T_{ear} - T_{air}/T_{core} - T_{ear}$ can be used as a relative indication of ear blood flow. When the average skin temperature of the rabbit's trunk is 37.5°C, the PO/AH temperature (T_{po}) must be warmed above 39.5°C before ear blood flow begins to increase. Blood flow increases linearly with an increase in T_{po} above this T_{set}. The hypothalamic thermosensitivity (α) can be determined by the slope or regression coefficient of this function. If the trunk skin is warmed to 38.5°C, not only does the T_{set} decrease but also the α decreases. Conversely, trunk-skin cooling to 35.7°C increases both the T_{set} and the α of this response. Thus, in the integration of central and peripheral temperatures for the control of SBF, skin temperature affects not only the hypothalamic T_{set} but also the effective hypothalamic thermosensitivity for this response (Boulant and Gonzalez, 1974).

D. Evaporative Heat Loss

When environmental temperature approaches or exceeds body temperature, evaporative heat loss becomes the body's only effective heat-loss response to keep core temperature from rising. Evaporative heat loss (EHL) occurs when a liquid vaporizes on the surface of the skin or respiratory tract. The heat necessary to vaporize the liquid comes from this surface, and thus vaporization removes heat from the body. A variety of physiological and behavioral EHL responses exist throughout the animal kingdom. Panting and sweating are two common physiological responses; the hypothalamic control of these two responses will be considered separately in the remainder of this section. Skin wetting or saliva spreading is an example of behavioral EHL responses and is discussed in the section on behavioral thermoregulation.

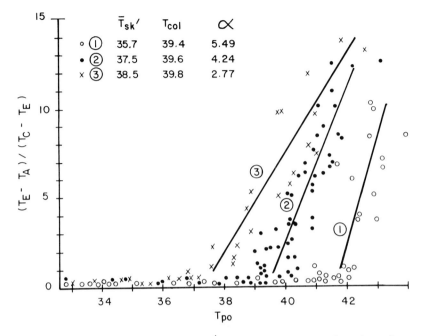

Figure 3 Relative skin blood flow ratio, SBF$'$, in the rabbit ear as a function of preoptic temperature, T_{po}. T_E, ear temperature; T_A, air temperature; T_C and T_{col}, core (colonic) temperature; $\overline{T}_{sk}{}'$, average trunk skin temperature. The ear was constantly maintained in a neutral environment. Each numbered line and symbol indicates a different clamped $\overline{T}_{sk}{}'$ and T_{col}. Thermosensitivity (α) refers to the slope of each line, in SBF$'$ per degree centigrade.

Regardless of the EHL response, it is obvious that the humidity of the air is an important limiting factor. When the ambient vapor pressure is high, less water is vaporized from the body surface. Thus, in a high humidity, central temperature and sweat rate may continue to increase, but EHL remains low. The temperature of the evaporative surface is another determinant of EHL. As mentioned previously, the hypothalamic control of blood flow to the skin and nasal mucosa is quite obviously a necessary component for effective vaporization during sweating and panting. In fact, Gonzalez et al. (1971) have shown that even in a nonsweating animal like the rabbit, 40% of the EHL comes from the moisture on the skin surface, which is closely associated with the thermal conductance of the skin. The other 60% of the EHL comes from the respiratory tract.

1. Panting

Panting is a means to promote evaporative heat loss from the respiratory tract. It is actually a complex of responses and has many variations in different animals. Polypnea, or an increased respiratory rate, is usually the most evident component of panting. In the rabbit, for example, an increase in hypothalamic temperature can increase breathing frequency from a normal 60 breaths per minute to as much as 500 breaths per minute (Boulant and Gonzalez, 1977). Panting in rabbits is usually considered to start at frequencies of 100-200 breaths per minute; above this level, there is a strong correlation between the amount of respiratory EHL and the level of breathing frequency (Gonzalez et al., 1971). In addition to the increased frequency, in many animals respiration becomes shallow and the mouth is often opened with the tongue protruding. In sheep, central or peripheral warming increases blood flow in the tongue and nasal mucosa (Hales and Iriki, 1975). In the dog, hypothalamic or peripheral warming increases the rate of salivation (Hammel and Sharp, 1971). Thus, an increased respiratory frequency increases air movement over the warm, moist surfaces of the nasal mucosa and tongue, and this promotes EHL.

a. PO/AH

In 1933, Hammouda showed that warming the optic thalami and other neural structures near the third ventricle elicited panting in urethane-anesthetized dogs. In an early thermode study in cats, Magoun et al. (1938) showed that localized PO/AH warming caused a marked polypnea accompanied by tongue and nostril movements and, occasionally, sweating from the foot pads. In that study also, the PO/AH appeared to be the most sensitive area in the brain for eliciting this response. Later, Lim and Grodins (1955) demonstrated that panting could be induced in anesthetized dogs by hypothalamic (i.e., carotid blood) warming alone but not by skin warming alone. They further showed that panting, induced by whole-body heating, could be abolished by cooling either the hypothalamus or the skin.

It is now well established that local warming of the rostral hypothalamus, especially the PO/AH, produces panting and EHL in a variety of species, including dogs (Fusco et al., 1961; Hellström and Hammel, 1967; Jessen and Mayer, 1971), cats (Chai et al., 1965; Jacobson and Squires, 1970), goats (Andersson et al., 1956; Jessen, 1976), oxen (Findlay and Ingram, 1961; Ingram and Whittow, 1962), and rabbits (Guieu and Hardy, 1970a; Boulant and Gonzalez, 1977). In some studies, a warm environment or warming selected areas of skin was necessary in order for PO/AH warming to produce panting (Fusco et al., 1961; Jacobson and Squires, 1970; Ingram and Legge, 1972). In other studies, however, sufficient PO/AH warming could elicit panting and EHL at neutral or even cool skin and ambient temperatures (Hellström and Hammel, 1967; Jessen and Mayer, 1971; Boulant and Gonzalez, 1977). Thus, skin thermoreceptors have some control over panting, and, in certain animals, peripheral cold receptors may eliminate the response to PO/AH warming. Conversely, local PO/AH cooling can also reduce or eliminate panting in response to peripheral warming (Ingram and Legge, 1972; Adair and Rawson, 1974).

In 1952, Forster and Ferguson studied the panting response in cats during peripheral warming. Five out of seven cats were termed "central panters." Despite warm ambient temperatures, these cats did not begin to pant until their hypothalamic temperatures began to rise. The other two cats were termed "reflex panters" because they began to pant when ambient temperature increased, with no change in hypothalamic temperature. Apparently, then, cutaneous warm receptors are capable of evoking panting; more often, however, central thermoreceptors must be stimulated in order to produce the response. In addition, since only hypothalamic temperature was measured in the Forster and Ferguson (1952) study, it is entirely possible that extrahypothalamic, deep-body thermoreceptors contributed to the response in the reflex panters.

It has been shown in cats and other animals possessing a carotid rete, that panting in a warm environment will preferentially cool the hypothalamus despite a rise in core temperature (Baker and Hayward, 1967, 1968; Baker, 1972). The carotid rete is a vascular network connecting the external carotid artery with the circle of Willis. The rete lies near the nasal cavity and respiratory passages. Thus, during respiratory EHL, blood in the rete is cooled before it reaches the circle of Willis and the brain. In these animals, EHL in the nasal cavity cools venous blood which in turn cools arterial blood passing through the carotid rete. This cools the rostral brainstem and possibly serves to protect this neural area from heat damage. During exercise or warm exposure, panting in such animals would allow hypothalamic temperature to remain below core temperature. If this is so, then other deep-body thermoreceptors (e.g., in the abdomen and spinal cord) must function to insure that panting continues; otherwise, panting would stop when hypothalamic temperature decreases. Recently, Kluger and D'Alecy (1975) showed that panting can also preferentially decrease hypothalamic temperature in the rabbit, an animal which lacks a carotid rete. These investigators suggest that various degrees of hypothalamic cooling may occur during panting in a wide range of species. The verification of this suggestion and its consequences on extrahypothalamic thermosensitivity await further study.

Certainly the importance of hypothalamic thermosensitivity in panting has been demonstrated in a variety of animals, animals both with and without a carotid rete. Using arrays of thermodes, Jessen (1976) has tested the thermosensitivity throughout the hypothalamus in the goat. He found that the region producing the greatest EHL response lies near the midline and contains the supraoptic and paraventricular nuclei. In the cat, Magoun et al. (1938) found the PO/AH to be the most thermosensitive region for panting. A region of lesser thermosensitivity apparently extends caudally from the PO/AH, occupying the dorsal and posterior parts of the hypothalamus and the periventricular gray matter of the midbrain. Local heating in these caudal areas still produces brief panting, but the frequency is less and the latency is longer. Magoun and co-workers could not produce panting by localized heating in other parts of the brain stem or telencephalon.

Early lesion studies suggested that the PO/AH is necessary for panting, but this view has since been modified. In 1933, Hammouda showed that removal of the cortex and corpora striata did not eliminate panting in dogs, but lesions in the optic thalami did eliminate panting, even during central warming. In that same year Bazett et al. (1933) indicated that cats having transections near the PO/AH, below the anterior hypothalamus, or in the midbrain and pons, did not show panting polypnea despite elevated core temperatures above 43°C. On the other hand, Squires and Jacobson (1968) have indicated that, in cats, PO/AH lesions do not eliminate panting but do cause a significant hyperthermia when the animals are exposed to warm temperatures. It is likely that PO/AH destruction leaves panting under the control of the less thermosensitive caudal parts of the hypothalamus and midbrain, described by Magoun et al. (1938). In this case, core and hypothalamic temperature would have to rise to a higher level before panting would be initiated. In addition, the panting might have a lower breathing frequency and would, therefore, be less effective in preventing a further rise in core temperature. A PO/AH lesion also removes some of the hypothalamic integration of peripheral thermoreceptor afferents, making it less likely that cutaneous warm receptors will initiate or contribute to panting. Thus, with PO/AH lesions, central and skin temperature must rise to higher levels before panting is initiated. Once panting is initiated, however, the EHL may be sufficient to maintain the core temperature constant at some hyperthermic level.

Other studies substantiate that, during warm exposure, animals having hypothalamic lesions at or below the PO/AH, allow their core temperatures to rise to higher levels before panting occurs. Keller and McClaskey (1964) have studied the effects of transections (at various levels) on the ability of dogs to thermoregulate in a 38°C environment. During a 6-hr heat exposure, dogs having transections rostral to the PO/AH show only a slight 0.5°C rise in core temperature before panting is initiated; thereafter, temperatures are maintained constant. If the transection is below the anterior commissure and preoptic area, effective panting does not begin until core temperature has risen 1 - 4°C. At this point, panting and cutaneous vasodilation prevent any further rise in core temperature. Similar effects occur in dogs having lesions in the preoptic area (i.e., "prechiasmal ablation") rostral to the anterior hypothalamus (Keller, 1963). Such lesions impair heat loss in a warm environment much more than they impair heat production in a cold environment.

b. Caudal hypothalamus and midbrain

Keller (1963) has also shown that discrete lesions to the ventral posterior hypothalamus (rostral to the mammalothalamic and habenular tracts) impair heat-production responses much more than heat-loss responses. In fact, heat production in the cold may be virtually eliminated, but evaporative heat loss in the warm environment is only moderately impaired by such lesions. Using Keller's dogs, Hardy and Hammel (1963) showed that removal of the posterior hypothalamus raises the threshold temperature for panting to about 1.5°C above normal and reduces the thermosensitivity (α) for panting by about 80%.

In the Keller and McClaskey study (1964), lesions which remove the rostral third of the midbrain (i.e., low-midbrain preparation) eliminated any effective panting and abolished thermoregulation in hot, cold, and even neutral environments. Lesions directly between the hypothalamus and midbrain (high-midbrain preparation) have similar effects and eliminate panting in undisturbed, warm-exposed dogs. However, as long as core temperature is at least 38°C, panting can still be evoked if these high-midbrain dogs are disturbed or display body movements. An earlier study by Keller (1933) also indicated that complete removal of the hypothalamus produces panting in response to nonthermal somatosensory stimulations and body movements, even at low core temperatures. Evidently, thermosensitive structures in the hypothalamus, especially in the PO/AH, contribute to panting when core temperature is elevated. When the preoptic region is removed, panting is controlled by higher-threshold, less thermosensitive structures, possibly in the more caudal hypothalamus and rostral midbrain. Complete removal of the hypothalamus drastically impairs the panting response. Not only does it eliminate hypothalamic thermosensitive inputs, but it apparently eliminates much of the contribution of peripheral warm receptors, which may or may not play a significant role in panting. Evidently, however, nonthermal somatosensory or proprioceptive inputs to the rostral midbrain can still elicit some panting in these lesioned animals.

Keller (1963) has shown in dogs that heat-loss responses (but not heat-production responses) can be impaired by removal of the dorsal half of the lower midbrain. In addition, partial heat production is retained (but heat loss is eliminated) by pontine transections, which leave only the pyramidal tracts intact. It appears, therefore, that a pathway controlling panting descends from the PO/AH, circumvents the posterior hypothalamus (either laterally or dorsally, as speculated by Hardy, 1973), and continues through the dorsal midbrain and pons to the respiratory areas of the pons and medulla.

c. Lower brain stem and spinal cord

Thermosensitive structures in the pons, medulla, and spinal cord undoubtedly have some control over panting. Unfortunately, there is conflicting evidence in different studies, making it difficult to assess the role of these lower areas. As just mentioned, Keller and McClaskey (1964) indicated that the caudal hypothalamus and rostral midbrain are structures necessary for effective panting in dogs. Bard and Macht (1958) reported, however, that decerebrate cats (having variety of transections ranging from just behind the hypothalamus to the lower midbrain and pons) display panting responses when rectal temperatures reach 41.2-44.0°C. For the lower transections, however, panting occurred only if the mouth was forced open. Although these investigators do not report breathing frequencies, they do indicate that the frequencies in all of the transected animals are not as high as those of normal, intact cats. Chai and Lin (1973) have reported the effects of warming the skin, the spinal cord, and the medulla on the respiratory frequency of normal and decerebrate rabbits. In intact rabbits, ambient warming to 35°C resulted in a panting polypnea of 307 breaths per minute after 1 hr. During this time, rectal temperature rose 0.9°C; therefore, it is impossible to distinguish separate effects of central and peripheral warm-receptors. In addition, before warm exposure, the average respiratory frequency was 162 breaths per minute, or about 100 breaths per minute above normal. This makes a comparison with warm exposure difficult. Despite these difficulties, midcollicular decerebration did eliminate the panting response to peripheral and central warming. After 1 hr in the 35°C environment, respiratory frequency increased only from 61 to 75 breaths per minute in decerebrate rabbits. It should be noted that Gonzalez et al. (1974) only consider a polypnea of 200 breaths per minute to be indicative of panting in rabbits.

In the Chai and Lin (1973) study, warming the medulla about 4°C slightly increased respiration in both intact and midcollicular decerebrate rabbits, i.e., to 126 and 111 breaths per minute, respectively. Again, it is debatable whether or not this polypnea should be considered panting. Equivalent warming of the spinal cord caused even less of an increased respiration, i.e., 117 breaths per minute in intact rabbits and 99 breaths per minute in decerebrate rabbits. In another study of cats, localized warming of the medullary reticular formation to 41-42°C also reportedly increased respiratory frequency, although exact values were not given (Holmes et al., 1960). Guieu and Hardy (1970a) have shown that spinal warming enhances panting in anesthetized rabbits, but only after panting has already been initiated by PO/AH warming. In addition, intense spinal cooling has no effect on panting frequency during PO/AH warming. Finally, Kosaka et al. (1969) have shown in rabbits that spinal warming to about 41.8°C slightly increases respiration to about 85 breaths per minute. A midcollicular decerebration abolishes this response, suggesting that hypothalamic and rostral midbrain structures may be important in even these minor increases in respiratory frequency. Based on this evidence alone, it would seem that without the hypothalamus and rostral midbrain, peripheral and extrahypothalamic warming produces, at best, only marginal increases in respiration. In addition, spinal and medullary thermoreceptors seem to have only a slight effect on breathing frequency, even in animals having an intact hypothalamus. Moreover, in the studies mentioned above, it is doubtful whether the observed changes in breathing frequency have a significant effect on respiratory EHL.

There is also the distinct possibility that localized warming of the lower brain stem may directly affect neurons and pathways of respiratory "centers" concerned with

Hypothalamic Control of Thermoregulation

breathing frequency. Chai and Lin (1972), for example, have shown that warming the medulla or spinal cord of monkeys increases respiration from about 40 to about 75 breaths per minute. Conversely, cooling these two areas decreases respiration by about nine breaths per minute. Thus, even in an animal that is not known to pant, it is possible to alter respiratory frequency by changing local brain-stem and spinal temperatures. There is also evidence in cats and monkeys that localized warming of the medulla actually decreases breathing frequency, while localized cooling increases breathing frequency (Chai et al., 1965; Chai and Wang, 1970). Such thermally inappropriate responses must lend credence to the possibility that these artificial changes in temperature are affecting some system other than the neural control of thermal panting.

Despite this dubious evidence for the extrahypothalamic control of panting, Jessen and Mayer (1971) and Clough and Jessen (1974) have presented the convincing evidence that, at least in the intact dog, spinal warming effectively increases respiratory evaporative heat loss. This occurs not only in a 30°C environment but at air temperatures of 18 and 24°C as well. During spinal warming, the magnitude of the EHL increase is comparable to that produced by a similar change in hypothalamic temperature. In addition, Rawson and Quick (1970) have found the existence of deep-body thermoreceptors in the abdomen of sheep. Localized warming of these receptors will increase respiratory EHL and preferentially decrease hypothalamic temperature due to heat exchange in the carotid rete. Thus, there is some evidence that, in addition to the hypothalamus, thermoreceptors in the skin and deep-body also contribute to the panting response.

Because of the conflicting evidence concerning the role of the spinal cord in panting, it would seem that further study in different species is justified. Moreover the "permissive" role of the PO/AH in allowing panting in response to extrahypothalamic warming should be closely examined.

d. Effect of extrahypothalamic temperature on hypothalamic control of panting

What is the effect of these extrahypothalamic thermoreceptors on the hypothalamic control of panting? In the conscious dog, when the spinal cord and the hypothalamus are warmed simultaneously, the amount of respiratory EHL is greater than it is for individual hypothalamic or spinal warming alone (Jessen and Ludwig, 1971). Spinal warming decreases the hypothalamic set-point temperature, T_{set}, for panting. In a 24°C environment, the hypothalamus must be warmed above 42°C to elicit an increased evaporative heat loss. With spinal warming, however, panting begins at a hypothalamic temperature of 40°C. The hypothalamic thermosensitivity (α) for EHL is the change in EHL for each degree change in hypothalamic temperature above T_{set}. The hypothalamic thermosensitivity does not increase during spinal warming; rather, it either decreases or remains about the same, depending on the ambient temperature.

Jessen and Simon (1971) have shown that spinal cooling has no effect on the EHL response to hypothalamic warming in conscious dogs. Similarly, hypothalamic cooling does not interfere with the increase in EHL during spinal warming. This would be important for those animals (with and without a carotid rete) in which panting preferentially decreases hypothalamic temperature. During exercise, for example, an increase in hypothalamic or spinal temperature could initiate panting. Even if this EHL cools the hypothalamus, panting would still continue due to efferents from spinal and possibly abdominal thermoreceptors. In addition, it is probable that nonthermal (proprioceptor) afferents from moving limbs tend to facilitate panting during exercise (Scott et al., 1980), just as body movements facilitated panting in the (high-midbrain) decerebrate dogs (Keller and McClaskey, 1964).

The effect of skin temperature on the hypothalamic control of panting has been mentioned. Certain studies have shown that a cool or even neutral environment may completely block the panting response to hypothalamic warming (Hemingway et al., 1940; Fusco et al., 1961; Jacobson and Squires, 1970). In dogs, however, Hellstrøm and Hammel (1967) indicated that skin temperature merely increases or decreases the hypothalamic T_{set} for panting, without changing the hypothalamic thermosensitivity for this response. Compared with cutaneous vasoconstriction and metabolic heat production, the hypothalamic control of panting is only slightly affected by changes in skin temperature. In one dog, for example, warming the skin from 33 to 37.4°C decreased the hypothalamic T_{set} for panting from 41.1 to 39.4°C (i.e., a T_{set} change of 1.7°C). This same change in skin temperature decreased the hypothalamic T_{set} for vasoconstriction and metabolism from above 40°C to 36°C or lower (Hellstrøm and Hammel, 1967). This suggests that in the neural integration of thermal signals, peripheral thermal afferents have relatively less effect on effector neurons controlling EHL than on effector neurons controlling metabolic heat production.

In addition to changing the apparent hypothalamic T_{set} for panting, it is likely that skin temperature also changes the effective hypothalamic thermosensitivity (α) for this response (Boulant and Gonzalez, 1977). The effect of skin temperature on the hypotha-

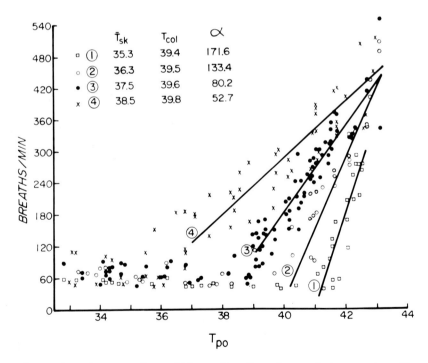

Figure 4 Breathing frequency in a rabbit as a function of preoptic temperature, T_{po}. Each numbered line and symbol indicates a different level of trunk skin ($\overline{T}_{sk}$) and colonic temperature (T_{col}). Thermosensitivity (α) refers to the slope of each line, in breaths per minute per degree centigrade. (From a study by Boulant and Gonzalez, 1974.)

lamic control of panting polypnea in the unanesthetized rabbit is illustrated in Figure 4. As noted, at any constant skin and core temperature, an increase in PO/AH temperature (T_{po}) past a particular T_{set} causes a proportional increase in respiratory rate. The hypothalamic thermosensitivity (α) of this response is merely the slope or regression coefficient of each thermoresponse line, that is, the change in breaths per minute for each degree of T_{po} above T_{set}. Warming the skin not only increases the respiratory rate at each T_{po} (i.e., it decreases T_{set}) but also decreases the α for this response. Conversely, cooling the skin increases both the hypothalamic T_{set} and the α for panting. Other studies have suggested that skin temperature produces similar changes in the effective hypothalamic thermosensitivity—in the rabbit for respiratory frequency (Boulant and Gonzalez, 1974, 1977; Gonzalez et al., 1974) and EHL (Stitt, 1976), and in the dog for salivation during exercise (Hammel and Sharp, 1971).

2. Sweating

In certain mammals, including humans, evaporative heat loss from the skin surface is greatly enhanced by the secretion of sweat from cutaneous glands. Bligh (1973) has reviewed some of the morphological and functional differences in the two types of sweat glands. Atrichial sweat glands open onto the skin surface, while epitrichial sweat glands secrete into the lumens of hair follicles. Both types are under sympathetic neural control, but the atrichial glands found in humans and cats involve cholinergic inputs, while the epitrichial glands in cattle involve adrenergic inputs.

a. Human studies: central control and effect of skin temperature

Most studies of the thermoregulatory control of sweating have been in humans. This accounts for the sparsity of information concerning the neural control of sweating, since in humans it is impossible to thermally stimulate selected sites in the hypothalamus and brain stem. It is only possible to increase whole-body temperature by exercise or peripheral heating. Also, specific hypothalamic and brain stem temperatures cannot be measured in humans. Instead, intracranial temperature must be approximated by measuring tympanic and nasopharyngeal temperatures (Benzinger and Taylor, 1963) or esophageal and right atrial temperatures (Wyss et al., 1974).

Even in humans, there is still a great deal of evidence that this internal or intracranial temperature is the primary determinant of the rate of sweating (just as it is for panting). Once internal temperature rises above 37.0°C, sweating begins, and the amount of sweating is proportional to the level of internal temperature above this T_{set} level (Benzinger et al., 1963; Nadel et al., 1971; Wyss et al., 1974). Benzinger et al. (1963) have further stated that, even during peripheral heating, the rate of sweating is determined only by the increase in internal temperature, not by the increase in skin temperature. Accordingly, the skin may be warmed anywhere from 33 to 39°C, and the internal T_{set} and thermosensitivity for sweating remain unchanged. These investigators do indicate, however, that cooling the skin below 33°C will progressively increase the internal T_{set} for sweating. For example, if the skin is 31°C, the internal T_{set} to elicit sweating is elevated to 37.2°C. If the skin is cooled to 29.5°C, the internal T_{set} is elevated even higher, to 37.4°C. This suggests that cutaneous warm receptors have little or no functional role in the neural regulation of sweating, but afferents from cutaneous cold receptors might inhibit central thermosensory signals, possibly in the hypothalamus.

Nadel et al. (1971) found that the rate of skin cooling correlated quite well with the amount of inhibition of sweating, but there was no comparable correlation between the rate of skin warming and the amount of sweating. This again suggests that peripheral cold receptors are more influential in altering sweat rate than peripheral warm receptors. Unlike the previous study of Benzinger and co-workers, Nadel and co-workers indicate that warming the skin above 33°C increases the sweat rate and, in effect, decreases the hypothetical internal T_{set} for sweating. Apparently, at any given esophageal temperature above 37°C, local sweating is greater at a skin temperature of 36°C than at a skin temperature of 35°C.

Nadel et al. (1971) have also shown that a local skin temperature has an effect on the local sweat rate, measured under sweat collection capsules. At a constant esophageal temperature, an increase in the average skin temperature causes a proportional increase in sweat rate at a given location. This occurs even if the local temperature under the sweat capsule remains constant. If the local skin (under the capsule) is then clamped at a warmer temperature, the local sweating increases and is even more sensitive to a change in the average temperature of the body surface. Thus, if the esophageal and average ("body") skin temperatures remain warm and constant, increasing just the local temperature under a sweat capsule increases the local sweat rate. This suggests that, in addition to the central integration of internal and peripheral temperature, there is also a local effect of temperature, either directly upon the sweat glands themselves or by way of a very precise reflex. Studies in humans also indicate that a small amount of reflex sweating does exist even when the spinal cord is completely isolated from the brain stem (Randall et al., 1966; Fuhrer, 1971). Such local cutaneous effects would have to be eliminated in studying the hypothalamic integration of central and peripheral thermal signals in the control of sweating; otherwise, it would be impossible to separate the local effect of skin temperature on the sweat glands from the effect of skin thermal afferents on the hypothalamic control of sweating. One way to eliminate the local skin response would be to use the sweat capsules so that the skin under the capsules could be maintained at a constant temperature, despite temperature changes in the rest of the skin surface.

Using constant-temperature sweat capsules, Downey et al. (1976) have recently reported the effects of independently changing skin and core temperatures in a human having a complete T_6 transection. Sweating was measured from the sentient skin innervated above the transection. At a constant oral temperature above 37.0°C, warming the sentient skin elicited local sweating. The core temperature was independently changed by warming or cooling the lower body below the transection. In this individual, warming the sentient skin to 36.3°C effectively decreased the internal T_{set} for sweating. Skin temperature apparently did not affect the internal thermosensitivity for sweating. As reported by others, sweating did not occur at internal (oral) temperature below 37.0°C, and, once induced, sweating could be rapidly inhibited by skin cooling. It should be noted, however, that, in this study, the only measurement of internal temperature was oral temperature, which is a less accurate indication of intracranial temperature. Some studies, therefore, indicate that sweat rate may be increased by skin warming, as well as decreased by skin cooling. Part of this is due to a local or reflex effect, and part may be due to peripheral and central thermal integration at higher neural levels.

Other recent studies, however, tend to support the earlier conclusions of Benzinger et al. (1963) that skin warming above 33°C does not influence the central control of sweating. Wyss et al. (1974) indicate that sweat rate is dependent solely on the level of internal temperature above 37.0°C and is independent of skin temperatures over the range of 34-40°C. These investigators measured right atrial temperature and feel that this approximates intracranial temperature better than esophageal temperature. They also found that the rate of skin cooling does substantially reduce sweating. Similarly, Chappuis et al. (1976) indicate that cutaneous EHL is directly related to tympanic temperature and is independent of ambient temperatures over the range of 20-30°C. At this point, it may be said that, in humans, internal temperature (probably hypothalamic temperature) is by far the principal determinant of sweating. Aside from a local cutaneous effect, skin cooling appears to have some part in the central integration and control of sweating and probably is more influential than skin warming.

b. Animal studies: hypothalamic control and effect of skin temperature

Only a few studies have considered the neural control of sweating in animals. In 1929, Hasama reported that unilateral electrical or thermal stimulation of the PO/AH produced bilateral sweating on the foot pads of cats. In their thermode study, Magoun et al. (1938) found that localized PO/AH warming in cats produced panting accompanied by foot-pad sweating. This sweating may require an intact hypothalamus, since no foot-pad sweating was observed in warm-exposed, decerebrate cats despite panting, cutaneous vasodilation, and rectal temperatures in the 41-44°C range (Bard and Macht, 1958).

Perhaps the most promising type of study to understand the neural control of sweating involves monkeys having implanted hypothalamic thermodes and sweat capsules to measure local sweat rates. Using this technique on rhesus monkeys, Smiles et al. (1976) have shown that at a skin temperature of 36.4°C, sweat rate is directly proportional to an increase in hypothalamic temperature above a T_{set} of about 38.7°C. If the skin temperature is warmed to 37.7°C, the hypothalamic T_{set} decreases to about 38.1°C, but the hypothalamic thermosensitivity (α) for sweating does not change. Similar shifts in the hypothalamic T_{set} for sweating have been observed in a single squirrel monkey at mean skin temperatures of 38.1, 38.5, and 39.0°C (Stitt et al., 1971). Undoubtedly, future thermode studies such as these are necessary to understand the central integration of hypothalamic and cutaneous thermal information and the possible role of other neural structures in the control of sweating. In these studies, hypothalamic thermosensitivity should be determined at several clamped skin temperatures in order to determine whether α does or does not remain constant.

3. Conservation of Body Water

As noted, an increase in hypothalamic temperature causes an increase in evaporative heat loss, and a natural consequence of this is a loss of body fluids. This loss of fluids elicits compensatory water-retention responses, notably an increase in the levels of antidiuretic hormone (ADH). The possibility exists that central and peripheral temperatures may have some direct effect on ADH release. However, since ADH release is known to be affected by volume receptors and osmoreceptors, it is more likely that nonthermal stimuli ultimately account for the thermally induced ADH and diuretic changes. Accordingly, central or peripheral warming increases peripheral blood flow and elicits compensatory responses regulating central blood pressure and volume. Facilitation of ADH release would result as a consequence of these shifts in vascular fluid. In addition, the extended depletion of body water due to EHL would also facilitate ADH release. Conversely, peripheral or central cooling would shift much blood to central venous pools and consequently inhibit ADH release (Hayward and Baker, 1968).

The evidence for hypothalamic thermal control of ADH is somewhat conflicting. Again further study is necessary to distinguish direct thermal effects from effects due to shifts in vascular volume. PO/AH warming in dogs increases the levels of plasma ADH and suppresses initiated water diuresis (Szczepanska-Sadowska, 1974). Forsling et al. (1976) have shown in pigs that warming either the skin or the preoptic region does not substantially or consistently elevate plasma ADH levels. Large increases in plasma ADH do occur, however, when rectal temperatures rise in hot environments. These elevated ADH levels can be occasionally inhibited by cervical spinal cooling but can be completely inhibited by preoptic cooling. Partial inhibition occurs during cooling of the supraoptic nucleus, but preoptic cooling alone does not consistently change ADH levels or promote water diuresis. Other studies in monkeys have shown that preoptic cooling does increase urine flow and decrease urine osmolality. Moreover, this diuresis can be blocked by ADH injections, despite elevated core temperatures (Hayward and Baker, 1968; Hayward, 1970). Similar responses occur in goats during hypothalamic cooling and less consistently during spinal cord cooling (Jessen and Simon-Oppermann, 1976). It is apparent, therefore, that more studies are needed to resolve, first, the effect that different thermosensitive areas have on ADH release, and second, whether the ADH release is a direct or an indirect consequence of thermal stimulation. One possible approach might be to produce maximal cardiovascular changes by skin warming and cooling and then to superimpose PO/AH warming and cooling. If the central thermal stimulus causes further changes in ADH release without further cardiovascular changes, this might be taken as evidence for a direct thermal effect on ADH release (J. T. Stitt, personal communication).

E. Thermoregulatory Behavior

In addition to physiological responses, body temperature is regulated to a large extent by behavior. In fact, thermoregulatory behavior is the principle means of thermoregulation in all forms of animal life, including mammals. Autonomic heat-production and heat-loss responses are employed when core temperatures, and sometimes skin temperatures, are altered above and below thermoneutrality. The purpose of these autonomic responses is to insure that core temperature remains constant. These autonomic responses do not regulate skin temperature, which often approaches ambient temperature. Bligh (1973) points out that thermoregulatory behavior is usually the first response to occur when skin temperature begins to change. In contrast to autonomic responses, the primary purpose of behavior is to insure that skin temperature remains at thermoneutrality. This would indirectly insure that core temperature remains constant. In a sense, behavioral responses predominate over and actually conserve autonomic responses. Behavioral responses are the first to be initiated when skin temperature changes, and as long as skin temperature can be maintained near thermoneutrality, autonomic responses will be minimal.

There are two basic types of thermoregulatory behavior: behavior that is inherent to a particular animal, and behavior that must be learned. Both types have received wide attention in thermoregulatory studies. Postural changes would be a good example of inherent thermoregulatory behavior. When an animal is cold, it assumes a huddled position, which decreases its exposed surface area and, thereby, decreases its heat loss to the environment. When the same animal is warm, it assumes a prone or spread-out position, which increases its exposed surface area and increases its heat loss. Grooming or saliva spreading is another inherent thermoregulatory behavior; in this case, to increase evaporative heat loss from the skin. In rats, the amount of saliva spreading increases with an increase in ambient temperature. In fact, it is impossible for heat-exposed desalivated rats to regulate their body temperatures (Hainsworth, 1967). A thermoregulatory behavior common to all animals is movement away from a harsh environment, toward a more thermoneutral environment. While this type of behavior is often associated with nonmammalian species, such as reptiles and fish, it can be equally important in mammals as well. The most thoroughly studied learned thermoregulatory behavior consists of training animals to bar press for short exposures of radiant heat or bursts of cold air. A cold-exposed rat, for example, will work or bar press for radiant heat, while a warm-exposed rat will bar press for bursts of cold air (Weiss, 1957). In addition, the amount of work or the frequency of bar pressing progressively increases the farther ambient temperature is from thermoneutrality (Weiss, 1957; Murgatroyd and Hardy, 1970).

1. Hypothalamic Control

As indicated, thermoregulatory behavior can be evoked by changes in skin temperature. These same types of behavior can also be evoked by changes in hypothalamic temperature. Thermode studies have shown that, when behavior is experimentally permitted, alterations in PO/AH temperature induce both inherent and learned behaviors. Thermodes, for example, have been implanted in the rostral brain stem of lizards and fish. When the PO/AH is cooled in these animals, they remain in warm environments longer than normal and incur body temperatures above normal. Conversely, PO/AH warming causes these animals to remain in cool environments longer and incur body temperatures that are below normal (Hammel et al., 1967, 1969; Myhre and Hammel, 1969; Crawshaw and Hammel, 1974). In these nonmammalian species, thermoregulatory behavior appears to be determined by a combination of skin, hypothalamic, and possibly other deep-body temperatures. Moreover, thermosensitive neurons have been found in and near the rostral hypothalamic area in both lizards (Cabanac et al., 1967) and fish (Greer and Gardner, 1970).

Hypothalamic thermal stimulation also induces inherent behaviors in mammals. In unanesthetized cats, localized cooling of the PO/AH produces a huddled posture with the tail wrapped close to the body. In these same animals, PO/AH warming causes a relaxed, stretched out position of the body, limbs, and tail (Freeman and Davis, 1959). A similar type of relaxed recumbent position has been observed in opossums and rats during localized warming at sites near the PO/AH. In these same animals, localized warming in other PO/AH sites results in grooming behavior (Roberts et al., 1969; Roberts and Mooney, 1974). Since these two different types of heat-loss behavior can be elicited from slightly different sites, it has been suggested that different hypothalamic thermosensitive neurons control different behaviors. Alternatively, the general state of the animal may determine which behavior predominates at any given time. In the rat, the stretched-out posture is elicited solely by warming PO/AH sites; the grooming is elicited primarily from sites in the posterior hypothalamus and rostral midbrain; and tail vasodilation results from warming selected sites extending from the preoptic region back into the pons (Roberts and Mooney, 1974).

Local changes in hypothalamic temperature also elicit thermoregulatory behavior which has been learned. In 1964, Satinoff showed that local PO/AH cooling causes rats to bar press for radiant heat and that, during PO/AH cooling, rats work harder in a 5°C environment than in a 24°C environment. Evidently, an integration of central and peripheral thermal signals occurs, even for this type of response. Murgatroyd and Hardy (1974) further showed that the frequency of bar pressing for heat is a function of the decrease in preoptic temperature (T_{po}). Rats begin to increase bar pressing when T_{po} drops below 36.5°C and subsequently work harder at each lower level of preoptic temperature, down to about 33°C. In addition, these same rats can be trained to bar press for bursts of cool air. When T_{po} is warmed past 39.5°C, the bar pressing frequency for peripheral cooling begins to increase. This work rate reaches a maximum at a preoptic temperature of about 40.5°C. Thus, when PO/AH temperature deviates from the 36.5-39.5°C range, rats will actively work for a change in peripheral temperature. Similar types of learned thermoregulatory responses have been demonstrated by changing hypothalamic temperature in baboons (Gale et al., 1970), squirrel monkeys (Adair et al., 1970), and pigs (Baldwin and Ingram, 1967). The chapter by Myers extensively describes the role of pyrogenic agents in fevers. Fevers can be produced experimentally by learned behavioral responses. After receiving intravenous pyrogen injections (Gale et al., 1970) or intrahypothalamic (i.e., PO/AH) prostaglandin E_1 injections (Crawshaw and Stitt, 1975), monkeys can become febrile by bar pressing for peripheral heat which drives core temperature to hyperthermic levels. They are, thus, behaviorally regulating their hypothalamic temperature by means of their skin temperature. Had behavioral thermoregulation not been available to these animals, the fevers would have been produced solely by shivering and cutaneous vasoconstriction. The behavioral response can, therefore, minimize the deployment of these autonomic responses.

2. Effect of Extrahypothalamic Temperature on Hypothalamic Control

Corbit (1969, 1973) has shown that rats will work to directly control their hypothalamic temperatures, and that the rate at which they work is determined by a combination of both skin and hypothalamic temperature, skin temperature being the more dominant stimulus. Rats having implanted hypothalamic thermodes can be trained to bar press for very brief periods of constant hypothalamic cooling. At a skin temperature of 36.2°C, "clamping" the hypothalamic temperature above about 39°C initiates bar pressing for these brief hypothalamic coolings. The greater the hypothalamic temperature above 39°C, the greater the bar-pressing rate. In fact, there is a fairly linear hypothalamic thermosensitivity for behaviorally "regulating" hypothalamic temperature (an admittedly difficult concept to envision). In addition, at any clamped hypothalamic temperature, an increase in skin temperature substantially increases the bar-pressing frequency. This is equivalent to decreasing the hypothalamic T_{set} for eliciting this response. (Actually, a change in skin temperature is a two to three times greater stimulus than a comparable change in hypothalamic temperature.) In effect, then, warming the skin decreases the hypothalamic T_{set} for this behavioral control of hypothalamic temperature. Corbit indicates that skin temperature does not alter the hypothalamic thermosensitivity (α); however, because of the small number of data points, this is difficult to determine with certainty. Thus, in the rat, at least, it is possible to demonstrate a behavior which controls hypothalamic temperature and a separate behavior which controls skin temperature. Corbit (1973) has shown that both of these behaviors are quantitatively similar in terms of their responsiveness to changes in hypothalamic and skin temperature.

Previous studies show that changes in either skin or hypothalamic temperature produce behavioral responses which affect skin temperature. The interaction between skin and hypothalamic temperature appears to be complex, and a consensus has not resulted from experiments to date. Corbit (1973) indicates that the effect of these two temperatures is additive. In one group of rats, an increase in hypothalamic temperature from 36 to 41°C causes a proportional increase in bar pressing for bursts of cool air. When this experiment is performed at ambient temperatures of 34, 36, and 38°C, the bar-pressing frequency at each hypothalamic temperature increases with an increase in ambient temperature; therefore, skin warming apparently decreases the hypothetical hypothalamic T_{set} for eliciting this behavioral control of skin temperature. Because of the limited data, it is difficult to determine whether ambient temperature also affects the hypothalamic thermosensitivity (α) for this response. Corbit suggests that this thermosensitivity remains constant. However, if a curve is fitted through the data points (i.e., rat T15), differences in sensitivity may, in fact, occur. At an ambient temperature of 34°C, the hypothalamic thermosensitivity appears to be greater at hypothalamic temperatures above 38°C, while at an ambient temperature of 38°C, the hypothalamic thermosensitivity appears greater at hypothalamic temperatures below 38°C.

A study by Adair (1977, see Fig. 5) suggests that skin and deep-body temperatures affect both the hypothalamic T_{set} and thermosensitivity for behavioral responses. It is known that prolonged alterations in hypothalamic temperature result in changes in core and skin temperature. For example, in the baboon, hypothalamic cooling can initiate bar pressing for radiant heat. During prolonged hypothalamic cooling, however, this bar pressing will decrease once the radiant heat has elevated skin and core temperature to some higher level (Gale et al., 1970). Similarly, prolonged hypothalamic warming causes a decrease in both skin and core temperature. By using this means to alter core temperature in squirrel monkeys, Adair (1977) has shown that extrahypothalamic temperature alters the hypothalamic thermosensitivity for behavioral responses (Fig. 5). When the hypothalamic temperature is clamped below normal, rectal temperature is increased. During this time, the changes in bar-pressing behavior are less for a given change in PO/AH temperature. When the hypothalamic temperature is clamped above normal, rectal temperature is decreased. During this time, the change in the behavioral response is much greater for a comparable change in PO/AH temperature. In other words, not only do hypothalamic and extrahypothalamic temperatures both influence learned behavior but extrahypothalamic cooling actually increases the hypothalamic thermosensitivity (α) for this response, while extrahypothalamic warming decreases the hypothalamic thermosensitivity for this response. This is a response quite similar to that described above for the PO/AH control of shivering and NST (see Fig. 2).

3. Hypothalamic Interactions with Brain Stem and Spinal Cord

While most thermode studies of learned behavioral responses have been confined to the PO/AH, some studies have described other neural areas that affect these responses. In squirrel monkeys, changes in posterior hypothalamic temperature can elicit the same type of behavior as changes in PO/AH temperature, although the thermosensitivity of the posterior hypothalamus is not as great (Adair, 1974). It might be argued, however, that

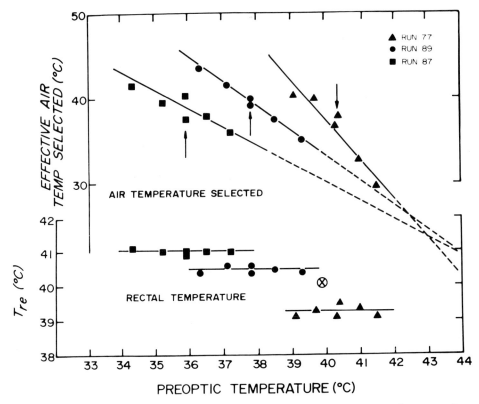

Figure 5 Behavioral responses in a squirrel monkey, trained to regulate environmental temperature by selecting between two incoming air temperatures, 10 and 50°C. Upper panel: Behavioral thermoregulation of environmental temperature as a function of preoptic temperature. Arrows denote clamped level of preoptic temperature in each of three experiments. Lower panel: Mean rectal temperatures corresponding to behavioral points plotted above. Symbol (X) denotes normal rectal temperature. (From Adair, 1977.)

in such a small hypothalamus, posterior hypothalamic heating also causes conductive heating of the PO/AH, which would explain the diminished response. Lipton (1971, 1973) has shown in intact rats that the medial medulla possesses the same thermosensitivity as the PO/AH for behavioral thermoregulation. Equivalent changes in either PO/AH or medulla temperature produce equivalent changes in bar-pressing behavior for peripheral cooling. More importantly, if the PO/AH is destroyed, the thermosensitivity of the medulla actually increases, so that medullary temperature becomes more effective in behavioral thermoregulation. This suggests, again, that in the hierarchical structure of the nervous system, thermosensitive PO/AH structures normally tend to dominate more caudal thermosensitive areas. If the PO/AH is removed, then some of these lower areas may be less restricted in exerting control over selected thermoregulatory responses. Additional confirmatory studies are necessary to substantiate this possibility.

Another example of hypothalamic control over a lower thermosensitive structure is shown in a study of pigs implanted with hypothalamic and spinal thermodes and trained to bar press for peripheral radiant heat (Carlisle and Ingram, 1973). Cooling either the hypothalamus or the cervical spinal cord causes an increase in work rate for radiant heat. Cooling the hypothalamus, however, elicits a substantially greater work rate than equivalent cooling in the spinal cord. Hypothalamic or spinal warming depresses bar pressing below control levels, but this time the spinal warming is slightly more effective. If the hypothalamic and spinal temperatures are changed in opposite directions, the bar-pressing behavior is determined solely by the hypothalamic temperature and is apparently not affected by the opposite change in spinal cord temperature. On the other hand, spinal cooling appears to be more effective than hypothalamic cooling in eliciting a "cold-defensive" prone position, which is an inherent behavior in this animal. This particular spinal domination over the hypothalamus is evident even when the two temperatures are changed in opposite directions. Consequently, it appears that, in the pig, the hypothalamus is most important for learned thermoregulatory behaviors, while the spinal cord may be most important for inherent thermoregulatory behaviors. The significance and duplication of this observation await further study.

In many cases, the PO/AH appears to have substantial control over thermoregulatory behaviors, even though these behaviors can also be evoked by other thermosensitive neural structures. If the PO/AH is lesioned, behavioral thermoregulation often increases, possibly to compensate for some loss in autonomic thermoregulation. In rats trained to bar press for radiant heat, PO/AH or septal lesions produce an increase in bar pressing despite possible decreases in rectal temperature (Lipton, 1968; Carlisle, 1969; Wakeman et al., 1970). The most likely explanation for this would be that, as in the previous study by Lipton (1973), removal of the PO/AH allows other thermosensitive areas, such as the medulla, to become effectively more sensitive to changes in core and possibly skin temperature. In cats, hypothalamic lesions, including the dorsomedial posterior hypothalamus, can abolish shivering but do not impair behavior thermoregulation. Such animals still huddle and actively seek out and move to warmer environments (Stuart et al., 1962a). In 1958, Bard and Macht reported that (midbrain) decerebrated adult cats showed either a huddled posture for increased movement during cold exposure. This occurs even though core temperature is not maintained. Bignall and Schramm (1974) suggest that a decerebrated kitten has essentially the same responses and core temperature as an unlesioned littermate. In addition to shivering and piloerection, a decerebrate kitten can actively select a thermoneutral environment and avoid both cold and hot environments. Continuation of similar studies will, it is hoped, bring to question the widely accepted view that thalamic and cortical structures are necessary for all forms of thermoregulatory behavior.

4. The Hypothalamus as a Limbic Structure

In the intact brain, the hypothalamus is an important structure in the limbic system, which is responsible for emotions and motivated behavior, as well as learning. In a sense, the hypothalamus connects the hippocampus, amygdala, and cingulate gyrus with the lower brain stem. Accordingly, the septum, PO/AH, mamillary bodies, and posterior hypothalamus are important structures in this link. Neurons in these areas receive, integrate, and contribute to information relayed in the major limbic pathways. These pathways include the median forebrain bundle, the fornix, and the stria terminalis. These same areas in and near the hypothalamus possess thermosensitive neurons and receive peripheral thermosensitive afferents. It is not surprising, then, that these areas strongly influence behavior affecting body temperature. It would also be anticipated that these same areas impart much of the emotional and subjective feelings of thermal comfort, with respect to both central and peripheral temperature. Just as these limbic areas determine the pleasantness or unpleasantness of cutaneous tactile and noxious stimuli, so too these same hypothalamic areas may determine whether or not a particular skin temperature is considered comfortable or uncomfortable. Quite likely, the activity of PO/AH thermosensitive neurons, at any particular time, contributes to this subjective determination. If a skin temperature is "emotionally" considered to be uncomfortable, then, by ascending and descending limbic signals, an appropriate behavior is employed to make skin temperature more comfortable. If this is not possible, then the appropriate autonomic responses are employed to insure that core temperature remains constant despite the change in skin temperature.

III. The Role of Hypothalamic Neurons in Thermoregulation

The preceding section has presented strong evidence that the hypothalamus, especially the PO/AH, is a thermosensitive structure having an important role in all thermoregulatory responses. We have seen that the neural areas in the brain stem and spinal cord are also important in temperature regulation, but by far the greatest evidence indicates that the hypothalamus is vital to the normal integration of central and peripheral thermal signals and the control of necessary and appropriate thermoregulatory responses. This section considers the thermosensitive and integrative characteristics of the PO/AH neurons. It is based primarily on single unit studies. Recent discussions of these neuronal characteristics include Hayward (1977) and Boulant (1980).

A. Neuronal Characteristics

1. Local Thermosensitivity

Beginning with the study of Nakayama et al. (1961), many single-unit studies have identified certain temperature-sensitive neurons in and near the hypothalamus. These thermosensitive neurons change their firing rates when local hypothalamic temperature is altered by implanted thermodes. The thermosensitivity of a neuron is defined by its change in firing rate for a particular change in hypothalamic temperature. Most studies use the slope (α) of the frequency-temperature response curve (see Fig. 1) as an indication of a neuron's thermosensitivity. This slope is also called the "thermal coefficient." Usually a neuron is considered to be warm sensitive if it has a positive thermal coefficient of at least 1.0 impulses per second per degree centigrade over a 3-4°C range of hypothalamic temperature (Hardy et al., 1964; Guieu and Hardy, 1970b; Hellon, 1972; Boulant and Bignall, 1973a). Such a warm-sensitive neuron, for example, would have to increase its firing rate by at least three impulses per second during a 3°C increase in PO/AH temperature. Other studies have used the Q_{10} (see Section I) of a neuron's discharge frequency as an indication of its thermosensitivity. Usually, neurons having a Q_{10} greater than 2.0 are considered to be warm sensitive (Eisenman and Jackson, 1967; Jell and Gloor, 1972). While there is not much difference between the types of neurons classified by either of these two criteria (Boulant and Bignall, 1973a), it is likely that more low-firing neurons would be considered thermosensitive based on their Q_{10}, because a small change in firing rate would represent a greater percent-change for the lower-firing neurons as compared with the higher-firing neurons. Cold-sensitive neurons increase their firing rate with decreases in hypothalamic temperature. Usually a neuron is considered to be cold sensitive if it has a negative thermal coefficient of at least -0.8 to -1.0 impulses per second per degree centigrade (Guieu and Hardy, 1970b; Hellon, 1972; Boulant and Bignall, 1973a). If a stable single unit does not meet these criteria for warm or cold sensitivity, then it is considered to be temperature insensitive.

Single-unit studies have indicated that thermosensitive neurons exist throughout the PO/AH (Nakayama et al., 1963; Hardy et al., 1964; Hellon, 1967; Boulant and Bignall, 1973a) as well as the septum (Cunningham et al., 1967; Eisenman, 1969; Beckman and Rozkowska-Ruttimann, 1974; Boulant and Hardy, 1974). In addition, neurons sensitive to local or central temperatures also exist in the posterior hypothalamus (Edinger and Eisenman, 1970; Wünnenberg and Hardy, 1972; Nutik, 1973a,b), the midbrain raphe nucleus (Weiss and Aghajanian, 1971; Cronin and Baker, 1976), the reticular formation of the midbrain (Cabanac and Hardy, 1969; Nakayama and Hardy, 1969; Nakayama and Hori, 1973), and the reticular formation of the pons and medulla (Wünnenberg and Brück, 1968; Boulant and Hardy, 1972; Inoue and Murakami, 1976).

The reported proportions of the various types of neurons vary among different studies and possibly different locations. In every study of the rostral hypothalamic area, the cold-sensitive neurons represent the smallest population and are usually outnumbered by warm-sensitive neurons by a ratio of three or four to one. Many neuronal studies include single units which, for one reason or another, were held for only short periods of time. Apparently the inclusion of these units affects the reported proportions of the different types of units. It has been shown, for example, that among those PO/AH single units that could be observed for only 15-75 min, 19% were warm sensitive, 18% were cold sensitive, and 63% were temperature insensitive. However, among more stable units that could be observed for at least 90 min (and averaging about 4 hr), 53% were warm sensitive, 16% were cold sensitive, and 31% were temperature insensitive (Boulant and Bignall, 1973a).

2. The "Shape" of Thermoresponse Curves

The neuronal model in Figure 1 describes the activity of both thermoreceptor neurons (i.e., high-Q_{10}, warm-sensitive neurons) and set-point interneurons or effector neurons (i.e., cold-sensitive and warm-sensitive neurons having an apparent T_{set}). Single unit studies (Eisenman and Jackson, 1967; Hellon, 1967; Jell and Gloor, 1972) have also observed certain thermosensitive neurons which have characteristics similar to thermoreceptors and set-point interneurons. Set-point interneurons include those units having nonlinear thermoresponse curves. These units either increase or decrease their firing rates only when PO/AH temperature is changed above or below a certain T_{set}; otherwise, firing rate remains constant. In contrast, thermoreceptor neurons have linear thermoresponse curves and are thermosensitive over a wide range of PO/AH temperatures.

Neuronal models of Hammel (1965) and Hardy (1972) propose the most likely roles for the nonlinear neurons. That is, those neurons which are sensitive only in the hyperthermic PO/AH temperature range may control heat-loss responses (e.g., panting and sweating) which function in this temperature range. Conversely, those neurons which are sensitive only in the hypothermic range may control heat-production responses which function in this temperature range. Eisenman and Jackson (1967) state that the activity of the nonlinear "interneurons" is depressed by barbiturates which apparently might block their synaptic inputs. These investigators indicate, however, that the linear "thermodetectors" are only slightly depressed by barbiturates, therefore suggesting that they are relatively free from synaptic inputs and have inherent spontaneous firing rates.

This concept of an independent receptor controlling the activity of a "nonreceptive" interneuron may be applicable to spinal reflexes but probably not to the hypothalamus. It is likely that thermosensitivity is a common, rather than uncommon, occurrence. Moreover, the thermosensitivity of synaptic events (i.e., the effect of temperature on the release, diffusion, binding, breakdown, and reuptake of neurotransmitters) or the thermosensitivity of local reverberating circuits (in which the moderate thermosensitivity of individual neurons in a pool of mutual excitatory neurons can render each neuron highly thermosensitive) may be more important than the local thermosensitivity of an individual neuron, free from synaptic inputs. Finally, evidence is presented below to indicate that the "shape" or linearity of a thermoresponse curve is no indication of whether a neuron is a thermodetector or T_{set}-interneuron. This arbitrary separation of central neurons as receptors and effectors or interneurons is probably unrealistic and is not necessary to explain the PO/AH control of thermoregulation.

3. Natural and Induced Changes in Neuronal Activity

The thermosensitivity of temperature-sensitive and even temperature-insensitive neurons is not fixed. Rather, when observed for long periods, the firing rate and thermosensitivity of most PO/AH units tends to fluctuate over time (Boulant and Bignall, 1973b). Figure 6 shows a warm-sensitive PO/AH neuron which slowly increases and decreases its firing rate and thermosensitivity. Such fluctuations occur even though all other central and peripheral temperatures remain constant; they are seen in both anesthetized and decerebrate, unanesthetized animals. Recent studies even demonstrate fluctuations in units recorded in vitro from hypothalamic tissue slices (Kelso et al., 1980).

Most neurons display only minor fluctuations in their firing rates. A warm-sensitive unit showing a large change in firing rate is shown in Figure 7. While this is an extreme example, it does indicate the typical thermosensitivity change that accompanies the firing-rate changes. When this unit was first observed (1), it was nonlinear and sensitive primarily to PO/AH temperatures above 37°C. Over time, its level of firing rate increased (2) and its thermoresponse curve became linear. Finally, its firing rate increased still further (3), its thermoresponse curve became nonlinear, and its greatest thermosensitivity occurred in the extreme hypothermic range. Such thermosensitivity changes in individual neurons suggest that the linearity or nonlinearity of a neuron's thermoresponse curve is not an adequate criterion to classify a neuron as either a thermoreceptor or a T_{set}-interneuron, as was suggested by earlier single-unit studies. Depending on its level of firing rate, the neuron in Figure 7, for example, might first be considered to be a hyperthermic heat-loss effector (1), then a linear thermoreceptor (2), and finally a hypothermic heat-production effector (3). Unrelated changes in firing rate and thermosensitivity also occur in cold-sensitive neurons (Fig. 8). and temperature-insensitive neurons (Boulant and Bignall, 1973b).

These changes in neuronal characteristics emphasize some important points regarding neuronal models and single-unit studies. First, it stresses that in neuronal models, each neuron type represents the average characteristics of an entire population of neurons. At any particular period of time, an individual neuron in a population may have thermosensitive characteristics which are quite different from the rest of its population. Individual neurons should be observed over long periods of time in order to determine fully their thermoresponse characteristics. Second, it emphasizes the need for rigid criteria when testing neuronal responsiveness to factors known to affect thermoregulation. Several single-unit studies have observed changes in hypothalamic neuronal activity in response to controlled variables. Some studies, for example, have demonstrated the effects of changes in skin and spinal temperatures on unit activity (Wit and Wang, 1968a; Hellon, 1969, 1970, 1971, 1972b; Guieu and Hardy, 1970b; Boulant and Hardy, 1972, 1974; Wünnenberg and Hardy, 1972; Boulant and Bignall, 1973c; Knox et al., 1973b; Nutik, 1973a; Jell, 1974). Other studies have shown the effects of injected anesthetics (Eisenman and Jackson, 1967; Murakami et al., 1967; Jell and Gloor, 1972), metabolites (Boulant, 1971; Boulant and Bignall, 1971), neurotransmitters (Cunningham et al., 1967; Knox et al., 1973a), and pyrogenic agents (Cabanac et al., 1968; Wit and Wang, 1968b; Eisenman, 1969, 1970; Schoener and Wang, 1975). Some of these same substances have also been applied locally by microiontophoresis (Beckman and Eisenman, 1970; Hori and Nakayama, 1973; Murakami, 1973; Beckman and Rozkowska-Ruttimann, 1974; Ford, 1974; Jell, 1974; Stitt and Hardy, 1975). In addition, other studies have described hypothalamic neuronal responses to electrical stimulation in hypothalamic, brain-stem, and limbic areas (Nutik, 1973b; Boulant and Demieville, 1977). This extensive list of studies illustrates the variety of single-unit experiments that have been undertaken. In every case, an alteration in neuronal firing rate is considered the primary indication of a unit's responsiveness to a particular variable. All too frequently, however, adequate controls are not used to insure a causal relationship between a manipulated variable and a neuron's activity. It is quite likely that some of these "induced" changes in neuronal activity would have occurred even if the variable had not been presented. The previous study (Boulant and Bignall, 1973b) emphasizes the need for adequate control measurements to insure that changes in neuronal activity are in response to the stimulus presented and are not simply unrelated, "naturally occurring" changes in neuronal activity. At the very least, an "affected" unit must be shown to return toward control activity when the manipulated variable is removed. Repeatable unit responses would be better proof of a causal relationship.

4. Effect of Firing Rate on Neuronal Thermosensitivity

We have seen that the shape of a unit's thermoresponse curve is not an adequate criterion for the distinction between receptors and interneurons, because the thermoresponse of an individual neuron may display linearity and different temperature ranges of nonlinearity depending on its level of firing rate. The warm-sensitive unit in Figure 7 is an extreme example of the effect of a substantial change in spontaneous firing rate on the thermoresponse characteristics of a neuron. Most single units show only minor changes in spontaneous firing rate and minor alterations in local thermosensitivity. Still, as seen in Figure 7, these changes in the firing rate of a neuron's thermosensitivity suggest some interesting possibilities about neuronal populations. If the characteristics of individual neurons can be applied to entire neuron populations, then, possibly the low-firing, warm-sensitive neurons have a different range of thermosensitivity compared with the normally high-firing, warm-sensitive neurons. For warm-sensitive neurons, the low-firing population

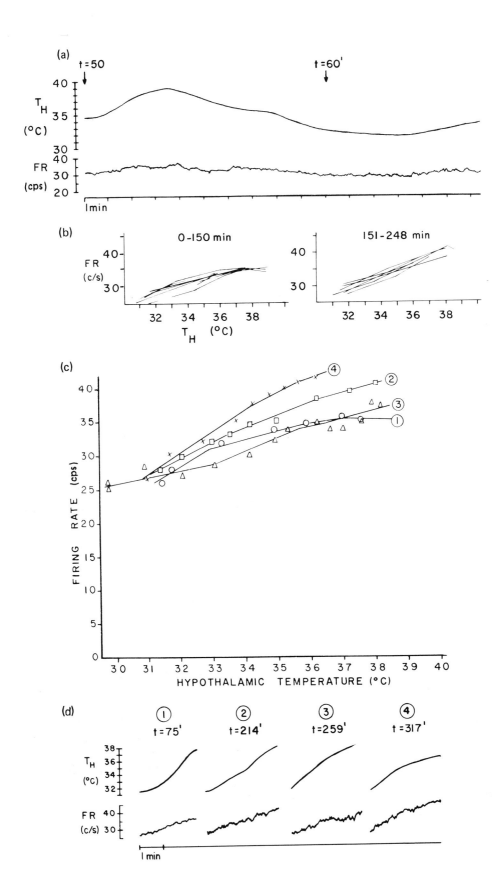

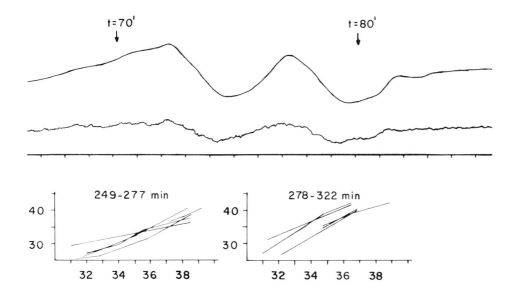

Figure 6 Firing rate responses of a PO/AH warm-sensitive neuron observed over 322 min in a ground squirrel. (a) Record showing typical changes in firing rate (FR) to slow and rapid changes in hypothalamic temperature (T_H). (b) All thermoresponse curves for each increase or decrease in T_H. Curves are grouped into four periods of time to illustrate consistency of responses during each period. (c) One representative curve from each of four periods shown in B. (d) Records for each of the four curves shown in (c). (From Boulant and Bignall, 1973b.)

might be more sensitive in the hyperthermic range, while the high-firing population might be more sensitive in the hypothermic range. This range of thermosensitivity suggests the most likely thermoregulatory role for these neurons. The most likely role of a neuron thermosensitive only in the hyperthermic range is controlling a heat-loss response, which actually functions in this range. The most likely role of a neuron sensitive only to hypothermic temperatures is controlling a heat-production response, which functions in this lower temperature range. Similarly, neurons equally sensitive both above and below thermoneutrality might function in heat-retention responses, such as the control of skin blood flow and thermoregulatory behavior, which are most active immediately above and below thermoneutrality.

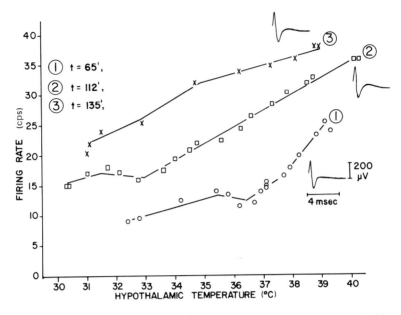

Figure 7 Thermoresponse curves of one PO/AH warm-sensitive neuron recorded in an anesthetized ground squirrel. Each response curve refers to the time (in minutes) after the single unit was first observed. Photographs of oscilloscope traces indicate a constant waveform over this time period. (From Boulant, 1971.)

Boulant (1974) and Boulant and Demieville (1977) have shown that substantial differences occur in the ranges of PO/AH thermosensitivity between neuronal populations having different spontaneous firing rates. Figure 9 shows the combined data from these two studies for 86 warm-sensitive neurons. These warm-sensitive units are divided into five groups based on their level of firing rate at 38°C ($FR_{38°}$). The average thermoresponse curve for each group is determined by plotting the average firing rate at 1°C intervals of preoptic temperature. The regression coefficient or thermosensitivity is determined over three ranges of preoptic temperature: 34-37°C, 37-40°C, and 40-42°C. As noted in Figure 9, the lowest-firing (i.e., zero to five impulses per second) group of warm-sensitive neurons have their greatest thermosensitivity in the hyperthermic range and are fairly insensitive to temperature changes in the other two ranges. The most likely role of these low-firing, hyperthermic neurons is in the control of heat-loss responses. In contrast, the two groups of high-firing neurons (i.e., 15-25 and 25-60 impulses per second) are most thermosensitive in the hypothermic range and are essentially temperature insensitive in the hyperthermic range. Accordingly, the most likely role of these high-firing, hypothermic neurons is in the control of heat-production responses. Finally, the average thermoresponse curve of the medium-firing neurons tends to be linear. This is especially true for those warm-sensitive neurons having spontaneous firing rates between 10 and 15 impulses per second. The most likely role of these medium-firing, linear neurons is in the control of heat-retention responses, which must function both above and below thermoneutrality.

A similar type of analysis for groups of cold-sensitive neurons shows that the higher-firing, cold-sensitive neurons are more thermosensitive in all three ranges of preoptic

Hypothalamic Control of Thermoregulation

temperature (Boulant, 1974). This is especially true in the hyperthermic range, where the thermosensitivity of the highest-firing group is almost three times greater than the thermosensitivity of the lowest-firing group. In the hyperthermic range, therefore, the level of spontaneous firing rate has the opposite effect on warm- and cold-sensitive neurons. In warm-sensitive neurons, hyperthermic thermosensitivity decreases among the higher-firing groups, but in cold-sensitive neurons, hyperthermic thermosensitivity increases among higher-firing groups.

5. Temperature-Insensitive Neurons

Temperature-insensitive PO/AH neurons do not display a wide range of firing rates but, instead, are limited almost entirely to very low spontaneous firing rates of five impulses per second or less (Boulant and Bignall, 1973a; Boulant and Hardy, 1974). In many neuronal models (e.g., Fig. 1), these temperature-insensitive neurons serve as necessary "reference" inputs for T_{set} interneurons. In reality, there is very little evidence that these insensitive neurons have any role in thermoregulation. Actually, considering the changes in the thermoresponse curves of warm- and cold-sensitive neurons, it is doubtful that temperature-insensitive neurons are even necessary in hypothetical neuronal models. Moreover, it is possible that these temperature-insensitive neurons are primarily the "glucoreceptors, osmoreceptors, and estrogen-receptors," etc., that have been found in the PO/AH and which are proposed to control other nonthermal homeostatic systems.

While some studies differ, in many PO/AH single-unit studies, stimuli that often affect temperature-sensitive neurons do not consistently affect temperature-insensitive neurons. Murakami et al. (1967) report that anesthetics do not affect low-firing preoptic neurons. As mentioned, most low-firing neurons are temperature insensitive (Boulant and Bignall,

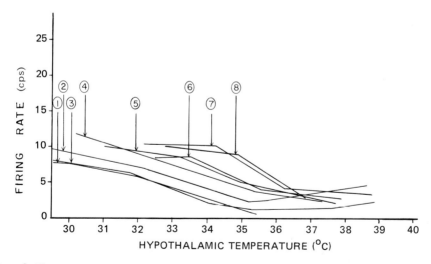

Figure 8 Thermoresponse curves of one PO/AH cold-sensitive neuron recorded in an anesthetized ground squirrel. The eight curves refer to different times after the single unit was first observed. These times range from (1) 105 min to (8) 179 min. (From Boulant and Bignall, 1973b.)

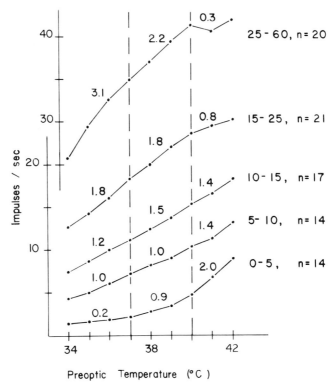

Figure 9 Averaged thermoresponse curves for 86 warm-sensitive neurons grouped into five ranges of firing rate at $38°C$ ($FR_{38°}$). The n refers to the number of neurons in each group. The points indicate the average firing rates of each group at $1°C$ intervals. The values above each section are the average thermosensitivity (impulses per second per degree centigrade) for that temperature range. (Derived from data from Boulant, 1974, and Boulant and Demieville, 1977.)

1973a; Boulant and Hardy, 1974). Jell (1973) indicates that most neurons that are unresponsive to the microiontophoresis of different neurotransmitters are also insensitive to local temperature. Other studies show that temperature-insensitive neurons are essentially unaffected by pyrogen injections (Cabanac et al., 1968; Wit and Wang, 1968b; Eisenman, 1969). Only a very small proportion of the PO/AH temperature-insensitive neurons are affected by changes in spinal cord temperature (Boulant and Hardy, 1974), see Figure 10(a), or by brain-stem or hippocampal electrical stimulation (Boulant and Demieville, 1977), see Figure 10(b-c) and 11. Wit and Wang (1968a) and Boulant and Hardy (1974) did not find any PO/AH temperature-insensitive neurons that were affected by changes in skin temperature. Therefore, there is little to indicate that these insensitive neurons receive much synaptic input from peripheral thermoreceptors (as suggested in Fig. 1).

Since temperature-insensitive neurons uniformly have very low firing rates, it is possible that they receive very little synaptic input compared with the thermosensitive neurons. Certainly some of the evidence just mentioned would support this. Alternatively, Boulant and Bignall (1973a) have proposed that cell size can account for some of the differences between temperature-sensitive and -insensitive neurons. It is suggested that thermosensitive neurons are smaller and thus have lower membrane thresholds. This would account for the higher spontaneous firing rates among thermosensitive neurons and might also provide clues concerning their greater susceptibility to local temperature changes. Anatomical support for this concept comes from a study in which capsaicin desensitization of apparent thermodetectors caused mitochondrial changes only in the smaller neurons in the PO/AH (Szolcsanyi et al., 1971). Continued study is obviously needed to substantiate these suggestions and to elaborate the basis of neuronal warm sensitivity, cold sensitivity, and temperature insensitivity. Intracellular labelling with horseradish peroxidase would identify the size, shape, and dentritic arborization of these various cell types (D.O. Nelson, personal communication).

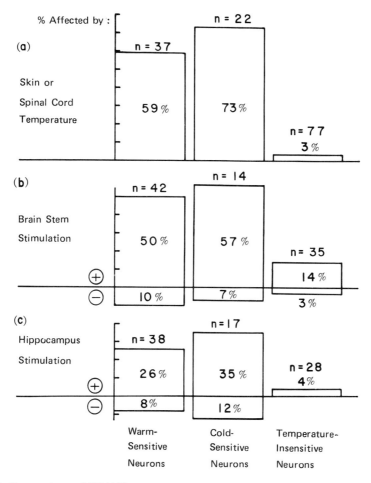

Figure 10 Proportions of PO/AH warm-sensitive, cold-sensitive, and temperature-insensitive neurons affected by (a) thermal stimulation of the skin or spinal cord (Boulant and Hardy, 1974); (b) electrical stimulation of brain-stem sites (RF, LL, ST as shown in Fig. 11) along the ascending anterolateral pathway; and (c) electrical stimulation of the hippocampus (HP as shown in Fig. 11). The n refers to the total number of neurons in each group. (+), Excitatory responses; (−), inhibitory responses. (From Boulant and Demieville, 1977.)

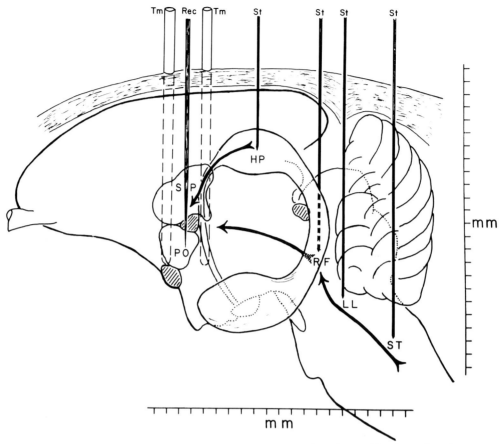

Figure 11 Sagittal view of the rabbit's brain showing the locations of implanted probes employed in the study described in Figure 10. Tm, thermode; Rec, recording microelectrode; St, electrode for electrical stimulation; SP, septum; PO, preoptic region; HP, hippocampus; RF, midbrain reticular formation; LL, lateral lemniscus; and ST, the lateral spinothalamic tract. (From Boulant and Demieville, 1977.)

B. Neuronal Integration of Central and Peripheral Temperatures

Some single-unit studies have shown that certain PO/AH neurons are affected by temperature changes in the skin (Knox et al., 1973b; Jell, 1974; Reaves and Heath, 1975) and midbrain (Cabanac and Hardy, 1969). More importantly, other studies indicate that it is primarily the thermosensitive PO/AH and septal neurons (see Fig. 10) that receive the thermal afferent input from the skin (Wit and Wang, 1968a; Hellon, 1970, 1972b; Boulant, 1971; Boulant and Bignall, 1973c; Boulant and Hardy, 1974), from the spinal cord (Guieu and Hardy, 1970b; Boulant and Hardy, 1974), and from the midbrain (Nakayama and Hardy, 1969). This supports the contention that the PO/AH is an important site for the integration of peripheral and central thermal signals. Furthermore, most of the neurons that are sensitive to both hypothalamic and extrahypothalamic temperatures have the same type of thermosensitivity (or thermal coefficient) for both temperatures. PO/AH neurons that are excited by increases in spinal or skin temperature are also excited by increases in PO/AH temperature. PO/AH neurons that are excited by decreases in spinal or skin temperature are also excited by decreases in PO/AH temperature. This indicates that there is some "logical" order involved in the afferent input to the different types of neurons. It would be of immense importance to understand how such a logical order develops. How, for example, does a cutaneous warm-receptor pathway develop excitatory synapses only to the warm-sensitive PO/AH neurons and not to the temperature-insensitive or cold-sensitive neurons? This suggests that the type of afferent input a PO/AH neuron receives may eventually determine that neuron's local thermosensitivity and ultimate function in thermoregulation.

The extent of this extrahypothalamic input to PO/AH neurons varies considerably among different single-unit studies. Earlier studies, for example, did not find any PO/AH single units that were affected by peripheral temperatures (Hardy et al., 1964; Murakami et al., 1967). At the other extreme, Boulant and Hardy (1974) indicate that most PO/AH neurons that are sensitive to local PO/AH temperature are also sensitive to skin or spinal temperature (see Fig. 10). Again, most of these neurons have the same type of thermal coefficient for both hypothalamic and extrahypothalamic temperatures. Very few studies have examined these characteristics of neural integration. There is a substantial need to examine further the ability of hypothalamic neurons to integrate local temperature information with thermal afferents from one, but preferably more, locations.

The neural integration of central and peripheral thermal information is more complex than simply the summation of afferents from different locations as suggested by Hammel's model (Fig. 1). The following sections consider the effect of extrahypothalamic temperature on the activity of warm-sensitive and cold-sensitive hypothalamic neurons.

1. Warm-Sensitive Neurons

Studies of warm-sensitive neurons show that changes in skin and spinal temperature affect not only the level of spontaneous firing rate but also the sensitivity of these neurons to changes in hypothalamic temperature (Hellon, 1972a,b; Boulant and Bignall, 1973c; Boulant and Hardy, 1974). Most of these changes in hypothalamic thermosensitivity are similar to the "unrelated" thermoresponse changes seen previously in the neuron described in Figure 7. Figure 12 shows the responses of two different warm-sensitive neurons; one neuron also responds to spinal temperature, and the other neuron also responds to skin temperature. The effect of extrahypothalamic temperature on the hypothalamic thermosensitivity is similar in both units. Extrahypothalamic cooling not only decreases the spontaneous firing rate ($FR_{38°}$) but also increases the hypothalamic thermosensitivity (α) at PO/AH temperatures above 37-38°C. Extrahypothalamic warming not only increases the spontaneous firing rate toward some maximal level but also decreases the local thermosensitivity at PO/AH temperatures above 37°C. It should be noted that these single units were recorded in rabbits, animals whose normal core temperature usually exceeds 39°C.

The effect of peripheral temperature on warm-sensitive neurons is quite similar to the effect of peripheral temperature on the hypothalamic control of panting (Fig. 4) and skin blood flow (Fig. 3). Consequently, it is quite likely that certain warm-sensitive neurons display all the integrative characteristics needed to explain the hypothalamic control of heat-loss responses, like panting (Fig. 13). Not only does peripheral warming increase both neuronal firing rate and EHL, but it also decreases the hypothalamic thermosensi-

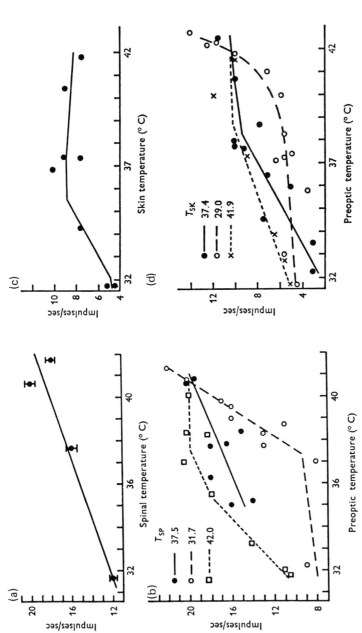

Figure 12 Responses of two different PO/AH warm-sensitive neurons recorded in rabbits during changes in extrahypothalamic temperature. (a) The spinal thermoresponse curve of one neuron at a preoptic temperature of 38.0°C. (b) The preoptic thermoresponse curves of the same neuron (a) at three different clamped spinal temperatures, T_{SP}. (c) The skin thermoresponse curve of another neuron at a preoptic temperature of 38.0°C. (d) The preoptic thermoresponse curves of this second neuron (c) at three different clamped skin temperatures, T_{SK}. (From Boulant and Hardy, 1974.)

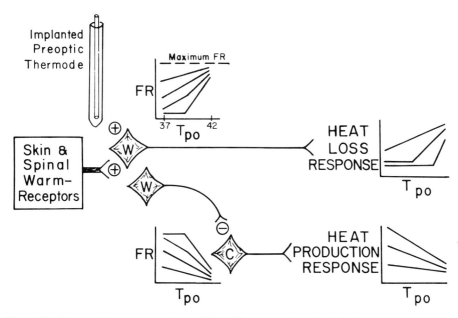

Figure 13 Schematic representation of PO/AH neuronal and whole-body responses to changes in both PO/AH and extrahypothalamic temperature. This indicates the possible functional role of hypothalamic neurons in thermoregulation. W, warm-sensitive neuron; C, cold-sensitive neuron; (+), excitatory input; (−), inhibitory input. The graphs near the two types of neurons summarize the effect of skin temperature on neuronal firing rate (FR) and hypothalamic thermosensitivity. T_{po} indicates preoptic temperature. The graphs to the right describe the changes in hypothalamic thermosensitivity for different thermoregulatory responses during changes in skin temperature. (From Boulant and Gonzalez, 1977; Boulant, 1980.)

tivity (α) for both the neuronal response and the heat loss response. At any hypothalamic temperature peripheral cooling decreases both neuronal firing rate and the level of EHL. In addition, peripheral cooling increases the hypothalamic thermosensitivity of both the warm-sensitive neuron and the heat-loss response.

As depicted by a (+) in Figure 13, each warm-sensitive neuron receives excitatory synaptic inputs from ascending thermoreceptive pathways, as well as local excitatory inputs which may be synaptic or nonsynaptic in origin. These local inputs contribute to the neuron's local thermosensitivity. The nonsynaptic excitatory factors include all of the endogenous factors in a cell's immediate environment that could affect its firing rate (e.g., ions, metabolites, osmotic pressure, endocrines, etc.). The effect of each of these factors on the neuronal membrane is potentially temperature dependent.

In Figure 13, the thermoresponse curves for the two warm-sensitive neurons summarize the effect of spinal and skin temperature on most warm-sensitive neurons observed in previous studies (Boulant and Bignall, 1973c; Boulant and Hardy, 1974). When extrahypothalamic temperature causes neuronal discharge to increase towards some maximal level, the neuron's local thermosensitivity tends to decrease, at least for hypothalamic temperatures above 37-38°C. This suggests that local excitatory afferent synapses not only summate temporally but also compete with one another for determining neuronal discharge. Presumably, the closer a warm-sensitive neuron is to its maximal firing rate, the greater the competition between excitatory inputs.

When the skin is cool, the excitatory synaptic input from cutaneous warm receptors is low. Under these conditions, neuronal discharge is determined solely by the local, temperature-dependent, excitatory factors in the neuron's immediate environment. For example, suppose that local conditions determine the frequency or amplitude of local changes in the neuronal membrane potential or of excitatory postsynaptic potentials (EPSPs) due to local synapses. Suppose, also, that these conditions are highly temperature dependent in these warm-sensitive neurons. Therefore, during skin cooling, the spontaneous firing rate might be low, but the local thermosensitivity could be quite high.

When the skin is warm, peripheral warm-receptor afferents increase the synaptic release of excitatory transmitters on these warm-sensitive neurons. This increases the frequency of EPSPs in the warm-sensitive neuron. The EPSPs from the local inputs (or the temperature-dependent membrane potential) and the EPSPs from the afferent inputs all summate temporally, and the resultant increased firing rate may reflect this summation. To a degree, however, these different EPSPs tend to compete with each other for determining action potentials. Consequently, during skin warming, the firing rate may be high, but the neuron may be much less sensitive to local excitatory factors. Thus, the local thermosensitivity is reduced. This competition is greatest when the two different excitatory inputs are both high (e.g., local warming plus skin warming).

How can this competition between excitatory inputs occur? Suppose, hypothetically, that one EPSP produces one neuronal action potential. If the local input causes one EPSP per second and the afferent input causes one EPSP per second, then neuronal firing rate would be about two impulses per second, as long as these two EPSPs do not occur simultaneously. If the two different EPSPs do occur simultaneously, then the neuronal firing rate would only be one impulse per second, because only one EPSP is necessary for an action potential. As long as the two different inputs remain low, the chances of simultaneous EPSPs would be minimal; therefore, firing rate would reflect the summation of these two inputs. When either or both of these inputs increase, however, the occurrence of simultaneous EPSPs becomes more common. Thus, neuronal firing rate increases, but it no longer reflects the summation of the two separate inputs. In reality, a neuron's membrane potential is constantly fluctuating, and many simultaneous EPSPs are often necessary to reach threshold. This concept of competition between EPSPs may still occur, however. Once threshold is reached, only one action potential occurs in a given period of time. Any additional EPSPs during this time period do not contribute to neuronal firing rate. Consequently, there is competition between excitatory inputs for producing this action potential.

2. Cold-Sensitive Neurons

Only one single-unit study (Boulant and Hardy, 1974) has described the repeatable effect of skin and spinal temperatures on the preoptic thermosensitivity of cold-sensitive neurons. Most of these thermosensitivity changes are similar to the "unrelated" changes in thermosensitivity shown previously in the neuron described in Figure 8. Figure 14 shows the thermoresponse curves for two different cold-sensitive neurons. Both of these neurons also respond to changes in skin temperature; one has a negative thermal coefficient for skin temperature (a and b) and the other has a positive thermal coefficient for skin

Hypothalamic Control of Thermoregulation

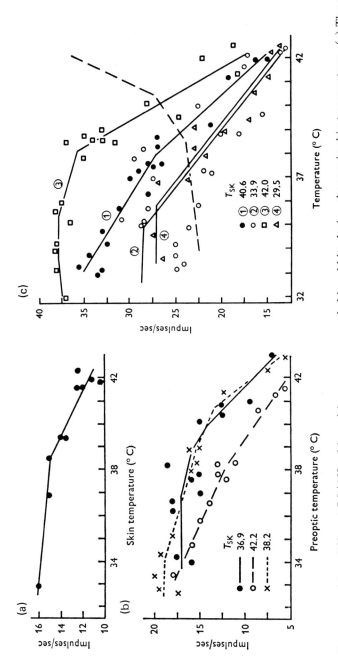

Figure 14 Responses of two different PO/AH cold-sensitive neurons recorded in rabbits during changes in skin temperature. (a) The skin thermoresponse curve of one neuron at a preoptic temperature of $38.4°C$, and the preoptic thermoresponse curves (solid lines) at four different clamped skin temperatures. (b) The preoptic thermoresponse curves of the same neuron (a) at three different clamped skin temperatures. (c) The skin thermoresponse curve (dashed line) of another neuron at a preoptic temperature of $38.0°C$. (From Boulant and Hardy, 1974.)

temperature (c). In both cases, an increase in spontaneous firing rate ($FR_{38°}$) is associated with an increase in hypothalamic thermosensitivity at PO/AH temperatures above 37-38°C. Most of the cold-sensitive neurons that respond to either skin or spinal temperature are excited by extrahypothalamic cooling and inhibited by extrahypothalamic warming.

Figure 13 summarizes the general effect of extrahypothalamic temperature on the thermoresponse curve for many cold-sensitive neurons. That is, as the level of firing rate increases due to extrahypothalamic temperature, the local hypothalamic thermosensitivity also increases, and a higher maximum firing rate is often observed. These changes in neuronal activity are the inverse of the changes observed in warm-sensitive neurons. This suggests that cold-sensitive neurons are not independently thermosensitive; rather, their cold sensitivity could be attributed to synaptic inhibition from nearby, warm-sensitive neurons. Otherwise, if cold-sensitive neurons were completely independent, one might expect to see the same type of responses observed in warm-sensitive neurons. That is, the same concept of competition of excitatory inputs might be expected for the cold-sensitive neurons, such that there would be only one apparent maximal firing rate, and as the $FR_{38°}$ increased and approached the maximum firing rate, the local cold sensitivity would be expected to decrease. Since these possibilities are not observed, the simplest explanation is that cold-sensitive neurons receive inhibitory inputs from nearby warm-sensitive neurons.

As indicated in Figure 13, a warm-sensitive neuron could inhibit a neuron having some tonic firing rate. Accordingly, when hypothalamic temperature is changed, this inhibited neuron would appear to be cold sensitive. A decreased hypothalamic temperature would decrease the firing rate in the inhibitory warm-sensitive neuron, which would increase the firing rate in the apparent "cold-sensitive" neuron. Thus, a tonically firing neuron, with little or no inherent thermosensitivity, could appear to be cold sensitive due to this tonic inhibition from a nearby warm-sensitive neuron. Peripheral cooling would decrease the spontaneous firing in the warm-sensitive neuron, which in turn would increase the activity in the "cold-sensitive neuron" which it inhibits. Accompanying this increased activity in the cold-sensitive neuron, an increase in hypothalamic thermosensitivity would be expected, since the inhibiting warm-sensitive neuron increased its own thermosensitivity when its firing rate decreased. Thus, the thermoresponse curves of cold-sensitive neurons are simply mirror-images of those in warm-sensitive neurons.

Note the similarities between the responses of cold-sensitive neurons and those thermoregulatory responses that increase with PO/AH cooling, such as metabolic heat production (Fig. 2) and certain behavioral responses (Fig. 5). Peripheral cooling not only increases heat-production and thermoregulatory behavior but also increases the hypothalamic thermosensitivity (α) for these responses. This strongly suggests that the PO/AH cold-sensitive neurons have an important integrative role in the control of these responses. Thus, many heat-production and heat-loss responses conform to the representation shown in Figure 13. Peripheral cooling tends to increase the hypothalamic thermosensitivity for both types of responses, and peripheral warming tends to decrease the hypothalamic thermosensitivity for both types of responses. Hypothalamic thermoresponse curves for the thermosensitive neurons and the thermoregulatory responses all tend to converge at high PO/AH temperatures. The simplest explanation for this conformity of responses is that PO/AH warm-sensitive neurons provide the basis for the neuronal integration of central and peripheral temperatures, and that excitatory inputs to these neurons not only summate but also compete with one another for determining neuronal excitation. Therefore, the concept of competition of excitatory inputs to warm-sensitive neurons provides

Hypothalamic Control of Thermoregulation

the basis of the convergence seen in the thermoresponse curves of hypothalamic neurons and thermoregulatory responses.

C. Neuronal Control of Thermoregulation

The neuronal model in Figure 13 explains how individual hypothalamic neurons may integrate central and peripheral thermal signals to control thermoregulatory responses. It does not explain, however, which neurons would be most likely to have specific functional roles in particular thermoregulatory responses. It does not specify, for example, which warm-sensitive neurons would be more likely to control heat-loss responses and which warm-sensitive neurons would be more likely to control heat-production responses. Earlier (in Figure 9), we saw that the level of spontaneous firing rate determines the PO/AH temperature range in which a warm-sensitive neuron is most thermosensitive. The range of greatest thermosensitivity suggests the most likely thermoregulatory role of each neuron; that is, neurons sensitive only to high PO/AH temperatures would be expected to function primarily in heat-loss responses, while neurons sensitive only to low PO/AH temperatures would be expected to function primarily in heat-production responses. This is the basis for the neuronal model in Figure 15, which is a simplified version of an earlier model (Boulant, 1974, 1977). As noted (Figs. 9 and 15), the low-firing, warm-sensitive neurons are thermosensitive only in the hyperthermic range of PO/AH temperatures. It is proposed, therefore, that these neurons have a role in the control of heat-loss responses, such as panting and sweating, which actually function in this temperature range. While

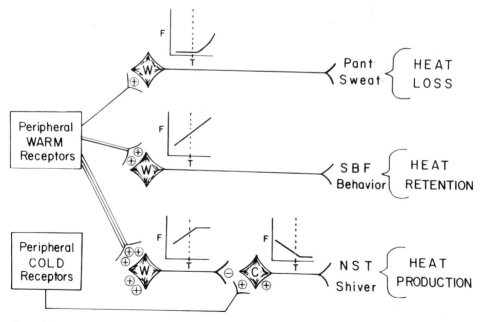

Figure 15 A neuronal model for thermoregulation. W, warm-sensitive neuron; C, cold-sensitive neuron; (+), excitatory input; (−), inhibitory input. The hypothalamic thermoresponse curves of various neuronal types are shown next to those neurons. F, firing rate; T, hypothalamic temperature; SBF, to skin blood flow; and NST, nonshivering thermogenesis.

the reason for the low firing rates is now known, it is suggested that these neurons receive relatively little excitatory synaptic input from extrahypothalamic thermoreceptors. An earlier model (Boulant, 1974) suggested that these low-firing neurons might receive a greater amount of inhibitory synaptic input from ascending cold-receptor afferents. Recent studies indicate, however, that the relative amount of inhibitory input to warm-sensitive neurons is very small, as in Figure 10(b) and is uniformly low for neurons in all ranges of spontaneous firing rate (Boulant and Demieville, 1977); see Figure 16(b). Therefore, inhibitory inputs probably synapse on all types of warm-sensitive neurons, but for the sake of simplicity, they are not included in the model in Figure 15.

Heat-retention or heat-conserving responses, like skin blood flow (SBF) and thermoregulatory behavior, are most active near thermoneutrality and are the first responses to change when PO/AH temperature is altered above or below thermoneutrality. It would seem likely, therefore, that these responses are controlled by neurons which are equally sensitive to temperature changes above and below thermoneutrality. This includes primarily the warm-sensitive PO/AH neurons having medium levels of firing rate, i.e., 10-15 impulses per second (Fig. 9). Figures 16 and 17 suggest that these particular neurons receive a moderate amount of excitatory input from extrahypothalamic warm receptors, thus accounting for their medium firing rates. This is supported by evidence from studies employing thermal stimulation of the skin and spinal cord, as in Figures 16(a) and 17(a) and electrical stimulation of the ascending anterolateral pathway, as in Figures 11 and 16(b).

Although not shown in Figure 15, it is probable that some of these medium-firing, warm-sensitive neurons inhibit and thus determine the thermosensitivity of nearby cold-sensitive neurons. It is therefore likely that some of the cold-sensitive neurons function in heat-retention or even heat-loss responses.

Support for the behavioral role of the medium-firing, warm-sensitive neurons comes from a recent study employing electrical stimulation in the hippocampus (Fig. 11), an important limbic structure associated with behavior, attention-directing, spatial orientation, and memory. Such hippocampal stimulation selectively excited most of the warm-sensitive neurons having firing rates of 10-15 impulses per second, as in Figure 16(c). Hippocampal stimulation affected only a small proportion of warm-sensitive neurons having higher or lower firing rates (Boulant and Demieville, 1977). Therefore the hippocampus (a structure closely linked with behavior and memory) has a specific input to the medium-firing, warm-sensitive neurons, i.e., the neuronal population most likely to have a role in heat-retention thermoregulatory behavior.

As noted in Figure 9, the high-firing (i.e., 15-25 and 25-60 impulses per second), warm-sensitive neurons are not very sensitive to PO/AH warming. Rather, these high-firing, warm-sensitive neurons have their greatest thermosensitivity in the hypothermic range—i.e., they substantially decrease their firing rates only when the PO/AH is cooled below normal. Consequently, it is proposed in Figure 15 that these neurons control heat production responses by inhibiting and, thus, determining the "thermosensitivity" of most cold-sensitive neurons. These cold-sensitive neurons might regulate shivering and nonshivering thermogenesis (NST). It is suggested that these hypothermic, warm-sensitive neurons receive the greatest amount of excitatory input, either synaptically from extrahypothalamic warm-receptors or from local factors. As indicated below, support for these two possibilities is presented in Figures 16(a,b) and 17.

In addition to inhibitory inputs from warm-sensitive neurons, the PO/AH cold-sensitive neurons probably also receive tonic excitation from local inputs, as well as from

Hypothalamic Control of Thermoregulation

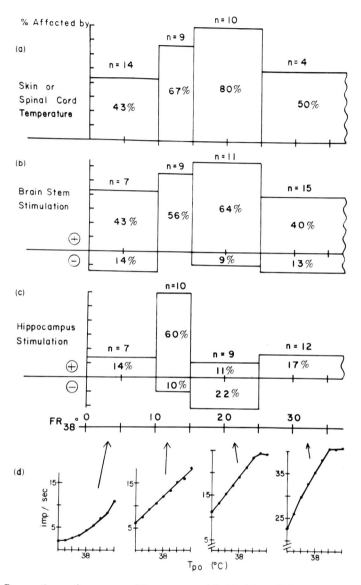

Figure 16 Proportions of warm-sensitive neurons (divided into firing rate groups as in Fig. 9) affected by (a) thermal stimulation of the skin or spinal cord (Boulant and Hardy, 1974); (b) electrical stimulation of brain-stem sites (RF, LL, and ST as shown in Fig. 11) along the ascending anterolateral pathway; and (c) electrical stimulation of the hippocampus (HP as shown in Fig. 11). The n refers to the total number of neurons in each group. The curves in (d) are the average thermoresponse curves for each firing-rate group of warm-sensitive neurons in (b) and (c). (From Boulant and Demieville, 1977.)

extrahypothalamic cold receptors (Fig. 15). As noted in Figure 10, cold-sensitive neurons represent the smallest population of PO/AH neurons, yet they receive the greatest proportion of synaptic input from cutaneous and spinal thermal afferents. Furthermore electrical stimulation along the anterolateral afferent pathway indicates that most of this input is excitatory to these cold-sensitive neurons (Boulant and Demieville, 1977).

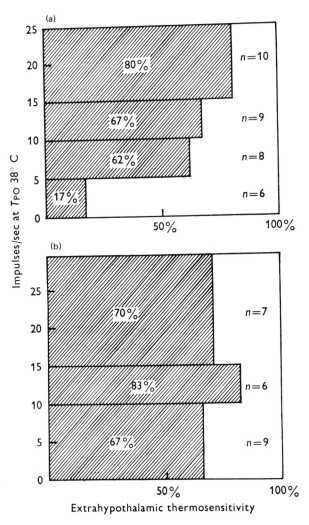

Figure 17 The incidence of extrahypothalamic thermosensitivity among populations of (a) warm-sensitive PO/AH neurons and (b) cold-sensitive PO/AH neurons. Neurons in each population are grouped according to their firing rates at 38°C preoptic temperature, T_{po}. The n refers to the number of neurons. In each group, the proportion of neurons sensitive to either spinal or skin temperature is indicated by the shaded area. (From Boulant and Hardy, 1974.)

The neuronal model (Fig. 15) is further based on the fact that many PO/AH single-unit studies indicate that warm-sensitive neurons outnumber cold-sensitive neurons by a ratio of about three or four to one (Hardy et al., 1964; Eisenman and Jackson, 1967; Hellon, 1967; Wit and Wang, 1968a; Guieu and Hardy, 1970b; Boulant and Bignall, 1973a). This same ratio is maintained if most of the cold-sensitive neurons are determined primarily by the high-firing, hypothermic, warm-sensitive neurons.

Figure 15 suggests that the amount of excitatory afferents from peripheral thermoreceptors is the primary determinant of spontaneous firing rate in warm-sensitive neurons.

Figures 16(a,b) and 17 indicate that this is quite probable for warm-sensitive neurons firing between 0 and 25 impulses per second. Warm-sensitive neurons are grouped according to their firing rates at 38°C. These figures show the proportion of neurons in each group which are also sensitive to changes in skin or spinal temperature or to electrical stimulation of afferent pathways. The low-firing, warm-sensitive neurons receive little thermal afferent input, while 80% of these neurons firing between 15 and 25 impulses per second are affected by an extrahypothalamic temperature. Almost identical proportions of these groups of warm-sensitive neurons are excited by electrically stimulating peripheral thermoreceptor afferent pathways as they ascend through the brain stem (Boulant and Demieville, 1977). Consequently, for warm-sensitive neurons firing between 0 and 25 impulses per second, the amount of excitatory synaptic input from peripheral warm receptors may determine the level of the firing rate, and thus, the range of local thermosensitivity. This concept may not hold true for warm-sensitive neurons having firing rates greater than 25 impulses per second. Only about one-half of these neurons are affected by extrahypothalamic temperatures (Boulant and Hardy, 1974) or brain-stem electrical stimulation (Boulant and Demieville, 1977). It may be that these neurons are affected to a greater extent by local excitatory factors, accounting for their very high firing rates.

In certain animals (e.g., rat) that rely heavily on nonshivering thermogenesis, these differences between warm-sensitive neurons (firing at 15-25 impulses per second and 25-60 impulses per second) may determine the functional differences between shivering and nonshivering thermogenesis. Even among the high-firing, hypothermic, warm-sensitive neurons, it is possible that two neuronal groups exist having different firing rates, different threshold temperatures, and different amounts of local and afferent inputs. The highest-firing (25-60 impulses per second; Fig. 9) neurons may control shivering and may have the lowest hypothalamic T_{set}. These neurons may also be affected more by local inputs than by peripheral afferent inputs. The hypothermic neurons firing between 15 and 25 impulses per second may control NST. These may have a somewhat higher T_{set} and receive a great amount of excitatory input from peripheral thermoreceptors. Consequently, in some animals, NST is the first heat-production response to be elicited during peripheral or PO/AH cooling, and shivering is evoked only during intense PO/AH cooling. In other animals this process may be reversed, such that the hypothermic neurons, firing between 15 and 25 impulses per second, may control shivering, and the highest-firing hypothermic neurons may control NST.

Figure 15 predicts that extrahypothalamic temperature affects most of the PO/AH cold-sensitive neurons. Not only do these neurons receive input from peripheral cold receptors, but they are inhibited by the high-firing, warm-sensitive neurons, which receive the greatest proportion of peripheral warm-receptor input. Figures 10 and 17 support this concept. Approximately 70-80% of all cold-sensitive neurons are affected by skin or spinal temperature. This is about the same proportion of extrahypothalamic thermosensitivity observed in warm-sensitive neurons having firing rates between 10 and 25 impulses per second. However, it should be noted again that most of the input to cold-sensitive neurons is excitatory rather than inhibitory. It is quite likely that both excitatory and inhibitory inputs occur in cold-sensitive neurons and that the excitatory response is predominant during indiscriminant electrical stimulation.

The model in Figure 15 does not depend on T_{set} interneurons receiving mutually antagonistic inputs from PO/AH neurons having different thermosensitivities. Rather, the amount of excitatory input determines a neuron's level of firing rate and range of local thermosensitivity. This range of thermosensitivity indicates the most likely thermoregulatory role of a particular neuron. Thus, developmentally, the amount of afferent input a neuron receives may determine that neuron's role in thermoregulation. This would explain many previously mentioned examples indicating that peripheral temperatures have a great influence on the hypothalamic control of metabolic heat production but have only a small influence on the hypothalamic control of heat-loss responses, like sweating and panting (Benzinger et al., 1963; Hellstrøm and Hammel, 1967).

Finally, recent single-unit studies of thermosensitive neurons in hypothalamic tissue slices provide further support for the models proposed in Figures 13 and 15 (Kelso et al., 1980). With the removal of afferent synaptic input, in vitro warm-sensitive neurons all have very low firing rates and are primarily sensitive only to hyperthermic temperatures. The in vitro cold-sensitive neurons are also sensitive only in the hyperthermic range; this would be expected if their thermosensitivity were due to inhibition from nearby, warm-sensitive neurons. Such single-unit work in isolated hypothalamic tissue offers a powerful tool for future studies to test and revise neuronal models.

D. Concluding Remarks

It seems appropriate to end this chapter with some personal comments regarding hypothalamic neurons involved in thermoregulation.

First of all, regarding neuronal models: A primary purpose of a model should be to facilitate experiments designed to prove the model false, rather than true. As noted earlier in this chapter, neuronal models and single-unit studies often tend to become mutually reinforcing. The model dictates experimental procedures and predicts experimental results. The single-unit data are then interpreted as they fit the model's predictions. This process cannot be avoided; it can be limited, however, by devising alternative hypotheses to test and by giving equal emphasis to the conflicting and often confusing data.

Second, regarding hypothalamic thermosensitivity (or α): Certain figures have been presented throughout this chapter to stress the point that peripheral thermal afferents can change the hypothalamic thermosensitivity for both whole-body responses and neuronal responses. These responses include metabolic heat production (Fig. 2), skin blood flow (Fig. 3), panting (Fig. 4), thermoregulatory behavior (Fig. 5), and the firing rates of warm- and cold-sensitive PO/AH neurons (Figs. 12 and 14). On the other hand, there is also evidence indicating that the peripheral afferent effect is simply additive and does not alter the hypothalamic thermosensitivity for these responses. This would be equivalent to a shift in the hypothalamic T_{set} as described in Figure 1. In terms of whole-body thermoregulation, it isn't really that important whether peripheral thermal afferents change the hypothalamic thermosensitivity or the hypothalamic T_{set}. As long as one or both of these parameters is changed, the appropriate thermoregulatory response could probably be made to maintain a constant core temperature.

Why then, are the changes in hypothalamic thermosensitivity (α) emphasized in this chapter? The importance of these slope or sensitivity changes is that they provide clues to PO/AH neuronal characteristics and the organization of neuronal networks. If a change in skin temperature shifts the hypothalamic T_{set} but does not change the hypothalamic α of neurons or whole-body responses, then there is no reason to speculate that the thermosensitivity of cold-sensitive neurons is due to inhibitory inputs from nearby warm-sensitive neurons. If, however, peripheral afferents do change the value of α as suggested in Fig. 13, then a functional link between warm- and cold-sensitive neurons is substantiated. Moreover, the concept of competition between different excitatory inputs to warm-sensitive neurons then becomes the basis for the slope changes observed in the neuronal and whole-body hypothalamic thermoresponse curves.

A third comment also regards neuronal characteristics and neuronal models. In some ways, models often postulate neuronal networks that are more complicated than need be, in order to explain regulatory control. In other ways, however, all models necessarily oversimplify the complicated interaction of synaptic and nonsynaptic electrical events of the neuronal membrane. As noted above, for a given neuron, the summation of one excitatory input with another excitatory input does not necessarily equal the equivalent of two excitatory inputs. Moreover, one excitatory input plus one inhibitory input does not necessarily equal zero. An EPSP is not the exact opposite of an inhibitory postsynaptic potential, a fact that is overlooked by all of the models presented in this chapter. Synaptic or neurochemical interactions, therefore, have not been realistically considered in this chapter. The neurochemical basis of thermoregulation is presented in the following chapter. If the models proposed above (Figs. 13 and 15) are valid, then a given neurotransmitter could have a substantially different effect on neurons having different afferent inputs and different functional roles. An excitatory neurochemical, for example, might have a completely different effect on a slow firing, warm-sensitive neuron compared with a warm-sensitive neuron that is already firing near its maximal rate. The reader should consider these possibilities when analyzing the effects of various natural and artificial central stimulations, whether chemical, electrical, or thermal.

Finally, regarding neuronal specificity. It should be obvious to the reader that not all of the hypothalamic neurons can function in the specific roles suggested throughout this text. In this chapter, for example, it is stated that about one-half of the PO/AH neurons are temperature sensitive. Does this mean the temperature-sensitive neurons control thermoregulation and the temperature-insensitive neurons control the other regulatory functions of the PO/AH? Of course not! Undoubtedly, other chapters in this text will report that certain proportions of PO/AH neurons act as receptors of various neurohormones. Still other chapters will report that a certain proportion of the PO/AH neurons are glucoreceptors or osmoreceptors, etc. Before long, we quickly run out of PO/AH neurons necessary to control the various systems in which the PO/AH acts. The point is, it would be naive to think that all temperature-sensitive neurons are involved only in thermoregulation and not in the other homeostatic systems controlled by the PO/AH. It would be equally naive to believe that some thermosensitive PO/AH neurons could not be influenced by endogenous glucose, progesterone, estrogen, osmotic pressure, etc. How much, for example, is thermoregulation affected by alterations in body metabolites, body water, or endocrine levels, or better yet, by simultaneous alterations in all three parameters? Conversely, how much is osmoregulation affected by alterations in body temperature, body metabolites, and endogenous endocrines? In the opening paragraph of this chapter, it was noted that a vast interrelated neuronal network regulates the body's internal environment, both autonomically and behaviorally, and that the separation of the thermoregulatory system from all other homeostatic systems was a separation contrived for simplification. It is important, therefore, that we continue to recognize this fact and that we view each hypothalamic neuron as it relates to the entire homeostatic system and not just as it relates to the specific system we happen to be studying.

Acknowledgments

The author would like to thank Dr. Robert Myers and Dr. Jack Rall for critical review of this chapter.

Much of the author's research cited in this chapter has been supported by grants from NIH and the American Heart Association.

References

Abdel-Sayed, W. S., Abboud, F. M., and Calvelo, M. G. (1970). Effect of local cooling on responsiveness of muscular and cutaneous arteries and veins. *Am. J. Physiol. 219,* 1772-1778.

Abrahams, V. C., Hilton, S. M., and Zbrozyna, A. (1960). Active muscle vasodilation produced by stimulation of the brainstem: its significance in the defense reaction. *J. Physiol. 154,* 491-513.

Adair, E. R. (1974). Hypothalamic control of thermoregulatory behavior. Preoptic-posterior hypothalamic interaction. In *Recent Studies of Hypothalamic Function,* K. Lederis and K. E. Cooper (Eds.). Karger, Basel, pp. 341-358.

Adair, E. R. (1977). Skin, preoptic and core temperatures influence behavioral thermoregulation. *J. Appl. Physiol. 42,* 559-564.

Adair, E. R., and Rawson, R. O. (1974). Autonomic and behavioral temperature regulation unilateral vs. bilateral preoptic thermal stimulation. *Pfluegers Arch. Gesamte Physiol. 352,* 91-103.

Adair, E. R., Casby, J. U., and Stolwijk, J. A. J. (1970). Behavioral temperature regulation in the squirrel monkey: changes induced by shifts in hypothalamic temperature. *J. Comp. Physiol. Psychol. 72,* 17-27.

Andersson, B. (1957). Cold defense reactions elicited by electrical stimulation within the septal area of the brain in goats. *Acta Physiol. Scand. 41,* 90-100.

Andersson, B., and Persson, N. (1957). Pronounced hypothermia elicited by prolonged stimulation of the "heat loss center" in unanesthetized goats. *Acta Physiol. Scand. 41,* 277-282.

Andersson, B., Ekman, L., Gale, C. C., and Sundsten, J. W. (1962). Thyroidal response to local cooling of the pre-optic "heat loss center." *Life Sci. 1,* 1-11.

Andersson, B., Grant, R., and Larsson, S. (1956). Central control of heat loss mechanisms in the goat. *Acta Physiol. Scand. 37,* 261-280.

Andersson, B., Ekman, L., Gale, C. C., and Sundsten, J. W. (1963a). Control of thyrotrophic hormone (TSH) secretion by the "heat loss center." *Acta Physiol. Scand. 59,* 12-33.

Andersson, B., Gale, C. C., Hökfelt, B., and Ohga, A. (1963b). Relation of preoptic temperature to the function of the sympathetico-adrenomedullary system and the adrenal cortex. *Acta Physiol. Scand. 61,* 182-191.

Baker, M. A. (1972). Influence of the carotid rete on brain temperature in cats exposed to hot environments. *J. Physiol. 220,* 711-728.

Baker, M. A., and Hayward, J. N. (1967). Carotid rete and brain temperature of cat. *Nature* (London) *216,* 139-141.

Baker, M. A., and Hayward, J. N. (1968). The influence of the nasal mucosa and the carotid rete upon hypothalamic temperature in sheep. *J. Physiol. 198,* 561-579.

Baldwin, B. A., and Ingram, D. L. (1967). The effect of heating and cooling the hypothalamus on behavioral thermoregulation in the pig. *J. Physiol. 191,* 375-392.

Banet, M., and Hensel, H. (1976). Nonshivering thermogenesis induced by repetitive hypothalamic cooling in the rat. *Am. J. Physiol. 230,* 522-526.

Bard, P., and Macht, M. B. (1958). The behavior of chronically decerebrate cats. In *Symposium on the Neurological Basis of Behavior*–Ciba Symposium, London, Little, Brown, Boston, pp. 55-71.

Bard, P., Woods, J. W., and Bleier, R. (1970). The effects of cooling, heating and pyrogen on chronically decerebrate cats. In *Physiological and Behavioral Temperature Regulation,* J. D. Hardy, A. P. Gagge, and J. A. J. Stolwijk (Eds.). Thomas, Springfield, Ill., pp. 519-545.

Bazett, H. C., Alpers, B. J., and Erb, W. H. (1933). Hypothalamus and temperature control. *Arch. Neurol. Psychiat. 30,* 728-748.

Beckman, A. L., and Eisenman, J. S. (1970). Microelectrophoresis of biogenic amines on hypothalamic thermosensitive cells. *Science 170,* 334-336.

Beckman, A. L., and Rozkowska-Ruttimann, E. (1974). Hypothalamic and septal neuronal responses to iontophoretic application of salicylate in rats. *Neuropharmacology 13,* 393-398.

Benzinger, T. H., and Taylor, G. W. (1963). In *Temperature–Its Measurement and Control in Science and Industry,* Vol. 3, Pt. 3, *Biology and Medicine,* J. D. Hardy (Ed.). Reinhold, New York, pp. 111-120.

Benzinger, T. H., Kitzinger, C., and Pratt, A. W. (1963). The human thermostat. In *Temperature–Its Measurement and Control in Science and Industry,* Vol. 3, Pt. 3, *Biology and Medicine,* J. D. Hardy (Ed.). Reinhold, New York, pp. 637-665.

Bignall, K. E., and Schramm, L. (1974). Behavior of chronically decerebrated kittens. *Exp. Neurol. 42,* 519-531.

Birzis, L., and Hemingway, A. (1956). Descending brain stem connections controlling shivering in cat. *J. Neurophysiol. 19,* 37-43.

Birzis, L., and Hemingway, A. (1957a). Shivering as a result of brain stimulation. *J. Neurophysiol. 20,* 91-99.

Birzis, L., and Hemingway, A. (1957b). Efferent brain discharge during shivering. *J. Neurophysiol. 20,* 156-166.

Bligh, J. (1973). *Temperature Regulation in Mammals and Other Vertebrates.* North-Holland, Amsterdam, pp. 61-63, 68-70, 192-213.

Boulant, J. A. (1971). *Time-Dependent Thermosensitive Characteristics of Preoptic and Anterior Hypothalamic Neurons* (Dissertation). University of Rochester, New York.

Boulant, J. A. (1974). The effect of firing rate on the local thermosensitivity of preoptic neurones. *J. Physiol. (London) 240,* 661-669.

Boulant, J. A. (1977). A hypothalamic neuronal model for thermoregulation. In *Selected Topics in Environmental Biology,* B. Bhatia, G. S. Chhina, and B. Singh (Eds.). Pergamon, Oxford, England, Chapter 7, pp. 41-44.

Boulant, J. A. (1980). Hypothalamic mechanisms in thermoregulation. *Fed. Proc.* In press.

Boulant, J. A., and Bignall, K. E. (1971). Time-dependent characteristics of hypothalamic temperature-sensitive neurons. *Fed. Proc. 30,* 319.

Boulant, J. A., and Bignall, K. E. (1973a). Determinants of hypothalamic neuronal thermosensitivity in ground squirrels and rats. *Am. J. Physiol. 225,* 306-310.

Boulant, J. A., and Bignall, K. E. (1973b). Changes in the thermosensitive characteristics of hypothalamic units over time. *Am. J. Physiol. 225*, 311-318.

Boulant, J. A., and Bignall, K. E. (1973c). Hypothalamic neuronal responses to peripheral and deep-body temperatures. *Am. J. Physiol. 225*, 1371-1374.

Boulant, J. A., and Demieville, H. N. (1977). Responses of thermosensitive preoptic and septal neurons to hippocampal and brain stem stimulation. *J. Neurophysiol. 40*, 1356-1368.

Boulant, J. A., and Gonzalez, R. R. (1974). The effect of extrahypothalamic temperature on preoptic control of thermoregulation. *Fed. Proc. 33*, 457.

Boulant, J. A., and Gonzalez, R. R. (1977). The effect of skin temperature on the hypothalamic control of heat loss and heat production. *Brain Res. 120*, 367-372.

Boulant, J. A., and Hardy, J. D. (1972). Hypothalamic and medullary neuronal responses to deep-body, peripheral and local temperatures. *Int. J. Biometerology 16* (Suppl.), 38-39.

Boulant, J. A., and Hardy, J. D. (1974). The effect of spinal and skin temperatures on the firing rate and thermosensitivity of preoptic neurones. *J. Physiol. 240*, 639-660.

Brück, K., and Schwennicke, H. P. (1971). Interaction of superficial and hypothalamic thermosensitive structures in the control of non-shivering thermogenesis. *Int. J. Biometeorology 15*, 156-161.

Brück, K., and Wünnenberg, W. (1970). "Meshed" control of two effector systems: non-shivering and shivering thermogenesis. In *Physiological and Behavioral Temperature Regulation*, J. D. Hardy, A. P. Gagge, and J. A. J. Stolwijk (Eds.). Thomas, Springfield, Ill., pp. 562-580.

Brück, K., Wünnenberg, W., and Zeisberger, E. (1969). Comparison of cold-adaptive metabolic modifications in different species, with special reference to the miniature pig. *Fed. Proc. 28*, 1035-1041.

Burton, H. (1975). Responses of spinal cord neurons to systematic changes in hindlimb skin temperatures in cats and primates. *J. Neurophysiol. 38*, 1060-1079.

Cabanac, M., and Hardy, J. D. (1969). Réponses unitaires et thermorégulatrices lors de réchauffements et refroidissements localisés de la région préoptique et du mésencéphak chez le lapin. *J. Physiol. (Paris) 61*, 331-347.

Cabanac, M., Hammel, H. T., and Hardy, J. D. (1967). *Tiligua scincoides:* Temperature-sensitive units in lizard brain. *Science 158*, 1050-1051.

Cabanac, M., Stolwijk, J. A. J., and Hardy, J. D. (1968). Effect of temperature and pyrogens on single unit activity in the rabbit's brain stem. *J. Appl. Physiol. 24*, 645-652.

Calaresu, F. R., and Mogenson, G. J. (1972). Cardiovascular responses to electrical stimulation of the septum in the rat. *Am. J. Physiol. 223*, 777-782.

Carlisle, H. J. (1969). Effect of preoptic and anterior hypothalamic lesions on behavioral thermoregulation in the cold. *J. Comp. Physiol. Psychol. 69*, 391-402.

Carlisle, H. J., and Ingram, D. L. (1973). The effects of heating and cooling the spinal cord and hypothalamus on thermoregulatory behaviour in the pig. *J. Physiol. 231*, 253-364.

Chai, C. Y., and Lin, M. T. (1972). Effects of heating and cooling the spinal cord and medulla oblongata on thermoregulation in monkeys. *J. Physiol. 225*, 297-308.

Chai, C. Y., and Lin, M. T. (1973). Effects of thermal stimulation of medulla oblongata and spinal cord on decerebrate rabbits. *J. Physiol. 234*, 409-419.

Chai, C. Y., and Wang, S. C. (1970). Cardiovascular and respiratory responses to cooling of the medulla oblongata of the cat. *Proc. Soc. Exp. Biol. Med. 134*, 763-767.

Chai, C. Y., Mu, J. Y., and Brobeck, J. R. (1965). Cardiovascular and respiratory responses from local heating of the medulla oblongata. *Am. J. Physiol. 209*, 301-306.

Chambers, W. W., Koenig, H., and Windle, W. F. (1949). Site of action in the central nervous system of a bacterial pyrogen. *Am. J. Physiol. 159*, 209-216.

Chambers, W. W., Seigel, M. S., Liu, J. C., and Liu, C. N. (1974). Thermoregulatory responses of decerebrate and spinal cats. *Exp. Neurol. 42*, 282-299.

Chappius, P., Pittlet, P., and Jequier, E. (1976). Heat storage regulation in exercise during thermal transients. *J. Appl. Physiol. 40*, 384-392.

Chowers, I., Hammel, H. T., Stromme, S. B., and McCann, S. M. (1964). Comparison of effect of environmental and preoptic cooling on plasma cortisol levels. *Am. J. Physiol. 207*, 577-582.

Clarke, N. P., and Rushmer, R. F. (1967). Tissue uptake of ^{86}Rb with electrical stimulation of hypothalamus and midbrain. *Am. J. Physiol. 213*, 1439-1444.

Clough, D. P., and Jessen, C. (1974). The role of spinal thermosensitive structures in the respiratory heat loss during exercise. *Pfluegers Arch. Gesamte Physiol. 347*, 235-248.

Conner, J. D., and Crawford, I. L. (1969). Hyperthermic in midpontine lesioned cats. *Brain Res. 15*, 590-593.

Coote, J. H., Nilton, S. M., and Zbrożyna, A. W. (1973). The ponto-medullary area integrating the defense reaction in the cat and its influence in muscle blood flow. *J. Physiol. 229*, 257-274.

Corbit, J. D. (1969). Behavioral regulation of hypothalamic temperature. *Science 166*, 256-258.

Corbit, J. D. (1973). Voluntary control of hypothalamic temperature. *J. Comp. Physiol. Psychol. 83*, 394-411.

Crawshaw, L. I., and Hammel, H. T. (1974). Behavioral regulation of internal temperature in the brown bullhead, *Ictalurus nebulosus*. *Comp. Biochem. Physiol. 47*, 51-60.

Crawshaw, L. I., and Stitt, J. T. (1975). Behavioural and autonomic induction of prostaglandin E_1 fever in squirrel monkeys. *J. Physiol. 244*, 197-206.

Cronin, M. J., and Baker, M. A. (1976). Heat-sensitive midbrain raphe neurons in the anesthetized cat. *Brain Res. 110*, 175-181.

Crosby, E. C., Humphrey, T., and Lauer, E. W. (1962). *Correlative Anatomy of the Nervous System*, Macmillan, New York, p. 125.

Cunningham, D. J., Stolwijk, J. A. J., Murakami, N., and Hardy, J. D. (1967). Responses of neurons in the preoptic area to temperature, serotonin, and epinephrine. *Am. J. Physiol. 213*, 1570-1581.

Denny-Brown, D. (1966). *The Cerebral Control of Movement*. Thomas, Springfield, Ill.

Dickenson, A. H. (1976). Neurones in raphé nuclei of the rat responding to skin temperature. *J. Physiol. 256*, 110 P.

Downey, J. A., Huckaba, C. E., Kelley, P. S., Tam, H. S., Darling, R. C., and Cheh, H. Y. (1976). Sweating responses to central and peripheral heating in spinal man. *J. Appl. Physiol. 40*, 701-706.

Dworkin, S. (1930). Observations on the central control of shivering and of heat regulation in the rabbit. *Am. J. Physiol. 93*, 227-244.

Edinger, H. M., and Eisenman, J. S. (1970). Thermosensitive neurons in tuberal and posterior hypothalamus of cats. *Am. J. Physiol. 219*, 1098-1103.

Eisenman, J. S. (1969). Pyrogen-induced changes in the thermosensitivity of septal and preoptic neurons. *Am. J. Physiol. 216*, 330-334.

Eisenman, J. S. (1970). The action of bacterial pyrogen on thermoresponsive neurons. In *Physiological and Behavioral Temperature Regulation,* J. D. Hardy, A. P. Gagge, J. A. J. Stolwijk, (Eds.). Thomas, Springfield, Ill., pp. 507-518.

Eisenman, J. S., and Jackson, D. C. (1967). Thermal response patterns of septal and preoptic neurons in cats. *Exp. Neurol. 19*, 33-45.

Evans, S. E., and Ingram, D. L. (1974). The significance of deep-body temperature in regulating the concentration of thyroxine in the plasma of the pig. *J. Physiol. 236*, 159-170.

Findlay, J. D., and Ingram, D. L. (1961). Brain temperature as a factor in the control of thermal polypnea in the ox (*Bos taurus*). *J. Physiol. 155*, 72-85.

Folkow, B., Johansson, B., and Öberg, B. (1959). A hypothalamic structure with a marked inhibitory effect on tonic sympathetic activity. *Acta Physiol. Scand. 47*, 262-270.

Ford, D. M. (1974). A selective action of prostaglandin E_1 on hypothalamic neurones in the cat which respond to brain cooling. *J. Physiol. 242*, 142-143.

Forsling, M. L., Ingram, D. L., and Stanier, M. W. (1976). Effects of various ambient temperatures and of heating and cooling the hypothalamus and cervical spinal cord on antidiuretic hormone secretion and urinary osmolality in pigs. *J. Physiol. 257*, 673-686.

Forster, R. E., II, and Ferguson, T. B. (1952). Relationship between hypothalamic temperature and thermoregulatory effectors in unanesthetized cat. *Am. J. Physiol. 169*, 255-269.

Forsyth, R. P. (1970). Hypothalamic control of the distribution of cardiac output in the unanesthetized rhesus monkey. *Circ. Res. 26*, 783-794.

Freeman, W. J., and Davis, D. D. (1959). Effect on cats of conductive hypothalamic cooling. *Am. J. Physiol. 197*, 145-148.

Fuhrer, M. J. (1971). Skin conductance responses mediated by the transected human spinal cord. *J. Appl. Physiol. 30*, 663-669.

Fuller, C. A., Horwitz, B. A., and Horowitz, J. M. (1975). Shivering and nonshivering thermogenic responses of cold-exposed rats to hypothalamic warming. *Am. J. Physiol. 228*, 1519-1524.

Fusco, M. M., Hardy, J. D., and Hammel, H. T. (1961). Interaction of Central and peripheral factors in physiological temperature regulation. *Am. J. Physiol. 200*, 572-580.

Gale, C. C., Jobin, M., Proppe, D. W., Notter, D., and Fox, H. (1970). Endocrine thermoregulatory responses to local hypothalamic cooling in unanesthetized baboons. *Am. J. Physiol. 219*, 193-201.

Gale, C. C., Mathews, M., and Young, J. (1970). Behavioral thermoregulatory responses to hypothalamic cooling and warming in baboons. *Physiol. Behav. 5*, 1-6.

Ganong, W. F. (1975). *Review of Medical Physiology.* Lange, Los Altos, Calif., pp. 241, 264, 269-270, 278.

Gebber, G. L., and Snyder, D. W. (1970). Hypothalamic control of baroreceptor reflexes. *Am. J. Physiol. 218*, 124-131.

Gonzalez, R. R., Kluger, M. J., and Hardy, J. D. (1971). Partitional calorimetry of the New Zealand White rabbit at temperatures 5-35°C. *J. Appl. Physiol. 31,* 728-734.

Gonzalez, R. R., Kluger, M. J., and Stolwijk, J. A. J. (1974). Thermoregulatory responses to thermal transients in the rabbit. *Am. J. Physiol. 227,* 1292-1298.

Greer, G. L., and Gardner, D. R. (1970). Temperature-sensitive neurons in the brain of brook trout. *Science 169,* 1220-1223.

Guieu, J. D., and Hardy, J. D. (1970a). Effects of preoptic and spinal cord temperature in control of thermal polypnea. *J. Appl. Physiol. 28,* 540-542.

Guieu, J. D., and Hardy, J. D. (1970b). Effects of heating and cooling of the spinal cord on preoptic unit activity. *J. Appl. Physiol. 29,* 675-683.

Hainsworth, F. R. (1967). Saliva spreading, activity, and body temperature regulation in the rat. *Am. J. Physiol. 212,* 1288-1292.

Hales, J. R. S. (1973). Effects of exposure to hot environments on the regional distribution of blood flow and on cardiorespiratory function in sheep. *Pfluegers Arch. Gesamte Physiol. 344,* 133-148.

Hales, J. R. S., and Iriki, M. (1975). Integrated changes in regional circulatory activity evoked by spinal cord and peripheral thermoreceptor stimulation. *Brain Res. 87,* 267-279.

Hales, J. R. S., Fawcett, A. A., and Bennett, J. W. (1975). Differential influences of CNS and superficial body temperatures on the partition of cutaneous blood flow between capillaries and arteriovenous anastomoses (AVA's). *Pfluegers Arch. Gesamte Physiol. 361,* 105-106.

Hammel, H. T. (1965). Neurons and temperature regulation. In *Physiological Controls and Regulations,* W. S. Yamamoto and J. R. Brobeck (Eds.). Philadelphia, pp. 71-97.

Hammel, H. T., and Sharp, F. (1971). Thermoregulatory salivation in the running dog in response to preoptic heating and cooling. *J. Physiol. (Paris) 63,* 260-263.

Hammel, H. T., Hardy, J. D., and Fusco, M. M. (1960). Thermoregulatory responses to hypothalamic cooling in anesthetized dogs. *Am. J. Physiol. 198,* 481-486.

Hammel, H. T., Jackson, D. C., Stolwijk, J. A. J., Hardy, J. D., and Strømme, S. B. (1963). Temperature regulation by hypothalamic proportional control with an adjustable set point. *J. Appl. Physiol. 18,* 1146-1154.

Hammel, H. T., Caldwell, F. T., and Abrams, R. M. (1967). Regulation of body temperature in the blue-tongued lizard. *Science 156,* 1260-1262.

Hammel, H. T., Strømme, S. B., and Myhre, K. (1969). Forebrain temperature activates behavioral thermoregulatory response in Arctic sculpins. *Science 165,* 83-85.

Hammouda, M. (1933). The central and reflex mechanism of panting. *J. Physiol. 77,* 319-336.

Hardy, J. D. (1972). Peripheral inputs to the central regulatory for body temperature. In *Advances in Climatic Physiology,* S. Itoh, K. Ogata, and H. Yashimura (Eds.). Igaky Shoin Ltd., Tokyo, and Springer-Verlag, New York, 3-21.

Hardy, J. D. (1973). Posterior hypothalamus and the regulation of body temperature. *Fed. Proc. 32,* 1564-1571.

Hardy, J. D., and Hammel, H. T. (1963). Control system in physiological temperature regulation. In *Temperature–It's Measurement and Control in Science and Industry,* C. M. Herzfeld and J. D. Hardy (Eds.). Reinhold, New York, pp. 613-625.

Hardy, J. D., Hellon, R. F., and Sutherland, K. (1964). Temperature-sensitive neurones in the dog's hypothalamus. *J. Physiol. 175,* 242-253.

Hasama, B. (1929). Pharmakologische and physiologische Studien über die Schweisszentren. II. Über den Einfluss der direkten mechanischen, thermischen and elektrischen Reizung auf die Schweiss-sowie Wärmezentren. *Arch. Exp. Pathol. Pharmakol. 146*, 129-161.

Hayward, J. N. (1970). Central neural regulation of antidiuretic hormone release and unit activity in the supraoptic nucleus of the behaving rhesus monkey. *Am. J. Anat. 129*, 203-206.

Hayward, J. N. (1977). Functional and morphological aspects of hypothalamic neurons. *Physiol. Rev. 57*, 574-658.

Hayward, J. N., and Baker, M. A. (1968). Diuretic and thermoregulatory responses to preoptic cooling in the monkey. *Am. J. Physiol. 214*, 843-850.

Hellon, R. F. (1967). Thermal stimulation of hypothalamic neurones in unanesthetized rabbits. *J. Physiol. 193*, 381-395.

Hellon, R. F. (1969). Environmental temperature and firing rate of hypothalamic neurones. *Experientia 25*, 610.

Hellon, R. F. (1970). The stimulation of hypothalamic neurones by changes in ambient temperature. *Pfluegers Arch. Gesamte Physiol. 321*, 56-66.

Hellon, R. F. (1971). Central thermoreceptors and thermoregulation. In *Handbook of Sensory Physiology*. Vol. 3, Pt. 1, *Enteroceptors*. E. Neil (Ed.). Springer, Berlin, pp. 161-186.

Hellon, R. F. (1972a). Central transmitters and thermoregulation. In *Essays on Temperature Regulation*, J. Bligh and R. E. Moore (Eds.). North-Holland, Amsterdam, pp. 71-85.

Hellon, R. F. (1972b). Temperature-sensitive neurons in the brain stem: their responses to brain temperature at different ambient temperatures. *Pfluegers Arch. Gesamte Physiol. 335*, 323-334.

Hellon, R. F. (1975). Monoamines, pyrogens and cations: their actions on central control of body temperature. *Pharmacol. Rev. 26*, 289-321.

Hellon, R. F., and Misra, N. K. (1973a). Neurones in the dorsal horn of the rat responding to scrotal skin temperature changes. *J. Physiol. 232*, 375-388.

Hellon, R. F., and Misra, N. K. (1973b). Neurones in the rat thalamus responding to scrotal skin temperature changes. *J. Physiol. 232*, 389-399.

Hellon, R. F., Misra, N. K., and Provins, K. A. (1973). Neurones in the somatosensory center of the rat responding to scrotal skin temperature changes. *J. Physiol. 232*, 401-411.

Hellstrøm, B., and Hammel, H. T. (1967). Some characteristics of temperature regulation in the unanesthetized dog. *Am. J. Physiol. 213*, 547-556.

Hemingway, A., and Stuart, D. G. (1963). Shivering in man and animals. In *Temperature—Its Measurement and Control in Science and Industry*, Vol. 3, Pt. 3, *Biology and Medicine*. J. D. Hardy (Ed.). Reinhold, New York, pp. 407-427.

Hemingway, A., Rasmussen, T., Wikoff, H., and Rasmussen, A. T. (1940). Effects of heating hypothalamus of dogs by diathermy. *J. Neurophysiol. 3*, 329-338.

Hemingway, A., Forgrave, P., and Birzis, L. (1954). Shivering suppression by hypothalamic stimulation. *J. Neurophysiol. 17*, 375-386.

Hensel, H. (1973). Neural processes in thermoregulation. *Physiol. Rev. 53*, 948-1017.

Hilton, S. M., and Spyer, K. M. (1971). Participation of the anterior hypothalamus in the baroreceptor reflex. *J. Physiol. 218*, 271-293.

Holmes, R. L., Newman, P. P., and Wolstencroft, J. H. (1960). A heat-sensitive region in the medulla. *J. Physiol. 152*, 93-98.

Horeyseck, G., Jänig, W., Kirchner, F., and Thämer, V. (1976). Activation and inhibition of muscle and cutaneous postganglionic neurones to hindlimb during hypothalamically induced vasoconstriction and atropine-sensitive vasodilation. *Pfluegers Arch. Gesamte Physiol. 361*, 231-240.

Hori, T., and Nakayama, T. (1973). Effects of biogenic amines on central thermoresponsive neurones in the rabbit. *J. Physiol. 232*, 71-85.

Ingram, D. L., and Legge, K. F. (1972). The influence of deep body and skin temperatures on thermoregulatory responses to heating of the scrotum in pigs. *J. Physiol. 224*, 477-487.

Ingram, D. L., and Whittow, G. C. (1962). The effect of heating the hypothalamus on respiration in the ox (*Bos taurus*). *J. Physiol. 163*, 200-210.

Inoue, S., and Murakami, N. (1976). Unit responses in the medulla oblongata of rabbit to changes in local and cutaneous temperature. *J. Physiol. 259*, 339-356.

Iriki, M., and Kozawa, E. (1976). Patterns of differentiation in various sympathetic efferents induced by hypoxic and by central thermal stimulation in decerebrated rabbits. *Pfluegers Arch. Gesamte Physiol. 362*, 101-108.

Jacobson, F. H., and Squires, R. D. (1970). Thermoregulatory responses of the cat to preoptic and environmental temperatures. *Am. J. Physiol. 218*, 1575-1582.

Jell, R. M. (1973). Responses of hypothalamic neurones to local temperature and to acetylcholine, noradrenaline and 5-hydroxy-tryptamine. *Brain Res. 55*, 123-134.

Jell, R. M. (1974). Responses of rostral hypothalamic neurones to peripheral temperature and to amines. *J. Physiol. 240*, 295-307.

Jell, R. M., and Gloor, P. (1972). Distribution of thermosensitive and nonthermosensitive preoptic and anterior hypothalamic neurones in unanesthetized cats, and effects of some anesthetics. *Canad. J. Physiol. Pharmacol. 50*, 890-901.

Jessen, C. (1976). Two-dimensional determination of thermosensitive sites within the goat's hypothalamus. *J. Appl. Physiol. 40*, 514-520.

Jessen, C., and Ludwig, O. (1971). Spinal cord and hypothalamus as core sensors of temperature in the conscious dog. II. Addition of signals. *Pfluegers Arch. Gesamte Physiol. 324*, 205-216.

Jessen, C., and Mayer, E. Th. (1971). Spinal cord and hypothalamus as core sensors of temperature in the conscious dog. I. Equivalence of responses. *Pfluegers Arch. Gesamte Physiol. 324*, 189-204.

Jessen, C., and Simon, E. (1971). Spinal cord and hypothalamus as core sensors of temperature in the conscious dog. III. Identity of functions. *Pfluegers Arch. Gesamte Physiol. 324*, 217-226.

Jessen, C., and Simon-Oppermann, Ch. (1976). Effects of cooling hypothalamus or spinal cord on urine formation in conscious goats. *Pfluegers Arch. Gesamte Physiol. 362* (Suppl.), R24.

Jessen, C., Simon, E., and Kullmann, R. (1968). Interaction of spinal and hypothalamic thermodetectors in body temperature regulation of the conscious dog. *Experientia 24*, 694-695.

Keller, A. D. (1933). Observations on the localization in the brain-stem of mechanisms controlling body temperature. *Am. J. Med. Sci. 185*, 746-748.

Keller, A. D. (1935). The separation of the heat loss and heat production mechanisms in chronic preparations. *Am. J. Physiol. 113,* 78-79.

Keller, A. D. (1963). Temperature regulation disturbances in dogs following hypothalamic ablations. In *Temperature–Its Measurement and Control in Science and Industry,* Vol. 3, Pt. 3, *Biology and Medicine.* J. D. Hardy (Ed.). Reinhold, New York, pp. 571-584.

Keller, A. D., and McClaskey, E. B. (1964). Localization, by the brain slicing method, of the level or levels of the cephalic brainstem upon which effective heat dissipation is dependent. *Am. J.Phys. Med. 43,* 181-211.

Kelso, S. R., Perlmutter, M. N., and Boulant, J. A. (1980). Temperature-sensitive neurons in hypothalamic tissue slices. *Fed. Proc. 39,* 989.

Kluger, M. J., and D'Alecy, L. G. (1975). Brain temperature during reversible upper respiratory bypass. *J. Appl. Physiol. 38,* 268-271.

Kluger, M. J., Gonzalez, R. R., and Stolwijk, J. A. J. (1973). Temperature regulation in the exercising rabbit. *Am. J. Physiol. 224,* 130-135.

Klussman, F. W. (1969). Der Einfluss der Temperatur auf die afferente und efferente motorische Innervation des Rückenmarks. I. Temperaturabhängigkeit der afferenten und efferenten Spontantatigkeit. *Pfluegers Arch. Gesamte Physiol. 305,* 295-315.

Klussmann, F. W., and Henatsch, H. D. (1969). Der Einfluss der Temperatur auf die afferente und efferente motorische Innervation des Rückenmarks. II. Temperaturabhängigkeit der Muskelspindelfunktion. *Pfluegers Arch. Gesamte Physiol. 305,* 316-339.

Knigge, K. M., and Bierman, S. M. (1958). Evidence of the central nervous system's influence upon cold induced acceleration of thyroidal [131] iodine release. *Am. J. Physiol. 192,* 625-630.

Knox, G. V., Campbell, C., and Lomax, P. (1973a). The effects of acetylcholine and nicotine on unit activity in the hypothalamic thermoregulatory centers of the rat. *Brain Res. 51,* 215-223.

Knox, G. V., Campbell, C., and Lomax, P. (1973b). Cutaneous temperature and unit activity in the hypothalamic thermoregulatory centers. *Exp. Neurol. 40,* 717-730.

Kosaka, M., and Simon, E. (1968). Der zentralnervöse spinal Mechanismus der Kättezitterns. *Pfluegers Arch. Gesamte Physiol. 302,* 357-373.

Kosaka, M., Simon, E., Thauer, R., and Walther, O. E. (1969). Effect of thermal stimulation of spinal cord on respiratory and cortical activity. *Am. J. Physiol. 217,* 858-863.

Kullman, R., Schonung, W., and Simon, E. (1970). Antagonistic changes in blood flow and sympathetic activity in different vascular beds following central thermal stimulation. I. Blood flow in skin, muscle and intestine during spinal cord heating and cooling in anesthetized dogs. *Pfluegers Arch. Gesamte Physiol. 319,* 146-161.

Lim, P. K., and Grodins, F. S. (1955). Control of thermal panting. *Am. J. Physiol. 180,* 445-449.

Lindsley, D. B., Schreiner, L. H., and Magoun, H. W. (1949). An electromyographic study of spasticity. *J. Neurophysiol. 12,* 197-205.

Lipton, J. M. (1968). Effects of preoptic lesions on heat-escape responding and colonic temperature in the rat. *Physiol. Behav. 3,* 165-169.

Lipton, J. M. (1971). Thermal stimulation of the medulla alters behavioral temperature regulation. *Brain Res. 26,* 439-442.

Lipton, J. M. (1973). Thermosensitivity of medulla oblongata in control of body temperature. *Am. J. Physiol. 224,* 890-897.

Lipton, J. M., Dwyer, P. E., and Fossler, D. E. (1974). Effects of brainstem lesions on temperature regulation in hot and cold environments. *Am. J. Physiol. 226*, 1356-1365.

Lutherer, L. O., Fregly, M. J., and Anton, A. H. (1969). An interrelationship between theophylline and catecholamines in the hypothyroid rat acutely exposed to cold. *Fed. Proc. 28*, 1238-1242.

McLennan, H., and Miller, J. J. (1974). The hippocampal control of neuronal discharges in the septum of the rat. *J. Physiol. 237*, 607-624.

Magoun, H. W., Harrison, F., Brobeck, J. R., and Ranson, S. W. (1938). Activation of heat loss mechanisms by local heating of the brain. *J. Neurophysiol. 1*, 101-114.

Martin, J. B., and Reichlin, S. (1970). Thyrotropin secretion in rats after hypothalamic electrical stimulation or injection of synthetic TSH-releasing factor. *Science 168*, 1366-1368.

Meurer, K. A., Jessen, C., and Iriki, M. (1967). Kältezittern während isolierter Kühlung des Rückenmarks nach Durchschneidung der Hinterwurzeln. *Pfluegers Arch. Gesamte Physiol. 293*, 236-255.

Meyer, H. H. (1913). Theorie des Fiebers und seiner Behandlung. *Verhandl. Deuts. Bes Inner. Med. 30*, 15-25.

Murakami, N. (1973). Effects of iontophoretic application of 5-hydroxytryptamine, noradrenaline and acetylcholine upon hypothalamic temperature-sensitive neurones in rats. *Jpn. J. Physiol. 23*, 435-446.

Murakami, N., Stolwijk, J. A. J., and Hardy, J. D. (1967). Responses of preoptic neurons to anesthetics and peripheral stimulation. *Am. J. Physiol. 213*, 1015-1024.

Murgatroyd, D., and Hardy, J. D. (1970). Central and peripheral temperatures in behavioral thermoregulation of the rat. In *Physiological and Behavioral Temperature Regulation*, J. D. Hardy, A. P. Gagge, and J. A. J. Stolwijk (Eds.). Thomas, Springfield, Ill., pp. 874-891.

Myers, R. D., and Veale, W. L. (1970). Body temperature: possible ionic mechanism in the hypothalamus controlling the set point. *Science 170*, 95-97.

Myers, R. D., and Veale, W. L. (1971). The role of sodium and calcium ions in the hypothalamus in the control of body temperature of the unanesthetized cat. *J. Physiol. 212*, 411-430.

Myers, R. D., and Yaksh, T. L. (1969). Control of body temperature in the unanaesthetized monkey by cholinergic and aminergic systems in the hypothalamus. *J. Physiol. 202*, 483-500.

Myers, R. D., Simpson, C. W., Higgins, D., Nattermann, R. A., Rice, J. C., Redgrave, P., and Metcalf, G. (1976). Hypothalamic Na^+ and Ca^{++} ions and temperature set-point: new mechanisms of action of a central or peripheral thermal challenge and intrahypothalamic 5-HT,NE,PGE_1 and pyrogen. *Brain Res. Bull. 1*, 301-327.

Myhre, K., and Hammel, H. T. (1969). Behavioral regulation of internal temperature in the lizard *Tiligua scincoides*. *Am. J. Physiol. 217*, 1490-1495.

Nadel, E. R., Bullard, R. W., and Stolwijk, J. A. J. (1971). Importance of skin temperature in the regulation of sweating. *J. Appl. Physiol. 31*, 80-87.

Nakayama, T., and Hardy, J. D. (1969). Unit responses in the rabbits brain stem to changes in brain and cutaneous temperature. *J. Appl. Physiol. 27*, 848-857.

Nakayama, T., and Hori, T. (1973). Effects of anesthetic and pyrogen on thermally sensitive neurons in the brainstem. *J. Appl. Physiol. 34*, 351-355.

Nakayama, T., Eisenman, J. S., and Hardy, J. D. (1961). Single unit activity of anterior hypothalamus during local heating. *Science 134,* 560-561.

Nakayama, T., Hammel, H. T., Hardy, J. D., and Eisenman, J. S. (1963). Thermal stimulation of electrical activity of single units of the preoptic region. *Am. J. Physiol. 204,* 1122-1126.

Necker, R. (1975). Temperature-sensitive ascending neurons in the spinal cord of pigeons. *Pfluegers Arch. Gesamte Physiol. 353,* 275-286.

Ninomiya, I., Judy, W. V., and Wilson, M. F. (1970). Hypothalamic stimulus effects on sympathetic nerve activity. *Am. J. Physiol. 218,* 453-462.

Nutik, S. L. (1973a). Posterior hypothalamic neurons responsive to preoptic region thermal stimulation. *J. Neurophysiol. 36,* 238-249.

Nutik, S. L. (1973b). Convergence of cutaneous and preoptic region thermal afferents on posterior hypothalamic neurons. *J. Neurophysiol. 36,* 250-257.

Pierau, Fr.-K., Klee, M. R., and Klussmann, F. W. (1976). Effect of temperature on postsynaptic potentials of cat spinal motoneurones. *Brain Res. 114,* 21-34.

Pinkston, J. O., Bard, P., and Rioch, D. McK. (1934). The responses to changes in environmental temperature after removal of portions of the forebrain. *Am. J. Physiol. 109,* 515-531.

Poletti, C. E., Kinnard, M. A., and Maclean, P. D. (1973). Hippocampal influence on unit activity of hypothalamus, preoptic region and basal forebrain in awake, sitting squirrel monkeys. *J. Neurophysiol. 36,* 308-324.

Proppe, D. W., and Gale, C. C. (1970). Endocrine thermoregulatory responses to local hypothalamic warming in unanesthetized baboons. *Am. J. Physiol. 219,* 202-207.

Randall, W. C., Wurster, R. D., and Lewin, R. J. (1966). Responses of patients with high spinal transection to high ambient temperatures. *J. Appl. Physiol. 21,* 985-993.

Ranson, S. W. (1940). Regulation of body temperature. In *The Hypothalamus and Central Levels of Autonomic Function,* Hafner, New York, pp. 342-399.

Rawson, R. O., and Quick, K. P. (1970). Evidence of deep-body thermoreceptor response to intra-abdominal heating of the ewe. *J. Appl. Physiol. 28,* 813-820.

Reaves, T. A., and Heath, J. E. (1975). Interval coding of temperature by CNS neurones in thermoregulation. *Nature (London) 257,* 688-690.

Roberts, W. W., and Mooney, R. D. (1974). Brain areas controlling thermoregulatory grooming, prone extension, locomotion and tail vasodilation in rats. *J. Comp. Physiol. Psychol. 86,* 470-480.

Roberts, W. W., Bergquist, E. H., and Robinson, T. C. L. (1969). Thermoregulatory grooming and sleep-like relaxation induced by local warming of preoptic area and anterior hypothalamus in opossum. *J. Comp. Physiol. Psychol. 67,* 182-188.

Satinoff, E. (1964). Behavioral thermoregulation in response to local cooling of the rat brain. *Am. J. Physiol. 206,* 1389-1394.

Schoener, E. P., and Wang, S. C. (1975). Leukocytic pyrogen and sodium acetylsalicylate on hypothalamic neurons in the cat. *Am. J. Physiol. 229,* 185-190.

Schönung, W., Jessen, C., Wagner, H., and Simon, E. (1971a). Regional blood flow antagonism induced by central thermal stimulation in conscious dogs. *Experientia 27,* 1291-1292.

Schönung, W., Wagner, H., Jessen, C., and Simon, E. (1971b). Differentiation of cutaneous and intestinal blood flow during hypothalamic heating and cooling in anesthetized dogs. *Pfluegers Arch. Gesamte Physiol. 328,* 145-154.

Scott, I. M., Toner, M. M., and Boulant, J. A. (1980). The effect of passive limb movement on hypothalamic-evoked thermoregulatory responses in rabbits. *Soc. Neurosci. (Abstr.) 6.* In press.

Simon, E. (1971). Regional differentiation of vasomotor activity underlying thermoregulatory adjustments of blood flow. *Int. J. Biometeor. 15,* 219-224.

Simon, E. (1972). Temperature signals from skin and spinal cord converging on spinothalamic neurons. *Pfluegers Arch. Gesamte Physiol. 337,* 323-332.

Simon, E., and Iriki, M. (1971). Sensory transmission of spinal heat and cold sensitivity in ascending spinal neurons. *Pfluegers Arch. Gesamte Physiol. 328,* 103-120.

Smiles, K. A., Elizondo, R. S., and Barney, C. C. (1976). Sweating responses during changes of hypothalamic temperature in the rhesus monkey. *J. Appl. Physiol. 40,* 653-657.

Spence, R. J., Rhodes, B. A., and Wagner, H. N., Jr. (1972). Regulation of arteriovenous anastomotic and capillary blood flow in the dog leg. *Am. J. Physiol. 222,* 326-332.

Squires, R. D., and Jacobson, F. H. (1968). Chronic deficits of temperature regulation produced in cats by preoptic lesions. *Am. J. Physiol. 214,* 549-560.

Stitt, J. T. (1976). The regulation of respiratory evaporative heat loss in the rabbit. *J. Physiol. 258,* 157-171.

Stitt, J. T., and Hardy, J. D. (1975). Microelectrophoresis of PGE_1 onto single units in the rabbit hypothalamus. *Am. J. Physiol. 229,* 240-245.

Stitt, J. T., Adair, E. R., Nadel, E. R., and Stolwijk, J. A. J. (1971). The relation between behavior and physiology in the thermoregulatory response of the squirrel monkey. *J. Physiol. (Paris) 63,* 424-472.

Stitt, J. T., Hardy, J. D., and Stolwijk, J. A. J. (1974). PGE_1 fever: its effect on thermoregulation at different low ambient temperatures. *Am. J. Physiol. 227,* 622-629.

Stuart, D. G., Kawamura, Y., and Hemingway, A. (1961). Activation and suppression of shivering during septal and hypothalamic stimulation. *Exp. Neurol. 4,* 485-506.

Stuart, D. G., Kawamura, Y., Hemingway, A., and Price, W. M. (1962a). Effects of septal and hypothalamic lesions on shivering. *Exp. Neurol. 5,* 335-347.

Stuart, D. G., Freeman, W. J., and Hemingway, A. (1962b). Effects of decerebration and decortication on shivering in the cat. *Neurology 12,* 99-107.

Sundsten, J. W. (1967). Effects of steady and cyclic hypothalamic thermal stimulation in unanesthetized cats. *J. Appl. Physiol. 22,* 1129-1134.

Szczepanska-Sadowska, E. (1974). Plasma ADH increase and thirst suppression elicited by preoptic heating in the dog. *Am. J. Physiol. 226,* 155-161.

Szolcsanyi, J., Joó, F., and Jancsó-Gábor, A. (1971). Mitochondrial changes in preoptic neurones after capsaicin desensitization of the hypothalamic thermodetectors in rats. *Nature (London) 229,* 116-117.

Tanche, M., and Therminarias, A. (1969). Thyroxine and catecholamines during cold exposure in dogs. *Fed. Proc. 28,* 1257-1260.

Thauer, R. (1970). Thermosensitivity of the spinal cord. In *Physiological and Behavioral Temperature Regulation,* J. D. Hardy, A. P. Gagge, and J. A. J. Stolwijk (Eds.). Thomas, Springfield, Ill., pp. 472-492.

Thomas, S., and Anand, B. K. (1970). Effect of electrical stimulation of the hypothalamus on thyroid secretion in monkeys. *J. Neuro-Visceral Relations 31,* 399-408.

Von Euler, C., and Holmgren, B. (1956). The role of hypothalamo-hypophyseal conversions in thyroid secretion. *J. Physiol. 131,* 137-146.

Vanhoutte, P. M., and Shepherd, J. T. (1970). Effect of temperature on reactivity of isolated cutaneous veins of the dog. *Am. J. Physiol. 218,* 187-190.

Wakeman, K. A., Donovick, P. J., and Burright, R. G. (1970). Septal lesions increase bar pressing for heat in animals maintained in the cold. *Physiol. Behav. 5,* 1193-1195.

Walther, O. E., Iriki, M., and Simon, E. (1970). Antagonistic changes in blood flow and sympathetic activity in different vascular beds following central thermal stimulation. II. Cutaneous and visceral sympathetic activity during spinal cord heating and cooling in anesthetized rabbits and cats. *Pfluegers Arch. Gesamte Physiol. 319,* 162-184.

Webb-Peploe, M. M., and Shepherd, J. T. (1968). Responses of the superficial limb veins of the dog to changes in temperature. *Circ. Res. 22,* 737-746.

Weiss, B. (1957). Thermal behavior of the sub-nourished and pantothenic-acid-deprived rat. *J. Comp. Physiol. Psychol. 50,* 481-485.

Weiss, B. L., and Aghajanian, G. K. (1971). Activation of brain serotonin metabolism by heat: role of midbrain raphe neurons. *Brain Res. 26,* 37-48.

Wenger, C. B., Roberts, M. F., Nadel, E. R., and Stolwijk, J. A. J. (1975a). Thermoregulatory control of finger blood flow. *J. Appl. Physiol. 38,* 1078-1082.

Wenger, C. B., Roberts, M. F., Stolwijk, J. A. J., and Nadel, E. R. (1975b). Forearm blood flow during body temperature transients produced by leg exercise. *J. Appl. Physiol. 38,* 58-63.

Wit, A., and Wang, S. C. (1968a). Temperature-sensitive neurons in preoptic/anterior hypothalamic region: effects of increasing ambient temperature. *Am. J. Physiol. 215,* 1151-1159.

Wit, A., and Wang, S. C. (1968b). Temperature-sensitive neurons in preoptic/anterior hypothalamic region: actions of pyrogen and acetylsalicylate. *Am. J. Physiol. 215,* 1160-1169.

Woods, J. W. (1964). Behavior of chronic decerebrate rats. *J. Neurophysiol. 27,* 635-644.

Wünnenberg, W., and Brück, K. (1968). Single unit activity evoked by thermal stimulation of the cervical spinal cord in the guinea pig. *Nature (London) 218,* 1268-1269.

Wünnenberg, W., and Brück, K. (1970). Studies on the ascending pathways from the thermosensitive region of the spinal cord. *Pfluegers Arch. Gesamte Physiol. 321,* 233-241.

Wünnenberg, W., and Hardy, J. D. (1972). Response of single units of the posterior hypothalamus to thermal stimulation. *J. Appl. Physiol. 33,* 547-552.

Wyss, C. R., Brengelmann, G. L., Johnson, J. M., Rowell, L. B., and Niederberger, M. (1974). Control of skin blood flow, sweating, and heart rate: role of skin vs. core temperature. *J. Appl. Physiol. 36,* 726-733.

Zanick, D. C., and Delaney, J. P. (1973). Temperature influences on arteriovenous anastomoses. *Proc. Soc. Exp. Biol. Med. 144,* 616-620.

2
Hypothalamic Control of Thermoregulation
Neurochemical Mechanisms *

R. D. Myers / University of North Carolina School of Medicine, Chapel Hill, North Carolina

I. Introduction

Unique humoral mechanisms in the hypothalamus underlie the control of one's body temperature. Each has at its heart a distinct set of neurochemical events. This fact stems from a rather recent scientific drama that has been as exciting as it is profound. Today, it still possesses two surprising facets: first, none of the characters, although clearly recognized, are fully developed; second, other participants have not as yet made their appearance on stage.

As is commonplace in all of the diverse fields covered in the volumes of this handbook, many reviews have been written in recent years which are devoted to the intricacies of the physiology of temperature regulation per se (Benzinger, 1969; Hardy, 1972a; Bligh, 1973; Hensel, 1973; Hellon, 1975). In the preceding chapter, Boulant has dealt mainly with the difficult neurophysiological questions as they pertain to the hypothalamus; but the counterpoint presented in this chapter attacks the neurochemical issues of thermoregulation, particularly with reference to the activity of an endogenous substance such as a neurotransmitter, hormone, cation, or pyrogen. In stark contrast to the vantage point of the earlier treatises, the footing of this review is anchored almost exclusively within the neuroanatomical context of one structure, the hypothalamus. Although this strategy has not been adopted without some misgivings, the focus is necessarily centered on the neurohumoral properties of the anterior, posterior, or other hypothalamic areas which mediate the body's temperature control system. The reasons for this follow.

First, pathological damage to any of several structures within the hypothalamus may cause a profound disturbance in an individual's thermoregulatory capacity (Boulant, 1979, this volume). Even though the timeworn, all-pervasive concept of a functional "center"

---

*This chapter is dedicated to the memory of Professor Philip Bard. In his Laboratory at Johns Hopkins, the spark of interest in temperature regulation was irrevocably kindled in this author.

of thermal control is now rejected, one cannot eschew one important piece of clinical knowledge. Survival itself is threatened after an accidental lesion of the hypothalamus that is of a sufficiently disseminated dimension to interfere with the maintenance of body temperature. A pernicious ablation to other parts of the neuraxis ordinarily does not cause such an impairment in thermoregulation, unless, of course, the vital tissue that supports cardiac or respiratory function is concurrently damaged.

Second, the hypothalamus, without any question, does contain neurons which are physically reactive to a perturbation in body temperature. These cells are thus classified as thermosensitive (J. Boulant, this volume). Of great significance is the fact that these neurons not only react to a change in the local temperature of the hypothalamus but also respond equally to the disturbances in skin temperature that are characteristic of the thermal challenge which must be met almost continuously in an environment in which an ambient temperature approaching euthermia is the exception rather than the rule.

Third, certain parts of the hypothalamus exhibit a differential sensitivity to a neurochemical change of a most subtle nature. Some areas are responsive to one substance, such as a bacterial pyrogen. Yet another region will react to an ever-so-slight fluctuation in the ionic milieu of the interstitial fluid which bathes it. Thus, within a morphologically separate area, compartmentalized chemoreactivity of the cells is unique.

Fourth, these selfsame hypothalamic regions liberate or release substances differentially, after an external or internal thermal stimulus. Since the characteristics of their cellular elaboration depends on the specificity of the direction and intensity of the thermal load, this sort of observation takes on considerable functional significance. As described in this chapter, the physiological meaning of these findings with respect to the control of body temperature is exceedingly complex.

A. Neurochemical Aspects of Thermoregulation

On the surface, one could easily make the simple mistake that the neuropharmacological or neurochemical experiment on temperature regulation is straightforward. After all, as a result of an injection of a drug, the temperature of the animal will either rise, fall, or remain unchanged. Indeed, there is an element of truth resting in the simplicity of the response, but the unraveling of the precise mechanism which is responsible for either a change in temperature or an inherent tendency toward its stability is unimaginably difficult. The main reason for this is that so many actions of the autonomic nervous system are brought into play as secondary responses when an animal's body temperature is modified (Feldberg, 1968). Then again, an autonomic "storm" or even a change in a single autonomic process can produce a great effect on body temperature.

To illustrate, the rapid respiratory rate that accompanies an epileptic episode typically causes a remarkable dissipation of body heat which in turn eventuates in a seizure-induced, dramatic decline in body temperature (Beleslin and Myers, 1973). Yet, the pathological manifestations of a convulsion have nothing whatsoever to do with the "controller" system in the central nervous system (CNS) for thermoregulation, i.e., an epileptic attack is not utilized functionally as a thermoregulatory response in the body's defense against heat. Other well-known factors clearly affect body temperature secondarily, including generalized muscular activity, the drinking of an iced beverage, a cardiovascular response, an emotional reaction, the consumption of hot food, the plasma level of a hormone, exercise, alcohol or other drugs, and a physiological stressor. The point here is that a central stimulus acting primarily to influence one of these factors or processes may cause a secondary or even tertiary response which will help to initiate a change in body temperature (Myers and Waller, 1977). But the change in itself is not at all a regulatory adjustment in response to a cold or hot environmental temperature.

B. Which Thermoregulatory Function Does a Neurochemical Event Serve?

In every vertebrate's nervous system, a biochemical event is the handmaiden of the physical input to the neurons. Such a chemical event can be exemplified in many ways: by a change in ion flux, transmitter release, or by some aspect of metabolic activity of the neuron. Each change provides the physiological basis of the cell's response. In the case of temperature control, two types of input alter the chemical activity of the neurons within restricted anatomical regions of the animal's hypothalamus.

First, within the anterior hypothalamic, preoptic area, a change in the temperature of the blood circulating locally within the capillary beds affects the depolarization of the local cells whose rate and property of discharge is characteristic of thermosensitive neurons (Hellon, 1967).

Second, the temperature of the skin constitutes an equally vital input to the anterior hypothalamic, preoptic area which in turn emits the neural impulses which call for heat gain or loss (Boulant, 1979, this volume). Each action potential is contingent upon the reversal of the resting transmembrane current established intrinsically by extracellular sodium ions and intracellular potassium ions. When the depolarizing potential reaches the nerve ending, the release of a chemical transmitter at the synapse subsequently depolarizes an adjacent neuron and its respective axon fiber. Thus, either a local alteration in the temperature of the blood in the hypothalamus or a cutaneous afferent stimulation may serve to enhance and/or concurrently inhibit the transmitter activity within distinct pools of contiguous synapses. Ultimately, a succession of chemically mediated synaptic transmissions in this hypothalamic area leads to the activation of pathways morphologically delegated either to peripheral processes for heat production or heat dissipation. Experimental evidence indicates that the signals carried by the respective pathways are mediated similarly, at anatomical points along the effector system, by the presynaptic release of specific transmitter substances (Myers, 1974a). Thus, a thermal balance is achieved by these synaptic and membrane events so that the defense of the "set-point" temperature is maintained.

A third important characteristic of the neurons in the rostral hypothalamus can best be thought of as the property of "chemorecognition." In a sense, this feature is just as life-saving as that of the temperature sensor-effector mechanism and it may be independent of the sensing elements that react to temperature per se. Here, we refer to the remarkable capacity of cells in the anterior hypothalamic, preoptic area to react to the presence of (1) bacterial and other pyrogens, (2) cellular material such as leukocytes, (3) toxins, and (4) drugs such as anesthetics (see Myers, 1974a; Hellon, 1975). In hibernating species, for example, there is now evidence to suggest that the season of the year, the combination of a change in light and temperature over a prolonged period, and the presence of a "trigger factor" of high molecular weight all may influence the activity of the neurons in the anterior hypothalamic, preoptic area (see South et al., 1972; Dawe et al., 1970). Again, the presence of a physicochemical factor in this region is translated into a series of physiological responses by virtue of neurochemical mediators that function within the hypothalamus.

C. Hypothalamic Neurohumoral Model: Rationale

In 1963, the monoamine theory of thermoregulation was proposed (Feldberg and Myers, 1963). Since then a relatively complex neurochemical model of thermoregulation has gradually evolved. This model incorporates as a focal point the putative synaptic transmitters that are present endogenously in the hypothalamus, i.e., serotonin (5-HT), norepinephrine, and acetylcholine, as well as other substances which may influence synaptic activity. In addition, nontransmitter factors that are also found endogenously, such as the prostaglandins and sodium and calcium ions, are also incorporated by many investigators into the general schema involving neurohumoral control of thermoregulation.

As the details of this model unfold in the succeeding pages, one fact will become clear. The comprehension of the hypothalamic control system requires an anatomical foundation whereby separate physiological mechanisms can be clearly distinguished. Within one region, a given endogenous substance exerts a specific action on body temperature, whereas another substance does not. In a different region, a shift in ambient temperature may release a given hypothalamic factor and, at the same time, inhibit the release of another. Just as there are marked differences anatomically in the inherent thermosensitivity of neurons, clear-cut distinctions also exist in the chemosensitivity of the selfsame tissue. To understand the hypothalamic control mechanism, a model derived from results obtained with a compound injected systemically cannot be considered simply because all of the brain, spinal cord, the autonomic and peripheral neurons are affected by the substance (see Section II.A).

Even though some sort of an inference can often be drawn from an experiment in which a compound is administered by the intraventricular route and subsequently alters temperature, the result is, at best, provisionary. It may, in fact, even be misleading (Myers, 1974). Three illustrations are presented below which compare the findings with intraventricular and intrahypothalamic injections.

First, when 5-HT is injected into the lateral cerebral ventricle of a cat or in a much smaller dose into the anterior hypothalamus, there is little change in respiratory rate, and the animal's temperature rises (Feldberg and Myers, 1964, 1965a). On the other hand, when 5-HT is injected in a very high dose into the ventricle of this or another animal such as the ox, its respiratory rate increases sharply, the animal may become sedated, and body temperature declines precipitously (Findlay and Robertshaw, 1967). Since it is not known if these responses would occur if 5-HT were applied in a small dose directly to the thermosensitive zone of the ox's anterior hypothalamus, we are faced with the question of whether other autonomic effects could mask and, in effect, cancel out the hypothalamic role of 5-HT in temperature control of this animal.

Second, when nicotine is infused into the cerebral ventricle of a monkey, its temperature declines; however, if nicotine is microinjected into the posterior hypothalamus, the primate's temperature rises (Hall and Myers, 1971, 1972). It is probable that the fall in temperature after intraventricular nicotine could actually be due to an alternative action of nicotine within the anterior hypothalamus. In fact, the microinjection of nicotine at an adrenergic-sensitive site lowers temperature in an identical way to that produced by norepinephrine (Hall and Myers, 1972), presumably because of nicotine's action in releasing catecholamines from nerve tissue (Hall and Turner, 1972). Thus, nicotine's central action is diverse and cannot be delineated by a morphologically nonspecific intraventricular injection.

Third, after thyroid-releasing hormone (TRH) is injected in a nanogram dose into the cerebral ventricle of the cat, hypothermia is produced (Metcalf, 1974); however, the response is characterized readily by a host of concomitant autonomic changes including tachypnea and salivation. But a microinjection of a similarly tiny dose into either the anterior or posterior hypothalamus fails to cause any change in the cat's body temperature. On the other hand, if the tripeptide is infused into the mesencephalon at a site and dosage that causes the animal to pant (exceeding 200 respirations per minute), then the temperature of the animal declines. This hypothermia is apparently due solely to the tachypnea (Myers et al., 1977a) and not to any effect or function of TRH in the temperature-controlling mechanism of hypothalamic neurons (Myers, 1976).

From these and other illustrations, a scientific picture of remarkable clarity begins to emerge. An extrapolation based on an intracerebroventricular injection may be false, partially correct, or entirely reasonable. In spite of a series of findings with an antagonist and an agonist, as well as the consistent responses of an animal tested at several ambient temperatures, a precise deduction about a hypothalamic mechanism will continue to be difficult to make until the neurochemical components within the anterior or posterior parts are examined. Today, many researchers are attempting to isolate the thermoregulatory part of a given neurochemical process from a more generalized autonomic response by quantitating the animal's respiratory rate, cutaneous blood flow, oxygen consumption, and even cortical electroencephalographic (EEG) activity.

D. Hypothalamic Intervention and Manipulation: Caution on Methods

A valid experiment on the neurochemical basis of thermoregulation, which provides a set of consistent and unequivocal observations, is exceedingly difficult to undertake. Among the technical difficulties which have plagued this field over the years are those pertaining to the measurement of body temperature in an unrestrained animal, the confounding effect of an anesthetic, and the length of time typically required for a change in temperature to take place.

A number of variables are likewise involved in the chemical manipulation of a portion of brain tissue of diminutive anatomical dimension. For example, an injection of a neuroactive drug into the hypothalamus of a rat can cause a response which will depend upon the site of application, the dose given, the volume of the carrier vehicle, and the physiological state of the animal. Each of these factors is crucial to an interpretation of a temperature response. If the volume injected in the rat's hypothalamus exceeds 0.5 μl, the number of nuclei or fiber tracks reached by the injected fluid may be so disseminated (Myers and Hoch, 1978) that the response is inordinately complex and the subsequent interpretation dubious (Myers, 1974a). Correspondingly, if the dose is too high and the receptor sites are subsequently "swamped," a given neuronal response may be entirely blocked. Thus, a pathological hypothermia analogous functionally to that caused by a sharp thump on the skull may occur (Myers and Waller, 1977). A clue to nonspecific responses is found in the animal's consequent inactivity or other signs of malaise.* Indeed,

---

*Interestingly, acute truncation of the monkey's brain stem just caudal to the hypothalamus causes a rapid loss of body heat without shivering, or pilomotor or vasomotor responses (Liu, 1979).

an unwarranted confusion about the neurochemical property of a substance may arise simply because the necessity of generating a dose-response curve, with intrahypothalamic microinjections, is not yet adequately recognized.

A common property shared by many biologically active compounds is that one dose evokes a certain reaction whereas another dose given by the same route causes a diametrically opposite response (Chase and Murphy, 1973). As shown by Feldberg and Vartiainen (1934), a small amount of acetylcholine excites the sympathetic ganglion, but, in a large dose, the synaptic transmitter paralyzes the ganglion cells. Thus, an excitatory action of a transmitter substance at a synapse is converted into a paralyzing one if an excessive quantity is applied; this principle may extend to compounds other than transmitters as well (Banerjee et al., 1968a).

Another factor centers on the vehicle which is used to carry the test substance. For example, if a solution is not pyrogen-free, the natural and specific chemoreactivity of the particular region of the hypothalamus under study often will be overridden by the presence of bacteria. Fever then ensues. Similarly, if a pocket of neuroglia accumulates around the shaft of the cannula or if a mass of leukocytes is present at the tip of the cannula, a microliter injection through the lumen in either case can dislodge a fragment of tissue. If this tissue touches upon reactive neurons, an intense hyperthermia ensues (Myers et al., 1974). Again, such a pyrogen fever could either exacerbate or attenuate the actual effect which the substance ordinarily would exert at that site.

When a microinjection or a push-pull perfusion of a circumscribed region of the hypothalamus is undertaken, the size of the cannula system is a major consideration. In a small animal such as the rat, guinea pig, or squirrel, a test of a compound's efficacy on an isolated region is difficult because of the significant amount of diffusion of the injected solution into the parenchyma (Myers, 1971a). Therefore, careful anatomical "charts" in terms of the latency and magnitude of the temperature response must be plotted with the concentration of the test solution held constant. In this way, the anatomical region of greatest sensitivity to the chemical substance can be demarcated and pinpointed (Myers et al., 1974).

As alluded to by Boulant (this volume), the interpretation of an experiment which uses the method of iontophoresis can be formidable. Jell (1974) has enunciated a few of the important problems associated with the technique: (1) the extrusion of an amine from the capillary pipette does not insure that the amine will reach a membrane that contains receptors that are specially sensitive to that amine; (2) the solution may reach the soma or axon which possesses a low density receptor; (3) the iontophoresed solution will reach nonthermoresponsive neurons which can bias the categorization of neurons according to frequency and direction of change; and (4) since electronic coupling between the cell and electrode is necessary for the adequate recording of an action potential, the effect of iontophoresis could be greater on the soma than on the dendritic processes.

II. Serotonergic Pathways

Once the initial observations had been made in the 1950s that 5-HT given intraperitoneally alters body temperature (Bächtold and Pletscher, 1957; Hoffman, 1958), a host of speculations emerged as to the possible role that this indole-alkylamine played in the central control of thermoregulation. Ironically, the origin of several current concepts about 5-HT and the CNS control mechanism was derived from results following its peripheral administration, even though the amine does not penetrate the blood-brain barrier. As reviewed recently (Myers and Waller, 1977), a systemic injection of 5-HT or the serotonin precursor, 5-HTP, produces a syndrome of often intense symptoms characterized by respiratory distress, cardiovascular disturbance, profound perturbations of the alimentary canal, lethargy, alterations in water balance, and other marked autonomic effects (Svanes, 1968).

Of special significance to this discussion is the difficulty of disentangling the specific responses of cells in the hypothalamus that could actually be involved in temperature control from a more generalized serotonergic poikilothermia. According to Shemano and Nickerson (1958), a systemic injection of 5-HT as well as a plethora of other compounds, including the metabolic precursors of serotonin, causes a nonspecific change in body temperature which is exclusively related to the animal's environmental temperature. The dilemma posed by their finding is equally applicable to studies in which a compound such as parachlorophenylalanine (pCPA), morphine, imipramine hydrochloride, a monoamineoxidase (MAO) inhibitor, or other drug is administered. In some instances, a part of a drug's action may or may not be on the central nervous system. Even then, however, one cannot conclude from the systemic administration of a drug that the alteration in an animal's colonic temperature arises from the influence of the compound on neurons in the anterior or posterior hypothalamus or other region of the brain stem (Lin, 1978).

A parallel situation also exists in relation to the myriad of investigations in which 5-HT or a related compound is infused directly into the fluid spaces of the brain. Whether given by the cisternal route or into the lateral or third cerebral ventricle, the multitude of effects that an indoleamine could have on many structures which anatomically form the walls of the ventricular system is considerable. This is not to gainsay the importance of the ventricular approach as the first approximation to an understanding of the effect of 5-HT on the hypothalamic thermoregulatory mechanism. Nevertheless, the intricacies of its specific role cannot be understood until an anatomical approach, such as a direct intrahypothalamic microinjection or perfusion, is undertaken.

A. 5-HT and the Control of Heat Production

Following the classical report of Horita and Gogerty (1958), a hallmark paper by Canal and Ornesi (1961) revealed that 5-HT or 5-HTP injected into the cisterna magna of the rabbit evokes a striking hyperthermia. Moreover, the 5-HT antagonist, cyproheptadine, given either systemically or by the same intracisternal route, blocks the indoleamine-induced rise in the rabbit's temperature (see Jacob and Girault, 1974). Later it was discovered that 5-HT infused into the lateral ventricle of the cat produces a similar sort of rise in body temperature (Feldberg and Myers, 1963). Since shivering and vasoconstriction accompany the cat's hyperthermia, and because norepinephrine infused similarly antagonizes 5-HT hyperthermia, a new theory was put forth that body temperature is regulated by the delicate balance in the presynaptic release of 5-HT and norepinephrine within the "thermoregulatory centers" of the hypothalamus (Feldberg and Myers, 1964).

1. 5-HT and the Thermosensitive Zone: Cat and Monkey

This early set of observations set the stage for a crucial question: does 5-HT actually mediate a change in body temperature by altering the local function within the anterior hypothalamic, preoptic area, which had already been implicated as a vital region for thermoregulatory control, or does 5-HT act on the more posterior integrative region of the hypothalamus?

In order to localize 5-HT's action, the amine was microinjected directly into the hypothalamus in doses of as much as a hundred-times less than that in the ventricle. When applied to the anterior hypothalamus, a rise in the cat's body temperature is evoked after a short latency. Again, the cat shivers, its ear vessels constrict, and the over-all response of thermogenesis is found to be dose dependent (Feldberg and Myers, 1965a). Of significance is the fact that 5-HT has no effect whatsoever on the cat's temperature when microinjected similarly into the posterior part of its hypothalamus. This observation coincided with the discovery made at about the same time, that cell bodies of 5-HT neurons within the brain stem send some of their terminal projections to the rostral hypothalamic area with collaterals touching also upon cells in this region (Fuxe et al., 1968; Battista et al., 1972; Jouvet, 1972). On the horizon, there was evidence of a neuroanatomical-neurochemical link for the temperature control function.

Subsequently, the localized effect of 5-HT on cells of the anterior hypothalamus was verified anatomically in the primate, among others. When 5 μg of 5-HT is infused in a 1.0-μl volume into the anterior hypothalamic, preoptic region of the rhesus monkey, the animal's temperature rises (Myers, 1968). However, if a larger dose of 5-HT is injected at the same anatomical locus, a transient hypothermia precedes the increase in temperature (Myers, 1968), an effect described also in the rat and other species (Crawshaw, 1972; Komiskey and Rudy, 1975). The precise site of action of 5-HT in the primate's hypothalamus, as delineated by means of a morphological "mapping" of injection loci, is a circumscribed region ventral to the anterior commissure and nucleus accumbens but dorsal to the optic chiasm (Myers and Yaksh, 1969). This region of 5-HT receptivity, which is depicted in Figure 1, corresponds identically to the area demonstrated in the primate's hypothalamus to be particularly thermosensitive (Hayward and Baker, 1968).

In other species of monkey, such as the pigtailed macaque, anatomical and functional concordances are again seen, with the rostral hypothalamus being the region of greatest sensitivity to 5-HT in terms of the evocation of thermogenesis (Myers and Waller, 1975a). In the baboon, a sustained elevation in temperature is similarly produced by the infusion of 15.0 μg of 5-HT into the rostral hypothalamus, but this occurs only after the transient fall typically observed after the local injection of a high dose of the monoamine (Toivola and Gale, 1970).

Figure 1 Anatomical mapping at seven coronal planes, extending from AP 10.0 to AP 16.0, of sites distributed throughout the hypothalamus of the unanesthetized nemestrina monkey at which a microinjection of 5-HT produces hyperthermia (▲) or no change in brain temperature (○). The doses ranged from 2.5 to 10.0 μg and were given in a volume of 1.0-1.5 μl. Anatomical abbreviations are: a, anterior commissure; al, ansa lenticularis; c, caudate nucleus; cd, nucleus centralis; ci, internal capsule; d, dorsomedial nucleus; f, fornix; ff, fields of Forel; g, globus pallidus; h, anterior hypothalamus; hp, paraventricular nucleus; m, mammillary body; o, nucleus of oculomotor nerve; p, posterior hypothalamus; pp, cerebral peduncle; r, reticular nucleus; re, nucleus reuniens; s, subthalamic nucleus; sn, substantia nigra; to, optic tract; v, ventromedial nucleus; va, ventral nucleus of thalamus; vl, lateral ventricle; z, zona incerta; 3, third ventricle. (From Myers and Waller, 1975a, reprinted with permission from *Comp. Biochem. Physiol. 51A*, 639-645, Pergamon Press, Ltd.)

Neurochemical Mechanisms in Thermoregulation

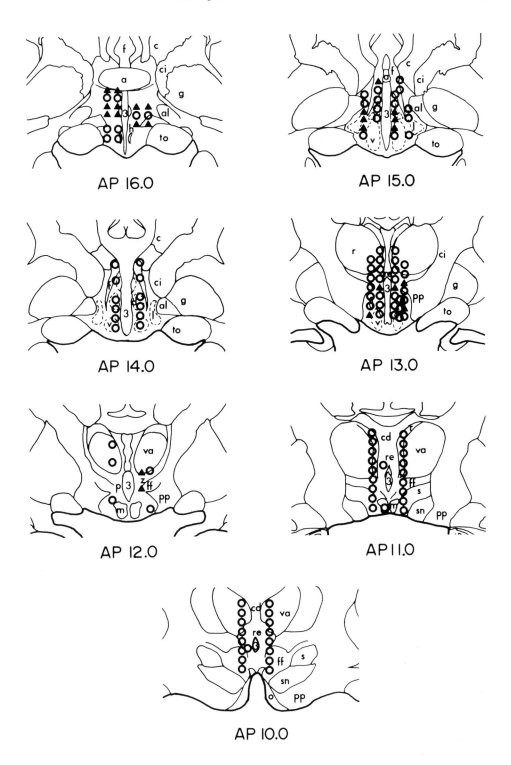

2. 5-HT in the Rat and Other Animals

In an animal having a relatively small brain, the use for an injection of a cannula assembly that exceeds the dimension of the rostral hypothalamus continues to be a persistent technical problem. Nevertheless, in 1969, Kubikowski and Rewerski first showed in the laboratory rat that 5-HT microinjected in a volume of 1.0 µl into the anterior hypothalamic area produces hyperthermia (Kubikowski and Rewerski, 1969). Other independent investigations on the local action of 5-HT on the hypothalamus of this species have served to clear up the inconsistent findings pertaining to an intracerebroventricular injection of the amine (see review of Myers and Waller, 1977). Whereas several authors have found that 5-HT given by this ventricular route produces hyperthermia in the rat; other workers have described a hypothermic response; and still others have reported mixed effects (Feldberg and Lotti, 1967; Myers and Yaksh, 1968; Bruinvels, 1970; Jacob et al., 1972; Kruk and Brittain, 1972; Francesconi and Mager, 1976).

In 1972, Crawshaw at last demonstrated unequivocally that 5-HT elicits a dose-dependent rise in the temperature of the rat in spite of the relatively large 1-µl volume employed. As seen in the monkey and cat, a higher dose of 5-HT given in the ventricle also evokes a fall in the rat's temperature (Crawshaw, 1972). This led to the conclusion that an intraventricular injection of 5-HT does not necessarily reflect the occurrence of a thermoregulatory response within the temperature-sensitive cells of the anterior hypothalamus. Moreover, this would, of course, explain the hypothermia after an intrahypothalamic injection of 5-HT in a 20-µl volume (Boros-Farkas and Illei-Donhoffer, 1969) which surely must leak out into the lumen of the third ventricle (Myers et al., 1971d). When a small volume (0.5 µl) is used for an infusion of the amine into the anterior hypothalamus of the rat, distinct dose-response curves for 5-HT can be generated in terms of the temporal characteristics of the temperature rise (Myers, 1975a). As portrayed in Figure 2, a dose of 1.25 µg of 5-HT causes a rise of about $0.8°C$. Although double the dose elevates the rat's temperature by more than $2.0°C$, the rise is delayed; in fact, a slight hypothermia precedes this latter rise, possibly because of a transitory overloading of the receptor sites (Myers, 1974a) which does not occur with the lower dose of 5-HT. Thus, in this rodent, 5-HT in the animal's thermosensitive zone would seem to subserve thermogenesis.

Extensive research has been carried out on the chicken by Marley and his co-workers. Infused in a volume of 1.0 µl, 5-HT evokes behavioral and EEG sleep and a concomitant hypothermia when the amine is given in a small dose but a mild hyperthermia at a larger dose (Marley and Whelan, 1975). The position of the cannula is reportedly important; further, different salts of 5-HT exert variable effects on the bird's behavior and body temperature. It is interesting that the hyperthermia evoked in the adult chicken by an intrahypothalamic infusion of tryptamine or 5-HT is prevented by a prior intrahypothalamic or intraventricular infusion of methysergide, a 5-HT receptor blocking agent (Nistico et al., 1976). This observation suggests the presence of 5-HT receptors which are anatomically demonstrable. Their antagonism, of course, would be expected to impair 5-HT thermogenesis. And this corresponds to the report that methysergide, given peripherally, also interferes with the initial rise in body temperature produced by 5-HT injected into the anterior hypothalamus of the cat (Komiskey and Rudy, 1975).

In the adult pigeon, a relatively high dose of 5.0 or 10.0 µg of 5-HT generally produces a slight decline in body temperature when the amine is microinjected in the bird's anterior hypothalamus; however, 40 min after an injection is given, the bird begins to

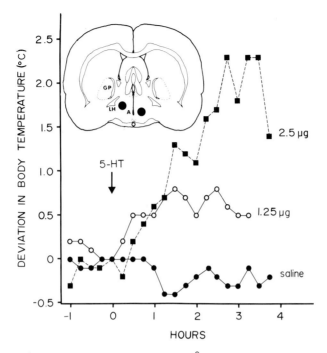

Figure 2 Changes from the base-line temperature (°C) of a representative rat after a 0.5-μl injection of 5-HT at zero hour. The two doses indicated and the saline control were given at weekly intervals. Sites of injection are designated by the dots on the histological map. Abbreviations: A, anterior hypothalamic area; GP, globus pallidus; LH, lateral hypothalamus; O, optic chiasm. (From Myers, 1975.)

shiver and its temperature increases (Hissa and Rautenberg, 1975). Unfortunately, the sustained rise in body temperature which is typically seen after a 5-HT microinjection was not recorded in these investigations. At a high ambient temperature (38°C), 5-HT produced transient inhibition of pigeon panting (Pyörnilä et al., 1978); recently, however, this group found sites caudal to the anterior hypothalamus in which 5-HT evoked a rise of up to 2°C. In the same birds, 5-HT injected more anteriorly produced hypothermia when administered in a very high, 10 μg-dose.

When a small dose of 5-HT is infused into the rostral hypothalamus of the ground squirrel, the body temperature of this hibernator is increased if the animal is in torpor. This occurs when the animal's temperature is less than 10°C or if the squirrel is awake and active with a core temperature of approximately 37°C (Beckman and Satinoff, 1972). In another hibernator, the hamster, an intrahypothalamic injection of 5-HT reportedly has little or no effect on temperature when the animal is normothermic, possibly because of the volume or dose used (Riegle and Wolfe, 1974).

A dose of 5-HT as high as 40.0 μg induces a fall in the rabbit's body temperature, as might be expected, when the amine is injected intrahypothalamically (Cooper et al., 1965; Gonzalez, 1971). As is discussed more fully in Section III.A.2, this observation became a part of the initial belief that profound species differences exist in the responses to monoamines purportedly because of the different "neurohumoral coding" of hypothalamic processes (Myers and Sharpe, 1968). However, according to Preston (1975), two crucial factors could ultimately resolve the enigma concerning the thermal effects of a monoamine on the hypothalamus: (1) the requirement of an anatomically differentiated injection volume, and (2) the maintenance of the animal at its own thermoneutral temperature.

At last, the enigma appears to be solved concerning the role of 5-HT in the rabbit. When Borsook et al. (1977) infused 0.8-5.0 µg of 5-HT into the rabbit's anterior hypothalamus, they were able to generate a family of dose-response curves of hyperthermia. Even the phenomenon of the transient fall in temperature preceding the subsequent rise was replicated. Figure 3 portrays the data which provide the basis for the resolution of the historical question of serotonergic thermogenesis mediated by 5-HT-sensitive sites in the lagomorph's hypothalamus. As in the pigeon, sites have been identified in the cat's preoptic area in which 5-HT evoked a dose-dependent hypothermia (Komiskey and Rudy, 1977) which was partially antagonized by methysergide, a serotinergic receptor blocker. Recently it was found that phentolamine, an α-adrenergic antagonist, inhibits a 5-HT-induced fall in temperature when both substances are administered in the preoptic region of the cat (W. E. Ruwe and R. D. Myers, unpublished observation, 1980). Therefore, the question of whether or not 5-HT neurons within this area are functionally involved in the heat-loss mechanism will not be answered until the demonstration of its presynaptic release from preoptic neurons in response to heat stress.

B. Action of 5-HT on Thermosensitive Neurons

Thermosensitive neurons located within the anterior hypothalamic, preoptic region comprise a substantial portion of the anatomical region in which a gross deflection in the local temperature causes a compensatory thermoregulatory response (Hardy, 1972b; Hensel, 1973; Hellon, 1975). On the basis of a series of microinjection studies, one logical assumption is that 5-HT receptors are located on neurons that are either cold sensitive or are influenced by neurons that react to the cold stimulus. Several observations uphold this assumption, but others do not.

When Cunningham et al. (1967) infused 5-HT into the ventricle of the anesthetized dog in a dose of 200-500 µg, they found that the firing rate of both temperature sensitive and insensitive neurons recorded in the preoptic area was considerably depressed. In applying 5-HT from a microcapillary tube directly onto a thermosensitive neuron by the method of iontophoresis, Beckman and Eisenman uncovered two new facts. First, most thermosensitive units are unresponsive to the microiontophoresis of 5-HT onto their cell membranes. Second, in the rat or cat, a few cells that were identified as cold sensitive tended to decrease their firing rate upon the localized iontophoretic ejection of 5-HT (Beckman and Eisenman, 1970b).

Different results have been obtained in other studies. In the rat, the indoleamine applied by iontophoresis excited 11 of 22 warm-sensitive neurons and depressed the firing-rate in only two of eight cold-sensitive units (Murakami, 1973). Over-all, one-third of the temperature-sensitive neurons in the rat's anterior hypothalamus failed to respond to 5-HT or norepinephrine. Somewhat surprising is the parallelism between one result of 5-HT infused into the cerebral ventricle of the rat (Myers and Yaksh, 1968) and that obtained in this study of single-unit activity. This is perlexing, particularly since hyperthermia is produced when 5-HT is microinjected into the rat's anterior hypothalamus. Another

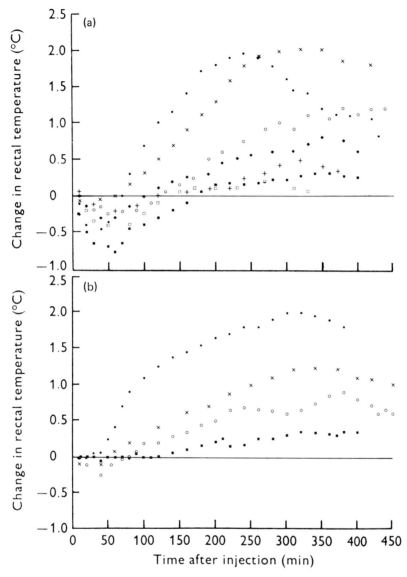

Figure 3 Effects of intrahypothalamic injections of (a) 28 nmol and (b) 14 nmol 5-HT in the rabbit. Each curve represents the rectal temperature change from pre-injection levels (zero on temperature scale) for a single rabbit. (Modified after Borsook et al., 1977.)

investigation devoted to the issue of unit activity has revealed that 5-HT increases the firing rate of nearly every warm-sensitive neuron recorded in the preoptic area of the rabbit (Hori and Nakayama, 1973). Conversely, in all but one of the cold-sensitive units identified in the same area, the iontophoresis of 5-HT depresses the firing rate of each cell (Hori, 1977). This would seem to be a paradoxical finding physiologically in view of the clear-cut results of Borsook et al. (1977) on 5-HT hyperthermia in the rabbit.

Another set of single-unit experiments has been carried out by Jell (1973, 1974) who recorded the firing pattern of several hundred spontaneously firing neurons in the anterior, preoptic area of the unanesthetized cat and who reported still different results. When the local temperature of these units was varied by means of an implanted thermode through which cold or hot water is passed, the excitatory or inhibitory response after the iontophoresis of 5-HT onto a single neuron was independent of the thermosensitivity of that neuron. Thus, no clear-cut correlation is ascribed to the response of a given thermosensitive neuron and the application of 5-HT or, for that matter, norepinephrine or acetylcholine (Jell, 1973). Similarly, there is no consistent relationship between amine responsiveness and the thermal characteristics of each cell as delineated by altering the temperature of the air blown onto the cat's face.

Of the cogent points made by Jell to account for the inexplicable results, a particular one stands out. The mediating effect of a monoamine in terms of its local synaptic function could be at an amine-sensitive receptor site that is not located on the soma of the neuron (Jell, 1974). Since microiontophoresis itself does not necessarily involve the cell's processes as a solution is ejected from the microcapillary tube but rather has the soma as its target, this explanation seems plausible. The limitations of the micropipette fluid-ejection procedure (Bloom, 1974a) thus would seem to hinder interpretation about the functional role of the temperature synapses. Moreover, from the results with the rat or rabbit in which the iontophoresis of an amine frequently inhibits the firing rate of the cell, an equal concern rests with the concentration of amine reaching receptor sites involved in the alteration of the cells' firing pattern. Again, a troublesome concentration-response analysis is encountered, since the method of iontophoresis, given the present technology, does not provide the experimenter with dose-response information. Other methodological problems may also help to account for the discrepancies and inconsistencies in the iontophoresis studies: interaction of the neuron's response with the anesthetic, the ambient temperature, and the influence of other, local neurons on the firing pattern of the recorded cell.

C. Endogenous Activity of 5-HT During Thermoregulation

Ever since the discovery was made that 5-HT is present in relatively high concentrations in the hypothalamus, the content of the diencephalic amine has been correlated with every conceivable functional change associated with this structure (see review of Barchas and Usdin, 1973). As would be anticipated, thermoregulation did not escape this scrutiny, and, indeed, many attempts have been made to relate the content or turnover of the indoleamine with the level of an animal's body temperature. The interpretation of the simple measure of the content of an amine or one of its metabolites in the whole brain or even the hypothalamus is generally subject to serious pitfalls (Myers, 1968). For example, a 5-HT content assay fails to separate the synaptosomal from other subcellular fractions. In addition, if the content of the indoleamine is elevated in the tissue sample, this could reflect a concomitant decline in the release of 5-HT from synaptic vesicles (Myers, 1974a). Unless a tissue-punch technique for excising nuclei anatomically (Jacobowitz et al., 1975) or other morphologically restricted approach is employed, a change in 5-HT activity within the hypothalamus may or may not relate to a specific function.

1. Temperature and Hypothalamic Content of 5-HT

Following the classic paper of Toh (1960), a large number of studies were published which showed that the content of 5-HT may be reduced in the brain of the rat, mouse, or other animal in response to a high ambient temperature (e.g., Garattini and Valzelli, 1958; Sochański and Samek, 1962) or may be elevated when rectal temperature is low (Friedman and Walker, 1968; Novotná and Janský, 1976) or remain entirely unchanged regardless of the displacement in ambient temperature (Werdinius, 1962; Laverty, 1963; Ingenito and Bonnycastle, 1967; Gál et al., 1968; Shellenberger and Elder, 1967). A complex result was reported by Harri and Tirri (1969) who found that the content of 5-HT in the whole brain of the mouse first increases and then falls during a continuous exposure of the animal to cold air; on exposure to warm temperature, the content value of brain 5-HT then reverses.

A midbrain lesion, which reduces the amount of 5-HT in the whole brain of the ground squirrel, disrupts the entry of this animal into hibernation (Spafford and Pengelley, 1971). Corresponding to the heat-production role postulated for 5-HT generally, the content of the indoleamine in the whole brain of the squirrel declines significantly as the hibernator enters into torpor. However, other assays to measure 5-HT content in the whole hypothalamus of the arctic ground squirrel reveal that the level of the amine apparently remains relatively unchanged during the animal's active season as well as during periods of hibernation and arousal (Feist and Galster, 1974).

Simmonds (1970) has dissected out the hypothalamus of the rat after its exposure to different environmental temperatures. In assuming that the turnover of 5-HT can be estimated by the degree of disappearance of the radiolabeled amine from brain tissue, Simmonds has found that the activity of 5-HT, particularly in the preoptic area, increases as the environmental temperature of the rat is elevated. Whether or not these postmortem measures do reflect directly the presynaptic activity of 5-HT in the living neurons of a thermoregulating animal cannot be answered with certainty at the present time (Preston and Schönbaum, 1976).

2. 5-HT Release During Thermoregulation

Concerning the nature of amine activity in neurons involved in temperature control, Hensel (1973) has raised a salient point: the release of a putative neurotransmitter must occur at hypothalamic synapses which are likely to be activated by a thermoregulatory challenge. Thus, the designation of an endogenous substance in a given aspect of the control of body temperature must be paralled by the functional demonstration of its endogenous hypothalamic release (Feldberg and Myers, 1964) in the living animal. In this context, the postmortem preparation and analysis of hypothalamic tissue is useless.

In 1969, it was shown that thermal stimulation by exposure of an unanesthetized animal to a cold or warm environment does, in fact, alter the resting release of a monoamine. After peripheral cooling of the conscious monkey, nanogram quantities of 5-HT are detectable in samples of push-pull perfusates collected at circumscribed sites in the hypothalamus (Myers et al., 1969). The region in which the cold challenge enhances the release of 5-HT encompasses the anterior hypothalamic, preoptic area and is homologous to the area sensitive to the local application of 5-HT. Figure 4 portrays several of the localized sites at which 5-HT activity is increased by cooling the monkey. The fact that

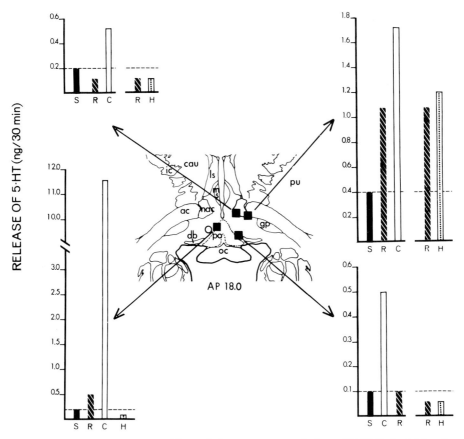

Figure 4 Individual sites in rostral hypothalamus at coronal plane AP 18.0 at which release of 5-HT changed (■) or was not detected (○) in push-pull perfusates (inset). Output of 5-HT is expressed in nanograms per 30-min perfusion interval, for control or resting release (R), cooling (C), or heating (H) of monkey. Minimum sensitivity (S) of isolated organ strip is denoted by a solid bar and horizontal dashed lines for each assay. (From Myers and Beleslin, 1971.)

warming the primate fails to augment the output of 5-HT, and even in some instances suppresses it, gives further support to 5-HT's postulated synaptic role in the heat-maintenance mechanism. When the anatomy of the releasing sites was examined, it was found that the greatest enhancement of 5-HT activity after cooling occurs close to the midline in the classical thermosensitive zone between the anterior commissure and the optic chiasm (Myers and Beleslin, 1971).

Remarkably similar results have been obtained with the cat. Isaac (1973) indicates that 5-hydroxyindoleacetic acid (5-HIAA) activity in the anterior hypothalamus is enhanced after the cat is exposed to a cold environment of 5°C. This suggests that not only is the actual release of 5-HT altered within the rostral hypothalamus but also the metabolism of the amine in the neurons of this structure is affected by the animal's exposure to cold. In this connection, the release of norepinephrine from the homologous anatomical region in the hypothalamus is unchanged when the cat is exposed to cold air (Myers and Chinn, 1973).

From these studies, one could conclude that the 5-HT released endogenously within other areas of the hypothalamus and under different conditions probably has little bearing on the functional role of the amine in thermoregulation (Myers and Beleslin, 1971). This would explain why an assay of the entire hypothalamus, or indeed the whole brain, often yields conflicting results. Actually, the presynaptic activity of 5-HT-containing neurons within the anterior hypothalamic thermosensitive region could be masked or canceled out by the discharge of alternative groups of serotonergic neurons in response to sleep, stress, arousal, hunger, or other physiological states (Myers, 1974a).

D. Destruction of 5-HT Neurons

The prior pieces of evidence supporting the concept that 5-HT is the principal mediator on the heat-production side in the thermosensitive zone are based on (1) anatomical localization, (2) 5-HT's pharmacological action, and (3) the enhanced release of endogenous 5-HT during cold. There is, however, a fourth piece of evidence. When a chemical lesion is placed in the anterior hypothalamic, preoptic area so that 5-HT neurons are destroyed, thermoregulation is severely impaired. In two species tested thus far, the rat and monkey, a neurotoxin such as 5,6-dihydroxytryptamine (5,6-DHT) or 5,7-dihydroxytryptamine (5,7-DHT), which may be relatively selective depending upon its use, was microinjected at a site already identified as being reactive to exogenously applied 5-HT.

When a bilateral microinjection of as little as 1.25 μg of 5,6-DHT was made into the anterior hypothalamic, preoptic area of the unanesthetized rat, a marked thermoregulatory deficit ensued, which was evident almost immediately (Myers, 1975a). After the animal's peripheral exposure to either cold or warm air, its temperature rose or fell, respectively, to these challenges. A larger dose of 5,6-DHT (2.5 μg) produced an even greater impairment in temperature regulation. However, in both cases, the rat did recover its capacity to thermoregulate in the warm within three weeks after a microinjection of 5,6-DHT but failed to do so in response to cold. This clearly indicates that the impairment produced by a localized 5,6-DHT lesion is functionally selective to the pathway for heat production (Myers, 1977, 1978a).

With respect to the primate, a small dose of the same 5-HT neurotoxin injected into the anterior hypothalamic, preoptic area of the rhesus monkey also produces the same sort of marked deficit in thermoregulation. Moreover, the impairment can be so severe that within 24-48 hr after the serotonergic lesion, the monkey may succumb due to a deep hypothermia. An example of this is depicted in Figure 5. Such a terminal decline in body temperature can follow a temporary 1-2-hr period of hyperthermia, during which time an agonistic action of the 5,6-DHT is seen as the neurotoxin presumably binds to serotonergic receptors within the anterior hypothalamus (Waller et al., 1976).

A fascinating aspect of such a neurotoxin lesion is the sparing of the animal's capacity to thermoregulate behaviorally. Even though the monkey fails to respond to the cold by the typical physiological mechanisms of shivering, vasoconstriction, and increased metabolic heat production after such a 5,6-DHT lesion, some monkeys nevertheless attempt to keep warm. Although its behavior depends on the site and extent of the lesion, the monkey persistently pulls a lever with its forepaw to obtain warmth from a heat lamp as the

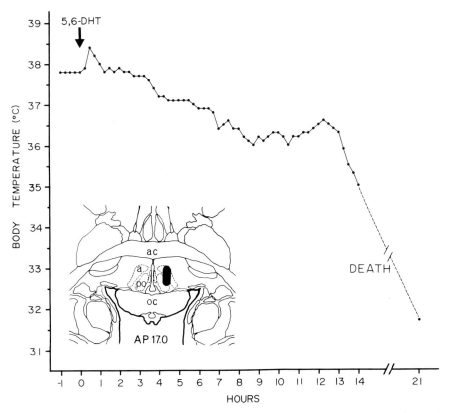

Figure 5 Temperature of a monkey after a single microinjection at zero hour (arrow) of 2.7 μg 5,6-DHT into the area designated by the black dot in the histological inset. Warm air was blown into the chair chamber between hours 9 and 12. (From Waller et al., 1976; reprinted with permission from *Neuropharmacology 15*, 61-68, Pergamon Press, Ltd.)

trunk of its body is exposed to cold. The heat generated in this way by the lamp is sufficient to prevent the animal's temperature from falling even though physiological processes for maintaining heat are inoperative (Waller et al., 1976). Since 5,6-DHT does not exert its neurotoxic effect on temperature regulation when it is injected at other anatomical sites within the hypothalamus, the concept that intact 5-HT nerve terminals form a vital link in the network of neurons mediating the heat-production pathway (Myers, 1974a) is strengthened further.

A summary of the over-all anatomical, pharmacological, and physiological evidence for the theory that 5-HT in the anterior hypothalamic, preoptic area subserves heat production is presented in Table 1. Note that this table is based principally on experiments in which each functional event has been localized anatomically.

III. Catecholaminergic Pathways

Shortly after the turn of this century, the adrenal principal (epinephrine, suprarenin) became implicated in various facets of thermoregulation (von Euler, 1961). Since then, the

Table 1 Summary of Anatomical Evidence for the Role of Endogenous 5-HT in the Mediation of the Anterior Hypothalamic Mechanism for Heat Production

A. Indirect Evidence
   1. 5-HTP administered peripherally evokes hyperthermia (Horita and Gogerty, 1958).
   2. Intracisternal or intraventricular 5-HT evokes hyperthermia (Canal and Ornesi, 1961; Feldberg and Myers, 1964).
   3. A 5-HT receptor blocking agent antagonizes a 5-HT-induced hyperthermia (Canal and Ornesi, 1961; Girault and Jacob, 1979).

B. Direct Evidence
   1. 5-HT microinjected into the anterior hypothalamus evokes a dose-dependent hyperthermia, under an appropriate experimental condition in the rat, rabbit, cat, monkey, and bird (see Myers, 1974a; Borsook et al., 1977).
   2. Cooling of the body evokes the local release of 5-HT from the anterior hypothalamus (Myers and Beleslin, 1971).
   3. Peripheral cooling enhances 5-HT turnover within the anterior hypothalamus (Isaac, 1973).
   4. 5-HT microinjected into the anterior hypothalamus evokes the release of acetylcholine in caudal sites along the heat-production pathway as body temperature rises (Myers and Waller, 1973).
   5. Bacterial pyrogen given centrally evokes the release of 5-HT from the anterior hypothalamic, preoptic area as fever develops (Myers, 1971b, 1977).
   6. Lesioning of 5-HT-containing neurons by 5,6-DHT neurotoxin microinjected into the anterior hypothalamus impairs heat conservation during the state of normothermia as well as heat production in response to a cold challenge (Waller et al., 1976; Myers, 1976a, 1978a).
   7. 5-HT microinjected into the anterior hypothalamus causes the efflux of Ca^{2+} from the posterior hypothalamus and a consequent increase in body temperature (Myers et al., 1976b).

role of both epinephrine and norepinephrine in the peripheral response to cold stress has been elucidated (Brück, 1976). For example, the increased secretion of these two catecholamines which occurs as an animal acclimates to the cold is well known (Carlson, 1960; Leduc, 1961; Baldwin et al., 1969). In fact, catecholamine-induced peripheral thermogenesis is a major response to cold stress (Chaffee and Roberts, 1971; Fregly et al., 1977). Nonshivering thermogenesis and the calorigenic response to norepinephrine appear to be related to the presence of brown adipose tissue (see review by Chaffee and Roberts, 1971).

Parenthetically, the importance of the increase in metabolism after the exogenous elevation of the concentration of norepinephrine in the blood stream cannot be overemphasized (Myers, 1980). One reason for this lies in the possibility that a catecholamine injected centrally, if in a sufficiently high dose, may exert its effect on body temperature by way of the peripheral mechanism for nonshivering thermogenesis (Schmidt, 1963) rather than through a synaptic process in the hypothalamus.

A. Norepinephrine and the Control of Heat Dissipation

The history of the part played by the catecholamines in the brain's control of body temperature is sprinkled with fascinating accounts. In connection with their experiments on "heat-puncture" fever, Jacobj and Roemer first reported in 1912 that epinephrine injected into the ventricle produces a fall in temperature. In another classic paper, Barbour and Wing (1913) infused a solution of epinephrine or other drug in a large volume directly into the brain substance of the rabbit through a cannula implanted aseptically and affixed to the animal's skull. Each of the single experiments elaborated upon revealed that epinephrine causes a fall in the rabbit's temperature. Barbour and Wing concluded that epinephrine has an antipyretic action analagous to that of quinine, antipyrin, and chloral hydrate infused similarly.

Of great importance is the initial insight of Barbour and Wing who inferred that this catecholamine has a specific effect on nerve tissue: "Such an action is very important and may be the real cause; furthermore, the work of Cow (1911) and others indicates that epinephrine dilates rather than constricts the cerebral vessels and herein one would find also an explanation for reduction in body temperature by the increase in brain temperature which would result from the improved circulation." This statement is in direct contradiction to the idea of Jacobj and Roemer (1912) who believed that there is a general irritative condition on the ventricle produced by the catecholamine. Even in that era, over 60 years ago, it is clear that controversy flared as it does today.

In 1954, Vogt showed that catecholamines are distributed unevenly in the brain but are present in the hypothalamus in a relatively high concentration. Literally hundreds of investigators have documented this fact in the mammal and in other species. Further, widespread fiber pathways, projection systems, and the CNS distribution of the catecholamines have been delineated for the rat, cat, primate, and others (Fuxe et al., 1970; Olson and Fuxe, 1971; Maeda and Shimizu, 1972; Ungerstedt, 1971; Palkovits et al., 1974). A topographic atlas of catecholamine-containing neurons has been constructed depicting structures within the morphological confines of the telencephalon, diencephalon, and other regions (Jacobowitz and Palkovits, 1974). In the rat, cat, and monkey, the organization of discrete catecholamine pathways and cell bodies has also been described (Garver and Sladek, 1975; Hubbard and DiCarlo, 1974).

Of special importance to the thermoregulatory system of the diencephalon is the well-established fact that catecholaminergic nerve bundles are juxtaposed with serotonergic fiber pathways which project from nuclei located in the mesencephalon (Ungerstedt, 1971; Morgane and Stern, 1972). This point becomes pivotal to the understanding of the interrelationship between 5-HT and the catecholamines in the anterior hypothalamus with reference to the mechanism which initiates the loss or gain of body heat.

1. Hypothermia in the Primate and Cat

Coinciding with the experiments in which a rise in temperature is produced by 5-HT injected in the ventricles, Feldberg and Myers (1964) showed that both epinephrine and norepinephrine not only antagonize the hyperthermic action of 5-HT but also lower normal body temperature when either catecholamine is injected into the cerebral ventricle of the normothermic cat. In the initial attempt to localize the hypothalamic action of a catecholamine on body temperature, it was found that a microinjection of either norepinephrine or epinephrine into the anterior hypothalamus induces a profound, dose-dependent fall in the cat's body temperature (Feldberg and Myers, 1965a). The doses of 0.5-5.0 μg required to produce these changes in the animal's colonic temperature are as much as a hundred times less than those required to evoke an equivalent change when the compounds are administered by the ventricular route. Again, the two catecholamines exert no effect on the cells of the posterior hypothalamus or the pathways that traverse this area, an observation confirmed by Saxena (1973).

In the rhesus monkey, species continuity has been demonstrated in that norepinephrine or epinephrine causes the same sort of hypothermia when either catecholamine is microinjected at homologous sites (Myers, 1968; Myers and Sharpe, 1968; Toivola and Gale, 1970; Myers and Waller, 1975a). An anatomical examination of the specific sites in the hypothalamus at which norepinephrine evokes hypothermia reveals an interesting phylogenetic relationship. In the rhesus macaque, a dose as low as 1.0 μg infused in the anterior hypothalamus, below the anterior commissure and dorsal to the optic chiasm, is effective in producing a transient hypothermia (Myers and Yaksh, 1969). Within the same circumscribed region of the cat's hypothalamus, norepinephrine possesses the same sort of high potency in evoking a decline in the animal's temperature. Again, the catecholamine has no effect when microinjected in the posterior or other areas outside of the rostral hypothalamus (Metcalf and Myers, 1978). An anatomical analysis of the precise location of the rostral hypothalamic region where norepinephrine produces a decline in body temperature reveals that the tissue juxtaposed to the preoptic area is most reactive. With a microinjection volume of 0.5 μl and a constant dose of 5.0 μg, the magnitude of hypothermia correlates significantly with the proximity to this anatomical position (Metcalf and Myers, 1978).

Within a small percentage of sites, the catecholamine simultaneously elicits a sudden and voracious ingestion of food pellets, even though the animal had been fed previously. Generally, however, the norepinephrine-induced hypothermia and feeding response are independent of each other and contingent usually upon the specific site of catecholamine injection (Yaksh and Myers, 1972). In other species of primate, the same sort of hypothermic response is clearly evident when norepinephrine is applied in an area dorsal to the optic chiasm and ventral to the commissure. An infusion of norepinephrine at a rate of 1.0 μg/min induces an almost immediate cutaneous vasodilation and a fall in the baboon's brain temperature, heart rate, and blood pressure (Toivola and Gale, 1970). Similarly, a bilateral injection of norepinephrine in the anterior region of the pigtailed macaque results in a synergism, i.e., in that the magnitude of the hypothermia is much greater than that elicited by a unilateral injection (Myers and Waller, 1975a).

A notable observation has been made in the cat by Cooper et al. (1976a) who demonstrated that norepinephrine applied to the anterior hypothalamic, preoptic area produces a decline in temperature of similar intensity regardless of the ambient temperature at which the animal is kept, i.e., cool (10°C), neutral (22°C), or warm (35°C). However, the type of physiological response associated with the hypothermia is related directly to the ambient temperature. In the cold, the cat's typical heat production responses are simply inhibited, whereas in the warm environment, the mechanisms for heat loss are activated, which include panting, vasodilatation, and the adoption of a posture conducive to heat loss with limbs apart and the cat's body sprawled out. Figure 6 shows the hypothermic responses to different doses of the catecholamine at this high ambient temperature (Cooper et al., 1976a).

The properties of the pharmacological receptors which mediate the heat-loss response of norepinephrine have also been studied; thus far, the findings favor an action of the

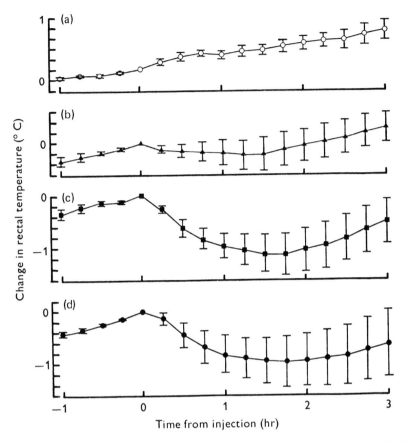

Figure 6 Mean changes in body temperature from the temperature at the time of bilateral injection of norepinephrine (NE) into the anterior hypothalamic, preoptic area of cats at an ambient temperature of 35°C. (a) 0.5 µl saline (n = 6); (b) 2.5 µg NE (n = 6); (c) 5.0 µg NE (n = 4); (d) 10.0 µg NE (n = 6). Vertical bars equal ± SEM. (From Cooper et al., 1976a.)

catecholamine on α-adrenergic receptors. Saxena (1973) has shown that isoprenaline, a β-adrenergic agonist, microinjected into the anterior hypothalamus of the cat, has no effect on the animal's body temperature. Further, a β-receptor blocker does not prevent the hypothermic action of norepinephrine in this same species (Burks, 1972). On the other hand, an injection into the hypothalamus of phenylephrine, an α-receptor agonist, evokes a similar sort of hypothermia as that produced by norepinephrine but a weaker response than that following epinephrine (Rudy and Wolf, 1971). In the latter study, pretreatment of the hypothalamic microinjection site with phentolamine, an α-receptor antagonist, reduces the usual hypothermia produced by the catecholamine. Phentolamine infused into the hypothalamus also attenuates the norepinephrine-induced decline in pigeons' temperature, with propanolol having no antagonistic effect on the hypothermic response (Hissa and Pyörnilä, 1977).

2. Responses of the Rabbit

Cooper and his colleagues (1965) first raised the intriguing possibility of an inherent species difference with respect to the hypothalamic role of the monoamines in body temperature. After 10.0 μg of norepinephrine was injected into the anterior hypothalamus in a 2.0- to 10.0-μl volume, the rabbit's temperature increased rather than fell. In one-half of the experiments undertaken, however, the rabbit's temperature failed to change with that dose. In addition, doses of 2.0-5.0 μg of the catecholamine, which lower the temperature in other species such as the cat or monkey, seemed to exert little effect on the rabbit's temperature. Conversely, Gonzalez has reported that norepinephrine injections into the hypothalamus of the rabbit in the same dose range but in a volume of 5.0 μl cause an immediate decrease in rectal temperature and a concomitant rise in hypothalamic temperature (Gonzalez, 1971). Although this inverse relationship is consistent, the absolute changes in temperature are very small.

A coherent set of observations has been made by Preston (1975) concerning the role of norepinephrine in the rabbit's anterior hypothalamic, preoptic area. In contrast to the hyperthermia produced by a large dose injected into the cerebral ventricle of the rabbit, norepinephrine infused in the rostral hypothalamus generally causes a marked vasodilatation and a fall in temperature. When the volume of the microinjection is held to only 0.5 μl, the decline in body temperature averages between 0.2 and 0.5°C. This fall is accompanied by vasodilatation as illustrated in Figure 7. A significant point over and above the injection technique is that Preston has carried out the experiments with rabbits kept at an environmental temperature of 15°C. This ambient temperature approaches that of

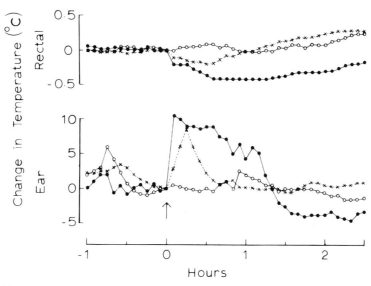

Figure 7 Decrease in rectal temperature and accompanying rise in ear temperature after microinjection of a total of 6 (x- - -x) or 20 (●——●) μg norepinephrine into the anterior hypothalamic, preoptic area on separate occasions in same conscious rabbit. Control injection of NaCl solution (o——o) was without effect. (From Preston, 1975.)

thermal neutrality for this species. In view of this, Preston draws the conclusion that "... no support is provided for the concept that in the AH/POA [anterior hypothalamic, preoptic area] of the rabbit, the effect of exogenously applied norepinephrine and the role of the endogenous catecholamine is to activate temperature raising mechanisms."

One current explanation of norepinephrine hyperthermia in the rabbit is that, after its injection into the ventricle, it causes thermogenesis because of its diffusion or dispersion to other brain-stem structures or its subsequent exit from the ventricular cavities and the resultant action on a peripheral or central locus other than the hypothalamic temperature controller. If this is so, then the issue of species difference among mammals and birds could be on its route to resolution. Monotremes and other animals pose yet another question with regard to the part played by monoamines in their thermoregulatory process (Baird et al., 1974).

3. Responses of the Rat

An injection of norepinephrine into the rostral hypothalamus of the rat usually causes a dose-dependent hypothermia which varies from 0.4 to 0.8°C (Lomax et al., 1969). When a small, 0.5-μl volume of the catecholamine is infused into the anterior hypothalamic area, the same sort of hypothermic effect on the rat's body temperature is seen (Avery, 1971). At a high-ambient temperature, a similar microinjection of the catecholamine does not entirely prevent the typical heat-induced elevation in the animal's temperature but the magnitude is far less than that following a control injection (Avery, 1972).

A dose of 10.0 μg of norepinephrine injected into the anterior hypothalamus induces a hyperthermia which is greater than that produced by a control injection of water (Kubikowski and Rewerski, 1969). In consideration of this finding with an animal having a brain as small as that of the rat, it is possible that the size of the lesion produced by the injection or the position of the microinjection cannula itself could be important factors. For example, Veale and Whishaw (1976) showed that hypothermia is produced by a microinjection of norepinephrine in a low dose of 5.0 μg into the anterior hypothalamus. On the other hand, 25.0 μg injected similarly induces a sharp rise in the rat's temperature which, as will be discussed later, may well be due to hyperactivity (Day et al., 1979). If the position of the cannula is moved slightly caudalward and encroaches upon the lateral hypothalamus, the dose of 25.0 μg is far more efficacious in evoking a prolonged hypothermia, which corresponds to the response after intraventricular infusion of the amine (Fukushima and Itoh, 1975). This is probably because of the more gradual diffusion of norepinephrine onto the rostral hypothalamic receptors involved in the mediation of heat loss in the rodent.

If catecholamine content is acutely depleted in the hypothalamus by a 32 to 64 μg microinjection of guanethidine, body temperature rises (Armstrong et al., 1973). This is due possibly to the imbalance in 5-HT-catecholamine release in the anterior hypothalamus with the serotonergic pathway predominating following norepinephrine reduction.

In investigations on behavioral thermoregulation, Beckman (1970) has found that an intrahypothalamic injection of norepinephrine in a volume larger than that used by Avery causes an increase in the animal's lever-pressing responses to obtain heat-lamp reinforcement. Therefore, the rat's behavior could counteract a noradrenergically induced condition indicative of heat loss, hypothermia, and the animal's apparent sensing of cold. Thus, the animal's temperature rises with repeated bursts of heat from the lamp.

A corresponding type of experiment by Avery and Penn (1973) shows conclusively that, as norepinephrine lowers body temperature after its injection into the anterior hypothalamus, the rat temporarily reduces its responding to escape from externally applied heat. This result presumably is due to the fact that the skin temperature is elevated through norepinephrine's induction of peripheral vasodilatation. Interestingly, when the anterior hypothalamic area is lesioned, the fall in the rat's temperature typically produced by an intraventricular injection of norepinephrine may be even more intense (Satinoff and Cantor, 1975; Cantor and Satinoff, 1976). This result would favor the idea that the catecholamine, acting from the ventricular lumen, possesses a secondary action on structures other than the hypothalamus. Alternatively, it could reflect a denervation supersensitivity of the amine receptors still intact in the rostral area.

It is indeed likely that any rise in temperature following intrahypothalamic norepinephrine is due to enhanced motor activity (Day et al., 1979); restraint abolishes this and, as a result, dose-dependent falls in temperature only are evoked by the monoamine. It is possible that in the diminutive brain, norepinephrine in a large dose spills onto receptors of the nearby nucleus accumbens which may be involved in the heightened locomotion and a resultant hyperthermia.

4. Guinea Pig, Chicken, and Other Species

Changes in body temperature are brought about by a catecholamine applied to the hypothalamus of other species as well. Norepinephrine injected in a dose of 20 μg into the anterior hypothalamus of the newborn sheep produces an immediate fall in body temperature (Pittman et al., 1977a), which is in contrast to what is seen when the catecholamine is infused into the sheep's ventricle (Bligh et al., 1971). The guinea pig's rostral hypothalamus in a region caudal to the thermosensitive area reacts to norepinephrine in that a shift in the threshold for heat production is induced by the amine (Zeisberger and Brück, 1971a,b). A single 2.0-μg dose of norepinephrine causes a biphasic temperature response and an increase in oxygen consumption of as much as 50%. Also, the threshold for shivering is raised by norepinephrine microinjected into the rostral hypothalamus of either the newborn or cold-adapted guinea pig (Zeisberger and Brück, 1976). Interestingly, a slight hypothermia at several of the thermocouples measuring various parts of the pig's body temperature occurs when the ambient temperature approximates that of the rabbit as described in Preston's experiment (1975), i.e., 15°C. However, when the pig's environmental chamber is rewarmed, its temperature increases to the normal level.

When a very high dose (10-100 μg) of the α-adrenergic antagonist, phentolamine, is infused into the hypothalamus of the guinea pig kept at an ambient temperature of 14°C, not only is body heat lost, but its oxygen consumption is reduced as well. Zeisberger and Brück (1976) suggested the idea that "catecholamines act post-synaptically at the inhibitory neurons" in the hypothalamus. Although their results support such a suggestion in this species, information would be useful on (1) a dose-response analysis, (2) evidence of catecholamine release within the guinea pig's hypothalamus, and (3) the anatomical distribution of sites sensitive to norepinephrine. Interestingly, intrahypothalamically infused phentolamine also blocks thermogenesis produced by electrical stimulation of the pig's lower brainstem (Szelényi et al., 1977). Thus, these data, taken in toto, could represent a formidable argument in favor of a true species difference in the role of hypothalamic norepinephrine in thermoregulation.

Marley and his co-workers have published a series of elegant papers on the hypothermic actions of catecholamines when they are infused into the hypothalamus of the unanesthetized chicken (Marley and Stephenson, 1970; Marley and Nistico, 1972; Nistico and Marley, 1972). As reviewed by Nistico and Marley (1974), the sites at which catecholamines exert a hypothermic action are localized to the hypothalamus in both the adult and young fowl. Although the dose required to produce a fall in body temperature is one-fifth to one-twentieth of that required when the intraventricular route is used, norepinephrine also produces sleep and a marked reduction in activity, which could account partially for the decline in the bird's temperature (Marley and Nistico, 1972). Marley and Stephenson (1975) find that norepinephrine infused into the hypothalamus of the chicken kept in a cold environment inhibits the vasoconstrictor response but not the pilomotor or behavioral reaction. Further, Nistico et al. (1976) have characterized the receptors for catecholamine hypothermia and conclude that either α- or β-adrenergic receptors subserve the temperature fall, since blockers of both classes of drug are equally effective in attenuating or abolishing the catecholamine-induced hypothermia.

Species continuity has been demonstrated in the bird by Hissa and Rautenberg (1974) who found that epinephrine injected into the anterior hypothalamic, preoptic area of the pigeon likewise causes a sharp fall in core temperature; however, these authors report that, whereas the pretreatment with a microinjection of phentolamine prevents the hypothermia, propranolol given at the same site has no effect on the norepinephrine-induced fall in the bird's temperature. This suggests that adrenergic receptors of the α class mediate the hypothalamic system for heat loss in the pigeon (Hissa and Rautenberg, 1975).

When norepinephrine is infused into the anterior hypothalamus of the ground squirrel, its arousal from hibernation is initiated. At a normothermic temperature, this catecholamine injected similarly reportedly produces either hypothermia or hyperthermia (Beckman and Satinoff, 1972), the inconsistencies of which are not readily understandable at this time.

B. Iontophoresis of Norepinephrine onto Thermosensitive Neurons

As described succinctly by Carette and Fessard (1976), the iontophoresis of norepinephrine onto a neuron in the rostral hypothalamus can exert either an excitatory or inhibitory effect on the cell's discharge rate, or no effect at all. Thus, the experiments on the iontophoresis of catecholamines, as those with 5-HT, are fraught with the same sort of difficulties in interpretation. As a forerunner of the iontophoretic investigations, Cunningham et al. (1967) showed that epinephrine infused into the cerebral ventricle of a dog in a dose of 50 μg reduces the discharge rate of all thermosensitive neurons; however, one warm-sensitive unit increased its activity after the injection of 10 μg of epinephrine into the third cerebral ventricle. The firing rate of thermally insensitive units in the hypothalamus is either depressed or enhanced by the catecholamine to a variable degree.

In 1970, Beckman and Eisenman announced that in the cat, the iontophoresis of norepinephrine onto a warm-sensitive cell raises its firing rate but lowers the firing rate of a cold-sensitive neuron (Beckman and Eisenman, 1970a). Later that same year, they unexplainably retracted this observation and reported in another paper that warm-sensitive neurons are instead inhibited by norepinephrine, and the firing-rate of cold-sensitive units is accelerated by this catecholamine (Beckman and Eisenman, 1970b). In the rat,

Murakami also found that the microiontophoresis of norepinephrine reduces the activity in 6 out of 20 warm-sensitive units, whereas the discharge rate in 3 of 10 cold-sensitive units is increased (Murakami, 1973).

As found characteristically in the single-unit studies, most neurons within the thermoregulatory pathway of the rat's hypothalamus are unresponsive to the iontophoretic application of the catecholamine or other amine for that matter, with only a small proportion generally affected by the chemical (Bloom, 1974b). In an extensive study with the rabbit, Hori and Nakayama (1973) were able to identify a somewhat greater proportion of thermosensitive cells that respond to the cellular application of amines. Their results agree with the earlier experiments on iontophoresis in that norepinephrine ejected onto a warm-sensitive neuron attenuates its firing rate, but on a cold-sensitive unit the catecholamine augments the discharge pattern. However, the findings diverge from that which one would expect after the microinjection of the amine into the rabbits' hypothalamus (Preston, 1975; Borsook et al., 1977).

Although Jell (1973) was able to find a relatively large number of neurons that were excited by iontophoretically applied 5-HT (37 of 222) and inhibited by norepinephrine (38 of 222), their individual responses to a change in local temperature produced by an implanted thermode are essentially uncorrelated with the distribution of firing patterns seen in response to the amine. Jell proposes that, since the number of neurons sensitive to either amine is considerable, the possibility exists that the unresponsive units may be the "true thermodetectors" while the other cells are suspected "interneurons." He speculates further that a rostral hypothalamic neuron could easily possess multiple receptor sites which are differentially sensitive to each of the different amines. Since 29% of the neurons respond to 5-HT or norepinephrine, as reflected by their altered discharge rate in an opposite direction, Jell contends that the iontophoretic data could perhaps support an amine model of thermoregulation. When the experiments were repeated using peripheral thermal stimulation onto the face of the anesthetized cat to identify thermosensitive neurons, again no consistent relationship was observed between the response of the single unit to an amine and facial temperature. By contrast, Jell (1974) did find some neurons which respond in a way which could be predicted from the amine model of thermoregulation based either on neuroanatomical experiments with the cat and primate (Myers, 1974a) or on ventricular injections in the sheep (Bligh et al., 1971).

The concordance of responses to iontophoresis of the amines in the cat, rabbit, and rat, when thermosensitive and amine-sensitive units are indeed concurrently identified, generally appear to be opposite to the results of experiments on (1) localized microinjection of the amines, (2) analysis of monoamine release by the technique of push-pull perfusion, and (3) the deficits in thermoregulation following neurotoxin lesions of the anterior hypothalamus. The reasons for this apparent discrepancy are similar to those discussed at the conclusion of Section I.D and of Section II.B on 5-HT.

C. Hypothalamic Content and Release of Catecholamines During Thermal Stress

To correlate the endogenous level of activity of the catecholamines within the brain during thermoregulation or after thermal stress has been the goal of many investigators. However, as we have seen with the studies of 5-HT activity, such analytical efforts again are plagued often by inconsistency in observation. Nevertheless, one would expect that in view of the marked changes that occur in catecholamine activity in the periphery in the cold-exposed animal, some variation in the stores of the amines most likely would occur in the CNS.

1. Catecholamine Levels and Thermal Stress

In pioneering work done in the early 1960s, Strzoda (1962) found that the content of norepinephrine in the diencephalon increases while that in the rhombencephalon decreases in the mouse exposed to a high temperature (40°C). Strzoda hypothesized that elevated temperature enhances the diencephalic metabolism of the catecholamines in relation to the "vasomotor center's resistance to heat." However, in determining postmortem the hypothalamic content of norepinephrine in a thermally stressed rat, Ingenito and Bonnycastle (1967) were unable to correlate the amount of the amine with exposure to a high or low temperature. This result with hypothalamic tissue per se was in sharp contrast to the more generalized increase in norepinephrine turnover occurring in the whole brain of the rat exposed either to a hot or cold environmental temperature (Costa and Neff, 1966; Gordon et al., 1966; Corrodi et al., 1967; Duce et al., 1968; Reid et al., 1968). The latter findings could be due to the nonspecific stress of a thermal challenge which naturally would affect all systems in the mammal's brain. Unfortunately, other researchers working during this time also failed to demonstrate a clear-cut and reliable change in the postmortem content of norepinephrine in the whole brain of the rat after its exposure to extreme environmental temperatures (Maynert and Levi, 1964; Beauvallet et al., 1967).

Of considerable interest is the finding of Shellenberger and Elder (1967) who demonstrated that the level of norepinephrine in the brain stem of the rabbit exposed to a temperature of 42°C was reduced to 65% of the normal level. This thermal stress failed to change the content of 5-HT or dopamine measured similarly. This result coincides with the fact that rabbits pretreated systemically with α-methyltyrosine, which reduces the level of brain stem norepinephrine, are virtually unable to cope with an induced hyperthermia produced by peripheral exposure to heat.

In an early study done with the rat in which histochemical and biochemical approaches were combined, Corrodi et al. (1967) found that, in norepinephrine-containing neurons in the rat's hypothalamus, the rate of depletion of this amine was markedly increased when the rat was exposed to external heat up to 40°C. Since cold exposure to 3°C has no effect on the noradrenergic neurons, Corrodi et al. concluded that the activity of hypothalamic noradrenergic neurons increases as the rat dissipates heat, but not when it gains heat. Most interesting is the fact that none of the hypothalamic dopamine-containing nerve terminals are influenced by either heat or cold.

In the whole brain of the mouse, a corresponding decline in the content of norepinephrine occurs again, but only if the rodent is exposed to a high ambient temperature of 40-42°C (Harri and Tirri, 1969). In a follow-up study, the analysis of the hypothalamus of the mouse again revealed that the exposure of the animal to a warm environmental temperature causes a depletion in the content of norepinephrine in this structure (Tirri, 1970). Figure 8 exemplifies the sharp fall in the catecholamine's level 1 hr after exposure to 40°C (Harri and Tirri, 1969).

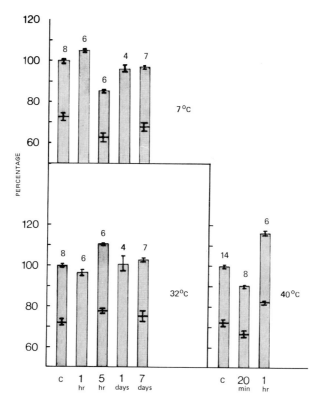

Figure 8 The level percentage of control values of 5-HT in mouse brain after exposure of animals to various ambient temperatures. The control values = 0.67 µg/g at 7 and 32°C, and 0.66 µg/g at 40°C. The number of experiments is given on the top of each column. The duration of the exposure is indicated at the base of the columns. The upper part of divided columns represents the "free" form and the lower part the "bound" form of NE. C, Control animals, kept at 22°C. (Modified after Harri and Tirri, 1969.)

In examining the rate of turnover of norepinephrine in the rat's hypothalamus, Simmonds and Iversen (1969) found that the radioactivity of tritiated norepinephrine injected intracisternally declines after the rat is warmed to 32°C. This is consistent with Corrodi's supposition (1967) that hypothalamic norepinephrine may regulate heat loss in the rat under a normal physiological condition. In a subsequent experiment, Simmonds (1971) found that intraventricular atropine inhibits the disappearance of the catecholamine if the rat is cooled to 9°C but not when warmed to 32°C. When the hypothalamus is dissected into anterior and posterior parts and the preoptic area, after the rat had been exposed to a high or low environmental temperature, the rate of disappearance of the catecholamine from either area of the hypothalamus is much faster than when the rat is kept at a neutral temperature (Simmonds, 1969). An interpretation of this result is difficult to make in view of other results obtained with tissue assays. Possibly sufficient anatomical resolution is not achieved by the method of dissection used; hence, the "whole-structure assay" may not be valid (Myers, 1969b).

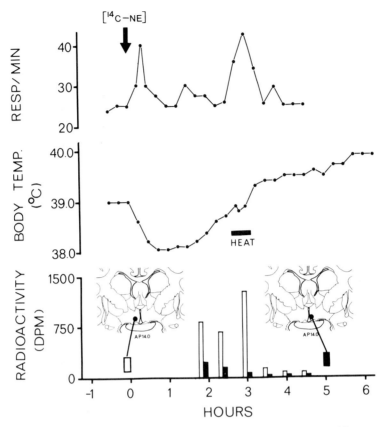

Figure 9 Changes in respiratory rate, body temperature, and output of [^{14}C] NE and metabolites in a representative cat before, during, and after warming cat's chamber to 40°C. At zero hour, 25 μCi [^{14}C] NE were injected into lateral ventricle as indicated by arrow. Interval of heating is indicated by horizontal bar (—). Local perfusion sites at AP 14.0 are indicated in insets. (From Myers and Chinn, 1973.)

2. Evoked Release of Norepinephrine

The necessity of understanding the actual in vivo dynamics of norepinephrine activity in the anterior hypothalamus of the conscious animal as it thermoregulates against a warm or cold temperature is self-evident. In an experiment undertaken to uncover these changes, the anterior hypothalamic, preoptic area of the cat was labeled with radioactive norepinephrine, then perfused slowly with an artificial cerebrospinal fluid (CSF) at the same time that the animal was exposed to a warm or cold ambient temperature (Myers, 1970). Successive perfusates were analyzed for the quantity of norepinephrine activity. At anatomical sites identified within the anterior hypothalamic area dorsal to the optic chiasm, norepinephrine release is sharply enhanced while the cat is exposed to an ambient temperature of 40°C (Myers and Chinn, 1973). The time course of such a release is illustrated in Figure 9. On the other hand, when the cat is placed in a cold ambient temperature of 10°C, there is no change in the output of the catecholamine from the anterior hypothalamus. In general, this finding with the cat corresponds to the attenuated hypothermia seen in the cat given desmethylimipramine (DMI) after the hypothalamus is depleted of norepinephrine by αMpT (Cranston et al., 1972).

Taken together, these findings would suggest that, in the face of a warm thermal challenge, norepinephrine is released presynaptically from noradrenergic nerve terminals located within the thermosensitive zone of the cat's anterior hypothalamus (Myers and Chinn, 1973). Such a presynaptic release of the catecholamine would be predicted on the basis of the pharmacological experiments in which the microinjection of norepinephrine in this same anatomical region of the cat as well as other species evokes a relatively shortlasting but significant hypothermia (Metcalf and Myers, 1978). If the amine is indeed acting on postsynaptic receptors, then the two sets of coalescent observations indicate that the catecholamine mediates the neuronal process underlying the cellular mechanism for the control of heat dissipation.

D. Dopamine and Thermoregulation

The metabolic precursor of norepinephrine, dopamine, was first implicated in the mechanism for thermoregulation in 1966. When dopamine was injected into the third ventricle of the unanesthetized monkey in doses ranging from 25 to 300 μg, the primate's temperature declined (Myers, 1966). Subsequent analyses revealed that the dose of dopamine required to evoke hypothermia is 50% higher than that of norepinephrine; moreover, the temperature fall induced by dopamine was far less intense. Thus, its role in a neurochemical mode of thermoregulation has been difficult to justify until recently (Myers and Sharpe, 1968).

The lack of potency of dihydroxyphenylalanine (L-Dopa) or dopamine is likewise seen when either is infused into the lateral cerebral ventricle of the rat. Again, only very minor fluctuations are produced in the animal's colonic temperature, i.e., less than 0.5°C in either direction (Myers and Yaksh, 1968). In the conscious mouse, an intraventricular injection of dopamine in an exceedingly high dose, relative to the size and weight of this rodent's brain, does cause its temperature to decline (Brittain and Handley, 1967). Since these early studies, the hypothermic action of very high doses of dopamine given by the intraventricular route has been confirmed repeatedly in several species including the pigeon (Chawla et al., 1974), cat (Kennedy and Burks, 1974), and again in the rat (Bruinvels, 1970; Kruk, 1972).

Because of an incongruency in the required dose of dopamine, in comparison to the potency of norepinephrine, the results with ventricular injections tend to dispel any potential role for dopamine in the hypothalamic control of body temperature. Somewhat more substantial support for this notion has been obtained with an anatomical analysis in the monkey. After dopamine is microinjected in the same range of doses as norepinephrine in the anterior hypothalamic, preoptic region, the maximum decline in temperature is only 0.3°C (Myers and Yaksh, 1969). In the rat, a 10 μg dose of dopamine injected into the preoptic region lowered body temperature by about 1.0°C; this fall was antagonized by prior injection of pimozide, a dopamine-receptor blocking agent (Cox and Lee, 1977). In tissue slices of anterior hypothalamus, incubation with dopamine augmented 3,5-cyclic AMP synthesis which was blocked by haloperidol (Cox et al., 1978) suggesting the existence of dopaminergic receptors in this portion of the hypothalamus.

In a similar type of experiment in the unanesthetized cat, Quock and Gale (1974) showed that dopamine given in a dose of 10 µg causes a fall in temperature but again with a maximum fall of only 0.4°C within 15 to 30 minutes following its infusion. They also found that haloperidol given systemically or intrahypothalamically in a high dose of 25 µg does attenuate the slight hypothermia brought about by dopamine. This result corresponds to that of Kennedy and Burks (1974) who showed that the hypothermia following intraventricularly-given dopamine is likewise antagonized by haloperidol infused by the same route.

In an anatomical study of 113 microinjection sites carried out in the unrestrained cat, Ruwe and Myers (1978) demonstrated that dopamine given in somewhat smaller doses can elicit a hypothermia which is dose-dependent. This effect is seen, however, only when dopamine is injected at very circumscribed sites in the cat's anterior hypothalamic, preoptic area. Of significance is the fact that some of the dopamine-reactive sites do not overlap with those that are reactive to norepinephrine. This suggests that there are at least two types of catecholaminergic receptors in the anterior hypothalamus that activate heat loss or, alternatively, inhibit heat production (Ruwe and Myers, 1978). These receptors could mediate different physiological components of the mechanism responsible for a fall in body temperature. An example of a hypothermia induced by dopamine at sites insensitive to norepinephrine is presented in Figure 10.

Dopaminergic hypothermia is clearly antagonized by a prior microinjection of either haloperidol or butaclamol given at the same hypothalamic locus (Ruwe and Myers, 1978). This result suggests that dopamine may be involved in some component of heat loss subserved by synapses which differ from noradrenergic terminals. For example, the presynaptic release of dopamine could be involved in only one of the motorial components of heat loss, for example, the activation and maintenance of cutaneous vasodilatation. In relation to this is the report that apomorphine, a dopamine-receptor agonist, when injected into the nucleus accumbens also evokes a fall in the body temperature (Grabowska and Andén, 1976). Because the large 2.0 µl volume of the apomorphine solution used undoubtedly reached the anterior hypothalamus, a portion of the dopamine agonist solution could have acted on amine receptors which mediate a hypothermic response.

When dopamine is delivered by microiontophoresis to thermosensitive neurons, the monoamine depresses the firing rate mainly of cold-responsive cells (Sweatman and Jell, 1977). This would indicate that dopamine release at the synapse may suppress the heat production pathway by inhibiting the activity of cold-sensitive neurons. Unfortunately, dopamine applied locally by the iontophoretic technique tends to depress all neurons apparently in a nonspecific manner. Moreover, a precise distinction between the effect of norepinephrine versus dopamine on single units in the hypothalamus has not been delineated (Whitehead and Ruf, 1974). There is some indirect evidence based on intraventricular studies with apomorphine, pimozide, and other compounds that also implicates dopamine in the hypothalamic mechanism for heat loss. This will be discussed in Section VIII.D.

Finally, in a double isotope label study, it was shown that warming of a cat evoked the release of [^{14}C] dopamine from sites at which [^{3}H] norepinephrine was not necessarily liberated (Ruwe and Myers, 1978). This again strongly suggests a multiple system of catecholamine pathways intermingled within the same preoptic, anterior hypothalamic area, which is responsible for the activation of heat loss responses or, alternatively, the blockade of heat production, or both.

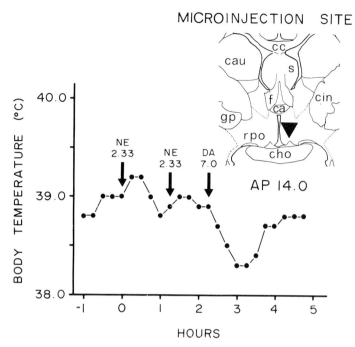

Figure 10 Colonic temperature after a microinjection into the site denoted in the histological inset (▼): at zero hour, 2.33 μg NE; 2.33 μg NE 75 min later; and 7.0 μg DA 135 min after zero hour. Anatomical abbreviations are: ca, anterior commissure; cau, caudate nucleus; cc, corpus callosum; cho, optic chiasm; cin, internal capsule; f, fornix; gp, globus pallidus; rpo, preoptic area; s, septal region. (From Ruwe, 1977.)

1. 6-Hydroxydopamine (6-OHDA) and Hypothermia

In species in which the chemoreactivity of the anterior hypothalamus has been anatomically elucidated (except perhaps the guinea pig), it is apparent that the catecholamines mediate hypothermia. When 6-OHDA, a neurotoxin which is known to release the catecholamines from their respective nerve terminals (see review of Kostrzewa and Jacobowitz, 1974), is injected intraventricularly in the rat, a fall in the body temperature is produced (Simmonds and Uretsky, 1970; Breese and Howard, 1971; Nakamura and Thoenen, 1971; Breese et al., 1972; Hansen and Whishaw, 1973). In the cat, the same sort of hypothermia is seen after the catecholamine neurotoxin is injected similarly (Howard and Breese, 1974).

An injection of 6-OHDA given intraventricularly or directly into the hypothalamus of the rat reportedly has little or no effect on the rat's physiological response to cold or heat when tested approximately one week after the 6-ODHA is given (Van Zoeren and Stricker, 1976); however, behavioral thermoregulation may be impaired by a large 6-OHDA lesion of the hypothalamus (Van Zoeren and Stricker, 1977).

In experiments in which 6-OHDA has been microinjected in hypothalamic tissue in a punctate volume of less than 1.0 μl, a profound impairment in the rat's thermoregulatory capacity is seen. If the animal is exposed to a warm ambient temperature several days

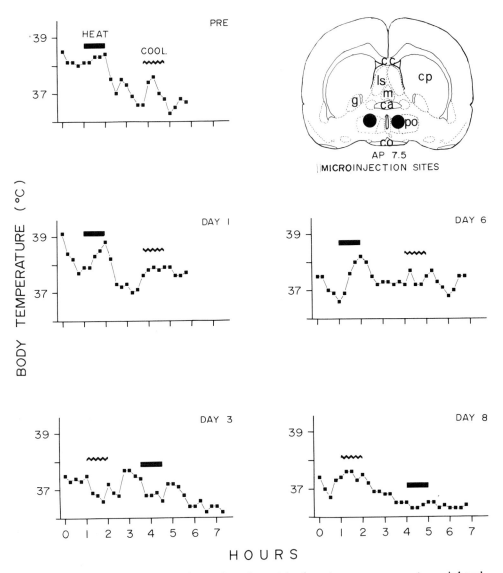

Figure 11 Transitory thermoregulatory impairment in the rat upon exposure to peripheral thermal challenges. A single microinjection of 3.0 μg of 6-OHDA in 0.75 μl CSF was administered bilaterally into the AH/POA sites as in the inset (●). Rats were alternatively exposed to cold, 8.0°C (∼∼), or to heat, 35.0°C (━━), for a 1-hr interval before 6-OHDA, 24 hr after microinjection of the neurotoxin, and at 2-3 day intervals thereafter. Anatomical abbreviations are: ca, anterior commissure; cc, corpus callosum; co, optic chiasm; cp, caudate-putamen nucleus; g, globus pallidus; ls, lateral septal nucleus; m, medial septal nucleus, po, preoptic area. (From Myers and Ruwe, 1978.)

after the catecholamine neurotoxin had been infused in a low dose at a highly circumscribed site in the anterior hypothalamic, preoptic area, the rat's body temperature follows the ambient and rises (Myers and Ruwe, 1978). Depending upon the specific hypothalamic site of the microinjection, the time course of the impairment to heat regulation can endure for less than 2 days or not become fully evident until 10 days or even longer. However, with the passage of time, this thermoregulatory deficit does abate, and the animal subsequently recovers its capacity to withstand the heat challenge. Figure 11 portrays the characteristics of temperature impairment caused over time by 6-OHDA's action on the anterior hypothalamus.

An over-all summary is presented in Table 2 of the anatomical evidence that the catecholamine-containing neurons within the anterior hypothalamic, preoptic thermosensitive region mediate the mechanism subserving heat loss. Again, it should be noted that the

Table 2 Summary of Anatomical Evidence for the Role of Catecholamines in the Mediation of the Anterior Hypothalamic Preoptic Mechanism for Heat Loss, Blockade of Heat Production, or Both Simultaneously

A. Indirect Evidence
  1. Catecholamine-containing nerve terminals are present in the anterior hypothalamic, preoptic area (Andén et al., 1966).
  2. Intraventricular injections of norepinephrine or dopamine induce hypothermia (Feldberg and Myers, 1964; Kennedy and Burks, 1974).

B. Direct Evidence
  1. Dose-dependent hypothermia is evoked by catecholamines and is localized to the thermosensitive area of the anterior hypothalamus (Feldberg and Myers, 1964).
  2. Blockade by α-adrenergic antagonist applied intrahypothalamically of the catecholamines' hypothermic effect on the anterior hypothalamus (Rudy and Wolf, 1971).
  3. Magnitude of norepinephrine induced hypothermia is, within limits, independent of the ambient temperature (Cooper et al., 1976a).
  4. Depletion of catecholamine occurs in norepinephrine-containing hypothalamic neurons when the animal is exposed to external heating. (Corrodi et al., 1967).
  5. Neuronal release of endogenous norepinephrine from the anterior hypothalamus is enhanced when the conscious animal is exposed to heat but not to cold (Myers and Chinn, 1973).
  6. Lesioning of catecholamine-containing nerve terminals in the anterior hypothalamus by 6-hydroxydopamine neurotoxin impairs thermoregulation against the heat (Myers and Ruwe, 1978).
  7. Microinjection of dopamine in the anterior hypothalamus induces hypothermia (Quock and Gale, 1974).
  8. Neuronal release of dopamine from the anterior hypothalamic, preoptic area in the animal exposed to heat (Ruwe, 1977).
  9. Blockade of dopamine-induced decline in body temperature pretreatment of anterior hypothalamic, preoptic area with dopaminergic antagonists (Ruwe and Myers, 1978).

points of evidence are based mainly on experiments in which an anatomical localization of the functional response to the amines was completed.

IV. Cholinergic Pathways Mediating Thermogenesis

In 1968, it was suggested that acetylcholine (ACh) is a likely transmitter at the central synapses which mediate the effector pathways involved in heat production (Myers and Yaksh, 1968). The idea arose from the observation that ACh in a mixture with the anticholinesterase, physostigmine, or the anticholinesterase itself, evokes a dose-dependent hyperthermia when the compound is infused into the lateral cerebral ventricle of the unrestrained rat (Myers and Yaksh, 1968). At that time, it was believed that the cholinergic synapses for heat production are located just caudal to or at the border of the anterior hypothalamus (Myers, 1969a). Earlier, Hulst and De Wied (1967) had found that carbachol, a cholinomimetic which is not readily hydrolyzed, causes a rise in the rat's temperature when it is deposited in crystalline form in the ventral thalamus. However, when carbachol was applied similarly in other regions including the rostral hypothalamus and septum, the hydrated (water-replete) rat suddenly drank a large volume of water, ranging from approximately 10 to 13 ml. This polydipsic response was accompanied by a fall in the rat's core temperature of varying magnitude.

In other species, ACh induces a rise in body temperature when this neurotransmitter is applied centrally (see Tangri et al., 1974). For example, Bligh and Maskrey (Bligh and Maskrey, 1969; Maskrey and Bligh, 1971) showed that a marked rise in temperature follows the injection of ACh into the lateral cerebral ventricle of the sheep. In addition to the increase in body temperature produced by an intraventricular injection of an ACh-physostigmine mixture, the sheep's oxygen consumption and plasma free-fatty acid rise simultaneously (Darling et al., 1974). The concomitant shivering and change in the sheep's respiratory quotient, following intraventricular ACh, suggest also that the metabolism of fat increases as a result of a central effect on cholinergic synapses.

A. Anatomical Localization of Cholinergic Hyperthermia

Although the role of ACh in the anterior hypothalamic, preoptic area was initially unclear, eventually it was found that carbachol, ACh alone, and ACh mixed in a solution with physostigmine to retard its degradation are hyperthermic substances. All of these cholinomimetic substances produce a transient but intense rise in the colonic temperature of the unanesthetized rhesus monkey when they are microinjected at sites distributed throughout the primate's hypothalamus (Myers and Yaksh, 1969). As shown in Figure 12, ACh injected in the caudal part of this structure caused a sharp rise in temperature of short duration; at the same site, both 5-HT and norepinephrine are without effect.

An anatomical mapping of the cholinergic thermogenesis, however, shows that, in a relatively small percentage of sites localized to a circumscribed region between the mesencephalon and posterior hypothalamus, the cholinergic receptor agonists do induce a fall in the monkey's temperature. Based on an anatomical mapping of cholinoceptive sites in the hypothalamus, the anatomical pathway for heat production is relatively diffuse (Myers

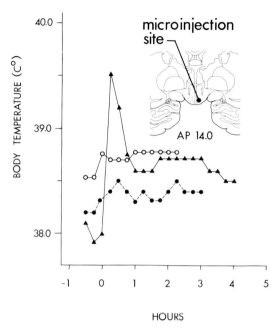

Figure 12 Temperature responses of two monkeys after zero-hour microinjection in the posterior part of the ventromedial nucleus (●) at AP 14.0 (inset) of 25 μg 5-HT (●--●); of 12 μg norepinephrine (○——○); and of 6 μg acetylcholine-eserine mixture (ACH) (▲——▲). 5-HT and norepinephrine were given to one monkey, and acetylcholine to the other. (From Myers and Yaksh, 1969.)

and Yaksh, 1969). As a result, the conclusion was reached that ACh acts as a transmitter subserving the heat-production pathway which projects caudalward from the rostral hypothalamus through the lateral and posterior areas and on into the mesencephalon (Myers, 1970). A cholinergically mediated heat-loss pathway of much smaller dimension apparently arises in the caudal portion of the hypothalamus and it again projects caudally into the mesencephalon. However, there is no evidence based on anatomical analysis by mapping of microinjection sites that this efferent pathway originates within the anterior hypothalamus and traverses the posterior region (Myers, 1976a).

Carbachol microinjected into the anterior hypothalamus of the conscious cat generally produces a marked hyperthermic response at sites located just dorsal to the optic chiasm (Rudy and Wolf, 1972). This corresponds to the thermogenesis brought about by cholinomimetics infused into the ventricle of this species. If the dose of carbachol given intrahypothalamically is increased, the cat's temperature may fall. Thus, the idea is supported that a depolarizing blockade of the animal's heat production mechanism caused by a high dose of the ACh analogue infused into the rostral hypothalamus may well account for the cholinergic hypothermia (Myers, 1974a).

Another experiment bears this out. A clear-cut, dose-dependent elevation in the rectal temperature of the unanesthetized golden hamster results from an intrahypothalamic injection of carbachol. However, as the dose is systematically raised, the temperature-

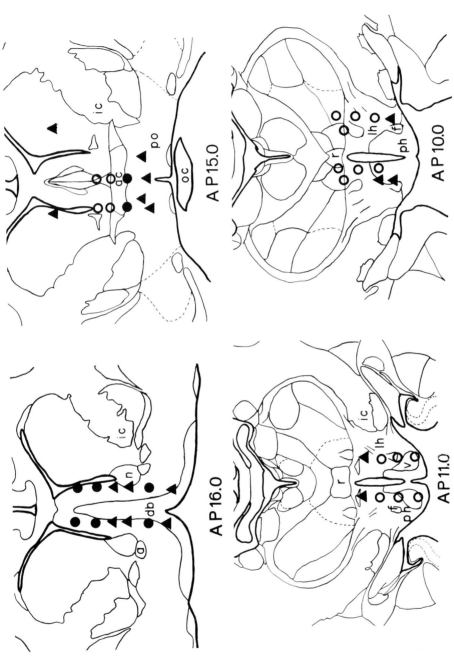

Figure 13 Representative frontal sections of cat brain illustrating the anatomical sites at which push-pull perfusions of RX 72601 produce a rise in temperature. ▲, Sites where push-pull perfusions of 1.0 μg/min RX 72601 produce hyperthermia of 0.4°C or greater; ○, sites where doses of drug 0.125–0.5 μg/min are ineffective; ●, sites where 1.0 μg/min is ineffective. Key to labeled structures: a, nucleus accumbens; ac, anterior commissure; c, caudate nucleus; db, diagonal band of Broca; f, fornix; ic, internal capsule; lh, lateral hypothalamus; n, nucleus of anterior commissure; oc, optic chiasm; ph, posterior hypothalamus; po, preoptic area; r, nucleus reuniens; v, ventromedial hypothalamus. (From Metcalf et al., 1975.)

raising response is reversed, and a significant hypothermia follows (Reigle and Wolf, 1975). In the rabbit, carbachol injected into the preoptic area and anterior or posterior hypothalamus similarly evokes a dose-dependent rise in temperature (Sharpe et al., 1979) which corresponds to that produced by ACh-eserine mixtures. An injection of ACh into the hypothalamus of the hibernating ground squirrel causes arousal and thermogenesis. The same response is seen when the animal is kept at 5°C ambient but not at 22°C (Stanton and Beckman, 1977).

The hypothalamic application of an anticholinesterase, which serves to raise the level of ACh endogenously, also affects temperature. When RX 72601, a potent anticholinesterase, is perfused at sites extending from the preoptic area to the mammillary bodies, a hypothermic response of 0.4°C or greater occurs in the cat (Metcalf et al., 1975). The anatomical loci mediating this response are illustrated in Figure 13.

1. Cholinergic Thermogenesis in the Rat

The hyperthermia produced by ACh delivered intraventricularly in the rat was duplicated in tissue studies by Avery (1970), who microinjected the cholinomimetic, carbachol, into the anterior hypothalamic, preoptic area. Examples of the cholinomimetic's potent effect on thermogenesis are presented in Figure 14. Following this observation, Avery generated a dose-response curve for the carbachol-induced hyperthermia and found that when the dose is raised from 5.0 to 8.0 μg, the magnitude of temperature rise is reduced. In these experiments, Avery injected a small 0.5-μl volume of the solution and kept each of the experimental rats unrestrained throughout the experiments. These two factors are hypothesized by Avery to account for the discrepancy between his observations and the opposite response of hypothermia reported by Lomax and Jenden (1966) who had not only injected a much larger volume of carbachol solution in the animal's hypothalamus

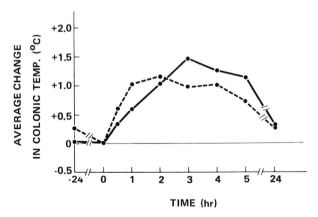

Figure 14 Hyperthermic responses to two doses of the cholinomimetic, carbachol, microinjected into rat rostral hypothalamus in a volume of 0.5 μl. At zero hour, 5 μg of carbachol (——) or 3 μg of carbachol (– – –) were injected. (From Avery, 1970; reprinted with permission from *Neuropharmacology 9*, 175-178, Pergamon Press, Ltd.)

but also kept the rat restrained. In other words, a larger volume could spread to such an extent that a depolarizing blockade of the heat-production pathway could have occurred as receptor sites were swamped (Myers, 1974a).

At an ambient temperature of either 5 or 24°C, carbachol injected into the anterior preoptic area causes hyperthermia. If the rat is kept at a high temperature of 35°C at the time of the carbachol microinjection, the temperature of the rat does not rise further but rather remains stable (Avery, 1972). This would be expected, since metabolic heat production is not necessary physiologically if the ambient temperature is already high. Similarly, cholinergic stimulation of the anterior hypothalamus leads also to more frequent bar-holding responses by the rat to escape the heat generated by a 250-W lamp above its chamber (Avery and Penn, 1973).

2. Cholinergic Hypothermia

Following the initial observation of Lomax and Jenden (1966) that oxotremorine or carbachol elicits a fall in body temperature of the restrained rat when either drug is injected into its preoptic area, Lomax and colleagues found that pilocarpine produces a long-lasting hypothermia when given similarly in a single, high dose, i.e., 40 μg (Lomax et al., 1969). A sharp fall in temperature occurs when high doses (from 20 to 100 μg) of carbachol or 15 μg of ACh are applied electrophoretically to the rat's rostral hypothalamus (Kirkpatrick and Lomax, 1970).

A similar result occurs after an intrahypothalamic injection of ACh is given in the guinea pig (Zeisberger and Brück, 1971b) as well as in the pigeon infused with a 1.0 μl-volume (Hissa and Rautenberg, 1975). In the latter case, the core temperature of the bird declines when it is kept at an ambient temperature of 15°C, but not at a higher or lower ambient temperature. Further, atropine can prevent the carbachol-induced hypothermia in the pigeon (Pyörnilä et al., 1977). Carlisle and his group have replicated the high-dose hypothermic effect of ACh. In addition, they showed that the rat's bar-pressing for heat is interrupted by an ACh injection (Beckman and Carlisle, 1969) but then resumes at a relatively high rate as the animal's temperature begins to rise above its base line.

After a bilateral microinjection of a 100-μg dose of ACh into the preoptic area of the rat, transitory hypothermia is followed by a continuous rise in body temperature of over 1.4°C (Crawshaw, 1973). Laudenslager and Carlisle (1976), injecting the same range of doses of ACh into the rat's hypothalamus, find that the animal depresses a lever to turn off a "heat gun" and thereby escape heat. However, the increase in behavioral responding, which is analogous to that described by Avery and Penn (1973), is transient and corresponds to the action of ACh in producing the short-lived fall in temperature.

Poole and Stephenson (1979a,b) have also shown that both carbachol and norepinephrine infused into the rat's rostral hypothalamus induce a decline in temperature. Atropine and phentolamine pretreatment of the injection site antagonized, respectively, the action of the two compounds. Muscarine produced a hypothermia similar to carbachol, which is consistent with that elicited by oxotremorine given in the same site (Cox and Lee, 1978). Other workers have verified the carbachol hypothermia with high doses in a large microinjection volume (Netherton et al., 1977).

In general, ACh may be a transmitter that mediates hypothermia in the rodent; however, several types of experimental observations would be helpful to substantiate its role:

(1) a complete anatomical mapping of thermal responses to much lower doses of a cholinomimetic infused into anterior, posterior or other hypothalamic areas; (2) usage of a microinjection volume of less than 0.5 μl that would permit a precise morphological localization of ACh's effects (Myers and Hoch, 1978); (3) a measure of metabolic heat production so that indices of active or passive heat loss could be obtained; (4) quantitation of activity, malaise, cardiovascular changes, or other factors which may be partially involved in the rat's hypothermia; and (5) ruling out ACh's local effect in releasing norepinephrine from noradrenergic synapses in the rostral hypothalamus (Hall and Myers, 1972).

B. Nicotinic and Muscarinic Receptor Mechanisms

Relatively few investigators have devoted their efforts to the problem of characterizing the cholinergic receptors in the hypothalamus involved in heat production. As Brimblecombe (1973) has reviewed the situation, a pharmacological analysis of cholinoceptive sites or cholinoreceptors has been hampered somewhat by the vagaries in the temperature response produced by a peripherally-acting cholinolytic or cholinomimetic compound. Again, we find that the dose of an anticholinergic agent administered systemically, and probably centrally, seems to determine the nature of the temperature change. The main reason for this is the multifaceted action of each of the powerful drugs tested. To illustrate, as an antagonist is chemically "recognized" and taken up by its appropriate receptors, an agonistic effect may be produced temporarily because of the cholinomimetic's property. By the same token, the compound may initially possess a toxic action; this is one of the well-known aspects of the literature on atropine (Lomax, 1970).

Generally, an intraventricular injection of a muscarinic blocking agent such as atropine tends to impede the temperature-raising mechanism of the animal, although this is not always seen (Tangri et al., 1975). For example, after atropine is injected into the ventricle of the rabbit, its body temperature declines whether or not the rabbit is in a thermoneutral environment or is in a state of fever (Preston, 1974). Observation by Cooper's group is in agreement with this. When given by the same route to a rabbit exposed to a cold environment, atropine inhibits shivering, decreases oxygen consumption, and attenuates a fever induced by leukocyte pyrogen or prostaglandin microinjected into the anterior hypothalamus (Cooper et al., 1976b). An intraventricular injection of the cholinomimetic, methacholine, also induces hyperthemia in the cat, but this rise is attenuated or blocked by central pretreatment with intraventricular atropine (Baird and Lang, 1973).

In an instance when ACh or a cholinomimetic drug produces a hypothermia after it is injected into the hypothalamus in a large dose, the response can usually be antagonized by an agent that blocks muscarinic receptors. For example, a microinjection of atropine into the hypothalamus of the rat blocks the fall in temperature ordinarily produced by oxotremorine applied in a high dose at the same site (Lomax and Jenden, 1966; Staib and Andreas, 1970; Poole and Stephenson, 1979a,b). A similar muscarinic blocker, DKJ-21, exerts the same antagonistic effect (Kirkpatrick et al., 1967). Although atropine itself, injected intrahypothalamically, can cause a fall in temperature (Myers et al., 1976), it reportedly elevates the body temperature of the rat on its own (Kirkpatrick and Lomax, 1967). A dose-response relationship plus anatomical analysis of this atropine hyperthermia have not yet been reported.

Nicotine introduced into the cerebral ventricle of the cat and monkey causes a marked hypothermia (Hall and Myers, 1971; Hall, 1973), but, in other species, such as the chicken, nicotine reportedly has very little effect on body temperature (Marley and Seller, 1974). In 1972, a study was published which revealed the extraordinary complexity of the nicotinic agonist action of nicotine itself. When microinjected into the anterior hypothalamus of the conscious rhesus monkey, nicotine elicits a hypothermia which is similar to that seen after the intraventricular injection of the alkaloid (Hall and Myers, 1972). Conversely, when nicotine is injected into the posterior hypothalamus, an intense rise in the primate's body temperature occurs. The further characterization of the remarkable anatomical specificity of nicotine's action, depicted in Figure 15, is as follows.

If the same dose of norepinephrine is applied at the same site in the anterior hypothalamus, the hypothermia is virtually identical to that produced by nicotine in terms of latency, magnitude, and duration. However, when ACh is mixed with eserine and microinjected in the same dose as nicotine into the posterior hypothalamus, the curve of hyperthermia can be superimposed upon that following a nicotine microinjection at the same caudal locus. Nicotine's hypothermic action on the anterior hypothalamus may be interpreted in light of nicotine's norepinephrine-releasing property (Hall and Turner, 1972). However, within the caudal hypothalamus, nicotine apparently occupies cholinergic receptor sites that mediate heat production (Hall and Myers, 1972; Myers and Waller, 1973).

Interestingly, the infusion of nicotine into the hypothalamus of the rat also evokes a transient hypothermia followed by a sharp rise in the rodent's temperature (Knox and Lomax, 1972). However, if the nicotine microinjection is repeated in the rat 5 and 13 days later, the initial hypothermic response is lost; instead, the rat's temperature increases in the same way as that observed in the monkey. This result may be explained by the possibility that some noradrenergic synapses in the rostral hypothalamus are lesioned by the injection; thus, nicotine dispersing to somewhat more caudal nicotinic receptor sites could mediate heat production. In the anesthetized rat, the local hypothalamic application of nicotine produces an initial increase in the frequency of firing of heat-sensitive neurons recorded in the anterior hypothalamus (Knox et al., 1973). However, the pattern of initial excitation is followed by a subsequent inhibition, which is also reported in experiments in which nicotine is administered systemically.

Over-all, both muscarinic and nicotinic synapses are implicated in the mechanisms underlying heat loss and heat gain. Nevertheless, from the anatomical studies on ACh completed in the primate, it would appear that this endogenous transmitter substance mediates impulses along a heat-production pathway from the anterior to the posterior hypothalamic area. More direct evidence of this action of ACh is found in the experiments on the endogenous release of the amines.

C. Temperature-evoked Release of ACh

If the delicate balance in monoamine release within the anterior hypothalamus controls the activation of the heat-production and heat-loss mechanisms, the synaptic information emanating from the thermosensitive zone must be transmitted through the neuroaxis. The hypothesis that ACh serves this transmitter function is based not only on microinjection studies cited above but also on the fact that the activity of ACh in the diencephalon changes as the animal regulates its body temperature (Myers and Waller, 1973).

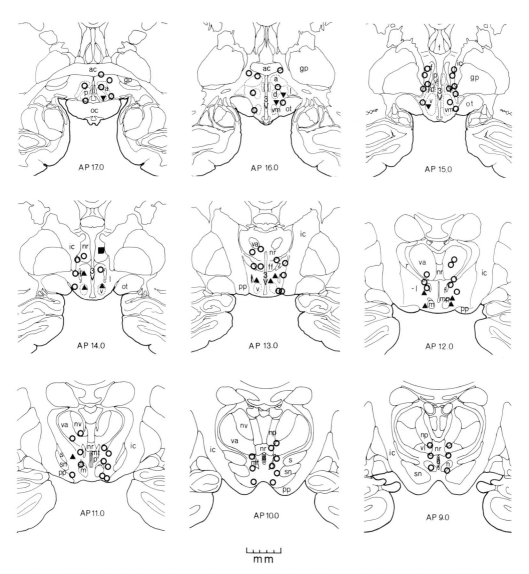

Figure 15 Anatomical mapping at nine coronal levels of sites in the hypothalamus of the unanesthetized rhesus monkey at which 5 μg nicotine injected in a volume of 1 μl evoked no change (○), a fall (▼), or a rise (▲) in body temperature of 0.4°C or greater, with a latency of 15 min. ■, A biphasic change in temperature. Abbreviations for anatomical sites are as follows: a, anterior hypothalamus; ac, anterior commissure; do, dorsomedial nucleus of the hypothalamus; f, fornix; ff, fields of Forel/zona incerta; gp, globus pallidus; ic, internal capsule; l, lateral hypothalamus; m, mamillary body; mp, mamillary fasciculus princeps; mt, mamillothalamic tract; np, nucleus paracentralis of the thalamus; nr, reuniens nucleus of the thalamus; nv, ventrolateral nucleus; oc, optic chiasm; ot, optic tract; p, posterior hypothalamus; pp, cerebral peduncles; r, reticular nucleus of the thalamus; s, nucleus subthalamicus; sn, substantia nigra; va, anteroventral nucleus of the thalamus; vl, ventrolateral medial nucleus of the thalamus; vm, ventromedial nucleus of the hypothalamus; 3v, third ventricle. (From Hall and Myers, 1972.)

That the origin of the cholinergic pathway could be in the anterior hypothalamic, preoptic area is supported by the histochemical studies revealing the presence of ACh in this region (Shute and Lewis, 1967; Lewis and Shute, 1967; Jacobowitz and Palkovits, 1974). Insofar as ACh content is concerned, Jacobowitz and Goldberg (1977) have shown that the medial preoptic and anterior hypothalamic areas possess the highest content of ACh per sample of protein of any of the structures that they have examined. In a corresponding anatomical study, in which the technique of tritiated amino acid autoradiography was employed, it was revealed that short-axon projections from these two medial regions extend directly to the mammillary region of the posterior hypothalamus of the rat (Conrad and Pfaff, 1975). Finally, electrophysiological experiments performed by Nutik (1971) indicate that 72% of the single neurons recorded in the posterior hypothalamus change their firing rate when the local temperature of the preoptic area is cooled; however, when the same anterior area is warmed, the firing rate of only 28% of the posterior units changes. Taken together, these observations provide the backdrop for studies of the actual characteristics of ACh release during thermal stress of the unanesthetized animal.

In the conscious rhesus monkey, 101 sites in the hypothalamus and mesencephalon were perfused by means of push-pull cannulae, and samples of perfusate were assayed for their content of ACh. When the primate is exposed to a cold ambient temperature, the output of ACh from the anterior hypothalamic, preoptic area is markedly enhanced at 88% of the active ACh-releasing sites (Myers and Waller, 1973). Conversely, peripheral warming of the monkey reduces the release of ACh at 80% of these same perfusion loci. Caudal to the anterior region, ACh output increases at about one-half of the active ACh-releasing sites when the monkey is exposed to cold air, but the release of ACh is inhibited at nearly every locus when the monkey is warmed. In the mesencephalon, ACh output in two-thirds of all sites is augmented by peripheral cooling, the magnitude of which is illustrated in Figure 16. Conversely, ACh activity is diminished by hot-air warming of the monkey; ACh release is unaffected within the remaining third.

Overall, this anatomical analysis of the endogenous activity of ACh along both the heat-production and heat-loss pathways shows that ACh would appear to serve as a transmitter delegated principally to the pathways for heat gain rather than for heat loss. In considering the "linking" function of the anatomically separate midportion of the hypothalamus, warming the monkey by hot air fails to augment the increase of ACh at any site; rather, it inhibits its output at every perfusion site tested (Myers and Waller, 1973). This is summarized in Table 3.

In a follow-up study, the hypothesis was tested directly as to whether 5-HT in the anterior preoptic area, given to simulate a functional release of the amine, would actually activate the cholinergic synapses of the postulated heat-production pathway (Myers and Yaksh, 1969); therefore, the indoleamine was microinjected into the anterior, preoptic area of the monkey, and, immediately thereafter, samples of ACh were collected at sites caudal to the injection locus in the diencephalon and mesencephalon by the same push-pull perfusion method. As the 5-HT hyperthermia progressed, the output of ACh was unchanged at 17 of 36 loci but enhanced at 19 of the remaining sites located within the thalamus and mesencephalon (Myers and Waller, 1975b).

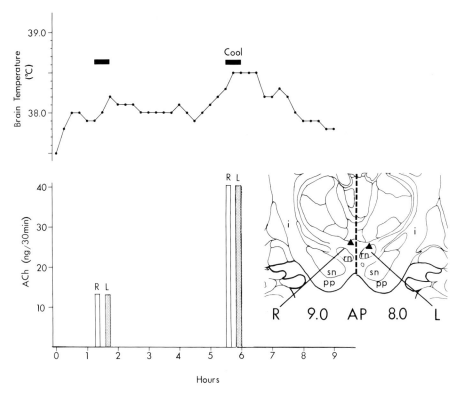

Figure 16 ACh release in nanograms per 30-min perfusion interval from two mesencephalic sites in a single monkey as indicated on the anatomical reconstruction in the inset. The perfusate contained neostigmine, 1.0 μg/ml. The control perfusion was followed 3.5 hr later by a second perfusion during which the monkey's ambient temperature was lowered to 0°C. Anatomical abbreviations are: i, internal capsule; pp, cerebral peduncle; rn, nucleus ruber; sn, substantia nigra. (From Myers and Waller, 1973.)

If norepinephrine is injected instead into the same anterior region, the typical fall in body temperature of the monkey is accompanied by a simultaneous decline in ACh output at the more caudal homologous sites. Several exceptions to these two sets of observations occurred. In some cases the release of ACh was inhibited by 5-HT but enhanced by norepinephrine, during the respective rise or fall in the monkey's temperature. Therefore, one could deduce that there are four types of cholinergic synapses in the thermal effector fiber systems concerned with the temperature function: synapses that transmit signals for (1) heat production; (2) the inhibition of heat production; (3) heat loss; and (4) inhibition of heat loss (Myers and Waller, 1975b). Thus, the activity of efferent cholinergic synapses in the hypothalamus and mesencephalon of the primate would depend upon the stimulus for either heat gain or an inhibition of active heat loss, or the converse of both of these responses.

Table 3 Percentage Enhancement or Suppression of ACh Release in Four Separate Regions as a Result of Raising or Lowering the Monkey's Ambient Temperature (from Myers and Waller, 1973)

Anatomical region	Increase in ACh release (%)		Suppression of ACh release (%)	
	Cooling	Warming	Cooling	Warming
Anterior preoptic area (AP 16.0-19.0)	70	20	30	80
Midhypothalamus (AP 13.0-15.0)	40	0	60	100
Mamillary region (AP 10.0-12.0)	54	25	46	75
Mesencephalon (AP 6.0-9.0)	65	33	35	67

Although the original model suggesting that ACh-containing neurons simply mediate the excitatory pathway delegated to maintaining an animal's temperature (Myers, 1974a) must now be modified, one interesting fact still stands. If the animal is cold and heat production is called for, release of ACh is enhanced at sites that outnumber by 2 to 1 those loci which release ACh during heat dissipation following a heat stimulus. This 2 to 1 relationship is strikingly similar to that found in the thermally-evoked ACh release study (Myers and Waller, 1973) as well as by Nutik in his study of single units. In the posterior hypothalamus, cells which fired in response to anterior hypothalamic cooling outnumbered by 2 to 1 those which discharged to heating of the same rostral locus (Nutik, 1971).

The survival value of the 2 to 1 relationship of anatomical sites releasing ACh to cold would seem to rest in the functional property of the thermoregulatory system itself. The primate and most other mammals must continually defend their set-point temperature of 37°C (or thereabouts) against an environmental temperature which is much lower than its own. In other words, a condition of cold "load" exists continually except perhaps in a tropical climate. Therefore, since a greater proportion of afferent input from the skin is delegated to heat production (Boulant, 1976; this volume), it is easily envisaged that the greater proportion of the effector synapses in the hypothalamus which mediate the thermoregulatory response should be delegated principally to heat production. Parenthetically, if one postulates that the role for ACh is exclusively that subserving heat loss, then a transmitter factor which carries "messages" or output signals for heat production from anterior through to posterior hypothalamus would have to be identified. No such substance has been found nor even proposed.

Hierarchically, the individual responses for heat production in close order of their functional activation are: (1) behavioral, such as seeking warmth, lying huddled and compact; (2) the redistribution of cutaneous blood flow, such as vasoconstriction; (3) a pilomotor reaction in many species; (4) an increase in metabolic rate involving the liver and brown fat; and (5) enhanced activity of muscle groups as noted in the shivering response. Concerning the latter, the relative contribution of shivering in the normal maintenance of body heat can be debated readily. Rarely does shivering ever occur, in humans at least, even when one takes a plunge into a pool containing ice-cold water. Only perhaps in the case of a fever does the human actually shiver, and then primarily during the rising phase of this pathological reaction.

Table 4 presents a summary of morphological evidence that implicates ACh as a transmitter in effector pathways for the heat-production mechanisms, primarily, and in active

Table 4 Summary of Anatomical Evidence for the Role of ACh as a Transmitter at Synapses Located Along the Effector Pathway for Heat Production in the Diencephalon and Mesencephalon

A. Indirect Evidence
  1. Passage of efferent cholinergic pathways through hypothalamus (Lewis and Shute, 1967).
  2. High content of ACh in rostral hypothalamus (Jacobowitz and Goldberg, 1977) with axon projections from this region extending directly to the mammillary area of the posterior hypothalamus (Conrad and Pfaff, 1975).
  3. Centrally applied ACh, carbamylcholine, or other cholinomimetic given intraventricularly evokes hyperthermia (Myers and Yaksh, 1968; Bligh and Maskrey, 1969), an increase in O_2 consumption and an elevation in plasma FFA (Darling et al., 1974).
  4. Intraventricular atropine reduces a normal or febrile temperature (Preston, 1974) or attenuates a cholinergic hyperthermia (Baird and Lang, 1973).

B. Direct Evidence
  1. ACh, an anticholinesterase, or a mixture of the two, given in the hypothalamus evokes a dose-dependent hyperthermia at microinjection sites distributed diffusely in this structure (Myers and Yaksh, 1969).
  2. Nicotine microinjected in the caudal hypothalamus evokes a hyperthermia having a response curve precisely congruent with that produced by a similar injection of ACh and eserine (Hall and Myers, 1972).
  3. Atropine injected in the caudal hypothalamus elicits a fall in temperature or exacerbates Ca^{2+}-induced hypothermia (Myers, 1975).
  4. Atropine inhibits shivering, O_2 consumption, and the febrile response induced by pyrogen or a prostaglandin microinjected into the anterior hypothalamus (Cooper et al., 1976).
  5. Cooling of the body causes a significantly greater release of ACh at sites within the hypothalamus than does the similar exposure of an animal to a warm ambient temperature (Myers and Waller, 1973).
  6. 5-HT injected into the anterior hypothalamus enhances the release of ACh along the caudal heat production pathways as temperature rises (Myers and Waller, 1975b); norepinephrine, which induces heat loss, often inhibits or fails to augment ACh release within homologous sites.

heat-loss fiber systems to a much lesser degree. These neuronal pathways are believed to originate anatomically in the anterior hypothalamus, traverse the structure caudalward, and send projections rhombencephalically to the spinal cord.

V. Role of Cyclic AMP, Peptides, and Other Humoral Factors

Ever since the astute observation of Shemano and Nickerson (1958), it has been clear that any one of a number of substances, each possessing a unique chemical property, can affect body temperature when given systemically. One common pharmacological attribute of an exceedingly large number of compounds is their potential to produce an incapacitating poikilothermia. Moreover, certain factors that have been isolated and identified as occurring endogenously within the central nervous system may exert an equally profound and deleterious effect on body temperature when administered exogenously. Whether or not a specific endogenous substance actually subserves the sensor-effector mechanism for thermoregulation, which involves the anterior and posterior parts of the hypothalamus, cannot be ascertained purely on the basis of a pharmacological experiment.

A. Adenine Nucleotides

In 1969, Breckenridge and Lisk deposited crystals of the cyclic nucleotide adenosine-3',5'-monophosphate (cyclic AMP), as well as dibutyrl (db) cyclic AMP, into the anterior hypothalamus of the rat. Applied to this region, cyclic AMP produces a marked hyperthermia. Cyclic AMP is believed to be involved in the release of various transmitter substances, and it exerts effects on many other of the cellular functions of the neuron. Since monoamines such as norepinephrine and 5-hydroxytryptamine can influence the synthesis of the nucleotides in brain tissue (Kakiuchi and Rall, 1968), its involvement in thermoregulation is a distinct possibility. This prospect was strengthened by the finding that db cyclic AMP infused into the cerebral ventricle of the cat produces a relatively long-lasting hyperthermia lasting several hours. This rise in temperature may or may not be preceded by a transient hypothermia (Varagić and Beleslin, 1973; Clark et al., 1974).

In other anatomical studies, db cyclic AMP microinjected in the anterior hypothalamus of the rabbit also raises the animal's temperature (Laburn et al., 1975). Given in only a 1.0-μg dose, cAMP elevates the temperature up to 0.8°C with a latency of onset of about 4 min (Woolf et al., 1975; Willies et al., 1976a). In the squirrel monkey, 500-750 ng of db cyclic AMP injected intrahypothalamically causes a short-term hypothermia followed by a slight hyperthermia or no significant change in body temperature (Lipton and Fossler, 1974).

In attempting to extend these observations to the feline species, Dascombe and Milton (1975) have completed an exacting pharmacological assessment of cyclic AMP's effect on the anterior hypothalamic and other regions of the unanesthetized cat. Generally, both cyclic AMP and its dibutyrl derivative evoke a dose-dependent hypothermia when infused in concentrations of 50-500 μg, with the dibutyrl molecule being more potent. However,

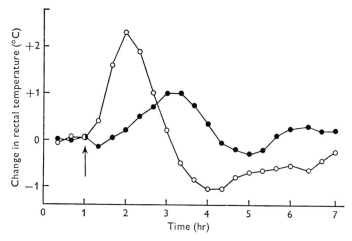

Figure 17 Recordings of rectal temperature in two unanesthetized cats. At the arrow, 0.96 μmol cyclic AMP was injected into the posterior hypothalamus of one cat (○) and into the preoptic anterior hypothalamus of the other cat (●). Drug solutions applied to the preoptic anterior hypothalamus of the latter animal had access to the third ventricle. The rise in temperature in each animal was associated with increased motor activity. (From Dascombe and Milton, 1975.)

in some instances, db cyclic AMP causes a complex series of changes, as witnessed by Lipton and Fossler in the monkey. A transient hyperthermia is followed by a fall in the cat's temperature, then a secondary rise in body temperature, and finally defervescence. Further, when the nucleotide is microinjected into the posterior hypothalamus, an intense rise in body temperature of over 2.0°C often ensues. As shown in Figure 17, the hyperthermic spike is reminiscent of a cholinergic rise (Fig. 12). At least on the surface, it would appear that cAMP itself could have an essential role in two or more of the thermoregulatory mechanisms within the hypothalamus for heat production, heat dissipation, or the inhibition or excitation of both. Because of its wide-ranging action in relation to the physiological activity of several transmitter candidates in the brain stem, this assumption was not unreasonable initially.

In cleverly conceived follow-up experiments, Dascombe and Milton (1976) tested directly the idea that cyclic AMP is specifically involved in thermoregulation per se. In the unanesthetized cat in which samples of CSF were collected at given intervals, the exposure of the cat to a very high or low ambient temperature (0 or 45°C for as much as 3½ hr) was used as a thermal challenge. When Dascombe and Milton performed subsequent assays on the cat's CSF, they found that neither cold nor heat stress has any effect whatsoever on the concentration of cyclic AMP in the CSF samples. Moreover, even though the level of cyclic AMP increases in the CSF of the cat in association with a bacterially induced fever, the subsequent administration of an antipyretic, which abolishes the cat's fever, fails to reduce this elevated concentration of cyclic AMP. Finally, a cat undergoing collapse with a manifest hypothermia, following an intense bacterial loading, has a level of cyclic AMP in its CSF which is nevertheless just as high as that seen in the absence of the toxemic condition.

From these latter observations, the conclusion can be drawn that cAMP in the brain stem really does not mediate the rise in temperature which is produced by a pyrogen or other substance. Further, it is not at all likely that this nucleotide has any involvement in the diencephalic control of an animal's body temperature, at least as reflected rather directly by the endogenous output of cyclic AMP in the cat's CSF. Caution should be raised at once, however, since Willies et al. (1976b) have demonstrated that a microinjection into the rabbit's hypothalamus of an exotoxin bacillus, which inhibits adenylate cyclase, does prevent a pyrogen fever. Nevertheless, the experiments of Dascombe and Milton illustrate a point worthy of special emphasis. In spite of the presence of a specific endogenous compound in the brain and in spite of its action on body temperature when administered intracerebrally, the compound may not have physiological significance which could well be attributed to it on the basis of a single pharmacological observation. Indeed, only the demonstration of an enhanced release or inhibition of its release from an anatomically circumscribed population of neurons, during a cold or warm challenge, provides the necessary functional evidence for its role (see Hensel, 1973; Myers, 1976a).

B. Peptides in the Brain

In 1974, Metcalf first announced that a tripeptide found in the hypothalamus, TRH, evokes a profound fall in the body temperature of the unanesthetized cat after the hormone is injected into the lateral cerebral ventricle (Metcalf, 1974). Later, it was shown that TRH is approximately 1000 times more potent than norepinephrine and 27,000 times more potent than calcium ions when either of these two substances is given by the same intracerebral route (Metcalf and Myers, 1976). Of significance is the fact that TRH injected in this manner produces salivation, tachypnea, growling, vasodilatation, and other autonomic symptoms, the most important of which is an exceptionally rapid respiratory rate analogous to hyperventilation.

In the conscious rabbit, TRH given intraventricularly or into the hypothalamus also evokes a pronounced tachypnea, piloerection, increase in locomotor activity, and general alertness, but, in this case, a rise in temperature occurs which is characteristic of a state of arousal (Horita et al., 1976; Carino et al., 1976). On the other hand, TRH and another hypothalamic peptide, neurotensin, evoke hypothermia in the mouse when either is given intracisternally (Bissette et al., 1976). As a result of these observations, three questions arise. What is the site of action of TRH? What is its mechanism of action? Is it involved in the thermoregulatory processes?

In order to determine whether TRH actually acts on the thermosensitive zone in the anterior or posterior hypothalamus or other site in the brain stem, the peptide's effect has been subjected to an anatomical analysis (Myers et al., 1977b). At 347 sites scattered throughout the hypothalamus and mesencephalon, TRH has been microinjected in an 0.5-μl volume while the body temperature of the unanesthetized cat was monitored. Surprisingly, TRH fails to alter the animal's temperature when infused at anterior hypothalamic sites at which norepinephrine produces its characteristic hypothermia (Myers et al., 1977b). Further, TRH applied similarly to the posterior hypothalamus exerts no notable

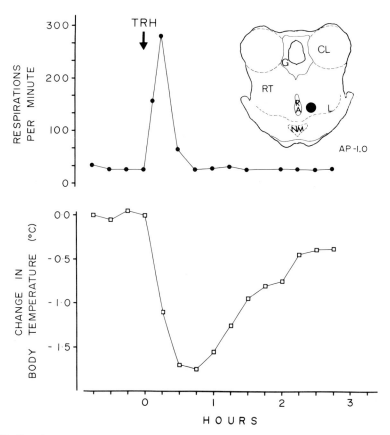

Figure 18 Respiratory rate and changes in base-line temperature after a 0.5-μl injection of TRH in a dose of 10 ng. The site of the microinjection in the mesencephalon at AP − 1.0 is denoted in the histological inset by the dot (●). Anatomical abbreviations are: CL, claustrum; G, griseum centrale; L, leminscus medialis; NM, nucleus pontis medialis; RA, raphe nuclei; RT, substantia reticularis mesencephalica. (From Myers et al., 1977.)

effect on the cat's temperature. However, after its microinjection into certain loci in the mesencephalon, TRH does evoke an intense tachypnea in the cat, profuse salivation, vasodilatation, and a consequent acute hypothermia of marked intensity. The respiratory and temperature changes produced by TRH are illustrated in Figure 18.

Because the fall in the cat's temperature always followed the onset of rapid breathing (200 times per minute), it follows that the tripeptide exerts a principal action on the autonomic mechanism in the mesencephalon underlying respiration. Thus, the hypothermic response in all likelihood is simply a secondary side effect of tachypnea (Myers et al., 1977b) which would contribute to evaporative heat loss. A finding in another species upholds this contention. When injected into the hypothalamus of the chicken, TRH evokes a non-dose-related rise in temperature which is associated with stereotyped head-neck movements, "wet-dog shakes," and other locomotor activity (Nistico et al., 1978). Again, this suggests that the TRH induced alteration in temperature is not a thermoregulatory response but rather a secondary result of intense motor activity.

Other peptides, including vasoactive intestinal peptide (VIP), neurotensin, bombesin, and the endogenous opiod, enkephalin, can cause effects on body temperature when they are injected intraventricularly (Clark, 1977). Although it is unlikely that the opiate-like pentapeptide is a neuronal candidate for the thermoregulatory mechanism (Clark, 1977), its action on anterior hypothalamic tissue has been tested in the cat (Myers and Ruwe, 1980). When microinjected in a volume of 1.0 μl into sites reactive to PGE, 5-HT, and endotoxin, leu-enkephalin at a low dose has no effect; however, a dose of 5.0 μg produces a rise of only 0.5°C after 1 hr. A 20-μg dose of enkephalin also evokes a very gradual increase in the cat's temperature, dissimilar to that produced by the aforementioned hyperthermic substances (Myers and Ruwe, 1980). Thus, it is most unlikely that hypothalamic enkephalin is involved in the thermoregulatory defense of an animal's set-point temperature.

Just as in the case of cyclic AMP, the evidence favoring a role of a peptide in the anterior hypothalamic, preoptic control of body temperature is not on a sound footing. Once again, the requisite demonstration of a peptide's functionally induced release from hypothalamic tissue, in association with a thermal challenge, has yet to be presented.

C. Effect of Histamine

Still other substances that are normally present in the brain may also affect body temperature when applied directly to diencephalic tissue. For example, histamine has been tested with the view that its endogenous presence could indicate some sort of a physiological role (Lomax and Green, 1975). In experiments with the rabbit, 12 μg of histamine infused in a volume of 12 μl into the rostral hypothalamus did not alter the animal's body temperature (Cooper et al., 1965). In addition, an injection of 20 μg of histamine into the homologous region of the monkey's anterior hypothalamus also had no effect on the primate's temperature (Myers, 1968). However, Shaw (1971) reported that histamine injected into the cerebral ventricle of the mouse, in a dose of 1-10 μg, elicits a dose-dependent decline in the rodent's temperature. In the cat, the brief hypothermia following intraventricular histamine is followed by a rise in the animal's temperature (Clark and Cumby, 1976a).

When microinjected in a dose of 2.5-5.0 μg into the rostral hypothalamus of the rat, histamine causes a fall in temperature which lasts for more than 20 min (Brezenoff and Lomax, 1970). An antihistaminic agent, chlorcyclazine, given systemically 1 hr before the intracerebral injection, can antagonize the histamine-induced hypothermia. In a follow-up to this observation, it was found that histidine, the metabolic precursor of histamine, infused directly into the rostral hypothalamus of the rat also causes a fall of approximately 1.2°C in the rodent's core temperature (Green et al., 1975a). Although the hypothermic effect of histidine might be attributed to an increased formation of histamine in the brain, it would be necessary to document the latency, duration, and other temporal characteristics of the hypothermic response. In this connection, the antagonists of histamine H-1 and H-2 receptors, pyrilamine and burimamide, respectively, injected into the anterior hypothalamus of the rat fail to block the hypothermic response to a systemic injection of histidine. However, burimamide, the H-2 receptor-blocker, did antagonize the hypothermia evoked by histidine but only when infused into the third cerebral ventricle.

Thus, there may be both H-1 and H-2 receptors that are affected by histamine in producing an increase in heat loss in the rat (Green et al., 1975b, 1976).

As alluded to earlier, many endogenous and synthetic compounds exert profound effects on the autonomic nervous system, which can result in a change in body temperature as a secondary side effect (Maickel, 1970). Just as caution should be exercised in interpreting a result obtained with cAMP, a tripeptide, or amino acid such as taurine (Sgaragli et al., 1975; Harris and Lipton, 1977), certain physiological questions would have to be answered before histamine could be assigned a role in the thermoregulatory mechanism within the anterior hypothalamic, preoptic area. For example, the demonstration of the in vivo release of histamine within this rostral area, as well as the enhancement or inhibition of its activity by a cold or warm thermal challenge, would be a logical next step in the analysis of its function. This is particularly important in view of the results of Sweatman and Jell (1977) who showed that the iontophoresis of histamine onto thermosensitive neurons in the rostral hypothalamus of the cat either excites or depresses the cells. No correlation whatsoever is evident between the sensitivity of a neuron to histamine and its thermoresponsiveness following a local deflection in temperature.

VI. Hypothalamic Ions and the Concept of a Temperature Set Point

Early in this century, physiologists discovered that a systemic injection of calcium or other cation causes an abrupt change in the body temperature of an animal (Freund, 1911; Schütz, 1916). By the 1930s, experimental technology had advanced to such an extent that Hasama (1930) was able to examine the effect of a solution with an aberrant cation concentration injected directly into the hypothalamus. Given in this way, Hasama found that excess sodium, potassium, and barium produce hyperthermia in the unanesthetized cat, whereas excess calcium ions given similarly cause an intense fall in temperature. This pioneering work exemplified the antagonistic action of substances acting directly within the central nervous system.

Shortly thereafter, Kym (1934) published a remarkable set of observations on the rabbit. An intrahypothalamic injection of a sodium chloride solution causes a rise in temperature that is similar to that produced by ergotoxin administered intravenously. When a solution containing excess calcium ions is injected into the same region, the pyrexia is prevented, thus revealing the antipyretic action of this cation. Although an injection of excess potassium ions has an inconsistent effect, Kym found that, in some cases, this cation does cause a rise in temperature. Many years later, the hyperthermic action of the cation infused into the anterior hypothalamus of the same species was reaffirmed (Cooper et al., 1965). In 1940, Rosenthal confirmed Kym's observations but showed, in addition, that an intrahypothalamic injection of excess calcium ions potentiates the decrease in the rabbit's core temperature produced by picrotoxin infused by the same route (Rosenthal, 1941).

A. A Physiological Basis for the Set Point

According to one point of view, the term set point, as evolved from engineering and physical control theory (Hardy, 1965), is essentially a mathematical abstraction which describes the values of input and output in a control system (Mitchell et al., 1970). In physiological terms, a set-point temperature is derived presumably from a functional mechanism which has a temperature-dependent property and is unique in terms of its neuroanatomical and neurophysiological characteristics (Atkins, 1964; Benzinger, 1969). As recounted by Boulant in the first chapter in this volume, the precise definition of a set point is not agreed upon. Nevertheless, some physiologists consider the set-point mechanism to be inherently established as an intrinsic, built-in reference temperature of approximately 37°C in many mammals (Myers, 1978b). Around this set level, regulatory adjustments take place continually in defense of this metabolically efficient and life-sustaining temperature.

The fact that there are internal "clocks" that determine certain bodily rhythms and cycles (Richter, 1975), coupled with the existence of a conglomeration of physiological constants around which bodily processes are adjusted, lends credence to the general concept of the set point for body temperature (Myers, 1976a). Equally as compelling is a wide range of experimental evidence which favors the concept. For example, disseminated bilateral ablation of the posterior hypothalamus destroys the animal's capacity to maintain its body at a stable temperature in either a cold or warm environment (Keller, 1933). To a certain extent, the animal is rendered poikilothermic, as if it had lost its set-point mechanism or a part of it. Because of the perturbation of the set temperature following a discrete lesion placed in the posterolateral hypothalamus of the cat or monkey (Ranson, 1940), one could imagine that the mechanism for the defense of the reference temperature rests within this structure. Such a diminutive lesion can result in a prolonged displacement in the animal's body temperature after which the animal then thermoregulates anew. Another piece of experimental evidence is the fact that the laboratory animal or human patient exposed to heat or cold defends the elevated set-point temperature that occurs during bacteria-caused fever by normal regulatory processes including vasomotor responses (Cooper, 1972). In consideration of these observations, a basic set of assumptions about the nature of a set-point function has been developed as follows (Myers, 1974b; Myers, 1978b).

First, the site of the set-point mechanism in the central nervous system is conceived to be anatomically separate from the neuronal locus involved in the regulatory process. That is, neurons responsible for holding a set-point temperature would have to be different from those which sense temperature and control the regulatory reactions. The thermosensitive neurons would alter their firing pattern to compensate for a rise or fall in a set temperature and thus could not be one and the same with set-point neurons.

Second, the set-point temperature can be elevated or lowered either normally or pathologically. Such a change is thought to occur during sleep, fever, hibernation, and exercise.

Third, when the set point is actually deflected, regulation around the new level of temperature takes place normally by means of the vasomotor, pilomotor, metabolic, and other bodily functions ordinarily utilized when the reference value is 37°C.

Fourth, a functioning set point in terms of a reference temperature such as 37°C is present inherently at parturition in most mammals. However, the capacity to thermoregulate and defend 37°C against cold or heat often is not inborn.

Fifth, because of its universal nature, the set-point mechanism in the mammal should be governed by a fundamental and intrinsic property of a select population of neurons in the brain stem which are in close anatomical proximity and share connectivity with temperature-sensing neurons.

Sixth, the set-point temperature, which is normally an invariant value, can be shifted ordinarily without serious or permanent pathological consequences. However, such a shift can occur only within certain upper and lower limits of body temperature or death ensues.

B. Na^+-Ca^{2+} Balance in the Posterior Hypothalamus

Upon finding that calcium serves to block the elevation in a conscious cat's temperature induced by excess sodium ions perfused through the cerebral ventricles, Feldberg et al. (1970) speculated that the temperature's set point could have its origin in a balance between these cations. Within a few years, it was surprisingly found that there are no species differences nor inconsistancy in the hyper- and hypothermic effects which are evoked by Na^+ and Ca^{2+} ions, respectively, given by this route. In fact, in animals tested at a euthermic temperature, these ions exert their hyper- or hypothermic effects after each is infused or perfused within the ventricles of the rabbit (Feldberg and Saxena, 1970), golden hamster (Myers and Buckman, 1972), monkey (Myers et al., 1971a), rat (Myers and Brophy, 1972), dog (Greenleaf et al., 1974), pigeon (Saxena, 1976), and chicken (Denbow and Edens, 1980).

In 1970, an attempt to localize the ionic phenomenon within the anterior hypothalamic, preoptic area failed (see Myers, 1974b). Instead, the site at which each cation evoked its dramatic effect on body temperature was found to lie in the posterior part of the hypothalamus rather than in the anterior region (Myers and Veale, 1970). If a circumscribed area of the posterior hypothalamus is perfused by means of push-pull cannulae with a solution of one of several salts of sodium, shivering, vasoconstriction, piloerection, and a sharp hyperthermia occur in the cat. A concentration of less than 10.0 mM excess Na^+ ions is effective. In corresponding experiments, a solution containing excess calcium, from 1.2 to 10.0 mM in excess of CSF, perfused within the same posterior site causes vasodilatation and a sharp fall in body temperature, which again is concentration-dependent. Figure 19 illustrates these changes and the anatomical sites at which the cations are active. Interestingly, neither potassium nor magnesium ions, increased in the same proportion as Na^+ or Ca^{2+} above their normal values, have any effect when they are perfused identically at these same sites (Myers and Veale, 1971).

Shortly thereafter, the fact was confirmed that the disturbance to the natural balance between Na^+ and Ca^{2+} was the cause of the given temperature perturbation (Myers and Yaksh, 1971). For example, the localized chelation of Ca^{2+} ions with ethyleneglycol-bis-[β-amino-ethyl ether] N,N'-tetra-acetic acid (EGTA), delivered by push-pull cannulae, causes an intense rise in the monkey's body temperature presumably because Na^+ ions predominate within the local ionic milieu of the diencephalic parenchyma. Distinguishing between thermoregulatory and set-point functions, the unanesthetized monkey's temperature can be reset for an extended period of time (more than one-half day) simply by repeating the push-pull perfusions of either Na^+ or Ca^{2+} ions at intermittent intervals once the animal's temperature is elevated or lowered. Indeed, there is no tachyphylaxis or decline in tissue sensitivity to either cation, unlike that seen with a similar infusion or injection of most substances given by the intracerebral or intrahypothalamic route.

A most surprising finding, however, is the monkey's retention of the capacity to regulate its body temperature around the new set-point level, high or low, established earlier

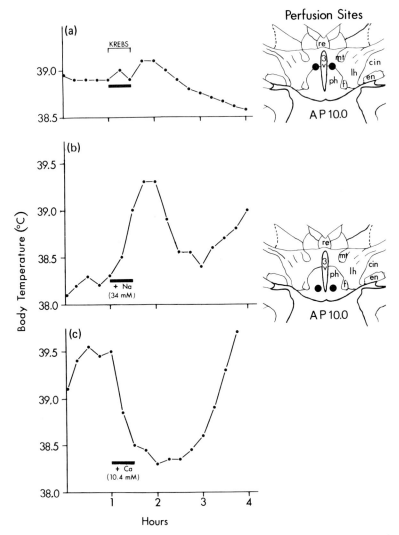

Figure 19 Temperature records of an unanesthetized cat in response to the local perfusions for 30 min of: (a) Krebs solution alone; (b) Krebs solution plus 34 mM excess sodium; and (c) Krebs solution plus 10.4 mM excess calcium. The sites of the bilateral perfusions are indicated by the filled circles (●) in the inset. Anatomical abbreviations are: cin, internal capsule; en, endopeduncular nucleus; f, fornix; lh, lateral hypothalamic area; mt, mamillothalamic tract; ph, posterior hypothalamic area; re, nucleus reuniens of the thalamus; 3v, third ventricle. (From Myers and Veale, 1971.)

by the intermittent perfusion of a solution of excess Na^+ or Ca^{2+} ions. When the newly stabilized temperature of the monkey is challenged by a warm or cold stimulus, provided in the form of hot or cold water given intragastrically, normal thermoregulatory responses including shivering and an appropriate vasomotor or pilomotor reaction result.

That Ca^{2+} ions could merely cause a nonspecific depressive effect or that Na^+ ions have a nonspecific excitatory action (Hanegan and Williams, 1973; Nielsen and Greenleaf,

1977), so as to impair temperature regulation, does not now appear to be a valid assumption. Should Ca^{2+} ions simply depress the thermoregulatory system involving the heat production pathway, then the animal undoubtedly could neither shiver nor vasoconstrict in response to a cold thermal load. In relation to this issue, Beleslin et al. (1974) have reported that several salts of sodium infused into the cerebral ventricle evoke a sharp rise in the temperature of the unanesthetized cat without any evidence of hyperactivity or hyperexcitability. Differential effects of excess Ca^{2+} ions on EEG and states of arousal are also clearly separated in the rat (Tószeghi et al., 1978). Additional observations listed in Table 5 seem to repudiate the notion of the nonspecificity of cation activity in the posterior hypothalamus.

The specific response to an ionic disturbance has been localized to the posterior hypothalamus in two other species. A push-pull perfusion of a solution containing excess Ca^{2+} ions in the posterior hypothalamus of the rat causes a concentration-dependent fall in temperature (Myers et al., 1976a). The amount of the cation in the perfusate required to produce the hypothermia is much below that which must be either microinjected at the same site in the rat or infused into the animal's lateral cerebral ventricle. The Ca^{2+}-induced fall in a rat's temperature is not prevented by an intraventricular infusion of mecamylamine; however, the Ca^{2+} hypothermia is often accentuated by atropine given similarly. This suggests that muscarinic receptors may be involved in the pathways through which Ca^{2+} ions exert their effect of lowering the set-point temperature.

Using the rabbit, Veale and colleagues have locally perfused excess Na^+ or Ca^{2+} ions within the posterior hypothalamus by means of push-pull cannulae. Again, a similar rise or fall in the rabbit's temperature is seen which is concentration- and site-dependent (Veale and Jones, 1976; Veale et al., 1977a). Of significance is their result that the artificial disturbance in the normal ionic balance within this caudal hypothalamic area can be overridden if glucose is added to the perfusion medium (Jones et al., 1978). The carbohydrate greatly lessens or even abolishes the responses evoked by the altered ratio between Na^+ and Ca^{2+} ions, most likely because of the extra availability of utilizable energy substrate (Veale et al., 1977a). The specificity of this effect is borne out by the lack of effect of sucrose perfused in the same molar concentration (Jones et al., 1978).

In relation to the issue of functional specificity, it is notable that a monkey feeds itself normally, following an intrahypothalamic injection of norepinephrine into the anterolateral hypothalamus, during the course of deep hypothermia produced by repeated push-pull perfusions of excess Ca^{2+} ions within the posterior hypothalamus (Myers and Yaksh, 1972).

C. Set Point and Exercise

During submaximal or maximal exercise, one's body temperature rises sharply just to a given level and remains there as long as exercise continues (Nielsen, 1969). It is generally conceded that the hyperthermia observed so commonly is a regulated rise. This is ordinarily independent of a fairly wide range of ambient temperatures (Nielsen, 1938; Greenleaf et al., 1975). In the animal that is running on a treadmill, the elevation in central temperature during exercise has been demonstrated to be proportional to the work intensity during this steady-state activity (Young et al., 1959).

A rat trained to run on a small-rodent treadmill often exhibits a rise in temperature of up to 2.0°C during this exercise. When the posterior hypothalamus is perfused with an artificial CSF, there is no effect on the rat's elevated temperature or on its work output during a 90-min interval (Gisolfi et al., 1976). However, when Ca^{2+} ions slightly in excess of that in CSF are perfused by push-pull cannulae at the same hypothalamic locus, the exercising rat continues to run at the same rate as before. Astonishingly, the rat's temperature at the same time declines steadily in response to Ca^{2+} perfusion (Gisolfi et al., 1976). By contrast, excess Na^+ ions infused in the same locus intensify the exercise-induced rise in the rat's temperature (C. V. Gisolfi, unpublished observation). EGTA perfused through the same region, so as to chelate the Ca^{2+} ions, has the same exacerbating effect on the running rat's temperature (Wilson et al., 1978). These two findings indicate that the exercise-related rise in temperature, often hypothesized to be due to an upward deflection of the set-point temperature, may perhaps be related to a subtle alteration in the balance between Na^+ and Ca^{2+} ions which could occur during exercise (Gisolfi et al., 1976).

After the rhesus monkey is trained to exercise strenuously in a special rowing machine, an infusion of excess Ca^{2+} ions into its cerebral ventricle likewise ameliorates the typically intense rise in this primate's body temperature (Myers et al., 1977b,c). If the concentration of the Ca^{2+} solution is not too high, the suppression of the exercise-induced rise in the monkey's set-point temperature has little effect on the over-all work output of the monkey, as shown in Figure 20. That is to say, the monkey exercises at nearly the same rate even though the usual rise in temperature is prevented by excess intracerebral calcium. This demonstrates further that the rise in core temperature produced by muscular activity can be entirely overridden by Ca^{2+} ions acting on the animal's diencephalon.

D. Ambient Temperature and Cations

When the golden hamster is placed in a chamber with an ambient temperature of $+5°$ or $-10°C$, it is able to thermoregulate quite satisfactorily over a short term of several hours. If excess calcium is injected into the cerebral ventricle of this hibernator, the fall in temperature is not only concentration dependent but also is considerably intensified by the cold ambient temperature. From this result, the conclusion could be reached that a disturbance to the balance of cations within the hypothalamus may be ultimately dependent on the environmental temperature of the animal (Myers and Buckman, 1972). In addition, if calcium acted solely to "anesthetize" the hypothalamus, then the temperature of the animal should fall passively during a period of cold exposure. This possibility would raise serious questions about the establishment and maintenance of a set-point mechanism by the ratio of the two cations in the hypothalamus. A series of experiments has dealt with this vital issue.

Two factors seem to determine the relationship between ambient temperature and the central effect of either Na^+ or Ca^{2+} ions: (1) the interval during which the animal is exposed to a warm or cold temperature; and (2) the anatomical region in the brain stem at which the alteration in the ratio of the two cations is introduced (Myers, 1974b). If the ambient temperature of an animal is altered at precisely the same time that the cerebral ventricle or hypothalamus is perfused with a solution containing excess cations, the temperature response is markedly different from that seen when the ambient challenge is given prior to perfusion. For example, when the posterior hypothalamus of the cat is perfused with excess Na^+ or Ca^{2+} ions at the same time as the animal is exposed to air warmed to $38°C$, or cooled to $6°C$, the magnitude and duration of the respective hyper-

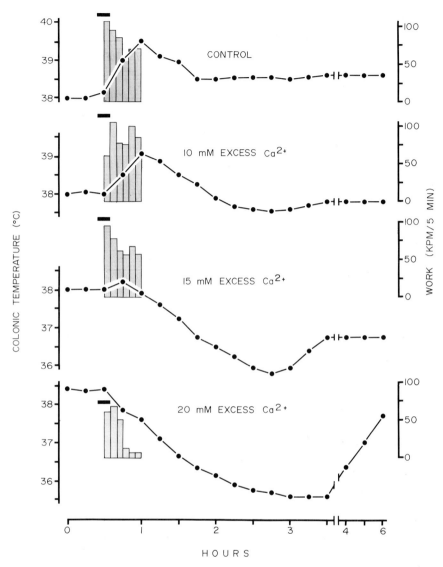

Figure 20 Colonic temperature (°C) of a chaired macaque monkey recorded in four separate experiments before, during, and after exercise on a "rowing machine." Work rate, recorded in kilocalories per minute (kpm) per 5 min is denoted by vertical shaded bars. The 10-min interval of infusion of control CSF solution or of Ca^{2+} ions 10, 15, or 20 mM in excess of CSF into the cerebral ventricle is indicated by the black bar. (From Myers et al., 1977c.)

or hypothermia are virtually identical to that seen when the cat is kept at a neutral environmental temperature of 22°C (Myers et al., 1976b). The change in temperature produced by either cation is accentuated by neither warm nor cold peripheral challenges. However, when the cat is placed in either a warm or cold environment before the cation concentration of the posterior hypothalamus is artificially altered by perfusion, a

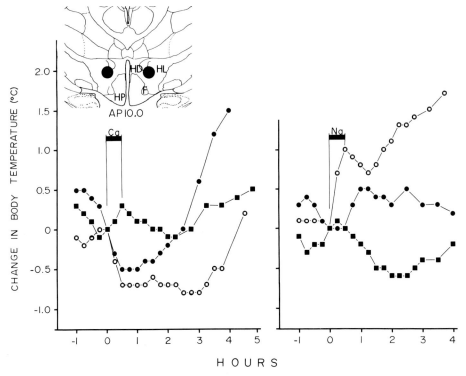

Figure 21 Deviations in body temperature of the unanesthetized cat produced by push-pull perfusion at 50 μl/min of 10.4 mM excess Ca^{2+} ions (left panel) or 13.6 mM excess Na^+ ions (right panel). The interval of perfusion is denoted by the black bars and the sites of perfusion are designated by the dots in the histological inset. The temperature of the cat's cage was held at 22°C (o——o), or elevated (38°C, ■——■), or lowered (6°C, ●——●) during a 30-min interval before start of the perfusion and throughout the course of perfusion with excess cations. Anatomical abbreviations are: F, fornix; HD, dorsal hypothalamic area; HL, lateral hypothalamus; HP, posterior hypothalamic area. (From Myers et al., 1976b.)

compensatory reaction against the abnormal Na^+ or Ca^{2+} concentration occurs. In fact, the typical change in body temperature is counteracted so effectively that the temperature of the animal may rebound in a direction opposite to that which normally ensues after a perfusion with Na^+ or Ca^{2+} ions. As illustrated in Figure 21, raising the ambient temperature to 38°C or lowering it to 6°C, 30 min before a hypothalamic perfusion with excess Na^+ or Ca^{2+} ions, greatly attenuates the typical Na^+- or Ca^{2+}-induced shifts in body temperature. Actually, when the animal is kept at the high temperature of 38°C, the Ca^{2+} hypothermia is delayed until after the perfusion of excess Ca^{2+} ions is terminated. Similarly, the high environmental temperature retards the rise in body temperature which always follows the local perfusion with excess Na^+ ions in the hypothalamus. In this instance, a hypothermia occurs after the termination of the Na^+ perfusion.

Thus, the postulated cation set-point mechanism actually withstands an aberrant physiological condition. This is more evidence that a cation could not simply cause a poikilothermia because of a nonspecific Na^+-ion excitation of the cells in the posterior hypothalamus or Ca^{2+}-ion depression of these neurons (Hanegan and Williams, 1975). Two alternative hypotheses are suggested. First, as the posterior hypothalamic set-point mechanism is challenged by an external thermal load in the form of a cold or hot stimulus to the skin, the set point reacts actively in its own defense. Hence, if the environment is already cold, a downward shift in the set point caused ordinarily by an excess in Ca^{2+} ions certainly would not be called for. Similarly, if the environment is already warm, an upward shift in the set point customarily caused by an excess in Na^+ ions would likewise not be necessary. In both instances, a displacement in the set-point temperature in either direction would be counter to the animal's own survival.

Second, the effect of a cation in a circumscribed region of the posterior hypothalamus is apparently functionally specific. One could arrive at the supposition that there are actual changes in membrane binding or unbinding as well as extracellular transport or intracellular metabolism of the given cation (e.g., Baker, 1972; Cheung, 1980). These shifts presumably would occur endogenously in the posterior hypothalamus as the set point for body temperature is actively defended under extreme thermal stress or is displaced during a fever or other atypical state. This latter presumption is considered in the following section.

E. Endogenous Activity of Na^+ and Ca^{2+} Ions

In the late 1960s, it was discovered that an endotoxin administered systemically produces a peculiar perturbation in the profile of serum electrolytes: the concentration of Ca^{2+} ions is reduced (Skarnes, 1968). Not to come until several years later was the first substantive evidence that the cerebral ratio of cations could be related to the febrile temperature.

In the unanesthetized cat, either $^{45}Ca^{2+}$ or $^{22}Na^+$ injected into the cerebral ventricle radiolabels the endogenous stores of calcium and sodium in the hypothalamus, respectively. Upon continuous perfusion of the third cerebral ventricle of the unanesthetized cat, there is a steady-state efflux of both of these cations. However, when a bacterial pyrogen such as *Salmonella typhosa* is administered, the resultant rates of Na^+ and Ca^{2+} efflux suddenly change so as to parallel the course of the ensuing bacterially induced fever (Myers and Tytell, 1972). As the temperature of the cat rises, the egress of Ca^{2+} into the third ventricle increases sharply, whereas the Na^+ ions are retained within the animal's diencephalon. These two observations offer relatively direct evidence that a bacteria must act on the brain stem to shift the balance in the activity of Na^+ and Ca^{2+} ions during an interval which corresponds to the deviation in the cat's set-point temperature.

In a reciprocal fashion, the efflux of the cations reverses completely during antipyresis following the administration of acetaminophen systemically. In an experiment done in the febrile cat pretreated earlier with bacteria, the efflux of $^{22}Na^+$ into the third ventricle is greatly augmented after an antipyretic is administered systemically (Myers, 1976b). In contrast, the efflux of $^{45}Ca^{2+}$ ions declines as the animal's temperature falls following the antipyretic. These findings conform to the pharmacological effects of an excess in Na^+ or Ca^{2+} ions delivered by the same intraventricular route. When heat production is required, Na^+ ions are retained in the diencephalon while Ca^{2+} ions are extruded. Conversely, the heat loss of antipyresis is accompanied by a retention of Ca^{2+} ions while Na^+ ions are released from this brain stem structure.

1. Anatomical Localization of Ion Flux

Somewhat more definitive anatomical information has been provided in an investigation consisting of five separate studies. In the experiments, the dynamics in the activity of Ca^{2+} ions in the posterior hypothalamus have been examined during both thermoregulation and the defense of the set-point temperature. After a Ca^{2+}-sensitive locus in the posterior hypothalamus of the cat is first identified by an anatomical exploration using push-pull perfusions with calcium, the site is then prelabeled with $^{45}Ca^{2+}$. As soon as the cat is subjected to a cold environmental temperature of 0-5°C, the efflux of $^{45}Ca^{2+}$ ions into the push-pull perfusate collected in the posterior hypothalamus is enhanced (Myers et al., 1976b); however, exposure of the cat to a warm ambient temperature of 35°C causes a retention of the cation within the same hypothalamic area. Figure 22 illustrates the opposite changes in Ca^{2+} efflux in response to the extremes in ambient temperature.

The magnitude of the change in $^{45}Ca^{2+}$ activity in the posterior hypothalamus depends entirely upon the intensity of the thermal challenge. In contrast to the absence of any change in Ca^{2+} flux during a small deflection in ambient temperature, a severe thermal disturbance is required to produce any quantitative deviation in Ca^{2+} efflux. Thus, it would seem that a regulatory mechanism, as distinct from a functioning set-point mechanism, can accomodate the typical moment-to-moment variations in an animal's ambient temperature. But when the ambient temperature is drastically displaced and the set point level itself is severely challenged, then the cation mechanism in the posterior hypothalamus seems to come into full play (Myers et al., 1976b).

With regard to the cellular functioning of the set point, the stability of Ca^{2+} ion flux in the posterior hypothalamus is thought to be determined principally by two factors: (1) the thermal input that is conveyed from the periphery to the anterior hypothalamic, preoptic area, and (2) the local temperature of the rostral hypothalamus. The experimental evidence is as follows. When this anterior region is cooled by thermodes, the efflux of Ca^{2+} ions within the posterior hypothalamus is enhanced. Alternatively, warming of the anterior hypothalamus causes the Ca^{2+} ions to be retained in the posterior hypothalamic area (Myers et al., 1976b).

A local anesthetic, procaine, microinjected into the anterior hypothalamus temporarily prevents any modification in the flux of Ca^{2+} ions in the cat's caudal hypothalamus when the animal is placed in a cold or warm ambient temperature. In addition, if a hyperthermia-inducing substance such as a bacterial pyrogen, 5-HT, or a prostaglandin is microinjected directly into the anterior hypothalamic area so that the heat production mechanism is immediately stimulated, the efflux of Ca^{2+} is once again greatly increased in the posterior hypothalamus at the same time that the cat's body temperature rises.

These results would seem to provide almost unequivocal evidence that the posterior hypothalamus receives its physiologically significant signals by way of neuronal input from the anterior hypothalamus (Myers, 1978). Whether the impulses are derived from neurons in the rostral hypothalamus that are locally thermosensitive and chemosensitive or originate by way of the cold or heat sensor pathway emanating from the periphery, the net effect on ionic activity of the posterior hypothalamus is identical. Once a certain threshold of neuronal activity is reached, impulses are transmitted along pathways caudalward from the anterior-preoptic region. Subsequently, the molecular properties of the posterior hypothalamus seem to be drastically realigned in terms of the binding, uptake, metabolism, or transport of Ca^{2+} ions (Cheung, 1980). Such a resultant change in the

Neurochemical Mechanisms in Thermoregulation

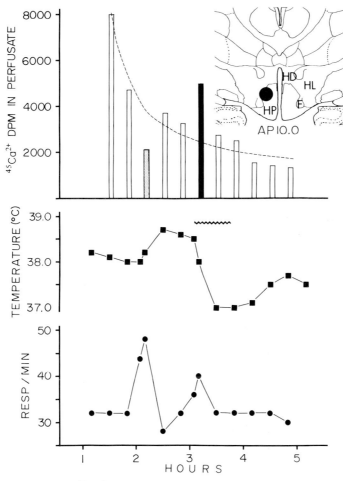

Figure 22 Efflux (top) of $^{45}Ca^{2+}$ in successive push-pull perfusates collected at a rate of 50 µl/min from the perfusion site denoted by the dot in the inset. The site had been labeled with 1.0 µCi $^{45}Ca^{2+}$, 18 hr earlier. The chamber temperature of the cat was raised to 40°C (open bars) or lowered to 0°C (black bar) just preceding and during the third and sixth perfusion, respectively, as denoted by the bars. Colonic temperature (middle) and respiratory rate (bottom) were recorded continuously. Shivering is designated by the zig-zag line (middle). Anatomical abbreviations are the same as in Figure 21. (From Myers et al., 1976b.)

ionic milieu apparently either facilitates or retards the transmission of efferent signals for heat production or heat dissipation to the respective organs in the periphery which are crucially involved in vasomotor responses, respiration, shivering, sweating, piloerection, nonshivering thermogenesis, and behavioral reactions.

Table 5 presents a summary of the three sets of evidence for the involvement of hypothalamic Na^+ and Ca^{2+} ions in the establishment of a postulated set-point mechanism for body temperature.

Table 5 Summary of Evidence for the Ionic Theory of the Temperature Set-Point[a]

A. Indirect Evidence
   1. Evolution of a universal mechanism for ratio of cations in body tissue (Stevens, 1973).
   2. Intrinsic stability of the Na^+ to Ca^{2+} ratio in the extracellular fluid spaces under normal conditions.
   3. Universality of effect on temperature of excess Na^+ and Ca^{2+} or reduction by chelating agent.

B. Pharmacological Evidence
   1. Anatomical separation of the cations' effects on posterior hypothalamus from the anterior hypothalamic, preoptic area of thermosensitivity.
   2. Na^+ and Ca^{2+} ions act specifically in terms of the relative activity of other essential cations including K^+ and Mg^{2+}.
   3. Set-point temperature can be reset for a prolonged period at high or low temperatures, within limits, without pathological sequelae.
   4. There is no tachyphylaxis to repeated local perfusions of excess cations, i.e., temperature set-point can be driven to the brink of death in either direction.
   5. Retention of the capacity to thermoregulate normally around a new temperature established by an altered ratio of Na^+ to Ca^{2+} ions; no evidence of nonspecific depression or excitation of neurons.
   6. An exercise-induced rise in set-point temperature is overcome by elevating Ca^{2+} in the posterior hypothalamus while the animal continues to exercise.
   7. Peripheral cold or warm challenge blocks a typical Na^+-induced rise or Ca^{2+}-induced shift in set-point temperature.
   8. Behavioral responding to obtain heat during regulatory adjustment fails to occur during a change in temperature induced by cations perfused in the posterior hypothalamus.
   9. When the cation ratio is shifted in other diencephalic structures, the cation-evoked response to feeding, arousal, sleep, drinking, etc., is functionally specific.
   10. Normality of EEG and the capacity to feed are retained after the set-point temperature is shifted artificially by excess Na^+ or Ca^{2+} in the posterior hypothalamus.

C. Physiological (Endogenous) Evidence
   1. Bacteria administered systemically or intracerebrally evoke a reciprocal shift in the ratio of endogenous Na^+ to Ca^{2+} ions in the diencephalon as fever develops.
   2. An antipyretic given systemically reverses reciprocally the induced shift in the endogenous diencephalic ratio of the two cations during defervescence.
   3. As the set-point temperature rises during exercise, the efflux of Ca^{2+} ions from the diencephalon is enhanced.
   4. A prostaglandin or bacteria injected into the anterior hypothalamus evokes a consequent efflux of Ca^{2+} ions within the posterior hypothalamus as the set-point temperature rises.
   5. A mild thermal stress of heat or cold, within a normal thermoregulatory band width, fails to affect the endogenous ratio of Na^+ to Ca^{2+} ions in the posterior hypothalamus, i.e., the cation ratio remains unchanged during thermoregulation.

Table 5 (Continued)

6. A severe peripheral challenge to the set-point temperature in terms of an intense cold or heat stressor causes an immediate compensatory shift in the activity of endogenous Ca^{2+} ions in the posterior hypothalamus.
7. A sharp deflection of the anterior hypothalamic, preoptic temperature sufficient to activate thermosensitive neurons enhances the efflux or retention of endogenous Ca^{2+} ions within the posterior hypothalamus.

[a] According to this concept, the set-point is determined by a stable extracellular concentration of sodium and calcium in the posterior hypothalamus. This ionic milieu theoretically establishes a steady-state firing rate of efferent neurons in the caudal hypothalamic pathways for heat production and heat loss (for specific references see reviews of Myers, 1974a,b, 1976b, 1978b).

VII. Fever and Pyrogens

By the turn of this century, the idea had already been promulgated that bacteria somehow have a direct effect on the temperature "centers" of the brain stem. For example, in 1910 Grafe attributed fever to the irritating action of the decomposition products of bacteria upon a so-called "heat center" which was presumably located at the base of the brain. That a neural mechanism is essential to fever was proposed by Citroni and Leschke (1913) and by Judah (1916), who discovered independently that an infectious agent fails to alter body temperature after an extensive lesion of the midbrain region. Ranson et al. (1939) concluded that a disseminated lesion of the hypothalamus interferes with the development of a typical fever. However, a more localized ablation of the preoptic region may leave the febrile reaction to a systemic endotoxin intact (Veale and Cooper, 1975; Lipton and Trzcinka, 1976). In the human patient with a relatively restricted sarcoidosis lesion of the rostral hypothalamus, febrile episodes develop even though a considerable thermolability to warm or cold room temperature is evident (Lipton et al., 1977).

The first evidence that bacteria can act directly on the tissue of the brain was published in a remarkable account by Hashimoto in 1915. He showed that an intense febrile response is evoked by a bacterial pyrogen injected directly into the so-called temperature centers of the brain of a rabbit or guinea pig. As early as this time, Hashimoto was even able to demonstrate a functional tachyphylaxis to an intracerebrally injected chemical complex, horse serum, which he attributed to an "immunizing reaction" that developed after two or three weeks upon repeated injections. Just as significant was the further observation that the amount of pyrogen required to produce a rise in the animal's temperature was substantially smaller when given by the intracerebral route as compared to the intravenous route (Hashimoto, 1915a,b). This fact has been "rediscovered" by many researchers ever since.

The concept that a fever of bacterial origin is a controlled elevation in body temperature was proposed by Meyer in 1913. Generally, it was a well-accepted viewpoint even in that era. Here it is appropriate to consider a memorable passage written 60 years ago (Barbour, 1921): "In fever, it is said that the body thermostat is 'set at a higher level.' A normal temperature becomes apparently interpreted as 'cold' by the temperature centers, while one of 40° perhaps feels 'neutral'. Similarly, the skin nerves seem hypersensitive to cold, or hypersensitive to heat, whence the subjective 'chill.' "

Thirty years later, the stage was set for the pivotal study of Fox and McPherson (1954). In a patient who had unwittingly developed a fever during the course of an exercise study (but nevertheless wished to carry on the experiment), the pattern of the work-induced rise in temperature was identical to that as seen when the patient was afebrile. That is, the patient regulated his body temperature normally around a new set level, independent of the existing fever. Clinically, this has been witnessed on numerous occasions. An experiment with the rabbit demonstrated that a pyrogen does act to shift the common reference temperature, whereas a change in ambient temperature does not (Hardy, 1976; Stitt, 1976). In addition, an elevated ambient temperature seems to cause a progressive loss of hypothalamic thermosensitivity, although a pyrogen may not necessarily affect the thermosensitivity of this structure (Stitt, 1976).

A. Hypothalamic Locus of Pyrogen Action

Whether or not bacteria act directly on the cells of the hypothalamus (Cooper et al., 1975) is an issue which is still actively debated today because of findings both pro and con (Milton, 1976). In spite of the observation that radiolabeled bacteria are not necessarily detectable in the central nervous system, a classic paper published by Morgan (1938) documented lucidly the pathological insult which systemically administered bacteria exert upon nuclei of the hypothalamus. In this investigation, Morgan (1938) showed that an intravenous injection in the dog or rabbit of typhoid bacillus or Bronchisepticus toxin produces severe cytological damage to three major clumps of cells: those dorsal to the optic chiasm, within the region of the tuber cinereum and those extending from the fornix to the mammillary bodies. Actually, a marked chromatolysis of these hypothalamic cell groups occurred within a few hours after an injection of either one of these two types of bacteria. This finding reveals that if the plasma level of a bacterial organism is sufficiently high, pathological damage is instigated which may destroy the anatomical integrity of the blood-brain barrier.

By using the typhoid organism, *Salmonella typhosa*, Bennett and his colleagues were able to induce a reliable fever in the dog by injecting a suspension of the bacteria directly into the cerebral ventricle (Bennett et al., 1957). Later, Sheth and Borison (1960) demonstrated the same sort of hyperpyrexia in the cat after an intraventricular injection of the bacterial pyrogen. In a brief abstract, Grant et al. (1955) reported that 50 μl of endotoxin injected into the hypothalamus failed to alter the temperature of the rabbit. Thus, the search began in the early 1960s for a possibly specific anatomical site of action of a pyrogen in the brain.

1. Bacterial Pyrogen

In the unanesthetized cat, it was discovered that bacteria induce an intense fever when injected directly into the anterior hypothalamus in a 1- to 2-μl volume (Villablanca and Myers, 1964). A concentration of *S. typhosa* of only 1/1000 of the intravenous threshold dose or less was required to produce the pyrexic response. The latency of the onset of hyperthermia was only 5-20 min after the intrahypothalamic microinjection of the endotoxin (Villablanca and Myers, 1965). If a bacterial pyrogen was infused into either the lateral or posterior parts of the cat's hypothalamus, a much longer latency was recorded, and the febrile episode was often less intense.

The extraordinary reactivity of the anterior hypothalamus to *S. typhosa* as well as to *S. dysenteriae* organisms was confirmed, also in the same species, by Feldberg and Myers (1965a); however, in their experiments, the injection of the typhoid suspension in a dilution of 1:500 evoked an almost immediate rise in the cat's body temperature. In both cases, these responses were accompanied by vigorous shivering, vasoconstriction, and piloerection, the normal components of the heat-producing response of a fever. The pyrexia caused by an endotoxin applied to the anterior but not to other areas of the cat's hypothalamus was later confirmed (Jackson, 1967).

In general, the infrahuman primate is much less responsive than the human to a bacterial pyrogen that is administered systemically (see Myers, 1971a). However, in the unanesthetized rhesus monkey, each of three endotoxins, *S. dysenteriae, s. typhosa*, or *E. coli* microinjected into the anterior hypothalamus causes a severe and long-lasting fever (Myers et al., 1971b). The hyperthermic reaction is characterized by intermittent shivering, piloerection, and huddling behavior, which are typical of the monkey's heat conservation process. In some primates, an endotoxin given systemically elicits a biphasic or two-peaked fever rather than a simple monophasic rise in temperature. Such a biphasic hyperthermia is equally reproducible if one of several potent endotoxins is microinjected directly into the anterior hypothalamic, preoptic area of the monkey (Myers et al., 1973). In fact, the similarity in the biphasic nature of the temperature rise indicates that the second phase of a fever, typically evoked by endotoxin given peripherally, could be due to a localized release of endogenous pyrogen, or perhaps a sequestration of leukocytes directly within the hypothalamus itself.

A careful mapping of the diencephalic area of the monkey that is sensitive to killed bacteria reveals an interesting positive relationship between anatomical loci and the febrile response. When the linear distance of an injection site from a coronal plane formed by the optic chiasm and anterior commissure is plotted against the latency and magnitude of hyperthermia as well as the over-all fever index (area under the temperature curve), the resultant correlation coefficients are significantly high (Myers et al., 1974). Thus, all of the features of a fever correlate morphologically with the reactivity of the thermosensitive zone demarcated within the monkey's hypothalamus (Hayward and Baker, 1968). This region is portrayed in Figure 23.

Since tolerance to repeated microinjections of the endotoxin into the anterior, preoptic area does not develop (Myers et al., 1974), the first peak of a fever may be caused by a direct action of the pyrogen on the hypothalamus. Alternatively, the second peak of the biphasic response may be brought about by the release, within this rostral region, of an intermediary substance that induces thermogenesis, such as 5-HT (Myers et al., 1974) or a protein factor (Ruwe and Myers, 1979). Of further interest is the fact that the thermosensitivity of the monkey's anterior, preoptic area, when heated or cooled locally by a thermode, is considerably reduced during a fever caused by typhoid bacteria microinjected into this rostral area (Lipton and Kennedy, 1979). This suggests that a dampening down of hypothalamic reactivity to thermal stimuli may be due to a predominance of input from peripheral thermal sensors which trigger regulatory responses around an elevated temperature.

When a bacterial pyrogen is injected into the medial, preoptic region of the rabbit, a correspondingly long-lasting fever develops which is contingent upon the concentration of organisms in the suspension of the injected pyrogen (Repin and Kratskin, 1967). In several more recent studies, the pyrexic response following an intrahypothalamic injection of

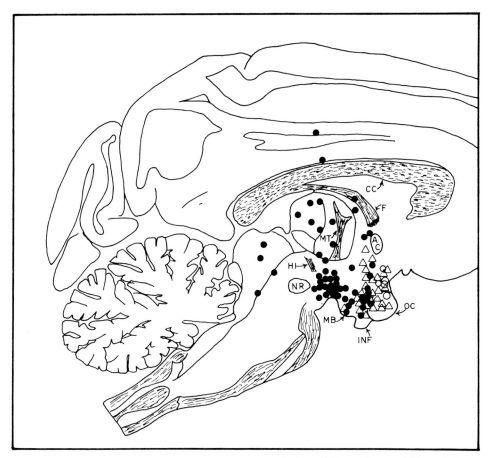

Figure 23 Composite anatomical map in a parasagittal plane 2.5 mm from midline of sites in monkey brain at which a suspension of a 1:2 dilution of *E. coli* was microinjected. In terms of the onset of a fever, an open circle (○) indicates a latency of 30 min or less, a triangle (△) a latency of 31-60 min, and a filled circle (●) a latency greater than 60 min or no effect. AC, anterior commissure; CC, corpus callosum; F, fornix, HI; habenulointerpeduncular tract; INF, infundibulum; MB, mamillary body; MT, mamillothalamic tract; NR, red nucleus; OC, optic chiasm. (From Myers et al., 1974.)

bacterial pyrogen into one of several regions of the hypothalamus has verified the finding that the anterior hypothalamic, preoptic area is indisputably the site most sensitive to an endotoxin. The species in which this has been demonstrated include the rat (Lipton et al., 1973; Splawinski et al., 1977), the lamb (Pittman et al., 1975), and the chicken (Pittman et al., 1976a). Today, it is still not known why the original experiment of Grant et al. (1955) failed (vide supra).

2. Leukocyte Pyrogen

To explore the theory that the rise in body temperature during a fever is due to the direct action of a leukocyte-derived or endogenous pyrogen on cells in the brain, several investigators have attempted to identify the locus of action of this factor (see Cooper et al., 1976b). In 1967, Cooper and his colleagues found that the pyrogen fraction obtained following the incubation of cells from rabbit peritoneal exudate has its major effect on the same region as the endotoxin, the anterior hypothalamus. Within a latency period of 10 to 25 min, the intrahypothalamic microinjection of leukocyte pyrogen produces shivering, an increase in respiration, and a sharp rise in temperature (Cooper et al., 1967). Curiously, the interval to the onset of the febrile response is nearly identical to or perhaps slightly longer than the latency after a similar injection of endotoxin in the cat (Villablanca and Myers, 1965). Nonetheless, an intrahypothalamic injection of *Salmonella abortus equi* in the rabbit causes a fever with a latency of about 25 min.

Polymorphonuclear leukocytes microinjected into the anterior hypothalamus of the cat likewise induce a sharp rise in temperature reportedly within 5 min after the injection (Jackson, 1967). This latency corresponds to the shortest interval of onset reported for *Salmonella* endotoxin injected into the hypothalamus of the cat (Villablanca and Myers, 1965; Feldberg and Myers, 1965a). Unfortunately, the large 6.0 μl volume of the pyrogen infused (Jackson, 1967) does not preclude a disseminated action of the leukocytes on a very large number of cells outside of this rostral region. Since the latency data may be confounded by the volume effect in hypothalamic tissue, a more quantitative study of latency effects is required.

When Rosendorff and his co-workers used a somewhat smaller 2.0 μl volume, leukocyte pyrogen applied to several areas of the hypothalamus and brain stem of the unanesthetized rabbit exerted differential effects. Only when the pyrogen was injected into the anterior hypothalamus did a fever with a short latency develop (Rosendorff et al., 1970). Although a comparable injection into other areas of the brain stem also produces a hyperthermia in many of the rabbits (Rosendorff and Mooney, 1971), the much longer latency to thermogenesis indicates that the pyrogenic material probably diffuses to an alternate site, probably the rostral hypothalamus (Myers et al., 1974).

An important observation has been made by Cooper and his colleagues (1976b) who have injected plasma collected from febrile patients into the anterior hypothalamic, preoptic area of the unanesthetized rabbit. The resulting pyrexia seen in the animal strongly suggests that the rostral hypothalamus, due to its extreme sensitivity, would provide an excellent assay for the pyrogenic properties of plasma obtained during a pathological condition. Human leukocyte pyrogen infused into the preoptic area of the monkey also produces an intense hyperpyrexia accompanied by augmented oxygen consumption (Lipton et al., 1979). Both results indicate that the cells of the rostral hypothalamus are certainly reactive to heterologous pyrogen and that there is no species-specificity to endogenous pyrogen after all.

After the anterior hypothalamic area of the rabbit is ablated, leukocyte pyrogen injected intravenously evokes a similar kind of febrile response to that seen when the animal's hypothalamus is intact (Cooper and Veale, 1974). However, if the lesion is placed in the posterior hypothalamus, as illustrated in Figure 24, a fever does not develop after the same sort of intravenous leukocyte challenge. Whether the posterior hypothalamus constitutes an essential anatomical link in the central pathway responsible for the mediation of a febrile response has been a controversial point over the years (Bard et al., 1970; Chambers et al., 1949, 1974). However, this key result of Cooper and Veale favors its crucial role in the hypothalamic mechanism for fever (Myers et al., 1976b).

Although the precise way by which leukocyte pyrogen is inactivated within the hypothalamus is presently unknown, it is possible that pyrogen which reaches the anterior

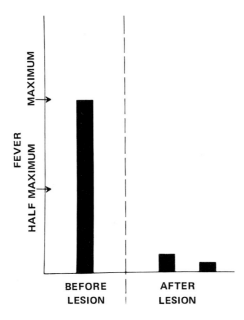

Figure 24 Maximum fever heights due to standard doses of leukocyte pyrogen injected intravenously in rabbit no. 2 before and after posterior hypothalamic lesioning. [From Cooper et al., 1976b, reprinted from *Brain Dysfunction in Infantile Febrile Convulsions*, M. A. B. Brazier and F. Coceani (Eds.), Raven Press, New York.]

hypothalamus is eliminated through its egress into the cerebral ventricle. In an important experiment done by Cooper and Veale (1972), inert mineral oil was kept in the cerebral ventricle during a pyrogen fever. As long as the oil prevented the exit of leukocytes from the hypothalamus, the fever was sustained. Upon removal of the oil from the ventricle, the febrile response dissipated.

3. Action of Pyrogen on Single Neurons

In order to determine whether bacteria could alter the discharge characteristics of temperature-sensitive cells in the hypothalamus, a number of investigators have attempted to correlate neuronal firing rate with the onset and development of a pyrogen fever. In a hallmark study, Wit and Wang (1968) revealed that *Pseudomonas* polysaccharide endotoxin given systemically has no effect on the temperature-insensitive neurons identified in the anterior hypothalamic, preoptic area of the cat. However, these investigators did find that the reactivity of warm-sensitive neurons in this region is significantly depressed or abolished by the endotoxin when the cat's brain temperature is held in the range of 37-39°C. However, these warm sensors continue to respond to an environmental challenge in the range of 39-41°C. This favors the view that the set point for body temperature during a pyrogen fever is newly reset at a higher level. In this same study, the systemic administration of aspirin after the suppression of firing of the warm-sensitive units results in a return of their normal thermal reactivity toward the prepyrogen level (Wit and Wang, 1968).

When neurons were tested which show a decrease in discharge frequency after the animal's peripheral exposure to a warm stimulus (Fig. 25), leukocyte pyrogen injected systemically excites these depressed neurons (Schoener and Wang, 1975). In contrast, the administration of salicylate, as shown in Figure 25, causes a return of the discharge pattern of these excited neurons to their control firing rate. Again we see that an antipyretic offsets the aberrant firing pattern of a thermosensitive neuron. These observations have two noteworthy implications. First, aspirin may reinstate the sensitivity of heat-loss neurons to their normal level of discharge. This in turn would enable the hypothalamic set-point mechanism to return body temperature to within a normal range. Second, the locus of the antipyretic action of aspirin appears to be in the central nervous system (see Section VIII).

Cabanac et al. (1968) have reported a similar finding in the rabbit. The systemic injection of typhoid vaccine suppresses the activity of single neurons identified as being warm sensitive. Later, Eisenman (1969) revealed that neurons whose discharge rate is augmented after hypothalamic warming also exhibit a decline in thermosensitivity after the systemic injection of a bacterial pyrogen. Supplementing the research on the anesthetized preparation, Eisenman (1974) has extended these observations advantageously to the unanesthetized animal. Of importance is the corresponding change in the depolarization rate of identified units in the reticular formation of the urethane-anesthetized rabbit (Nakayama and Hori, 1973); that is, after the peripheral injection of a bacterial endotoxin, some cold-sensitive neurons increase their rate of spontaneous discharge, whereas the firing rate of warm-sensitive neurons declines. Since this part of the rabbit's brain stem exhibits

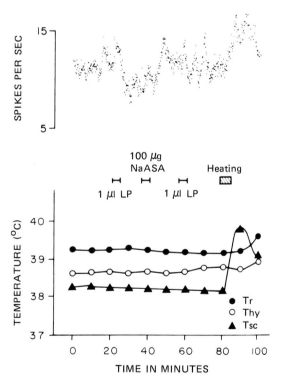

Figure 25 Effects of pyrogen (LP), sodium-salicylate (NaASA), and pyrogen consecutively on a thermopositive PO/AH neuron in cat. Thermal responsiveness was tested before and after drug administrations; first test not shown. First injection of pyrogen near neuron caused depression of discharge. While rate of discharge was still below control, injection of NaASA in same area caused reversal of pyretic effect with an overshoot. Second injection of pyrogen after NaASA had minimal effect. (From Schoener and Wang, 1975.)

little sensitivity to bacteria, as deduced from the volume and dose analysis of Rosendorff and Mooney (1971), one might infer that these neurons mirror the activity of their cellular counterparts within the anterior hypothalamus.

B. Content and Release of Amines

When a rabbit is pretreated with reserpine, a systemic injection of *E. coli* endotoxin fails to evoke a fever of typical intensity (Kroneberg and Kurbjuweit, 1959; Yasuda, 1962). In fact, after the endotoxin injection, the temperature of the reserpine-treated animal may actually decline rather than rise (Des Prez et al., 1966). Along the same lines, Göing (1959) and Cooper and Cranston (1966) have found that a fever produced by an intravenous injection of leukocyte pyrogen is substantially potentiated in the rabbit after it is pretreated with a monoamine oxidase inhibitor.

As originally described by Takashima (1962) and later confirmed by others (Rašková et al., 1966; Takagi and Kuruma, 1966), the content of 5-HT in the brain of an animal undergoing a pyrogen fever declines significantly. The reduction in the level of this indoleamine may reflect an increased presynaptic release, and hence loss of 5-HT, or its utilization by nerve cells that are involved in the pyrexic response. In agreement with this, Gardey-Levassort et al. (1970) and Olive et al. (1971) were not able to detect any increase in the rate of accumulation of 5-HT in the whole hypothalamus or brain stem of a febrile rabbit given a bacterial pyrogen earlier. However, if the rabbit was pretreated with pargyline, an endotoxin challenge did cause an elevation in the content of hypothalamic and brain-stem 5-HIAA during fever.

A microinjection of *p*-chlorophenylalanine (pCPA), into the rabbit's anterior hypothalamus potentiates the febrile response to leukocyte pyrogen given systemically (Teddy, 1969); contrary to this, α-methyl-metatyrosine (aMMT) given by the same route attenuates the rabbit's fever. This finding represents a replication, within the CNS, of the study of Giarman et al. (1968), who administered pCPA and aMMT to the rabbit peripherally. Apparently, an exceedingly complex action results from the treatment with a drug that can cause a depletion, a rebound elevation, or perhaps a paradoxical change in the activity of 5-HT, norepinephrine, or dopamine within the brain (Myers and Waller, 1977). The question of the validity of the systemic approach to depleting monoamines in the CNS is borne out by the fact that the 30-40% of the amine remaining in the hypothalamis is probably sufficient to mediate thermogenesis (Carruba and Bächtold, 1976).

With respect to the actual function of an amine in the hypothalamus per se, Mašek et al. (1973) have further elucidated the phenomenon of the localized turnover of 5-HT and norepinephrine. After a fever is induced by streptococcal mucopeptide, the turnover rate of 5-HT within the hypothalamus is significantly enhanced whereas that of norepinephrine is unchanged. In relation to this, 5-hydroxytryptophan (5-HTP) injected intravenously likewise enhances the pyrexic response to the streptococcal organism (Mašek et al., 1968). Over-all, it appears that hypothalamic 5-HT in the rat is used in the genesis of a fever produced by a bacterial organism. As in Section II.C, the presynaptic release of 5-HT seemingly represents a part of the sequential mechanism underlying the neuronal stimulation of heat production.

Corresponding to the change in 5-HT is the alteration in the catecholamine activity within the hypothalamus of the febrile rabbit. Lechat and Gardey-Levassort (1975) report that norepinephrine accumulates in the hypothalamus during the course of a bacterial fever, whereas the level of dopamine remains unchanged. On the other hand, the concentration of homovanillic acid and 3,4-dihydroxyphenylacetic acid are suppressed in the hypothalamus. Concerning 5-HT metabolism, only 5-HIAA is increased in the CSF of the febrile rabbit given either *E. coli* or gonococcus organisms (Gardey-Levassort et al., 1977). Thus, a reciprocal shift in the endogenous activity of 5-HT and norepinephrine during the course of a fever may occur in the hypothalamus of both the rat and rabbit. As the activity of the indoleamine increases, that of the catecholamine apparently decreases possibly in relation to their respective release, synthesis, or turnover. This result would fit the general schema of the monoamine theory as it relates to pyrogen fever (Feldberg and Myers, 1964).

It is of interest that the in vivo release of 5-HT is markedly elevated from locally perfused sites within the anterior hypothalamus of both the unanesthetized monkey and cat as a pyrogen fever develops (Myers, 1971c; Myers, 1977). The enhancement in the serotonergic activity of the rostral hypothalamus, as portrayed in Figure 26, correlates directly with the rising phase in temperature during the febrile reaction. These observations on the intact, thermoregulating animal provide additional physiological evidence for the view that a subtle perturbation in the release of a given monoamine may well comprise a part of a final neurohumoral pathway that is so profoundly affected by a lipopolysaccharide or other pyrogenic molecule.

C. Postulated Role of Prostaglandins

The prostaglandins are a family of biologically active derivatives of fatty acids which are synthesized within a large number of different animal tissues (Horton, 1969). They are extremely potent, anatomically ubiquitous, with a major source of synthesis within the seminal vesicles (Bergström et al., 1968).

In the last few years, the speculation has arisen that a prostaglandin of the E series (PGE) may be an intermediary substance in the hypothalamic genesis of a fever of bacterial origin. In 1966, an unidentified contractile substance having prostaglandin-like activity was detected in the cerebrospinal fluid of the anesthetized cat (Feldberg and Myers, 1966). A few years later, Milton and Wendlandt (1970) found that PGE_1 injected in only nanomolar doses into the third cerebral ventricle of the cat evokes an intense fever. Later that year, it was reported that effluents collected by means of push-pull perfusion cannulae from 17 of 24 hypothalamic sites examined in three monkeys contain a prostaglandin-like substance, whose activity increased either during the rising phase or after the fever has reached a plateau (Myers, 1971c). Then, Flower and Vane (1972) announced that the antipyretic, acetaminophen, inhibits PG formation. The idea gradually evolved that an aspirin-like drug possesses an antipyretic effect because of a reduction in PGE activity (see reviews of Feldberg, 1975; Milton, 1976).

1. Hypothalamic Action of PGE

The site of PGE's powerful action has been examined in a wide variety of species. As portrayed in Figure 27, the microinjection into the hypothalamus of 100 ng or less of PGE_1 or PGE_2 generally produces an intense rise in body temperature in the cat (Feldberg

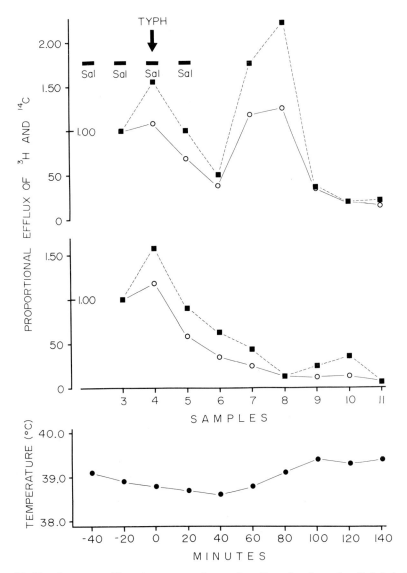

Figure 26 Simultaneous efflux, into successive push-pull perfusates, of radiolabeled 5-HT (■---■) and PGE_2 (○——○) from the anterior (top) and anterodorsal (middle) hypothalamus of the unrestrained cat. Sodium salicylate (SAL) was added (5 μg/min) to four consecutive perfusates as indicated by the black bars. *Salmonella typhosa* (TYPH) was added to the third perfusate (1:10 dilution) at zero hour. Colonic temperature (bottom) in degrees centigrade. (R. D. Myers, unpublished observation.)

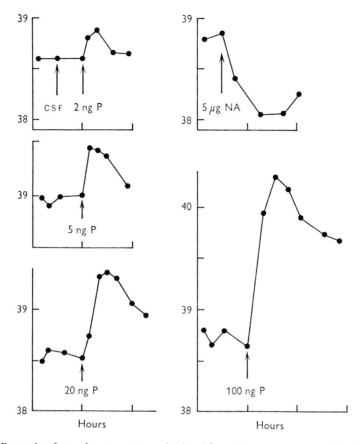

Figure 27 Records of rectal temperature obtained from two unanesthetized cats, each with an indwelling microinjection cannula in the left anterior hypothalamus (site of injection: 1.5 mm lateral to midline; 13.5 mm anterior to and 6.5 mm above interaural zero line of Snider and Niemer, 1961). Left records and upper right record obtained from same cat on different days. Each arrow indicates a microinjection of 1 μl fluid of artificial CSF or of artificial CSF containing 2, 5, 20, or 100 ng PGE_1 (P) or 5 μg noradrenaline (NA). (From Feldberg and Saxena, 1971.)

and Saxena, 1971), as well as rat (Lipton et al., 1973; Spławiński et al., 1978), rabbit (Veale and Cooper, 1973; Stitt, 1973), monkey (Waller and Myers, 1973; Crawshaw and Stitt, 1975), and chicken (Nistico and Marley, 1973, 1976; Pittman et al., 1976a). Anatomical mapping studies done in several species including the rat, rabbit, and monkey (Veale and Cooper, 1975; Stitt, 1973; Simpson et al., 1977; Williams et al., 1977) have revealed that the region of maximum sensitivity to PGE is ventral to the anterior commissure and dorsal to the optic chiasm. Morphologically, this area is identical to that within which 5-HT, norepinephrine and a bacterial pyrogen all exert their respective neuropharmacological actions. Further, it would appear that, in some species, PGE and pyrogen-sensitive sites are anatomically concordant (Veale and Cooper, 1975; Cooper et al., 1976b). Since the latency of the febrile response after an injection into the anterior hypothalamus of PGE_1 is usually somewhat shorter than that of a leukocyte pyrogen (Stitt, 1973), a pyrogen reaching this region could cause a local release of prostaglandin within the nerve tissue. The possibility that endoperoxide precursors of PGE could be responsible for the hyperthermia is at the moment ruled out. The reason is that stable analogues of the cyclic endoperoxides injected into PG-sensitive sites in the hypothalamus are without effect on the body temperature of the rat (Hawkins and Lipton, 1977). Arachidonic acid, thromboxane A_2, and prostaglandin ($F1\alpha$) also seem to be eliminated as possible mediators of a pyrogen fever, since there is little correlation with their hypothalamic site of action and that of endotoxin or PGE_2 (Spławiński et al., 1978).

Perhaps the most significant boost given to the "pyrogen-prostaglandin-fever" theory was an experiment of Feldberg and Gupta (1973); they found that, as measured by bioassay, PGE-like activity in the CSF obtained from the ventricle of the febrile cat is much higher than when the animal is normothermic. Representative patterns of PGE output are presented in Figure 28. Further, an antipyretic drug such as acetaminophren not only abolishes the fever of the unanesthetized cat but also reduces the PGE-like activity of CSF (Fig. 28) collected in the same way (Feldberg et al., 1973). A thin-layer chromatographic analysis and other assays revealed that the biologically active fraction contained in the CSF is probably PGE_2.

Phillip-Dormston and his colleagues have demonstrated independently that the level of PGE is relatively high in a rabbit's CSF after a fever is produced either by an endotoxin or leukocyte pyrogen (Phillip-Dormston and Siegert, 1974). Actually, both cyclic AMP and PGE levels are double their normal value in the CSF samples collected from the febrile rabbit (Phillip-Dormston, 1976). Veale and Cooper (1974b) found that radioactive prostaglandin delivered to the anterior hypothalamic, preoptic area escapes readily into the cerebrospinal fluid of the rabbit. Thus, this route of egress may be a physiologically significant mode of inactivation of a prostaglandin that is present in the hypothalamus (Veale and Cooper, 1974a). In relation to this, if the degradation of intrahypothalamic PGE_2 is disrupted by systemically given polyphoretin phosphate, an inhibitor of 15-hydroprostaglandin dehydrogenase, a PGE or endotoxin fever is prolonged significantly (Spławiński et al., 1979).

In contrast to PGE's possible role in the development of the fever, little evidence has been forthcoming that a prostaglandin is involved in the thermoregulatory mechanism per se. Hales and his colleagues have infused several prostaglandins into the ventricle of the sheep at different ambient temperatures; whether the animal is kept at a high, low, or thermoneutral temperature, the magnitude of the PGE_1 hyperthermia is the same (Hales et al., 1973). Essentially, this same result is seen in the rabbit (Hori and Harada, 1974). Local cooling of the anterior hypothalamus can enhance a PGE hyperthermia, whereas warming of this structure tends to attenuate the rise (Stitt, 1973; Hori and Harada, 1974).

If PGE_1 is infused directly into the rabbit's rostral hypothalamus, the thermosensitivity of this region is retained. Again, at three levels of ambient temperature, PGE_1 increases metabolic rate, as expressed as an upward displacement of the hypothalamic temperature threshold for stimulating metabolism (Stitt et al., 1974). At an ambient temperature of 32°C, an injection of PGE_1 into the anterior hypothalamus of the squirrel monkey produces a fever as a result of an increased vasoconstriction and, to a lesser degree, a change in basal metabolism (Crawshaw and Stitt, 1975). In the rhesus monkey, the degree of sweating and metabolic rate during a PGE_1 induced fever is, however, influenced

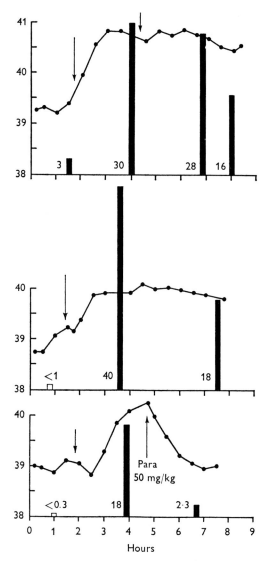

Figure 28 Records of rectal temperature from three unanesthetized cats. The height of the columns and the values beside them show PGE_1-like activity in nanograms per milliliter of cisternal CSF; the position of the columns refers to the time but not the duration of the CSF collection. All arrows, except the second one in the bottom record, indicate injections of 150 ng pyrogen into the cisterna magna. The second arrow in the bottom record indicates an intraperitoneal injection of paracetamol 50 mg/kg (Para). (From Feldberg et al., 1973.)

by the ambient temperature at which the primate is held (Barney and Elizondo, 1978). Veale and Whishaw (1976) also showed that PGE_1 infused directly into the anterior hypothalamus of the unanesthetized rat raises its temperature for the same duration whether the animal is kept at an ambient temperature of 5 or 35°C.

Unlike the correlation between prostaglandin-like activity in the cat's CSF during a bacterial fever, there is no evidence of increased PGE synthesis as reflected in CSF activity of the cat placed in the cold (Cranston et al., 1975a). This corresponds with other findings, which indicate that PGE is not involved in the anterior hypothalamic mechanism for thermoregulation (Veale and Whishaw, 1976). However, Jell and Sweatman (1977) do find a high proportion of single neurons in the anesthetized cat that are both thermoresponsive and sensitive to locally applied PGE_1 and PGE_2 but not $PGE_{2\alpha}$. Such a high incidence of concordant responses raises the possibility of a regulatory involvement of PGE.

A 10-ng dose of PGE_1 injected into the hypothalamus of a rhesus monkey alters its thermoregulatory behavior. If the animal is given the opportunity to obtain warmth from a heat lamp by pulling a switch, its rate of responding to obtain warmth increases sharply as the PGE fever develops (Myers, 1974b; Waller, 1975). The maximal rate of responding for heat reward correlates closely with the rate of the temperature rise. As shown in Figure 29, this is sharply contrasted to the relatively low rate of lever-responding to obtain heat by a monkey recovering from calcium-induced hypothermia. This result indicates that a PGE_1 hyperthermia affects behavioral regulation in terms of responses to conserve body heat, enhance exposure to warmth, minimize metabolic demand, and reduce the discomfort associated with shivering. The behavioral responding to obtain heat has been confirmed in the squirrel monkey also given intracerebral PGE (Crawshaw and Stitt, 1975).

2. PGE and Single-Unit Activity

The iontophoresis of a PGE onto neurons in the hypothalamus exerts an undifferentiated excitatory effect on all cells independent of the area of application (Poulain and Carette, 1974). Surprisingly, the period of activation outlasts the duration of iontophoresis, with tachyphylaxis often occurring. Jell (1975) has reported that of nine thermoresponsive neurons in the anterior hypothalamus whose firing rate changes after the iontophoresis of PGE_1 or PGE_2, five of the units were cold-sensitive and four were warm-sensitive cells. Moreover, when PGE_1 is applied by iontophoresis to single units of the rostral hypothalamus of the unanesthetized rabbit, less than 9% of the neurons tested show any response to the fatty acid (Stitt and Hardy, 1975). In addition, none of these neurons responsive to PGE_1 could be selectively designated as being thermally sensitive or insensitive. Thus, the proportion of units affected by PGE_1 is approximately equal for warm-sensitive, cold-sensitive, or thermally insensitive neurons.

A prostaglandin synthetase inhibitor such as fenoprofen applied by iontophoresis to the rostral hypothalamus generally causes a depression of the depolarization rate of single neurons (Jell and Sweatman, 1976). The response of these cells to either norepinephrine or 5-HT applied by iontophoresis ordinarily is unaffected by prior application of the synthetase inhibitor. Thus, the neuron's response to a microiontophoretically applied monoamine may not depend upon a prostaglandin.

When PGE_1 is injected by infusion into the rostral hypothalamus of the cat, at a point remote from the recording electrode, the identified "thermopositive" units (warm-sensitive) are depressed by PGE_1, whereas the "thermonegative" units (cold-sensitive) are

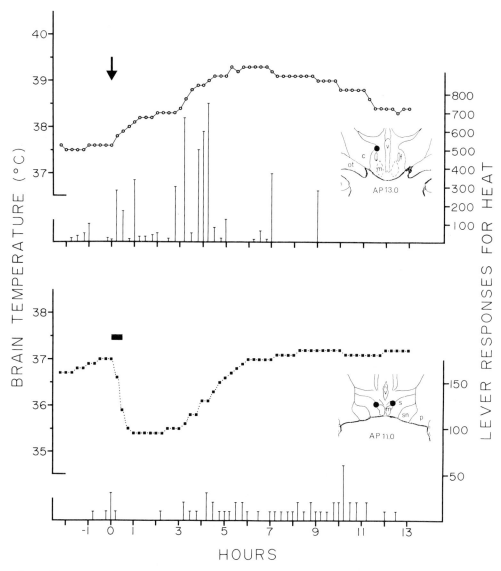

Figure 29 Lever responses (right axis) for a 4-sec burst of a heat lamp, portrayed by the vertical bars, emitted by a *Macaca nesmestrina* monkey seated in a restraining chair. Top: At arrow, 10 ng prostaglandin E_1 was injected in a 1-μl volume at the site indicated by the dot in the inset. Bottom: At bar, bilateral 30-min push-pull perfusion at 50 μl/min of the sites indicated in the inset of an artificial CSF containing Ca^{2+} ions 10.4 mM in excess of the normal value. c, Internal capsule; f, fornix; m, mamillary body; ot, optic tract; p, cerebral peduncle; s, subthalamic nucleus; sn, substantia nigra; v, third ventricle. (From Myers, 1974b.)

excited (Schoener and Wang, 1976). A higher dose of PGE_1 causes a biphasic alteration in the discharge pattern since the units usually excited are first depressed, whereas the initially depressed units are excited. Schoener and Wang (1976) have suggested that PGE, because of its powerful vasodilator action and the subsequent enhanced loss of tissue heat, may ultimately depress warm thermoreceptors and excite cold receptors. According to them, "it is possible that the neuronal effects of prostaglandin are partially due to vasomotor actions of the substance" (Schoener and Wang, 1976).

3. Does Prostaglandin Mediate a Pyrogen Fever?

The main evidence to support the idea that a prostaglandin functions in the hypothalamus to mediate a fever caused by bacteria has been reviewed comprehensively by Veale and Cooper (1974), Feldberg (1975), and Milton (1976). It is twofold: (1) hyperthermia is evoked in most species when PGE is applied to the anterior hypothalamus; (2) PGE-like activity increases in CSF during an endotoxin fever and is reduced following an antipyretic. Several experimental observations now indicate that a functional disassociation may well exist between the pharmacological effect of PGE on the hypothalamus in raising body temperature and the intrinsic biological activity of a bacterial pyrogen in producing fever. They are enumerated as follows:

First, a prostaglandin injected into the cerebral ventricle of the lamb fails to evoke a fever; however, in the same animal, a bacterial pyrogen given by the same route produces the typical febrile response (Pittman et al., 1975). Injected directly into the rostral hypothalamus of the newborn lamb, PGE_1 and PGE_2 fail to cause fever even though the same hypothalamic loci are highly reactive to bacterial pyrogen, which does evoke pyrexia (Pittman et al., 1977a,b).

Second, a potent inhibitor of prostaglandin synthesis in brain media, indomethacin, does not inhibit a fever produced by an intraventricular injection of PGE_1 in the cat. Nevertheless, indomethacin either attenuates or abolishes the febrile response to a leukocyte pyrogen given similarly (Clark and Cumby, 1975). A fever produced by bacterial pyrogen injected into the chicken's hypothalamus is unaffected by indomethacin pretreatment (Artunkal et al., 1977). Further, PGE infused in the bird kept below thermoneutrality causes a decline in body temperature, whereas bacteria have no such effect. In the chicken, an antipyretic abolishes or prevents a fever elicited by intrahypothalamic endotoxin, but fails to affect a PGE hyperthermia (Nistico and Rotiroti, 1978). When administered systemically to the rabbit, a pharmacological antagonist of prostaglandin, SC-19220, does not prevent a fever induced by a pyrogen (Sanner, 1974).

Third, after the anterior hypothalamic, preoptic area of the rabbit is subjected to a massive lesion, leukocyte pyrogen injected into the ventricle still causes its characteristic fever. However, a prostaglandin applied by the same intraventricular route has no effect on temperature.

Fourth, no clear-cut relationship has been established between the discharge pattern of a single neuron in the rostral hypothalamus, which responds to a prostaglandin delivered by microiontophoresis, and its identification as a thermosensitive unit that reacts specifically to either a cold or warm stimulus (Stitt and Hardy, 1975; Jell and Sweatman, 1976).

Fifth, when sodium salicylate is infused intravenously in the rabbit in a dose that does not prevent the development of a fever caused by intravenous leukocyte pyrogen, the concentration of prostaglandin in its CSF does not rise after all but instead remains constant (Cranston et al., 1975). Therefore, the relationship between an enhanced synthesis or release of prostaglandin in brain tissue, as measured in CSF, and the induction of a fever by a pyrogen is not a causal one.

Sixth, if the anterior hypothalamic, preoptic area is radiolabeled simultaneously with [^{3}H] 5-HT and [^{14}C] PGE$_2$, a bacterial pyrogen delivered by a push-pull perfusion within this region causes a concurrent release of both 5-HT and the prostaglandin (Myers, 1977). However, when sodium salicylate is perfused simultaneously within the anterior hypothalamus in a low dose (5.0 μg/μl) that does not prevent a fever after pyrogen perfusion, the enhancement in PGE$_2$ release is somewhat reduced in spite of the ensuing hyperthermia caused by the bacteria. At the same time, as shown in Figure 26, the efflux of 5-HT from the anterior hypothalamus nevertheless persists at a high level. Therefore, the local activity within the anterior hypothalamus of 5-HT, but not necessarily prostaglandin, correlates with the development of a bacterial fever (Myers, 1978a).

Seventh, the simultaneous intraventricular injection of leukocyte pyrogen plus the PG antagonist, SC-19220, produces a marked fever in the rabbit. On the other hand, the simmultaneous injection of the same antagonist plus the prostaglandin attenuates the fever (Cranston et al., 1976).

In summary, the role of PGE in fever presently constitutes a most "interesting enigma" (Veale et al., 1977b). Concerning the possible action of a PGE on the hypothalamus, the local injection of PGE to the anterior hypothalamic, preoptic area could alter the release of the two monoamines, 5-HT and norepinephrine. However, this possibility seems to be discounted. If PGE$_1$ is perfused locally within the rostral hypothalamus of the unanesthetized monkey, the steady-state pattern of efflux of either 5-HT or norepinephrine within that region remains unchanged (Myers and Waller, 1977). Similarly, neither 5-HT nor norepinephrine perfused within the rostral hypothalamus alters the characteristics of PGE release within the anterior hypothalamus (Myers and Waller, 1976).

Another possibility is that PGE may act by chelating or displacing calcium ions, which would thereby result in an increase in sodium ion conductance (Clark and Coldwell, 1973). Alternatively, PGE could serve as a calcium ionophore (Kirtland and Baum, 1972). If this were the case, however, PGE$_1$ would produce hyperthermia when applied to calcium reactive sites in the posterior hypothalamus, which it clearly does not do (Veale and Whishaw, 1976). PGE also could interact with cAMP at the hypothalamic level, since the concentration of both substances increases twofold during a pyrogen fever (Siegert et al., 1976). If this were true, there should be an exacting correlation between the level of cAMP in the CSF and fever; again this also has not been demonstrated (Dascombe and Milton, 1976).

Yet another possibility is that PGE attaches to 5-HT receptor sites in the anterior hypothalamus to activate the 5-HT-cholinergic heat-production pathway (Myers, 1974b; Myers et al., 1974). Although experimental evidence from peripheral pharmacological studies is not yet clear, the inhibition of 5-HT by drugs such as pCPA can reduce a PGE-induced fever (Kandasamy et al., 1975; Milton and Harvey, 1975) just as a hypothermia is caused in the human patient with carcinoid syndrome (Vaidya and Levine, 1971). Interestingly, the microinjection of methysergide, a 5-HT receptor blocker, into the anterior hypothalamus of the unanesthetized monkey, attenuates a fever produced by PGE$_1$ given later at the same site (Simpson et al., 1977). The antagonistic action of methysergide is similar to its action on 5-HT hyperthermia, when the indoleamine is microinjected in the hypothalamus of the cat (Komiskey and Rudy, 1975). In addition, it appears that PGE could antagonize the catecholaminergic neurons in the hypothalamus, which would serve to potentiate the hyperthermia (Milton and Harvey, 1975; Simpson et al., 1977). Possibly, PGE alters the release of ACh or affects cholinergic receptors in the hypothalamus (Rudy and Viswanathan, 1975).

Although PGE does not seem to interfere with the presynaptic release of either 5-HT or a catecholamine within the anterior hypothalamus (Myers and Waller, 1976), one current supposition (see above) is that PGE attaches to postsynaptic 5-HT receptor sites when it is artificially applied to the hypothalamus. What is so perplexing and difficult to explain is the increase in PGE level seen in the CSF during fever. This augmented activity of PGE apparently is not due to the neuronal release of the fatty acid, because there is no morphological basis for classifying a hypothalamic cell as a "prostaglandinergic neuron."

Perhaps molecules of PGE are sequestered in the capillary beds or on membranes of neuroglia in the hypothalamus or other region having direct access to the ventricular system. Then, during fever, they are unbound from the anterior hypothalamus (Myers, 1977). Along with the change in activity of other substances such as cAMP and amino acids, the localized mobilization of PGE could reflect a more generalized increase in neuronal activity or glial metabolism within the hypothalamus during fever. In any case, Veale et al. (1977b) underscored the complexity of the PGE issue with a cogent statement: "What was a few years ago thought to be the ultimate answer to the central mechanism of fever has now opened up once again the need for critical experimental re-evaluation."

D. Hypothalamic Protein Mediator of Fever

Cycloheximide, which is a strong inhibitor of the in vivo synthesis of new protein, has the capacity to interfere with the bacterial induction of a fever (Siegert et al., 1976). Thus, a protein or enzyme factor of an unknown chemical nature could be a cellular mediator of the febrile response resulting from an endotoxin challenge. Because of the generalized toxicity of cycloheximide, the possibility of nonspecific interference with thermogenesis has tempered the interpretation of the initial observation.

In the unrestrained cat, experiments have been undertaken with an alternative antibiotic, anisomycin, which possesses three advantageous properties: low cytotoxicity, short-lived action, and reversibility of its effect on protein synthesis inhibition (Myers and Ruwe, 1980). Given systemically, anisomycin prevents or attenuates the endotoxin fever produced by *Salmonella* microinjected into the anterior hypothalamic, preoptic area (Myers and Ruwe, 1980). This finding supports the contention that synthesis of new protein material may be required centrally for the thermogenic response caused by bacteria.

Anatomical evidence has also been collected which favors this conclusion and pinpoints the site of the protein factor's action. The microinjection of a 10- to 25-μg dose of anisomycin directly into the cat's rostral hypothalamus up to an hour before a similar microinjection of typhoid bacteria into the same locus, blocks or delays the pyrogen fever (Ruwe and Myers, 1980). Of special note is the further observation that the antiobiotic does not impair the hypothalamic mechanism for thermogenesis nonspecificially. That is, 5-HT or PGE_2 infused into the anterior hypothalamus after the anisomycin pretreatment produces its typical hyperthermic response with an identical latency and magnitude. Thus, the prevention of a newly synthesized protein factor locally within the hypothalamus does interfere with the development of a fever. This suggests that an as-yet-unknown intermediary substance, involved in the pathogenesis of a fever, constitutes a vital neurohumoral link between the bacterial insult and the efferent pathway of neurons in the diencephalon which activate heat production (Ruwe and Myers, 1980).

VIII. Hypothalamic Effect of Drugs That Alter Body Temperature

Many drugs, as well as numerous compounds not used as pharmacological agents, exert a profound effect on body temperature when administered systemically. Given in a sufficiently high dose, virtually any compound may exert a toxic or sedative side effect (Lessin and Parkes, 1957) and thus possess the capacity to induce a fall in temperature. Such a nonspecific action has been viewed epigrammatically as being tantamount to a "blow on the head" (Myers and Waller, 1977). The loss in body heat may be due to a poikilothermic rather than to a hypothermic response per se, involving active heat-dissipating functions.

Alternatively, a substance which acts primarily to increase an animal's activity, arouse its autonomic nervous system, or alter the distribution of cutaneous blood flow might well produce a rise in temperature (Cremer and Bligh, 1969; Cox and Lomax, 1977). But again, the thermoregulatory specificity of such a hyperthermia, in terms of the maintenance of body temperature in the cold or heat, would be highly suspect (Myers and Waller, 1977).

Even though a given substance does exert a direct action on the temperature-controlling mechanism of the anterior hypothalamus, the same compound may also exert quite a different effect on certain organ systems in the periphery. For example, epinephrine circulating systemically is a potent thermogenic hormone but when acting on the hypothalamus this monoamine typically inhibits heat production. Whether the effect of a substance is predominantly on a central or peripheral set of "receptors" actually determines the nature of the response, i.e., hyper- or hypothermia. In this section, drugs that can affect elements in the thermosensitive region of an animal's hypothalamus will be examined.

A. The Antipyretics

In 1944, Guerra and Brobeck discovered that a lesion placed in the anterior hypothalamus of the monkey interferes with the antipyretic action of aspirin. Following from this classic work, the suspicion has grown that an antipyretic works through an unknown but direct effect on the central nervous system. Of course, aspirin and other antipyretics do exhibit a wide spectrum of effects on peripheral organ systems (Collier, 1969) including antiinflammatory and analgesic properties (Ziel and Krupp, 1976). One clever suggestion is that a salicylate reduces a pyrogen-induced fever because it prevents the pyrogen which circulates in the blood stream from entering the central nervous system (Cooper et al., 1968). Although a salicylate given in a therapeutic but nontoxic dose exerts virtually no effect on the normal body temperature of an animal or human patient (Barbour, 1921), one can lower the temperature of an afebrile animal by giving a very large, almost toxic dose of sodium salicylate (e.g., Chai and Lin, 1975) particularly if the treated subject is kept in a cold environment (Francesconi and Mager, 1974).

Of considerable bearing on the mechanism of sodium salicylate's action is the fact that it sequesters differentially in the central nervous system of the febrile or afebrile rabbit (Rawlins et al., 1973). In blood samples collected from anterior or posterior hypothalamic tissue of a rabbit undergoing a fever produced by leukocyte pyrogen, the content of salicylate is nearly double the normal value of that in the afebrile animal. When a fever is generated by bacterial endotoxin injected directly into the cerebral ventricle of the cat, the peripheral administration of a salicylate or acetaminophen reduces the animal's temperature significantly (Milton and Wendlandt, 1968; Clark, 1970). The same holds true for a fever induced by leukocyte pyrogen. In a cat pretreated systemically with salicylate before an injection of leukocyte pyrogen into the lateral cerebral ventricle, the onset of the fever is substantially delayed and the peak of the consequent hyperthermia is materially reduced (Schoener and Wang, 1974). Salicylate injected into the cerebral ventricle of the rabbit or cat diminishes a leukocyte pyrogen fever (Cranston et al., 1970a; Clark and Alderdice, 1972), possibly because the antipyretic competes with leukocytes for receptor sites in the brain (Clark and Coldwell, 1972).

To test the hypothesis that salicylate reduces a fever by preventing circulating pyrogen from entering into the CNS, several bacterial pyrogens have been microinjected directly into the anterior hypothalamus of the monkey. When sodium salicylate is given intragastrically during the plateau phase of the pyrexic response, the hypothalamically induced fever is reduced or abolished by the antipyretic (Myers et al., 1971c). This suggests that salicylate antipyresis is mediated by a direct antagonism of the effect of a pyrogen on cellular elements within the anterior hypothalamus (Myers et al., 1974). As expected, salicylate has no effect on the body temperature of the afebrile monkey. In the chicken, salicylate or indomethacin given systemically or intraventricularly also abolishes fever elicited by *Shigella* infused intrahypothalamically (Nistico and Rotiroti, 1978).

1. Effect of an Intrahypothalamic Antipyretic

Cortisone is often prescribed clinically in the case of an inflammatory response in which a high fever is a cardinal symptom. The first clue that a substance could exert an antipyretic effect when applied directly to the hypothalamus was provided by an experiment with centrally applied cortisol (Chowers et al., 1968). Just before a rabbit was given *Pseudomonas* polysaccharide complex either intravenously or intracerebrally, hydrocortisone was microinjected into the animal's preoptic area. In contrast to a control injection, this steroid partially attenuated the typical rise in body temperature produced by the pyrogen. Although an intraventricular injection of a small dose of salicylate given 4 hr after the start of an intravenous infusion of leukocyte pyrogen causes a dose-dependent defervescence, the antipyretic acting by this route (Cranston et al., 1970b) has no effect on the temperature of the afebrile rabbit.

When a dose of 6 to 30 μg of salicylate is microinjected directly into the hypothalamus or other area of the brain, a distinct antipyretic response can be observed. After the induction of a fever by an intravenous infusion of leukocyte pyrogen, salicylate reduces a rabbit's body temperature after it is microinjected into the preoptic area or the mesencephalon (Cranston and Rawlins, 1972). The sites at which the drug acted to lower a febrile temperature are depicted in Figure 30. When salicylate is infused very slowly into the same preoptic area, a dose-dependent defervescence is observed (Vaughn et al., 1979). In the latter report, the sites of salicylate sensitivity correlated well with those that showed sensitivity to PGE_1.

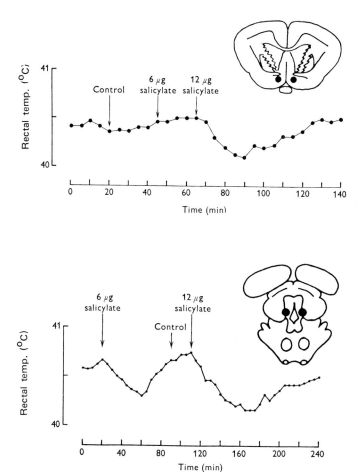

Figure 30 The effect of bilateral (10-μl) microinjections of artificial cerebrospinal fluid alone and containing 6 μg and 12 μg sodium salicylate on rectal temperature in rabbits with pyrogen-induced fevers. The accompanying diagrams show the injection sites which were in the preoptic hypothalamus (above) and the periaqueductal gray matter of the midbrain (below). (From Cranston and Rawlins, 1972.)

Ordinarily, the rat is not subject to pyrogen fever, but an intraperitoneal injection of *Pseudomonas* polysaccharide consistently produces a fever if a sufficiently high dose of this pyrogen is employed. Subsequently, a microinjection of 5.0 μg of salicylate into the anterior hypothalamus blocks the pyrogen-induced fever, whereas a similar injection into the posterior hypothalamic area does not (Avery and Penn, 1974). Again, salicylate does not affect the body temperature of the afebrile rat when the antipyretic is given in this manner.

Avery and Penn (1976) have further elucidated the mechanism by which sodium salicylate might affect neurons in the anterior hypothalamus. In the rat in which 5-HT applied to the anterior hypothalamus can induce hyperthermia, the intragastric administration of salicylate blocks the rise in temperature produced by the indole-amine. However, the cholinergic hyperthermia induced by carbachol, injected in the same way, is not affected by systemically administered salicylate. Thus, an antipyretic may compete for 5-HT receptor sites postsynaptically and thereby prevent the action of the normally released 5-HT, which apparently occurs within the anterior hypothalamic area during the genesis of a bacterial fever (Myers, 1971b, 1977).

Since salicylate does not normally affect thermoregulatory responses, unless given in near-lethal doses, it is surprising that the iontophoresis of salicylate to a warm-sensitive neuron in the anterior preoptic region of the nonfebrile rat generally will excite the cell (Beckman and Rozkowska-Ruttiman, 1974). Perhaps salicylate has a twofold mechanism of action: it could (1) selectively alter the membrane potential or characteristics of depolarization of a thermosensitive neuron, or (2) alter the balance in activity of transmitter substances in rostral hypothalamic nerve terminals. The first alternative is not, however, supported by the observation that salicylate, infused into the anterior preoptic area fails to alter the thermoresponsiveness of this region (Lipton and Kennedy, 1979).

2. Mechanism of Action of Antipyretics

These studies raise a major question pertaining to the mechanism of an antipyretic's action. By what neuronal processes in the anterior hypothalamus or other region does salicylate or acetaminophen work? In this section, factors including the monoamines, cyclic AMP, prostaglandins, and the ratio of sodium to calcium ions will be considered.

When salicylate is given in a dose above the therapeutic range, tryptophan and 5-HIAA levels in the whole brain of the rat are elevated; under this circumstance, however, 5-HT content remains unchanged (Guerinot et al., 1974). This observation could indicate that salicylate alters the synthesis or turnover of cerebral 5-HT. Alternatively, the effect may be nonspecific since other areas of the brain including the hippocampus exhibit the same sort of change in indoleamine activity. That a catecholamine within the anterior hypothalamus may also mediate the antipyretic effect of salicylate or acetaminophen is also a reasonable assumption at present. For example, norepinephrine microinjected into a prostaglandin-sensitive site in the anterior hypothalamus is a far more potent antagonist of a PGE fever than an antipyretic itself (Waller and Myers, 1974). Further, the anterior hypothalamic site at which a microinjection of PGE evokes hyperthermia is clearly concordant with the most sensitive locus at which norepinephrine induces its characteristic hypothermia (Veale and Whishaw, 1976).

Laburn et al. (1975) have shown that a series of intrahypothalamic injections of 6-hydroxydopamine, 150 μg given on three to five successive days, will attenuate the febrile response to an intravenous injection of pyrogen. However, the high dose of 6-OHDA used could conceivably destroy all of the neurons, including serotonergic and noradrenergic cells (Evans et al., 1975) which are postulated to be involved in the pyrexic response. In agreement with this, prostaglandin injected intrahypothalamically is relatively ineffective in inducing hyperthermia after such a 6-OHDA lesion. In a succeeding experiment, Laburn et al. (1975) found that an intrahypothalamic injection of 50 μg of phenoxybenzamine also reduces a pyrogen- or prostaglandin-induced fever. Again however, the α-adrenergic antagonist given in this does may act nonspecifically as a local anesthetic.

In relation to the premise upon which these experiments with the rabbit are based, Preston and Cooper (1976) documented the fact that a dose of 2.0 or 8.0 μg of norepinephrine injected in a small volume into the anterior hypothalamic area of the rabbit fails to elevate the animal's body temperature. On the other hand, PGE_1 injected at the same site causes an intense fever when a dose of only 75 ng is used. Therefore, it is most unlikely that, within the anterior hypothalamus of the rabbit, norepinephrine serves as a mediator of normal thermogenesis or of a pathologically related increase in the lagomorph's body temperature (Preston and Cooper, 1976). Parenthetically, a demonstration of catecholamine release evoked within the anterior hypothalamus of the rabbit exposed to high and low environmental temperatures is now clearly necessary.

Cyclic AMP represents another candidate as a neurochemical link between an antipyretic drug and the neuronal mechanism responsible for the alleviation of a fever. Clark and his colleagues have found that salicylate, acetaminophen, or indomethacin all serve to inhibit a fever induced in the cat by an injection of cholera enterotoxin (Clark et al., 1974). Moreover, these antipyretics block the hyperthermia induced by the dibutyrl form of cAMP injected into the cat's cerebral ventricle. A corresponding view concerning the possible function of cyclic AMP is supported by the work of Rosendorff and his colleagues. Using the rabbit, they found that an intrahypothalamic injection of cyclic AMP produces a fever of rapid latency (Woolf et al., 1975). Theophylline, a nucleotide phosphodiesterase inhibitor, potentiates the fever which is elicited either by an intrahypothalamic microinjection of a prostaglandin or by bacteria administered intravenously (Woolf et al., 1975). However, sodium salicylate given systemically fails to ameliorate the cyclic AMP-induced hyperthermia. The similar failure of pretreating the rabbit with the antipyretic so as to alter the induction of a central prostaglandin fever is consistent with this finding (Willies et al., 1976a).

The somewhat puzzling observation that the systemic administration of an antipyretic fails to prevent the development of a prostaglandin hyperthermia but yet abolishes a fever evoked by leukocyte or bacterial pyrogen has been noted as well in the cat and chicken (Milton and Wendlandt, 1971; Schoener and Wang, 1974; Nistico, 1975). One explanation for this is that, although the salicylate may inhibit the local hypothalamic synthesis of prostaglandin after an injection of leukocyte pyrogen, it might have virtually no effect on the prostaglandin which is already present in this structure. Nevertheless, once a prostaglandin fever has developed, aspirin or acetaminophen is a potent antagonist of the fever, causing a fall in the animal's temperature toward the basal level (Waller and Myers, 1974). Interestingly, Clark and Cumby (1976b) have shown that an antipyretic antagonizes the hyperthermia produced by the prostaglandin precursor, sodium arachidonate. Because of this latency of antipyresis, it is possible that the drug may exert its action at a step just before PGE synthesis. Since indomethacin can attenuate the elevated temperature caused by massive mechanical damage to the rostral hypothalamus of the rat (Rudy et al., 1977), cellular or capillary or other vascular damage may release or sequester a PGE locally in the lesioned tissue.

Calcium ions may also serve as a "brake" in the genesis of a pyrogen fever by retarding the synthesis of the prostaglandin (Feldberg and Milton, 1973). In fact, excess calcium ions infused into the cerebral ventricle of the unanesthetized animal are an exceedingly potent antipyretic agent (Rosenthal, 1941) when used to combat fever produced either by a centrally given prostaglandin or by a bacterial pyrogen (Dey et al., 1974). In this connection, an antipyretic clearly shifts the ratio of Ca^{2+} to Na^+ ions within the cerebral ventricle of the cat during a bacterial fever (Myers, 1976b). As described earlier (Section VI.E), the ratio of efflux of the two cations, Na^+ and Ca^{2+}, is opposite to that seen when the fever of bacterial origin develops.

In any case, an antipyretic such as sodium salicylate does not interfere with the normal body temperature of a rabbit, even when the animal is placed in an environment of 10°C, if the amount of salicylate administered is matched to the dose that is efficacious against a fever produced by a leukocyte pyrogen (Pittman et al., 1976b). With respect to its amelioration of fever, salicylate may indeed be harmful to an animal in which pathogenic bacteria are introduced. In fact, the mortality rate may rise following bacterial challenge if a fever is prevented by the drug (Bernheim and Kluger, 1976).

In summary, an antipyretic could act within the diencephalon, particularly the anterior hypothalamic, preoptic area, in one or more of the following ways: (1) decrease the regional synthesis of a prostaglandin; (2) alter the synaptic influence of cyclic AMP; (3) augment the release of norepinephrine presynaptically; (4) retard the presynaptic release of 5-HT; (5) compete with or block acetylcholine receptors postsynaptically in the cholinergic pathway for heat production; or (6) reestablish the normal balance in the extracellular ratio of Na^+ to Ca^{2+} ions within the posterior hypothalamus. Presumably, a muliplicity of neuronal events occur simultaneously as an antipyretic, with all of its myriad effects, manifests its restorative property on a pathologically elevated body temperature.

B. Amphetamines, Phenothiazines, and Related Compounds

The CNS stimulant, amphetamine, and many of its analogues exert a complex set of effects on body temperature. As one might predict, any substance which increases locomotor or other activity ought to elevate an animal's body temperature by means of the muscles' generation of body heat. As reviewed by Horita and Hill (1973), a peripherally administered amphetamine ordinarily elevates body temperature. Some workers believe that this hyperthermia is caused by amphetamine's systemic action (Gessa et al., 1969), whereas others propose that amphetamine disturbs normal temperature through an effect on a central control system (Frey, 1975). In both the rat and cat, the systemic or intraventricular injection of an amphetamine may produce either a hyper- or hypothermic response (Cox and Tha, 1975). Generally speaking, the direction of the temperature change depends on the size of the dose employed (Jellinek, 1971; Terwelp et al., 1973). Recently, evidence has been presented that 5-HT in the hypothalamus may play a significant role in the amphetamine-induced rise in temperature; after one of several amphetamines is administered subcutaneously, the hypothalamic turnover of 5-HT is enhanced (Frey, 1975a, b). Further, destruction of serotinergic nerve terminals by intraventricular 5,6-DHT inhibits amphetamine and apomorphine hyperthermia (Carruba et al., 1978).

A phenothiazine or other tranquilizer exerts a multiplicity of numerous effects on body temperature in much the same way as amphetamine. Some of these could be on the hypothalamic control system. Initially, Lotti (1965) reported that the core temperature of the rat rises rather than falls after a direct injection of chlorpromazine into its rostral hypothalamus. Although, in another study, chlorpromazine microinjected into the anterior hypothalamus was found to evoke a long-lasting rise in temperature, either hyper- or hypothermia occurs when the phenothiazine is injected into the posterior part of the hypothalamus or other regions of the rat's brain stem (Rewerski and Jori, 1968). In addition, pretreatment with α-methyltyrosine, which depletes the central stores of catecholamines, does not prevent the chlorpromazine-induced rise in temperature (Rewerski and Gumulka,

1969). This coincides with the view that a catecholaminergic mechanism mediates heat loss. Recently, the phenothiazine-induced rise has been verified but not in a dose-response manner (Trzcinka et al., 1978).

In confirming the central hyperthermic effect of chlorpromazine, Kirkpatrick and Lomax (1971) also showed that its quaternary analogue, n-methylchlorpromazine, causes an identical sort of hyperthermia. The rostral part of the hypothalamus again is the principal site of sensitivity of the analogue. One exception to this series of findings on chlorpromazine has been published by Reigle and Wolf (1971). They found that chlorpromazine injected into the anterior hypothalamus of the hamster evokes a dose-dependent hypothermia. Two differences, however, may explain the discrepancy in these results. First, the latter experiment was conducted on the hamster maintained in an ambient temperature of 10°C. Second, the doses of chlorpromazine employed were much larger than those of the other investigators. This raises the possibility that a nonspecific poikilothermia could be caused by chlorpromazine since it may induce a local depolarizing blockade of neurons in the rostral hypothalamus (Myers, 1974a).

Temperature changes are also produced by the antidepressant, desmethylimipramine (DMI) which prevents the reuptake of a catecholamine into its respective nerve terminal. This drug elicits a moderate increase in body temperature when it is microinjected directly into the anterior hypothalamus of the lightly anesthetized rat (Rewerski and Jori, 1968). Since the rise is somewhat independent of the dose of DMI and is almost as intense when DMI is injected into either the posterior hypothalamus or nuclei of the thalamus, it is presently difficult to ascribe a given specificity to DMI's action on a monoamine system within the hypothalamus.

Sympathomimetic compounds whose properties have been attributed to their affinity for both peripheral and central α-adrenergic receptors also affect body temperature. When microinjected into the rostral hypothalamus of the restrained rat, two imidazoline derivatives, naphazoline and tetrahydrozoline, evoke a hyperthermia which can be blocked by a prior intracerebral injection of the α-adrenergic antagonist, tolazoline (Lomax and Foster, 1969). On the other hand, another imidazoline derivative, clonidine, infused into the hypothalamus of the chicken, causes a fall in the bird's temperature as well as a flushing of its comb (Marley and Nistico, 1975). These responses are illustrated in Figure 31. Since clonidine also produces tachypnea and other significant autonomic signs before a change in body temperature, the results of this and other studies should be interpreted very cautiously with respect to their specificity to thermoregulation per se.

C. Anesthetics and Toxic Substances

A compound that possesses an anesthetic action or produces a depolarizing blockade of nerve cells, would logically be expected to produce a poikilothermia, with a subsequent decline in body temperature, if applied directly to the thermoregulatory control region. Just as metrazol or other convulsant compound causes a decline in temperature of the cat when administered systemically (Beleslin and Myers, 1973), substances with similar properties have the same effect when given centrally. As early as 1941, it was shown that picrotoxin or aconitine given in minute doses in the hypothalamus of the unanesthetized rabbit produces a sharp decline in the animal's body temperature (Rosenthal, 1941).

A general anesthetic also can exert a potent effect on cells of the hypothalamus. For example, chloralose injected in doses of 7-32 μg into the anterior hypothalamus of the

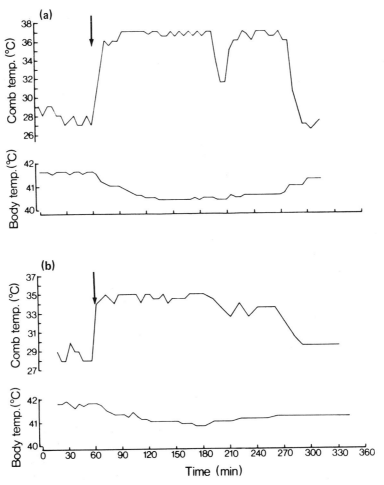

Figure 31 Records of comb and body temperatures of two unanesthetized fowls (a and b) to illustrate effects of 0.1 μM clonidine infused into the third cerebral ventricle (a) and hypothalamus (b), with ambient temperature set at 22°C in both experiments. Both infusions promptly elevated comb temperature but lowered body temperature slowly. Comb and body temperatures recovered *pari passu*. (From Marley and Nistico, 1975.)

lightly anesthetized cat blocks shivering, lowers the animal's respiratory rate, and induces a dose-dependent fall in body temperature (Feldberg and Myers, 1965b). Because of the localized nature of the response, the neuronal processes subserving shivering and respiration are certainly influenced by such an anterior hypothalamic injection of the anesthetic (Feldberg and Myers, 1965b). In the restrained rat, hypothermia occurs also after sodium pentobarbital is injected in a dose of 45 μg into the hypothalamus, but the barbiturate has very little effect at a lower dose (Lomax, 1966). In this experiment, a sodium chloride solution injected into the same site in the same microliter volume often elicited a similar sort of temperature fall. In the unrestrained rat, an unspecified dose of sodium pentobarbital injected into the hypothalamus evokes varied changes in temperature depending primarily on the locus of application (Humphreys et al., 1976).

Capsaicin, the pungent ingredient of red pepper, produces a prompt fall in the body temperature of the rat when it is injected into the anterior hypothalamus (Jancso-Gabor et al., 1970). Successive injections of capsaicin seem to desensitize the anterior hypothalamus, since the rat loses its ability to thermoregulate against an environmental thermal stress. Thus, the pepper product produces poikilothermia. An electron microscopic analysis of sections taken through the capsaicin-treated area of the hypothalamus reveals severe pathology to the mitochondria of the smaller neurons (Szolcsányi et al., 1971). This coincides with the proposition by Boulant and Bignall (1973a) that a thermosensitive neuron, based on its electrophysiological characteristics, should be smaller than the larger, temperature-insensitive cell.

The cardiac glycosides, acetylstrophanthidin and deslanoside, injected into the hypothalamus of the rhesus monkey, induce a marked hypothermia (Chai et al., 1971). Again, the most sensitive locus of action is the preoptic area. Because of a rebound rise in temperature that occurs in the monkey, the necrotizing, possibly poikilothermic, action of the glycoside is apparently a transient one. In related studies, 2-deoxy-D-glucose (2-DG), infused into the lateral ventricle of the rat, causes a hypothermia that is potentiated by intraventricular 5,6-DHT but antagonized by amphetamine injected peripherally (Müller et al., 1974). The fall in body temperature induced by the major constituent of marijuana, Δ-9-tetrahydrocannabinol, is also hypothesized to have its locus of action in the preoptic region of the rat (Schmeling and Hosko, 1976); however, it is not clear whether the drug is a poikilothermic agent or truly exerts a hypothermia action.

The powerful solvent and degreasing agent, tetrahydronaphthylamine, has been known since 1909 to alter body temperature. Given either systemically or directly into the hypothalamus, this hydrocarbon causes a rise in temperature (Sacharoff, 1909; Barbour and Wing, 1913). In a carefully done study, Bruinvels and Kemper (1971) demonstrated that an intense rise in temperature follows the direct application of tetrahydronaphthylamine to the medial preoptic area of the lightly anesthetized rat. If the animal is pretreated with a monoamine oxidase inhibitor, the tetrahydronaphthylamine hyperthermia is potentiated. Since the pretreatment of the rat with pCPA can counteract the effect of this compound, Bruinvels and Kemper suggest that the hyperthermia is mediated through a change in the activity of 5-HT in the rat's rostral hypothalamus. This idea corresponds to the finding of Peters et al. (1972) who relate the hyperpyrexia caused by the insecticide, DDT, to a change in the turnover of brain-stem 5-HT. Similarly, the inhibition of 5-HT synthesis by p-chloroamphetamine markedly enhances the typical hypothermia produced by ethanol given parenterally (Pohorecky et al., 1976). Again, the involvement of 5-HT in thermogenesis of the rat is apparent.

D. Opiate Alkaloids

Today, it is broadly recognized that an opiate alkaloid such as morphine most likely alters body temperature through an action on the central nervous system (Cox and Lomax, 1977). Generally speaking, it would not be unexpected that morphine produces a hypothermia in much the same way as any substance which interferes with the diencephalic heat-regulating mechanism (Domino, 1962). Whether or not it induces a poikilothermia in the same way as an anesthetic is not known.

As early as 1915, Hashimoto had proposed that morphine in high doses acts on the heat production center by abolishing its function (Hashimoto, 1915a,b). Then in 1965,

Lotti and others showed that the hypothermic effect of morphine is localized to the rostral hypothalamus. If morphine is injected into the posterior hypothalamus, hyperactivity and a subsequent rise in temperature are seen (Lotti et al., 1965a). Since morphine given systemically possesses a dual effect on temperature—hyper- followed by hypothermia—the local action of the alkaloid on the hypothalamus could clearly depend on the site of injection. One reason why morphine produces hypothermia is the drug's reduction of the rat's metabolic heat production; in support of this, oxygen consumption declines simultaneously with the body temperature of the animal (Lotti et al., 1966a).

When the morphine antagonist, nalorphine, is injected into the anterior hypothalamus, the morphine-induced temperature decline is antagonized (Lotti et al., 1965b). Thus, nalorphine's antagonistic action is conceivably exerted on receptors located within the chemosensitive area of the rostral hypothalamus. Quite interesting is the fact that repeated injections of morphine into the rat's anterior hypothalamus cause the same magnitude of hypothermia but only if the doses of the opiate are increased up to fourfold (Lotti et al., 1966b). Alternatively, repeated injections of n-methylmorphine into the same area cause a fall in temperature, the response of which shows no apparent evidence of developing tolerance (Foster et al., 1967). These results indicate that an acute tolerance to morphine may occur either within the hypothalamic neurons or the glial tissue surrounding the nerve cells, or both.

If the unrestrained rat is kept at an ambient temperature of 24°C, a single high dose of 50 μg of morphine injected into the rostral hypothalamus has little or no effect on the animal's body temperature according to Paolino and Bernard (1968). However, if the rat's environmental temperature is lowered to 5°C, a profound fall in body temperature occurs in response to morphine. Conversely, when the environmental temperature is held at 32°C, the same intrahypothalamic injection of morphine produces a hyperthermia. From these results, it would appear that morphine, instead of altering a hypothalamic set point for body temperature (Lomax, 1967), really may interfere with the synaptic transmission of impulses for both heat production and heat loss within the anterior hypothalamic, preoptic area. The net effect of this, of course, is a functional poikilothermia.

That a hyperthermia may be evoked by a lower dose of morphine acting within the rostral hypothalamic area has been confirmed by Cox et al. (1976a), using only 4 μg of the opiate in a 1.0-μl volume. But the issue of restraint, as it determines an animal's temperature response, continues to persist. Morphine infused into the hypothalamus of the freely moving rat kept at 22°C caused a dose-dependent rise in temperature (Martin and Morrison, 1978), which was prevented by restraint. Another study carried out with rats maintained at an undisclosed ambient temperature provided similar results except that the condition of restraint and a 500-μg dose infused into the anterior, preoptic area caused an intense decline in body temperature (Trzcinka et al., 1977). Overall, morphine could reestablish or reset the set point for body temperature at a new level. However, if the set point were truly changed, the animal would thermoregulate around the new temperature and defend this level in heat or cold. Such a demonstration has not been reported.

It has been suspected that the action of morphine on the hypothalamus could well be on 5-HT or other monoamine systems (Banerjee et al., 1968b; Medon et al., 1969; Medon and Blake, 1973). For example, a lesion of the midbrain raphe system abolishes the hyper- and hypothermia elicited by both a low and high dose of morphine given intraperitoneally (Samanin et al., 1972). In this connection, morphine-elicited hypothermia is potentiated by pretreatment with an inhibitor of serotonin uptake (Fuller and Baker, 1974), which could simulate the hypothermia following the intrahypothalamic injection of a high dose of the indoleamine in the rostral hypothalamus (Crawshaw, 1972). Recently, Warwick and Schnell (1976) found that both naloxone and morphine inhibit the reuptake of 5-HT in hypothalamic nerve endings, even though naloxone itself attenuates the hypothermic action of morphine. Thus, the hypothermic action of the alkaloid on the rostral hypothalamus is probably unrelated to the inhibition of the serotonergic reuptake mechanism.

In the morphine-dependent rat undergoing naloxone-induced withdrawal hypothermia, the dopamine receptor antagonist, pimozide, prevents the fall in temperature when it is injected into the rostral hypothalamus (Ary and Lomax, 1976). Although this result resembles that seen when pimozide is administered systemically (Cox et al., 1975, 1976b), it is possible that dopaminergic neurons in the hypothalamus are involved in the heat dissipation which typically accompanies not only opiate withdrawal but also the pharmacological responses to other drugs such as clonidine, amphetamine, and ouabain (Doggett, 1973; Horita and Quock, 1975; Reid et al., 1975).

IX. Conclusion

Without question, the multitude of neurochemical or neurohumoral events which continually take place during the course of thermoregulation is truly astounding. The physiological responses that are under the influence of a temperature control system are fascinating in themselves. To a certain degree their complexity may even appear bewildering, a situation candidly acknowledged by some reviewers (e.g., Cooper, 1972). Nonetheless, some of the views held commonly until the mid-1960s concerning the functions of the rostral and caudal areas of the hypothalamus are now no longer valid.

Based on his clinical impressions, Hans Meyer in 1913 drew together much of the theory and observation of that time to propose that the anterior hypothalamus contains a "heat-loss" center" whereas the posterior hypothalamus consists of a "heat-maintenance center" (Meyer, 1913). The historical features of this concept have been carefully examined (Benzinger, 1969, 1976, 1977). In spite of the continued promulgation of Meyer's theory in some quarters of physiology, we know today that the anterior and posterior regions of the hypothalamus both participate in the mediation of heat production and heat loss in a dual fashion. As long ago as 1930, the "center" theory of Meyer began to run into difficulty when Keller and his colleagues discovered that a dog exhibits poikilothermia following an extensive surgical lesion of its hypothalamus (e.g., Keller and Hare, 1932; Keller and McClaskey, 1964). Astonishingly, the more recent interpretation of their studies has not always been entirely clear, even though the results obtained with this sort of preparation tend to abrogate the possibility that the posterior hypothalamus exclusively comprises the heat-production center (Myers and Yaksh, 1971). To quote from the penetrating analysis of Hardy (1973), "... from today's vantage point, we can safely say that Hans Meyer was wrong and that his theory is no longer tenable."* Of course, it must be remembered that a large and disseminated hypothalamic ablation destroys afferent pathways as well as the efferent system indiscriminately.

---

*The lag time for pedagogically updating an academic curriculum with new knowledge is staggering. A casual perusal of some of the commonly used textbooks of physiology for medical and graduate students will reveal that this incorrect view is still espoused.

The authenticity of the dual physiological nature of both the anterior and posterior areas of the hypothalamus is derived essentially from three experimental sources: (1) investigations in which the response properties of single units in both hypothalamic areas are analyzed; (2) the functional delineation by direct stimulation with chemical substances of the two areas either by the method of microinjection or by push-pull perfusion; and (3) differentiated patterns of the temperature-evoked efflux or release of Ca^{2+} ions, monoamines, acetylcholine, and other neurohumoral factors from these two disparate regions. This chapter concludes with a brief analysis and an interpretation of the facts, as we now have them, relating a given neurochemical event to the anatomically circumscribed mechanisms for the control of body temperature.

A. Neurohumoral Coding in the Anterior Hypothalamus

If the anterior hypothalamic, preoptic area is not the heat-loss center, then what is its function? How does it operate?

From the abundant amount of evidence accumulated both at the neurophysiological level (Hardy, 1973; Boulant, 1976, and this volume) as well as through the neurochemical investigations presented herein, the anterior hypothalamus seems to play the role of a sensory integrator for thermal information. The anterior hypothalamus receives coded impulses from the periphery which signal the respective condition of both the cutaneous thermodetectors and those located in internal structures. Coupled with this peripheral sensing of either a cold or warm environmental or systemic temperature is another source of input, the localized thermosensitivity of the anterior hypothalamus. This remarkable physical property of the anterior hypothalamus reveals that the temperature of the blood coursing through the capillary beds and supplying the cells of this region is monitored locally.

Of significance, in addition, is another attribute of the rostral hypothalamus, namely, a special chemoreactivity to a pathogen such as a bacterial organism, leukocyte, or endogenous pyrogen. If it operates as a sensory integrator, the anterior hypothalamic area must possess such a receptor capacity, particularly in view of the fact that a febrile response may be beneficial if not essential to an animal's recovery from an illness caused by the presence of a pathogenic pyrogen in the body (Bernheim and Kluger, 1976; Kluger et al., 1977). Finally, cells of the anterior hypothalamus are equally reactive to the presence of toxins, poisons, drugs, and perhaps other factors that could endanger the life of the organism. Such chemosensitivity is essential for the over-all activation of the autonomic nervous system and, hence, the animal's survival.

The neurons of the anterior hypothalamus continuously integrate incoming sensory signals in an inordinately complicated manner, the precise details of which are still unknown. There is already evidence indicating the existence of four or more different categories of thermosensitive neurons (Hellon, 1967). Each of these could react in a different way to the presence of a pyrogen or other chemical substance. Nevertheless, the anterior hypothalamus seems to be responsible for integrating both types of sensory information: that received from the periphery as well as that generated within the blood supply to its own thermoreactive neurons.

Synaptically, the integrator neurons seem to be chemically coded by means of the presynaptic release of monoamines and acetylcholine. As has been explained lucidly by Cox and Lomas (1977), the specific mode of an amine's action is not universally agreed upon.

However, as hypothesized in the 1960s, the concept that a balance in the release within the anterior hypothalamus of 5-HT, norepinephrine, (probably dopamine), and acetylcholine, as a result of a peripheral thermal disturbance or a bacterial pyrogen, has continued to receive experimental support. It is envisaged that the end result of the integration of the sensory information in the anterior region is rapidly translated into successive depolarizations of individual neurons. The mediation of these neuronal discharges is accomplished presumably by the release of a monoamine onto a cholinergically coupled neuron (Myers, 1974a). In turn, ACh serves to transmit the efferent information caudalward to activate pathways either for heat production or heat dissipation. The synaptic action of a given amine can thereby modify, facilitate, or retard the transmission of impulses along the respective efferent pathway (Myers, 1978b).

B. Neurohumoral Coding in the Posterior Hypothalamus

If the posterior hypothalamus is not the heat-maintenance center, then what is its function? How does it operate?

Several critical features distinguish the posterior from the anterior hypothalamus. First, the supramamillary region contains few monoamine terminals; as one might expect, the neurons in this area are entirely insensitive to the presence of an exogenously applied monoamine, i.e., 5-HT, norepinephrine, or dopamine, at least insofar as a temperature response is concerned (Myers, 1975b). Second, although some thermosensitive units have been identified in this area (Hardy, 1973), the posterior hypothalamus is, on the whole, a structure which is insensitive to a thermal displacement. Third, the posterior region does not react to the presence of pyrogens or other potent chemical substances including the prostaglandins, which do exert a profound effect on the neurons of the anterior hypothalamus. Thus, the possibility that the function of the posterior hypothalamus is one devoted to the integration of sensory input may be somewhat limited.

From the preponderance of results in which lesions, the effect of electrical stimulation, and the local application of both ACh and cations have been investigated, the posterior hypothalamus would seem to operate to mediate convergent efferent responses. Signals received from the anterior hypothalamus could be integrated with impulses arising from extrahypothalamic structures including the spinal cord, septum, and subcortical motor pathways. The posterior hypothalamic neurons would conceivably process this convergent input and subsequently translate it into a set of appropriate effector responses, including vasomotor, pilomotor, sudomotor, and metabolic (Myers et al., 1976b).

Although the source of input and the type of signal is crucial to the type of response made by neurons in the posterior hypothalamus, the nature of their origin seems to be irrelevant. That is, if the incoming impulses from the anterior hypothalamic area call for heat production, specific neuronal elements in the posterior hypothalamus signal the activation of the efferent pathways for thermogenesis which triggers shivering, vasoconstriction, and an increase in oxygen consumption. In view of the uniform extrusion of Ca^{2+} ions, it does not seem to matter whether the signal is derived from a fall in the temperature of the blood pumping through the rostral hypothalamus, or stimulation of the cold thermodetectors in the skin, or from the presence of a foreign substance. Regardless of what comprises the stimulus, the specific neurons within the posterior hypothalamus that are delegated to thermogenesis ostensibly react in virtually the same way (Myers et al., 1976b).

The neurohumoral coding in the posterior hypothalamus through which the specific efferent impulses are transmitted is now believed to be accomplished through the presynaptic release of acetylcholine. Generally, the acetylcholine-releasing sites for heat production seem to outnumber those for heat loss by about 2 to 1. This ratio would be expected because of the fact that the body temperature of the mammal is usually much higher than the environmental temperature, which necessitates nonshivering thermogenesis continually. This proportion corresponds also to the unit activity in this structure recorded in an animal exposed to cold and warm temperatures (Nutik, 1971), which could partially reflect afferent input from the periphery (Boulant and Bignall, 1973b).

1. Set-Point Mechanism

Apart from the concept of a set point which is based on principles of physics or engineering control systems, the set point has been defined in physiological terms as that reference temperature of about 37°C, established at birth in most mammals, around which body temperature is regulated (e.g., Myers, 1974b). As described in Section VI.A, the steady-state firing rate of hypothalamic neurons involved in the maintenance of vasomotor tone and metabolic heat production at euthermia is postulated as the mechanism whereby the set temperature is stabilized. Should a physiological mechanism for the set point be located in the posterior hypothalamus, certain neurons in this region, by virtue of a steady-state firing pattern presumably required for a set point, could also process the input arising from the anterior hypothalamus. Since the majority of day-to-day thermoregulatory responses do not involve a resetting or shifting of the set point for body temperature (see Section VI.A), efferent pathways subserving thermoregulation per se are most likely anatomically distinct from the network of neurons delegated principally to the maintenance of a set temperature $\simeq 37°C$. Nevertheless, the firing rate of a collection of set-point neurons would necessarily have to be integrated in this area of motor convergence with thermoregulatory signals arising from the anterior hypothalamus, as is demonstrated by the changes in the activity of endogenous calcium ions (Myers, 1978b).

Conceptually, it is envisaged that the defense of the set temperature at euthermia or at a higher level (e.g., fever) or lower level (e.g., hibernation) is achieved by independent impulses that are transmitted, probably via cholinergic neurons, from the anterior hypothalamic area through to the posterior hypothalamus. To illustrate this idea as deduced from the data presented in Section VI.A, if an endogenous pyrogen should reach and bathe neurons of the rostral hypothalamus, or if a given thermal challenge (e.g., cold exposure) is extreme, these thermosensitive and/or chemoreactive neurons would (1) change their firing pattern, (2) concomitantly release ACh, which in turn would (3) stimulate the cholinergic pathway for heat production, (4) simultaneously shift the normal pattern of calcium (and possibly sodium) extrusion within the posterior hypothalamic region, in order to (5) mobilize a consequent shift in the set-point temperature. The kinetics of cation movement is quite clearly anatomically compartmentalized to the posterior hypothalamic area in that a thermal disturbance unexpectedly fails to alter calcium activity in the anterior hypothalamus. Therefore, it is difficult to entertain a concept of "gating" of efferent impulses to explain the concordance of ion efflux and set-point shift.

The ionic milieu in the caudal hypothalamus, which is notably stable under normal circumstances, would seem to provide an ideal mechanism whereby a reasonable stability in temperature could be achieved. The remarkable anatomical separation between thermosensitive neurons and nonreactive neurons would seem to underscore the earlier theoretical concept that a set-point element would have to be an independent morphological entity. The reason for this rests in an otherwise impossible physiological situation as follows. If a set-point neuron and a thermosensitive neuron were one and the same, then the deflection in the firing rate of a thermosensor upon thermal challenge would shift its own reference value. As a consequence, whenever the set-point temperature underwent a displacement, each thermosensitive neuron would necessarily alter its own firing rate. Thus, during a pyrogen fever, for example, cold-sensitive neurons, underlying thermogenesis, would reduce their firing so as to counter the thermal challenge of a rising temperature which would, of course, be caused by the febrile response.

C. Current Status and Future Directions

With each passing year, a "new" chemical substance (or old one) is discovered to exert an effect on body temperature when it is administered intracerebrally. If the substance is present endogenously in the brain, a scientist's immediate temptation is to ascribe the resultant hyper- or hypothermic action to some sort of "transmitter" property of the compound. In view of such a speculation, it would be appropriate at this stage of knowledge to shift the research emphasis from the simple pharmacological application of the compound to a physiological demonstration of its intrinsic activity within a circumscribed region of the anterior or posterior hypothalamus or other part of the brain stem.

To illustrate, if one finds that TRH, histamine, cyclic AMP, GABA, or other endogenous substance infused intrahypothalamically induces heat loss, there are at least three steps to be taken to ascertain its function in temperature control. First, if a receptor antagonist exists, its infusion into the hypothalamus is called for to block the local effect on temperature-regulating pathways. The next step is to demonstrate that the substance is not causing a nonspecific poikilothermia, so that the temperature of the animal falls at laboratory temperature or simply rises if the animal's chamber is warm. The last step is to determine whether the activity of that substance actually changes in the hypothalamus of the animal as it is exposed to a warm environment. The great bonus of this latter physiological approach is the demonstration of the relationship between the temperature change and the endogenous activity of the substance within the thermoregulating cells of the hypothalamus. A "temperature transmitter," as any transmitter, must interact extracellularly with adjacent neurons (Myers, 1976a). Therefore, unless the extracellular release of the endogenous substance can be functionally demonstrated as a response to a thermal challenge, it cannot be considered as a neurotransmitter (Myers, 1974a).

With respect to the controversy and debate over inconsistent experimental findings, clear-cut technical problems are today a major concern in determining the role played by a specific substance in thermoregulation (Frens, 1975). Postulated differences between species in the thermoregulatory process have been one major stumbling block hampering a coherent explanation of each neurochemical mechanism. Unfortunately, some of the discrepancies reported each year may have been artificially created, unwittingly in some cases, mainly because of one or more procedural irregularities: an excessive, perhaps lesioning, volume injected into the brain's parenchyma; the temporal slot assigned the animal in an experimental series; a paralyzing dose inadvertently given; use of a pyrogen-contaminated solution; primary autonomic effects; a nonanatomical approach; restraint of the animal; uncontrolled ambient temperature in the laboratory; and/or a lack of reliability of the injection procedure.

An inconsistency in a result may also be due to the passage of the compound from the brain's substance into the CSF, and then onto an ependymal structure whose function may override the temperature system. Or alternatively, the substance may pass from CSF into the systemic circulation. Indeed, it is conceivable that some species may have a "leaky" brain-to-blood barrier or cerebrospinal fluid-blood barrier. As a result, a substance injected either into the CSF spaces or into the parenchyma itself could readily exit into the capillary network and thence into the systemic circulation. Already, it is known that a tritiated monoamine injected into the ventricle of the rat is detected within minutes in the animal's urine (Glowinski et al., 1966; Martin and Myers, 1975). In our laboratory, we have found that a monoamine microinjected even in a volume of 0.5 μl into the brain stem dissipates so quickly that less than 5% of the total dose applied actually reaches the vicinity of neuronal receptors at the site of application (Myers and Hoch, 1978).

Nevertheless, the exceptions to a reasonably well documented observation together with its theoretical veiwpoint must be taken into account. For example, in considering instances of 5-HT-induced hypothermia or dopaminergic hyperthermia, receptor properties of the region of infusion become just as important to the interpretation as the procedures (see above) used to study the effect. To illustrate, the fall in temperature produced by a microinjection of 5-HT into the preoptic area can be blocked by pretreating the infusion site with phentolamine. Thus, it could be envisaged that in an area where serotonergic terminals are sparsely scattered, 5-HT attaches to noradrenergic receptors, and, in so doing, activates heat loss (W. E. Ruwe and R. D. Myers, unpublished observation). The ultimate proof, however, of the existence of a serotonergic system of thermoregulatory synapses cannot be obtained indirectly by a study of the action of antagonists. Instead, the physiologically evoked release of the amine by a specific thermal stimulus is required.

A creative, sagacious, yet realistic approach to an understanding of hypothalamic control functions in all their intricacies is difficult to come by. In the years to come, the difficulties will become even more formidable. Signs already on the horizon intimate that each mechanism involved in the regulation of body temperature operates according to a profile of neurochemical activity. Several neurohumoral factors located within the anterior and posterior parts of the hypothalamus ostensibly act simultaneously and in synchrony with one another. Research in the next decade should in all probability focus upon the contribution made by the respective chemical factors within a morphologically defined region. In this way, the complex neurohumoral "code" (Myers, 1974a) by which a given chemical substance could be responsible for mediating a specific afferent impulse or a differentiated efferent response may be delineated anatomically.

In the last part of this century there will be great challenges to the scientist working in the fields of thermoregulation and hypothalamic physiology. From the neurochemical vantage point, there are acknowledged biases both for and against the acceptance of an amine theory of thermoregulation, the role of a prostaglandin in fever, the function of cations in the set-point mechanism, and the influence of peptides in temperature control. Partialities are easily understood not only because of the relative newness of each of the ideas, but also because of the expected difficulty in interpreting a new scientific concept in light of a more traditional viewpoint. This is a healthy hallmark for the whole of the scientific process. An inexplicable or exceptional result does not represent a hindrance to the scientific progress in this field. Rather it leads to a more thorough examination of the particular hypothalamic mechanism that has been, and continues to be, so difficult to unravel.

The documentation of the role of a given factor in temperature regulation should be considered only in light of the accumulated data, viewed in its entirety, and not on the basis of one observation or one species. For example, the pieces of evidence presented in Tables 1 through 5 represent just that which has accrued to the present time with respect to a hypothesis of a given thermoregulatory function. In the future, the idea that a monoaminergic neuron in the hypothalamus functions in temperature control may well have to be modified. But that is the rewarding marvel of science: the ultimate truth of a mechanism eventually will be revealed. As long as the scientist approaches the question from a perspective that is not entrenched in bias and narrowness of mind, forward steps can be taken, and a "red herring" that may impede such an endeavor can be eliminated.

Looking ahead objectively, the essential role of the neurochemical "characters" alluded to at the beginning of this chapter, not yet fully developed, will be determined exclusively by how carefully they are manipulated at the laboratory bench. Indeed, the ultimate understanding of the hypothalamic control system for thermoregulation rests not upon the quantity of observation taken; rather, it depends upon the anatomical precision and physiological meticulousness with which each new experiment on the hypothalamus is completed.

References

Andén, N.-E., Dahlström, A., Fuxe, K., Larsson, K., Olson, L., and Ungerstedt, U. (1966). Ascending monoamine neurons to the telencephalon and diencephalon. *Acta Physiol. Scand. 67*, 313-326.

Armstrong, S., Burnstock, G., Evans, B., and Singer, G. (1973). The effects of intrahypothalamic injections of guanethidine on catecholamine fluorescence, food intake and temperature regulation in the rat. *Pharmacol. Biochem. Behav. 1*, 307-312.

Artunkal, A. A., Marley, E., and Stephenson, J. D. (1977). Some effects of prostaglandins E_1 and E_2 and of endotoxin injected into the hypothalamus of young chicks: dissociation between endotoxin fever and the effects of prostaglandins. *Br. J. Pharmac. 61*, 39-46.

Ary, M., and Lomax, P. (1976). Dopaminergic sites involved in morphine withdrawal hypothermia. *Proc. West. Pharmacol. Soc. 19*, 290-294.

Ary, M., Lomax, P., and Cox, B. (1977). Apomorphine hypothermia in the rat: central sites and mechanisms. *Neuropharmacol. 16*, 731-735.

Atkins, E. (1964). Elevation of body temperature in disease. *Ann. N. Y. Acad. Sci. 121*, 26-29.

Avery, D. D. (1970). Hyperthermia induced by direct injections of carbachol in the anterior hypothalamus. *Neuropharmacology 9*, 175-178.

Avery, D. D. (1971). Intrahypothalamic adrenergic and cholinergic injection effects on temperature and ingestive behavior in the rat. *Neuropharmacology 10*, 753-763.

Avery, D. D. (1972). Thermoregulatory effects of intrahypothalamic injections of adrenergic and cholinergic substances at different environmental temperatures. *J. Physiol. 220*, 257-266.

Avery, D. D., and Penn, P. E. (1973). Effects of intrahypothalamic injections of adrenergic and cholinergic substances on behavioral thermoregulation and associated skin temperature levels in rats. *Pharmacol. Biochem. Behav. 1,* 159-165.

Avery, D. D., and Penn, P. E. (1974). Blockage of pyrogen induced fever by intrahypothalamic injections of salicylate in the rat. *Neuropharmacology 13,* 1179-1185.

Avery, D. D., and Penn, P. E. (1976). Interaction of salicylate and body temperature changes caused by injections of neurohumours into the anterior hypothalamus: possible mechanisms in salicylate antipyresis. *Neuropharmacology 15,* 433-438.

Bächtold, H., and Pletscher, A. (1957). Einfluss von Isonikotinsäurehydraziden auf den Verlauf der Korpertemperatur nach Reserpin, Monoaminer, und Chlorpromazin. *Experientia 13,* 163-165.

Baird, J., and Lang, W. J. (1973). Temperature responses in the rat and cat to cholinomimetic drugs injected into the cerebral ventricles. *Eur. J. Pharmacol. 21,* 203-211.

Baird, J. A., Hales, J. R. S., and Lang, W. J. (1974). Thermoregulatory responses to the injection of monoamines, acetylcholine and prostaglandins into a lateral cerebral ventricle of the echidna. *J. Physiol. 236,* 539-548.

Baker, P. F. (1972). Transport and metabolism of calcium ions in nerve. *Prog. Biophys. Mol. Biol. 24,* 177-223.

Baldwin, B. A., Ingram, D. L., and LeBlanc, J. (1969). The effects of environmental temperature and hypothalamic temperature on excretion of catecholamines in the urine of the pig. *Brain Res. 16,* 511-516.

Banerjee, U., Burks, T. F., and Feldberg, W. (1968a). Effect on temperature of 5-hydroxytryptamine injected into the cerebral ventricles of cats. *J. Physiol. 195,* 245-251.

Banerjee, U., Feldberg, W., and Lotti, V. J. (1968b). Effect on body temperature of morphine and ergotamine injected into the cerebral ventricles of cats. *Br. J. Pharmacol. Chemother. 32,* 523-538.

Barbour, H. G. (1921). The heat-regulating mechanism of the body. *Physiol. Rev. 1,* 295-326.

Barbour, H. G., and Wing, E. S. (1913). The direct application of drugs to the temperature centers. *J. Pharmacol. 5,* 105-147.

Barchas, J., and Usdin, E. (1973). *Serotonin and Behavior,* J. Barchas and E. Usdin (Eds.). Academic, New York.

Bard, P., Woods, J. W., and Bleier, R. (1970). The effects of cooling, heating, and pyrogen on chronically decerebrate cats. *Comm. Behav. Biol. 5,* 31-50.

Barney, C. C., and Elizondo, R. S. (1978). Effect of ambient temperature on development of prostaglandin E_1 hyperthermia in the rhesus monkey. *J. Appl. Physiol. 44,* 751-758.

Battista, A., Fuxe, K., Goldstein, M., and Ogawa, M. (1972). Mapping of central monoamine neurons in the monkey. *Experientia 28,* 688-689.

Beauvallet, M., Legrand, M., and Fugazza, J. (1967). Hyperthermie et teneur en noradrenaline, dopamine et 5-hydroxytryptamine du cerveau chez le rat. *C. R. Seanc. Soc. Biol. 161,* 1291-1292.

Beckman, A. L. (1970). Effect of intrahypothalamic norepinephrine on thermoregulatory responses in the rat. *Am. J. Physiol. 218,* 1596-1604.

Beckman, A. L., and Carlisle, H. J. (1969). Effect of intrahypothalamic infusion of acetylcholine on behavioural and physiological thermoregulation in the rat. *Nature (London) 221*, 561-562.

Beckman, A. L., and Eisenman, J. S. (1970a). Responsiveness of temperature-sensitive hypothalamic neurons to microelectrophoretically applied amines in rats and cats. *Fed. Proc. 29*, Abstr. 1548.

Beckman, A. L., and Eisenman, J. S. (1970b). Microelectrophoresis of biogenic amines on hypothalamic thermosensitive cells. *Science 170*, 334-336.

Beckman, A. L., and Rozkowska-Ruttimann, E. (1974). Hypothalamic and septal neuronal responses to iontophoretic application of salicylate in rats. *Neuropharmacol. 13*, 393-398.

Beckman, A. L., and Satinoff, E. (1972). Arousal from hibernation by intrahypothalamic injections of biogenic amines in ground squirrels. *Am. J. Physiol. 222*, 875-879.

Beleslin, D. B., and Myers, R. D. (1973). Observations on the hypothermic effect of metrazol in the conscious cat. *Pharmacol. Biochem. Behav. 1*, 727-729.

Beleslin, D. B., Dimitrijević, M., and Samardzić, R. (1974). Hyperthermic effect of palmitate sodium, stearate sodium and oleate sodium injected into the cerebral ventricles of conscious cats. *Neuropharmacology 13*, 221-223.

Bennett, I. L., Jr., Petersdorf, R. G., and Keene, W. R. (1957). Pathogenesis of fever: evidence for direct cerebral action of bacterial endotoxins. *Assoc. Am. Phys. Trans. 70*, 64-73.

Benzinger, T. H. (1969). Heat regulation: homeostasis of central temperature in man. *Physiol. Rev. 49*, 671-759.

Benzinger, T. H. (1976). Role of thermoreceptors in thermoregulation. In *Sensory Functions of the Skin*, Vo. 27, Y. Zotterman, (Ed.), Pergamon, New York, pp. 379-397.

Benzinger, T. H. (1977). *Benchmarks in Physiology*, Dowden Hutchinson and Ross, Stroudsburg, Pa.

Bergström, S., Carlson, L. A., and Weeks, J. R. (1968). The prostaglandins: a family of biologically active lipids. *Pharmacol. Rev. 20*, 1-48.

Bernheim, H. A., and Kluger, M. J. (1976). Fever: effect of drug-induced antipyresis on survival. *Science 193*, 237-239.

Bissette, G., Nemeroff, C. B., Loosen, P. T., Prange, A. J., Jr., and Lipton, M. A. (1976). Hypothermia and intolerance to cold induced by intracisternal administration of the hypothalamic peptide neurotensin. *Nature (London) 262*, 607-609.

Bligh, J. (1973). *Temperature Regulation in Mammals and Other Vertebrates*, North-Holland, Amsterdam.

Bligh, J., and Maskrey, M. (1969). A possible role of acetylcholine in central control of body temperature in sheep. *J. Physiol. (Lond.) 203*, 55-57 P.

Bligh, J., Cottle, W. H., and Maskrey, M. (1971). Influence of ambient temperature on the thermoregulatory responses to 5-hydroxytryptamine, noradrenaline and acetylcholine injected into the lateral cerebral ventricle of sheep, goats and rabbits. *J. Physiol. 212*, 377-392.

Bloom, F. E. (1974a). Dynamics of synaptic modulation: Perspectives for the future. In *The Neurosciences: Third Study Program*, F. O. Schmitt and F. G. Worden (Eds.). MIT Press, Cambridge, Mass., pp. 989-999.

Bloom, F. E. (1974b). To spritz or not to spritz: the doubtful value of aimless iontophoresis. *Life Sci. 14*, 1819-1834.

Boros-Farkas, M., and Illei-Donhoffer, A. (1969). The effect of hypothalamic injections of noradrenaline, serotonin, pyrogen, γ-amino-*n*-butyric acid, triodothyronine, triiodothyroacetic acid, DL-phenylalanine, DL-alanine and DL-γ-amino-β-hydroxybutyric acid on body temperature and oxygen consumption in the rat. *Acta Physiol. Acad. Sci. Hung. 36*, 105-116.

Borsook, P., Laburn, H. P., Rosendorff, C., Willies, G. H., and Woolf, C. S. (1977). A dissociation between temperature regulation and fever in the rabbit. *J. Physiol. 266*, 423-433.

Boulant, J. (1976). A hypothalamic neuronal model for thermoregulation. In *Selected Topics in Environmental Biology*. B. Bhatia, G. S. Chhina, and B. Singh (Eds.). Interprint, New Delhi, pp. 41-44.

Boulant, J. A., and Bignall, K. E. (1973a). Determinants of hypothalamic neuronal thermosensitivity in ground squirrels and rats. *Am. J. Physiol. 225*, 306-310.

Boulant, J. A., and Bignall, K. E. (1973b). Changes in thermosensitive characteristics of hypothalamic units over time. *Am. J. Physiol. 225*, 311-318.

Breckenridge, B. M., and Lisk, R. D. (1969). Cyclic adenylate and hypothalamic regulatory functions. *Proc. Soc. Exp. Biol. Med. 131*, 934-935.

Breese, G. R., and Howard, J. L. (1971). Effect of central catecholamine alterations on the hypothermic response to 6-hydroxydopamine in desipramine treated rats. *Br. J. Pharmacol. 46*, 671-674.

Breese, G. R., Moore, R. A., and Howard, J. L. (1972). Central actions of 6-hydroxydopamine and other phenylethylamine derivatives on body temperature in the rat. *J. Pharmacol. Exp. Ther. 180*, 591-602.

Brezenoff, H. E., and Lomax, P. (1970). Temperature changes following microinjection of histamine into the thermoregulatory centers of the rat. *Experientia 26*, 51-52.

Brimblecombe, R. W. (1973). Effects of cholinomimetic and cholinolytic drugs on body temperature. In *The Pharmacology of Thermoregulation*, E. Schönbaum and P. Lomax (Eds.). Karger, Basel, pp. 182-193.

Brittain, R. T., and Handley, S. L. (1967). Temperature changes produced by the injection of catecholamines and 5-hydroxytryptamine into the cerebral ventricles of the conscious mouse. *J. Physiol. 192*, 805-813.

Brück, K. (1976). Temperature regulation and catecholamines. *Israel J. Med. Sci. 12*, 924-933.

Bruinvels, J. (1970). Effect of noradrenaline, dopamine and 5-hydroxytryptamine on body temperature in the rat after intracisternal administration. *Neuropharmacol. 9*, 277-282.

Bruinvels, J., and Kemper, G. C. M. (1971). Role of noradrenaline and 5-hydroxytryptamine in tetrahydronaphthylamine-induced temperature changes in the rat. *Br. J. Pharmac. 43*, 1-9.

Burks, T. F. (1972). Central alpha-adrenergic receptors in thermoregulation. *Neuropharmacol. 11*, 615-624.

Cabanac, M., Stolwijk, J. A. J., and Hardy, J. D. (1968). Effect of temperature and pyrogens on single-unit activity in the rabbit's brain stem. *J. Appl. Physiol. 24*, 645-652.

Canal, N., and Ornesi, A. (1961). Serotonina encefalica e ipertermia da vaccino. *Atti del. Accad. Med. Lomb. 16*, 69-73.

Cantor, A., and Satinoff, E. (1976). Thermoregulatory responses to intraventricular norepinephrine in normal and hypothalamic-damaged rats. *Brain Res. 108*, 125-142.

Carette, M. B., and Fessard, M. A. (1976). Iontophorèse de la noradrénaline sur des neurones des aires septale, préoptique et hypothalamique antérieure. *C. R. Acad. Sc. Paris 282,* 1877-1880.

Carino, M. A., Smith, J. R., Weick, B. G., and Horita, A. (1976). Effects of thyrotropin-releasing hormone (TRH) microinjected into various brain areas of conscious and pentobarbital-pretreated rabbits. *Life Sci. 19,* 1687-1692.

Carlson, L. D. (1960). Nonshivering thermogenesis and its endocrine control. *Fed. Proc. 19* (Suppl. 5), 25-30.

Carruba, M. O., and Bächtold, H. P. (1976). Pyrogen fever in rabbits pretreated with p-chlorophenylalanine or 5,6-dihydroxytryptamine. *Experientia 32,* 729-730.

Carruba, M. O., Tofanetti, O., Picotti, G. B., and Mantegazza, P. (1978). Involvement of serotoninergic neurons in the hyperthermic response to dopaminergic agonists. *Pharmac. Res. Comm. 10,* 357-370.

Chaffee, R. R. J., and Roberts, J. C. (1971). Temperature acclimation in birds and mammals. *Ann. Rev. Physiol. 33,* 155-202.

Chai, C. Y., and Lin, M. T. (1975). Hypothermic effect of sodium acetylsalicylate on afebrile monkeys. *Br. J. Pharmacol. 54,* 475-479.

Chai, C. Y., Chen, H. I., and Yin, T. H. (1971). Central sites of action and effects of acetylstrophanthidin on body temperature in monkeys. *Exp. Neurol. 33,* 618-628.

Chambers, W. W., Koenig, H., Koenig, R., and Windle, W. F. (1949). Site of action in the central nervous system of a bacterial pyrogen. *Am. J. Physiol. 159,* 209-216.

Chambers, W. W., Seigel, M. S., Liu, J. C., and Liu, C. N. (1974). Thermoregulatory responses of decerebrate and spinal cats. *Exp. Neurol. 42,* 282-299.

Chase, T. N., and Murphy, D. L. (1973). Serotonin and central nervous system function. *Ann. Rev. Pharmacol. 13,* 181-197.

Chawla, N., Johri, M. B. L., Saxena, P. N., and Singhal, K. C. (1974). Effects of catecholamines on thermoregulation in pigeons. *Br. J. Pharmacol. 51,* 497-501.

Cheung, W. Y. (1980). Calmodulin plays a pivotal role in cellular regulation. *Science 207,* 19-27.

Chowers, I., Conforti, N., and Feldman, S. (1968). Local effect of cortisol in the preoptic area on temperature regulation. *Am. J. Physiol. 214,* 538-542.

Citroni, J., and Leschke, E. (1913). *Zeitschr. Exp. Pathol. 14,* 3, 379 (cited by Barbour, 1921).

Clark, W. G. (1970). The antipyretic effects of acetaminophen and sodium salicylate on endotoxin-induced fevers in cats. *J. Pharmacol. Exp. Ther. 175,* 469-475.

Clark, W. G. (1977). Emetic and hyperthermic effects of centrally injected methionine-enkephalin in cats. *Proc. Soc. Exp. Biol. Med. 154,* 540-542.

Clark, W. G., and Alderdice, M. T. (1972). Inhibition of leukocytic pyrogen-induced fever by intracerebroventricular administration of salicylate and acetaminophen in the cat. *Proc. Soc. Exp. Biol. Med. 140,* 399-403.

Clark, W. G., and Coldwell, B. A. (1972). Competitive antagonism of leukocytic pyrogen by sodium salicylate and acetaminophen. *Proc. Soc. Exp. Biol. Med. 141,* 669-672.

Clark, W. G., and Coldwell, B. A. (1973). The hypothermic effect of tetrodotoxin in the unanesthetized cat. *J. Physiol. 230,* 477-492.

Clark, W. G., and Cumby, H. R. (1975). The antipyretic effect of indomethacin. *J. Physiol. 248,* 625-638.

Clark, W. G., and Cumby, H. R. (1976a). Biphasic changes in body temperature produced by intracerebroventricular injections of histamine in the cat. *J. Physiol. 261*, 235-253.

Clark, W. G., and Cumby, H. R. (1976b). Antagonism by antipyretics of the hyperthermic effect of a prostaglandin precursor, sodium arachidonate in the cat. *J. Physiol. 257*, 581-595.

Clark, W. G., Cumby, H. R., and Davis, H. E., IV. (1974). The hyperthermic effect of intracerebroventricular cholera enterotoxin in the unanaesthetized cat. *J. Physiol. 240*, 493-504.

Collier, H. O. J. (1969). A pharmacological analysis of aspirin. In *Advances in Pharmacology and Chemotherapy*, Vol. 7, S. Garattini, A. Goldin, F. Hawking, and I. J. Kopin (Eds.). Academic, New York, pp. 333-405.

Conrad, L. C. A., and Pfaff, D. W. (1975). Axonal projections of medial preoptic and anterior hypothalamic neurons. *Science 190*, 1112-1114.

Cooper, K. E. (1972). Central mechanisms for the control of body temperature in health and febrile states. In *Modern Trends in Physiology*, Vol. 1, C. B. B. Downman (Ed.). Butterworths, London, pp. 33-54.

Cooper, K. E., and Cranston, W. I. (1966). Pyrogens and monoamine oxidase inhibitors. *Nature (London) 210*, 203-204.

Cooper, K. E., and Veale, W. L. (1972). The effect of injecting an inert oil into the cerebral ventricular system upon fever produced by intravenous leucocyte pyrogen. *Can. J. Physiol. Pharmacol. 50*, 1066-1071.

Cooper, K. E., and Veale, W. L. (1974). Fever, an abnormal drive to the heat-conserving and -producing mechanisms? In *Recent Studies of Hypothalamic Function*, K. Lederis and K. E. Cooper (Eds.). Karger, Basel, pp. 391-398.

Cooper, K. E., Cranston, W. I., and Honour, A. J. (1965). Effects of intraventricular and intrahypothalamic injection of noradrenaline and 5-HT on body temperature in conscious rabbits. *J. Physiol. 181*, 852-864.

Cooper, K. E., Cranston, W. I., and Honour, A. J. (1967). Observations on the site and mode of action of pyrogens in the rabbit brain. *J. Physiol. 191*, 325-337.

Cooper, K. E., Grundman, M. J., and Honour, A. J. (1968). Observations on sodium salicylate as an antipyretic. *Proc. Physiol. Soc. J. Physiol. 196*, 56-57.

Cooper, K. E., Pittman, Q. J., and Veale, W. L. (1975). Observations on the development of the "fever" mechanism in the fetus and newborn. In *Temperature Regulation and Drug Action*. J. Lomax, E. Schönbaum, and J. Jacob (Eds.). Karger, Basel, pp. 43-50.

Cooper, K. E., Jones, D. L., Pittman, Q. J., and Veale, W. L. (1976a). The effect of noradrenaline, injected into the hypothalamus, on thermoregulation in the cat. *J. Physiol. 261*, 211-222.

Cooper, K. E., Veale, W. L., and Pittman, Q. J. (1976b). The pathogenesis of fever. In *Brain Dysfunction in Infantile Febrile Convulsions*, M. A. B. Brazier, and F. Coceani (Eds.). Raven, New York, pp. 107-115.

Corrodi, H., Fuxe, K., and Hökfelt, T. (1967). A possible role played by central monoamine neurones in thermoregulation. *Acta Physiol. Scand. 71*, 224-232.

Costa, E., and Neff, N. H. (1966). The dynamic process for catecholamine storage as a site for drug action. International Congress of Neuro-psychopharmacology, Washington, D.C., pp. 757-764.

Cow, D. (1911). *J. Physiol. 42*, 125 (cited by Barbour and Wing, 1913).

Cox, B., and Lee, T. F. (1977). Do central dopamine receptors have a physiological role in thermoregulation? *Br. J. Pharmacol. 61*, 83-86.

Cox, B., and Lee, T. F. (1978). Is acetylcholine involved in a dopamine receptor mediated hypothermia in mice and rats? *Br. J. Pharmacol. 62*, 339-347.

Cox, B., and Lomax, P. (1977). Pharmacologic control of temperature regulation. *Ann. Rev. Pharmacol. Toxicol. 17*, 341-53.

Cox, B., and Tha, S. J. (1975). The role of dopamine and noradrenaline in temperature control of normal and reserpine-treated mice. *J. Pharmaceut. Pharmacol. 27*, 242-247.

Cox, B., Ary, M., and Lomax, P. (1975). Dopaminergic mechanisms in withdrawal hypothermia in morphine dependent rats. *Life Sci. 17*, 41-42.

Cox, B., Ary, M., Chesarek, W., and Lomax, P. (1976a). Morphine hyperthermia in the rat: an action on the central thermostats. *Eur. J. Pharmacol. 36*, 33-39.

Cox, B., Ary, M., and Lomax, P. (1976b). Dopaminergic involvement in withdrawal hypothermia and thermoregulatory behavior in morphine dependent rats. *Pharmacol. Biochem. Behav. 4*, 259-262.

Cox, B., Kerwin, R., and Lee, T. F. (1978). Dopamine receptors in the central thermoregulatory pathways of the rat. *J. Physiol. 282*, 471-483.

Cranston, W. I., and Rawlins, M. D. (1972). Effects of intracerebral microinjection of sodium salicylate on temperature regulation in the rabbit. *J. Physiol. 222*, 257-266.

Cranston, W. I., Luff, R. H., Rawlins, M. D., and Rosendorff, C. (1970a). The effects of salicylate on temperature regulation in the rabbit. *J. Physiol. 208*, 251-259.

Cranston, W. I., Hellon, R. F., Luff, R. H., Rawlins, M. D., and Rosendorff, C. (1970b). Observations on the mechanism of salicylate-induced antipyresis. *J. Physiol. 210*, 593-600.

Cranston, W. I., Hellon, R. F., and Mitchell, D. (1975a). A dissociation between fever and prostaglandin concentration in cerebrospinal fluid. *J. Physiol. 253*, 583-591.

Cranston, W. I., Hellon, R. F., and Mitchell, D. (1975b). Is brain prostaglandin synthesis involved in responses to cold? *J. Physiol. 249*, 425-434.

Cranston, W. I., Hellon, R. F., Luff, R. H., and Rawlins, M. D. (1972). Hypothalamic endogenous noradrenaline and thermoregulation in the cat and rabbit. *J. Physiol. 223*, 59-67.

Cranston, W. I., Duff, G. W., Hellon, R. F., Mitchell, D., and Townsend, Y. (1976). Evidence that brain prostaglandin synthesis is not essential in fever. *J. Physiol. 259*, 239-249.

Crawshaw, L. I. (1972). Effects of intracerebral 5-hydroxytryptamine injection on thermoregulation in rat. *Physiol. Behav. 9*, 133-140.

Crawshaw, L. I. (1973). Effect of intracranial acetylcholine injection on thermoregulatory responses in the rat. *J. Comp. Physiol. Psych. 83*, 32-35.

Crawshaw, L. I., and Stitt, J. T. (1975). Behavioural and autonomic induction of prostaglandin E_1 fever in squirrel monkeys. *J. Physiol. 244*, 197-206.

Cremer, J. E., and Bligh, J. (1969). Body-temperature and responses to drugs. *Br. Med. Bull. 25*, 299-306.

Cunningham, D. J., Stolwijk, J. A. J., Murakami, N., and Hardy, J. D. (1967). Responses of neurons in the preoptic area to temperature, serotonin, and epinephrine. *Am. J. Physiol. 213*, 1570-1581.

Darling, K. F., Findlay, J. D., and Thompson, G. E. (1974). Effect of intraventricular acetylcholine and eserine on the metabolism of sheep. *Pfluegers Arch. Gesamte Physiol. 349*, 235-245.

Dascombe, M. J., and Milton, A. S. (1975). The effects of cyclic adenosine $3',5'$-monophosphate and other adenine nucleotides on body temperature. *J. Physiol. 250*, 143-160.

Dascombe, M. J., and Milton, A. S. (1976). Cyclic adenosine $3',5'$-monophosphate in cerebrospinal fluid during thermoregulation and fever. *J. Physiol. 263*, 441-463.

Dawe, A. R., Spurrier, W. A., and Armour, J. A. (1970). Summer hibernation induced by cryogenically preserved blood "trigger." *Science 168*, 497-498.

Day, T. A., Willoughby, J. O., and Geffen, L. B. (1979). Thermoregulatory effects of preoptic area injections of noradrenaline in restrained and unrestrained rats. *Brain Res. 174*, 175-179.

Denbow, D. M., and Edens, F. W. (1980). Effects of intraventricular injections of sodium and calcium on body temperature in the chicken. *Am. J. Physiol.*, in press.

Des Prez, R., Helman, R., and Oates, J. A. (1966). Inhibition of endotoxin fever by reserpine. *Proc. Soc. Exp. Biol. Med. 122*, 746-749.

Dey, P. K., Feldberg, W., Gupta, K. P., Milton, A. S., and Wendlandt, S. (1974). Further studies on the role of prostaglandin in fever. *J. Physiol. 241*, 629-646.

Doggett, N. S. (1973). Possible involvement of a dopaminergic pathway in the depressant effects of ouabain on the central nervous system. *Neuropharmacology 12*, 213-220.

Domino, E. F. (1962). Sites of action of some central nervous system depressants. *Ann. Rev. Pharmacol. 2*, 215-250.

Duce, M., Crabai, F., Vargiu, L., Piras, L., Adamo, F., and Gessa, G. L. (1968). Effeto del'α-metil-*m*-tirosina sulle catecolamine di animali esposti af freddo. *Boll. Soc. Ital. Biol. Sper. 43*, 1607-1609.

Eisenman, J. S. (1969). Pyrogen-induced changes in the thermosensitivity of septal and preoptic neurons. *Am. J. Physiol. 216*, 330-334.

Eisenman, J. S. (1974). Depression of preoptic thermosensitivity by bacterial pyrogen in rabbits. *Am. J. Physiol. 227*, 1067-1073.

Evans, B. K., Armstrong, S., Singer, G., Cook, R. D., and Burnstock, G. (1975). Intracranial injection of drugs: comparison of diffusion of 6-OHDA and guanethidine. *Pharmacol. Biochem. Behav. 3*, 205-217.

Feist, D. D., and Galster, W. A. (1974). Changes in hypothalamic catecholamines and serotonin during hibernation and arousal in the arctic ground squirrel. *Comp. Biochem. Physiol. 48A*, 623-662.

Feldberg, W. (1968). The monoamines of the hypothalamus as mediators of temperature responses. In *Recent Advances in Pharmacology*, J. M. Robson and R. S. Stacey (Eds.). Churchill, London, pp. 349-397.

Feldberg, W. (1975). The Ferrier Lecture. Body temperature and fever: changes in our views during the last decade. *Proc. R. Soc. Lond. B 191*, 199-229.

Feldberg, W., and Gupta, K. P. (1973). Pyrogen fever and prostaglandin-like activity in cerebrospinal fluid. *J. Physiol. 228*, 41-53.

Feldberg, W., and Lotti, V. J. (1967). Temperature responses to monoamines and an inhibitor of MAO injected into the cerebral ventricles of rats. *Br. J. Pharmacol. Chemother. 31*, 152-161.

Feldberg, W., and Milton, A. S. (1973). Prostaglandin fever. In *The Pharmacology of Thermoregulation,* E. Schönbaum and P. Lomax (Eds.). Karger, Basel, pp. 302-310.

Feldberg, W., and Myers, R. D. (1963). A new concept of temperature regulation by amines in the hypothalamus. *Nature (London) 200,* 1325.

Feldberg, W., and Myers, R. D. (1964). Effects on temperature of amines injected into the cerebral ventricles. A new concept of temperature regulation. *J. Physiol. 173,* 226-237.

Feldberg, W., and Myers, R. D. (1965a). Changes in temperature produced by microinjections of amines into the anterior hypothalamus of cats. *J. Physiol. 177,* 239-245.

Feldberg, W., and Myers, R. D. (1965b). Hypothermia produced by chloralose acting on the hypothalamus. *J. Physiol. 179,* 509-517.

Feldberg, W., and Myers, R. D. (1966). Appearance of 5-hydroxytryptamine and an unidentified pharmacologically active lipid acid in effluent from perfused cerebral ventricles. *J. Physiol. 184,* 837-855.

Feldberg, W., and Saxena, P. N. (1970). Mechanism of action of pyrogen. *J. Physiol. 211,* 245-261.

Feldberg, W., and Saxena, P. N. (1971). Further studies on prostaglandin E_1 fever in cats. *J. Physiol. 219,* 739-745.

Feldberg, W., and Vartiainen, A. (1934). Further observations on the physiology and pharmacology of a sympathetic ganglion. *J. Physiol. 83,* 103-128.

Feldberg, W., Myers, R. D., and Veale, W. L. (1970). Perfusion from cerebral ventricle to cisterna magna in the unanaesthetized cat. Effect of calcium on body temperature. *J. Physiol. 207,* 403-416.

Feldberg, W., Gupta, K. P., Milton, A. S., and Wendlandt, S. (1973). Effect of pyrogen and antipyretics on prostaglandin activity in cisternal C.S.F. of unanaesthetized cats. *J. Physiol. 234,* 279-303.

Findlay, J. D., and Robertshaw, D. (1967). The mechanism of body temperature changes induced by intraventricular injections of adrenaline, noradrenaline and 5-hydroxytryptamine in the ox (*Bos taurus*). *J. Physiol. 189,* 329-336.

Flower, R. J., and Vane, J. R. (1972). Inhibition of prostaglandin synthetase in brain explains the anti-pyretic activity of Paracetamol (4-acetamidophenol). *Nature (London) 240,* 410-411.

Foster, R. S., Jenden, D. J., and Lomax, P. (1967). A comparison of the pharmacologic effects of morphine and *n*-methyl morphine. *J. Pharmacol. Expl. Therp. 157,* 185-195.

Fox, R. H., and MacPherson, R. K. (1954). The regulation of body temperature during fever. *J. Physiol. 125,* 21-22P.

Francesconi, R. P., and Mager, M. (1974). L-Tryptophan: effects on body temperature in rats. *Am. J. Physiol. 227,* 402-405.

Francesconi, R., and Mager, M. (1976). Thermoregulatory effects of monoamine potentiators and inhibitors in the rat. *Am. J. Physiol. 231,* 148-152.

Fregly, M. J., Field, F. P., Nelson, E. L., Jr., Tyler, P. E., and Dasler, R. (1977). Effect of chronic exposure to cold on some responses to catecholamines. *J. Appl. Physiol. 42,* 349-354.

Frens, J. (1975). Thermoregulation set-point changes during lipopolysaccharide fever. In *Temperature Regulation and Drug Action,* P. Lomax, E. Schönbaum, and J. Jacob (Eds.). Karger, Basel, pp. 59-64.

Freund, H. (1911). Über das Kochsalzfieber. *Arch. Exp. Pathol. Pharmakol.* 65, 225-238.

Frey, H.-H. (1975a). Central tryptaminergic mediation of the hyperthermia induced by amphetamine and halogenated analogues of it? *Pol. J. Pharmacol. Pharmaceut.* 27, 89-90.

Frey, H.-H. (1975b). Hyperthermia induced by amphetamine, p-chloroamphetamine and fenfluramine in the rat. *Pharmacology* 13, 163-176.

Friedman, A. H., and Walker, C. A. (1968). Circadian rhythms in rat mid-brain and caudate nucleus biogenic amine levels. *J. Physiol.* 197, 77-85.

Fukushima, N., and Itoh, S. (1975). Role of adrenergic receptors in the central thermoregulatory mechanism of the rat. *Jpn. J. Physiol.* 25, 621-631.

Fuller, R. W., and Baker, J. C. (1974). Further evidence for serotonin involvement in thermoregulation following morphine administration from studies with an inhibitor of serotonin uptake. *Chem. Pathol. Pharmacol.* 8, 715-718.

Fuxe, K., Hokfelt, T., and Ungerstedt, U. (1968). Localization of indolealkylamines in CNS. In *Advances in Pharmacology*, Vol. 6, S. Garattini and P. A. Shore (Eds.). Academic, New York, pp. 235-251.

Fuxe, K., Hokfelt, T., and Ungerstedt, U. (1970). Morphological and functional aspects of central monoamine neurons. In *International Review of Neurobiology*, Vol. 13, C. C. Pfeiffer and J. R. Smythies (Eds.). Academic, New York, pp. 93-126.

Gál, E. M., Heater, R. D., and Millard, S. A. (1968). Studies on the metabolism of 5-hydroxytryptamine (serotonin). VI. Hydroxylation and amines in cold-stressed reserpinized rats. *Proc. Soc. Exp. Biol.* 128, 412-415.

Garattini, S., and Valzelli, L. (1958). Researches on the mechanism of reserpine sedative action. *Science* 128, 1278-1279.

Gardey-Levassort, C., Olive, G., Fontagne, J., Szafranowa, H., and Lechat, P. (1970). Réponse fébrile du lapin aux pyrogènes bactériens et teneur de l'hypothalamus en sérotonine et noradrénaline apres un pré-traitement aux I.M.A.O. *J. Pharmacol. (Paris)* 1, 57-64.

Gardey-Levassort, C., Tanguy, O., and Lechat, P. (1977). Brain concentrations of biogenic amines and their metabolites in two types of pyrogen-induced fever in rabbits. *J. Neurochem.* 28, 177-182.

Garver, D. L., and Sladek, J. R., Jr. (1975). Monoamine distribution in primate brain I. Catecholamine-containing perikarya in the brain stem of *Macaca speciosa*. *J. Comp. Neurol.* 159, 289-304.

Gessa, G. L., Clay, G. A., and Brodie, B. B. (1969). Evidence that hyperthermia produced by d-amphetamine is caused by a peripheral action of the drug. *Life Sci.* 98, 135-141.

Giarman, N. J., Tanaka, C., Mooney, J., and Atkins, E. (1968). Serotonin, norepinephrine, and fever. *Adv. Pharmacol. Chemotherap.* 6A, 307-317.

Girault, J.-M. T., and Jacob, J. J. (1979). Serotonin antagonists and central hyperthermia produced by biogenic amines in conscious rabbits. *Eur. J. Pharmacol.* 53, 191-200.

Gisolfi, C. V., Wilson, N. C., Myers, R. D., and Phillips, M. I. (1976). Exercise thermoregulation: hypothalamic perfusion of excess calcium reduces elevated colonic temperature of rats. *Brain Res.* 101, 160-164.

Gisolfi, C. V., Mora, F., and Myers, R. D. (1977). Diencephalic efflux of calcium ions in the monkey during exercise, thermoregulation and feeding. *J. Physiol.* 273, 617-630.

Glowinski, J., Axelrod, J., and Iversen, L. L. (1966). Regional studies of catecholamines in the rat brain. IV. Effects of drugs on the disposition and metabolism of H^3-norepinephrine and H^3-dopamine. *J. Pharmacol. Exp. Ther. 153*, 30-41.

Göing, V. H. (1959). Beeinflussung der Fieber erzeugenden Wirkung bakterieller Pyrogene durch Iproniacid, Reserpin und Dibenamin. *Arzn.-Forsch. 9*, 793-794.

Gonzalez, R. R. (1971). Heat production and hypothalamic temperature changes to norepinephrine and 5-hydroxytryptamine in cold- and warm-adapted rabbits. *Proc. Soc. Exp. Biol. 137*, 1126-1130.

Gordon, R., Spector, S., Sjoerdsma, A., and Udenfriend, S. (1966). Increased synthesis of norepinephrine and epinephrine in the intact rat during exercise and exposure to cold. *J. Pharmacol. Exp. Ther. 153*, 440-447.

Grabowska, M., and Andén, N. E. (1976). Apomorphine in rat nucleus accumbens: effects on synthesis of 5-hydroxytryptamine and noradrenaline, motor activity and the body temperature. *J. Neural. Trans. 38*, 1.

Grafe, E. (1910). *Deutsch. Arch. F. Klin. Med. 101*, 209 (cited by Barbour, 1921).

Grant, R., Lewis, J., and Ahrne, I. (1955). Effects of intrahypothalamic injections of pyrogens. *Fed. Proc. 14*, 61.

Green, M. D., Simon, M. L., and Lomax, P. (1975a). Histidine induced hypothermia in the rat. *Life Sci. 16*, 1293-1300.

Green, M. D., Cox, B., and Lomax, P. (1975b). Histamine H_1- and H_2-receptors in the central thermoregulatory pathways of the rat. *J. Neurosci. Res.*, in press.

Green, M. D., Cox, B., and Lomax, P. (1976). Sites and mechanisms of action of histamine in the central thermoregulatory pathways of the rat. *Neuropharmacol. 15*, 321-324.

Greenleaf, J. E., Kozlowski, S., Nazar, K., Kaciuba-Uscilko, H., and Brzezinska, Z. (1974). Temperature responses of exercising dogs to infusion of electrolytes. *Experentia 30*, 769-770.

Greenleaf, J. E., Kozlowski, S., Nazar, K., Kaciuba-Uscilko, H., and Brzezinska, Z. (1975). Temperature responses to infusion of electrolytes during exercise. In *Temperature Regulation and Drug Action*, P. Lomax, E. Schönbaum, and J. Jacob (Eds.). Karger, Basel, pp. 352-360.

Guerinot, F., Poitou, P., and Bohuon, C. (1974). Serotonin synthesis in the rat brain after acetylsalicyclic acid administration. *J. Neurochem. 22*, 191-192.

Guerra, F., and Brobeck, J. R. (1944). The hypothalamic control of aspirin antipyresis in the monkey. *J. Pharmacol. Exp. Ther. 80*, 209-216.

Hales, J. R. S., Bennett, J. W., Baird, J. A., and Fawcett, A. A. (1973). Thermoregulatory effects of prostaglandins E_1, E_2, $F_{1\alpha}$ and $F_{2\alpha}$ in sheep. *Pflügers Arch. 339*, 125-133.

Hall, G. H. (1973). Effects of nicotine on thermoregulatory systems in the hypothalamus. In *The Pharmacology of Thermoregulation*, E. Schönbaum and P. Lomax (Eds.). Karger, Basel, pp. 244-254.

Hall, G. H., and Myers, R. D. (1971). Hypothermia produced by nicotine perfused through the cerebral ventricles of the unanaesthetized monkey. *Neuropharmacol. 10*, 391-398.

Hall, G. H., and Myers, R. D. (1972). Temperature changes produced by nicotine injected into the hypothalamus of the conscious monkey. *Brain Res. 37*, 241-251.

Hall, G. H., and Turner, D. M. (1972). Effects of nicotine on the release of ^{3}H-noradrenaline from the hypothalamus. *Biochem. Pharmacol. 21*, 1829-1838.

Hanegan, J. L., and Williams, B. A. (1973). Brain calcium: role in temperature regulation. *Science 181*, 663-664.

Hanegan, J. L., and Williams, B. A. (1975). Ca^{2+} induced hypothermia in a hibernator (*Citellus beechyi*). *Comp. Biochem. Physiol. 50A*, 247-252.

Hansen, M. G., and Wishaw, I. Q. (1973). The effects of 6-hydroxydopamine, dopamine and *dl*-norepinephrine on food intake and water consumption, self-stimulation, temperature and electroencephalographic activity in the rat. *Psychopharmacol. 29*, 33-44.

Hardy, J. D. (1965). The "set-point" concept in physiological temperature regulation. In *Physiological Controls and Regulations*, W. S. Yamamoto and J. R. Brobeck (Eds.). W. B. Saunders, Philadelphia, pp. 98-116.

Hardy, J. D. (1972a). Models of temperature regulation: a review. In *Essays on Temperature Regulation*. J. Bligh and R. E. Moore (Eds.), North-Holland, Amsterdam, pp. 163-186.

Hardy, J. D. (1972b). Peripheral inputs to the central regulator for body temperature. In *Advances in Climatic Physiology*, S. Ito, K. Ogata, and H. Yoshimura (Eds.). Igaku Shoin, Tokyo, pp. 3-21.

Hardy, J. D. (1973). Posterior hypothalamus and the regulation of body temperature. *Fed. Proc. 32*, 1564-1571.

Hardy, J. D. (1976). Fever and thermogenesis. *Israel J. Med. Sci. 12*, 942-950.

Harri, M., and Tirri, R. (1969). Brain monoamines in the temperature acclimation of mice. *Acta Physiol. Scand. 75*, 631-635.

Harris, W. S., and Lipton, J. M. (1977). Intracerebroventricular taurine in rabbits: effects on normal body temperature, endotoxin fever and hyperthermia produced by PGE_1 and amphetamine. *J. Physiol., 266*, 397-410.

Hasama, B. (1930). Pharmakologische und physiologische Studien über die Schweisszentren. *Arch. Exp. Pathol. Pharmakol. 153*, 291-308.

Hashimoto, M. (1915a). Mitteilung: Über die spezifische Überempfindlichkeit des Warmzentrums an sensibilisierten Tieren. *Arch. Exp. Pathol. Pharmakol. 70*, 370-393.

Hashimoto, M. (1915b). Mitteilung: Über den Einfluß unmittelbarer Erwärmung und Abkuhlung des Wärmzentrums auf die Temperaturwirkungen von verschiedenen pyrogenen und antipyretischen Substanzen. *Arch. Exp. Pathol. Pharmakol. 70*, 394-425.

Hawkins, M., and Lipton, J. M. (1977). Analogs of endoperoxide precursors of prostaglandins: failure to affect body temperature when injected into primary and secondary central temperature controls. *Prostaglandins 13*, 209-217.

Hayward, J. N., and Baker, M. A. (1968). Diuretic and thermoregulatory responses to preoptic cooling in the monkey. *Am. J. Physiol. 214*, 843-850.

Hellon, R. F. (1967). Thermal stimulation of hypothalamic neurones in unanaesthetized rabbits. *J. Physiol. 193*, 381-395.

Hellon, R. F. (1975). Monoamines, pyrogens and cations: their actions on central control of body temperature. *Pharmacol. Rev. 26*, 289-321.

Hensel, H. (1973). Neural processes in thermoregulation. *Physiol. Rev. 53*, 948-1017.

Hissa, R., and Pyörnilä, A. (1977). Effect of intrahypothalamic phentolamine on hypothermia produced by peripheral noradrenaline in the pigeon. *Br. J. Pharmac. 61*, 163-166.

Hissa, R., and Rautenberg, W. (1974). The influence of centrally applied noradrenaline on shivering and body temperature in the pigeon. *J. Physiol. 238*, 421-435.

Hissa, R., and Rautenberg, W. (1975). Thermoregulatory effects of intrahypothalamic injections of neurotransmitters and their inhibitors in the pigeon. *Comp. Biochem. Physiol. 51*, 319-326.

Hoffman, R. A. (1958). Temperature response of the rat to action and interaction of chlorpromazine, reserpine and serotonin. *Am. J. Physiol. 195*, 755-758.

Hori, T. (1977). Single-unit studies on thermo-sensitive neurons. *Bull. Inst. Const. Med. Kumamoto Univ. 27* (Suppl.), 17-34.

Hori, T., and Harada, Y. (1974). The effects of ambient and hypothalamic temperatures on the hyperthermic responses to prostaglandins E_1 and E_2. *Pflügers Arch. 350*, 123-134.

Hori, T., and Nakayama, T. (1973). Effects of biogenic amines on central thermoresponsive neurones in the rabbit. *J. Physiol. 232*, 71-85.

Horita, A., and Gogerty, J. H. (1958). The pyretogenic effect of 5-hydroxytryptophan and its comparison with that of LSD. *J. Pharmacol. Exp. Ther. 122*, 195-200.

Horita, A., and Hill, H. F. (1973). Hallucinogens, amphetamines and temperature regulation. In *The Pharmacology of Thermoregulation*, E. Schönbaum and P. Lomax (Eds.). Karger, Basel, pp. 417-431.

Horita, A., and Quock, R. M. (1975). Dopaminergic mechanisms in drug-induced temperature effects. In *Temperature Regulation and Drug Action*, J. Lomax, E. Schönbaum, and J. Jacob (Eds.). Karger, Basel, pp. 75-84.

Horita, A., Carino, M. A., and Weick, B. G. (1976). Effects of thyrotropin releasing hormone (TRH) microinjected into various areas of conscious and pentobarbital-pretreated rabbits. *Proc. West. Pharmacol. Soc. 19*, 212-213.

Horton, E. W. (1969). Hypotheses on physiological roles of prostaglandins. *Physiol. Rev. 49*, 122-161.

Howard, J. L., and Breese, G. R. (1974). Physiological and behavioral effects of centrally-administered 6-hydroxydopamine in cats. *Pharmacol. Biochem. Behav. 2*, 651-661.

Hubbard, J. E., and Di Carlo, V. (1974). Fluorescence histochemistry of monoamine-containing cell bodies in the brain-stem of the squirrel monkey (*Saimir sciureus*). II. Catecholamine-containing groups. *J. Comp. Neurol. 153*, 369-384.

Hulst, S. G. T., and De Wied, D. (1967). Changes in body temperature and water intake following intracerebral implantation of carbachol in rats. *Physiol. Behav. 2*, 367-371.

Humphreys, R. B., Hawkins, M., and Lipton, J. M. (1976). Effects of anesthetic injected into brainstem sites on body temperature and behavioral thermoregulation. *Physiol. Behav. 17*, 667-674.

Ingenito, A. J., and Bonnycastle, D. D. (1967). The effect of exposure to heat and cold upon rat brain catecholamine and 5-hydroxytryptamine levels. *Can. J. Physiol. Pharmacol. 45*, 733-743.

Isaac, L. (1973). Temperature alteration of monoamine metabolites in cerebrospinal fluid. *Nat. New Biol. 243*, 269-271.

Jackson, D. L. (1967). A hypothalamic region responsive to localized injection of pyrogens. *J. Neurophysiol. 30*, 586-602.

Jacob, J. J., and Girault, J.-M. T. (1974). The influence of cyproheptadine and of D-lysergamide on the rise in temperature induced by intracerebroventricular 5-hydroxytryptamine, noradrenaline and dopamine in conscious rabbits. *Eur. J. Pharmacol. 27*, 59-67.

Jacob, J., Sauaudeau, C., and Michaud, G. (1971). Actions de la Noradrénaline, de la dopamine, de l'isopropylnoradrénaline et de la 5-hydroxytryptamine administrées par voie intracisternale et sous-cutanée sur la température du rat éveillé. *J. Pharmacol. 2,* 401-422.

Jacob, J., Girault, J. M., and Peindaries, R. (1972). Actions of 5-hydroxytryptamine and 5-hydroxytryptophan injected by various routes on the rectal temperature of the rabbit. *Neuropharmacol. 11,* 1-16.

Jacobj, C., and Roemer, C. (1912). *Arch. Exp. Pathol. 70,* 109 (cited by Barbour, 1921).

Jacobowitz, D. M., and Goldberg, A. M. (1977). Determination of acetylcholine in discrete regions of the rat brain. *Brain Res. 122,* 575-577.

Jacobowitz, D. M., and Palkovits, M. (1974). Topographic atlas of catecholamine and acetylcholinesterase-containing neurons in the rat brain. 1. Forebrain (telencephalon, diencephalon). *J. Comp. Neurol. 157,* 13-20.

Jacobowitz, D. M., Roizen, M. F., Suttora, N. L., and Muth, E. A. (1975). A method for the discrete removal of a segment of the dorsal noradrenergic bundle and its application to analysis of catecholamine and dopamine-β-hydroxylase. *Brain Res. 98,* 377-382.

Jancso-Gabor, A., Szolcsanyi, J., and Jancso, N. (1970). Stimulation and desensitization of the hypothalamic heat-sensitive structures by capsaicin in rats. *J. Physiol. 208,* 449-459.

Jell, R. M. (1973). Responses of hypothalamic neurones to local temperature and to acetylcholine, noradrenaline and 5-hydroxytryptamine. *Brain Res. 55,* 123-134.

Jell, R. M. (1974). Responses of rostral hypothalamic neurones to peripheral temperature and to amines. *J. Physiol. 240,* 295-307.

Jell, R. M. (1975). Amine-prostaglandin modulation of activity of thermoregulatory neurones. In *Temperature Regulation and Drug Action,* P. Lomax, E. Schönbaum, and J. Jacob (Eds.). Karger, Basel, pp. 119-123.

Jell, R. M., and Sweatman, P. (1976). Action of prostaglandin synthetase inhibitors on rostral hypothalamic neurones: thermoregulation and biogenic amines. *Can. J. Physiol. Pharmacol. 54,* 161-166.

Jell, R. M., and Sweatman, P. (1977). Prostaglandin sensitive neurones in cat hypothalamus: relationship to thermoregulation and to biogenic amines. *Can. J. Physiol. Pharmacol., 55,* 560-567.

Jellinek, P. (1971). Dual effect of dexamphetamine on body temperature in the rat. *Eur. J. Pharmacol. 15,* 389-392.

Jones, D. L., Veale, W. L., and Cooper, K. E. (1978). Perfusions of the posterior hypothalamus of cats with various ions and saccharides: effects on body temperature. *Can. J. Physiol. Pharmacol. 56,* 571-577.

Jouvet, M. (1972). The role of monoamines and acetylcholine-containing neurons in the regulation of the sleep-waking cycle. In *Ergebnisse der Physiologie,* Springer-Verlag, Heidelberg, pp. 166-307.

Judah, L. J. (1916). *J. Hyg. 15,* 169 (cited by Barbour, 1921).

Kakiuchi, S., and Rall, T. W. (1968). The influence of chemical agents on the accumulation of adenosine $3',5'$-phosphate in slices of rabbit cerebellum. *Mol. Pharmacol. 4,* 367-378.

Kandasmay, B., Girault, J.-M., and Jacob, J. (1975). Central effects of a purified bacterial pyrogen, prostaglandin E_1 and biogenic amines on the temperature in the awake rabbit. In *Temperature Regulation and Drug Action,* P. Lomax, E. Schönbaum, and J. Jacob (Eds.). Karger, Basel, pp. 124-132.

Keller, A. D. (1933). Observations on the localization in the brain-stem of mechanisms controlling body temperature. *Am. J. Med. Sci. 185*, 746-748.

Keller, A. D., and Hare, W. K. (1932). The hypothalamus and heat regulation. *Proc. Soc. Expl. Biol. Med. 29*, 1069-1070.

Keller, A. D., and McClaskey, E. B. (1964). Localization, by the brain slicing method, of the level or levels of the cephalic brainstem upon which effective heat dissipation is dependent. *Am. J. Phys. Med. 43*, 181-213.

Kennedy, M. S., and Burks, T. F. (1974). Dopamine receptors in the central thermoregulatory mechanism of the cat. *Neuropharmacology 13*, 119-128.

Kirkpatrick, W. E., and Lomax, P. (1967). The effect of atropine on the body temperature of the rat following systemic and intracerebral injection. *Life Sci. 6*, 2273-2278.

Kirkpatrick, W. E., and Lomax, P. (1970). Temperature changes following iontophoretic injection of acetylcholine into the rostral hypothalamus of the rat. *Neuropharmacology 9*, 195-202.

Kirkpatrick, W. E., and Lomax, P. (1971). Temperature changes induced by chlorpromazine and N-methyl-chlorpromazine in the rat. *Neuropharmacol. 10*, 61-66.

Kirkpatrick, W. E., Jenden, D. J., and Lomax, P. (1967). The effect of N-(4-diethylamino-2-butynyl)-succinimide (DKJ 21) on oxotremorine induced hypothermia in the rat. *Int. J. Neuropharmacol. 6*, 273-277.

Kirtland, S. J., and Baum, H. (1972). Prostaglandin E_1 may act as a "calcium ionophore." *Nat. New Biol. 236*, 47-49.

Kluger, M. J., Bernheim, H. A., Vaughn, L. K., Foster, M. A., and D'Alecy, L. G. (1977). Evolution and adaptive value of fever. In *Drugs, Biogenic Amines and Body Temperature*. K. E. Cooper, P. Lomax, and E. Schönbaum (Eds.). Karger, Basel.

Knox, G. V., and Lomax, P. (1972). The effect of nicotine on thermosensitive units in the rostral hypothalamus. *Proc. West. Pharmacol. Soc. 15*, 179-183.

Knox, G. V., Campbell, C., and Lomax, P. (1973). The effects of acetylcholine and nicotine on unit activity in the hypothalamic thermoregulatory centers of the rat. *Brain Res. 51*, 215-223.

Komiskey, H. L., and Rudy, T. A. (1975). The involvement of methysergide-sensitive receptors and prostaglandins in the hyperthermia evoked by 5-HT in the cat. *Res. Comm. Chem. Pathol. Pharmacol. 11*, 195-208.

Komiskey, H. L., and Rudy, T. A. (1977). Serotonergic influences on brain stem thermoregulatory mechanisms in the cat. *Brain Res. 134*, 297-315.

Kostrzewa, R. M., and Jacobowitz, D. M. (1974). Pharmacological actions of 6-hydroxydopamine. *Pharmacol. Rev. 26*, 199-287.

Kroneberg, G., and Kurbjuweit, H. G. (1959). Die Beeinflussung von experimentellem Fieber durch Reserpin und Sympathicolytica am Kaninchen. *Arzn.-Forsch. 9*, 556-558.

Kruk, Z. L. (1972). The effect of drugs acting on dopamine receptors on the body temperature of the rat. *Life Sci. 11*, 845-850.

Kruk, Z. L., and Brittain, R. T. (1972). Changes in body core and skin temperature following intracerebroventricular injection of substances in the conscious rat: interpretation of data. *J. Pharmaceut. Pharmacol. 24*, 835-837.

Kubikowski, P., and Rewerski, W. (1969). Biogenic amines and body temperature regulation in the rats. *Diss. Pharmaceut. Pharmacol. 21*, 207-212.

Kym, O. (1934). Die Beeinflussung des durch verschiedene fiebererzeugende Stoffe erregten Temperaturzentrums durch lokale Applikation von Ca, K und Na. *Arch. Exp. Pathol. Pharmakol. 176,* 408-424.

Laburn, H. P., Rosendorff, C., Willies, G., and Woolf, C. (1974). A role for noradrenaline and cyclic AMP in prostaglandin E_1 fever. *J. Physiol. 240,* 49-50P.

Laburn, H., Woolf, C. J., Willies, G. H., and Rosendorff, C. (1975). Pyrogen and prostaglandin fever in the rabbit. II. Effects of noradrenaline depletion and adrenergic receptor blockade. *Neuropharmacology 14,* 405-411.

Laudenslager, M. L., and Carlisle, H. J. (1976). Heat-escape behavior in the rat following intrahypothalamic injection of acetylcholine. *Pharmacol. Biochem. Behav. 4,* 369-373.

Laverty, R. (1963). The influence of cold environmental temperature and histamine treatment on the effect of reserpine in the rat. *J. Neurochem. 10,* 151-154.

Lechat, P., and Gardey-Levassort, C. (1975). Brain levels of noradrenaline, dopamine, homovanillic acid and 3,4-dihydroxyphenylacetic acid in rabbits made tolerant to a bacterial pyrogen. In *Temperature Regulation and Drug Action,* P. Lomax, E. Schönbaum, and J. Jacob (Eds.). Karger, Basel, pp. 143-149.

Leduc, J. (1961). Effect of exposure to cold on the production and release of catecholamines. *Acta Physiol. Scand. 53* (Suppl. 183), 18-38.

Lessin, A. W., and Parkes, M. W. (1957). The relation between sedation and body temperature in the mouse. *Br. J. Pharmacol. 12,* 245.

Lewis, P. R., and Shute, C. C. D. (1967). The cholinergic limbic system: projections to hippocampal formation, medial cortex, nuclei of the ascending cholinergic reticular system, and the subfornical organ and supra-optic crest. *Brain 90,* 521-540.

Lin, M. T. (1978). Effects of specific inhibitors of 5-hydroxytryptamine uptake on thermoregulation in rats. *J. Physiol. 284,* 147-154.

Lipton, J. M., and Fossler, D. E. (1974). Fever produced in the squirrel monkey by intravenous and intracerebral endotoxin. *Am. J. Physiol. 226,* 1022-1027.

Lipton, J. M., and Kennedy, J. I. (1979). Central thermosensitivity during fever produced by intra-PO/AH and intravenous injections of pyrogen. *Brain Res. Bull. 4,* 23-34.

Lipton, J. M., and Trzcinka, G. P. (1976). Persistence of febrile response to pyrogens after PO/AH lesions in squirrel monkeys. *Am. J. Physiol. 231,* 1638-2648.

Lipton, J. M., Welch, J. P., and Clark, W. G. (1973). Changes in body temperature produced by injecting prostaglandin E_1, EGTA and bacterial endotoxins into the PO/AH region and the medulla oblongata of the rat. *Experientia 29,* 806-808.

Lipton, J. M., Kirkpatrick, J., and Rosenberg, R. N. (1977). Hypothermia and persisting capacity to develop fever. Occurrence in a patient with sarcoidosis of the central nervous system. *Arch. Neurol. 34,* 498-504.

Lipton, J. M., Dinarello, C. A., and Kennedy, J. I. (1979). Fever produced in the squirrel monkey by human leukocytic pyrogen (40463). *Proc. Soc. Exp. Biol. Med. 160,* 426-428.

Liu, J. C. (1979). Tonic inhibition of thermoregulation in the decerebrate monkey (*Saimiri sciureus*). *Exptl. Neurol. 64,* 632-648.

Lomax, P. (1966). The hypothermic effect of pentobarbital in the rat: sites and mechanisms of action. *Brain Res. 1,* 296-302.

Lomax, P. (1967). Investigations on the central effects of morphine on body temperature. *Arch. Biol. Med. Exp. 4,* 119-124.

Lomax, P. (1970). Drugs and body temperature. *Int. Rev. Neurobiol. 12,* 1-43.

Lomax, P., and Green, M. D. (1975). Histamine and temperature regulation. In *Temperature Regulation and Drug Action*, P. Lomax, E. Schönbaum, and J. Jacob (Eds.). Karger, Basel, pp. 85-94.

Lomax, P., and Jenden, D. J. (1966). Hypothermia following systematic and intracerebral injection of oxotremorine in the rat. *Int. J. Neuropharmacol. 5,* 353-359.

Lomax, P., and Foster, R. S. (1969). Temperature changes induced by imidazoline sympathomimetics in the rat. *J. Pharmacol. Exp. Ther. 167,* 159-165.

Lomax, P., Foster, R. S., and Kirkpatrick, W. E. (1969). Cholinergic and adrenergic interactions in the thermoregulatory centers of the rat. *Brain Res. 15,* 431-438.

Lotti, V. J. (1965). The hypothermic effect of morphine in the rat: site of mechanism of action. *Doct. Diss. UCLA,* pp. 166-187.

Lotti, V. J., Lomax, P., and George, R. (1965a). Temperature responses in the rat following intracerebral microinjection of morphine. *J. Pharmacol. Exp. Ther. 150,* 135-139.

Lotti, V. J., Lomax, P., and George, R. (1965b). N-Allylnormorphine antagonism of the hypothermic effect of morphine in the rat following intracerebral and systemic administration. *J. Pharmacol. Exp. Ther. 150,* 420-425.

Lotti, V. J., Lomax, P., and George, R. (1966a). Heat production and heat loss in the rat following intracerebral and systemic administration of morphine. *Int. J. Neuropharmacol. 5,* 75-83.

Lotti, V. J., Lomax, P., and George, R. (1966b). Acute tolerance to morphine following systemic and intracerebral injection in the rat. *Int. J. Neuropharmacol. 5,* 35-42.

Maeda, T., and Shimizu, N. (1972). Projections ascendantes du locus coeruleus et d'autres neurones aminergiques pontiques au niveau de prosencéphale du rat. *Brain Res. 36,* 19-35.

Maickel, R. P. (1970). Interaction of drugs with autonomic nervous function and thermoregulation. *Fed. Proc. 29,* 1973-1979.

Marley, E., and Nistico, G. (1972). Effects of catecholamines and adenosine derivatives given into the brain of fowls. *Br. J. Pharmacol. 46,* 619-636.

Marley, E., and Nistico, G. (1975). Central effects of clonidine 2-(2,6-dichlorophenylamino)-2-imidazoline hydrochloride in fowls. *Br. J. Pharmacol. 55,* 459-473.

Marley, E., and Seller, T. J. (1974). Effects of nicotine given into the brain of fowls. *Br. J. Pharmacol. 51,* 335-346.

Marley, E., and Stephenson, J. D. (1970). Effects of catecholamines infused into the brain of young chickens. *Br. J. Pharmacol. 40,* 639-658.

Marley, E., and Stephenson, J. D. (1975). Effects of noradrenaline infused into the chick hypothalamus on thermoregulation below thermoneutrality. *J. Physiol. 245,* 289-303.

Marley, E., and Whelan, J. E. (1975). Some central effects of 5-hydroxytryptamine in young chickens at and below thermoneutrality. *Br. J. Pharmacol. 53,* 37-41.

Martin, G. E., and Morrison, J. E. (1978). Hyperthermia evoked by the intracerebral injection of morphine sulphate in the rat: the effect of restraint. *Brain Res. 145,* 127-140.

Martin, G. E., and Myers, R. D. (1975). Evoked release of [^{14}C-]norepinephrine from the rat hypothalamus during feeding. *Am. J. Physiol. 229,* 1547-1555.

Mašek, K., Rašková, H., and Rotta, J. (1968). The mechanism of the pyrogenic effect of streptococcus cell wall mucopeptide. *J. Physiol. 198,* 345-353.

Mašek, K., Kadlecová, O., and Rašková, H. (1973). Brain amines in fever and sleep cycle changes caused by streptococcal mucopeptide. *Neuropharmacology 12*, 1039-1047.

Maskrey, M., and Bligh, J. (1971). Interactions between the thermoregulatory responses to injections into a lateral cerebral ventricle of the Welsh mountain sheep of putative neurotransmitter substances and of local changes in anterior hypothalamic temperature. *Int. J. Biometeor. 15*, 129-133.

Maynert, E. W., and Levi, R. (1964). Stress-induced release of brain norepinephrine and its inhibition by drugs. *J. Pharmacol. Exp. Ther. 143*, 90-95.

Medon, P. J., and Blake, D. E. (1973). Temperature effects of intraventricular serotonin, norepinephrine and pilocarpine in the morphine-tolerant rat. *Life Sci. 13*, 1395-1402.

Medon, P. J., Haubrich, R. D., and Blake, D. E. (1969). Modification of the hypothermic action of morphine after the depletion of brain serotonin or catecholamines. *Pharmacologist 11*, 258.

Metcalf, G. (1974). TRH: a possible mediator of thermoregulation. *Nature (London) 252*, 310-311.

Metcalf, G., and Myers, R. D. (1976). A comparison between the hypothermia induced by intraventricular injections of thyrotropin releasing hormone, noradrenaline or calcium ions in unanaesthetized cats. *Br. J. Pharmacol. 58*, 489-495.

Metcalf, G., and Myers, R. D. (1978). Precise location within the preoptic area where noradrenaline produces hypothermia. *Eur. J. Pharmacol. 51*, 47-53.

Metcalf, G., Myers, R. D., and Redgrave, P. C. (1975). Temperature and behavioural responses induced in the unanaesthetized cat by the central administration of RX72601, a new anticholinesterase. *Br. J. Pharmacol. 55*, 9-15.

Meyer, H. H. (1913). Theorie des Fiebers und seiner Behandlung. *Verhandl. Deuts. Bes. Inner. Med. 30*, 15-25.

Milton, A. S. (1976). Modern views on the pathogenesis of fever and the mode of action of antipyretic drugs. *J. Pharmaceut. Pharmacol. 28*, 393-399.

Milton, A. S., and Harvey, C. A. (1975). Prostaglandins and monoamines in fever. In *Temperature Regulation and Drug Action*, P. Lomax, E. Schönbaum, and J. Jacob (Eds.). Karger, Basel, pp. 133-142.

Milton, A. S., and Wendlandt, S. (1968). The effect of 4-acetamindophenol in reducing fever produced by the intracerebral injection of 5-hydroxytryptamine and pyrogen in the conscious cat. *Br. J. Pharmacol. 34*, 215P.

Milton, A. S., and Wendlandt, S. (1970). A possible role for prostaglandin E_1 as a modulator for temperature regulation in the central nervous system of the cat. *Proc. Physiol. Soc. 207*, 16-17.

Milton, A. S., and Wendlandt, S. (1971). Effects on body temperature of prostaglandins of the A, E and F series on injection into the third ventricle of unanaesthetized cats and rabbits. *J. Physiol. 218*, 325-336.

Mitchell, D., Snellen, J. W., and Atkins, A. R. (1970). Thermoregulation during fever: change of set-point or change of gain. *Pflügers Arch. 321*, 293-302.

Morgan, L. O. (1938). Cell-changes in some of the hypothalamic nuclei in experimental fever. *J. Neurophysiol. 1*, 281-285.

Morgane, P. J., and Stern, W. C. (1972). Relationship of sleep to neuroanatomical circuits, biochemistry and behavior. *Ann. N.Y. Acad. Sci. 193*, 95-111.

Müller, E. E., Cocchi, D., Jalanbo, H., Udeschini, G., Peruzzi, G., and Mantegazza, P.

(1974). Central hypothermia by 2-deoxy-D-glucose: antagonism by α-adrenergic activation. *Eur. J. Pharmacol. 26*, 243-255.

Murakami, N. (1973). Effects of iontophoretic application of 5-hydroxytryptamine, noradrenaline and acetylcholine upon hypothalamic temperature-sensitive neurones in rats. *Jpn. J. Physiol. 23*, 435-446.

Myers, R. D. (1966). Temperature regulation in the conscious monkey: chemical mechanisms in the hypothalamus. *Int. Biometerol. Congr. Proc. 125*,

Myers, R. D. (1968). Discussion of serotonin, norepinephrine, and fever. *Adv. Pharmacol. 6*, 318-321.

Myers, R. D. (1969a). Temperature regulation: neurochemical systems in the hypothalamus. In *The Hypothalamus*, W. Haymaker, E. Anderson, and W. Nauta (Eds.). Thomas, Springfield, Ill., pp. 506-523.

Myers, R. D. (1969b). Thermoregulation and norepinephrine. *Science 165*, 1030-1031.

Myers, R. D. (1970). The role of hypothalamic transmitter factors in the control of body temperature. In *Physiological and Behavioral Temperature Regulation*, J. D. Hardy (Ed.). Thomas, Springfield, Ill., pp. 648-666.

Myers, R. D. (1971a). Methods for chemical stimulation of the brain. In *Methods in Psychobiology*, Vol. I, R. D. Myers (Ed.). Academic, London, pp. 247-280.

Myers, R. D. (1971b). Primates. In *Comparative Physiology of Thermoregulation*, Vol. 2, C. C. Whittow (Ed.), Academic, New York, pp. 283-326.

Myers, R. D. (1971c). Hypothalamic mechanisms of pyrogen action in the cat and monkey. In *Ciba Foundation Symposium on Pyrogens and Fever*, G. E. W. Wolstenholme and J. Birch (Eds.). Churchill, London, pp. 131-153.

Myers, R. D. (1973). The role of hypothalamic serotonin in thermoregulation. In *Serotonin and Behavior*, J. Barchas and E. Usdin (Eds.). Academic, New York, pp. 293-302.

Myers, R. D. (1974a). Temperature regulation. In *Handbook of Drug and Chemical Stimulation of the Brain.* Von Nostrand-Reinhold, New York, pp. 237-285.

Myers, R. D. (1974b). Ionic concepts of the set-point for body temperature. In *Recent Studies of Hypothalamic Function*, K. Lederis and K. E. Cooper (Eds.). Karger, Basel, pp. 371-390.

Myers, R. D. (1975a). Impairment of thermoregulation, food and water intakes in the rat after hypothalamic injections of 5,6-dihydroxytryptamine. *Brain Res. 94*, 491-506.

Myers, R. D. (1975b). An integrative model of monoamine and ionic mechanisms in the hypothalamic control of body temperature. In *Temperature Regulation and Drug Action*, J. Lomax, E. Schönbaum, and J. Jacob (Eds.). Karger, Basel, pp. 32-42.

Myers, R. D. (1976a). Chemical control of body temperature by the hypothalamus: a model and some mysteries. *Proc. Austral. Physiol. Pharmacol. Soc. 7*, 15-32.

Myers, R. D. (1976b). Diencephalic efflux of $^{22}Na^+$ and $^{45}Ca^{2+}$ ions in the febrile cat: effect of an antipyretic. *Brain Res. 103*, 412-417.

Myers, R. D. (1977). New aspects of the role of hypothalamic calcium ions, 5-HT and PGE during normal thermoregulation and pyrogen fever. In *Drugs, Biogenic Amines and Body Temperature.* K. E. Cooper, P. Lomax, and E. Schönbaum (Eds.). Karger, Basel, pp. 51-53.

Myers, R. D. (1978a). Hypothalamic actions of 5-HT neurotoxins: feeding, drinking and body temperature. *Ann. N. Y. Acad. Sci., 305*, 556-575.

Myers, R. D. (1978b). Hypothalamic mechanisms underlying physiological set points. In *Current Studies of Hypothalamic Function,* Vol. 2, K. Lederis and W. L. Veale (Eds.). Karger, Basel, 17-28.

Myers, R. D. (1980). Catecholamines and the regulation of body temperature. In *Adrenergic Activators and Inhibitors,* L. Szekeres (Ed.). Springer, Berlin, pp. 549-567.

Myers, R. D., and Beleslin, D. B. (1971). Changes in serotonin release in hypothalamus during cooling or warming of the monkey. *Am. J. Physiol. 220,* 1746-1754.

Myers, R. D., and Brophy, P. D. (1972). Temperature changes in the rat produced by altering the sodium-calcium ratio in the cerebral ventricles. *Neuropharmacology 11,* 351-361.

Myers, R. D., and Buckman, J. E. (1972). Deep hypothermia induced in the golden hamster by altering cerebral calcium levels. *Am. J. Physiol. 223,* 1313-1318.

Myers, R. D., and Chinn, C. (1973). Evoked release of hypothalamic norepinephrine during thermoregulation in the cat. *Am. J. Physiol. 224,* 230-236.

Myers, R. D., and Hoch, D. B. (1978). ^{14}C-dopamine microinjected into the brain-stem of the rat: dispersion kinetics, site content and functional dose. *Brain Res. Bull. 3,* 601-609.

Myers, R. D., and Ruwe, W. E. (1978). Thermoregulation in the rat: deficits following 6-OHDA injections in the hypothalamus. *Pharmacol. Biochem. Behav. 8,* 377-385.

Myers, R. D., and Ruwe, W. D. (1980). Fever: intermediary neurohumoral factors serving the hypothalamic mechanism underlying hyperthermia. In *International Symposium on Fever,* J. Lipton (Ed.). Raven, New York, pp. 99-110.

Myers, R. D., and Sharpe, L. G. (1968). Intracerebral injections and perfusions in the conscious monkey. In *Use of Non-Human Primates in Drug Evaluation,* H. Vagtborg (Ed.). Texas Press, Austin, pp. 450-465.

Myers, R. D., and Tytell, M. (1972). Fever: reciprocal shift in brain sodium to calcium ratio as the set-point temperature rises. *Science 178,* 765-767.

Myers, R. D., and Veale, W. L. (1970). Body temperature: possible ionic mechanism in the hypothalamus controlling the set point. *Science 170,* 95-97.

Myers, R. D., and Veale, W. L. (1971). The role of sodium and calcium ions in the hypothalamus in the control of body temperature of the unanaesthetized cat. *J. Physiol. 212,* 411-430.

Myers, R. D., and Waller, M. B. (1973). Differential release of acetylcholine from the hypothalamus and mesencephalon of the monkey during thermoregulation. *J. Physiol. 230,* 273-293.

Myers, R. D., and Waller, M. B. (1975a). Species continuity in the thermoregulatory responses of the pigtailed macaque to monoamines injected into the hypothalamus. *Comp. Biochem. Physiol. 51A,* 639-645.

Myers, R. D., and Waller, M. B. (1975b). 5-HT- and norepinephrine-induced release of ACh from the thalamus and mesencephalon of the monkey during thermoregulation. *Brain Res. 84,* 47-61.

Myers, R. D., and Waller, M. B. (1976). Is prostaglandin fever mediated by the presynaptic release of hypothalamic 5-HT or norepinephrine. *Brain Res. Bull. 1,* 47-56.

Myers, R. D., and Waller, M. B. (1977). Serotonin and thermoregulation. In *Serotonin in Health and Disease,* W. B. Essman (Ed.). Spectrum, New York, pp. 1-67.

Myers, R. D., and Yaksh, T. L. (1968). Feeding and temperature responses in the unrestrained rat after injections of cholinergic and aminergic substances into the cerebral ventricles. *Physiol. Behav. 3,* 917-928.

Myers, R. D., and Yaksh, T. L. (1969). Control of body temperature in the unanaesthetized monkey by cholinergic and aminergic systems in the hypothalamus. *J. Physiol. 202*, 483-500.

Myers, R. D., and Yaksh, T. L. (1971). Thermoregulation around a new "set-point" established in the monkey by altering the ratio of sodium to calcium ions within the hypothalamus. *J. Physiol. 218*, 609-633.

Myers, R. D., and Yaksh, T. L. (1972). The role of hypothalamic monoamines and ions in hibernation and hypothermia. In *Hibernation and Hypothermia: Perspectives and Challenges*, F. E. South, J. P. Hannon, J. R. Willis, E. T. Pengelley, and N. R. Alpert (Eds.). Elsevier, Amsterdam, pp. 551-575.

Myers, R. D., Kawa, A., and Beleslin, D. B. (1969). Evoked release of 5-HT and NEFA from the hypothalamus of the conscious monkey during thermoregulation. *Experientia 25*, 705-706.

Myers, R. D., Veale, W. L., and Yaksh, T. L. (1971a). Changes in body temperature of the unanaesthetized monkey produced by sodium and calcium ions perfused through the cerebral ventricles. *J. Physiol. 217*, 381-392.

Myers, R. D., Rudy, T. A., and Yaksh, T. L. (1971b). Fever in the monkey produced by the direct action of pyrogen on the hypothalamus. *Experientia 27*, 160-161.

Myers, R. D., Rudy, T. A., and Yaksh, T. L. (1971c). Effect in the rhesus monkey of salicylate on centrally-induced endotoxin fevers. *Neuropharmacol. 10*, 775-778.

Myers, R. D., Tytell, M., Kawa, A., and Rudy, T. (1971d). Microinjection of ^{3}H-acetylcholine, ^{14}C-serotonin and ^{3}H-norepinephrine into the hypothalamus of the rat: diffusion into tissue and ventricles. *Physiol. Behav. 7*, 743-751.

Myers, R. D., Rudy, T. A., and Yaksh, T. L. (1973). Evocation of a biphasic febrile response in the rhesus monkey by intracerebral injection of bacterial endotoxins. *Neuropharmacology 12*, 1195-1198.

Myers, R. D., Rudy, T. A., and Yaksh, T. L. (1974). Fever produced by endotoxin injected into the hypothalamus of the monkey and its antagonism by salicylate. *J. Physiol. 243*, 167-193.

Myers, R. D., Melchior, C. L., and Gisolfi, C. V. (1976a). Feeding and body temperature in the rat: diencephalic localization of changes produced by excess calcium ions. *Brain Res. Bull. 1*, 33-46.

Myers, R. D., Simpson, C. W., Higgins, D., Nattermann, R. A., Rice, J. C., Redgrave, P., and Metcalf, G. (1976b). Hypothalamic Na^+ and Ca^{++} ions and temperature set-point: new mechanisms of action of a central or peripheral thermal challenge and intrahypothalamic 5-HT, NE, PGE_1 and pyrogen. *Brain Res. Bull. 1*, 301-327.

Myers, R. D., Metcalf, G., and Rice, J. C. (1977a). Identification by microinjection of TRH-sensitive sites in the cat's brainstem that mediate respiratory, temperature and other autonomic changes. *Brain Res. 126*, 105-115.

Myers, R. D., Gisolfi, C. V., and Mora, F. (1977b). Calcium levels in the brain underlie temperature control during exercise in the primate. *Nature (London) 266*, 178-179.

Myers, R. D., Gisolfi, C. V., and Mora, F. (1977c). Role of brain Ca^{2+} in central control of body temperature during exercise in the monkey. *J. Appl. Physiol. 43*, 617-630.

Nakamura, K., and Thoenen, H. (1971). Hypothermia induced by intraventricular administration of 6-hydroxydopamine in rats. *Eur. J. Pharmacol. 16*, 46-54.

Nakayama, T., and Hori, T. (1973). Effects of anesthetic and pyrogen on thermally sensitive neurons in the brainstem. *J. Appl. Physiol. 34*, 351-355.

Netherton, R. A., Lee, P. S., and Overstreet, D. H. (1977). Are there 2 cholinergic thermoregulatory centres in rats? *Experientia 33*, 1463-1464.

Nielsen, B. (1969). Thermoregulation in rest and exercise. *Acta Physiol. Scand. 323*, 1-74.

Nielsen, B., and Greenleaf, J. E. (1977). Electrolytes and thermoregulation. In *Drugs, Biogenic Amines and Body Temperature*, K. E. Cooper, P. Lomax, and E. Schönbaum (Eds.). Karger, Basel, pp. 39-47.

Nielsen, M. (1938). Die Regulation der Korpertemperatur bei Muskelarbeit. *Skand. ARch. Physiol. 79*, 193-230.

Nistico, G. (1975). Prostaglandine, pirogeno e febbre. *Rend. Atti Accad. Sci. Med. Chir. 129*, 1-15.

Nistico, G., and Marley, E. (1972). Central effects of amines in adult fowls. *Arch. Int. Pharmacod. Ther. 196*, 136-137.

Nistico, G., and Marley, E. (1973). Central effects of prostaglandin E_1 in adult fowls. *Neuropharmacology 12*, 1009-1016.

Nistico, G., and Marley, E. (1974). Neurotrasmettitori e termoregolazione. *Acta Neurol. 33*, 5-51.

Nistico, G., and Marley, E. (1976). Central effects of prostaglandins E_2, A_1 and $F_{2\alpha}$ in adult fowls. *Neuropharmacology 15*, 737-741.

Nistico, G., and Rotiroti, D. (1978). Antipyretics and fever induced in adult fowls by prostaglandins E_1, E_2 and O somatic antigen. *Neuropharmacology 17*, 197-203.

Nistico, G., Marley, E., and Preziosi, P. (1976). Receptor involvement in monoamine control of thermoregulatory mechanisms in adult fowls. *Pharmacol. Res. Commun. 8*, 53-63.

Nistico, G., Rotiroti, D., de Sarro, A., and Stephenson, J. (1978). Behavioural electrocortical and body temperature effects after intracerebral infusion of TRH in fowls. *Eur. J. Pharmacol. 50*, 253-260.

Novotná, R., and Janský, L. (1976). Effect of different environmental temperatures on the serotonin concentration and turnover in the brain stem of a hibernator. *Physiol. Bohemoslov. 25*, 37-42.

Nutik, S. L. (1971). Effect of temperature change of the preoptic region and skin on posterior hypothalamic neurons. *J. Physiol. (Paris) 63*, 368-370.

Olive, G., Gardey-Levassort, C., and Lechat, P. (1971). Détermination du turn-over de la serotoniné dans l'hypothalamus et le tronc cérébral du lapin a l'acmé de la fievre provoquée par un pyrogène bactérien. *J. Pharmacol. 2*, 61-70.

Olson, L., and Fuxe, K. (1971). On the projections from the locus coeruleus noradrenaline neurons: the cerebellar innervation. *Brain Res. 28*, 165-171.

Palkovits, M., Brownstein, M., Saavedra, J. M., and Axelrod, J. (1974). Norepinephrine and dopamine content of hypothalamic nuclei of the rat. *Brain Res. 77*, 137-149.

Paolino, R. M., and Bernard, B. K. (1968). Environmental temperature effects on the thermoregulatory response to systemic and hypothalamic administration of morphine. *Life Sci. 7*, 857-863.

Peters, D. A. V., Hrdina, P. D., Singhal, R. L., and Ling, G. M. (1972). The role of brain serotonin in DDT-induced hyperpyrexia. *J. Neurochem. 19*, 1131-1136.

Philipp-Dormston, W. K. (1976). Evidence for the involvement of adenosine $3'$, $5'$-cyclic monophosphate in fever genesis. *Pflügers Arch. 362*, 223-227.

Philipp-Dormston, W. K., and Siegert, R. (1974). Prostaglandins of the E and F series in rabbit cerebrospinal fluid during fever induced by Newcastle disease virus, *E. coli*-endotoxin, or endogenous pyrogen. *Med. Microbiol. Immunol. 159*, 279-284.

Philipp-Dormston, W. K., and Siegert, R. (1975). Adenosine $3',5'$-cyclic monophosphate in rabbit cerebrospinal fluid during fever induced by *E. coli*-endotoxin. *Med. Microbiol. Immunol. 161*, 11-13.

Pittman, Q. J., Veale, W. L., and Cooper, K. E. (1975). Temperature responses of lambs after centrally injected prostaglandins and pyrogens. *Amer. J. Physiol. 228*, 1034-1038.

Pittman, Q. J., Veale, W. L., Cockeram, A. W., and Cooper, K. E. (1976a). Changes in body temperature produced by prostaglandins and pyrogens in the chicken. *Am. J. Physiol. 230*, 1284-1287.

Pittman, Q. J., Veale, W. L., and Cooper, K. E. (1976b). Observations on the effect of salicylate in fever and the regulation of body temperature against cold. *Can. J. Physiol. Pharmacol. 54*, 101-106.

Pittman, Q. J., Veale, W. L., and Cooper, K. E. (1977a). Effect of prostaglandin, pyrogen and noradrenaline, injected into the hypothalamus, on thermoregulation in newborn lambs. *Brain Res. 128*, 473-483.

Pittman, Q. J., Veale, W. L., and Cooper, K. E. (1977b). Absence of fever following intrahypothalamic injections of prostaglandins in sheep. *Neuropharmacology 16*, 743-749.

Pohorecky, L. A., Brick, J., and Sun, J. Y. (1976). Serotoninergic involvement in the effect of ethanol on body temperature in rats. *J. Pharmaceut. Pharmacol. 28*, 157-159.

Poole, S., and Stephenson, J. D. (1979a). Effects of noradrenaline and carbachol on temperature regulation of rats. *Br. J. Pharmacol. 65*, 43-51.

Poole, S., and Stephenson, J. D. (1979b). Effects of noradrenaline and carbachol on temperature regulation of cold-stressed and cold-acclimated rats. *Br. J. Pharmacol. 66*, 307-315.

Poulain, P., and Carette, M. B. (1974). Iontophoresis of prostaglandins on hypothalamic neurons. *Brain Res. 79*, 311-314.

Preston, E. (1974). Central effects of cholinergic-receptor blocking drugs on the conscious rabbit's thermoregulation against body cooling. *J. Pharmacol. Exp. Ther. 188*, 400-409.

Preston, E. (1975). Thermoregulation in the rabbit following intracranial injection of norepinephrine. *Am. J. Physiol. 229*, 676-682.

Preston, E., and Cooper, K. E. (1976). Absence of fever in the rabbit following intrahypothalamic injections of noradrenaline into PGE_1-sensitive sites. *Neuropharmacology 15*, 239-244.

Preston, E., and Schönbaum, E. (1976). Monoaminergic mechanisms in thermoregulation. In *Brain Dysfunction in Infantile Febrile Convulsions*. M. A. B. Brazier and F. Coceani (Eds.). Raven, New York, pp. 75-87.

Pyörnilä, A., and Hissa, R. (1979). Opposing temperature responses to intrahypothalamic injections of 5-hydroxytryptamine in the pigeon exposed to cold. *Experientia 35*, 59-60.

Pyörnilä, A., Lahti, H., and Hissa, R. (1977). Thermoregulatory changes induced by intrahypothalamic injections of cholinomimetic substances in the pigeon. *Neuropharmacology 16*, 737-741.

Pyörnilä, A., Hissa, R., and Jeronen, E. (1978). The influence of high ambient temperature on thermoregulatory response to intrahypothalamic injections of noradrenaline and serotonin in the pigeon. *Pflügers Arch. 377*, 51-55.

Quock, R. M., and Gale, C. C. (1974). Hypothermia-mediating dopamine receptors in the preoptic anterior hypothalamus of the cat. *Arch. Pharmacol. 285*, 297-300.

Ranson, S. W. (1940). Regulation of body temperature. In *The Hypothalamus and Central Levels of Autonomic Function*, Williams & Wilkins, Baltimore, pp. 342-399.

Ranson, S. W., Jr., Clark, G., and Magoun, H. W. (1939). The effect of hypothalamic lesions on fever induced by intravenous injection of typhoid-paratyphoid vaccine. *J. Lab. Chem. Med. 25*, 160-168.

Rašková, H., Jiřička, A., Mašek, K., Smetana, R., Švihouec, J., and Vaněček, J. (1966). The release of biogenic amines by bacteria toxins. In *Mechanisms of Release of Biogenic Amines*, U. S. von Euler, S. Rosell, and B. Uvnas (Eds.). Pergamon, London, pp. 349-354.

Rawlins, M. D., Luff, R. H., and Cranston, W. I. (1973). Regional brain salicylate concentrations in afebrile and febrile rabbits. *Biochem. Pharmacol. 22*, 2639-2642.

Reid, J. L., Lewis, P. J., and Myers, M. G. (1975). Role of central dopaminergic mechanisms in piribedil and clonidine induced hypothermia in the rat. *Neuropharmacology 14*, 215-220.

Reid, W. D., Volicer, L., Smookler, H., Beaven, M. A., and Brodie, B. B. (1968). Brain amines and temperature regulation. *Pharmacology 1*, 329-344.

Reigle, T. G., and Wolf, H. H. (1971). The effects of centrally administered chlorpromazine on temperature regulation in the hamster. *Life Sci. 10*, 121-132.

Reigle, T. G., and Wolf, H. H. (1974). Potential neurotransmitters and receptor mechanisms involved in the central control of body temperature in golden hamsters. *J. Pharmacol. Exp. Ther. 189*, 97-109.

Reigle, T. G., and Wolf, H. H. (1975). A study of potential cholinergic mechanisms involved in the central control of body temperature in golden hamsters. *Neuropharmacology 14*, 67-74.

Repin, I. S., and Kratskin, I. L. (1967). Hypothalamic mechanisms of fever. *Fiziol. Zh. SSSR 53*, 1206-1211.

Rewerski, W. J., and Jori, A. (1968). Effect of desipramine injected intracerebrally in normal or reserpinized rats. *J. Pharmaceut. Pharmacol. 20*, 293-296.

Rewerski, W. J., and Gumulka, W. (1969). The effect of α-MT on hyperthermia induced by chlorpromazine. *Int. J. Neuropharmacol. 8*, 389-391.

Richter, C. P. (1975). Deep hypothermia and its effect on the 24-hour clock of rats and hamsters. *Johns Hopkins Med. J. 136*, 1-10.

Rosendorff, C., and Mooney, J. J. (1971). Central nervous system sites of action of a purified leucocyte pyrogen. *Am. J. Physiol. 220*, 597-603.

Rosendorff, C., Mooney, J. J., and Long, C. N. H. (1970). Sites of action of leucocyte pyrogen in the genesis of fever in the conscious rabbit. *Fed. Proc. 29*, Abstr. 1547.

Rosenthal, F. E. (1941). Cooling drugs and cooling centres. *J. Pharmacol. Exp. Ther. 71*, 305-314.

Rudy, T. A., and Viswanathan, C. T. (1975). Effect of central cholinergic blockade on the hyperthermia evoked by prostaglandin E_1 injected into the rostral hypothalamus of the rat. *Can. J. Physiol. Pharmacol. 53*, 321-324.

Rudy, T. A., and Wolf, H. H. (1971). The effect of intrahypothalamically injected sympathomimetic amines on temperature regulation in the cat. *J. Pharmacol. Exp. Ther. 179*, 218-235.

Rudy, T. A., and Wolf, H. H. (1972). Effect of intracerebral injections of carbamylcholine and acetylcholine on temperature regulation in the cat. *Brain Res. 38*, 117-130.

Rudy, T. A., Williams, J. W., and Yaksh, T. L. (1977). Antagonism by indomethacin of neurogenic hyperthermia produced by unilateral puncture of the anterior hypothalamic/preoptic region. *J. Physiol. 272*, 721-736.

Ruwe, W. D. (1977). Diencephalic mediation of thermoregulation and feeding in the cat by a dopaminergic mechanism. Masters thesis, Purdue University, Lafayette, Ind.

Ruwe, W. D., and Myers, R. D. (1978). Dopamine in the hypothalamus of the cat: pharmacological characterization and push-pull perfusion analysis of sites mediating hypothermia. *Pharmacol. Biochem. Behav. 9*, 65-80.

Ruwe, W. D., and Myers, R. D. (1980). Fever produced by intrahypothalamic pyrogen: effect of protein synthesis inhibition by anisomycin. *Brain Res. Bull. 4*, 741-745.

Sacharoff, G. P. (1909). Ueber die Wirkung des Tetrahydro-β-Naphthylamins auf die Körpertemperatur und den Blutkreislauf. *Zeit. Exp. Pathol. Ther. 7*, 225-241.

Samanin, R., Kon, S., and Garattini, S. (1972). Abolition of the morphine effect on body temperature in midbrain raphe lesioned rats. *J. Pharmaceut. Pharmacol. 24*, 374-377.

Sanner, J. H. (1974). Substances that inhibit the actions of prostaglandins. *Archs. Intern. Med. 133*, 133-146.

Satinoff, E., and Cantor, A. (1975). Intraventricular norepinephrine and thermoregulation in rats. In *Temperature Regulation and Drug Action,* P. Lomax, E. Schönbaum, and J. Jacob (Eds.). Karger, Basel, pp. 103-110.

Saxena, P. N. (1973). Mechanism of hypothermic action of catecholamines in the cat. *Ind. J. Pharmacol. 5*.

Saxena, P. N. (1976). Sodium and calcium ions in the control of temperature set-point in the pigeon. *Brit. J. Pharm. 56*, 187-192.

Schmeling, W. T., and Hosko, M. J. (1976). Hypothermia induced by Δ^9-tetrahydrocannabinol in rats with electrolytic lesions of the preoptic region. *Pharmacol. Biochem. Behav. 5*, 79-85.

Schmidt, J. (1963). Die Abhängigkeit der Temperaturbeeinflussung von Ratten durch biogene Amine von der Applikationsart und der Umgebungstemperatur. *Acta Biol. Med. Germ. 10*, 350-356.

Schoener, E. P., and Wang, S. C. (1974). Sodium acetylsalicylate effectiveness against fever induced by leukocytic pyrogen and prostaglandin E_1 in the cat. *Experientia 30*, 383.

Schoener, E. P., and Wang, S. C. (1975). Leukocytic pyrogen and sodium acetylsalicylate on hypothalamic neurons in the cat. *Am. J. Physiol. 229*, 185-190.

Schoener, E. P., and Wang, S. C. (1976). Effects of locally administered prostaglandin E_1 on anterior hypothalamic neurons. *Brain Res. 117*, 157-162.

Schütz, J. (1916). Zur Kenntnis der Wirkung des Magnesiums auf die Körpertemperatur. *Arch. Exp. Pathol. Pharmakol. 79*, 285-290.

Sgaragli, G. P., Pavàn, F., and Galli, A. (1975). Is taurine-induced hypothermia in the rat mediated by 5-HT? *N.-S. Arch. Pharmacol. 288*, 179-184.

Sharpe, L. G., Garnett, J. E., and Olsen, N. S. (1979). Thermoregulatory changes to cholinomimetics and angiotensin II, but not to the monoamines microinjected into the brain stem of the rabbit. *Neuropharmacology 18*, 117-125.

Shaw, G. G. (1971). Hypothermia produced in mice by histamine acting on the central nervous system. *Br. J. Pharmacol. 42*, 205-214.

Shellenberger, M. K., and Elder, J. T. (1967). Changes in rabbit core temperature accompanying alterations in brain-stem monoamine concentrations. *J. Pharmacol. Exp. Ther. 158*, 219-226.

Shemano, I., and Nickerson, M. (1958). Effect of ambient temperature on thermal responses to drugs. *Can. J. Biochem. Physiol. 36*, 1243-1249.

Sheth, U. K., and Borison, H. L. (1960). Central pyrogenic action of *Salmonella typhosa* lipopolysaccharide injected into the lateral cerebral ventricle in cats. *J. Pharmacol. Exp. Ther. 130*, 411-417.

Shute, C. C. D., and Lewis, P. R. (1967). The ascending cholinergic reticular system: neocortical, olfactory and subcortical projections. *Brain 90*, 497-520.

Siegert, R., Philipp-Dormston, W. K., Radsak, K., and Menzel, H. (1976). Mechanism of fever induction in rabbits. *Am. Soc. Microbiol. 14*, 1130-1137.

Simmonds, M. A. (1969). Effect of environmental temperature on the turnover of noradrenaline in hypothalamus and other areas of rat brain. *J. Physiol. 203*, 199-210.

Simmonds, M. A. (1970). Effect of environmental temperature on the turnover of 5-hydroxytryptamine in various areas of rat brain. *J. Physiol. 211*, 93-108.

Simmonds, M. A. (1971). Inhibition by atropine of the increased turnover of noradrenaline in the hypothalamus of rats exposed to cold. *Br. J. Pharmacol. 42*, 224-229.

Simmonds, M. A., and Iversen, L. L. (1969). Thermoregulation: Effects of environmental temperature on turnover of hypothalamic norepinephrine. *Science 163*, 473-474.

Simmonds, M. A., and Uretsky, N. J. (1970). Central effects of 6-hydroxydopamine on the body temperature of the rat. *Br. J. Pharmacol. 40*, 630-638.

Simpson, C. W., Ruwe, W. D., and Myers, R. D. (1977). Characterization of prostaglandin sensitive sites in the monkey hypothalamus mediating hyperthermia. In *Drugs, Biogenic Amines and Body Temperature,* K. E. Cooper, P. Lomax, and E. Schönbaum (Eds.). Karger, Basel, pp. 142-144.

Skarnes, R. C. (1968). *In vivo* interaction of endotoxin with a plasma lipoprotein having esterase activity. *J. Bacteriol. 95*, 2031-2034.

Sochánski, R., and Samek, D. (1962). The concentration of 5-hydroxytryptamine in the hindbrain of mice exposed to high temperatures. *Acta Physiol. Polon, 13*, 351-355.

South, F. E., Hannon, J. P., Willis, J. R., Pengelley, E. T., and Alpert, N. R. (Eds.) (1972). *Hibernation-Hypothermia: Perspectives and Challenges.* Elsevier, New York.

Spafford, D. C., and Pengelley, E. T. (1971). The influence of the neurohumor serotonin on hibernation in the golden-mantled ground squirrel, *Citellus lateralis. Comp. Biochem. Physiol. 38A*, 239-250.

Spławiński, J. A., Górka, Z., Zacny, E., and Kałuża, Z. (1977). Fever produced in the rat by intracerebral *E. coli* endotoxin. *Pflügers Arch. 368*, 117-123.

Spławiński, J. A., Gorka, Z., Zacny, E., and Wojtaszek, B. (1978). Hyperthermic effects of arachidonic acid, prostaglandin E_2 and $F_{2\alpha}$ in rats. *Pflügers Arch. 374*, 15-21.

Spławiński, J. A., Wojtaszek, B., and Swies, J. (1979). Endotoxin fever in rats: is it triggered by a decrease in breakdown of prostaglandin E_2? *Neuropharmacology 18*, 111-116.

Staib, A. H., and Andreas, K. (1970). Untersuchungen zur Lokalisation cholinerger

Rezeptoren mit Thermoregulatorischer Aktivität im Rattenhirn. *Acta Biol. Med. Germ.* 25, 605-611.

Stanton, T. L., and Beckman, A. L. (1977). Thermal changes produced by intrahypothalamic injections of acetylcholine during hibernation and euthermia in *Citellus lateralis*. *Comp. Biochem. Physiol.* 58A, 143-150.

Stevens, E. D. (1973). The evolution of endothermy. *J. Theor. Biol.* 38, 597-611.

Stitt, J. T. (1973). Prostaglandin E_1 fever induced in rabbits. *J. Physiol.* 232, 163-179.

Stitt, J. T. (1976). Preoptic anterior hypothalamic thermosensitivity during fever production by prostaglandin E_1. *Israel J. Med. Sci.* 12, 1056-1059.

Stitt, J. T., and Hardy, J. D. (1975). Microelectrophoresis of PGE_1 onto single units in the rabbit hypothalamus. *Am. J. Physiol.* 229, 240-245.

Stitt, J. T., Hardy, J. D., and Stolwijk, J. A. J. (1974). PGE_1 fever: its effect on thermoregulation at different low ambient temperatures. *Am. J. Physiol.* 227, 622-629.

Strzoda, L. (1962). Influence of elevated ambient temperature on the central nervous system—changes in the noradrenaline and adrenaline concentrations in the brain stem of animals subjected to hyperthermia. *Acta Physiol. Polon.* 13, 217-225.

Svanes, K. (1968). Microcirculatory changes and general responses to 5-hydroxytryptamine in normothermic and hypothermic mice. *Acta Physiol. Scand.* 72, 404-411.

Sweatman, P., and Jell, R. M. (1977). Dopamine and histamine sensitivity of rostral hypothalamic neurones in the cat: possible involvement in thermoregulation. *Brain Res.* 127, 173-178.

Szelényi, Z., Zeisberger, E., and Brück, K. (1977). A hypothalamic alpha-adrenergic mechanism mediating the thermogenic response to electrical stimulation of the lower brainstem of the guinea pig. *Pflügers Arch.* 370, 19-23.

Szolcsányi, J., Joo, F., and Jancsó-Gábor, A. (1971). Mitochondrial changes in preoptic neurones after capsaicin desensitization of the hypothalamic thermodetectors in rats. *Nature (London)* 229, 116-117.

Takagi, H., and Kuruma, I. (1966). Effect of bacterial lipopolysaccharide on the content of serotonin and norepinephrine in rabbit brain. *Jpn. J. Pharmacol.* 16, 478-479.

Takashima, H. (1962). Relation between pyrexia and amine levels in brain with special reference to the action of monoamine oxidase inhibitor. *Nip. Yakurigaku Zasshi* 58, 437-448.

Tangri, K. K., Bhargava, A. K., and Bhargava, K. P. (1974). Interrelation between monoaminergic and cholinergic mechanisms in the hypothalamic thermoregulatory centre of rabbits. *Neuropharmacology* 13, 333-346.

Tangri, K. K., Bhargava, A. K., and Bhargava, K. P. (1975). Significance of central cholinergic mechanism in pyrexia induced by bacterial pyrogen in rabbits. In *Temperature Regulation and Drug Action*, J. Lomax, E. Schönbaum, and J. Jacob (Eds.). Karger, Basel, pp. 65-74.

Teddy, P. J. (1969). The effects of alterations in hypothalamic monoamine content on fever in the rabbit. *J. Physiol.* 204, 140-141.

Terwelp, D. R., Kennedy, M. S., and Burks, T. F. (1973). Temperature responses to intraventricular amphetamine in unanesthetized cats. *Chem. Pathol. Pharmacol.* 6, 795-804.

Tirri, R. (1970). Brain monoamines in the control of body temperature in mice. *Ann. Zool. Fennici* 7, 323-328.

Toh, C. C. (1960). Effects of temperature on the 5-hydroxytryptamine (serotonin) content of tissues. *J. Physiol. 151*, 410-415.

Toivola, P., and Gale, C. C. (1970). Effect on temperature of biogenic amine infusion into hypothalamus of baboon. *Neuroendocrinol. 6*, 210-219.

Tószeghi, P., Tobler, I., and Borbely, A. A. (1978). Cerebral ventricular infusion of excess calcium in the rat: effects on sleep states, behavior and cortical EEG. *Eur. J. Pharmacol. 51*, 407-416.

Trzcinka, G. P., Lipton, J. M., Hawkins, M., and Clark, W. G. (1977). Effects on temperature of morphine injected into the preoptic/anterior hypothalamus, medulla oblongata, and peripherally in unrestrained and restrained rats. *Proc. Soc. Exp. Biol. Med. 156*, 523-526.

Trzcinka, G. P., Lipton, J. M., Hawkins, M., and Clark, W. G. (1978). Differential effects on temperature of chlorpromazine injected into PO/AH and medulla oblongata of the rat. *Arch. Int. Pharmacodyn. Ther. 232*, 111-116.

Ungerstedt, U. (1971). Histochemical studies on the effect of intracerebral and intraventricular injections of 6-hydroxydopamine on monoamine neurons in the rat brain. In *6-Hydroxydopamine and Catecholamine Neurons*, T. Malmfors and H. Thoenen (Eds.). North-Holland, Amsterdam, pp. 101-127.

Vaidya, A. B., and Levine, R. J. (1971). Hypothermia in a patient with carcinoid syndrome during treatment with para-chlorophenylalanine. *New Engl. J. Med. 284*, 255-257.

Van Zoeren, J. G., and Stricker, E. M. (1976). Thermal homeostasis in rats after intrahypothalamic injections of 6-hydroxydopamine. *Am. J. Physiol. 230*, 932-939.

Van Zoeren, J. G., and Stricker, E. M. (1977). Effects of preoptic, lateral hypothalamic, or dopamine-depleting lesions on behavioral thermoregulation in rats exposed to the cold. *J. Comp. Physiol. Psychol. 91*, 989-999.

Varagić, V. M., and Beleslin, D. B. (1973). The effect of cyclic N-2-O-dibutyryl-adenosine-$3',5'$-monophosphate, adenosine triphosphate and butyrate on the body temperature of conscious cats. *Brain Res. 57*, 252-254.

Vaughn, L. K., Veale, W. L., and Cooper, K. E. (1979). Sensitivity of hypothalamic sites to salicylate and prostaglandin. *Can. J. Physiol. Pharmacol. 57*, 118-123.

Veale, W. L., and Cooper, K. E. (1972). Fever produced by intravenous pyrogen following ablation of the anterior hypothalamic preoptic area. *Can. J. Physiol. 3*, 52.

Veale, W. L., and Cooper, K. E. (1974a). Evidence for the involvement of prostaglandins in fever. In *Recent Studies of Hypothalamic Function*, K. Lederis and K. E. Cooper (Eds.). Karger, Basel, pp. 359-370.

Veale, W. L., and Cooper, K. E. (1974b). Prostaglandin in cerebrospinal fluid following perfusion of hypothalamic tissue. *J. Appl. Physiol. 37*, 942-945.

Veale, W. L., and Cooper, K. E. (1975). Comparison of sites of action of prostaglandin E and leucocyte pyrogen in brain. In *Temperature Regulation and Drug Action*. J. Lomax, E. Schönbaum, and J. Jacob (Eds.). Karger, Basel, pp. 218-226.

Veale, W. L., and Jones, D. L. (1976). Alterations in hypothalamic ions: influence on limbic function. *Proc. Int. Neuro-Psychopharmacol.*

Veale, W. L., and Whishaw, I. Q. (1976). Body temperature responses at different ambient temperatures following injections of prostaglandin E_1 and noradrenaline into the brain. *Pharmacol. Biochem. Behav. 4*, 143-150.

Veale, W. L., Benson, M. J., and Malkinson, T. (1977a). Brain calcium in the rabbit: site of action for the lateration of body temperature. *Brain Res. Bull. 2*, 67-69.

Veale, W. L., Cooper, K. E., and Pittman, Q. J. (1977b). The role of prostaglandins in fever and temperature regulation. In *Prostaglandins*, Vol. III, P. Ramwell (Ed.). Plenum, New York, pp. 145-167.

Villablanca, J., and Myers, R. D. (1964). Fever production by endotoxin injections in the cat. *Arch. Biol. Med. Exp. 1*, 102.

Villablanca, J., and Myers, R. D. (1965). Fever produced by microinjection of typhoid vaccine into hypothalamus of cats. *Am. J. Physiol. 208*, 703-707.

Vogt, M. (1954). The concentration of sympathin in different parts of the central nervous system under normal conditions and after the administration of drugs. *J. Physiol. 123*, 451-481.

von Euler, C. (1961). Physiology and pharmacology of temperature regulation. *Pharmacol. Rev. 13*, 361-398.

Waller, M. B. (1975). The relationship between prostaglandin E and the amines in the control of body temperature. Doctoral dissertation, Purdue University, Lafayette, Ind.

Waller, M. B. and Myers, R. D. (1973). Hyperthermia and operant responding for heat evoked in the monkey by intrahypothalamic prostaglandin. *Soc. Neurosci. 118*.

Waller, M. B., and Myers, R. D. (1974). Mechanism of action on the hypothalamus of an antipyretic drug during the hyperthermia evoked by an endotoxin, a prostaglandin, 5-HT or 5,6-DHT. *Soc. Neurosci., 463*.

Waller, M. B., Myers, R. D., and Martin, G. E. (1976). Thermoregulatory deficits in the monkey produced by 5,6-dihydroxytryptamine injected into the hypothalamus. *Neuropharmacology 15*, 61-68.

Warwick, R. O., and Schnell, R. C. (1976). Studies relating morphine hypothermia with serotonin reuptake in the rat hypothalamus. *Eur. J. Pharmacol. 38*, 329-335.

Werdinius, B. (1962). Effect of temperature on the action of reserpine. *Acta Pharmacol. Toxicol. 19*, 43-46.

Whitehead, S. A., and Ruf, K. B. (1974). Responses of antidromically identified preoptic neurones in the rat to neurotransmitters and to estrogen. *Brain Res. 79*, 185-198.

Williams, J. W., Rudy, T. A., Yaksh, T. L., and Viswanathan, C. T. (1977). An extensive exploration of the rat brain for sites mediating prostaglandin-induced hyperthermia. *Brain Res. 120*, 251-262.

Willies, G. H., Woolf, C. J., and Rosendorff, C. (1976a). The effect of sodium salicylate on dibutyryl cyclic AMP fever in the conscious rabbit. *Neuropharmacology 15*, 9-10.

Willies, G. H., Woolf, C. J., and Rosendorff, C. (1976b). The effect of an inhibitor of adenylate cyclase on the development of pyrogen, prostaglandin and cyclic AMP fevers in the rabbit. *Pflügers Arch. 367*, 177-181.

Wilson, N. C., Gisolfi, C. V., and Phillips, M. I. (1978). Influence of EGTA on an exercise-induced elevation in the colonic temperature of the rat. *Brain Res. Bull. 3*, 97-100.

Wit, A., and Wang, S. C. (1968). Temperature-sensitive neurons in preoptic/anterior hypothalamic region: actions of pyrogen and acetylsalicylate. *Am. J. Physiol. 215*, 1160-1169.

Woolf, C. J., Willies, G. H., Laburn, H., and Rosendorff, C. (1975). Pyrogen and prostaglandin fever in the rabbit. I. Effects of salicylate and the role of cyclic AMP. *Neuropharmacology 14,* 397-403.

Yaksh, T. L., and Myers, R. D. (1972). Hypothalamic "coding" in the unanesthetized monkey of noradrenergic sites mediating feeding and thermoregulation. *Physiol. Behav. 8,* 251-257.

Yasuda, M. (1962). Effect of reserpine on febrile responses induced by pyrogenic substances. *Jpn. J. Pharmacol. 11,* 114-125.

Young, D. R., Mosher, R., and Erve, P., and Spector, H. (1959). Body temperature and heat exchange during treadmill running in dogs. *J. Appl. Physiol. 24,* 839-843.

Zeisberger, E., and Brück, K. (1971a). Central effects of noradrenaline on the control of body temperature in the guinea-pig. *Pflügers Arch. 322, 152-166.*

Zeisberger, E., and Brück, K. (1971b). Comparison of the effects of local hypothalamic acetylcholine and RF-heating on non-shivering thermogenesis in the guinea pig. *Int. J. Biometeor. 15,* 305-308.

Zeisberger, E., and Bruck, K. (1976). Alteration of shivering threshold in cold- and warm-adapted guinea pigs following intrahypothalamic injections of noradrenaline and of an adrenergic alpha-receptor blocking agent. *Pflügers Arch. 362,* 113-119.

Ziel, R., and Krupp, P. (1976). Mechanisms of action of antipyretic drugs. In *Brain Dysfunction in Infantile Febrile Convulsions,* M. A. B. Brazier and F. Coceani (Eds.). Raven, New York, pp. 141-152.

3
The Role of the Hypothalamus in Energy Homeostasis

Terry L. Powley,* Charles A. Opsahl, James E. Cox,† and
Harvey P. Weingarten‡ / Yale University, New Haven, Connecticut

I. Introduction

Both experimental manipulations and pathological changes of the hypothalamus produce dramatic effects on food intake and energy metabolism. These alterations are so striking—they can lead to extreme obesity or to inanition and death—that early experimental observations were interpreted in terms of reciprocally organized feeding and satiety centers located in the lateral and ventromedial hypothalamus, respectively. It is these observations, the interpretations that resulted from them, the more recent results that have been obtained, and further reinterpretations that are the subject of the present chapter.

The various hypothalamic disturbances in energy homeostasis that have been studied in depth offer potential insight into the hypothalamic control of basic regulatory physiology, behavior, and, perhaps, disease processes such as anorexia and obesity. Indeed, the literature that has resulted from the experimental interest in these hypothalamic feeding and metabolic disturbances is so vast that comprehensive reviews are probably no longer possible. The monographs by Stevenson (1969) and Morgane and Jacobs (1969) may represent the last attempts at complete reviews; most recent attempts at such coverage are published proceedings (Morgane, 1969; Novin et al., 1976; Wayner and Oomura, 1975) or collections of reviews (Code, 1967). Recognizing this problem, we have attempted to focus on certain issues and to provide only a selective review of the literature (completed December 1977). Many good reviews of particular issues exist, and in order to supplement our own coverage, we have identified these reviews at relevant points in the text.

Our decisions on which material to cover were based on a variety of specific considerations and orientations. While several of these orientations are discussed later in the text, the broader principles affecting the selection should be mentioned at the outset.

Present affiliation: Purdue University, West Lafayette, Indiana.
†*Present affiliation:* Bourne Behavioral Research Laboratory, New York Hospital-Cornell Medical Center, Westchester Division, White Plains, New York.
‡*Present affiliation:* McMaster University, Hamilton, Ontario, Canada.

First, the present chapter is concerned with the role of the hypothalamus in food intake. The choice of the chapter title is not fortuitous. It grows out of the conviction that the available work on the role of the hypothalamus in food intake can be evaluated properly only as part of the broader category of hypothalamic responses involved in energy homeostasis. The hypothalamic areas influencing food intake are neither simply feeding nor simply autonomic nor endocrine centers controlling metabolism. The richness of anatomical connections of the areas belies any simple unidimensional, functional role. The absence of "pure" behavioral disturbances or "pure" metabolic disturbances with hypothalamic manipulations also belies such a view. In a real sense, the hypothalamus appears to be one of the major levels of neural integration at which the behavioral and physiological dimensions of energy regulation come together. Both dimensions are examined in this chapter, and one of our central theses is that both facets of control must be considered together if one is to appreciate the role of the hypothalamus in energy balance. A corollary, developed below, is that for historical and largely irrelevant reasons the study of the role of the hypothalamus in energy homeostasis proceeded piecemeal; the extensive literature on the behavioral aspects of hypothalamic manipulations developed almost independently of the literature on the metabolic effects of such manipulations.

Second, we have only cursorily covered the effects of hypothalamic manipulations on the control of water balance. Problems of water and energy balance have frequently been confounded or inadvertently studied simultaneously in analyses of hypothalamic function. This has certainly been so in the case of the lateral hypothalamic feeding syndrome. Yet, the two types of balance are conceptually separable problems and are also at least partially dissociable from appropriate lesions or neurochemical manipulations. Practically, too, it is possible to test animals—for example, animals with lateral hypothalamic lesions—while experimentally controlling or eliminating challenges of water balance. Several good reviews of the hypothalamic roles in water intake and hydromineral metabolism are available (Stevenson, 1969; Mogenson, 1973; Epstein et al., 1973; Epstein, 1971).

Third, we have emphasized those experimental analyses of the role of the hypothalamus in energy homeostasis which have employed lesions, be they pathological or experimental, such as electrolytic, radio-frequency, knife-cut, or neurochemical lesions. We wish to emphasize, however, that this emphasis is a practical consideration, not a value judgment. The orientation reflects the fact that both electrophysiological and the electrical stimulation analyses are expertly reviewed elsewhere in this volume as well as in other recent reviews (e.g., Oomura, 1973; Rolls et al., 1976; Panksepp, 1974, 1975).

Fourth, we have emphasized consistently the historical antecedents and the chronological development of the problem area. This approach grows out of the conviction that analyses of an area are shaped as much by the existing conceptual tools as by the available physiological techniques. The prevailing thinking in the 1940s about "hunger" and "satiety" was probably as influential in the development of the subject as the advent of the Horsley-Clark stereotaxic instrument. We feel that the development of such conceptual tools as well as the impact they have had is most clearly seen in a chronological analysis. In addition, assumptions about the organization and functioning of both the brain and behavior frequently shape even the special vocabulary of the field, and the specialized vocabulary, in turn, often determines what questions are asked and even how those questions are asked.

Fifth, because of the last two considerations listed, we have focused on the two constellations of feeding disturbances classically called the ventromedial hypothalamic (VMH) syndrome and the lateral hypothalamic (LH) syndrome. Pragmatically, this organization is dictated by our decision to trace the development of work in the area. Most lesion analyses and the related literatures have recognized the two syndromes. Such analyses of brain function in terms of syndromes have obvious advantages and substantial limitations. Identification of syndromes—at least in the early stages of an analysis—has tremendous heuristic value. Syndrome analyses permit recognition of major constellations of symptoms. They also make possible the assumption that measurement of a few aspects will serve to identify the presence or absence of the whole set of disturbances. If every laboratory's lesion preparation were unique and produced a special combination of effects, it would be very difficult to study common phenomena in different laboratories. In the even more extreme case, every animal with brain lesions would be qualitatively unique. Analysis of brain function would be unable to proceed with general profiles of effects.

Syndrome analyses have several limitations, however. In the first place, the proposition that syndromes exist (at least as the proposition has developed in the feeding literature) seems to imply that the brain is organized into natural areas or "centers" that function to weld a variety of elements into integrated responses. Thus, the use of syndrome analyses tends to produce center explanations of brain function. In addition, identification of a syndrome implies that the constellation of effects has a unitary cause. Such conditions are possibilities—but unestablished ones—for the hypothalamic feeding disturbances. Further, the assumption that one need measure only one element of a syndrome, since the whole constellation of elements covaries, biases dramatically the results obtained. Potential dissociations of elements are apt to be overlooked. Yet other limitations of the syndrome concept are reviewed by Morgane (1975). In practice, our solution has been not only to review the various aspects of the material in terms of the classic lesion syndromes but also to recognize that various elements may not correlate well and, in fact, may be dissociated with more exhaustive experimentation. The latter possibility is explicitly examined at various points.

Sixth, we have emphasized experimentation on the ventromedial hypothalamic syndrome rather than work on the lateral hypothalamic syndrome. This focus reflects several factors. Historically, the VMH syndrome was recognized and studied first. A thorough review of this syndrome simultaneously provides a perspective on the factors that shaped analysis and experimentation on the lateral hypothalamus. In addition, there is simply much more work available on the VMH syndrome. The more extensive and multifaceted experimental results on the VMH syndrome make it feasible to examine more fully different dimensions of energy homeostasis; in contrast, analyses of the LH syndrome have been more sharply limited to examinations of the behavioral elements of the disturbance. And again, many of the principles and conclusions derived from a review of work on the ventromedial hypothalamus seem generally applicable to the lateral hypothalamus.

Finally, we have adopted the orientation that analyses of the etiology of hypothalamic disturbances of energy balance are central issues for illuminating the role of the hypothalamus in food intake, or, more generally, in energy homeostasis. A review of the general area might be limited to an anatomical and neurochemical analysis of the feeding syndromes, but energy homeostasis is fundamentally a *functional* issue to which the questions of etiology speak. To take an example, one might argue that the whole constellation of metabolic and behavioral effects produced by damage to the ventromedial hypothalamus are consequences of what many call "finickiness," a primary disturbance of sensory reactivity. In such a case, the effects of hypothalamic manipulations on energy balance would be secondary and indirect. Similarly, it has been argued that LH lesions produce their effects by causing arousal deficits which diminish an animal's responsiveness to homeostatic as well as other disturbances. In such a view, the effects of LH lesions on energy metabolism and feeding behavior are indirect manifestations of a deficit in arousal. Such views would imply that the direct role, or roles, of the hypothalamus in energy homeostasis are minimal. Such views, it seems to us, are not easily reconciled with the extensive amount of recording (e.g., Oomura, 1973; Cross, 1973; Rolls et al., 1976) and stimulation work (e.g., Frohman, 1971; Panksepp, 1974, 1975) which indicates that neurons in the lateral hypothalamus are selectively sensitive to both food-related stimuli and metabolites and, in turn, modulate both behavioral and physiological responses of energy homeostasis. Regardless of the controversies generated by specific proposals, however, the important point for the moment is that issues concerning the functional basis, or etiology, of hypothalamic disturbances of energy metabolism get at the heart of the question: what role does the hypothalamus play in energy homeostasis? Although the jury is still out, we have attempted to identify some of the major alternatives and essential strategies for bringing in a verdict.

II. Disturbances of Energy Homeostasis Produced by Ventromedial Hypothalamic Manipulations

A. The Hypothalamus and Energy Homeostasis: History

Historically, the first hypothalamic feeding disturbance to receive attention was the now-classic ventromedial hypothalamic (VMH) "syndrome." Early descriptions of this disturbance focused on the metabolic manifestations of VMH damage. More recent work has emphasized a behavioral, as opposed to a metabolic, analysis of VMH injury. This is a relatively modern development and represents a substantial reconceptualization of the etiology of the VMH syndrome. In view of some of the most recent (i.e., since 1970) rethinking of VMH disturbances in terms of primary metabolic components, it is worth reviewing the observations and analyses that led to the behavioral explanations of the VMH syndrome.

For 100 years, until the middle of the 1940s, clinical and experimental attention on the basomedial hypothalamus was focused on the adiposity that resulted from lesions in the area. Mohr (1840, cited in Brobeck, 1946) first described—in terms of obesity, or a metabolic explanation—the disturbance associated with a basomedial hypothalamic involvement of a pituitary tumor in a female patient. In the best-known analysis, Fröhlich (1901) characterized the human syndrome associated with ventral diecephalic lesions without paying particular note of food intake; he argued pituitary involvement was responsible. His description of the syndrome was sufficiently complete that the disturbance was subsequently referred to as the Fröhlich syndrome. In spite of occasional mentions in the clinical literature of voracious appetite (e.g., Bramwell, 1888), virtually none of the earlier papers report measurement of food consumption. Even the names dystrophia adiposogenitalis (Bartels, 1906) or adiposogenital syndrome, used as functional alternatives for Fröhlich's syndrome, focused attention on the classically recognized symptoms defining the syndrome; overeating was not one of them. In the years that followed, the adiposogenital syndrome was repeatedly observed in patients and an experimental animal analogue was identified. As in the clinical literature, the early experimental work on VMH damage in animals focused on adiposity, metabolism, and visceral function—not on food intake. The degree to which this orientation dominated thinking on the problem can be appreciated by reading the landmark symposium on hypothalamic research, *The Hypothalamus,* published in 1940 by the Association for Research in Nervous and Mental Disease. Food intake received no systematic treatment. Neither Long's chapter on "Evidence For and Against Control of Carbohydrate Metabolism by the Hypothalamus" nor that of Gildea and Man, "The Hypothalamus and Fat Metabolism," discusses the contribution of altered food intake to the metabolic changes produced by basomedial hypothalamic trauma or manipulations. Similarly, the seminal monograph of the time, *The Hypothalamus,* published in 1939 by Ranson and Magoun, contains sections on adiposity and carbohydrate metabolism but not food intake. In contrast, fully 15%—the chapter by Stevenson—of a more recent symposium on hypothalamic research (Haymaker et al., 1969) is devoted to food and water intake.

When attention was focused on the adiposity and metabolic disturbances that accompany VMH lesions, a major controversy concerned the critical site of the damage necessary to produce these effects. The crux of the issue was whether injury restricted to the tuberal hypothalamus, and thus not infringing on the pituitary, is sufficient to produce the classic obesity syndrome. This view was supported by Bramwell (1888) and Erdheim (1904). The popular alternative, that a pituitary dysfunction was responsible for the adiposogenital syndrome, was favored by Fröhlich (1901). This alternative was questioned by the experimental work of Smith (1927, 1930) and then, with the advent of the Horsley-Clark stereotaxic instrument, was convincingly rejected by Ranson and his colleagues (Hetherington and Ranson, 1939), who demonstrated that focal lesions in the region of the ventromedial hypothalamic nucleus (see Fig. 1), without concomitant pituitary damage, was sufficient to produce the characteristic obesity. Furthermore, hypophysectomy, which spared hypothalamic tissue, did not produce obesity (Hetherington and Ranson, 1939), and hypophysectomized animals could be rendered obese by subsequent VMH lesions (Hetherington, 1943). Finally, hypophysectomy did not modify obesity produced by prior hypothalamic damage (Hetherington and Ranson, 1942a). Since the brain damage necessary to produce obesity did not appear to depend on pituitary involvement, an alternative explanation in terms of other pathways was needed. Thus, the hypophysectomy studies cleared the way for a variety of subsequent hypotheses which sought to account for the elements of the ventromedial hypothalamic syndrome by relying on alternate constructs.

Food intake was the obvious behavioral candidate, and attention rapidly shifted to it as the primary process mediating hypothalamic obesity. Some observational data indicating that hypothalamic manipulations influenced food intake were already available. Indeed, as mentioned above, clinical reports had suggested that hypothalamic (or perhaps, more accurately, diencephalic) disturbances could produce "morbid appetite." Bramwell (1888) emphasized the tremendous appetite that accompanied tumors encroaching on the diencephalon, although he suggested that the voracious appetite was secondary to a general deterioration. In later clinical reports other physicians began to stress the importance of overeating in producing hypothalamic obesity (Newburgh, 1931; Wilder, 1932). For the most part, however, these clinical observations were largely anecdotal, not formal investigations of the role of overeating in hypothalamic obesity. In fact, in their recent review of hypothalamic obesity in man, Bray and Gallagher (1975) concluded that their own measurements of caloric intake in a single patient represent the first quantitative assessment of human hypothalamic hyperphagia.

In the experimental literature, Keller and his colleagues (Keller et al., 1932, 1936; Keller and Noble, 1935) first raised the possibility that food intake might play a signifi-

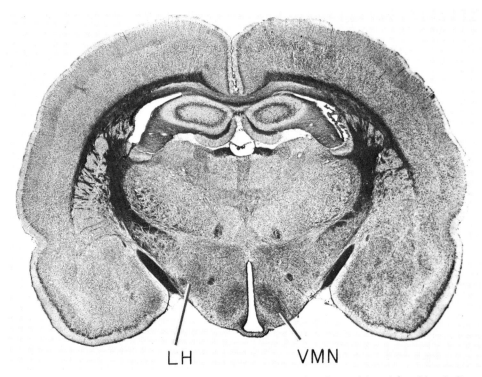

Figure 1 Coronal section of a rat brain stained with cresyl violet and luxol fast blue indicating location of ventromedial hypothalamus (VMH) and lateral hypothalamus (LH). (Reproduced from Keesey and Powley, 1975.)

cant role in the production of hypothalamic obesity when they informally reported increased food intake as well as a tendency to adiposity in dogs and cats with hypothalamic lesions. Keller and his group did not systematically measure food intake in their experiments, although they did report that, in a short consumption test, a dog and a kitten with basomedial hypothalamic damage consumed large quantities of milk but not of water. The classic series of papers which followed from Ranson's laboratory included the report of marked polyphagia in one monkey with ventromedial hypothalamic lesions (Ranson et al., 1938). In addition, increased food intake in VMH rats was reported by Hetherington (1941) and Hetherington and Ranson (1942b), although these authors stressed the importance of inactivity for the production of obesity and of diet palatability for the overeating response. Finally, in a landmark paper, dated 1943, the emphasis on food intake in the VMH syndrome was crystallized when Brobeck and co-workers argued that ventromedial hypothalamic lesions in rats produce a primary disturbance in food intake, a hyperphagia,* which in turn produces the metabolic conditions that characterize VMH obesity.

*In Brobeck's (1946) early review of the problem, he details the thinking that led to their selection of the term hyperphagia: "The word 'hyperphagia' was chosen because it does not have the subjective, psychological connotations of the terms 'hunger,' 'appetite,' 'satiety' and 'bulimia,' and because the word 'polyphagia,' which might have been adopted, implies 'omnivourousness; craving for all kinds of food' (Dorland). Hyperphagia is taken to mean simply increased eating" [p. 549]. Brobeck's excellent review (1946) should be consulted for a complete account of the early development of the problem.

They also reported an increased respiratory quotient (RQ) in VMH animals allowed to feed ad libitum (Tepperman et al., 1943). In contrast, however, when other rats sustaining hypothalamic damage were fasted, there was no elevation in the basal RQ. This widely cited paper offered an alternative account of hypothalamic obesity just at the time when hypophyseal explanations were being discounted, and it strongly influenced the direction of later research on the VMH syndrome. Indeed, in spite of several subsequent reports (see below) that indicated enhanced lipogenic processes even in pair-fed rats with VMH lesions, the general working assumption for the 30 years following the pivotal analysis by Brobeck and co-workers has been that hypothalamic destruction primarily affects behavioral controls over food intake, and metabolic controls are affected only as a consequence.

B. Elements of the Ventromedial Hypothalamic Syndrome

1. Hypothalamic Hyperphagia

a. Rats

Detailed and systematic measurement of food intake has come largely from experiments on rats with experimentally (and thus essentially instantaneously) produced lesions of the basomedial hypothalamus. Descriptions of the pattern of food intake that follows VMH destruction have variously recognized three or, more commonly, two successive stages.

The first, very transitory stage of effects of VMH lesions on food intake—the stage which Balagura and Devenport (1970) have called the "acute dynamic" stage—lasts for a matter of hours only and is often not distinguished in experimental accounts. It is characterized by frantic ingestive behavior. As the animals recover from the anesthetic, they begin to eat—often before they are able to stand and walk normally (Brooks et al., 1946; Brobeck et al., 1943). This persistent and voracious eating is often accompanied by extreme hyperkinesis (Hetherington and Ranson, 1942; Balagura and Devenport, 1970) and hyperreactivity (Brooks et al., 1946; Harrell and Remley, 1973). Immediately after the lesion procedure, the animals may take an immense meal (Becker and Kissileff, 1974) which may be up to 775% larger than any previous meal for that hour of the day and which may last for 6 hr (Harrell and Remley, 1973). In many respects, this overeating is not particularly stimulus-specific and may be characterized as bulimia or polyphagia. If food is unavailable, the animal may consume practically any nonnutritive material it is able to ingest together with tremendous amounts of water. The feeding in this stage also appears insensitive to normal gastrointestinal controls; there are reports of animals gorging themselves until they bloat or gag, (Harrell and Remley, 1973) or asphyxiate (Brobeck et al., 1943). As a preventive measure, many experimenters routinely deprive animals of food and water in this immediately postoperative period. For the most part, this transitory and indiscriminate polyphagia has not been the study of experimental analysis and has been attributed to the irritative stimulation and neural injury discharges characteristic of electrolytic lesions. Such an explanation receives support from the fact that nonelectrolytic lesions and knife-cut procedures apparently do not produce this initial, acute dynamic period.

In the context of this early phase, it is of interest to note that anesthetization of the VMH results in an immediate increase in food intake in both sated and deprived rats

(Epstein, 1960; Wagner and deGroot, 1963), as well as in chickens (Snapir et al., 1973). Furthermore, bilateral anesthetization of the VMH with procaine results, with a latency of only 16-120 sec, in performance of an operant task in order to obtain food (Larkin, 1975).

In the days after VMH lesions, an animal typically exhibits a characteristic pattern of hyperphagia that is generally called the dynamic phase. (Many accounts assume that the dynamic state begins at the moment of lesion.) The term dynamic phase—as well as the contrasting term for the next period, the static phase—was adapted to the ventromedial hypothalamic syndrome by Brooks and his co-workers (Brooks, 1946a,b; Brooks and Lambert, 1946; Brooks et al., 1946) from the monograph by Rony (1940) on the manifestations of essential obesity in neurologically intact humans. The dynamic phase is of varied duration which depends primarily on the extent of VMH damage (Reynolds, 1959) and on the taste and texture of the food. In the adult rat, this stage is characterized by a dramatic hyperphagia and weight increase. Figure 2 illustrates the hyperphagia and weight gain of an animal during a prolonged dynamic period; a shorter dynamic phase is illustrated in Figure 3. Animals overconsume standard laboratory diets (Brooks and Lambert,

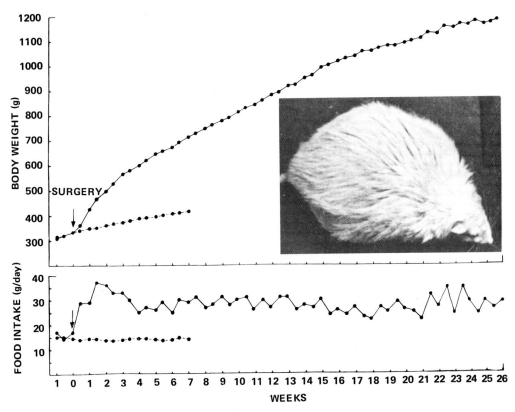

Figure 2 Hyperphagia and weight gain in a male rat maintained on a high-fat diet after parasagittal knife cuts between the VMH and LH. (Reproduced from Gold, 1970.) Copyright 1970 by the American Psychological Association. Reprinted by permission.

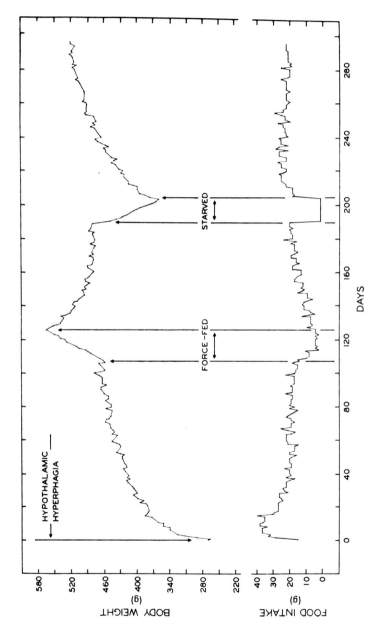

Figure 3 Hyperphagia and weight gain of a rat sustaining bilateral electrolytic lesions of the VMH. Note the rapid weight gain of the dynamic phase and the maintenance of a static body weight level in the face of body-weight challenges. (Reproduced from Hoebel and Teitelbaum, 1966.) Copyright 1966 by the American Psychological Association. Reprinted by permission.

1946) and eat prodigious amounts of certain "palatable" diets (Corbit and Stellar, 1964). During the initial period after hypothalamic destruction, this food intake can lead to average daily gains in gross body weight of 5% per day or more (Gold, 1970).

Continuous monitoring of the feeding behavior of rats after VMH lesions reveals a rather characteristic profile of changes in the pattern of food consumption. Tepperman et al., (1943) were among the first to emphasize the altered feeding pattern after VMH injury. They noted that if a VMH animal's food intake is restricted, it eats its food ration much more quickly than do control animals. In a situation where they are allowed ad libitum access to food, animals exhibit a substantial increase in the size of their individual meals; rats consume meals roughly 70% larger than normal in the dark and in the light during the first three days postlesion. Sclafani and Berner (1976) have reported almost threefold increases in meal size for rats given hyperphagia-inducing parasagittal hypothalamic knife cuts. Such (normally nocturnal) animals often show a slight increase in the number of meals as well (Becker and Kissileff, 1974), with a greater number of meals taken during the daylight phase. Thus, the normal circadian pattern of feeding practically disappears in the dynamic VMH rat.

The fact that the rather characteristic "VMH pattern" of hyperphagia is in great part accounted for by excessively large meals is the key observation supporting the initial proposal that the basomedial hypothalamus contains a "satiety center": once animals begin eating, they overeat because, in effect, they have a reduced behavioral or satiety brake (Brobeck, 1957; Brooks et al., 1946; Kennedy, 1950; Stellar, 1954).

In terms of this development of the satiety-center idea, it is instructive to add that the pattern of feeding of animals with VMH damage is flexible and accommodates a variety of experimental constraints; animals with VMH lesions will maintain a net hyperphagia even when they must adopt different eating strategies. When a VMH animal is prevented from taking large meals, it accommodates and maintains the positive energy balance by increasing its meal frequency (Thomas, 1971). Likewise, the VMH animal will overeat even if other temporal restraints (such as limited periods of access to food) are placed on its feeding (e.g., Gold et al., 1975; Peters, 1974; Smith et al., 1961). Thus, the relatively long popularity of the satiety-center concept notwithstanding, the hyperphagia of the VMH animal appears to be a flexible and integrated response, not the consequence of elimination or exaggeration of any single element of behavior.

Under a given set of experimental conditions, an animal's overconsumption eventually diminishes to a maintenance level, at which point abnormal weight gain no longer occurs. From this point on, the animal consumes control amounts—or just slightly more than control amounts—and its weight level remains at, or nearly at, a constant percentage of control weight. This period, which can last until senescence, has been characterized as the static phase (Brooks and Lambert, 1946).

While the animal in this phase may show only a very modest absolute hyperphagia, a fine-grained analysis of its behavior indicates that certain aspects of the disturbed meal patterns persist. In particular, even when the VMH animal takes only control amounts of food during the 24 hr, the disturbed diurnal pattern of meals still persists (Becker and Kissileff, 1974). This disturbed circadian pattern continues even if food intake of VMH animals is suppressed by reducing diet palatability (Sclafani and Berner, 1976).

Other evidence also indicates that the static phase represents not a recovery of temporarily lost control mechanisms but rather a new maintenance plateau at which the animal's physiological controls and behavioral responses reach a type of equilibrium. The animal's food intake varies so as to maintain the weight level established in the static phase, and, when animals are starved below this plateau, they exhibit renewed hyperphagia (or a second dynamic phase) until they have returned to the weight plateau (Brooks and Lambert, 1946; Hoebel and Teitelbaum, 1966). This effect can be seen in Figures 3 and 5. Similarly, if animals are forced above their static weight level by different hyperalimentation manipulations, they respond by reducing their food intake for as long as they remain above that level.

b. Other Species

Findings reported for other species seem generally congruent with the pattern identified for the rat. Among other related species, both intraperitoneal injections of gold thioglucose (GTG) (Mayer and Zighera, 1954; De Laey et al., 1974, 1975) and electrolytic lesions of the ventromedial hypothalamus (Mayer et al., 1955) have resulted in hyperphagia and marked weight gain in the mouse. Likewise, VMH lesions in the golden hamster produce weight gain and obesity (Marks and Miller, 1972). Parasagittal knife cuts increased food intake in guinea pigs by almost 50% and body-weight gain by more than 100% compared with noncut animals (Sclafani and Sperber, 1977). In rabbits, VMH lesions produce hyperphagia that can be five to seven times as high as preoperative food intake (Balinska, 1963); Romaniuk (1962) has reported that this increased ingestion subsides after two to three months. Lesions of the ventromedial hypothalamus in ground squirrels produce hyperphagia and obesity that result in an increase in the weight-gain phase of the body-weight cycle. These animals continue to exhibit normal circannual cycles of weight loss and gain around an elevated weight plateau (Mrosovsky, 1974, 1975).

In addition to the experiments cited above in which data mainly pertaining to hyperphagia and weight gain were collected, several species have been observed with an eye toward investigating the duration of the dynamic phase as well. Distinct periods of dynamic hyperphagia followed by near-normal static levels of food intake have been reported after VMH lesions in dogs (Rozkowska and Fonberg, 1971), rhesus monkeys (Hamilton and Brobeck, 1964), chickens (Lepkovsky and Yasuda, 1966), geese (Auffray and Blum, 1970; Snapir et al., 1976), and the white-throated sparrow (Kuenzel, 1972). A dynamic phase of hyperphagia and weight gain has been observed by two investigators (Auffray, 1969; Khalaf, 1969) to occur in swine after VMH lesions. No transition to a static phase was noted in either case, though the relatively short duration, eight weeks, of the studies might have been responsible.

In the case of humans, the food intake of Bray and Gallagher's (1975) patient was measured for 37 days, during which she averaged a net positive energy balance of roughly 400 cal (25%) in excess of her daily caloric need of 1600 cal. During this period, she gained weight at a rate commensurate with her overconsumption. Heldenberg et al., (1972) reported that a four-year-old girl with inflammatory lesions in the hypothalamus ate 4500-6000 cal per day and gained weight to the point of being almost unable to move. Reeves and Plum (1969) reported that their adult patient with a ventromedial hypothalamic neoplasm consumed 8000-10,000 cal per day.

Precise characterization of the time course or pattern of overconsumption in humans is not really possible. In most cases, the hypothalamic disturbance is the result of a developing disease process that gradually increases the extent of the lesion and the degree of hyperphagia. Rather than progressing to a stabilized static phase and then maintaining a stable weight plateau, the human patient often exhibits a progressive development that is interrupted by an eventual decline in the late stages of the disease (Bray and Gallagher, 1975). Further, the ingestional patterns of humans with hypothalamic obesity are rarely monitored, and, by the time any such monitoring is begun by the patient's physician or the hospital staff, it is more usually part of a program of enforced food restriction rather than a measure of ad libitum feeding.

2. Hypothalamic Hyperphagia Determined by Taste and Texture ("Finickiness")

a. Rats

One of the initial experimental observations made of food intake after VMH lesions was that the degree of hyperphagia an animal would exhibit was heavily—most researchers have concluded, inordinately—dependent upon the diet provided. Hetherington and Ranson (1942b) emphasized this point even though they were not primarily concerned with food intake per se. With the shift in focus to food intake after VMH injury, the dependence of hyperphagia and obesity on sensory aspects of the diet was clearly demonstrated. This phenomenon in VMH animals became one of the defining characteristics of the VMH syndrome and was termed finickiness.*

The exaggerated intake of VMH animals may be influenced in either direction by the food's sensory aspects. In the one case, the hyperphagia and subsequent obesity may be enhanced by a particular diet or dietary manipulation; in the other, the hyperphagia and obesity may be diminished or even eliminated by dietary manipulations. These effects have been considered positive finickiness (or finickiness to positive aspects of the diet) and negative finickiness (or finickiness to negative aspects of the diet), respectively, and it has been accepted that finickiness refers to the sensory aspects, as opposed to the caloric or postingestional consequences, of a diet.

The enhanced food intake of the VMH rat to palatable foods has been well documented. VMH rats consume approximately 100% more of a high-fat diet than do non-lesioned animals (see Figure 4) (Carlisle and Stellar, 1969; Corbit and Stellar, 1964; Strominger et al., 1953; Teitlebaum, 1955). In fact, many investigators routinely maintain their VMH animals on this greasy diet because of the dramatic hyperphagia and weight gain it supports. In a situation where VMH rats can select their diet from a variety of foodstuffs, they show an increased affinity for oily substances (Brobeck et al., 1943). Other studies have found that animals with VMH lesions overconsume dextrose solutions (Teitlebaum, 1955), dextrose mixed in powdered rat chow (Graff and Stellar, 1962; Teitlebaum, 1955), and sucrose solutions (Beatty, 1973).

VMH animals also typically overrespond to negative taste aspects of a diet. When the diet is laced with even small amounts of quinine, rats with VMH lesions (Miller et al.,

---

*The first use of the term finickiness in reference to the feeding habits of the VMH-lesioned rat seems to have been made by Teitlebaum (1955) when he noted that: "This combination of exaggerated reactions to the positive and negative qualities of the diet yields a picture of finickiness in the obese hyperphagic animals..." (p. 161). The term was then adopted by Graff and Stellar (1962) who incorporated it into the title of their paper, "Hyperphagia, Obesity, and Finickiness," which represented one of the major investigations of the contribution of the phenomenon to the VMH rat's hyperphagia.

Role of the Hypothalamus in Energy Homeostasis

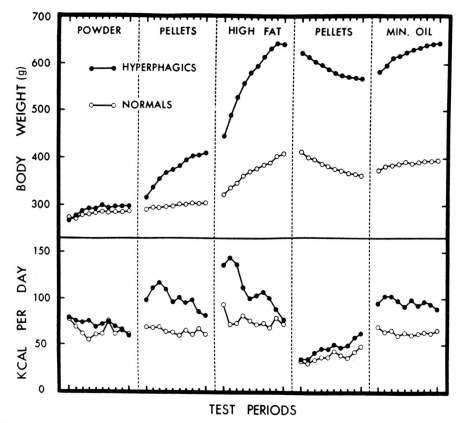

Figure 4 Food intake and body weight of normal and hypothalamic hyperphagic rats on successive exposures to different diets. (Reproduced from Corbit and Stellar, 1964.) Copyright 1964 by the American Psychological Association. Reprinted by permission.

1950; Brobeck et al., 1943; Teitelbaum, 1955) and parasagittal knife cuts in the basomedial hypothalamus (Sclafani and Berner, 1976) sharply reduce their food intake.* Typically, these effects are most evident when the VMH animal is overweight or is overeating. As the animal with lesions reduces its body weight to the level of nonlesioned animals, it begins to consume control amounts of the diet and to regulate its weight at the nonlesioned control level (Ferguson and Keesey, 1975; Kennedy, 1950; Sclafani, 1976). These data, according to some, indicate that negative finickiness is more a manifestation of an obese animal than of one sustaining VMH lesions (e.g., Franklin and Herberg, 1974). (For a more complete description of this issue, see Powley, 1977).

Although all of the above data are commonly taken as demonstrations of the finickiness of the VMH animal, they constitute only weak support for the notion that such animals select and overconsume solely on the basis of gustatory cues. All of these demonstrations confound changes in taste with changes in the nutritional content of the foodstuff. It remains conceivable that some or all of these effects which have been attributed

*In fact, maintaining VMH rats on a quinine-adulterated diet from the immediate postoperative period may eliminate completely the appearance of any hyperphagia (Ferguson and Keesey, 1975).

solely to taste may influence consumption by postingestional means. It is conceivable, for example, that a high-fat diet alters food consumption in the VMH animal by affecting some postingestional mechanism rather than acting through a preingestional palatability mechanism alone.

There are, however, other reports of finickiness that are not related to potential postingestional consequences of a diet, and we believe these demonstrations constitute the most unequivocal evidence for the existence of finickiness—i.e., animals with lesions of the VMH are, indeed, hyperresponsive to sensory aspects of diet.

A variety of experiments have shown that merely changing the textural quality of a diet, while holding its composition constant, affects the degree of hyperphagia following VMH lesions. Corbit and Stellar (1964) demonstrated that VMH animals fed powdered Purina rat chow showed little or no hyperphagia when compared with controls. When switched to the identical diet in pelleted form, however, they increased food intake over control levels and displayed a definite hyperphagia and weight gain. Similar findings have been reported with powdered vs. pelleted Purina Dog Chow (Brooks et al., 1946) and powdered vs. pelleted Gaines Meal (Teitelbaum, 1955). Changing the texture of a powdered diet by the addition of a nonnutritive oil (Carlisle and Stellar, 1969) dramatically increases the degree of hyperphagia in VMH rats. Animals with lesions also overconsume diets mixed with nonnutritive saccharin (Levison et al., 1973). In perhaps one of the most subtle demonstrations of the finickiness phenomenon, Brooks et al., (1946) demonstrated that merely supplying the VMH animal with fresh food on a frequent basis resulted in higher food intake than when the same amount of stale food was left available.

Data on the intake of high-carbohydrate foodstuffs by VMH animals are somewhat difficult to interpret in relation to the finickiness notion. If it is assumed that animals with VMH lesions will consume large quantities of palatable foods, one might predict that they would exhibit strong hyperphagia when offered sweet diets. Although VMH rats probably do consume larger quantities of a high-carbohydrate diet than of rat pellets (Teitelbaum, 1955; Graff and Stellar, 1962), this exaggeration appears to be very modest compared with their response to high-fat substances. Such observations might be viewed as not entirely consistent with the explanation that hypothalamic animals respond simply to some hedonic dimension of the food source, e.g., its palatability. On the other hand, the data on high-carbohydrate consumption by VMH rats only accentuate the conceptual problems surrounding the evaluation of finickiness in the VMH syndrome. First, none of the reports in the literature on high-carbohydrate consumption by the VMH animal involve independent manipulation of the gustatory and postingestional consequences of the diet (Beatty, 1973). Second, the definition of palatability has not been divorced from measures of food intake. And because the palatability of a high-carbohydrate food is unclear, it is difficult to evaluate observations on high-carbohydrate consumption. In sum, definitions of finickiness in this context lead to a problem of circularity which is described in Powley (1977).

b. Other Species

As in the case of diet quality, little systematic work is available for diet quantity in species other than the rat. Virtually no such observations on strictly nonnutritive manipulations of foodstuffs seem to be available, although there is some work on varied diets for VMH animals. While this work confounds the postingestional consequences of the different diets with their gustatory dimensions, it appears to describe a pattern of results that is generally analogous to that observed for rats. Rabbits with VMH lesions show a more extensive overconsumption of oats and potatoes after the lesions than before (Balinska, 1963). A clear postlesion preference developed for those foods as opposed to carrots. Hamsters show only normal food intake and weight gain after VMH lesions if given only Purina chow to eat. Hyperphagia and exaggerated weight gain will appear, however, if the diet is supplemented with sunflower seeds (Marks and Miller, 1972). Also, Rozkowska and Fonberg (1971) observed that, once their VMH dogs had reached the static phase, they refused to eat meatless meals and showed a clear preference for meat over other foods. In contrast, during the dynamic phase of the hyperphagia, their dogs readily accepted meatless and "tasteless" foods and even food adulterated with small quantities of quinine.

The reports on finickiness in humans are also sketchy. For the most part, anecdotes about feeding have stressed primarily a constant readiness to eat or preoccupation with food rather than special selectivity. Although it differs in detail from the patterns exhibited by rats with VMH damage, perhaps the closest human parallel with the finickiness observed in experimental animals is the frequent report that patients with hypothalamic (diencephalic would be more accurate) obesity have an exceptional desire for sweets. Kirschbaum (1951), for example, reported two cases of hyperphagia induced by cerebral trauma in which the patients exhibited an immense appetite for candy. This preference was so strong in one patient (case 9) that she would, for example, eat all the sugar in the sugar bowl and consume a whole can of condensed milk in a short time. Gallinek (1962) has also observed patients who excessively consumed sweets. Clinical reports on humans with such hypothalamic obesity give no real evidence for a preference for high-fat sources of calories.

While such clinical observations on diet preference do exist, it is also noteworthy that, in sharp contrast to the experimental work, other reports of diencephalic obesity in humans seem to stress a striking lack of sensitivity to food quality. Hyperphagic patients have been reported who gulp their food and swallow it without chewing (Kirschbaum, 1951). In these cases, the patients would eat indiscriminately (e.g., a bar of soap) and excessively. In one instance, right after breakfast, a patient was found eating a set of curtains which were soaking in the kitchen sink (Kirschbaum, 1951, case 6). Whether such accounts signal a substantial dissimilarity between the classic experimental hyperphagia and that seen in humans, or whether the apparent differences are due to additional neurological disturbances in the case of humans, is unclear at this point.

3. Food Intake of VMH Animals in Response to Feeding Challenges

A host of studies have investigated whether the physiological controls that influence food intake in the normal animal still operate after VMH injury. In this context, the food intake of VMH damaged animals has been observed after challenges that modulate the food intake of neurologically intact subjects. No class of regulatory responses is unequivocally absent in the VMH animal. Since it is probably impossible to match VMH animals with intact animals on all significant measures affecting energy balance and gustatory reactivity, the occasional reports of reduced or exaggerated responses of VMH animals to different challenges can, for the most part, be explained not as a primary effect of VMH damage but as secondary to differences in energy balance or finickiness. One of the few

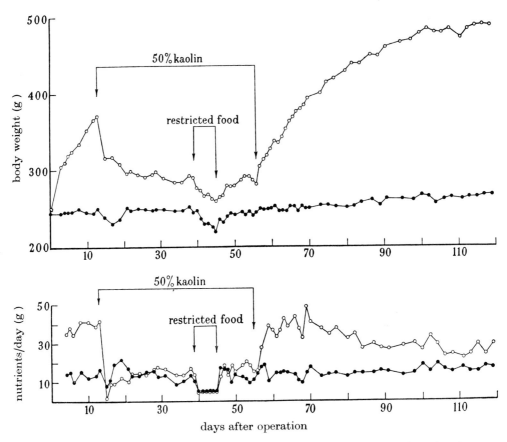

Figure 5 Food intake and body weight of a hypothalamic hyperphagic rat (○—○) and control rat (●—●) in response to 50% kaolin dilution of their basic diet. (Reproduced from Kennedy, 1950.)

unambiguous and permanent alterations in the VMH animal is that it maintains its weight at an abnormally high level—at least with palatable diets (see Figs. 3, 4, and 5).

a. Caloric Dilution Studies

Adolph (1947) demonstrated that rats, when presented with progressively diluted foodstuffs, increased their total food intake so as to maintain a relatively constant caloric intake. Rats would maintain this regulation until their diet was approximately 50% diluted, at which point they failed to ingest sufficient calories and lost weight. Kennedy (1950) was the first to test the ability of VMH rats to exhibit similar caloric compensation. His data indicated that VMH rats were slower to adapt to a 50% kaolin adulteration of their basic diet (see Fig. 5). The rats lost weight and required about four weeks before they began to exhibit signs of overconsumption. Control rats and dynamic-stage hyperphagics required only one week to adjust to the new feeding regimen and exhibited little or no body-weight loss. Strominger et al. (1953) repeated this experiment, showing that

VMH rats failed to compensate their food intake when their calf-meal diet was 50% diluted with nonnutritive cellulose—a compensation readily made by intact control animals. However, they also indicated that hypothalamic hyperphagics would show caloric compensation if their basic powder diet was diluted either 1:1, 1:2, or even 1:3 with water. A decreased ability to compensate calorically was also suggested by Teitelbaum (1955), who noted that static-stage VMH rats ate less of a powdered diet than did control rats when it was diluted 25% or more. However, static-stage rats were still hyperphagic relative to control animals at 15% dilutions and, similar to Kennedy's (1950) observation, lesioned rats still in the dynamic phase were virtually indistinguishable from control rats at all levels of diet adulteration. Levison et al. (1973) also indicated that the caloric compensation of VMH rats breaks down at lower dilutions than does that of control animals, and that the VMH rats' adjustment to a diluted diet is more sluggish than that of intact control rats. Singh and Meyer (1968) have suggested that this effect is more dramatic in the male VMH rat than in the female.

Although consistent in suggesting that VMH rats do not adjust their caloric intake as readily as control animals, the above data do not demonstrate a primary disturbance in caloric regulation after VMH damage. In a previous section of this paper, we reviewed the literature indicating that the VMH animal is inordinately responsive to such sensory aspects of diet as palatability and texture. In all the caloric regulation tests noted above, the influence of an alteration in finickiness might be sufficient to explain the animal's apparent inability to compensate calorically. The addition of nonnutritive substances may change the palatability of the diet and, as Teitelbaum (1955) notes, it certainly affects the texture. That finickiness may, in fact, account for the impaired ability of the VMH animal to ingest additional food to compensate for a diet's dilution is suggested by the fact that VMH rats exhibit no inability in compensating calorically on a palatable diet which is diluted even as much as 50% (Williams and Teitelbaum, 1959). Furthermore, the addition of nonnutritive saccharin to a kaolin-adulterated diet attenuates the apparent difficulty with caloric dilution in VMH rats (Levison et al., 1973). Fed on a Noyes pellet diet (as opposed to the generally used powder diet) which is adulterated up to 50%, VMH rats display no impairment in caloric compensation (Smutz et al., 1975). Finally, studies which systematically manipulate the palatability of the diet in dilution studies have identified palatability, especially greasiness, as a significant variable (Carlisle and Stellar, 1969).

b. Nutrient Preloads

Other studies, employing the technique of a premeal load of nutrients, have examined the response of VMH rats to a dietary (and caloric) challenge which bypasses the confusion of a finickiness mechanism. Neurologically intact rats suppress oral intake of a meal when it is preceded by exogenously delivered nutrients of sufficient magnitude (Wilson and Heller, 1975; Yin and Tsai, 1973). In general, rats make compensatory adjustments of their oral intake in the face of a nutrient infusion. The suppressive effect of premeal loading on subsequent oral intake has been extensively studied in animals sustaining VMH lesions, the consistent result being that these rats show an equal, if not superior, ability relative to controls in modulating oral intake after experimenter-administered nutrient infusions.

Using intragastric (IG) loads, Smith et al., (1961) measured 2-hr food intake after loads of saline, glucose, water, or kaolin. Hypothalamic hyperphagic rats and control animals showed a similar profile of oral-intake suppression after loads. Thomas and Mayer (1968) reported that VMH rats and control animals compensated nearly identically to a liquid diet after loads of the same diet. VMH rats displayed equal suppression of food intake as control rats to glucose, corn oil, or casein hydrosylate IG loads (Panksepp, 1971). Liu and Yin (1974) also demonstrated that control-like suppression of oral intake after loads is evident in both dynamic- and static-stage, hypothalamic hyperphagic animals. Monkeys rendered hyperphagic by parasagittal ventromedial hypothalamic disconnections also displayed an ability equal to that of control animals to suppress oral intake after IG nutrient loads (McHugh et al., 1975).

The ability of VMH animals to make caloric adjustments in oral intake after infused nutrient loads is also evident when the loads are administered intraperitoneally (Panksepp, 1974; Reynolds and Kimm, 1965; Russek and Morgane, 1963) or intravenously (Rowland et al., 1975). These data—that hypothalamic hyperphagic animals are able to adjust their food intake proportionately to prefeeding nutrient loads as well as control animals—have been taken as evidence that VMH lesions do not primarily impair a satiety mechanism.

c. Other Dietary Tests

The reactions of rats sustaining VMH lesions and normal rats to diets which are either unbalanced in or devoid of particular amino acids have been tested. Nassett et al., (1967) indicated that, whereas intact rats decreased food intake of an amino-acid-imbalanced diet, VMH rats overate these diets as much as control, nutritionally balanced diets. It is noteworthy, however, that these researchers tested rats that were still in the dynamic phase of obesity and appeared to be consuming both the unbalanced and control diets at a ceiling level. Rogers and Leung (1973) tested rats with VMH lesions on a more extensive battery of unbalanced or deficient diets and found that operated animals modulated their food intake in a manner similar to control animals. A similar profile of food intake in VMH and control rats has been observed with other amino-acid-imbalanced diets as well (Kraus and Mayer, 1965; Scharrer et al., 1970).

When given brief exposure to NaCl, glucose, or sucrose water of various concentrations, normal rats exhibit characteristic preference-aversion functions for each of the solutions. Rats with VMH lesions display preference-aversion functions to these diets that are similar to those of controls (Mook and Blass, 1970). Finally, the ability of VMH-lesioned rats to select a nutritionally adequate diet has also been investigated and found to be intact after lesions. Brobeck et al., (1943) found that, after bilateral lesions of the ventromedial hypothalamus, rats, although hyperphagic, maintained similar proportions of food intake as they had prelesion from the variety of foodstuffs that constituted their basic diet.

d. Environmental Challenges

The food intake of rats can be altered by changing the ambient temperature of their environment. The intake of VMH animals, even at a static stage of obesity, is apparently as sensitive to temperature changes as that of control animals. Kennedy (1953) demonstrated that VMH rats, like control animals, increased their food consumption in a cold environment and decreased it in a heated one. Hamilton (1963) also indicated that obese VMH rats displayed the same profile of food intake in response to changes in ambient temperature, both in an ad libitum feeding situation and when required to bar-press for their food.

Several investigators have suggested that the VMH rat reacts differently than control rats in restricted-feeding situations. Panksepp and colleagues (Panksepp, 1971; Panksepp and Dickinson, 1972) reported that VMH rats failed to eat as much as control rats when given only 1-5 hr access to food. These same rats were hyperphagic, relative to controls, on an ad libitum feeding situation. Epstein (1959) also reported that VMH rats that are hyperphagic on an ad libitum feeding schedule failed to eat more than controls when allowed only 6-hr exposure to a diet. Although van Putten et al., (1955) also reported that VMH rats were sluggish and exhibited greater difficulty than controls in adapting to two 1-hr feeding sessions per day, it is clear that the VMH rat can adjust its food intake so as to maintain a hyperphagia and weight gain in the face of constraints upon its eating. Furthermore, rats with parasagittal knife cuts increased their food intake in a 1-hr test meal with increasing levels of deprivation (Sclafani et al., 1976), and rats with VMH lesions maintained the correlation between meal size and subsequent intermeal interval (Thomas and Mayer, 1968).

e. Drug-Influenced Feeding

Several agents which have been documented as either increasing or decreasing food intake in neurologically intact rats have been tested on VMH rats. Amphetamine is one such anorexic agent, and its effect on VMH rats has been evaluated from the hypothesis that its central site of action might be the ventromedial hypothalamus. Stowe and Miller (1957) adulterated their rats' diet with amphetamine, or injected it subcutaneously, for one-week periods. Both hypothalamic hyperphagic and control rats reduced their food intake under such conditions, although the VMH animals exhibited a proportionately greater suppression. Epstein (1959) verified the hyperresponsiveness of the VMH rat to amphetamine injections; Reynolds (1959) demonstrated that amphetamine reduced the food intake of rats in a dose-related manner and that rats with lesions deprived of food for 24 hr exhibited greater suppression than did controls. The data on amphetamine from species other than the rat are somewhat more equivocal. McHugh et al. (1975) reported that monkeys with VMH disconnections are somewhat more resistant to the anorexic effect of amphetamine than controls, and Sharp et al. (1962) indicated a loss of responsiveness to amphetamine in cats after VMH lesions.

In terms of other agents which depress food intake in normal rats, it is noteworthy that synthetic cholecystokinin (CCK) reduced the food intake of both normal and VMH rats in a half-hour test meal following CCK injection (Kulkosky et al., 1976). Stern et al. (1976), however, reported a decrease in sensitivity in VMH animals to caerulin, a compound similar to CCK which suppresses food intake in intact rats.

The food-intake reaction of VMH rats has been examined to glucoprivic agents that cause intact animals to eat, with the results generally supporting the conclusion that insulin and 2-deoxy-D-glucose (2-DG) elicit overeating in lesioned animals as well as in controls. In the case of 2-DG, the response of sated VMH rats and controls to this drug administered intravenously is similar, both in terms of the degree of overeating and its latency after administration (Nicolaïdes and Meile, 1972). A similar correspondence in the response of lesioned animals and controls is seen when the drug is administered directly into the brain ventricles (Miselis and Epstein, 1971). King and Gaston (1977) reported that intraperitoneal injections of 350 mg/kg 2-DG caused an increase in food consumption in both static-phase VMH rats and controls but not in dynamic-phase hyperphagics. At higher doses (750 mg/kg), the drug-induced hyperphagia was not evident even in static obese

VMH rats, in spite of the fact that this dose increased the magnitude of the response in controls. Müller et al. (1974), using 750 mg/kg 2-DG, also failed to induce an increase in food intake in either static- or dynamic-phase hyperphagic rats sustaining very large VMH lesions. Recently, however, Houpt and Gold (1975) demonstrated that rats rendered hyperphagic by parasagittal knife cuts do elevate base-line responses to 2-DG administration. Furthermore, they provided explanations for the previous negative findings in the dynamic-phase hyperphagic and at the high-dose level. They noted that VMH rats in the dynamic phase are already eating as much as they would be if stimulated by a 2-DG injection, and therefore, presumably, the ineffectiveness of 2-DG in this rat may be explained by a variety of other competing factors in the animal.

Epstein and Teitelbaum (1967) monitored the food intake of VMH rats for 6 hr after insulin doses ranging from 8 to 16 U per rat, both static and dynamic hyperphagic rats increased food consumption relative to base-line measures during this period. This overconsumption, which was also apparent in controls, is obtained when the diet was either pellets or the more palatable eggnog. Sclafani et al., (1973) reported that rats rendered hyperphagic by parasagittal knife cuts either 1 or 2 mm lateral to the midline increased consumption of a powder diet in an acute 6-hr test after insulin injections. An increase in food intake both in animals sustaining knife cuts medial to the fornix and in control animals is also observed in doses ranging from 8 to 16 U per rat (Sclafani et al., 1975). King and Grossman (1977) have also reported that VMH animals display similar increases of food intake, and with equal latency, as control animals after insulin injections of 4 or 8 U per rat. Müller et al. (1974) found that VMH animals elevated their food intake only in response to an insulin dose of 8 U/kg body weight but not to a dose of 4 U/kg body weight. All these reports are consistent in suggesting that insulin injections increase feeding in the VMH rat in acute feeding tests. It should be noted that Panksepp and Nance (1972), however, reported that chronic administration of insulin to free-feeding animals elevates food intake in controls but not in VMH animals.

f. Operant Defense of Body Weight

One of the classic paradoxes surrounding the VMH animal's food intake is an apparent dissociation between its ad libitum food intake, i.e., hyperphagia, and formal measurements of its motivation to obtain food. Miller et al., (1950) were the first to document the phenomenon of an apparent "unwillingness" of the VMH rat to work for its food. It was their observation that VMH rats bar-pressed less than controls on a 5-min fixed-interval schedule, ran an alley slower than controls in order to obtain food, and pulled less weight down this alley in order to obtain food. Furthermore, lesioned rats reduced their food intake in the home cage when forced to manipulate a lid weighing 75 g to get to the food. Since, in unconstrained situations, VMH rats are markedly hyperphagic, their increased food intake coupled with Miller and colleagues' findings of their decreased motivation to obtain food seemed paradoxical. This apparent dissociation of food intake and food motivation was supported by Teitelbaum (1957), who reported that VMH rats pressed bars less than did control rats and therefore received fewer food pellets when required to make 16 or more bar-presses per pellet (fixed ratio [FR] $\geq$ 16), but that they bar-press as much or more than control rats when they had to press four or fewer times per pellet (FR $\leq$ 4). The reduction in operant responses in VMH rats at fixed-ratio schedules greater than approximately FR 4 has been reported by other experimenters (e.g., Sclafani and Kluge, 1974).

Subsequent studies on the willingness of VMH animals to work for food have revealed a variety of factors which affect the degree of the VMH animal's motivation to obtain food. The most prominent of these is the body weight at which the VMH rat is tested. In general, VMH rats tested at a percentage of an obese, static-phase body weight appear less willing to work for food. If tested at a percentage of prelesion body weight, however, they appear as motivated as intact control animals if not more so. Thus, for example, Marks and Remley (1972) indicated that rats sustaining bilateral damage to the VMH and tested at 90% of their static body weight manifested the typical bar-pressing decrement for food on a 2-min variable-interval (VI) schedule. Rats tested at 90% of their prelesion body weight exhibited no such performance decrement. Peters et al., (1973) report that VMH-lesioned rats maintained at a percentage of prelesion weight exhibit as many bar-presses as controls on a schedule as lean as FR-64. In fact, at 100% of prelesion body weight (i.e., mild deprivation), lesioned rats bar-press more than controls. Similar results are presented by Kent and Peters (1973) on a VI 1-min schedule. A number of other researchers have verified the dependence of the VMH animal's reduced willingness to work for food on its level of body weight (Porter and Allen, 1977; Falk, 1961; Wampler, 1973; Singh et al., 1974) and, in all cases, they indicated that rats tested at a percentage of preoperative body weight or of control body weight work as hard, or harder, than control animals for food. Sclafani and Kluge (1974) have presented a similar set of interactions for rats rendered hyperphagic by parasagittal knife cuts.

Body weight may not be the only factor that governs the existence of a VMH lesion-affected motivational deficit. In fact, the data of both Miller et al. (1950) and Teitelbaum (1957) are difficult to interpret with this single explanation, because in both their studies animals with lesions that were not currently overweight relative to control animals (the "low-feed" group of Miller and colleagues and the dynamic animals of Teitelbaum) exhibited a performance decrement as well. However, in the Teitelbaum study, dynamic-phase animals are not as unwilling to work for food as static-phase animals, and it may be of note that, in both reports, the lesion placements are not histologically verified.

Subsequent reports have identified other variables which affect the presence of a motivational deficit in animals with VMH lesions. Prelesion training of the operant response attenuates the performance decrement of VMH rats that press bars for food or water reinforcement. Even on a schedule as lean as FR-128, rats that were trained before the production of lesions increased their postlesion rate of response (Beatty et al., 1975b). Furthermore, the degree of improvement in the lesioned animals was positively correlated ($r = 0.91$) with their subsequent hyperphagia. King and Gaston (1973), as well as Porter and Allen (1977), also report an attenuation of a performance decrement in rats sustaining VMH lesions with prelesion training.

The palatability of the reinforcement also modulates the degree of response by the VMH rat in an operant situation. Rats sustaining lesions that receive a 32% sucrose solution as a reward, instead of the conventional Noyes pellet, may show a diminished or no performance decrement in a bar-pressing situation compared with controls (Beatty, 1973). Sclafani and Kluge (1974) also report an attenuation in performance decrement in parasagittal knife-cut rats rewarded with a palatable milk diet.

The temperature of the test environment influences the willingness to work for food in VMH rats. Such rats tested at 32°C reveal a performance decrement relative to controls which is eliminated by testing at temperatures of 27°C or lower (Colby and Smith, 1975).

Few reports exist on the operant performance of VMH animals other than the rat. The few data available, however, are consistent with the notion that animals increase food-reinforced behaviors after VMH injury. Hamilton and Brobeck (1964) report that static hyperphagic monkeys work as hard as controls on FR schedules for banana-flavored food pellets. Obese mice also increase their operant rate after VMH destruction (Anliker and Mayer, 1956). Rabbits rendered hyperphagic by lesions of the medial hypothalamus also increase the number of operants performed in order to obtain food (Balinska, 1963; Lewinska, 1964; Romaniuk, 1962). Furthermore, relative to prelesion extinction tests, the operant behavior of rabbits with medial hypothalamic lesions on a preferred diet is extinguished more slowly than that of controls (Balinska and Brutkowski, 1964).

4. The Ventromedial Hypothalamic Area and Obesity

As the previous discussion has indicated, hypothalamic hyperphagia is clearly associated with development of the obesity that originally identified the ventromedial hypothalamic syndrome. Animals with hypothalamic damage develop immense lipid stores and show a metabolic profile characteristic of active lipogenesis and a positive energy balance. Like the changes in food intake produced by hypothalamic lesions, these aspects of the syndrome have been most completely characterized in rodents.

a. Body Composition: Rodents

In general, body composition can be divided into three compartments: fat, fat-free solids, and water. By definition, hypothalamic obesity involves excessive accumulation of fat, and direct measurements indicate that the overwhelming bulk of the increase in body weight after basomedial hypothalamic damage is due to the increase in body fat (Hetherington and Weil, 1940; Kennedy, 1950). In the rat, increased carcass fat has been observed as early as 4 hr after VMH lesions (Slaunwhite et al., 1972). Although most methods of body-composition analysis preclude repeated measurements, it is clear that the fat accretion is progressive and apparently continues until the animal enters the static phase and its body weight reaches a plateau. Even when lesioned as adults, experimental animals with access to palatable foods can often more than triple their body weight (Brobeck et al., 1943; Gold, 1970), accrue sixfold increases in their carcass fat (Han and Young, 1963; Hetherington and Weil, 1940; Keesey and Powley, 1975), and reach a level at which ether-extractable lipid accounts for 45-60% of total body weight (Hetherington and Ranson, 1940; Montemurro and Stevenson, 1957a) (see Fig. 6). The added carcass fat can account for over 75% of the increase in body weight (DeLaey, 1975 [GTG-lesioned mouse]; Holm et al., 1973).

Since the precise anatomical locus of the VMH syndrome remains a matter of dispute, tight correlations between the degree of obesity and the amount of destruction are not possible. Still, observations indicate that there is a strong positive relationship between the two variables. In an experiment that varied lesion size by varying the number of small, punctate lesions, Bernardis and Goldman (1972) found that the more extensive lesions of the ventromedial area produced the largest increases in lipid content of the epididymal fat pads. In another experiment, Bernardis and Frohman (1970) reported that total carcass lipid in weanling rats with VMH lesions was positively correlated with lesion size. Similarly, McBurney et al. (1965) found that the GTG lesions destroying the greatest amounts of the ventromedial hypothalamic nuclei produced the fattest animals: the correlation between the amount of the VMH spared and lipid content of fat depots

Role of the Hypothalamus in Energy Homeostasis

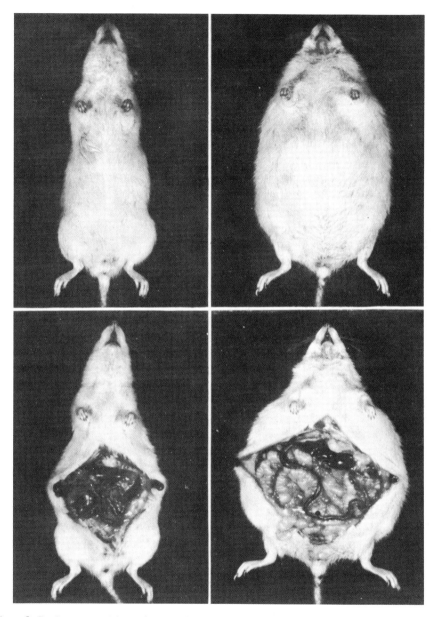

Figure 6 Body composition of control rat and obese rat 19 weeks after bilateral electrolytic lesions of the VMH. (Reproduced from Kennedy, 1950.)

was r = -0.86 for males and r = -0.76 for females. Finally, it is an accepted, though imprecise, working principle in research on the VMH syndrome that larger lesions produce fatter animals (see, for example, Brooks and Lambert, 1946; Gold, 1973).

There has been little investigation of possible differences of the accumulation of the excess fat in specific depots, but it appears that all depots are enlarged. Hetherington and Ranson (1940) and Brobeck et al. (1943) noted that their obese animals had large fat deposits subcutaneously, in the omentum and mesenteries, and in the retroperitoneal, perirenal, and pericardial regions. Hirsch and Han (1969) found cellular hypertrophy in both the epididymal and retroperitoneal fat pads after VMH lesions, and Johnson et al. (1971) found cellular hypertrophy in each of the three depots they examined: subcutaneous, gonadal, and retroperitoneal. Wise (1975) reports increases in abdominal and periovarian fat depots.

Recent work indicates that observed changes in the fat stores are typically the result of adipocyte hypertrophy, not hyperplasia. Hirsch and Han (1969) reported that VMH lesions in the rat can produce a fourfold increase in adipocyte size. These investigators also noted that this increase amounted to 2 μg of lipid stored in a cell and that the hypertrophic cells in their VMH rats were the largest adipocytes they had observed in any of the species they examined (human, rat, and mouse). Johnson and Hirsch (1972) found predominate hypertrophy of adipocytes in GTG-lesioned mice, although they detected a modest hyperplasia in the retroperitoneal fat pad in their animals. Wise (1975) has argued that the form of lipid storage in hypothalamic obesity is dependent on the time during the processes of maturation and fat-cell replication that lesions are produced: GTG (as well as bipiperidyl mustard) lesions of weanling mice induced adipocyte hyperplasia.

Though excessive food intake is required for the development of maximal obesity in the VMH animal, it has long been known that greater-than-control food consumption is not necessary for the occurrence of greater-than-control adiposity following lesions. Brobeck et al. (1943) and Brooks and Lambert (1946) reported that pair-feeding of VMH rats with controls did not prevent some of the experimental animals from outgaining the controls in body weight and becoming noticeably obese. These same experimenters also noticed that animals with lesions consumed their day's ration in much shorter order than did sham operates. Evidence that such engorgement results in increased lipid synthesis in normal animals (Tepperman et al., 1943; see section on lipid metabolism), led Brobeck and colleagues to conclude that the residual fattening of the pair-fed VMH animal is the result of its voracious pattern of feeding (subsequent research into the effects of "gorging" or "stuff and starve" patterns of food intake has established that they do, in fact, result in increased accumulation of carcass fat [Cohn, 1963; Fábry, 1967]). This claim was countered in part by the demonstration that VMH animals still often outgained controls even when the former were presented with control levels of food and in small portions (Brooks and Lambert, 1946). Holm and colleagues (1973) have recently found increased accumulation of ether-extractable lipids in VMH rats on a similar feeding regimen. Early attempts to control pattern of intake by requiring both controls and operated animals to consume their daily ration within a limited time yielded negative or equivocal results (Brooks and Lambert, 1946; van Putten et al., 1955). Subsequent experiments using such a regimen have revealed increased fat accumulation by VMH animals (Goldman et al., 1974; Han, 1967). Likewise, feeding by gavage has been used with the same result (Han and Liu, 1966).

There are three other classes of observations on adiposity in VMH animals with normal food intake, but, in these cases, the normophagia is voluntary. Weanling rats with hypothalamic lesions do not consume more food than controls and yet show increased carcass lipid (Han et al., 1965; Bernardis and Skelton, 1966). There have also been a number of observations of adult animals becoming obese after VMH lesions without exhibiting hyperphagia, possibly because of small or slightly misplaced lesions (Hetherington and Ranson, 1942b; Mu et al., 1968; Rabin, 1974). Finally, goldthioglucose lesions in mice produce obesity in the absence of excessive food intake if the animals are provided with a calorically dense diet (high fat) in warm (near thermal neutrality) rooms; similarly treated animals in cooler rooms will show increased food consumption (DeLaey et al., 1975).

Other body compartments, aside from carcass fat, also undergo changes after VMH lesions. For the most part, investigators have paid less attention to these changes because they are relatively modest in magnitude, they can be dissociated from changes in fat storage and metabolism, and in many cases, they appear secondary to excessive fat accretion.

Frequently, fat-free solids exhibit changes. Animals receiving lesions at weaning or during the juvenile periods of growth are generally stunted in terms of linear growth and show relative reductions in the levels of both total carcass fat-free solids (Bernardis and Skelton, 1966; Goldman et al., 1970; Han et al., 1965) and protein levels (Goldman et al., 1974). Dramatic reductions in these measures are also found in adult VMH animals whose food intake is not greater than that of controls, either voluntarily (Montemurro and Stevenson, 1957; Mu et al., 1968) or to restriction by the experimenter (Goldman et al., 1974; Han and Liu, 1966; Holm et al., 1973). Such reductions are found even though lipid accumulation is augmented in the lesioned animals. In cases where accumulation of lean body mass is not diminished or is even increased after VMH lesions, however, the relative increases in the fat compartment overshadow other tissue building (Chlouverakis et al., 1973; Montemurro and Stevenson, 1960; Mu et al., 1968). On the other hand, GTG lesions may produce increased linear growth as well as obesity (Powley and Plocher, 1980). The same can be said for parasagittal knife cuts in rats (Gold and Kapatos, 1975) and more extensive knife cuts that effectively produce basomedial hypothalamic "islands" or, perhaps more correctly, "disconnect" the basomedial hypothalamus from the rest of the brain (Mitchell et al., 1972; Palka et al., 1971).

When expressed as a percentage of total carcass weight, body water is reduced after VMH lesions. This reduction is in large part a consequence of the difference in body composition between experimental animals and controls: fat is stored in a very dehydrated state compared with other tissues, so that, as the amount of carcass fat increases, the proportion of total body weight that is water must decrease. However, work by Stevenson and his colleagues has indicated that, in addition to exhibiting this merely apparent deficit in water content, VMH rats are actually relatively dehydrated. Plasma sodium concentration in lesioned rats is increased relative to that in control rats, even after 24-hr food and water deprivation (Stevenson et al., 1950). Montemurro and Stevenson (1957, 1960) also found that, while absolute levels of extracellular fluid volume and total carcass water were increased in obese VMH rats, calculations taking into account the differences in body composition between experimental and control animals indicated that lesioned animals were actually deficient in fluids.

As might be predicted from the demands that adiposity and changes in water metabolism place on an animal's metabolism, several viscera exhibit size and structural changes in VMH animals. Brobeck and colleagues (1943) observed at autopsy that the gastrointestinal tracts of obese VMH rats were dilated and approximately twice the weight of those of controls. These authors considered this hypertrophy to be secondary to increased food intake. Other viscera showing increases in weight and hypothalamically obese animals are the heart, liver, and kidneys; the spleen and diaphragm are apparently unaffected (Brobeck et al., 1943; Brooks and Lambert, 1946; Mayer et al., 1955; Stevenson, 1949). Gonadal atrophy and dysfunction also typically occur in animals of both sexes after VMH lesions, though dissociations between the gonadal changes and obesity are quite common (Kennedy, 1969; Nance and Gorski, 1975; Stevenson, 1969).

The livers of hypothalamically obese rats may be twice the weight of those of controls (Brobeck et al., 1943; Kennedy and Pearce, 1958). A large part of this increase is fat, which can occupy twice the normal percentage of total liver weight. This percentage can be virtually returned to normal by 24-hr food deprivation (Brobeck et al., 1943). Kennedy and Pearce (1958) found that the greatest part of the increment in liver weight occurs within 24 hr of placement of hypothalamic lesions. However, this rapid increase is due to build-up of glycogen, which declines after 48 hr; by the fourth day postlesion, the excess glycogen has been replaced largely by fat. Total liver protein increases fairly steadily throughout the first week and has doubled by the end of that period. While this initial increase apparently represents hypertrophy only, there is evidence that, over a period of months, VMH animals show increases in hepatic cell number as well as in cell size (Kennedy, 1957; Kennedy and Pearce, 1958).

b. Body Composition: Species Other than Rodents

Basomedial hypothalamic lesions have been observed to produce increases in body weight and/or adipose tissue stores in virtually all mammals and birds studied to date. The early work of Bailey and Bremer (1921) showed that experimental punctures of the hypothalamus could lead to marked gain in body weight and obesity (as well as gonadal atrophy) in dogs. These findings were subsequently confirmed by a number of investigators, whether the neural damage accompanied removal of the pituitary (Graef et al., 1944; Keller and Noble, 1936) or was the result of attempts at producing discrete lesions within the hypothalamus alone (Biggart and Alexander, 1939; Heinbecker et al., 1944). During this early period, it was established that hypothalamic obesity could also be experimentally produced in the rhesus monkey (Brooks et al., 1942; Ranson et al., 1938) and the cat (Anand and Brobeck, 1951; Ingram, 1952; Wheatley, 1944). Other mammals that have shown exaggerated body weight after electrolytic VMH lesions are ground squirrels (Mrosovsky, 1974, 1975), golden hamsters (Marks and Miller, 1972), rabbits (Balinska, 1963), and swine (Auffray, 1969; Khalaf, 1969). Failure to produce increased body weight by electrolytic lesions of the VMH has been reported for guinea pigs (Joseph and Knigge, 1968), though, as mentioned earlier, parasagittal knife cuts have proved effective in this species (Sclafani and Sperber, 1977). Recent work has indicated that the VMH lesion effect generalizes to avian species, with increased body weight and lipid accumulation reported in chickens (Lepkovsky and Yasuda, 1966; Lepkovsky, 1973), geese (Snapir et al., 1976), and white-throated sparrows (Kuenzel and Helms, 1967).

Observations on the carcass distribution of the excess fat across the various species are very sketchy, but, within the limits of species variation, a pattern is seen similar to that in the rat. Thus, in the rhesus monkey, hypothalamic lesions result in large deposits of subcutaneous fat in the neck, thorax, and abdomen as well as heavy infiltration in the omentum and mesenteries. Pericardial and perirenal depots are also enlarged (Brooks et al., 1942; Poirier et al., 1962b; Ranson et al., 1938). Likewise, the fat accumulation following hypothalamic lesions in the dog has been described as "general" (Bailey and Bremer, 1921; Graef et al., 1944). Biggart and Alexander (1939) described large deposits of fat subcutaneously, in the abdomen, and "between the various muscle groups" (p. 414) in their obese dogs. In their report of a five-year-old girl who became obese as a result of leukemic infiltration of the hypothalamus, Heaney et al. (1954) noted that most of the excess fat accumulated in the face, trunk, and upper arms. Lepkovsky and Yasuda (1966) report that VMH lesions in chickens resulted in the development of large abdominal adipose tissue deposits, though they also state that fat accumulation was generally comparable throughout the carcass. The livers of some of the obese birds were elevated in weight and lipid content. Geese also show lesion-induced increases in abdominal adipose tissue (Snapir et al., 1976), but the most dramatic changes occur in the livers of these animals, which can, primarily through increases in lipid content, attain weights three to four times those of controls by several weeks postlesion (Auffray and Blum, 1970; Snapir et al., 1976).

Though quantitative analysis of carcass composition has only rarely been performed on nonrodents with hypothalamic obesity, body-weight data indicate that, in at least some of the species studied, the magnitude of the obesity can be virtually as impressive as it is in rodents. For example, Rozkowska and Fonberg (1971) found that their VMH dogs more than doubled their body weights within three weeks. Similarly, one of the two cats receiving basomedial hypothalamic lesions in a study performed by Anand and Brobeck (1951) increased its body weight from 2.2 to 5.8 kg within two months of the operation. Lesioned rhesus monkeys can also more than double their body weights over a period of several months (Brooks et al., 1942; Hamilton and Brobeck, 1964; Ranson et al., 1938).

In terms of body weight, however, hypothalamic obesity in several other species has appeared to be of rather modest magnitude. For example, golden hamsters have been reported to increase their body weights by roughly 30% in response to VMH lesions (Marks and Miller, 1972). Khalaf (1969) found that VMH-lesioned pigs increased their body weights only 13% more than controls over a seven-week period. (However, the animals used in this study were young—only eight weeks old at the time of the lesion—and still growing rapidly. It is possible that the rate of weight increase was at near-ceiling levels in the unlesioned animals—compare the lack of effect of hypothalamic lesions on body weight in weanling rats.) As Bray and Gallagher (1975) have summarized it, a review of cases of human hypothalamic obesity reveals that body weights over 140 kg, or roughly 310 pounds, are rare. Average weight gain is on the order of 20 kg. Such weight levels are relatively modest when compared with other types of human obesity (Bierman and Glomset, 1974). In birds, too, the effects of hypothalamic lesions on body weight have typically proved relatively minor. For example, Kuenzel and Helms (1967) found that white-throated sparrows increased their weight by only 30% after basomedial hypothalamic lesions. Comparably modest gains have been reported for chickens (Lepkovsky and Yasuda, 1966) and geese (Snapir et al., 1976). These cases may, of course, indicate important species differences in the basic mechanisms underlying hypothalamic obesity. For example, avian species apparently do not possess innervation of the pancreatic islets (Woods and Porte, 1974). It is possible that the apparently minor effects of VMH lesions in birds reflect this fact, in light of recent findings suggesting a critical role for neurally mediated insulin release in VMH obesity in mammals (see discussion below).

There is, however, one report of dramatic weight gain in birds after hypothalamic damage: Lepkovsky (1973) observed a more than twofold increase in the body weight of lesioned cockerels over a period of roughly four months after surgery. Several other observations also argue for the possibility, at least, of a robust VMH syndrome in birds: carcass fat can account for over 40% of body weight in white-throated sparrows with hypothalamic lesions (Kuenzel, 1972); the impressive effects of lesions on liver weights and lipid accumulation in geese are considerably greater in magnitude than the analogous changes in mammals; and, though the carcass fat in male chickens, for example, may not be particularly impressive (8-10 of abdominal fat per 100 body weight [Lepkovsky and Yasuda, 1966]) alongside that found in rats, it should be noted that normal male chickens contain very little fat (0.2 g of abdominal fat per 100 g body weight), so that increases expressed in proportion to control levels can actually be very dramatic. Thus, birds can show a rather impressive VMH obesity when differences in the levels of fat normally stored and in the mode of storage (e.g., greater stress on hepatic storage) of excess fat compared with mammals are taken into account. It should also be noted that an apparently smaller VMH effect in any of the species mentioned above might reflect, merely, the relatively small number of attempts to produce, or observations of, the obesity, which increases the likelihood that experimenters (or pathological processes) have not successfully pinpointed the effective neural locus.

The role of excessive food intake in the development of hypothalamic obesity in nonrodents appears to parallel that described for rats and mice. Wheatley (1944) noted that cats with hypothalamic lesions, which were later found to involve the ventromedial nuclei, showed above-normal weight gains though fed the same amount as control animals. (These cats also wolfed their food down in contrast to the leisurely pace of the control animals, just as was the case with the pair-fed VMH rats described by Brobeck et al., 1943). Poirier et al. (1962a, b) noted that some of their rhesus monkeys with medial hypothalamic lesions clearly outgained the control animals with which they were pair fed. Similarly, in a study performed by Lepkovsky and Yasuda (1966), several of the chickens to which they administered basomedial hypothalamic lesions showed normal food intake and weight gain but were found after sacrifice to have accumulated clearly more carcass fat than had control animals.

c. Metabolic Profile: Insulin

Hyperinsulinemia is a central, and perhaps defining, trait of the ventromedial hypothalamic syndrome. It has been observed in the rat after electrolytic lesions of the VMH (Hales and Kennedy, 1964) and parasagittal knife cuts (Tannenbaum et al., 1974) and in the mouse after electrolytic lesions of the basomedial hypothalamus (Chlouverakis et al., 1973; Coleman and Hummel, 1970) and GTG injections (Coleman and Hummel, 1970). Elevated plasma insulin has also been noted in human cases of hypothalamic obesity (Bray and Gallagher, 1975; Komorowski, 1977) and in rhesus monkeys after VMH lesions (Hamilton et al., 1972). Increased insulin secretion has been noted as early as 10 min after VMH damage (Berthoud and Jeanrenaud, 1977; Rohner et al., 1977). Bernardis and Frohman (1970) found that the degree of hyperinsulinemia correlated with VMH lesion size in weanling rats. In the same study there was a significant correlation between plasma insulin and amounts of excess carcass fat, though Han and Frohman (1970) found no such relation in VMH rats receiving control levels of food by gavage. However, the degree of hyperinsulinemia in lesioned rats pair-fed with control animals for 48 hr postlesion is a good predictor of subsequent weight gain on ad libitum food intake (Hustvedt and Løvø, 1972).

Although it is not yet possible to establish clearly whether both basal and stimulated phases of insulin secretion are initially potentiated by VMH destruction, or whether one is increased as a result of the other (see Powley, 1977, for a discussion of this issue), the data suggest on temporal grounds that stimulated insulin release is immediately exaggerated after lesions while augmented basal release develops over a period of days after the lesion. Basal plasma-insulin levels are not elevated in the first hour after lesions (Han and Frohman, 1970; Steffens et al., 1972), and there is evidence that insulin levels are normal at 48 hr postlesion if food has been withheld over that period (Louis-Sylvestre, 1971). In the latter experiment, subsequent ingestion by VMH rats of 1 ml of 50% glucose solution apparently resulted in an immediate and dramatic oversecretion of insulin. In fact, exaggeration of stimulated insulin release has been found 10 min postlesion to intravenous glucose (Berthoud and Jeanrenaud, 1977; Rohner et al., 1977) and within the first hour to food ingestion (Steffens et al., 1972). This heightened insulin responsiveness in rats is chronically maintained to a variety of challenges including food intake (Steffens, 1970), oral glucose administration (Louis-Sylvestre, 1976), intravenous glucose (Frohman et al., 1972a; Karakash et al., 1977), intraperitoneal glucose (Frohman et al., 1972b), or intravenous glibenclamide (Martin et al., 1974). Heightened insulin response to oral glucose is also present in cases of human hypothalamic obesity (Bray and Gallagher, 1975; though the finding of an increase might depend on how the data are scaled: a recent study by Komorowski (1977) indicates that, while the response may be increased over that of controls when expressed in absolute amounts, expressing the increase in plasma insulin as a percentage of basal levels eliminates the difference between the obese patients and the controls, owing to the increased basal levels of the former. Hamilton et al., (1972) found, however, that the rise in plasma insulin to intravenous glucose is considerably lower in obese rhesus monkeys with VMH lesions compared with controls.

Exaggerated basal-insulin levels do develop, but relatively slowly, and are probably influenced by, among other things, the nutritional state of the animal and the degree of obesity attained (Goldman et al., 1974; Martin et al., 1974; Hales and Kennedy, 1964). Actually, stimulated insulin levels are also probably influenced by states of nutrition and obesity as well (see Steffens, 1970). Most investigators have found basal-insulin levels to be normal on the second and third days after VMH lesions, insofar as food deprivation of from 3 to 24 hr results in equal plasma-insulin levels for control and experimental animals (Chikamori et al., 1977; Hales and Kennedy, 1964; Hustvedt et al., 1976). However, others have found basal-insulin levels to be already increased by that time (Hustvedt and Løvø, 1972; Tannenbaum et al., 1974 [parasagittal knife cuts]). Soon thereafter, by the fourth or fifth postoperative day (Han and Frohman, 1970; Slaunwhite et al., 1972), it has generally been found that food deprivation of up to 24 hr does not reverse the hyperinsulinemia of the VMH rat (Hales and Kennedy, 1964; Inoue and Bray, 1977; Han et al., 1972; Han and Frohman, 1970; Louis-Sylvestre, 1976; Slaunwhite et al., 1972; Steffens, 1970). However, Karakash and colleagues (1977) found that 24-hr food deprivation reduced the plasma insulin of VMH rats pair-fed with control animals to control levels up to 10 days postlesion. Also, fasts of 48 hr have been found to reverse the hyperinsulinemia of weanling rats with hypothalamic lesions (Frohman et al., 1971; Schnatz et al., 1971). Plasma insulin in hypothalamic obese humans has been found to remain elevated over the levels in patients with "essential" obesity for the first three days of a seven-day fast (Bray and Gallagher, 1975).

Restriction of the food intake of lesioned rats to control levels does not preclude the development of hyperinsulinemia (Han and Frohman, 1970; Han et al., 1972; Karakash et al., 1977), though maintaining VMH animals on ad libitum feeding results in an even greater exaggeration. For example, in lesioned animals pair-fed to controls, Hustvedt and Løvø (1972) found plasma-insulin increases of 60-280% 48 hr after the lesion, while experimental animals on an ad libitum intake showed increases on the order of 600% on the day after the lesion (Steffens et al., 1972, their Fig. 10). Steffens (1970) reported a similar pattern of results with regard to the meal-induced rise of plasma insulin. Also, weanling rats with basomedial hypothalamic lesions are hyperinsulinemic even though normophagic (Frohman and Bernardis, 1968). It is possible, however, to reduce the insulinemia of VMH rats to control levels by chronic maintenance on severely restricted food intake (Goldman et al., 1974; Inoue and Bray, 1977; Slaunwhite et al., 1972). Hamilton and Rabinowitz (1976) report dramatic improvements in the hyperinsulinemia of rhesus monkeys with VMH lesions by long-term restriction of food intake to 60-70% of control levels.

The composition of the diet also influences the degree of hyperinsulinemia exhibited by lesioned animals. A high-carbohydrate-low-fat diet is conducive to greater plasma insulin in VMH weanling rats than is a diet low in carbohydrate and high in fat (Frohman et al., 1971; Schnatz et al., 1971). Just the opposite effect was found in adult rats after hypothalamic lesions (Inoue et al., 1977b). However, this result is confounded by the greater weight levels and obesity and presumably greater food intake exhibited by the lesioned animals on high fat compared with those on a diet low in fat and high in carbohydrate. When experimental animals on both diets were pair-fed with controls, no effect of diet on insulinemia was observed.

As might be expected from the plasma-insulin levels, pancreatic islets of VMH animals exhibit changes indicative of overactivity: islet size is increased (Chikamori et al., 1973, cited in Chikamori et al., 1977; Coleman and Hummel, 1970; Han et al., 1970; Kennedy and Parker, 1963; Martin et al., 1974; Powley and Opsahl, 1976), and β cells demonstrate increases in nuclear (Sétáló, 1965) and granule size (Kennedy and Parker, 1963).

d. Metabolic Profile: Glucose Homeostasis

VMH animals tend to be nearly normoglycemic even in the face of the drastically altered insulin release. In fact, in spite of the hyperinsulinemia, deviations are about as likely to comprise increases as decreases in plasma glucose. The nature of the changes in blood sugar seems to vary considerably as a possible function of lesion placement and feeding conditions.

An immediate hyperglycemia can follow the production of VMH lesions when short-acting anesthetics (sodium hexobarbital or ether) are used (Han and Frohman, 1970; Steffens et al., 1972). Steffens and colleagues also observed a hyperglycemia of perhaps 15% 24 hr postlesion in animals on ad libitum food intake. Other experimenters have found normoglycemia: Bernardis and Frohman (1970) found that weanling VMH rats had normal blood glucose in the presence of hyperinsulinemia. Hales and Kennedy (1964) observed normoglycemia 2 days after hypothalamic lesions, hypoglycemia 7 days, and hyperglycemia 3 months after surgery. At 48 hr after VMH destruction, Martin et al.

(1974) found that basal blood-sugar levels were reduced in lesioned animals on restricted food intake but normal in those with free access to food. Parasagittal knife cuts were without effect on plasma glucose in a study performed by Tannenbaum et al., (1974). In mice, electrolytic lesions of the VMH have variously resulted in increased, decreased, and unaltered blood-sugar levels (Chlouverakis et al., 1973; Coleman and Hummel, 1970; and Mayer et al., 1955, respectively). GTG lesions also have little or no effect on blood sugar (Coleman and Hummel, 1970). Hypothalamic lesions leading to obesity also have minimal impact on glycemia in various other species, including humans (Komorowski, 1977), cats (Keller et al., 1932), dogs (Heinbecker et al., 1944), rabbits (Lewinska, 1964), and rhesus monkeys (Hamilton et al., 1972).

In terms of phasic, or stimulated, glycemic responses, Louis-Sylvestre (1976) found that the hypoglycemic response to oral administration of 1 ml of 50% glucose was exaggerated in VMH rats. Similarly, relative—but not absolute—hypoglycemia (i.e. lower levels than the increased levels of control animals) has been observed during the increased prandial-insulin release in animals with hypothalamic lesions (Hustvedt et al., 1976). Five weeks after VMH lesions, rats exhibit a slightly elevated (but similar in terms of time course) blood-sugar response to an oral-glucose tolerance test (Karakash et al., 1977).

The relatively effective maintenance of blood-sugar homeostasis despite dramatic hyperinsulinemia does not appear to be the result of insulin resistance. In vivo (Frohman et al., 1972; Hongslo et al., 1974) and in vitro (Han et al., 1972; Goldman et al., 1972) tests of insulin sensitivity indicate normal insulin resistance in both muscle and adipose tissue. Various data suggest that, instead, gluconeogenesis is increased in VMH animals: even on normal food intake, they display increased plasma urea (Goldman and Bernardis, 1975; Hustvedt et al., 1976) and a faster rise in urea levels after nephrectomy (Goldman and Bernardis, 1975), suggesting heightened deamination of amino acids. Nitrogen-balance studies reveal reduced nitrogen retention in rats after hypothalamic lesions (Holm et al., 1973). In vivo incorporation of label from bicarbonate and alanine into hepatic glycogen and plasma glucose is increased in VMH animals (Goldman and Bernardis, 1975). Both increased (Karakash et al., 1977) and decreased (Inoue et al., 1977b) plasma glucagon have been reported in animals with VMH lesions. Also consistent with an increase in gluconeogenesis are the decreases in growth and deposition of lean body mass and the hypertrophic changes in liver, possibly reflecting the increased anabolic demand, seen in VMH animals.

e. Metabolic Profile: Gastric Acid

In 1954, Heaney and colleagues described the case of a young girl who had exhibited the sudden onset of hyperphagia and rapid weight gain shortly before her death. At autopsy, she was found to possess, in addition to leukemic infiltration of the hypothalamus, an ulcer of the duodenum. This latter development suggests hypersecretion of gastric acid, which was, in fact, revealed a decade later in the VMH rat by Ridley and Brooks (1965). Subsequent investigation has confirmed the occurrence of increased basal gastric acid secretion in lesioned rats (Inoue and Bray, 1977; Powley and Opsahl, 1974) as well as demonstrating a heightened responsiveness to electrical stimulation of the vagus nerve (Inoue and Bray, 1977). However, VMH rats, in contrast to intact animals, apparently do not increase their secretion of gastric acid in response to insulin-induced hypoglycemia (Ridley and Brooks, 1965), which might indicate that neural mechanisms involved in the elaboration of that response have been destroyed. However, in the Ridley and Brooks study, the stimulated levels of secretion reached by the controls were no higher, and on some measures considerably lower, than those shown by the lesioned animals. Therefore, the failure of insulin hypoglycemia to increase gastric-acid secretion in animals with ventromedial hypothalamic lesions might reflect a ceiling effect or the operation of competing mechanisms.

Elevated acid secretion has been measured as early as several hours after lesion production (Ridley and Brooks, 1965). The increase is not appreciably altered by restriction of food intake to normal levels and is present after 24-hr food deprivation (Powley and Opsahl, 1974). Severe food restriction for several weeks can, however, reduce both basal secretion and the response to vagal stimulation to normal levels (Inoue and Bray, 1977).

f. Metabolic Profile: Lipid Metabolism

In line with the increased carcass lipid content and hyperphagia, VMH animals display increased lipogenesis and blood lipids. Early investigators found that, after food intake or intragastric glucose loads, rats with hypothalamic lesions showed larger increments in their respiratory quotients (RQ) than did control rats, which was thought to indicate augmented conversion of carbohydrate to fatty acids (Brooks, 1946a; Tepperman et al., 1943). Subsequently, numerous studies have shown increased incorporation of radioactively labeled substrates into lipid in VMH rats. In vivo uptake of labeled acetate into carcass and liver lipids is enhanced after electrolytic and GTG lesions (Bates et al., 1955a, c; Løvø and Hustvedt, 1973). VMH lesions also result in greater incorporation of label from intravenously injected [^{14}C] glucose into lipid in epididymal adipose tissue, the liver, and diaphragm (Frohman et al., 1972).

Similarly, in vitro studies have revealed increased incorporation of labeled glucose into liver and adipose-tissue lipid in rats with hypothalamic lesions (Bernardis and Goldman, 1972; Goldman et al., 1972b; Han et al., 1972). Christophe et al. (1961) found greater uptake of labeled carbon from acetate into adipose-tissue fatty acids in GTG-injected mice. Fatty acids from isolated hepatocytes show enhanced labeling after incubation with tritium-labeled water or [^{14}C] glucose in weanling rats with VMH lesions (Katz et al., 1977). Substrate-stimulated (by amino acids, glucose, or lactate) incorporation of tritium into fatty acids by in situ perfused liver is also elevated after hypothalamic lesions (Karakash et al., 1977). Karakash and colleagues also found greater stimulated (but not basal) activity of the hepatic lipogenic enzymes acetyl-CoA carboxylase and fatty-acid synthetase in the experimental animals.

Studies with the respiratory quotient suggested that exaggerated lipogenesis is not a direct consequence of VMH lesions and, in fact, requires two or three days to develop (Tepperman et al., 1943; Brooks, 1946a). However, incorporation studies have since revealed elevated lipid synthesis on the day of the lesion; in vivo uptake of tritium-labeled water into lipid in the paramedial fat pad is increased by 12 hr postlesion (Hustvedt and Løvø, 1973), and glucose incorporation into liver and adipose-tissue lipid in vitro has been reported to be elevated by 4 hr after VMH lesions (Slaunwhite et al., 1972).

Tepperman et al. (1943) found that the stimulated RQ of VMH rats remained abnormally high even if the food intake of the experimental animals was held at control levels. However, the voraciousness with which the animals with lesions consumed their reduced rations led these authors to train normal rats to eat their day's food in the space of a few hours, in imitation of the VMH animals' feeding patterns. Since the RQ responses of these neurologically intact, trained animals also mimicked those of pair-fed rats with hypothalamic lesions, Tepperman and colleagues concluded that the apparent elevation of lipogenesis of the VMH animal is secondary to its feeding behavior. Nonetheless, subsequent work has demonstrated that this lesion-produced edge in lipid synthesis is maintained even when both level and pattern of food intake are controlled, whether by spaced availability of portions of the daily ration (Brooks, 1946a; Karakash et al., 1977) or by tube feeding (Han et al., 1972; Slaunwhite et al., 1972). The weanling rat VMH syndrome also includes exaggerated lipogenesis (Frohman et al., 1972a).

In general, studies have found that fasts of 24-48 hr attenuate but do not completely abolish the exaggerated lipogenesis of VMH animals; Bates et al. (1955a) reported that 24-hr food deprivation abolished the increased uptake of labeled acetate into carcass lipid but not liver lipid in both VMH rats and GTG-injected mice. In their study of the RQ in lesioned animals, Tepperman and his colleagues (1943) routinely tested their animals after fasting them for 24 hr. Frohman et al. (1971) found that adipose-tissue lipid from weanling rats with VMH damage still showed increased uptake from glucose in vitro after 48-hr food deprivation. A 24-hr fast eliminates the increased lipogenesis but not the increased lipogenic enzyme activities in perfused liver in situ from VMH rats (Karakash et al., 1977). However, it appears that maintaining VMH rats (Goldman et al., 1974; Slaunwhite et al., 1972) and GTG-treated mice (Bates et al., 1955a) on severely restricted food intake can eliminate the exaggerated lipid synthesis in liver and adipose tissue.

As originally reported by Brobeck et al. (1943), animals with damage to the basomedial hypothalamus also exhibit increased blood-lipid levels. In that study, plasma fatty acids and cholesterol were elevated. Since that time, it has been well documented that damage to the VMH in rats can result in exaggerated blood triglyceride (Frohman et al., 1969; Han et al., 1972) and free-fatty-acid levels (Hales and Kennedy, 1964; Tannenbaum et al., 1974). There have also been other reports of increases in cholesterol levels, but the anatomical locus involved is not well defined; it may extend considerably farther anteriorly and laterally than the region classically considered the ventromedial hypothalamic area (Bernardis and Schnatz, 1971; Friedman et al., 1969). In addition, hypothalamic obesity is accompanied by elevated plasma triglycerides in humans (Bray and Gallagher, 1975) and rhesus monkeys (Hamilton et al., 1972) and supranormal blood levels of cholesterol in dogs (Heinbecker et al., 1944) and monkeys (Hamilton et al., 1972).

Increased plasma triglyceride has been reported as early as two days after hypothalamic lesions (Hustvedt et al., 1976), and increases in cholesterol have been noted at three days (Byers and Friedman, 1973). Hypertriglyceridemia is not dependent upon hyperphagia for its expression: it is found after VMH lesions in pair-fed adult rats (Han et al., 1972; Hustvedt et al., 1976; Karakash et al., 1977) and in nonhyperphagic weanlings (Frohman et al., 1969; Schnatz et al., 1971). However, the exaggerated triglyceride levels are dependent on diet insofar as they are eliminated by 24-hr food deprivation (Karakash et al., 1977) and maintenance on a fat-free diet (Schnatz et al., 1971), although Schnatz et al. found that, if the animals on the fat-free diet were fasted for 48 hr, those with VMH lesions had elevated plasma-triglyceride levels at the end of that period. Elevated plasma free fatty acids are not found in lesioned animals in the absence of increased food intake (Hustvedt et al., 1976; Tannenbaum et al., 1974). The hypercholesterolemia of lesioned rats is also abolished by a 24-hr fast (Brobeck et al., 1943). Hamilton and Rabinowitz (1976) found that severe food restriction of hypothalamic obese rhesus monkeys for several months resulted in a substantial decline of plasma-triglyceride levels while having little effect on cholesterol.

Lipoprotein electrophoretic patterns suggest that the hypertriglyceridemia of the VMH rat has both endogenous and exogenous components (Han et al., 1972; Schnatz et al., 1971). Karakash and colleagues (1977) obtained evidence of increased hepatic triglyceride synthesis after VMH damage, and it has also been reported that adipose tissue from lesioned rats shows reduced in vitro uptake of labeled plasma triglycerides relative to controls (Schnatz et al., 1971). Animals showing hypercholesterolemia after hypothalamic lesions are apparently normal in intestinal absorption and intestinal and hepatic synthesis of cholesterol but exhibit retarded hepatic uptake of cholesterol (Byers and Friedman, 1973).

A potentially related phenomenon is the increase in plasma cholesterol induced by lateral hypothalamic stimulation, which has been claimed to result from blockage of bile flow through the sphincter of Oddi into the duodenum (Gutstein et al., 1969). Since bile formation is a major catabolic pathway for cholesterol, blockade of this route could potentially account for the observed plasma increases through negative feedback acting at the liver (see Gutstein et al., 1970).

Excessive lipid accumulation could also result from an impairment in the mobilization of depot fat. Determination of lipid turnover in vivo after the addition of labeled palmitate to the food of VMH rats and mice with GTG-induced obesity indicated just such an impairment in both preparations (Bates et al., 1955b). However, lesioned rats subjected to long-term starvation apparently lose body weight and catabolize carcass fat at normal or even supranormal rates (Brooks and Lambert, 1946; Frohman et al., 1971; Montemurro and Stevenson, 1960; Stevenson and Montemurro, 1963). The RQ of fasting VMH rats is within the range shown by normal animals and interpreted as indicating fatty-acid utilization (Tepperman et al., 1943). The response of lesioned rats to severe food restriction is not clear: Haessler and Crawford (1967) found that lesioned animals lost body weight at roughly one-half the rate of controls in response to such a regimen. However, Bates et al. (1955a) and Slaunwhite et al. (1972) found that experimentals and controls lost weight about equally during severe food restriction.

Studies in vitro have indicated normal basal levels of release of free fatty acids (Haessler and Crawford, 1967) and glycerol (Kasemsri et al., 1972) from the adipose tissue of VMH rats. However, the free-fatty-acid response to the addition of epinephrine to the medium is considerably reduced in tissue from lesioned animals (Haessler and Crawford, 1967; Harrel and Remley, 1973; Kasemsri et al., 1972; Starr et al., 1966). In contrast, Kasemsri et al. (1972) report that glycerol release in response to epinephrine (when expressed per milligram of protein) and hormone-sensitive lipolytic activity in adipose tissue are not different in lesioned and control animals. They argue that the adipose tissue of VMH rats is not impaired in its stimulated release of fatty acids but that the subsequent reesterification and/or oxidation by adipose tissue is increased. Apropos of the latter point, Goldman and colleagues (1970) and Frohman and his co-workers (1971) found that palmitate oxidation by adipose tissue was actually reduced in weanling VMH rats; also, Haessler and Crawford (1967) reported that the ratio of glycerol to free fatty acids in the medium was the same for experimental and control animals, which implies that, since free fatty acids were reduced for the experimental animals, the same was true for glycerol. Finally, basal and epinephrine-stimulated glycerol release in vitro is comparable from adipose tissue of human patients with hypothalamic or "essential" obesities (Bray and Gallagher, 1975).

C. Etiology of the Ventromedial Hypothalamic Syndrome

The etiology of hypothalamic obesity has not been completely characterized and remains one of the central questions in the study of hypothalamic control of energy balance. Any of several elements of the VMH syndrome could conceivably produce the others. For example, work with intact subjects has established that hyperphagia and hyperalimentation will produce weight gain and obesity (Cohn and Joseph, 1962; Sims et al., 1973), as well as other aspects of the syndrome, such as reduced operant performance (Sclafani, 1976); or, alternatively, experimental administration of insulin can produce hyperphagia (May and Beaton, 1968), weight gain and obesity (Hoebel and Teitlebaum, 1966; Lotter and Woods, 1977), alterations in meal patterns (Panksepp, 1973), and inactivity (Campbell and Fibiger, 1970); or again, obesity is typically associated with hyperphagia (Steffens, 1975), hyperinsulinemia (Woods and Porte, 1974), and inactivity (Cohn and Joseph, 1962). Finally, inactivity may lead to increased weight gain (Mayer et al., 1954) and perhaps other elements of the classic VMH syndrome.

For perhaps 25 years after the 1943 paper, by Brobeck and colleagues, suggesting that hyperphagia was the basic disturbance produced by VMH lesions, the prevailing viewpoint was a behavioral one: the VMH was seen as a "satiety center" that inhibited food intake when it was stimulated by appropriate nutrient levels in the blood stream as well as afferent feedback from the viscera. As we reviewed above, this conclusion was supported as much by the theoretical views of the time as by the experimental analyses themselves. Such an explanation provided a central-nervous-system explanation of food intake (and hunger) that could replace some of the peripheral theories which had floundered between 1910 and 1930 (see Rosenzwieg, 1963). Further, the ventromedial nucleus, easily the most anatomically salient landmark in the tuberal hypothalamus, may have proved irresistible as a potential neural center onto which to map satiety.

In roughly the past decade a variety of physiological or metabolic explanations have been introduced. Much like the very early explanations of hypothalamic disturbances that had emphasized obesity and possible pituitary-mediated metabolic disturbances, these recent analyses have suggested that the metabolic defects that follow VMH lesions may be primary and produce, or at least contribute to, the hyperphagia and ultimate obesity that characterize the animal with VMH lesions. Such physiological analyses are based in part on a recognition of slowly accumulating evidence that behavioral explanations cannot account for the elements of the syndrome in some situations, especially those occurring in the absence of increased food intake.

Logically, of course, if one element is expressed unambiguously earlier than the other aspects of a syndrome, a reasonable hypothesis can be made that the element plays a primary role in the development of the syndrome. Or, if elimination of an element or interruption of a pathway involved in the expression of an element abolishes the other disturbances, it is possible that that element is a primary disturbance. Conversely, if the initial appearance of a component occurs later than that of other elements, or if the element can be eliminated without eliminating the others, one can reasonably conclude that that component represents a secondary symptom. A variety of experiments employing this logic have been performed on the VMH syndrome. Their results indicate that, while hyperphagia is not a necessary consequence of VMH lesions producing obesity, exaggerated insulin release (probably neurally mediated) may be necessary in the development of VMH obesity and possibly other components of the syndrome.

1. Temporal Dissociations

A number of experiments—many of them already described—have carefully detailed the onset of the disturbances that typically follow VMH lesions. Presumably, the disturbance occurring with the shortest latency might be primary. Hyperphagia has been observed within 30 min of lesion production (Becker and Kissileff, 1974); hyperinsulinemia (to intravenous glucose infusion) has been measured 10 min after VMH destruction (Rohner et al., 1977); and gastric hyperacidity is present within several hours of surgery (Ridley and Brooks, 1965). Practically, these times represent the earliest ones sampled to date. Other effects observed within the first 24 hr include altered glucose homeostasis (Han and Frohman, 1970), increased lipid synthesis (Hustvedt and Løvø, 1973), increased carcass lipid levels (Slaunwhite et al., 1972), as well as sensory reactivity (Marshall, 1975).

Unfortunately, unequivocal interpretation of these results in terms of possibly essential components of VMH lesion disturbances is very difficult. First, the separate elements do not appear singly on a sufficiently protracted time scale to make primacy decisions easy. Second, many of the disturbances occur so rapidly that they appear in the immediate postlesion interval, or "acute dynamic phase," during which the transitory side effects of lesion procedures (see Schoenfeld and Hamilton, 1977, for a recent review of these effects) complicate—and possibly even fundamentally distort—the pattern of disturbances seen. Finally, the problem of dissociating the components on this compressed time scale is further aggravated by problems of measurement: the present level of sophistication of the different techniques may permit us to detect modest changes in some responses or metabolites more readily than in other critically important responses or metabolites. A more slowly evolving set of symptoms would make attempts to dissociate the different elements far more feasible.

2. Metabolic Disturbances with Controlled Food Intake

As described above, one major group of experiments has sought, through the use of limited feeding strategies and preparations, to establish whether behavioral excesses, specifically the hyperphagia, are responsible for the increased lipid accumulation and other metabolic disturbances that characterize the VMH animal. Relevant experiments have repeatedly demonstrated that most of the metabolic parameters affected by hypothalamic lesions are still disturbed in animals prevented from consuming excess amounts of food.

The strongest conclusion from these observations—namely, that VMH lesions must directly alter metabolism in the absence of any behavioral disturbance—is controversial, however. Two major criticisms have been made repeatedly of experiments that have attempted to normalize the feeding behavior of VMH animals. First, at best, pair-feeding experiments have not provided completely normal patterns of food intake. Second, VMH lesions typically produce a decrease in locomotor activity (Hetherington and Ranson, 1942b; Brooks, 1946b) which presumably causes animals with lesions to expend fewer calories than their intact controls. Such unexpended calories would be available for fat synthesis and storage.

Each of these criticisms has been partially answered in experiments that have nevertheless found the typical VMH metabolic disturbances. In the case of patterning of food intake, several experiments have imposed identical patterns of tube feeding on experimentals and controls. In other experiments, the times of availability or delivery of food have been limited, either by requiring that both groups consume daily rations within a short time interval, thus making gorgers out of all animals, or by spacing the delivery of portions of food and, thus, minimizing gorging. While none of the above experimental strategies provides normal patterns of consumption, they certainly control feeding behavior. The results strongly suggest that the characteristic anabolic profile of the VMH animal does not depend upon aberrant feeding behavior for its expression. Furthermore, in recent experiments using new feeding paradigms that allow pair-feeding of obese, or potentially obese, animals with normal quantities and distributions of food—in contrast to the more traditional, limited-quantity paradigms—we found that the VMH rat (J. E. Cox, unpublished dissertation) and another rodent model of obesity, the diabetic (*db/db*) mouse (Cox and Powley, 1977), still fatten, and thus that pattern of food intake may not make major differences in outcome. The three classes of animals with hypothalamic damage that become obese even though voluntarily normophagic—weanling rats with surgical lesions, some adult rats with surgical lesions, and GTG-lesioned mice maintained in warm environs—also bolster the contention that altered feeding is not crucial to VMH disturbance.

The contribution of diminished locomotor activity to alterations in metabolism of the VMH rat is more difficult to evaluate. Hetherington and Ranson (1942b) considered the inactivity to be at least a contributing factor in the development of hypothalamic obesity, and Brooks (1946b) suggested that the residual fattening of the pair-fed VMH animal is due to their inactivity. One restricted-feeding experiment had controls for activity, however. Han (1968) tube-fed VMH and sham-operated rats in restriction cages to minimize the effects of differential activity. The striking result of this study was that comparison of restrained animals with groups in standard cages revealed that activity had no effect on body weight or fat accumulation in either VMH or sham-operated animals. Thus, animals with hypothalamic lesions which were matched with controls on amount of locomotor activity as well as quantity and pattern of food intake possessed, at time of killing, considerably more carcass than did controls.

In addition, it is generally conceded that the metabolic cost of activity is difficult to measure in terms of dietary calories and may, in fact, be negligible (see Collier et al., 1972). Finally, it is worth noting that a wide range of hypothalamic lesions produces comparable decreases in activity without the metabolic alterations exhibited by the VMH animal (Bernardis, 1972; Gladfelter, 1971; Gladfelter and Brobeck, 1962).

Although we have stressed that hyperphagia is not necessary for the expression of the major characteristics of the VMH syndrome, it is obvious that the increase in food intake plays a major role in determining the extent to which the disturbances develop. Dramatic weight gain and massive obesity occur only if lesioned animals consume excess calories. The more VMH animals eat, the heavier and fatter they become; the correlations between calories consumed and body weight can be very impressive (e.g., 0.90-0.98; Kemnitz et al., 1977). Limitation of food intake results in relatively modest disturbances of most elements of the syndrome. Outright starvation or chronic, severe food restriction can, in fact, often eliminate these disturbances.

In practical terms, these observations are the basis of the proposal that a disturbance in food intake is the primary basis of VMH obesity and its characteristic metabolic patterns. However, as documented above, attempts at rendering the feeding behavior of

VMH animals identical to that of controls have not yielded fundamental changes in the pattern of disturbances expressed—the disturbances merely become quantitatively less impressive. Relative magnitudes of metabolic disturbances with and without hyperphagia cannot be used to decide unequivocally on the causal primacy of the feeding disturbance. Hyperphagia may be necessary for extreme obesity, but necessity does not clearly imply primacy; it may merely indicate that the hypothalamic metabolic disturbances, even if causally primary, may depend upon adequate caloric substrates for their expression or degree of expression.

One line of experimentation, exemplified by a study performed by Hoebel and Teitelbaum (1966), suggests that VMH hyperphagia may, in fact, be secondary to the altered metabolic conditions of the animal. These authors found that animals made obese by insulin treatment did not exhibit the typical pattern of hyperphagia and rapid weight gain when they subsequently received hypothalamic lesions: the animals simply maintained approximately the same level of obesity. Similarly, if a VMH animal's weight is displaced in either direction from its static-phase level, the animal compensates by altering its food intake so as to cancel the displacement (Brooks and Lambert, 1946; Hoebel and Teitelbaum, 1966; Kennedy, 1953; see Fig. 3). These results clearly suggest that the VMH animal's feeding behavior is not an uncontrolled excess produced by the lesion but rather appears to be exquisitely sensitive to, or controlled by, the metabolic plateau or body-weight level resulting from the lesion.

The preceding observation finds some indirect support in other work. Several experiments indicate that animals with a variety of forms of obesity do not respond to VMH damage with an appropriate (i.e., commensurate with the neural damage produced) degree of hyperphagia and additional weight gain. For example, we have not succeeded in producing a strong, clear-cut VMH syndrome of weight gain and increased food intake in adult, already obese, Zucker "fatty" rats (T. L. Powley and C. A. Opsahl, unpublished results). Similarly, VMH lesions have only marginal effects on the body weight, carcass fat, and serum insulin of the already fat, obese hyperglycemic (*ob/ob*) mouse (Chlouverakis et al., 1973). Baile et al. (1970) also found that the *ob/ob* mouse does not show a clear weight gain after small (the animals were resistant to GTG) gold thioglucose lesions of the basomedial hypothalamus. Basomedial hypothalamic lesions aimed at the median eminence do not have any clear-cut effect on body weight (though they do ameliorate the hyperglycemia) in the sand rat, *Psammomys obesus* (Brodoff and Zeballos, 1970). Similar effects on body weight and blood chemistry have been observed in the diabetic (*db/db*) mouse, with medial hypothalamic lesions (both electrolytic and GTG), although this animal does show a lesion-induced exaggeration of its hyperinsulinemia (Coleman and Hummel, 1970). While such observations represent reinterpretations of experiments investigating different questions; while the results are consistent with other interpretations; and while a much more direct analysis of the conclusion is needed, in the aggregate the results with genetically obese rodents tend to support the contention that a prior obesity ameliorates the behavioral effects of VMH lesions.

Thus, considerable research indicates that hyperphagia is not crucial to the VMH syndrome and that perhaps the increased food intake is, in fact, secondary to the metabolic state of the animal. Instead of concluding, therefore, that altered feeding behavior represents the primary deficit of the VMH syndrome, it seems preferable to view the gluttony that follows hypothalamic lesions and, for example, the fattening of the pair-fed VMH animal as expressions of other primary disturbances produced by the lesion. One hypothesis in this field is that, owing to increased lipogenesis and an impairment of lipolysis, VMH animals tend to develop larger than normal fat deposits while, at the same time, suffering from a deficit in usable energy, all of which requires that they increase their food intake (cf. Harrel and Remley, 1973). (This particular hypothesis, though, is embarrassed by the immediacy of the hyperphagia that follows VMH lesions.)

3. Controlling Metabolic Responses and Monitoring Behavior

The other major class of experiments focused on identifying the primary element(s) have used different means of clamping or eliminating particular responses or pathways that might express the effects of VMH lesions. This type of experiment is epitomized by early studies which established that the development or maintenance of hypothalamic obesity may take place in the absence of the pituitary (Hetherington and Ranson, 1942a; Hetherington, 1943).

Since that time, similar experiments have demonstrated that many other elements of the VMH disturbance survive hypophysectomy, including hyperphagia (Kennedy and Parrot, 1958), increased plasma insulin and lipids, (Goldman et al., 1970; Han et al., 1972), islet hypertrophy (Han et al., 1970), and heightened adipose-tissue lipogenesis (Goldman et al., 1970; Han et al., 1972). What is more, some reports have indicated that VMH obesity is not at all diminished by pituitary removal (Goldman et al., 1970; Kennedy and Parrot, 1958). In the case of the GTG mouse, however, hypophysectomy blocks the (further) development of obesity, whether the operation is performed shortly after the injection (Liebelt and Perry, 1967; Redding et al., 1966) or well into the dynamic phase of weight gain (Powley and Plocher, 1980). In the latter case, the obesity is not reversed by the hypophysectomy. That the etiology of goldthioglucose-induced obesity might differ from that of the classic VMH obesity is an obvious possibility raised by these results.

The dispensability of the hypophysis for the expression of the disturbances produced by surgical lesions of the ventromedial hypothalamus, at least, makes it very difficult to argue for a critical effect of hypophyseal-endocrine mechanisms on the VMH disturbances. However, that one or more of these mechanisms might have a modulatory role in VMH obesity is suggested by several experiments. For example, VMH lesions result in reduced plasma levels of growth hormone in weanling rats (Frohman and Bernardis, 1968) and may have the same effect in adults (Martin et al., 1974; Tannenbaum et al., 1974). Studies of the effects of growth-hormone treatment have failed to reach a consensus, however, as repeated injections have been reported to have no effect on (Goldman et al., 1970), to diminish (Han, 1967), or to abolish (York and Bray, 1972) hypothalamic obesity.

Ventromedial hypothalamic damage has little or no effect on adrenal weights in VMH rats (Brooks and Lambert, 1946) or GTG mice (Redding et al., 1966) and, at most, a transient effect on plasma corticosterone in GTG mice (Liebelt et al., 1961). VMH obesity and hyperphagia are reportedly spared by adrenalectomy as well as by adrenalectomy with cortisone treatment (York and Bray, 1972). A study by Bernardis and Skelton (1965) suggests that adrenal demedullation might diminish hypothalamic hyperphagia in weanling rats, but the results are equivocal. Adrenalectomy and/or daily injections of adrenocorticotropic hormone (ACTH) are also without effect on the elevated plasma cholesterol which follows medial hypothalamic lesions (Friedman et al., 1972).

Thyroid weights may be increased (Brooks and Lambert, 1946), but thyroid secretion is reduced (Hinman and Griffith, 1973) in VMH rats, though secretion has been found to be normal in GTG mice (Schindler and Liebelt, 1967). Thyroxin injections have only slight effect on weight gain in hypothalamic obese rats (Brooks and Lambert, 1946). In addition, the hypercholesterolemia produced by hypothalamic lesions is little affected by thyroidectomy or injections of thyroid powder or thyroid-stimulating hormone (TSH) (Friedman et al., 1972).

In light of reports that VMH lesions can result in gonadal atrophy as well as disruptions of the estrus cycle (Brooks and Lambert, 1946; Hetherington and Ranson, 1940), there is the possibility that gonadal hypofunction plays a role in the development of hypothalamic obesity. If that is the case, then gonadectomy or injections of gonadal hormones should reduce the differences between VMH animals and neurologically intact animals. However, ovariectomy results in further increases in body weight and fat stores in VMH females as well as controls (Kemnitz et al., 1977; King and Cox, 1973), though the effect may be somewhat smaller in animals with hypothalamic lesions (Beatty et al., 1975a; Wright and Turner, 1973). Orchiectomy potentiates the effects on body weight of VMH lesions in male rats while producing a relative decrease in body weight in neurologically intact rats (Kemnitz et al., 1977). The reduction in body weight produced by injections of estradiol benzoate is no greater, and possibly less, in ovariectomized VMH rats than that produced in controls (Beatty et al., 1975; King and Cox, 1973; Nance, 1976). There has been one report, though, of oversensitivity (in terms of body-weight loss) of obese mice with electrolytic lesions of the hypothalamus to treatment with diethylstilbestrol, a synthetic estrogen (Montemurro, 1971). On the whole, the above results indicate that if lesioned-induced gonadal hypofunction plays any role in the production of VMH obesity, it is probably a rather small role.

VMH hyperphagia and increased body-weight gain on Purina pellets can apparently be reversed by a combination of adrenalectomy and ovariectomy (Mook et al., 1975); the elevations are to an extent reinstated if animals are maintained on eggnog or a high-fat diet. Whether this surgery works by the production of independent and/or nonspecific effects that compete with the mechanism of VMH lesion-induced weight gain is currently difficult to evaluate. Direct measurement of some of the other defining elements of the VMH disturbance would be useful for such an evaluation.

A set of experiments logically similar to those involving hypophysectomy has documented that insulin release is essential to the production of some of the disturbances resulting from VMH lesions. Diabetes produced by chemical (Friedman, 1972; Lazaris et al., 1976) or surgical (Young and Liu, 1965) means eliminates the excess weight gain and fat accumulation resulting from VMH lesions (but see Goldman et al., 1972b). Some authors have reported that insulin is necessary for VMH hyperphagia (Lazaris et al., 1976; Young and Liu, 1965), though hyperphagia in alloxan-diabetic VMH animals has been reported (Friedman, 1972). Insulin replacement in diabetic animals has been found to be effective in at least partially restoring VMH obesity (Friedman, 1972; Vilberg and Beatty, 1975; Young and Liu, 1965). An exception to the preceding statement is to be found in a study by York and Bray (1972). They found that streptozotocin injections prior to hypothalamic lesions precluded the expression of increased food intake and weight gain in experimental animals even though they were given daily insulin injections. However, diabetes (with insulin replacement), instated after VMH lesions, did not reduce body weights to normal. Hyperphagia was reversed in VMH animals already obese but not those in the early dynamic phase. (It is interesting that diabetic animals with hypothalamic lesions and receiving insulin treatments were found to be markedly hyperinsulinemic compared with their appropriate control animals.) Plasma cholesterol remains elevated after alloxan treatment (Friedman et al., 1972), though plasma triglyceride levels may be no higher than those of control animals after streptozotocin injection (Goldman et al., 1972b). Augmented adipose tissue lipogenesis has been reported to survive chemically induced diabetes (Goldman et al., 1972b), but anti-insulin serum has been found to restore hepatic lipogenesis to control levels in VMH rats (Karakash et al., 1977). Much as in the case of the pair-feeding experiments discussed above, observations on diabetes do not determine whether insulin serves in a merely permissive capacity in the production of some of the VMH disturbances or whether disturbed insulin release is a primary condition for the induction of the other disturbances produced by hypothalamic lesions.

There is yet another approach suggesting a possible pathway for the expression of VMH disturbances: the series of analyses employing vagotomy. In 1946, Brooks et al., attempted vagotomy on 10 animals with VMH lesions. Two animals—the only ones with apparently complete vagotomies—did not gain weight; however, the vagotomy procedures produced some general debilitation, and the effectiveness of the brain lesions was equivocal. More recently, we performed the same type of experiment and took pains to avoid the debilitating side effects that can accompany vagotomy (Powley and Opsahl, 1974). Complete subdiaphragmatic vagotomy performed on male rats, which had become fat after VMH lesions, eliminated the excess weight gain (Fig. 7) as well as the hyperphagia on pelleted rat chow. Similar results have been obtained by Inoue and Bray (1977) and

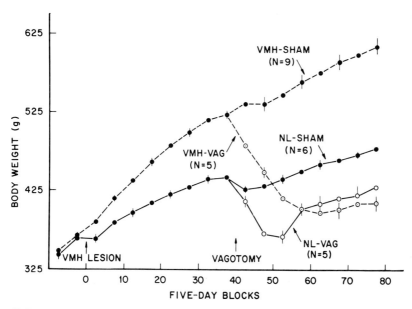

Figure 7 Body weights of male rats (grams ± standard error of the mean) that received bilateral electrolytic lesions of the VMH (or sham lesions) on day 0 and subdiaphragmatic vagotomy (or sham vagotomy) on day 40. NL, nonlesioned. (Reproduced from Powley and Opsahl, 1976.)

Sawchenko et al., (1977). In additional experiments, we have found that vagotomy also eliminates lesion-produced weight gain in females (Powley et al., 1974), as well as VMH hyperinsulinemia (Powley and Opsahl, 1976) and gastric hyperacidity (Powley and Opsahl, 1974), findings confirmed by Inoue and Bray (1977). Compared with intact controls, VMH animals that have been vagotomized still apparently manifest a somewhat exaggerated intake of both a high-fat diet (Powley and Opsahl, 1974) and a liquid eggnog diet (Powley et al., 1974).

Sawchenko et al. (1977) have now reported that interruption of the vagal celiac branch, which carries fibers to and from the pancreas, is the critical event responsible for the normalization of weight that occurs after vagotomy. Recently, Inoue et al., (1977a) placed VMH lesions in rats that had received streptozotocin injections followed by transplantation of donor pancreas. Though this noninnervated pancreatic tissue served to eliminate the diabetes produced by the streptozotocin injection, VMH lesions were without effect on the body weight and obesity (by the Lee index) and had only a very modest effect on the food intake of animals receiving this treatment. The possibly pivotal role of the celiac vagal fibers in innervating islet cells might account for the discrepant results of Chikamori and colleagues (1977), who found that their vagotomized VMH animals (treatment of the celiac branch unspecified) were nonetheless hyperinsulinemic and obese.

Finally, it has recently been reported that intestinal-bypass surgery performed on VMH rats reduces their body weights to normal, and reattachment of the bypassed intestinal segment reinstates the VMH body-weight elevation (Sclafani et al., 1977). In this case too, the specificity of the effect cannot be evaluated until further investigation reveals the effects of this surgery on other characteristics of the VMH disturbance.

D. VMH Hyperphagia and Metabolic Disturbances: Anatomical Substrates

1. The Two Major Models

In the 40 years of experimental work on hypothalamic obesity, essentially two major anatomical models have been suggested by research observations. One explanation emphasizes the role of a longitudinal system that runs through the hypothalamus and passes just ventrolateral to the ventromedial hypothalamic nucleus. The other model emphasizes the importance of the cells of the ventromedial nucleus and its adjacent neuropil and suggests a lateral projection from the area to the lateral hypothalamus. Both models are just that—models. Neither has been strongly substantiated in the sense that the appropriate anatomical pathways and connections have been identified (see chapters by Parent, Millhouse, Palkovits and his co-workers, and Morgane in these volumes). In addition, neither easily handles all of the available experimental observations.

Although we do not propose to include an exhaustive review of the potential anatomical substrates of hypothalamic obesity (see other chapters in the present volumes as well as recent analyses by Morgane, 1975; Hoebel, 1976; Coscina, 1977), several general observations on the problems are worth mentioning since they have a potential bearing on the functional issues just discussed. The two models are also of interest because they have influenced both the formulation of research and its evaluation. Finally, a brief review of the two proposed substrates and observations germane to each also emphasizes the kinds of information and anatomical analyses needed to identify the actual mechanisms involved in hypothalamic obesity.

In 1940 and 1942, soon after the pituitary had been eliminated as a major candidate as a pathway for the expression of hypothalamic obesity, Hetherington and Ranson (1942a) consolidated the early literature and their own work in a review that formulated the longitudinal explanation. They concluded:

> For herein seems to reside the heart of the data so far accumulated: Lesions which occupy the medial hypothalamus, particularly the region of the ventromedial hypothalamic nucleus, or are placed in the caudal hypothalamus in a position to interrupt a large number of the descending fibers from the hypothalamic cell groups (Kabat, Magoun, and Ranson, '35) cause a marked degree of obesity. It should be possible, theoretically at least, to sever these descending fibers at a yet more caudal point in their course and obtain the same effect.
>
> Obesity can be produced by fairly symmetrical lesions which destroy bilaterally: (1) most of the ventromedial hypothalamic nuclei together with some of the tissue immediately around them, especially on their lateral sides; (2) the caudal ends of the ventromedial hypothalamic nuclei, the premammillary area, and a considerable part of the lateral hypothalamic areas which lie dorsolateral to the mammillary body. These pairs of lesions appear merely to represent interruptions at successive levels of paired systems (one on each side of the hypothalamus, and each one capable of acting more or less independently).
>
> The systems in question seem to arise rostrally in, or in the neighborhood of the ventromedial hypothalamic nucleus and to proceed caudally into the midbrain in the company of the medial forebrain bundle. The descending pathway may of course consist of chains of short neurons [pp. 497-498].

Three particular aspects of the anatomical organization implied in this early work are worth noting. First, Hetherington and Ranson suggested that damage to cells of the hypothalamus lying in the vicinity of the anterior pole of the ventromedial nucleus or even as far anterior as the posterior pole of the paraventricular nucleus is responsible for hypothalamic obesity. They specifically excluded cells of the paraventricular nucleus itself, however (Hetherington, 1944). Second, they assumed that such anteriorly located cells involved in energy metabolism project their influence to systems in the caudal brain stem by way of a longitudinal descending pathway. Third, Hetherington and Ranson also inferred from the results they obtained with unilateral lesions and from accidental left-right asymmetries in lesion placement that the descending systems are uncrossed in the hypothalamus. Each of these general conclusions was consistent with observations available at the time, and each has also found considerable support in terms of recent work employing knife-cut procedures (e.g., Albert et al., 1971; Gold et al., 1972; Paxinos, 1974; Sclafani, 1971).

For most of the intervening years of research on the problem, however, a very different anatomical substrate was generally assumed. This alternative was suggested by Anand and Brobeck in 1951 when they proposed that the ventromedial nucleus per se represents the basic locus of the lesion effect and that the system involved a laterally oriented inhibitory projection from the ventromedial nucleus to the lateral hypothalamic nucleus which they recognized as a "feeding center" (see the quotation from Anand and Brobeck in Section III.A). They also suggested that the ventromedial nucleus itself—not the ventromedial hypothalamic area—inhibited this lateral hypothalamic feeding center by efferent fibers. By 1954, the ventromedial nucleus of the hypothalamus had become a satiety center (e.g., Stellar, 1954), and laterally projecting efferent connections were a widely accepted conclusion. In retrospect, part of the appeal of the ventromedial nucleus itself as a candidate for the basic mechanism appears to have been the fact that it complemented the early view that lesions produce their effects by destroying a satiety center. The ventromedial nucleus is the most prominent landmark in the tuberal hypothalamus (Fig. 1) Cajal (1966) called it the principal nucleus—and in theorizing about the classic symptoms, researchers have found it hard to resist. It is easier to envision a center as consisting of a coherent nucleus group rather than a cell-poor area or a fiber bundle.

This Anand-Brobeck anatomical model was widely accepted for about 20 years. Most recently, however, it has been strongly challenged on several important grounds (cf. Rabin, 1972; Sclafani et al., 1975a). Perhaps paradoxically, recent use of the knife-cut technique that was adopted in efforts to cut the postulated inhibitory efferents between the ventromedial nuclei and the lateral hypothalamic area has been largely responsible for the reassertion of the general view proposed by Hetherington and Ranson. In effect, in approximately a 20-year period, research and thinking on the anatomical substrate for hypothalamic obesity and hyperphagia has come nearly full circle.

2. Current Status of the Two Models

There are still some significant uncertainties about the locus of the basic substrate for hypothalamic obesity and hyperphagia. Recent work with knife-cut procedures has indicated that the cuts in the anterior hypothalamus are most effective if they are oriented transversely and vertically and pass through a ventral hypothalamic area ventromedial to the fornix. These transverse effects can be obtained with cuts as far anterior as the extreme caudal pole of the paraventricular nuclei and as far posterior as the mamillary nuclei (Clark, 1942; Gold et al., 1977). If the cuts are bilateral, different asymmetrical combinations of them are effective in producing the classic VMH obesity and hyperphagia (Gold et al., 1972, 1977).

This pattern of effective cuts suggests a longitudinally oriented bundle of uncrossed fibers that projects through the hypothalamus. At the level of the ventromedial nucleus, the system seems to course through the ventrolateral part of the nucleus itself as well as the area immediately ventrolateral to the nucleus proper. To date, lesions of the paraventricular nuclei or areas rostral and dorsal to them have not generated the usual adiposity and overeating (Hetherington, 1944; Gold et al., 1977). Attempts to follow the system caudally have met with some success, and transverse knife cuts as far posterior as the red nucleus have produced obesity (Paxinos and Bindra, 1973; Gold et al., 1977). Such a system is consistent with most of the effective lesion placements as well as with knife-cut work, and it is also consistent with the large literature that identifies an effective locus for catecholamine-induced feeding in the anterior hypothalamic region adjacent to the paraventricular nucleus (Leibowitz, 1976) and an effective site for electrically elicited suppression of feeding just dorsal to the ventrolateral pole of the VMH (Beltt and Keesey, 1975).

This hypothetical system does not correspond adequately to any well-known pathways or longitudinally organized cell groups; however, it seems likely that the host of sophisticated new procedures for mapping specific transmitters will make possible the identification of an appropriate substrate or substrates. Initial attempts to identify VMH obesity with the ventral noradrenergic pathway have not held up well, however. Recent demonstrations indicate that a syndrome of modest hyperphagia and weight gain can be produced by destruction of the ventral noradrenergic bundle (Ahlskog and Hoebel, 1973), but this syndrome is apparently fundamentally different from the VMH syndrome (Ahlskog, 1974; Ahlskog et al., 1975; Lorden et al., 1976).

The proposal for a longitudinal system such as the one just outlined also finds some support from another set of observations indicating that lesions of the ventromedial nucleus are not sufficient to produce major obesity or hyperphagia. Joseph and Knigge (1968) were unable to obtain weight gain in the guinea pig with lesions limited to the ventromedial nucleus. Similarly, earlier work with lesions performed before the satiety-center hypothesis had crystallized also did not find ventromedial lesions particularly effective (Biggart and Alexander, 1939; Heinbecker et al., 1944). Perhaps most dramatically, Gold (1973) argued that the ventromedial nucleus is merely the most prominent landmark in the general area of the basomedial hypothalamus and that lesions confined to the VMH do not produce substantial weight gain.

In contrast—at least superficially—to a longitudinal-fiber-bundle analysis of the VMH syndrome, there are several observations indicating that the ventromedial nucleus is intimately involved in energy metabolism. These challenge the conclusion that a longitudinal-fiber system passing just ventrolateral to the ventromedial nucleus and just ventromedial to the fornix constitutes the sole basic representation of the system in the hypothalamus.

First, electrophysiological studies have repeatedly confirmed that cells of the ventromedial nucleus proper respond to circulating nutrients (Oomura et al., 1975), gastrointestinal hormones (Oomura, 1973), and visceral afferent inputs (Anand and Pillai, 1967; Sharma et al., 1961). Second, karyometric analyses suggest that the cells of the ventromedial nucleus per se exhibit specific, selective changes in size to manipulation of nutritional status (Pfaff, 1969) or available insulin levels (Akmayev and Rabkina, 1976). Third, a number of tissue-metabolism studies have established that cells of the ventromedial nucleus or immediately adjacent ventrobasal cell bodies have a specialized and selective metabolism and are peculiarly sensitive to dynamic alterations of energy metabolism (Panksepp, 1974; Panksepp and Pilcher, 1973). Fourth, an impressive number of experiments have produced elements of the syndrome with circumscribed destruction of the VMH, and, further, many report an orderly dose-response relationship between the amount of VMH damage and the magnitude of the effect. While these experiments have not, for the most part, excluded the possibility that damage to the VMH is simply highly correlated with damage to the actual substrate, such a proposal seems strained. In addition, some of the experiments specifically avoid this type of criticism by employing very small lesions and grouping animals on the basis of the absence of damage to tissue outside of the ventromedial nuclei (e.g., Bernardis and Frohman, 1970).

The apparent inconsistencies between the studies supporting a longitudinal model as opposed to a VMH model may have several explanations, some of which, in turn, have implications for functional analyses.

First, it is possible that both models are partially correct. The hypothalamus may contain a complex system that can be manipulated differently at different points. The hypothalamic system must receive inputs from and, in turn, project outputs to extrahypothalamic structures. It is conceivable that different points in the system may be accessible to different techniques of analysis. Lesions and knife cuts must be most effective at points where fibers or cell bodies are spatially tightly organized. Recording procedures and measurements of cellular metabolism are most effective where cell bodies are located.

Second, it is also conceivable that the apparent discrepancy arises from the fact that different elements of the VMH syndrome have different anatomical substrates: some elements might be produced by destruction of a longitudinal pathway; others, by destruction of VMH neurons. Although it is typically assumed that the several elements of the syndrome have a covariant relationship, this assumption is not often tested. In this regard, it is worth noting that evidence for a longitudinal system has usually involved measurement of food intake and/or body weight, whereas evidence for a central role of the ventromedial nucleus has involved recording neural activity or measuring metabolic parameters. It is therefore conveivable—although not particularly parsimonious—that a longitudinal system is particularly central to food intake and substantial body-weight gain, while VMH neurons are particularly central to metabolic adjustments and reception of afferent inputs. Perhaps a more plausible alternative analysis is based on sensitivity: it could be argued that small lesions confined largely to the ventromedial nucleus produce subtle deficits in primary metabolic systems but do not produce the maximal degree of disturbance required for detection of large changes in food intake and body weight. In such a view, very considerable disruption of the system—e.g., interruption of its most coherent projections—might be necessary to produce major changes in food intake and body weight. Such a view would be consistent with the issues reviewed earlier on primacy: we know of no good evidence for VMH hyperphagia and weight gain occurring without metabolic disturbances, while the metabolic effects can apparently occur at a "lower threshold" than the behavioral effects.

III. Disturbances of Energy Homeostasis Produced by Lateral Hypothalamic Manipulations

A. The Lateral Hypothalamic Area and Energy Homeostasis: History

Current ideas about the role of the lateral hypothalamus (LH)—like those about the role of the VMH—have been shaped by a few key observations and concepts in the early experiments on the role of the lateral hypothalamus in energy homeostasis. These early experiments in turn were heavily influenced by the conceptual climate of their time and, in particular, by the view of the VMH disturbances that were then current.

Several early research attempts directed toward an anatomical and physiological analysis of hypothalamic obesity had incidentally observed decrements in or elimination of food intake after hypothalamic lesions (in the rat: Hetherington and Ranson, 1940; the cat: Clark et al., 1939, and Ingram et al., 1936; and the monkey: Ranson, 1939). A clinical literature which suggested that hypothalamic damage and disease could produce emaciation also existed (Brouwer, 1950; Haymaker and Anderson, 1955; Weinberger and Grant, 1941). These reports were even rarer—as might be predicted from anatomical hindsight—than those of hypothalamic obesity (White and Hain, 1959, estimated a ratio of 1:4) and, unlike the adipsogenital syndrome of Fröhlich, the characteristics described were not recognized as a distinct clinical entity or syndrome. In addition, the pattern of diecephalic emaciation and marked depletion of subcutaneous fat was not attributed to any particular behavioral or metabolic disorder.

Then, in 1951, Anand and Brobeck reported that they had pinpointed a lateral hypothalamic locus at which lesions in rats and cats effectively produced terminal cessations of feeding. These investigators, working within the conceptual framework of Brobeck's earlier conclusion that lesions of the ventromedial hypothalamus produced a primary feeding disturbance (Brobeck et al., 1943), assumed that the lateral hypothalamic effects were fundamentally behavioral and represented destruction of what they designated as a feeding center. Their conclusion contained most of the basic elements that informed perhaps 20 years of subsequent research:

> The behavior of these rats and cats suggests that the lesions have in some manner removed what Adolph as called the "urge to eat."
>
> It further appears from these observations that the more medial hypothalamic structures (where lesions cause hyperphagia) exert their influence on food intake through the "feeding center," possibly by means of inhibitory fibers running laterally into that center. Thus, an animal fails to eat following destruction of the "feeding centers" of both sides, while if the "feeding centers" (or even one of them) are intact and the more medial mechanism is removed by destruction of the ventromedial nuclei or their efferent fibers, the animal exhibits a state of hyperphagia. This concept is also in accord with the observation that the degree of hyperphagia and the rapidity with which obesity develops depends upon the size of the lesions in the more medial region, in other words, upon the number of inhibitory fibers damaged. On the other hand, combined bilateral lesions of the medial and lateral mechanisms, together, produce the same effect as lateral lesions alone, that is, failure to eat [p. 137].

Anand and Brobeck provided the idea that the lateral hypothalamus was a feeding center, and they also substantiated the idea that the ventromedial nucleus inhibited the LH, and that it did so by laterally projecting axons. (They, with proper caution, used the term "feeding center" exclusively in quotation marks, but the quotation marks were soon dropped by others.) Interestingly, their conclusion of lateral outflow, based wholly on circumstantial observations, rapidly supplanted the earlier conclusion, based on rather more direct evidence, that the VMH projections were predominantly longitudinal (see discussion above). Their conclusion also proposed that the feeding center elaborated an "urge to eat." No consideration was given to the possibility that the behavioral effects were mediated wholly or in part by metabolic and gastrointestinal events: the role of the hypothalamus in energy homeostasis was simply to determine consumption. In retrospect, too, since Anand and Brobeck provided food and water separately for their animals, their results were probably exaggerated by the adipsia produced by LH lesions.

Two later contributions played key roles in directing the experimentation that followed: Teitelbaum and Stellar (1954) described a recovery sequence for animals sustaining LH damage, and Montemurro and Stevenson (1957b) demonstrated a convincing dissociation of caloric and water deficits after LH lesions.

In their original report, Anand and Brobeck (1951) successfully tube-fed and maintained one cat with amygdaloid lesions for a week until it recovered spontaneous feeding and maintained itself. With this strategy of tube-feeding and special dietary measures, Teitelbaum and Stellar (1954) were able to maintain 14 rats with LH lesions until the animals recovered eating and drinking behavior (6-65 days). In 1962, Teitelbaum and Epstein published their classic formulation of the typical recovery sequence of animals maintained through the initial postlesion phase of aphagia and adipsia. On the basis of behavioral criteria, Teitelbaum and Epstein (1962) distinguished four distinct stages in which consummatory behavior is reinstated after LH lesions, and they suggested that the stages describe a lateral hypothalamic syndrome.

During this initial decade of analysis of the LH syndrome, almost no consideration was given to the possibility that LH lesions might produce primary alterations in "physiological energy homeostasis." Morgane (1961), in probably the first serious proposal that LH lesions might produce metabolic disturbances, described a possible "metabolic failure" in animals with far-lateral lesions which failed to survive even with tube-feeding. Stevenson and Montemurro (1963) detected an increased energy expenditure in LH animals at rest. For the most part, however, LH lesions were assumed to affect a behavioral "center." More recent work has indicated a number of primary metabolic changes that are produced by LH lesions.

B. Elements of the Lateral Hypothalamic Syndrome

1. The Lateral Hypothalamic Area and Food Intake

a. Rats

Well-placed LH lesions (see Fig. 8 for the area defined by Anand and Brobeck) of even modest size produce an initial aphagia and adipsia (see Anand and Brobeck, 1951; Teitelbaum and Stellar, 1954; Teitelbaum and Epstein, 1962). The duration of the aphagia, which is positively correlated with the size of the lesion (Teitelbaum and Epstein, 1962; Powley, 1970), can vary from virtually zero to many days. In extreme cases, there are reports of animals being totally aphagic for 60 or more days. Most observations, however, suggest that animals may typically be aphagic for perhaps three to five days (Epstein, 1971). Interpretation of the very long intervals is complicated by (1) the fact that massive ventral diencephalic lesions are usually employed in the extreme examples, (2) the possibility that postlesion complications or secondary insults may have occurred which perpetuate the behavioral disturbances, and/or (3) the fact that practices of tube-feeding may actually extend the period of aphagia and anorexia (Teitelbaum and Stellar, 1954; Keesey et al., 1976b). This period of aphagia—with the accompanying adipsia—is the interval that Teitelbaum and Epstein (1962) have identified as stage 1.

During the period of aphagia, animals with moderate-size LH lesions in the area identified by Anand and Brobeck (1951) show a very characteristic aversion to food, first described by Teitelbaum and Epstein (1962). Animals with such lesions avoid food placed in their cages and will actively reject food forced on them (Teitelbaum and Epstein, 1962; Epstein, 1971; Keesey et al., 1976b). This point may be particularly telling, because recent work with large electrolytic lesions which invade the far-lateral hypothalamus and the medial pole of the internal capsule or neurochemical lesions of the nigrostriatal pathway passing through the same area has identified a different pattern of behavior in which the animal ignores food (as well as almost all stimuli) entirely, exhibiting a characteristic "sensory neglect" (Marshall et al., 1971; Zigmond and Stricker, 1972, 1973). It seems more appropriate to consider this alternate response as a dysphagia rather than the active aphagia originally described by Teitelbaum and others (see discussion of this issue in Keesey et al., 1976b). During this period, LH animals also exhibit exaggerated and fragmented patterns of activity (Morrison, 1968; Rowland, 1977).

After an interval of aphagia, animals with LH lesions are temporarily anorexic; they eat some foods but not enough to maintain body weight. (In light of our analysis of one of the primary effects of LH lesions—lowering the level of body-weight regulation—it is instructive to note that anorexia is defined in terms of body-weight level). This anorexia, with concomitant adipsia, corresponds to the interval Teitelbaum and Epstein recognize

Role of the Hypothalamus in Energy Homeostasis

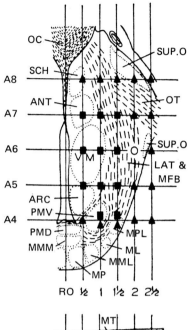

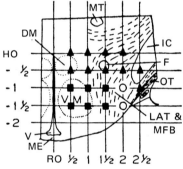

Figure 8 Upper: Horizontal section through rat brain indicating the effects of various lesion placements on feeding behavior. Lower: Coronal section through rat brain at level of ventromedial hypothalamus indicating the effects of various lesion placements on feeding behavior. In both cases, numbers refer to Horsley-Clarke coordinates. ▲, Normal food intake; ○, no feeding behavior; ■, hyperphagia. (Reproduced from Anand and Brobeck, 1951.)

as stage 2. During the period of anorexia, animals with LH lesions eat only wet foods (cf. the adipsia) and particularly what appear to be palatable foods such as cookie slurries and eggnog (Teitelbaum and Epstein, 1962; Rodgers et al., 1965; Powley and Keesey, 1970). As the animals' weight more closely approximates the plateau that they will subsequently defend, the animals accept a greater range of foods and will eat them in sufficient quantities to maintain this weight.

The emphasis on palatable foods for the anorexic LH animal is a potential source of confusion, however. In the first place, the subject of palatable diets for LH animals has all the ambiguities associated with the lack of clear definition or independent measurement discussed above for the VMH preparation. In the second place, though, this return of feeding (but only of palatable diets) is often taken to reflect the operation of mechanisms unique to the LH syndrome (see below). On the contrary, it is difficult to conceive of any insult to an organism that produces an anorexia selective for unpalatable foods, and it is almost as difficult to conceive of an anorexia that is totally insensitive to food quality. In all likelihood, any animal "off its feed" for any reason would first be induced to eat through offers of the most appetizing food. One has only to consult the suggestions for postoperative dietary therapy in texts of experimental surgery to appreciate how general such a sequence is. It seems to us that it is the occasional exception to the rule—e.g., an anorexic animal that will eat dry, adulterated food but not the most palatable food—that bears particular attention and may have diagnostic value.

After the period of anorexia, the animals begin consuming enough calories to maintain their body-weight level without further decline. (Again, this phase is defined in terms of body weight as well.) At this point, the animal has recovered feeding although, for a while longer, it typically remains adipsic. In the terminology of Teitelbaum and Epstein (1962), the animal is in stage 3 if it ingests adequate calories as long as it remains hydrated, which can be insured through such strategies as providing wet foods or flavored solutions for the animal to "eat." However, it is still adipsic and will develop a "secondary dehydration aphagia" if provided only with dry food and water separately. When the animal will again consume the water necessary to remain hydrated, it is in stage 4 or "recovery." The distinction between stage 3 and stage 4 turns on the problem of water balance, not caloric balance. Thus, for an analysis of the hypothalamic role in energy homeostasis, three phases of feeding behavior can be clearly distinguished after LH lesions: aphagia, anorexia, and normophagia. Animals maintained on wet diets that avoid confusing water-balance deficits with food-intake ones will observe only the three distinct phases (or fewer, with very small lesions) of feeding mentioned. The four stages of recovery identified by Teitelbaum and Epstein (1962) are more appropriate in cases where the deficits in hydromineral and caloric homeostasis are studied simultaneously. It is also important to note that the typical progression through the four successive stages is not invariant (cf. Morgane, 1975, p. 31).

In practice, not only the anorexia but the entire series of stages of recovery have been viewed as defining traits of the LH disturbance. Some investigators have reasoned that this sequence is a marker, and that any manipulation of the nervous system that produces a cessation of feeding followed by "recovery" similar to that shown by the LH animal must be affecting the LH mechanism. It is known, for example, that anterolateral decortication (Braun, 1975), vagotomy (Rezek et al., 1975), trigeminal deafferentation (Ziegler, 1975), interruption of central trigeminal structures (Zeigler and Karten, 1974), pallidal lesions (Levine et al., 1971), and nigrostriatal bundle damage (Marshall et al., 1974) produce feeding disturbances with sequences of feeding resumption similar to that following electrolytic damage to the lateral hypothalamus. It has also been observed that weanling rats go through a developmental sequence that resembles the LH sequence (Teitelbaum et al., 1969). Each of these disturbances has been conceptualized as the result of disabling a common feeding mechanism. It is at least as likely, however, that some of the similarities represent superficial resemblances based on a nearly ubiquitous and nondistinctive pattern of resumption (or assumption in the case of weanlings) of feeding. This dilemma of interpretation underscores the need for careful definition of the elements of each of the disturbances and for cautious evaluation of apparent similarities. Because of this concern, we have avoided assuming that the aphagia with active rejection of food initially identified with the lateral hypothalamic syndrome, the aphagia characterized by indifference or sensory neglect, and the aphagia characterized by an active approach to food coupled with a feeding "apraxia" (e.g., after trigeminal deafferentation) are all identical. Resolution of such issues will require additional detailed analyses of the elements of the feeding patterns with attention to the uniqueness (or lack of uniqueness) of the different components (see, for example, the recent experiments of Marshall and Teitelbaum, 1973, 1974; Stricker and Zigmond, 1974; and Zeigler, 1974).

LH animals that are feeding again and appear normophagic in terms of total food consumption still reveal characteristic disturbances of feeding behavior when subjected to a fine-grained microbehavioral analysis. Their meal patterns appear to be permanently disrupted; the animals are nibblers, i.e., they alternate between eating small amounts of food and pausing for short periods. Depending on the exact criteria set for the amount of food defining a meal and the time of intermeal intervals, this behavior can be recorded as a few extraordinarily long meals or as many short meals separated by brief intervals (Kissileff, 1970). This nibbling pattern is exaggerated with dry foods. In this case, the animal appears to alternate between a few bites of food and a small draught of water—the classical pattern of "prandial drinking" exhibited by the LH animal (Teitelbaum and Epstein, 1962; Epstein, 1971; Kissileff and Epstein, 1969). LH lesions, it should be noted, functionally desalivate animals by impairing neurally mediated release of saliva (Epstein, 1971) and apparently indirectly force an animal to drink as it eats dry food in order to chew and swallow effectively (Kissileff and Epstein, 1969; see Rowland, 1977, for an alternative account). When LH rats are maintained on a liquid diet, they exhibit nearly normal sizes and numbers of meals (Snowdon and Wampler, 1974).

The LH animal also tends to exhibit changes in the circadian pattern of feeding. On dry food and water, it eats almost exclusively during the night phase of the light cycle (Kissileff, 1970; Kakolewski et al., 1971). Apparently, this exaggerated nocturnality of feeding behavior may be diet-dependent since female LH rats exhibit a nearly normal circadian pattern of liquid-food intake (Snowdon and Wampler, 1974).

Animals with LH lesions (but not intraventricular 6-hydroxydopamine treatments) also exhibit an impaired ability to develop conditioned taste aversions (Roth et al., 1973; Stricker and Zigmond, 1974).

b. Other Species

In the original report of Anand and Brobeck (1951), two cats were described with lesions of the amygdala that spread into the lateral regions of the lateral hypothalamus. One animal stopped eating for seven days and the other never regained spontaneous feeding behavior. These results were replicated by Anand et al. (1955) in LH cats that actively rejected food even when it was placed inside their mouths. These same authors described a similar aphagia in monkeys after production of LH lesions. In dogs, Rozkowska and Fonberg (1970) produced LH aphagia which lasted anywhere from 2 to 16 days. After this period—during which the animals were maintained by tube-feeding—the amount of food spontaneously eaten was greatly reduced compared with prelesion amounts. A similar aphagia followed by a period of anorexia has been reported by Balinska and Brutkowski (1967) after lesions of the lateral hypothalamus in rabbits. Finally, LH destruction in swine (Khalaf, 1969) and in chickens (Feldman et al., 1957) has been reported to produce aphagia.

In humans, hypothalamic tumors in the region of the third ventricle have been associated with anorexia and extreme emaciation (Ingram, 1956). White and Hain (1959) and Kamalian et al. (1975) described such cases in adults with hypothalamic insult owing to cyst formation and demyelinating multiple sclerosis. Anorexia and severe inanition have been reported in children with third ventricular neoplasm (Kagan, 1958) and astrocytoma of the hypothalamus (Connors and Sheikholislam, 1977). In the only experimental study of lateral hypothalamic destruction in humans, Quadde and colleagues (1974) produced bilateral LH lesions in extremely obese adults to induce loss of appetite and reduction of body weight. Although the electrodes were positioned according to the patients' verbal reports of hunger when a stimulating current was passed through the tips, subsequent coagulation of this area produced only a transient decrease in food intake and little weight loss. Since histological material was not available from these patients, and, since adequate base lines for weight were not established, it is impossible to judge the exact size or location of the lesions or the precise similarity of the symptoms to those observed in animal models.

2. Exaggerated Responses to Food Quality ("Finickiness")

Although it is generally reported that LH animals are finicky, this proposal is difficult to substantiate except for some special uses of the term. The appellation has all of the problems associated with the use of the term for the VMH animal as well as some difficulties peculiar to the LH preparation. Three relatively independent usages of "finicky" have been associated with the LH syndrome. Two are very limited and appear to be quite different from the conventional use of the term. The third, while probably in keeping with conventional usage, finds little support.

First, many authors suggest that an LH animal is finicky when it is anorexic; it can be induced to eat only high-fat diets (Teitelbaum and Stellar, 1954), cookie slurries (Rodgers et al., 1965), or eggnog (Powley and Keesey, 1970) before it will take dry chows or standard laboratory formulas. As outlined above, though, this response pattern is probably nearly universal and not a specific trait of LH animals. Any physiological debilitation of neurologically intact animals would probably produce the same pattern of consumption. Certainly no experiment has yet shown that anorexic LH animals are more finicky than a control group of neurologically intact but anorexic animals. (For example, see Sclafani et al., 1976, on the anorexia of animals that have been fed excess calories.) In addition, this phase of anorexia-determined finickiness is transitory and ends when the LH animal reaches it chronically maintained weight plateau or when it recovers to stage 3.

The second limited use of finickiness with respect to the LH syndrome occurs when feeding reactions are confused with control of water intake. Anorexic LH animals (Teitelbaum and Epstein, stage 2) or LH animals that will consume normal amounts of food while still adipsic (Teitelbaum and Epstein, stage 3) are finicky in the sense that they will eat only liquid diets or wet mashes. Even the recovered LH animal exhibits certain responses in food consumption that are dependent on its altered regulation of body water. Animals with LH lesions may also have deficits in terms of their regulation of salt intake (see Wolf and Quartermain, 1967). The global deficits in the hydromineral metabolism of the animals with LH lesions frequently makes it a prandial drinker with minimal responses to osmotic and volemic challenges (see Epstein, 1971, for a comprehensive review). In such cases, the LH animal will not drink water when it is tainted (Teitelbaum and Epstein, 1962) and thus, in addition to the altered pattern of water consumption, will easily become aphagic or anorexic. Such a case is illustrated by Carlisle and Stellar's (1969) observation that LH animals evidenced poor caloric tracking on drier diet mixtures. This type of finickiness, however, is clearly secondary to problems of water consumption and does not represent a form of heightened reactivity to the sensory qualities of nutrients. In this sense the LH animal may be a finicky drinker, not a finicky eater.

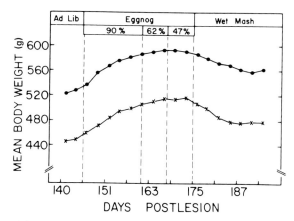

Figure 9 Body-weight changes of normal and LH rats to diet variation and diet dilution. Controls, 14 rats (●—●), 16 lesioned rats (∗—∗). (Reproduced from Powley and Keesey, 1970.) Copyright 1970 by the American Psychological Association. Reprinted by permission.

Finally, LH animals have been described as being finicky in a third sense. It has been suggested that LH animals that are taking enough food to maintain their weight can be made to lose weight—presumably by undereating—or gain weight—presumably by overeating—by manipulations of diet quality. Similar weight and feeding changes also occur in neurologically intact animals, however. A review of experiments that have systematically altered diet quality and composition and monitored body-weight changes of animals with lesions and of intact control animals suggests that, although LH animals have a different weight plateau, they adjust their weight and consumption proportionately to control animals. For example, the sequential varying of diets from rat chow to eggnog to diluted eggnog to wet mash produces comparable weight changes in animals with LH lesions and control animals (Powley and Keesey, 1970) (Fig. 9). Similar effects also occur for food intake (see Powley and Keesey, 1970). Other preferred diets such as high fat and sucrose-saccharin also produce parallel adjustments of food intake and body weight in LH-lesioned and control animals (Boyle and Keesey, 1975). Similarly, stage-4 rats show as rapid and as precise an adjustment to a one-to-one water dilution of a liquid diet as do control rats (Epstein, 1971). When maintained on 0.2% quinine-adulterated wet mash, rats with LH lesions also apparently decrease their weight and food intake by amounts that are comparable to intact-control levels (Keesey and Boyle, 1973) (Fig. 10). Leach and Braun (1976) have recently suggested that, although LH animals are not finicky in terms of a disproportionate weight loss to quinine-adulterated diets, these animals do show an exaggerated suppression of consumption on short-term quinine consumption tests. One of the few indications that animals with LH lesions might be more responsive, in terms of food intake and body weight, to diet quality than appropriate control animals (Mufson and Wampler, 1972) has proved difficult to replicate (Snowdon and Wampler, 1974).

In sum, altered patterns of food intake clearly exist in animals with LH lesions. On a purely descriptive level, finickiness occurs in LH animals in at least this limited sense. There are tremendous difficulties in evaluating this claim, however, if it is taken to mean that LH lesions produce a *primary* alteration in the reactivity of animals to nutrients. As

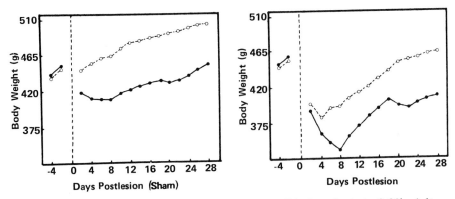

Figure 10 Body-weight changes in sham-lesioned and LH-lesioned rats to 0.2% adulteration of their diet with quinine. Control diet, ○—○; quinine, ●—●. (Reproduced from Keesey and Boyle, 1973.) Copyright 1973 by the American Psychological Association. Reprinted by permission.

in the case of research on the VMH syndrome, even if the circularity in the use of the term is ignored, most studies of LH animals on varying diets have ignored the possibility that the animals' different consumption patterns may reflect differences in postingestional consequences of the diets. In addition, virtually no experiments with LH animals have included neurologically intact, anorexic controls. Existing information also does not permit one to rule out the possibility that the altered consumption patterns are the result of a lack of saliva in the oropharynx and of the changes that such desalivation produces in the gustatory receptors. Any assumption that the finickiness of the LH animal has a primary basis is problematic.

3. LH Feeding and Regulatory Challenges

Although animals with LH lesions exhibit a wide spectrum of deficits in the regulation of body water (see Epstein, 1971), such animals show virtually normal compensation in a variety of challenges to caloric homeostasis. Some subtle deficits may still exist, but they are certainly far less comprehensive than originally assumed—particularly when increased experimental consideration is given to an adequate hydrational state, salivary secretory deficits, constraints imposed on LH animals by prandial drinking bouts, initial differences between controls and brain-damaged animals in body weight and energy stores, and so forth. The one clear exception to this generalization is the reduced capability of the LH animal to elaborate feeding responses to strong glucoprivic challenges.

The first two stages of the LH syndrome, aphagia and anorexia, have not been thoroughly examined for the presence of compensatory responses to regulatory challenges. The absence of any food intake during aphagia, the absence of a stable base-line of intake during anorexia, and the acute, rapidly evolving character of these phases all make measurement of the effects of regulatory challenges difficult. One type of observation indicates, however, that even during the initial stages of the syndrome, LH animals are sensitive to the state of their own energy stores. We have found that gradual reduction of an animal's weight prior to the production of LH lesions dramatically shortens and can even eliminate the typical period of aphagia and anorexia (Powley and Keesey, 1970). In the extreme case, animals reduced, by diet, to a weight level below that which they will defend after the production of LH lesions are actually hyperphagic after LH damage (Fig. 11). Similar effects of prelesion reduction of body energy stores have been seen in several additional experiments on both the duration of the aphagia and anorexia (Mufson and Wampler, 1972; Glick and Greenstein, 1972; Balagura and Harrell, 1974; Grijalva et al., 1976) and the duration of sensory neglect (Schallert, 1977). Myers and Martin (1973) obtained similar prestarvation results with some (but not other) 6-hydroxydopamine lesions of the lateral hypothalamus. In a reciprocal fashion, LH animals supplied with calories by a tube-feeding procedure during the first days after a lesion exhibit a dramatic protraction of anorexia (Keesey et al., 1976b). Similarly if animals are 120% of normal body weight when they receive LH lesions, they display an extended period of sensory neglect (Schallert, 1977). Finally, Balagura et al. (1973) have reported that the length of postlesion aphagia can be reduced or extended by administration of the glucodynamic hormones insulin and glucagon, respectively. Such observations suggest that the food intake, or the regulatory responses of the LH animal is still modified by its energy economy, even during the period of aphagia and anorexia.

More extensive observations on the control of caloric intake are available for the animal with LH lesions that has returned to normal or nearly normal levels of food intake (i.e., stage 3 or stage 4 "recovered laterals"). Such animals respond as vigorously and as completely as intact controls to several challenges. Rats with LH lesions show food intake increases in cold environments, and decreases in warm ones, that are as robust as those of control rats on the Greenstein diet, a liquid diet of average-to-low palatability (Epstein and Teitelbaum, 1967).

On the other hand, it has recently been demonstrated that, when LH animals with large lesions are shaved and then severely cold stressed (i.e., maintained in a 5°C

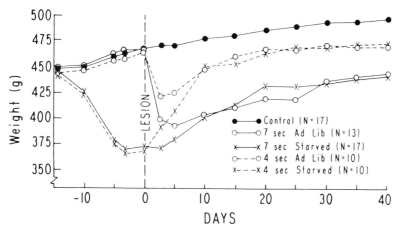

Figure 11 Body weights of rats after LH lesions (of varying current duration) or sham lesions, following prelesion manipulation of body weights. Note the weight gain of the two LH lesion-starved groups immediately after surgery and when put onto ad libitum feeding. (Reproduced from Powley and Keesey, 1970.) Copyright 1970 by the American Psychological Association. Reprinted by permission.

environment), they are less capable than control animals of surviving and increasing their food intake (Stricker et al., 1975). The investigators conclude that LH animals mobilize compensatory feeding responses to only a truncated range of stimulus intensities. Stricker and colleagues, however, did not match controls and LH animals for energy stores at the onset of the cold-stress test. Since LH animals have dramatically reduced energy reserves or body fat (Powley, 1971; Mitchell and Keesey, 1974; Keesey and Powley, 1975), it may be the case that, even though LH animals are capable of responding to intense thermal stimuli, they are unable, with their reduced endogenous stores of energy available for thermogenesis, to adequately redress their disadvantage of stores in such a severe test.

Rats that have stabilized their food intake and body weight after LH damage also exhibit apparently normal compensatory food-intake responses to both starvation and overfeeding. Epstein (1971) found that "recovered" laterals increase their intake of the Greenstein liquid diet by normal amounts during a 3-hr refeeding test after 21 hr of deprivation. In an analysis of the refeeding responses of LH animals subjected to more extreme deprivation, Mitchell and Keesey (1977) found that, when animals with LH damage and their controls were starved to 80% of their respective ad libitum weights, both groups consumed virtually identical amounts of wet mash over similar intervals in order to restore their body weights to their respective plateaus. When animals are matched in terms of hours of deprivation rather than percentage of body weight or fat stores, those with LH lesions—and thus lower levels of body weight (see Powley and Keesey, 1970)—eat with even shorter latencies than do controls (Devenport and Balagura, 1971). Finally, when an LH animal's body weight is experimentally increased above its stable postlesion level by intragastric hyperalimentation, the animal responds by decreasing food intake during the infusion period and exhibiting an anorexia when the hyperalimentation is discontinued (Keesey et al., 1976a).

When challenged with dietary dilutions or imbalances, the LH rat also appears to adjust its consumption as accurately as controls. Such animals respond with precision to progressive water dilution of a fortified eggnog diet (Powley and Keesey, 1970; see also Fig. 9). They also display as rapid and complete an adjustment of food intake when their diet (Greenstein liquid diet) is diluted to 50% with water (Epstein, 1971). LH animals also adjust their consumption to a water-diluted form of the same diet when they are self-administering the diet intragastrically (Williams and Teitelbaum, 1959). In addition, "recovered aphagics" adjust their consumption normally when maintained on high-leucine, amino-acid-imbalanced, and high-protein diets (Scharrer et al., 1970). In the only apparent exception, LH animals do not show a compensatory hyperphagia of the control diet after an interval of maintenance on an 80% casein diet.

Similarly, LH animals that have resumed feeding evidence caloric compensation by appropriately decreasing food intake after intragastric loads of glucose and vegetable oil (Yin and Liu, 1976).

The one clear case of reduced compensatory feeding responses on the part of the rat with LH brain damage is the feeding response to glucoprivic challenge. In 1967, Epstein and Teitelbaum reported that the recovered lateral animal did not eat in response to the glucoprivic challenge of insulin hypoglycemia (4-8 U administered subcutaneously). The deficit was virtually complete in LH animals and, as it appeared to be permanent, was recognized at the time as the loss of a major class of behavioral regulatory adjustments in such animals. Epstein and Teitelbaum (1967) reported one animal that still did not respond 503 days after lesion production. These observations were repeated and extended to other glucoprivic agents such as 2-deoxy-D-glucose (Smith and Epstein, 1969; Zigmond and Stricker, 1972; Marshall et al., 1974; Wayner et al., 1971). Although the feeding response of LH animals to glucoprivic feeding is sharply limited, more recent work indicates that it is not completely absent. Stricker and his colleagues (Stricker et al., 1975; Stricker and Zigmond, 1976a,b) replicated the previous results with strong, provocative doses of cytoglucoprivic agents and found that some LH animals do show compensatory food intake and weight gain when challenged by a graduated series of insulin doses (1-8 U) that begin at more moderate or physiological levels than those previously used. These investigators concluded that the deficits of the LH animal may be relative rather than absolute; the animals may respond to a truncated range of stimulus intensities. Although Stricker and his colleagues favor an "arousal" interpretation of such deficits, their observations with glucoprivic agents may mean simply that nonlethal and slowly debilitating doses of such agents provide sufficient time for the glucoprivation to be expressed by various metabolic pathways. These pathways may eventually produce indirect compensatory adjustments or feeding responses that are qualitatively different than the responses first mobilized by drastic provocations. These more moderate levels of glucoprivation also probably more closely approximate the more normal physiological operations by which glucose homeostasis operates on feeding (cf. Strubbe et al., 1977).

Further, it is not yet clear whether this deficiency in the LH animal amounts to the specific loss of a major homeostatic response or only a nonspecific reduced capacity to survive drastic and pharmacological disruptions of glucose homeostasis. Current analyses have suggested that blood-sugar levels may play far less direct roles in the determination of feeding behavior—at least in a short-run, meal-to-meal fashion (Strubbe et al., 1977; Steffens, 1969; Scharrer et al., 1974). Recent analyses have also questioned the physiological nature of glucoprivic feeding (Kraly and Blass, 1974; Blass and Kraly, 1974; Smith et al., 1972).

4. The Lateral Hypothalamic Area and Metabolic Responses

For the historical reasons outlined above, relatively little attention has been given to the metabolic and physiological concomitants of the LH syndrome. The experimentation available, however, points to major changes in energy metabolism in the animal.

a. Body Composition

Lesions in the lateral hypothalamus chronically reduce an animal's maintained body weight and this lower weight level reflects principally a reduction in total body fat.

In 1970 we observed that LH lesions permanently reduce an animal's weight to a fixed percentage of control weight (Powley and Keesey, 1970). Larger lesions produce lower body-weight set points. There is, in fact, a very regular linear relationship ($r = -0.85$) between the amount of a circumscribed area within the lateral hypothalamus that is destroyed and the body-weight level expressed as a percentage of control body weight at which an LH animal regulates (Fig. 12). This area within the hypothalamus was mapped with a specially developed computer-assisted sorting algorithm that treats the sampled brain areas as an array of cubes and recombines the destroyed cubes until the algorithm has identified the set of cubes, or "critical area," that best relates variation in neural damage to variation in a given dependent measure (Powley, 1970). The area identified

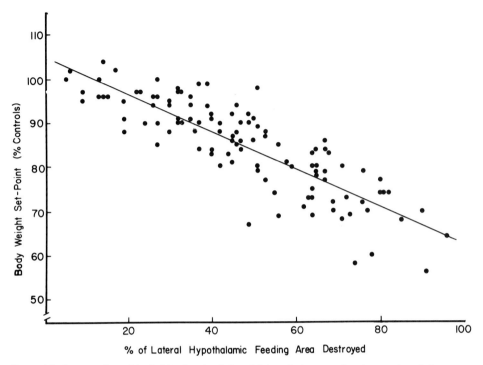

Figure 12 Scatterplot of individual animals' weight-maintenance levels or set points as a function of percentage of LH feeding area destroyed. The LH feeding area definition is based on the computer analysis described by Powley (1970). The rat body-weight set points are expressed as a percentage of control weights that have been corrected for the effects of sham operations. R, −0.85; y, −0.43x + 105; 110 subjects. (Reproduced from Powley, 1970.)

by this procedure (the shaded cubes in Fig. 13) corresponds well with that broad continuum classically known as the lateral hypothalamic area. This continuum also encompasses the areas previously identified as the effective LH feeding area (Anand and Brobeck, 1951; Morrison et al., 1958; Teitelbaum and Epstein, 1962, see Fig. 8).

The lowered levels of body weight regulation are achieved primarily by reduction in an animal's body fat or energy stores. This point has been confirmed by gravimetric (Powley, 1971) and ether-extraction (Keesey and Powley, 1975; Mitchell and Keesey, 1974) procedures. Although proportionately less dramatic, some reductions also occur in the water and fat-free solid compartments (Powley, 1971; Mitchell and Keesey, 1974). Recent analyses indicate that—at least for the genetically obese Zucker fatty rat—when LH lesions are produced during the period of on-going adipocyte hyperplasia, they effect the limitation of lipid stores in part by reducing the number of countable adipocytes (Stern and Keesey, 1978).

b. Metabolic Profile

The few reports available indicate that LH lesions may modify insulin release in a manner opposite to that of VMH lesions. The tentative conclusion can be made that LH

Role of the Hypothalamus in Energy Homeostasis

lesions diminish stimulated insulin release and reduce hyperinsulinemia but do not affect basal levels of insulin release. Steffens et al. (1972) described a group of recovered LH animals with the typical small-meal or nibbling pattern of intake that had appropriate basal-insulin levels and exhibited almost no meal-stimulated increase in insulin levels (see their Fig. 5). (Interpretation of these results is made difficult, though, because the investigators had another group of "recovered laterals" that ate uncharacteristically large meals and exhibited an insulin profile similar to that which one might expect after lesion encroachment on the ventromedial hypothalamus. The histological information does not permit a conclusive resolution of this possibility.) An additional analysis supporting the contention that elevated but not basal levels of insulin are reduced by LH damage is provided by Chlouverakis and Bernardis (1972), who report that LH lesions did not affect basal insulin levels in normal control mice but did lower the hyperinsulinemia of *ob/ob* mice. In a somewhat analogous comparison, Schnatz et al. (1973) found that LH lesions reduced the level of chronic hyperinsulinemia produced by prior VMH lesions.

As is the case after VMH damage, blood-glucose homeostasis is not dramatically affected by LH lesions. Steffens et al. (1972) reported relatively normal glucose values for

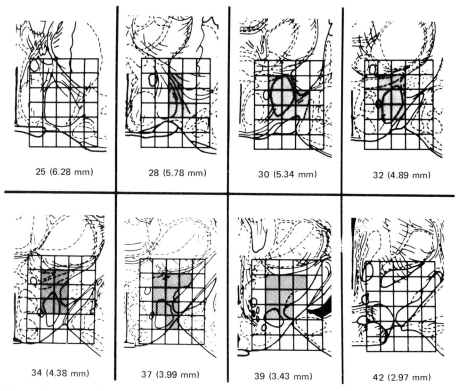

Figure 13 Summary of the final anatomical locus selected by the computer analysis as correlating with the reduction of rat body weight after LH lesions. The shaded cubes were included in the final definition of the set point area. Numbers in bold type refer to König and Klippel (1963) plate numbers. Distances, in parentheses, are expressed as millimeters anterior to the ear bars. (Reproduced from Powley, 1970.)

their LH animals, and Schnatz et al. (1973) reported no effect of LH lesions (when they are combined with prior VMH damage) on blood-glucose values. On the contrary, perhaps, LH lesions partially ameliorated the hyperglycemia of the *ob/ob* mouse, while they had no effect on blood-sugar levels of lean control mice (Chlouverakis and Bernardis, 1972).

The metabolic rate of animals with LH lesions is characterized by an initial acute increase followed by a chronic decrease to below control levels. In the initial postlesion period, the increased metabolic rate may last for 72 hr to one week or more (cf. Stevenson and Montemurro, 1963; Morrison, 1968; Davis, 1977); this response may account for the greater weight loss seen in LH animals compared with fasted control animals (Stevenson and Montemurro, 1963) and the "metabolic decay" of Morgane's (1961) far-lateral animals. During the same period, LH animals may also manifest increased core body temperatures (Harrell et al., 1975; Moore and Eichler, 1976). After this acute period, LH animals exhibit chronically decreased thyroid weights and plasma levels of thyroxine (Davis, 1977). Lateral animals also manifest elevated levels of creatinine excretion when aphagic and anorexic as well as when deprived of food and water after the resumption of feeding (Morrison, 1968).

LH animals also show several abnormal responses that are presumably mediated by autonomic and endocrine mechanisms. LH lesions acutely produce a higher-than-normal incidence of gastric pathologies (Lindholm et al., 1975). Gastric motility was disturbed in dogs during the period of aphagia following LH lesions (Glavcheva et al., 1970). Lesioned animals apparently exhibit a transitory hypersecretion of saliva immediately postlesion (Schallert, Leach, and Braun, unpublished results, cited in Schallert et al., 1977), which is followed by a chronically reduced release of saliva in some situations (Hainsworth and Epstein, 1966; Epstein, 1971). In chronic LH animals that have resumed feeding, fasting gastric-acid secretion rate is dramatically reduced, and the degree of the reduction correlates with the level of postlesion body-weight maintenance ($r = 0.69$, Opsahl and Powley, 1977). LH animals that have stabilized body weight and feeding and that are regulating at reduced body-weight levels also have heavier adrenals than controls (Opsahl, 1977).

C. Etiology of the Lateral Hypothalamic Syndrome

The logic of the etiological analyses of the VMH syndrome described above seems equally applicable to the LH disturbances. In our opinion, the need for such experimentation is at least as pressing in the case of the LH syndrome as in the case of the VMH syndrome. Corresponding to the relative paucity of work on the metabolic and physiological concomitants of the LH syndrome, however, is an almost complete dearth of experiments employing strategies designed to identify potential dissociations between elements of the disturbance that follows LH damage.

1. Temporal and Anatomical Dissociations

In general, the research documenting the correlations between the stages of resumption of feeding after LH lesions and the waxing or waning of specific behavioral and physiological components of the syndrome has not been sufficiently quantitative and precise to establish convincing correlations or dissociations. LH aphagia corresponds (at least in some, if not all, preparations) with the period of hyperactivity, sensory neglect, active weight reduction to a new weight level, increased metabolic rate, and elevated body temperature. All of these acute effects occur sufficiently soon after lesion to account for the aphagia, and none of them has clear temporal priority over the others. As in the case of the analyses of the VMH syndrome, the events appear to happen too close in time to permit a clear demonstration of chronological primacy.

Some dissociations of the elements of the LH syndrome in spatial dimensions have been identified, however. The area responsible for the aphagia and anorexia of the LH syndrome (see, for example, Figs. 8 and 13) and the area where lesions interfere with glucoprivic feeding are not anatomically coextensive: lesions of the anterior medial forebrain bundle interfere with glucoprivic feeding without producing the symptoms of aphagia, anorexia, and adipsia (Kraly and Blass, 1974; Blass and Kraly, 1974). Similarly, Schallert et al. (1977) have identified different foci for the lesion-produced gastric pathologies (more ventral and medial) and LH aphagia (more dorsal and lateral). Such results suggest that the classic behavioral elements of aphagia and anorexia are not the (simple) result of either disabling glucoprivic feeding mechanisms or generating gastrointestinal lesions.

2. Metabolic Disturbances with Controlled Intake

Several experiments have established that some of the specific metabolic responses during the initial aphagia of the LH animal cannot be produced by simple food and water deprivation of controls. Animals with LH lesions lose weight faster than totally deprived controls (Morgane, 1961; Stevenson and Montemurro, 1963; Harrell et al., 1975). Such animals also fail to gain weight at control rates when both groups are intragastrically fed equal quantities of food (Harrell et al., 1975). When controls and LH animals that have resumed feeding are deprived of food and water, the operated animals still exhibit an exaggerated creatinine excretion (Morrison, 1968). Some of these differences may be the results of the excessive activity that LH animals exhibit, particularly immediately postlesion (Morrison, 1968; Rowland, 1977), although activity disturbances apparently cannot explain all of the effects (Harrell et al., 1975; Davis, 1977).

3. Controlling Metabolic Responses and Monitoring Behavior

Manipulations of the LH animal's energy stores have been found to affect the time course of the aphagia-anorexia sequence. As reviewed above (see Fig. 11), we have demonstrated that prelesion dieting can reduce or even eliminate the characteristic period of aphagia and anorexia after LH lesions (Powley and Keesey, 1970). Other work has also indicated that body-weight levels (or energy stores, presumably) are determinants of the duration of the feeding interruption (Mufson and Wampler, 1972; Balagura and Harrell, 1974; Grijalva et al., 1976) as well as the duration of the sensory neglect (Schallert, 1977) and the extent of the gastric pathologies (Grijalva et al., 1976). Subsequent work has also suggested that use of a radical total-starvation method of weight reduction (as opposed to a partial food-restriction procedure) does not reduce and may even aggravate the postlesion recovery sequences (Grijalva et al., 1976; Myers and Martin, 1973; see also Chlouverakis and Bernardis, 1972).

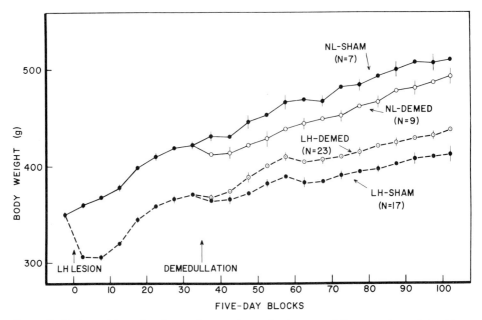

Figure 14 Body weights of male rats (grams ± standard errors of the mean) receiving bilateral electrolytic lesions of the LH (or sham lesions) on day 0 and adrenal demedullation (or sham surgery) on day 35. NL, nonlesioned. (Reproduced from Opsahl, 1977.)

In a related series of experiments, we have evaluated some of the potential roles of sympathetic and parasympathetic pathways in expressing elements of the LH syndrome. Though it decreases body-weight levels of nonlesioned controls, adrenal demedullation produces a significant increase in the body-weight maintenance level of LH animals (Fig. 14) (Opsahl, 1977). The results raise the possibility that some of the metabolic effects and body-weight loss produced by LH lesions may be mediated by an augmented secretion of catecholamines. Such a possibility is also consistent with the fact that LH animals manifest heavier adrenals (Opsahl, 1977). In additional experiments, we have also found that subdiaphragmantic vagotomy paradoxically increases fasting gastric-acid levels of LH animals (to nonlesioned vagotomized control levels) while it further decreases the body-weight maintenance level (Opsahl and Powley, 1977). This outcome suggests that the alterations in gastric-acid secretion in LH animals may involve the vagus, and, further, that the gastric secretory effect and the lowered body-weight levels can be experimentally dissociated.

D. LH Feeding and Metabolic Disturbances: Anatomical Substrates

Some of the relevant literature on the anatomical substrates of the LH disturbances has already been discussed in conjunction with the circuitry underlying the VMH syndrome and the history of the study of the lateral hypothalamic syndrome. As suggested for the VMH syndrome, some of the anatomical issues surrounding the lateral hypothalamic syndrome are particularly relevant to a functional analysis. These points are reviewed below. For a thorough review of the relevant anatomy, the reader should consult a number of chapters in the present volumes as well as reviews by Morgane (1975), Morgane and Stern (1974), and Nauta and Haymaker (1969).

Two major types of models have been developed to account for the elements of the LH syndrome. One emphasizes the role of the neurons that occupy the lateral hypothalamus, the bed nucleus of the medial forebrain bundle, and the fibers within the medial forebrain bundle. The original explanation proposed by Anand and Brobeck (1951) is such a model. The second type of explanation emphasizes the longitudinal fiber systems that travel through (and particularly just lateral to) the lateral hypothalamic area defined by cytoarchitectonic means. As was suggested earlier for the two types of models for the VMH syndrome, both anatomical explanations of the LH syndrome appear to have merit. It seems clear that the LH syndrome has a variety of elements and, in fact, consists of several variants (or perhaps even separate syndromes). As already discussed (see Section III.B.1.a), the conclusion that all ventral diencephalic aphagias and anorexias are necessarily reflections of damage to a single mechanism because of superficial similarities in the pattern of feeding disruption is questionable; that conclusion represents something of a carry-over from an earlier feeding-center orientation. In all likelihood, future progress in the study of hypothalamic role in energy homeostasis will require additional fractionation of the various elements of the lateral hypothalamic disturbance.

In the case of the first model (suggesting that the effective substrate for the lateral hypothalamic syndrome corresponds with the classically defined lateral hypothalamic area, which is occupied by the medial forebrain bundle and the bed nucleus of the bundle) several of the alterations in behavioral and metabolic responses produced by LH lesions seem to map reasonably well onto the lateral hypothalamic area. First, classic aphagia characterized by an active rejection of food has been consistently identified with this area (Anand and Brobeck, 1951; Morrison et al., 1958; Teitelbaum and Epstein, 1962; Powley, 1970; Powley and Keesey, 1970). Damage restricted primarily to this area—in contrast to the more lateral fiber systems—is characterized by a minimum of sensorimotor problems (Rodgers et al., 1965; Keesey et al., 1976), although it does produce the initial excessive and fragmentary locomotor activity (Morrison, 1968). In addition, Schallert (1977) has recently suggested that the lesions which produce this more active aphagia also tend to be more anterior within the lateral hypothalamus, whereas the lesions which produce a more passive aphagia with major components of sensory neglect and akinesis tend to be more posterior. Lesions restricted to the lateral hypothalamic area proper also reduce the characteristic body-weight regulation level (Powley, 1970; Powley and Keesey, 1970; see Figs. 12 and 13). This area also represents the major neural substrate whose destruction is responsible for the development of gastric pathologies seen in the early postlesion period (Schallert et al., 1977). It also appears to be the tissue primarily responsible for the decreased thyroxine levels, lighter thyroid weights, and lower basal metabolic rate after LH lesions (Davis, 1977). Other elements of the LH syndrome, including the wasting of dry food (presumably the result of reduced salivary release), disruption of taste-aversion learning, and diminished thermoregulation to heat challenges (wetting the coat with saliva), apparently are also the results of damage to this lateral hypothalamic area, since 6-hydroxydopamine disruption of the longitudinal dopaminergic fiber systems (which should not have major destructive effects on the nucleus itself) does not produce the effects (Stricker and Zigmond, 1974). Damage to the lateral hypothalamic area encompassed by the medial forebrain bundle and its bed nucleus also appear to be the primary basis for the disruption of the autonomic and endocrine defenses mobilized by glucoprivation, such as gastric acid secretion (Colin-Jones and Himsworth, 1969; Kadekaro et al., 1975), epinephrine secretion (Himsworth, 1970), and growth-hormone release (Himsworth et al., 1972).

The conclusion that a variety of the responses of energy homeostasis are modulated by lateral hypothalamic neurons and/or pathways is also strongly supported by both stimulation experiments and electrophysiological analyses of the lateral hypothalamus. The effective sites for eliciting stimulation-induced feeding extend through much of the lateral hypothalamic continuum and tend to cluster in an area immediately dorsolateral to the fornix (Keesey and Powley, 1968; Poschel, 1968; Valenstein et al., 1968; Valenstein and Phillips, 1970). Similarly, the extensive series of single-unit recording experiments has established that the activity of neurons in the area correlate well with physiological and behavioral events associated with energy homeostasis (Cross, 1964, 1973; Oomura, 1973; Rolls et al., 1976).

In contrast, other elements of the LH syndrome seem to be primarily the result of damage to the more lateral fiber systems traversing the lateral and far-lateral hypothalamus as well as the internal capsule and the subcapsular region. Major long-fiber systems implicated in some of the LH disturbances include the nigrostriatal pathway, pallidohypothalamic projections, and amygdalohypothalamic fibers. Electrolytic or neuropharmacological damage focused in these areas produces an aphagia-anorexia sequence marked by dramatic sensory, motor, and sensorimotor disabilities. Animals will exhibit sensory neglect (Marshall and Teitelbaum, 1974; Marshall et al., 1971; Ungerstedt, 1970), catalepsy (Balagura et al., 1969), somnolence (Levitt and Teitelbaum, 1975) and akinesis (Balagura et al., 1969; Levitt and Teitelbaum, 1975). As such disturbances would suggest, the aphagia and anorexia in this case are apparently more passive (Schallert, 1977).

As mentioned at the outset, it is also necessary to emphasize the spatial overlap of such mechanisms. Many of the elements of the syndrome can be produced by relatively selective damage within either of these major systems. It is not clear, however, that these systems can be effectively studied in isolation. As many have argued, the systems are so thoroughly interdigitated that lesion strategies—particularly the large-lesion strategies that have frequently been used in the study of the lateral hypothalamus—cannot satisfactorily manipulate the two systems independently. Finally, both models may well be correct: that is, they both may participate in many of the same neural systems responsible for the coordinated behavioral-physiological control of energy homeostasis. The anatomical dissociation may ultimately be relative, not absolute.

IV. Overview

A historical survey of the sort that appears in the preceding pages provides a useful perspective on the analysis of the hypothalamic role in energy balance. Several general conclusions seem appropriate, and some general directions for research can be identified. Our thinking about the hypothalamic disturbances of homeostasis has undergone a number of changes—this progress reflects advances in our understanding of hypothalamic anatomy, increased sophistication in the analysis of feeding behavior, new information on the physiological control of digestion and metabolism, and, also, a sharpening of our understanding of brain-behavior relationships. Each of these advances has contributed to our current comprehension of the role of the hypothalamus in energy homeostasis. Continued refinements in our understanding of each of these problems will undoubtedly continue to shape our views of the hypothalamic control of feeding behavior and energy homeostasis.

One major trend has been away from conceptualizing the hypothalamus as an amalgam of centers, each responsible for the execution of some behavioral construct—feeding and satiety in the case of the lateral and ventromedial hypothalamus, respectively. This trend has been stimulated, in large part, by the scarcity of data indicating the loss of particular classes of behavioral responses resulting from damage to either of these areas. One of the surprising conclusions suggested by a review of the literature is, in fact, that no, or practically no, major short-term regulatory capacities are completely disabled by hypothalamic lesions. The changes that do occur in the short-term defense of energy balance appear to reflect not the loss of unique elements of homeostatic control but rather the distortion of various acute defenses produced by the long-term disturbances that hypothalamic damage does produce. The point appears to hold for both the VMH and the LH animal. It also receives support from the work revealing the behavioral and physiological competencies of both "island" preparations in which the hypothalamus is isolated from the rest of the brain by knife cuts (Ellison, 1968; Ellison et al., 1970; Hefco et al., 1975; McHugh et al., 1975) and LH-VMH combination-lesion preparations (Carlisle and Stellar, 1969; Schnatz et al., 1973; Yin and Liu, 1976). On balance, the experimental observations better support the idea that the ventromedial and lateral hypothalamic areas participate in the modulation of behavioral and physiological responses of energy regulation that are largely organized at other sites in the nervous system.

The VMH preparation has been subjected to an exhaustive battery of challenges of its ability to respond to what are viewed as day-to-day exigencies in the maintenance of energy homeostasis and appears capable of responding to the full range of these acute regulatory challenges. The validity of this point becomes all the more apparent when it is recognized that the few indications of altered short-term regulatory responses in VMH animals may reflect experimental differences that are secondary to the few characteristic long-term alterations in metabolism or behavior (see below). Major difficulties are inherent in experimental attempts to equate completely VMH animals and nonlesioned controls for purposes of comparison. Ad libitum comparisons are vitiated by differences in energy reserves, basal hormone levels, patterns of feeding, and postlesion history of metabolic adaptations, all of which might influence regulatory tests. Comparisons of deprived animals are compromised by many of the same differences as well as by differences either in hours of deprivation or percentage of ad libitum body weight. Furthermore, though VMH animals exhibit several differences in feeding behavior indicated by microbehavioral analyses (such as increased meal sizes), the placing of constraints on the behavior of these animals, for example, by forcing VMH animals to eat small meals, have not had major effects on VMH energy balance. Instead, the VMH animal has shown itself capable of flexible, integrated, and appropriate adjustments in feeding behavior in the maintenance of its hyperphagia. Thus, proposed explanations of the VMH syndrome that have emphasized the absence of a single element of behavior or of a single acute regulatory response seem unlikely to be adequate.

In the case of the LH animal, similar conclusions seem reasonable, although less definite conclusions are warranted because of the sketchier nature of the available results.

Again, no convincing evidence seems to exist for a permanent loss of a unique behavioral response specific to the lateral hypothalamic animals. (That is not to say that it is impossible to extensively, and perhaps generally, debilitate an animal by massive ventral diencephalic lesions that unequivocally involve a multiplicity of neural systems. While such an experimental preparation may be an exception to the conclusion, it may be an uninteresting one in terms of understanding the particular homeostatic contributions of the lateral hypothalamus—or any other specific neural system.) With adequate controls for the major alterations in hydromineral metabolism produced by LH lesions, there is little evidence for the loss of an independent behavioral element. Similarly, relatively few deficiencies in regulatory responses to acute challenges seem to exist. The significance, extent, and specificity of the LH animal's one classic regulatory failure, that is, failure of feeding in response to glucoprivation, are now in doubt (also see Stricker and Zigmond, 1976b). As in the case of the VMH animal, the conclusion seems all the stronger for the LH animal if one recognizes that the partial deficits in short-term controls observed may be distortions that are secondary to some of the animal's chronic metabolic and behavioral adjustments. Experimentation on the LH syndrome has traditionally not employed controls for anorexia (e.g., neurologically intact, anorexic animals), altered feeding patterns, gastrointestinal pathologies, alterations in salivary competence, nutritional (or malnutritional) history, and levels of endogenous energy stores. Until such experimental checks are available, it seems most parsimonious to attribute the minor changes in regulatory capacity and behavioral responses of LH animals to such uncontrolled differences.

Obviously, of course, we do not mean to suggest that, after ventromedial or lateral hypothalamic lesions, animals are normal. What we are suggesting, though, is that many of their acute disturbances represent distortions and imbalances of still-existing regulatory responses and are, in large measure, the results of a few primary chronic changes produced by the lesions.

The VMH animal clearly manifests a few permanent changes. First, it maintains its body weight around an elevated weight plateau. This altered weight regulation typically involves a dramatic hyperphagia in the ad libitum feeding situation. Second, defense of the VMH animal's weight plateau is inordinately diet dependent (or, as traditionally expressed, the animal is finicky). Third, the VMH animal also manifests an anabolic profile of metabolism, including increased fat accumulation and insulin release. This profile develops immediately after the lesion (even when food intake is experimentally limited to control amounts) and then typically evolves to dramatically exaggerated levels over days or weeks. This anabolic profile is at least partially eliminated by autonomic (e.g., vagotomy) and endocrine manipulations. It remains to be established whether one of these three general disturbances is primary to the other two or whether the separate disturbances are all primary effects of damage to the ventromedial hypothalamus. Credible hypotheses have been advanced for each of the three elements being primary to the other two; see Powley (1977) for a review of such hypotheses.

Permanent changes also occur in LH animals. First, the animals typically reduce their level of body-weight maintenance. The newly established level—like that of the VMH animal—is maintained in part by appropriate adjustments in food intake. Second, LH animals characteristically exhibit alterations in energy utilization. This altered metabolic profile includes major endocrine and autonomic components. One component appears to be a reduction of (at least) stimulated and elevated phases of insulin release. Third, such animals exhibit a wide spectrum of disorders that are not disturbances of energy homeostasis per se but that often directly reduce the animal's ability to defend its energy balance. These different disorders are logically separable from those of energy homeostasis, they are in some cases anatomically dissociable and many of them can be manipulated independently with pharmacological agents. Nonetheless, because these disorders (e.g., deficits in hydromineral metabolism, salivary secretion, and, at least in some cases, arousal) have been confounded with tests of energy balance, they must also be recognized.

One of the exciting prospects for the immediate future of hypothalamic research is that more precise and complete explanations of the etiology of the VMH and LH disturbances of energy homeostasis will be generated. It seems clear that, as more progress is made on these issues, the hypothalamus will be appreciated not as an area with special executive functions for satiety or hunger or feeding or other unitary behavioral constructs but rather as one of many levels of integration that participate in the modulation of homeostatic responses.

References

Adolph, E. F. (1947). Urges to eat and drink in rats. *Am. J. Physiol. 151,* 110-125.

Ahlskog, J. E. (1974). Food intake and amphetamine anorexia after selective forebrain norepinephrine loss. *Brain Res. 82,* 211-240.

Ahlskog, J. E., and Hoebel, B. G. (1973). Overeating and obesity from damage to a noradrenergic system in the brain. *Science 182,* 166-169.

Ahlskog, J. E., Randall, P. K., and Hoebel, B. G. (1975). Hypothalamic hyperphagia: dissociation from hyperphagia following destruction of noradrenergic neurons. *Science 190,* 399-401.

Akmayev, I. G., and Rabkina, A. E. (1976). CNS-Endocrine pancreas system. I. The hypothalamus response to insulin deficiency. *Endokrinologie 68,* 211-220.

Albert, D. L., Storlien, L. H., Albert, J. G., and Mah, C. J. (1971). Obesity following disturbance of the ventromedial hypothalamus: a comparison of lesions, lateral cuts, and anterior cuts. *Physiol. Behav. 7,* 135-141.

Anand, B. K., and Brobeck, J. R. (1951). Hypothalamic control of food intake in rats and cats. *Yale J. Biol. Med. 24,* 123-140.

Anand, B. K., and Pillai, R. V. (1967). Activity of single neurons in the hypothalamic feeding centres: effect of gastric distension. *J. Physiol. 192,* 63-77.

Anand, B. K., Dua, S., and Shoenberg, K. (1955). Hypothalamic control of food intake in cats and monkeys. *J. Physiol. 127,* 143-152.

Anlicker, J., and Mayer, J. (1956). An operant conditioning technique for studying feeding-fasting patterns in normal and obese mice. *J. Appl. Physiol. 157,* 1574-1575.

Auffray, P. (1969). Effets des le'sions des noyaux ventro-médians hypothalamiques sur la prise d'aliment chez le porc. *Annal. Biol. Biochim. Biophys. 9,* 513-526.

Auffray, P., and Blum, J. C. (1970). Hyperphagic et steátose hépatique chez l'Oie après lésion du noyau ventromédian de l'hypothalamus. *C. R. Hebd. Seances Acad. Sci. 270,* 2362-2365.

Baile, C. A., Herrera, M. G., and Mayer, J. (1970). Ventromedial hypothalamus and hyperphagia in hyperglycemic obese mice. *Am. J. Physiol. 218,* 857-863.

Bailey, P., and Bremer, F. (1921). Experimental diabetes insipidus. *Arch. Intern. Med. 28,* 773-803.

Balagura, S., and Devenport, L. D. (1970). Feeding patterns of normal and ventromedial hypothalamic lesioned male and female rats. *J. Comp. Physiol. Psychol. 71*, 357-364.

Balagura, S., and Harrell, L. E. (1974). Lateral hypothalamic syndrome: its modification by obesity and leanness. *Physiol. Behav. 12*, 897-899.

Balagura, S., Wilcox, R. H., and Coscina, D. V. (1969). The effect of diencephalic lesions on food intake and motor activity. *Physiol. Behav. 4*, 629-633.

Balagura, S., Harrell, L., and Ralph, J. (1973). Glucodynamic hormones modify the recovery period after lateral hypothalamic lesions. *Science 182*, 59-60.

Balinska, H. (1963). Food preference and conditioned reflex type II activity in dynamic hyperphagic rabbits. *Acta Biol. Exp. 23*, 33-44.

Balinska, H., and Brutkowski, S. (1964). Extinction of food-reinforced responses after medial or lateral hypothalamic lesions. *Acta Biol. Exp. 24*, 213-217.

Balinska, H., and Brutkowski, S. (1967). The participation of the hypothalamus in food-reinforced performance and inhibition. *Acta Biol. Exp. (Warsaw) 27*, 289-295.

Bartels, M. (1906). Verber plattenepithelgeschwultse der hypophysengegend (des infundibulums). *Z. Augenheilkunde 16*, 407-438.

Bates, M. W., Mayer, J., and Nauss, S. F. (1955a). Fat metabolism in three forms of experimental obesity: acetate incorporation. *Am. J. Physiol. 180*, 304-308.

Bates, M. W., Mayer, J., and Nauss, S. F. (1955b). Fat metabolism in three forms of experimental obesity: Fatty acid turnover. *Am. J. Physiol. 180*, 309-312.

Bates, M. W., Zomzely, C., and Mayer, J. (1955c). Fat metabolism in three forms of experimental obesity. IV. 'Instantaneous' rates of lipogenesis in vivo. *Am. J. Physiol. 181*, 187-190.

Beatty, W. W. (1973). Influence of type of reinforcement on operant responding by rats with ventromedial lesions. *Physiol. Behav. 10*, 841-846.

Beatty, W. W., O'Briant, D. A., and Vilberg, T. R. (1975a). Effects of ovariectomy and estradiol injections on food intake and body weights in rats with ventromedial hypothalamic lesions. *Pharmacol. Biochem. Behav. 3*, 539-544.

Beatty, W. W., Vilberg, T. R., Shirk, T. S., and Siders, W. A. (1975b). Pretraining: effects on operant responding for food, frustration, and reactiveness to food-related ones in rats with VMH lesions. *Physiol. Behav. 15*, 577-584.

Becker, E. E., and Kissileff, H. R. (1974). Inhibitory controls of feeding by the ventromedial hypothalamus. *Am. J. Physiol. 226*, 383-396.

Beltt, B. M., and Keesey, R. E. (1975). Hypothalamic map of stimulation current thresholds for inhibition of feeding in rats. *Am. J. Physiol. 229*, 1124-1133.

Bernardis, L. L. (1972). Hypophagia, hypodipsia, and hypoactivity following dorsomedial hypothalamic lesions. *Physiol. Behav. 8*, 1151-1164.

Bernardis, L. L., and Frohman, L. A. (1970). Effect of lesion size in the ventromedial hypothalamus on growth hormone and insulin levels in the weanling rat. *Neuroendocrinology 6*, 319-328.

Bernardis, L. L., and Goldman, J. K. (1972). Effect of hypothalamic lesion localization and size on metabolic alterations in weanling rat adipose tissue. *J. Neuro-Visc. Relations 32*, 253-269.

Bernardis, L. L., and Schnatz, J. D. (1971). Localization in the ventromedial hypothalamic nuclei of an area affecting plasma triglyceride and cholesterol levels. *J. Neuro-Visc. Relations 32*, 90-103.

Bernardis, L. L., and Skelton, F. R. (1965). Failure to demonstrate a functional connection between the adrenal gland and ventromedial hypothalamic nuclei with regard to control of food intake. *Experientia 21,* 36-37.

Bernardis, L. L., and Skelton, F. R. (1966). Growth and obesity following ventromedial hypothalamic lesions placed in female rats at four different ages. *Neuroendocrinology 1,* 265-275.

Berthoud, H. R., and Jeanrenaud, B. (1977). Acute hypothalamic control of insulin secretion in the rat. Paper presented at the Sixth International Conference on the Physiology of Food and Fluid Intake, Paris, France.

Bierman, E. L., and Glomset, J. A. (1974). Disorders of lipid metabolism. In *Textbook of Endocrinology,* R. H. Williams (Ed.). Saunders, Philadelphia, pp. 890-937.

Biggart, J. H., and Alexander, G. L. (1939). Experimental diabetes insipidus. *J. Pathol. Bacteriol. 48,* 405-425.

Blass, E. M., and Kraly, F. S. (1974). Medial forebrain bundle lesions: specific loss of feeding to decreased glucose utilization in rats. *J. Comp. Physiol. Psychol. 86,* 679-692.

Boyle, P. C., and Keesey, R. E. (1975). Chronically reduced body weight in rats sustaining lesions of the lateral hypothalamus and maintained on palatable diets and drinking solutions. *J. Comp. Physiol. Psychol. 88,* 218-223.

Bramwell, B. (1888). *Intracranial Tumours,* Young J. Pentland.

Braun, J. J. (1975). Neocortex and feeding behavior in the rat. *J. Comp. Physiol. Psychol. 89,* 507-522.

Bray, G. A., and Gallagher, T. F. (1975). Manifestations of hypothalamic obesity in man: a comprehensive investigation of eight patients and a review of the literature. *Medicine 54,* 301-330.

Brobeck, J. R. (1946). Mechanism of the development of obesity in animals with hypothalamic lesions. *Physiol. Rev. 26,* 541-559.

Brobeck, J. R. (1957). Neural control of hunger, appetite, and satiety. *Yale J. Biol. Med. 29,* 566-574.

Brobeck, J. R., Tepperman, J., and Long, C. N. H. (1943). Experimental hypothalamic hyperphagia in the albino rat. *Yale J. Biol. Med. 15,* 831-853.

Brodoff, B. N., and Zeballos, G. (1970). Further studies on the effect of hypothalamic lesions in the sand rat (*Psammomys obesus*). *Diabetologia 6,* 366-370.

Brooks, C. McC. (1946a). A study of the respiratory quotient in experimental hypothalamic obesity. *Am. J. Physiol. 147,* 727-734.

Brooks, C. McC. (1946b). The relative importance of changes in activity in the development of experimentally produced obesity in the rat. *Am. J. Physiol. 147,* 708-716.

Brooks, C. M., and Lambert, E. F. (1946). A study of the effect of limitation of food intake and the method of feeding on the rate of weight gain during hypothalamic obesity in the albino rat. *Am. J. Physiol. 147,* 695-707.

Brooks, C. McC., Lambert, E. F., and Bard, P. (1942). Experimental production of obesity in the monkey (*Macaca mulatta*). *Fed. Proc. 1,* 11.

Brooks, C. M., Lockwood, R. A., and Wiggins, M. L. (1946). A study of the effect of hypothalamic lesions in the eating habits of the albino rat. *Am. J. Physiol. 147,* 735-741.

Brouwer, B. (1950). Positive and negative aspects of hypothalamic disorders. *J. Neurol. Neurosurg. Psychiat. 13,* 16-23.

Byers, S. O., and Friedman, M. (1973). Neurogenic hypercholesterolemia. III. Cholesterol synthesis, absorption, and clearance. *Am. J. Physiol. 226,* 1322-1326.

Cajal, S. Ramon y (1966). *Studies on the Diencephalon.* Thomas, Springfield, Ill.

Campbell, B. A., and Fibiger, H. C. (1970). Effects of insulin on spontaneous activity during food deprivation. *J. Comp. Physiol. Psychol. 71,* 341-346.

Carlisle, H. J., and Stellar, E. (1969). Caloric regulation and food preference in normal, hyperphagic, and aphagic rats. *J. Comp. Physiol. Psychol. 69,* 107-114.

Chikamori, K., Masuda, K., Izumi, H., Isaka, K., and Tezuka, U. (1977). Effect on vagotomy on hyperinsulinemia in obese rats with hypothalamic lesions. *Endocrinol. Japon. 24,* 251-258.

Chlouverakis, C., and Bernardis, L. L. (1972). Ventrolateral hypothalamic lesions in obese-hyperglycemic mice (*ob/ob*). *Diabetologia 8,* 179-184.

Chlouverakis, C., Bernardis, L. L., and Hojnicki, D. (1973). Ventromedial hypothalamic lesions in obese-hyperglycaemic mice (*ob/ob*). *Diabetologia 9,* 391-395.

Christophe, J., Jeanrenaud, B., Mayer, J., and Renold, A. E. (1961). Metabolism in vitro of adipose tissue in obese-hyperglycemic and goldthioglucose-treated mice. I. Metabolism of glucose. *J. Biol. Chem. 236,* 642-647.

Clark, G. (1942). Sexual behavior in rats with lesions in the anterior hypothalamus. *Am. J. Physiol. 137,* 746-749.

Clark, G., Magoun, H. W., and Ranson, S. W. (1939). Hypothalamic regulation of body temperature. *J. Neurophysiol. 2,* 61-80.

Code, C. F. (Ed.) (1967). Control of food and water intake. *Handbook Physiol.,* Sect. 6, *1.*

Cohn, C. (1963). Feeding frequency and body composition. *Annal. N. Y. Acad. Sci. 110,* 395-409.

Cohn, C., and Joseph, D. (1962). Influence of body weight and body fat on appetite of "normal" lean and obese rats. *Yale J. Biol. Med. 34,* 598-607.

Colby, J. J., and Smith, N. F. (1975). Ability of rats with ventromedial hypothalamic lesions to work: Effects of ambient temperature. *Physiol. Behav. 15,* 31-35.

Coleman, D. L., and Hummel, K. P. (1970). The effects of hypothalamic lesions in genetically diabetic mice. *Diabetologia 6,* 263-267.

Colin-Jones, D. G., and Himsworth, R. L. (1969). The secretion of gastric acid in response to a lack of metabolizable glucose. *J. Physiol. 202,* 97-109.

Collier, G., Hirsch, E., and Leshner, A. I. (1972). The metabolic cost of activity in activity-naive rats. *Physiol. Behav. 8,* 881-884.

Connors, M. H., and Sheikolislam, B. M. (1977). Hypothalamic symptomatology and its relationship to diencephalic tumor in childhood. *Child's Brain. 3,* 31-36.

Corbit, J. D., and Stellar, E. (1964). Palatability, food intake, and obesity in normal and hyperphagic rats. *J. Comp. Physiol. Psychol. 58,* 63-67.

Coscina, D. V. (1977). Brain amines in hypothalamic obesity. In *Anorexia Nervosa,* R. A. Vigersky (Ed.). Raven, New York, pp. 97-107.

Cox, J. E., and Powley, T. L. (1977). Development of obesity in diabetic mice pair-fed with lean siblings. *J. Comp. Physiol. Psychol. 91,* 347-358.

Cross, B. A. (1964). The hypothalamus in mammalian homeostasis. *Symp. Soc. Exp. Biol. 18,* 157-193.

Cross, B. A. (1973). Unit responses in the hypothalamus. *Frontiers Neuroendocrinol. 30,* 133-171.

Davis, J. R. (1977). Decreased metabolic rates contingent upon lateral hypothalamic lesion-induced body weight losses in male rats. *J. Comp. Physiol. Psychol. 91,* 1019-1031.

De Laey, P. (1975). Hypothalamic and nutritional obesities in mice. A comparative study of growth and body composition. *Arch. Internatl. Pharmacodyn. Ther. 216*, 288-314.

De Laey, P., Dent, C., Terry, A. C., and Quinn, E. H. (1974). Goldthioglucose, its toxicity, and efficacy in Charles River mice. *Arch. Internatl. Pharmacodyn. Ther. 211*, 341-362.

De Laey, P., Dent, C., Terry, A. C., and Quinn, E. H. (1975). Food and water intake in gold thioglucose-induced obese Charles River mice. *Arch. Internatl. Pharmacodyn. Ther. 213*, 145-162.

Devenport, L. D., and Balagura, S. (1971). Lateral hypothalamus: reevaluation of function in motivated feeding behavior. *Science 172*, 744-746.

Ellison, G. D. (1968). Appetitive behavior in rats after circumsection of the hypothalamus. *Physiol. Behav. 3*, 221-226.

Ellison, G. D., Sorenson, C. A., and Jacobs, B. L. (1970). Two feeding syndromes following surgical isolation of the hypothalamus in rats. *J. Comp. Physiol. Psychol. 70*, 173-188.

Epstein, A. N. (1959). Supression of eating and drinking by amphetamine and other drugs in normal and hyperphagic rats. *J. Comp. Physiol. Psychol. 52*, 37-45.

Epstein, A. N. (1960). Reciprocal changes in feeding behavior produced by intrahypothalamic chemical injections. *Am. J. Physiol. 199*, 969-974.

Epstein, A. N. (1971). The lateral hypothalamic syndrome: its implications for the physiological psychology of hunger and thirst. In *Progress in Physiological Psychology*, E. Stellar and J. Sprague (Eds.). Academic, New York, pp. 263-317.

Epstein, A. N., and Teitelbaum, P. (1967). Specific loss of the hypoglycemic control of feeding in recovered lateral rats. *Am. J. Physiology 213*, 1159-1167.

Epstein, A. N., Kissileff, H. R., and Stellar, E. (Eds.) (1973). *The Neuropsychology of Thirst: New Findings and Advances in Concepts*. Winston, New York.

Erdheim, J. (1904). Über Hypophysenganggeschwulste und Hirncholesteatome. *Sitzungsb. Akad. Wissensch. Wien Abt. III 113*, 537-726.

Fábry, P. (1967). Metabolic consequences of the pattern of food intake. *Handbook Physiol.*, Sect. 6, *3*, 31-49.

Falk, J. L. (1961). Comments of Dr. Teitelbaum's paper. In *Nebraska Symposium on Motivation*, Vol. 9, University of Nebraska Press, Omaha.

Feldman, S. E., Larsson, S., Dimick, M. K., and Lepkovsky, S. (1957). Aphagia in chickens. *Am. J. Physiol. 191*, 259-261.

Ferguson, N. B. L., and Keesey, R. E. (1975). Effect of a quinine-adulterated diet upon body weight maintenance in male rats with ventromedial hypothalamic lesions. *J. Comp. Physiol. Psychol. 80*, 478-488.

Franklin, K. B. J., and Herberg, L. J. (1974). Ventromedial syndrome: the rat's "finickiness" results from the obesity, not from the lesions. *J. Comp. Physiol. Psychol. 87*, 410-414.

Friedman, M. I. (1972). Effects of alloxan diabetes on hypothalamic hyperphagia and obesity. *Am. J. Physiol. 222*, 174-178.

Friedman, M., Elek, S. R., and Byers, S. O. (1969). Abolition of milieu-induced hyperlipidemia in the rat by electrolytic lesion in the anterior hypothalamus. *Proc. Soc. Exp. Biol. Med. 131*, 288-293.

Friedman, M., Byers, S. O., and Elek, S. (1972). Neurogenic hypercholesterolemia. II. Relationship to endocrine function. *Am. J. Physiol. 223,* 473-479.

Fröhlich, A. (1901). Ein Fall von Tumor der Hypophysis Cerebri ohne Akromegalie. *Wien. Klin. Rdsch. 15,* 883-886.

Frohman, L. A. (1971). The hypothalamus and metabolic control. In *Pathobiology Annual,* (Vol. 1), H. C. Ioachim (Ed.).

Frohman, L. A., and Bernardis, L. L. (1968). Growth hormone and insulin levels in weanling rats with ventromedial hypothalamic lesions. *Endocrinology 82,* 1125-1132.

Frohman, L. A., Bernardis, L. L., Schnatz, J. D., and Burek, L. (1969). Plasma insulin and triglyceride levels after hypothalamic lesions in weanling rats. *Am. J. Physiol. 216,* 1496-1501.

Frohman, L. A., Goldman, J. K., Schnatz, J. D., and Bernardis, L. L. (1971). Hypothalamic obesity in the weanling rat: effect of diet upon hormonal and metabolic alterations. *Metabolism 20,* 501-512.

Frohman, L. A., Goldman, J. K., and Bernardis, L. L. (1972a). Metabolism of intravenously injected ^{14}C-glucose in weanling rats with hypothalamic obesity. *Metabolism 21,* 799-805.

Frohman, L. A., Goldman, J. K., and Bernardis, L. L. (1972b). Studies of insulin sensitivity in vivo in weanling rats with hypothalamic obesity. *Metabolism 21,* 1133-1142.

Fulton, J. F. (Ed.). (1940). *The Hypothalamus.* Williams & Wilkins, Baltimore.

Gallinek, A. (1962). The Kleine-Levin syndrome: hypersomia, bulimia, and abnormal mental states. *World Neurol. 3,* 235-243.

Gladfelter, W. E. (1971). Hypothalamic lesions, general activity, and pituitary functioning in the rat. *Am. J. Physiol. 221,* 161-165.

Gladfelter, W. E., and Brobeck, J. R. (1962). Decreased spontaneous locomotor activity in the rat induced by hypothalamic lesions. *Am. J. Physiol. 203,* 811-817.

Glavcheva, L., Rozkowska, E., and Fonberg, E. (1970). The effect of lateral hypothalamic lesions on gastric motility in dogs. *Acta Neurobiol. Exp. 30,* 279-293.

Glick, S. D., and Greenstein, S. (1972). Facilitation of survival following lateral hypothalamic damage by prior food and water deprivation. *Psychonomic Sci. 28,* 163-164.

Gold, R. M. (1970). Hypothalamic hyperphagia: males get just as fat as females. *J. Comp. Physiol. Psychol. 71,* 347-356.

Gold, R. M. (1973). Hypothalamic obesity: the myth of the ventromedial nucleus. *Science 182,* 488-490.

Gold, R. M., and Kapatos, G. (1975). Delayed hyperphagia and increased body length after hypothalamic knife cuts in weanling rats. *J. Comp. Physiol. Psychol. 88,* 202-209.

Gold, R. M., Quackenbush, P. M., and Kapatos, G. (1972). Obesity following combination of rostrolateral to VMH cut and contralateral mammillary area lesion. *J. Comp. Physiol. Psychol. 79,* 210-218.

Gold, R. M., Sumpier, G., Ueberacher, H. M., and Kapatos, G. (1975). Hypothalamic hyperphagia despite imposed duirnal or nocturnal feeding and drinking rhythms. *Physiol. Behav. 14,* 861-865.

Gold, R. M., Jones, A. P., Sawchenko, P. E., and Kapatos, G. (1977). Paraventricular area: critical focus of a longitudinal neurocircuitry mediating food intake. *Physiol. Behav. 18,* 1111-1119.

Goldman, J. K., and Bernardis, L. L. (1975). Gluconeogenesis in weanling rats with hypothalamic obesity. *Horm. Metab. Res. 7,* 148-152.

Goldman, J. K., Schnatz, J. D., Bernardis, L. L., and Frohman, L. A. (1970). Adipose tissue metabolism of weanling rats after destruction of ventromedial hypothalamic nuclei: Effect of hypophysectomy and growth hormone. *Metabolism 19,* 995-1005.

Goldman, J. K., Bernardis, L. L., and Frohman, L. A. (1972a). Insulin responsiveness in vitro of diaphragm and adipose tissue from weanling rats with hypothalamic obesity. *Horm. Metab. Res. 4,* 328-331.

Goldman, J. K., Schnatz, J. D., Bernardis, L. L., and Frohman, L. A. (1972b). Effects of ventromedial hypothalamic destruction in rats with preexisting streptozotocin-induced diabetes. *Metabolism 21,* 132-136.

Goldman, J. K., Bernardis, L. L., and Frohman, L. A. (1974). Food intake in hypothalamic obesity. *Am. J. Physiol. 227,* 88-91.

Graef, I., Negrin, J., and Page, I. H. (1944). The development of hepatic cirrhosis in dogs after hypophysectomy. *Am. J. Pathol. 20,* 823-855.

Graff, H., and Stellar, E. (1962). Hyperphagia, obesity, and finickiness. *J. Comp. Physiol. Psychol. 55,* 418-424.

Grijalva, C. V., Lindholm, E., Schallert, T., and Bicknell, E. J. (1976). Gastric pathology and aphagia following lateral hypothalamic lesions in rats: effects of preoperative weight reduction. *J. Comp. Physiol. Psychol. 90,* 505-519.

Gutstein, W. H., Schneck, D. J., and Appleton, H. (1969). Mechanism of plasma lipid increases following brain stimulation. *Metabolism 18,* 300-310.

Gutstein, W. H., Schneck, D. J., Farrell, G., and Long, W. (1970). Hypothalamically induced hyperlipidemia: protection by pentaerythritol tetranitrate. *Metabolism 19,* 230-237.

Haessler, H. A., and Crawford, J. D. (1967). Fatty acid composition and metabolic activity of depot fat in experimental obesity. *Am. J. Physiol. 213,* 255-261.

Hainsworth, F. R., and Epstein, A. N. (1966). Severe impairment of heat-induced saliva-spreading in rats recovered from lateral hypothalamic lesions. *Science 153,* 1255-1257.

Hales, C. N., and Kennedy, G. C. (1964). Plasma glucose, non-esterified fatty acid and insulin concentrations in hypothalamic-hyperphagic rats. *Biochem. J. 90,* 620-624.

Hamilton, C. L. (1963). Interactions of food intake and temperature regulation in the rat. *J. Comp. Physiol. Psychol. 56,* 476-488.

Hamilton, C. L., and Brobeck, J. R. (1964). Hypothalamic hyperphagia in the monkey. *J. Comp. Physiol. Psychol. 57,* 271-278.

Hamilton, C. L., and Rabinowitz, J. L. (1976). Weight reduction and serum insulin levels in hypothalamic obese monkeys. *J. Med. Primatol. 5,* 276-283.

Hamilton, C. L., Kuo, P. T., and Feng, L. Y. (1972). Experimental production of syndrome of obesity, hyperinsulinemia, and hyperlipidemia in monkeys. *Proc. Soc. Exp. Biol. Med. 140,* 1005-1008.

Han, P. W. (1967). Hypothalamic obesity in rats without hyperphagia. *Trans. New York Acad. Sci. 30,* 229-243.

Han, P. W. (1968). Energy metabolism of tube-fed hypophysectomized rats bearing hypothalamic lesions. *Am. J. Physiol. 215,* 1343-1350.

Han, P. W., and Frohman, L. A. (1970). Hyperinsulinemia in tube-fed hypophysectomized rats bearing hypothalamic lesions. *Am. J. Physiol. 219,* 1632-1636.

Han, P. W., and Liu, A. C. (1966). Obesity and impaired growth of rats force fed 40 days after hypothalamic lesions. *Am. J. Physiol. 211,* 229-231.

Han, P. W., and Young, T. K. (1963). Adiposity and intestinal motility in hypothalamic hyperphagic rats. *Chinese J. Physiol. 19,* 99-105.

Han, P. W., Lin, C. H., Chu, K. C., Mu, J. Y., and Liu, A. C. (1965). Hypothalamic obesity in weanling rats. *Am. J. Physiol. 209,* 627-631.

Han, P. W., Yu, Y. K., and Chow, S. L. (1970). Enlarged pancreatic islets of tube-fed hypophysectomized rats bearing hypothalamic lesions. *Am. J. Physiol. 218,* 769-771.

Han, P. W., Feng, L. Y., and Kuo, P. T. (1972). Insulin sensitivity of pair-fed, hyperlipemic, hyperinsulinemic, obese-hypothalamic rats. *Am. J. Physiol. 223,* 1206-1209.

Harrell, E. H., and Remley, N. R. (1973). The immediate development of behavioral and biochemical changes following ventromedial hypothalamic lesions in rats. *Behav. Biol. 9,* 49-63.

Harrell, L. E., Decastro, J. M., and Balagura, S. (1975). A critical evaluation of body weight loss following lateral hypothalamic lesions. *Physiol. Behav. 15,* 133-136.

Haymaker, W., and Anderson, E. (1955). Disorders of the hypothalamus and pituitary gland. In *Clinical Neurology,* A. B. Baker (Ed.). Hoeber, p. 1160-1215.

Haymaker, W., Anderson, E., and Nauta, W. J. H. (Eds.) (1969). *The Hypothalamus.* Thomas, Springfield, Ill.

Heaney, R. P., Eliel, L. P., Joel, W., and Stout, H. (1954). Hyperphagia, obesity and duodenal ulcer associated with hypothalamic leukemic infiltration. *J. Clin. Endocrinol. Metab. 14,* 829-830.

Hefco, V., Rotenberg, P., and Jitarin, P. (1975). Carbohydrate metabolism in rats after partial or total isolation of the medral hypothalamus. *Endokrinologie 65,* 191-197.

Heldenberg, D., Tamir, I., Ashner, M., and Werbin, B. (1972). Hyperphagia, obesity, and diabetes insipidus due to hypothalamic lesion in a girl. *Helv. Paediatr. Acta (Basel) 27,* 489-494.

Heinbecker, P., White, H. L., and Rolf, D. (1944). Experimental obesity in the dog. *Am. J. Physiol. 141,* 549-565.

Hetherington, A. W. (1941). The relation of various hypothalamic lesions to adiposity and other phenomena in the rat. *Am. J. Physiol. 133,* 326-327.

Hetherington, A. W. (1943). The production of hypothalamic obesity in rats already displaying chronic hypopituitarism. *Am. J. Physiol. 140,* 89-92.

Hetherington, A. W. (1944). Non-production of hypothalamic obesity in the rat by lesions rostral or dorsal to the ventro-medial hypothalamic nuclei. *J. Comp. Neurol. 80,* 33-45.

Hetherington, A. W., and Ranson, S. W. (1939). Experimental hypothalamicohypophyseal obesity in the rat. *Proc. Soc. Exp. Biol. Med. 41,* 465-466.

Hetherington, A. W., and Ranson, S. W. (1940). Hypothalamic lesions and adiposity in the rat. *Anat. Rec. 78,* 149-172.

Hetherington, A. W., and Ranson, S. W. (1942a). Effect of early hypophysectomy on hypothalamic obesity. *Endocrinology 31,* 30-34.

Hetherington, A. W., and Ranson, S. W. (1942b). The spontaneous activity and food intake of rats with hypothalamic lesions. *Am. J. Physiol. 136,* 609-617.

Hetherington, A. W., and Ranson, S. W. (1942c). The relation of various hypothalamic lesions to adiposity in the rat. *J. Comp. Neurol. 76,* 475-499.

Hetherington, A. W., and Weil, A. (1940). The lipoid, calcium, phosphorous, and iron content of rats with hypothalamic and hypophyseal damage. *Endocrinology 26,* 723-727.

Himsworth, R. L. (1970). Hypothalamic control of adrenaline secretion in response to insufficient glucose. *J. Physiol. 206*, 411-417.

Himsworth, R. L., Carmel, P. W., and Frantz, A. G. (1972). The location of the chemoreceptor controlling growth hormone secretion during hypoglycemia in primates. *Endocrinology 91*, 217-226.

Hinman, D. J., and Griffith, D. R. (1973). Effects of ventromedial hypothalamic lesions on thyroid secretion rate in rats. *Horm. Metab. Res. 5*, 48-50.

Hirsch, J., and Han, P. W. (1969). Cellularity of rat adipose tissue: effect of growth, starvation, and obesity. *J. Lipid Res. 10*, 77-82.

Hoebel, B. G. (1976). Satiety: hypothalamic stimulation, anorectic drugs, and neurochemical substrates. In *Hunger: Basic Mechanisms and Clinical Applications*, D. Novin, W. Wyrwicka, and G. Bray (Eds.). Raven, New York, pp. 33-50.

Hoebel, B. G., and Teitelbaum, P. (1966). Weight regulation in normal and hypothalamic hyperphagic rats. *J. Comp. Physiol. Psychol. 61*, 189-193.

Holm, H., Hustvedt, B. E., and Løvø, A. (1973). Protein metabolism in rats with ventromedial hypothalamic lesions. *Metabolism 22*, 1377-1387.

Hongslo, C. F., Hustvedt, B. E., and Løvø, A. (1974). Insulin sensitivity in rats with ventromedial hypothalamic lesions. *Acta Physiol. Scand. 90*, 757-763.

Houpt, K. A., and Gold, R. M. (1975). Glucoprivic (2DG) eating in rats despite knife cut induced hyperphagia. *Pharmacol. Biochem. Behav. 3*, 583-588.

Hustvedt, B. E., and Løvø, A. (1972). Correlation between hyperinsulinemia and hyperphagia in rats with ventromedial hypothalamic lesions. *Acta Physiol. Scand. 84*, 29-33.

Hustvedt, B. E., and Løvø, A. (1973). Rapid effect of ventromedial hypothalamic lesions on lipogenesis in rats. *Acta Physiol. Scand. 87*, 28A-29A.

Hustvedt, B. E., Løvø, A., and Reichl, D. (1976). The effect of ventromedial hypothalamic lesions on metabolism and insulin secretion in rats on a controlled feeding regimen. *Nutr. Metab. 20*, 264-271.

Ingram, W. R. (1952). Brain stem mechanisms in behavior. *Electroencephal. Clin. Neurophysiol. 4*, 397-406.

Ingram, W. R. (1956). The hypothalamus. *Ciba Clin. Symp. 8*, 117-156.

Ingram, W. R., Barris, R. W., and Ranson, S. W. (1936). Catalepsy. An experimental study. *Arch. Neurol. Psychiat. 35*, 1175-1197.

Inoue, S., and Bray, G. A. (1977). The effects of subdiaphragmatic vagotomy in rats with ventromedial hypothalamic obesity. *Endocrinol. 100*, 108-114.

Inoue, S., Bray, G. A., and Mullen, Y. S. (1977a). Effect of transplantation of pancreas on development of hypothalamic obesity. *Nature (London) 266*, 742-744.

Inoue, S., Campfield, L. A., and Bray, G. A. (1977b). Comparison of metabolic alterations in hypothalamic and high fat diet-induced obesity. *Am. J. Physiol. 233*, R162-R168.

Johnson, P. R., and Hirsch, J. (1972). Cellularity of adipose depots in six strains of genetically obese mice. *J. Lipid Res. 13*, 2-11.

Johnson, P. R., Zucker, L. M., Cruce, J. A. F., and Hirsch, J. (1971). Cellularity of adipose depots in the genetically obese Zucker rat. *J. Lipid Res. 12*, 706-714.

Joseph, S. A., and Knigge, K. M. (1968). Effects of VMH lesions in adult and newborn guinea pigs. *Neuroendocrinology 3*, 309-331.

Kadekaro, M., Timo-laria, C., and Valle, L. E. R. (1975). Neural systems responsible for the gastric secretion provoked by 2-deoxy-D-glucose cytoglucopoenia. *J. Physiol.* 252, 565-584.

Kagan, H. (1958). Anorexia and severe inanition associated with a tumour involving the hypothalamus. *Arch. Dis. Child.* 33, 257-260.

Kakolewski, J. W., Deaux, E., Christensen, J., and Case, B. (1971). Diurnal patterns in water and food intake and body weight changes in rats with hypothalamic lesions. *Am. J. Physiol.* 221, 711-718.

Kamalian, N., Keesey, R. E., and Zurhein, G. M. (1975). Lateral hypothalamic demyelination and cachexia in a case of "malignant" multiple sclerosis. *Neurology* 25, 25-30.

Karakash, C., Hustvedt, B. E., Løvø, A., Le Marchand, Y., and Jeanrenaud, B. (1977). Consequences of ventromedial hypothalamic lesions on metabolism of perfused rat liver. *Am. J. Physiol.* 232, E286-E293.

Kasemsri, S., Bernardis, L. L., and Schnatz, J. D. (1972). Fat mobilization in adipose tissue of weanling rats with hypothalamic obesity. *Hormones* 3, 97-104.

Katz, J., Wals, P., Golden, S., Goldman, J. K., and Bernardis, L. L. (1977). Lipogenesis by hepatocytes of rats with hypothalamic obesity. *Horm. Metab. Res.* 9, 59-63.

Keesey, R. E., and Boyle, P. C. (1973). Effects of quinine adulteration upon body weight of LH-lesioned and intact male rats. *J. Comp. Physiol. Psychol.* 84, 38-46.

Keesey, R. E., and Powley, T. L. (1968). Enhanced lateral hypothalamic reward sensitivity following septal lesions in the rat. *Physiol. Behav.* 3, 557-562.

Keesey, R. E., and Powley, T. L. (1975). Hypothalamic regulation of body weight. *Am. Sci.* 63, 558-565.

Keesey, R. E., Boyle, P. C., Kenmitz, J. W., and Mitchel, J. S. (1976a). The role of the lateral hypothalamus in determining the body weight set point. In *Hunger: Basic Mechanisms and Clinical Implications,* D. Novin, W. Wyrwicks, and G. Bray (Eds.). Raven, New York, pp. 281-296.

Keesey, R. E., Powley, T. L., and Kenmitz, J. W. (1976b). Prolonged lateral hypothalamic anorexia by tube-feeding. *J. Comp. Physiol. Psychol.* 17, 367-371.

Keller, A. D., and Noble, W. (1935). Adiposity with normal sex function following extirpation of the posterior lobe of the hypophysis in the dog. *Am. J. Physiol.* 113, 79-80.

Keller, A. D., and Noble, W. (1936). Further observations on enhanced appetite with resultant adiposity following removal of the posterior lobe of the hypophysis. *Am. J. Physiol.* 116, 90-91.

Keller, A. D., Hare, W. K., and D'Amour, M. C. (1932). Ulceration in digestive tract following experimental lesions in brain stem. *Proc. Soc. Exp. Biol. Med.* 30, 772-775.

Keller, A. D., Noble, W., and Hamilton, Jr., J. W. (1936). Effects of anatomical separation of the hypophysis from the hypothalamus in the dog. *Am. J. Physiol.* 117, 467-473.

Kemnitz, J. W., Goy, R. W., and Keesey, R. E. (1977). Effects of gonadectomy on hypothalamic obesity in male and female rats. *Internat. J. Obesity* 1, 259-270.

Kennedy, G. C. (1950). The hypothalamic control of food intake in rats. *Proc. R. Soc. Ser. B 137,* 535-549.

Kennedy, G. C. (1953). The role of depot fat in the hypothalamic control of food intake in the rat. *Proc. R. Soc. Ser. B 140,* 578-592.

Kennedy, G. C. (1957). The effect of age on the somatic and visceral responses to overnutrition in the rat. *J. Endocrinol.* 15, 14-24.

Kennedy, G. C. (1969). The relation between the central control of appetite, growth, and sexual maturation. *Guys Hosp. Rep. 118*, 315-327.

Kennedy, G. C., and Parker, R. A. (1963). The islets of Langerhans in rats with hypothalamic obesity. *Lancet 2*, 981-982.

Kennedy, G. C., and Parrott, D. M. V. (1958). The effect of increased appetite and of insulin on growth in the hypophysectomized rat. *J. Endocrinol. 17*, 161-166.

Kennedy, G. C., and Pearce, W. M. (1958). The relation between liver growth and somatic growth in the rat. *J. Endocrinol. 17*, 149-157.

Kent, M. A., and Peters, R. H. (1973). Effects of ventromedial hypothalamic lesions on hunger-motivated behavior in rats. *J. Comp. Physiol. Psychol. 83*, 92-97.

Khalaf, F. (1969). Hyperphagia and aphagia in swine with induced hypothalamic lesions. *Res. Vet. Sci. 10*, 514-517.

King, B. M., and Gaston, M. G. (1973). The effects of pretraining on the bar-pressing performance of VMH-lesioned rats. *Physiol. Behav. 11*, 161-166.

King, B. M., and Grossman, S. P. (1977). Response to glucoprivic and hydrational challenges by normal and hypothalamic hyperphagic rats. *Physiol. Behav. 18*, 463-473.

King, J. M., and Cox, V. C. (1973). The effects of estrogens on food intake and body weight following ventromedial hypothalamic lesions. *Physiol. Psychol. 1*, 261-264.

Kirschbaum, W. R. (1951). Excessive hunger as a symptom of cerebral origin. *J. Nerv. Mental Dis. 113*, 95-114.

Kissileff, H. R. (1970). Free feeding in normal and "recovered lateral" rats monitored by a pellet-detecting eatometer. *Physiol. Behav. 5*, 163-173.

Kissileff, H. R., and Epstein, A. N. (1969). Exaggerated prandial drinking in the "recovered lateral" rat without saliva. *J. Comp. Physiol. Psychol. 67*, 301-308.

Komorowski, J. M. (1977). Blood sugar and immunoreactive insulin in women with hypothalamic, maternal and simple obesity. Part I. *Endokrinologie 70*, 182-191.

Kraly, F. S., and Blass, E. M. (1974). Motivated feeding in the absence of glucoprivic control of feeding in rats. *J. Comp. Physiol. Psychol. 87*, 801-807.

Krauss, R. M., and Mayer, J. (1965). Influence of protein and amino acids on food intake in the rat. *Am. J. Physiol. 209*, 479-483.

Kuenzel, W. J. (1972). Dual hypothalamic feeding system in a migratory bird, *Zonotrichia albicollis*. *Am. J. Physiol. 223*, 1138-1141.

Kuenzel, W. J., and Helms, C. W. (1967). Obesity produced in a migratory bird by hypothalamic lesions. *Bioscience 17*, 395-396.

Kulkosky, P. J., Breckenridge, C., Drinsky, R., and Woods, S. C. (1976). Satiety elicited by the C-terminal octapeptide of cholecystokinin-pancreozymin in normal and VMH-lesioned rats. *Behav. Biol. 18*, 227-234.

Larkin, R. P. (1975). Effect of ventromedial hypothalamic procaine injections on feeding, lever pressing, and other behavior in rats. *J. Comp. Physiol. Psychol. 89*, 1100-1108.

Lazaris, Y. A., Kasatkin, Y. N., Goldberg, R. S., and Smirnova, L. K. (1976). Role of the incretory function of the pancreatic islets in the mechanisms of development of hypothalamic obesity. Translated from *Byull. Eksper. Biol. Med. 82*, 1185-1187.

Leach, L. R., and Braun, J. J. (1976). Dissociation of gustatory and weight regulatory responses to quinine following lateral hypothalamic lesions. *J. Comp. Physiol. Psychol. 90*, 978-985.

Leibowitz, S. F. (1976). Brain catecholaminergic mechanisms for the control of hunger. In *Hunger: Basic Mechanisms and Clinical Applications*, D. Novin, W. Wyrwicka, and G. A. Bray (Eds.). Raven, New York, pp. 1-19.

Lepkovsky, S. (1973). Hypothalamic-adipose tissue interrelationships. *Fed. Proc. 32*, 1705-1708.

Lepkovsky, S., and Yasuda, M. (1966). Hypothalamic lesions growth, and body composition of male chickens. *Poultry Sci. 45*, 582-588.

Levine, M. S., Ferguson, N., Kreinick, C. J., Gustafson, J. W., and Schwartzbaum, J. S. (1971). Sensorimotor dysfunctions and aphagia and adipsia following pallidal lesions in rats. *J. Comp. Physiol. Psychol. 77*, 282-293.

Levison, M. J., Frommer, G. P., and Vance, W. B. (1973). Palatability and caloric density as determinants of food intake in hyperphagic and normal rats. *Physiol. Behav. 10*, 455-462.

Levitt, D. R., and Teitelbaum, P. (1975). Somnolence, akinesia, and sensory activation of motivated behavior in the lateral hypothalamic syndrome. *Proc. Natl. Acad. Sci. U. S. A. 72*, 2819-2823.

Lewinska, M. K. (1964). The effect of food deprivation on blood sugar level, food intake, and conditioning in rabbits with hypothalamic lesions. *Acta Biol. Exp. (Warsaw) 24*, 219-246.

Liebelt, R. A., and Perry, J. H. (1967). Action of goldthioglucose on the central nervous system. *Handbook Physiol.*, Sect. 6, *3*, 271-285.

Liebelt, R. A., Dear, W., and Guillemin, R. (1961). Goldthioglucose-induced hypothalamic lesion and ACTH release. *Proc. Soc. Exp. Biol. Med. 108*, 377-380.

Lindholm, E., Shumway, G. S., Grijalva, C. V., Schallert, T., and Ruppel, M. (1975). Gastric pathology produced by hypothalamic lesions in rats. *Physiol. Behav. 14*, 165-169.

Liu, C. M., and Yin, T. H. (1974). Caloric compensation to gastric loads in rats with hypothalamic hyperphagia. *Physiol. Behav. 13*, 231-238.

Lorden, J., Oltmans, G. A., and Margules, D. L. (1976). Central noradrenergic neurons: differential effects on body weight of electrolytic and 6-hydroxydopamine lesions in rats. *J. Comp. Physiol. Psychol. 90*, 144-155.

Lotter, E. C., and Woods, S. C. (1977). Injections of insulin and changes of body weight. *Physiol. Behav. 18*, 293-297.

Louis-Sylvestre, J. (1971). Increased insulin secretion in response to the food stimulus as a primary effect of ventromedial hypothalamic lesions in rats. Paper presented at the Fourth International Conference on the Regulation of Food and Water Intake, Cambridge, England.

Louis-Sylvestre, J. (1976). Preabsorptive insulin release and hypoglycemia in rats. *Am. J. Physiol. 230*, 56-60.

Løvø, A., and Hystvedt, B. E. (1973). Correlation between altered acetate utilization and hyperphagia in rats with ventromedial hypothalamic lesions. *Metabolism 22*, 1459-1465.

McBurney, P. L., Liebelt, R. A., and Perry, J. H. (1965). Quantitative relationships between food intake, lipid deposition, and hypothalamic damage in goldthioglucose obesity. *Texas Rep. Biol. Med. 23*, 737-752.

McHugh, P. R., Gibbs, J., Falasco, J. D., Moran, T., and Smith, G. P. (1975). Inhibitions on feeding examined in rhesus monkeys with hypothalamic disconnexions. *Brain 98*, 441-454.

Marks, H. E., and Miller, C. R. (1972). Development of hypothalamic obesity in the male golden hamster (*Mesocricetus auratus*) as a function of food preference. *Psychon. Sci.* 27, 263-265.

Marks, H. E., and Remley, N. R. (1972). The effects of type of lesion and percentage body weight loss on measures of motivated behavior in rats with hypothalamic lesions. *Behav. Biol.* 7, 95-111.

Marshall, J. F. (1975). Increased orientation to sensory stimuli following medial hypothalamic damage in rats. *Brain Res.* 86, 373-387.

Marshall, J. F., and Teitelbaum, P. (1973). A comparison of the eating in response to hypothermic and glucoprivic challenges after nigral 6-hydroxydopamine and lateral hypothalamic electrolytic lesions in rats. *Brain Res.* 55, 229-233.

Marshall, J. F., and Teitelbaum, P. (1974). Further analysis of sensory inattention following lateral hypothalamic damage in rats. *J. Comp. Physiol. Psychol.* 86, 375-395.

Marshall, J. F., Turner, B. H., and Teitelbaum, P. (1971). Sensory neglect produced by lateral hypothalamic damage. *Science* 174, 523-525.

Marshall, J. F., Richardson, J. S., and Teitelbaum, P. (1974). Nigrostriatal bundle damage and the lateral hypothalamic syndrome. *J. Comp. Physiol. Psychol.* 87, 808-830.

Martin, J. M., Konijnendijk, W., and Bouman, P. R. (1974). Insulin and growth hormone secretion in rats with ventromedial hypothalamic lesions maintained on restricted food intake. *Diabetes* 23, 203-208.

May, K. K., and Beaton, J. R. (1968). Hyperphagia in the insulin-treated rat. *Proc. Soc. Exp. Biol. Med.* 127, 1201-1204.

Mayer, J., and Zighera, C. Y. (1954). The multiple etiology of obesity: production of two types of obesity in littermate mice. *Science* 119, 96-97.

Mayer, J., Marshall, N. B., Vitale, J. J., Christensen, J. H., Mashayekhi, M. B., and Stare, F. J. (1954). Exercise, food intake and body weight in normal rats and genetically obese adult mice. *Am. J. Physiol.* 177, 544-548.

Mayer, J., French, R. G., Zighera, C. F., and Barrnett, R. J. (1955). Hypothalamic obesity in the mouse: production, description, and metabolic characteristics. *Am. J. Physiol.* 182, 75-83.

Miselis, R., and Epstein, A. N. (1971). Preoptic-hypothalamic mediation of feeding induced by cerebral glucoprivation. *Am. Zool.* 11, Abstr. 31.

Miller, N. E., Bailey, C. J., and Stevenson, J. A. F. (1950). Decreased "hunger" but increased food intake resulting from hypothalamic lesions. *Science* 112, 256-259.

Mitchel, J. S., and Keesey, R. E. (1974). The effects of lateral hypothalamic lesions and castration upon the body weight and composition of male rats. *J. Behav. Biol.* 11, 69-82.

Mitchel, J. S., and Keesey, R. E. (1977). Defense of a lowered weight maintenance level by lateral hypothalamically lesioned rats: evidence from a restriction-refeeding regimen. *Physiol. Behav.* 18, 1121-1125.

Mitchell, J. A., Smyrl, R., Hutchins, M., Schindler, W. J., and Critchlow, V. (1972). Plasma growth hormone levels in rats with increased naso-anal lengths due to hypothalamic surgery. *Neuroendocrinology* 10, 31-45.

Mogenson, G. (1973). Hypothalamic limbic mechanisms in the control of water intake. In *The Neuropsychology of Thirst: New Findings and Advances in Concepts*, A. N. Epstein, H. R. Kissilef, and E. Stellar (Eds.). John Wiley & Sons, New York, pp. 119-143.

Montemurro, D. G. (1971). Inhibition of hypothalamic obesity in the mouse with diethylstilbestrol. *Canad. J. Physiol. Pharmacol. 49*, 554-558.

Montemurro, D. G., and Stevenson, J. A. F. (1957a). Body composition in hypothalamic obesity derived from estimations of body specific gravity and extra cellular fluid volume. *Metabolism 6*, 161-168.

Montemurro, D. G., and Stevenson, J. A. F. (1957b). Adipsia produced by hypothalamic lesions in the rat. *Canad. J. Biochem. Physiol. 35*, 31-37.

Montemurro, D. G., and Stevenson, J. A. F. (1960). Survival and body composition of normal and hypothalamic obese rats in acute starvation. *Am. J. Physiol. 19*, 757-761.

Mook, D. G., and Blass, E. (1970). Specific hungers in hyperphagic rats. *Psychon. Sci. 19*, 34-35.

Mook, D. G., Fisher, J. C., and Durr, J. C. (1975). Some endocrine influences on hypothalamic hyperphagia. *Horm. Behav. 6*, 65-79.

Moore, R. Y., and Eichler, V. B. (1976). Central neural mechanisms in duirnal rhythm regulation and neuroendocrine response to light. *Psychoneuroendocrinology 1*, 265-279.

Morgane, P. J. (1961). Medial forebrain bundle and "feeding centers" of the hypothalamus. *J. Comp. Neurol. 117*, 1-25.

Morgane, P. J. (Ed.) (1969). Neural regulation of food and water intake. *Ann. New York Acad. Sci. 157*, 531-1216.

Morgane, P. J. (1975). Anatomical and neurobiochemical bases of the central nervous control of physiological regulations and behavior. In *Neural Integration of Physiological Mechanisms and Behavior* (Stevenson Memorial Volume), G. Mogenson and F. Calaresu (Eds.). Univ. Toronto Press, pp. 24-67.

Morgane, P. J., and Jacobs, H. L. (1969). Hunger and satiety. *World Rev. Nutr. Diet. 10*, 100-213.

Morgane, P. J., and Stern, W. C. (1974). Chemical anatomy of brain circuits in relation to sleep and wakefulness. In *Advances in Sleep Research*, Vol. 1, E. Weitzman (Ed.). Spectrum Publications, pp. 1-131.

Morrison, S. D. (1968). The relationship of energy expenditure and spontaneous activity to the aphagia of rats with lesions in the lateral hypothalamus. *J. Physiol. 197*, 325-343.

Morrison, S. D., Barrnett, R. J., and Mayer, J. (1958). Localization of lesions in the lateral hypothalamus of rats with induced adipsia and aphagia. *Am. J. Physiol. 193*, 230-234.

Mrosovsky, N. (1974). Hypothalamic hyperphagia without plateau in ground squirrels. *Physiol. Behav. 12*, 259-264.

Mrosovsky, N. (1975). The amplitude and period of circannual cycles of body weight in golden-mantled ground squirrels with medial hypothalamic lesions. *Brain Res. 99*, 97-116.

Mu, J. Y., Yin, T. H., Hami Hon, C. L., and Brobeck, J. R. (1968). Variability of body fat in hyperphagic rats. *Yale J. Biol. Med. 41*, 133-142.

Mufson, E. J., and Wampler, R. S. (1972). Weight regulation with palatable food and liquids in rats with lateral hypothalamic lesions. *J. Comp. Physiol. Psychol. 80*, 382-392.

Müller, E. E., Pecile, A., Cocchi, D., and Olgiati, V. R. (1974). Hyperglycemic or feeding response to glucoprivation and hypothalamic glucoreceptors. *Am. J. Physiol. 226,* 1100-1109.

Myers, R. D., and Martin, G. E. (1973). 6-OHDA lesions of the hypothalamus: interaction of aphagia, food palatability, set-point for weight regulation, and recovery of feeding. *Pharmacol. Biochem. Behav. 1,* 329-345.

Nance, D. M. (1976). Sex differences in the hypothalamic regulation of feeding behavior in the rat. In *Advances in Psychobiology,* Vol 3, G. Newton and A. H. Riesen (Eds.). John Wiley & Sons, New York, pp. 75-123.

Nance, D. M., and Gorski, R. A. (1975). Neurohumoral determinants of sex differences in the hypothalamic regulation of feeding behavior and body weight in the rat. *Pharmacol. Biochem. Behav. 3* (Suppl. 1), 155-162.

Nasset, E. S., Ridley, P. T., and Schenk, E. A. (1967). Hypothalamic lesions related to ingestion of an imbalanced amino acid diet. *Am. J. Physiol. 213,* 645-650.

Nauta, W. J., and Haymaker, W. (1969). Hypothalamic nuclei and fiberconnections. In *The Hypothalamus,* W. Haymaker, E. Anderson, and W. J. Nauta (Eds.). Thomas, Springfield, Ill., pp. 136-210.

Newburgh, L. H. (1931). The cause of obesity. *J. Am. Med. Assoc. 97,* 1659-1663.

Nicolaïdis, S., and Meile, M. J. (1972). Cartographie des lésions hypothalamiques supprimant la réponse alimentaire aux injections intracardiaques du 2-déoxy-d-glucose. *J. Physiol. (Paris) 65,* 151A.

Novin, D., Bray, G., and Wrywicka, W. (Eds.). (1976). *Hunger: Basic Mechanisms and Clinical Implications,* Raven, New York.

Oomura, Y. (1973). Central mechanisms of feeding. *Adv. Biophys. 5,* 65-142.

Oomura, Y., Sugimori, M., Nakamura, T., and Yamada, Y. (1975). Contribution of electrophysiological techniques to the understanding of central control systems. In *Neural Investigations of Physiological Mechanisms and Behavior,* G. Mogenson and F. R. Calaresu (Eds.). Univ. Toronto Press, pp. 375-395.

Opsahl, C. A. (1977). Sympathetic nervous system involvement in the lateral hypothalamic lesion syndrome. *Am. J. Physiol. 232,* R128-R136.

Opsahl, C. A., and Powley, T. L. (1977). Body weight and gastric acid secretion in rats with subdiaphragmatic vagotomy and lateral hypothalamic lesions. *J. Comp. Physiol. Psychol. 91,* 1284-1296.

Palka, Y., Liebelt, R. A., and Critchlow, V. (1971). Obesity and increased growth following partial or complete isolation of ventromedial hypothalamus. *Physiol. Behav. 7,* 187-194.

Panksepp, J. (1971). Is satiety mediated by the ventromedial hypothalamus? *Physiol. Behav. 7,* 381-384.

Panksepp, J. (1973). Reanalysis of feeding patterns in the rat. *J. Comp. Physiol. Psychol. 82,* 78-94.

Panksepp, J. (1974). Hypothalamic regulation of energy balance and feeding behavior. *Fed. Proc. 33,* 1150-1165.

Panksepp, J. (1975). Central metabolic and humoral factors involved in the neural regulation of feeding. *Pharmacol. Biochem. Behav. 3* (Suppl. 1), 107-119.

Panksepp, J., and Dickinson, A. (1972). On the motivational deficits after medial hypothalamic lesions. *Physiol. Behav. 9,* 609-614.

Panksepp, J., and Nance, D. M. (1972). Insulin, glucose, and hypothalamic regulation of feeding. *Physiol. Behav. 9,* 447-451.

Panksepp, J., and Pilcher, C. W. T. (1973). Evidence for an adipokinetic mechanism in the ventromedial hypothalamus. *Experientia 29*, 793.

Paxinos, G. (1974). The hypothalamus: neural systems involved in feeding, irritability, aggression, and copulation in male rats. *J. Comp. Physiol. Psychol. 87*, 110-119.

Paxinos, G., and Bindra, D. (1973). Hypothalamic and midbrain neural pathways involved in eating, drinking, irritibility, aggression, and copulation in rats. *J. Comp. Physiol. Psychol. 82*, 1-14.

Peters, R. H. (1974). Effects of ventromedial hypothalamic lesions on restricted feeding behavior in rats. *Physiol. Behav. 12*, 761-766.

Peters, R. H., Sensenig, L. D., and Reich, M. J. (1973). Fixed-ratio performance following ventromedial hypothalamic lesions in rats. *Physiol. Psychol. 1*, 136-138.

Pfaff, D. W. (1969). Histological differences between ventromedial hypothalamic neurones of well fed and underfed rats. *Nature (London) 223*, 77-78.

Poirer, L. J., Mouren-Mathieu, A. M., and Richer, C. L. (1962a). Neuroanatomical study of obese and non-obese hypothalamic monkeys in relation to food intake, locomotor activity, and temperature regulation. *Canad. J. Biochem. Physiol. 40*, 1185-1193.

Poirer, L. J., Mouren-Mathieu, A. M., and Richer, C. L. (1962b). Obesity in the absence of absolute hyperphagia in monkeys with hypothalamic lesions. *Rev. Canad. Biol. 21*, 127-134.

Porter, J. H., and Allen, J. D. (1977). Food-motivated performance in rats with ventromedial hypothalamic lesions: effects of body weight, deprivation, and preoperative training. *Behav. Biol. 19*, 238-254.

Poschel, B. P. H. (1968). Do biological reinforcers act via the self-stimulation areas of the brain? *Physiol. Behav. 3*, 53-60.

Powley, T. L. (1970). Reduction of body weight set-point by lateral hypothalamic lesions: implications for an analysis of the lateral hypothalamic feeding syndrome. Unpublished doctoral dissertation, University of Wisconsin, Madison.

Powley, T. L. (1971). Hypothalamic feeding centers control adipose tissue mass. Paper read at the Fourth International Conference on the Regulation of Food and Water Intake, Cambridge, England.

Powley, T. L. (1977). The ventromedial hypothalamic syndrome, satiety, and a cephalic phase hypothesis. *Psychol. Rev. 84*, 89-126.

Powley, T. L., and Keesey, R. E. (1970). Relationship of body weight to the lateral hypothalamic feeding syndrome. *J. Comp. Physiol. Psychol. 70*, 25-36.

Powley, T. L., and Opsahl, C. A. (1974). Ventromedial hypothalamic obesity abolished by subdiaphragmatic vagotomy. *Am. J. Physiol. 226*, 25-33.

Powley, T. L., and Opsahl, C. A. (1976). Autonomic components of the hypothalamic feeding syndromes. In *Hunger: Basic Mechanisms and Clinical Implications*, D. Novin, W. Wyrwicka, and G. Bray (Eds.). Raven, New York, pp. 313-326.

Powley, T. L., and Plocher, T. A (1980). Hypophysectomy blocks the weight gain and obesity produced by goldthioglucose lesions. *Behav. Neural Biol. 28*, 300-318.

Powley, T. L., Opsahl, C. A., and van den Pol, A. N. (1974). Vagotomy eliminates ventromedial hypothalamic obesity in female rats. Paper presented at the Fifth International Conference on the Physiology of Food and Fluid Intake, Kiriat, Anavim, Jerusalem, Israel.

Quaade, F., Vaernet, K., and Larsson, S. (1974). Stereotaxic stimulation and electrocoagulation of the lateral hypothalamus in obese humans. *Acta Neurochirurg. 30*, 111-117.

Rabin, B. M. (1972). Ventromedial hypothalamic control of food intake and satiety: a reappraisal. *Brain Res. 43*, 317-324.

Rabin, B. M. (1974). Independence of food intake and obesity following ventromedial hypothalamic lesions in the rat. *Physiol. Behav. 13*, 769-772.

Ranson, S. W. (1939). Somnolence caused by hypothalamic lesions in the monkey. *Arch. Neurol. Psychiat. 41*, 1-23.

Ranson, S. W., and Magoun, H. W. (1939). The hypothalamus. *Ergebn. Physiol. 41*, 56-163.

Ranson, S. W., Fisher, C., and Ingram, W. R. (1938). Adiposity and diabetes mellitus in a monkey with hypothalamic lesions. *Endocrinology 23*, 175-181.

Redding, T. W., Schally, A. V., and Bowers, C. Y. (1966). Effects of hypophysectomy on hypothalamic obesity in CBA mice. *Proc. Soc. Exp. Biol. Med. 121*, 726-729.

Reeves, A. G., and Plum, F. (1969). Hyperphagia, rage, and dementia accompanying a ventromedial hypothalamic neoplasm. *Arch. Neurol. 20*, 616-624.

Reynolds, R. W. (1959). The effects of amphetamine on food intake in normal and hypothalamic hyperphagic rats. *J. Comp. Physiol. Psychol. 52*, 682-684.

Reynolds, R. W., and Kimm, J. (1965). Effect of glucose on food intake in hypothalamic hyperphagic rats. *J. Comp. Physiol. Psychol. 60*, 438-440.

Rezek, M., Vanderweele, D. A., and Novin, D. (1975). Stages in the recovery of feeding following vagotomy in rabbits. *Behav. Biol. 14*, 75-84.

Ridley, P. T., and Brooks, F. P. (1965). Alterations in gastric secretion following hypothalamic lesions producing hyperphagia. *Am. J. Physiol. 209*, 319-323.

Rogers, Q. R., and Leung, P. M. (1973). The influence of amino acids on the neuroregulation of food intake. *Fed. Proc. 32*, 1709-1719.

Rodgers, W. L., Epstein, A. N., and Teitelbaum, P. (1965). Lateral hypothalamic aphagia: motor failure or motivational deficit. *Am. J. Physiol. 208*, 334-342.

Rohner, F., Dufour, A. C., Karakash, C., and Le Marchand, Y., Ruf, K. B., and Jeanrenaud, B. (1977). Immediate effect of lesion of the ventromedial hypothalamic area upon glucose-induced insulin secretion in anaesthetized rats. *Diabetologia 13*, 239-242.

Rolls, E. T., Burton, M. J., and Mora, F. (1976). Hypothalamic neuronal responses associated with the sight of food. *Brain Res. 111*, 53-66.

Romaniuk, A. (1962). The effects of lesions of the medial hypothalamus on the conditioned reflexes type II and emotional behavior. *Acta Biol. Exp. 22*, 59-67.

Rony, H. R. (1940). *Obesity and Leanness*. Lea & Febiger, Philadelphia.

Rosenzweig, M. R. (1963). The mechanisms of hunger and thirst. In *Psychology in the Making. Histories of Selected Research Problems*, L. Postman (Ed.). Knopf, New York, pp. 73-141.

Roth, S. R., Schwartz, M., and Teitelbaum, P. (1973). Failure of recovered lateral hypothalamic rats to learn specific food aversions. *J. Comp. Physiol. Psychol. 83*, 184-197.

Rowland, N. (1977). Fragmented behavior sequences in rats with lateral hypothalamic lesions: an alternative reason for intrameal prandial drinking. *J. Comp. Physiol. Psychol. 91*, 1039-1055.

Rowland, N., Meile, M. J., and Nicolaïdis, S. (1975). Metering of intravenously infused nutrients in VMH lesioned rats. *Physiol. Behav. 15*, 443-448.

Rozkowska, E., and Fonberg, E. (1970). The effects of lateral hypothalamic lesions on food intake and instrumental alimentary reflex in dogs. *Acta Neurolbiol. Exp. 30*, 59-68.

Rozkowska, E., and Fonberg, E. (1971). The effects of ventromedial hypothalamic lesions on food intake and alimentary instrumental conditioned reflexes in dogs. *Acta Neurobiol. Exp. 31*, 354-364.

Russek, M., and Morgane, P. J. (1963). Anorexic effect of peritoneal glucose in the hypothalamic hyperphagic. *Nature (London) 199*, 1004-1005.

Sawchenko, P. E., and Gold, R. M. (1977). Knife cut hypothalamic hyperphagia blocked by complete subdiaphragmatic vagotomy. Paper read at the annual meeting of the Eastern Psychological Association, Boxton, April, 1977.

Sawchenko, P. E., Eng, R., Gold, R. M., and Simson, E. L. (1977). Effects of selective subdiaphragmatic vagotomies on knife cut induced hypothalamic hyperphagia. Paper presented at the Sixth International Conference on the Physiology of Food and Fluid Intake, Paris, France.

Schallert, T. (1977). Changes in feeding behavior, cholinergic electroencephalographic activity, gastric pathology, sensory attention, and motor activity following lateral hypothalamic lesions in fattened and dieted rats. Paper presented at the Annual Meeting of the Eastern Psychological Association.

Schallert, T., Wishaw, I. Q., and Flannigan, K. P. (1977). Gastric pathology and feeding deficits induced by hypothalamic damage in rats: effects of lesion type, size, and placement. *J. Comp. Physiol. Psychol. 91*, 598-610.

Scharrer, E., Baile, C. A., and Mayer, J. (1970). Effect of amino acids and protein on food intake of hyperphagic and recovered aphagic rats. *Am. J. Physiol. 218*, 400-404.

Scharrer, E., Thomas, D. W., and Mayer, J. (1974). Absence of effect of intraoral glucose infusions upon spontaneous meals of rats. *Pfluegers Arch. Gesamte Physiol. 351*, 315-322.

Schindler, W. J., and Liebelt, R. A. (1967). Thyroid activity in mice with hypothalamic lesions induced by goldthioglucose. *Endocrinology 80*, 387-398.

Schnatz, J. D., Bernardis, L. L., Frohman, L. A., and Goldman, J. K. (1971). Hypertriglyceridemia in weanling rats with hypothalamic obesity. *Diabetes 20*, 655-663.

Schnatz, J. D., Frohman, L. A., and Bernardis, L. L. (1973). The effect of lateral hypothalamic lesions in weanling rats with lesions in the ventromedial hypothalamic nuclei. *Proc. Soc. Exp. Biol. Med. 142*, 256-257.

Schoenfeld, T. A., and Hamilton, L. W. (1977). Secondary brain changes following lesions: a new paradigm for lesion experimentation. *Physiol. Behav. 18*, 951-967.

Sclafani, A. (1971). Neural pathways involved in the ventromedial hypothalamic lesion syndrome in the rat. *J. Comp. Physiol. Psychol. 77*, 70-96.

Sclafani, A. (1976). Appetite and hunger in experimental obesity syndromes. In *Hunger: Basic Mechanisms and Clinical Implications*. D. Novin, W. Wyrwicka, and G. Bray (Eds.). Raven, New York, pp. 281-296.

Sclafani, A., and Berner, C. N. (1976). Influence of diet palatability on the meal taking behavior of hypothalamic hyperphagic and normal rats. *Physiol. Behav. 16*, 355-363.

Sclafani, A., and Kluge, L. (1974). Food motivation and body weight levels in hypothalamic hyperphagic rats: a dual lipostat model of hunger and appetite. *J. Comp. Physiol. Psychol. 86*, 28-46.

Sclafani, A., and Sperber, M. (1977). Hyperphagia and obesity in the guinea pig produced by hypothalamic knife cuts. *Behav. Biol. 19*, 394-400.

Sclafani, A., Berner, C. N., and Maul, G. (1973). Feeding and drinking pathways between the medial and lateral hypothalamus in the rat. *J. Comp. Physiol. Psychol. 85*, 29-51.

Sclafani, A., Berner, C. N., and Maul, G. (1975a). Multiple knife cuts between the medial and lateral hypothalamus: a reevaluation of hypothalamic feeding circuitry. *J. Comp. Physiol. Psychol. 88,* 210-217.

Sclafani, A., Gale, S. K., and Springer, D. (1975b). Effects of hypothalamic knife cuts on the ingestive responses to glucose and insulin. *Physiol. Behav. 15,* 63-70.

Sclafani, A., Koopmans, H. S., and Vasselli, J. R. (1977). Paper presented at the Sixth International Conference on the Physiology of Food and Fluid Intake, Paris, France.

Sclafani, A., Springer, D., and Kluge, L. (1976). Effects of quinine adulterated diets on the food intake and body weight of obese and non-obese hypothalamic hyperphagic rats. *Physiol. Behav. 16,* 631-640.

Sétáló, G. (1965). The mechanism of hypothalamic obesity in the rat. *Acta Physiol. Acad. Sci. Hung. 27,* 375-384.

Sharma, K. N., Anand, B. K., Dua, S., and Singh, B. (1961). Role of stomach in regulation of activities of hypothalamic feeding centres. *Am. J. Physiol. 201,* 593-598.

Sharp, J. C., Nielson, H. C., and Porter, P. B. (1962). The effect of amphetamine upon cats with lesions in the ventromedial hypothalamus. *J. Comp. Physiol. Psychol. 55,* 198-200.

Sims, E. A. H., Danforth, Jr., E., Horton, E. S., Bray, G. A., Glennon, J. A., and Salans, L. B. (1973). Endocrine and metabolic effects of experimental obesity in man. *Recent Prog. Horm. Res. 29,* 457-496.

Singh, D., and Meyer, D. R. (1968). Eating and drinking by rats with lesions of the septum and the ventromedial hypothalamus. *J. Comp. Physiol. Psychol. 65,* 163-166.

Singh, D., Lakey, J. R., and Sanders, M. K. (1974). Hunger motivation in gold thioglucose-treated and genetically obese female mice. *J. Comp. Physiol. Psychol. 86,* 890-897.

Slaunwhite, W. R., III, Goldman, J. K., and Bernardis, L. L. (1972). Sequential changes in glucose metabolism by adipose tissue and liver of rats after destruction of the ventromedial hypothalamic nuclei: effect of three dietary regimens. *Metabolism 21,* 619-631.

Smith, G. P., and Epstein, A. N. (1969). Increased feeding in response to decreased glucose utilization in the rat and monkey. *Am. J. Physiol. 217,* 1083-1087.

Smith, G. P., Gibbs, J., Stohmayer, A. J., and Stokes, P. E. (1972). Threshold doses of 2-deoxy-D-glucose for hyperglucemia and feeding in rats and monkeys. *Am. J. Physiol. 222,* 77-81.

Smith, M. H., Salisbury, R., and Weinberg, H. (1961). The reaction of hypothalamic-hyperphagic rats to stomach preloads. *J. Comp. Physiol. Psychol. 54,* 660-664.

Smith, P. E. (1927). The disabilities caused by hypophysectomy and their repair. *J. Am. Med. Assoc. 88,* 158-161.

Smith, P. E. (1930). Hypophysectomy and a replacement therapy in the rat. *Am. J. Anat. 45,* 205-256.

Smutz, E. R., Hirsch, E., and Jacobs, H. L. (1975). Caloric compensation in hypothalamic obese rats. *Physiol. Behav. 14,* 305-309.

Snapir, N., Robinzon, B., Godschalk, M., Heller, E. D., and Perek, M. (1973). The effect of intrahypothalamic administration of sodium pentobarbital on eating behavior and feed intake in chickens. *Physiol. Behav. 10,* 97-100.

Snapir, N., Yaakobi, M., Robinzon, B., Ravona, H., and Perek, M. (1976). Involvement of the medial hypothalamus and the septal area in the control of food intake and body weight in geese. *Pharmacol. Biochem. Behav. 5*, 609-615.

Snowdon, C. T., and Wampler, R. S. (1974). Effects of lateral hypothalamic lesions and vagotomy on meal patterns in rats. *J. Comp. Physiol. Psychol. 87*, 399-409.

Starr, S. E., Crawford, J. D., and Haessler, H. A. (1966). Dynamics of development of the metabolic and compositional alterations of depot fat in hypothalamic obese rats. *Metabolism 15*, 39-45.

Steffens, A. B. (1969). A method for frequent sampling of blood and continuous infusion of fluids in the rat without disturbing the animal. *Physiol. Behav. 4*, 833-836.

Steffens, A. B. (1970). Plasma insulin content in relation to blood glucose level and meal pattern in the normal and hypothalamic hyperphagic rat. *Physiol. Behav. 5*, 147-151.

Steffens, A. B. (1975). Influence of reversible obesity on eating behavior, blood glucose, and insulin in the rat. *Am. J. Physiol. 228*, 1738-1744.

Steffens, A. B., Mogenson, G. J., and Stevenson, J. A. F. (1972). Blood glucose, insulin, and free fatty acids after stimulation and lesions of the hypothalamus. *Am. J. Physiol. 222*, 1446-1452.

Stellar, E. (1954). The physiology of motivation. *Psychol. Rev. 61*, 5-22.

Stern, J., and Keesey, R. E. (1978). Panel discussion on "Control of food intake in genetic obesity." Paper presented at the Winter Conference on Brain Research. Keystone, Colorado, January, 1978.

Stern, J. J., Audillo, C. A., and Kruper, J. (1976). Ventromedial hypothalamus and short-term feeding suppression by caerulin in male rats. *J. Comp. Physiol. Psychol. 90*, 484-490.

Stevenson, J. A. F. (1949). Effects of hypothalamic lesions on water and energy metabolism in the rat. *Recent Progr. Horm. Res. 4*, 363-392.

Stevenson, J. A. F. (1969). Neural control of food and water intake. In *The Hypothalamus*, W. Haymaker, E. Anderson, and W. J. H. Nauta (Eds.). Thomas, Springfield, Ill., pp. 524-621.

Stevenson, J. A. F., and Montemurro, D. G. (1963). Loss of weight and metabolic rate of rats with lesions in the medial and lateral hypothalamus. *Nature (London) 198*, 92.

Stevenson, J. A. F., Welt, L. G., and Orloff, J. (1950). Abnormalities of water and electrolyte metabolism in rats with hypothalamic lesions. *Am. J. Physiol. 161*, 35-39.

Stowe, Jr., F. R., and Miller, Jr., A. T. (1957). The effect of amphetamine on food intake in rats with hypothalamic hyperphagia. *Experientia 13*, 114-115.

Stricker, E. M., and Zigmond, M. J. (1974). Effects on homeostasis of intraventricular injection of 6-hydroxydopamine in rats. *J. Comp. Physiol. Psychol. 86*, 973-994.

Stricker, E. M., and Zigmond, M. J. (1976a). Brain catecholamines and the lateral hypothalamic syndrome. In *Hunger: Basic Mechanisms and Clinical Implications*, D. Novin, W. Wywricka and G. Bray (Eds.). Raven, New York, pp. 19-32.

Stricker, E. M., and Zigmond, M. J. (1976b). Recovery of function after damage to central catecholamine-containing neurons: a neurochemical model for the lateral hypothalamic syndrome. *Progr. Psychobiol. Physiol. Psychol. 6*, 121-188.

Stricker, E. M., Friedman, M. I., and Zigmond, M. J. (1975). Glucoregulatory Feeding by rats after intraventricular 6-hydroxydopamine or lateral hypothalamic lesions. *Science 189*, 895-897.

Strominger, J. L., Brobeck, J. R., and Cort, R. L. (1953). Regulation of food intake in normal rats and in rats with hypothalamic hyperphagia. *Yale J. Biol. Med. 26*, 55-74.

Strubbe, J. H., Steffens, A. B., and De Ruiter, L. (1977). Plasma insulin and the time pattern of feeding in the rat. *Physiol. Behav. 18*, 81-86.

Tannenbaum, G. A., Paxinos, G., and Bindra, D. (1974). Metabolic and endocrine aspects of the ventromedial hypothalamic syndrome in the rat. *J. Comp. Physiol. Psychol. 86*, 404-413.

Teitelbaum, P. (1955). Sensory control of hypothalamic hyperphagia. *J. Comp. Physiol. Psychol. 48*, 156-163.

Teitelbaum, P. (1957). Random and food-directed activity in hyperphagic and normal rats. *J. Comp. Physiol. Psychol. 50*, 486-490.

Teitelbaum, P., and Epstein, A. N. (1962). The lateral hypothalamic syndrome: recovery of feeding and drinking after lateral hypothalamic lesions. *Psychol. Rev. 69*, 74-90.

Teitelbaum, P., and Stellar, E. (1954). Recovery from the failure to eat produced by hypothalamic lesions. *Science 120*, 894-895.

Teitelbaum, P., Cheng, M., and Rozin, P. (1969). Stages of recovery and development of lateral hypothalamic control of food and water intake. *Ann. New York Acad. Sci. 157*, 849-860.

Tepperman, J., Brobeck, J. R., and Long, C. N. H. (1943). The effects of hypothalamic hyperphagia and of alterations in feeding habits on the metabolism of the albino rat. *Yale J. Biol. Med. 15*, 855-874.

Thomas, D. W. (1971). Effects of altered meal size on the feeding behavior of normal and hyperphagic rats. Paper presented at the Fourth International Conference on the Regulation of Food and Water Intake, Cambridge, England.

Thomas, D. W., and Mayer, J. (1968). Meal taking and regulation of food intake by normal and hypothalamic hyperphagic rats. *J. Comp. Physiol. Psychol. 66*, 624-653.

Ungerstedt, U. (1970). Is interruption of the nigro-striatal dopamine system producing the "lateral hypothalamus syndrome"? *Acta Physiol. Scand. 80*, 35A-36A.

Valenstein, E. S., and Phillips, A. G. (1970). Stimulus-bound eating and deprivation from prior contact with food pellets. *Physiol. Behav. 5*, 279-282.

Valenstein, E. S., Cox, V. C., and Kakolewski, J. W. (1968). The motivation underlying eating elicited by lateral hypothalamic stimulation. *Physiol. Behav. 3*, 696-971.

van Putten, L. M., van Bekkum, D. W., and Querido, A. (1955). Influence of hypothalamic lesions producing hyperphagia, and of feeding regimens on carcass composition in the rat. *Metabolism 4*, 68-74.

Vilberg, T. R., and Beatty, W. W. (1975). Behavioral changes following VMH lesions in rats with controlled insulin levels. *Pharmacol. Biochem. Behav. 3*, 377-384.

Wampler, R. S. (1973). Increased motivation in rats with ventromedial hypothalamic lesions. *J. Comp. Physiol. Psychol. 84*, 275-285.

Wagner, J. W., and de Groot, J. (1963). Changes in feeding behavior after intracerebral injections in the rat. *Am. J. Physiol. 204*, 483-487.

Wayner, M. J., and Oomura, Y. (Eds.) (1975). Central neural control of eating and obesity. *Pharmacol. Biochem. Behav. 3* (Suppl. 1), pp. 1-177.

Wayner, M. J., Cott, A., Millner, J., and Tartaglione, R. (1971). Loss of 2-deoxy-D-glucose induced eating in recovered lateral rats. *Physiol. Behav. 7*, 881-884.

Weinberger, L. M., and Grant, F. C. (1941). Precocious puberty and tumor of hypothalamus, report of case in review of literature with pathophysiologic explanation of precocious sexual syndrome. *Arch. Intern. Med. 67*, 762-792.

Wheatley, M. D. (1944). The hypothalamus and affective behavior in cats. *Arch. Neurol. Psychiat. 52*, 296-316.

White, L. E., and Hain, R. F. (1959). Anorexia in association with a destructive lesion of the hypothalamus. *Arch. Pathol. 68*, 275-281.

Wilder, R. M. (1932). The regulation of the weight of the body. *Internatl. Clin. 1*, 30-41.

Williams, D. R., and Teitelbaum, P. (1959). Some observations on the starvation resulting from lateral hypothalamic lesions. *J. Comp. Physiol. Psychol. 52*, 458-465.

Wilson, W. H., and Heller, H. C. (1975). Elevated blood glucose levels and satiety in the rat. *Physiol. Behav. 15*, 137-143.

Wise, A. (1975). Adipocyte number and size in hypothalamic obesity induced in weanling mice by goldthioglucose and bipiperidyl mustard. *Nutr. Metab. 19*, 291-298.

Wolf, G., and Quartermain, D. (1967). Sodium chloride intake of adrenalectomized rats with lateral hypothalamic lesions. *Am. J. Physiol. 212*, 113-118.

Woods, S. C., and Porte, D. (1974). Neural control of the endocrine pancreas. *Physiol. Rev. 54*, 596-619.

Wright, P., and Turner, C. (1973). Sex differences in body weight following gonadectomy and goldthioglucose injections in mice. *Physiol. Behav. 11*, 155-159.

Yin, T. H., and Liu, C. M. (1976). Caloric compensation in rats with combined lesions in lateral and ventromedial hypothalamus. *Physiol. Behav. 16*, 461-469.

Yin, T. H., and Tsai, C. T. (1973). Effects of glucose on feeding in relation to routes of entry in rats. *J. Comp. Physiol. Psychol. 85*, 258-264.

York, D. A., and Bray, G. A. (1972). Dependence of hypothalamic obesity on insulin, the pituitary and the adrenal gland. *Endocrinology 90*, 885-894.

Young, T. K., and Liu, A. C. (1965). Hyperphagia, insulin and obesity. *Chinese J. Physiol. 19*, 247-253.

Zeigler, H. P. (1974). Feeding behavior in the pigeon: a neurobehavioral analysis. In *Birds: Brain and Behavior*, I. Goodman and M. Schein (Eds.). Academic, New York, pp. 101-132.

Zeigler, H. P. (1975). Trigeminal deafferentation and hunger in the pigeon (*Columba livia*). *J. Comp. Physiol. Psychol. 89*, 827-844.

Zeigler, H. P., and Karten, H. J. (1974). Central trigeminal structures and the lateral hypothalamic syndrome in the rat. *Science 186*, 636-637.

Zigmond, M. J., and Stricker, E. M. (1972). Deficits in feeding behavior after intraventricular injection of 6-hydroxydopamine in rats. *Science 177*, 1211-1214.

Zigmond, M. J., and Stricker, E. M. (1973). Recovery of feeding and drinking by rats after intraventricular 6-hydroxydopamine or lateral hypothalamic lesions. *Science 183*, 717-720.

4
Neurochemical Systems of the Hypothalamus
Control of Feeding and Drinking Behavior
and Water-Electrolyte Excretion

Sarah Fryer Leibowitz / The Rockefeller University, New York, New York

I. Introduction

The behaviors involved in the acquisition of food and water are clearly of vital importance for the survival of an organism. The motor responses underlying eating and drinking behavior are very basic and, thus, relatively simple to analyze. They are, however, merely the end points of complex biological control systems geared towards regulating the body levels of nutrients and energy and the intracellular and extracellular fluid compartments. Recent scientific investigations have significantly advanced our understanding of these control systems and have led to a clearer appreciation of the multitude of variables which may come to bear on the expression of ingestive behavior. The hypothalamus, as it lies at an interface between the brain and periphery, has through the years attracted interest from many points of view and, for example, has been demonstrated through lesion and electrical stimulation studies, to have an essential function in energy and water homeostasis. Clearly, the hypothalamus does not operate autonomously in this function but as part of a whole-brain circuitry which is organized at various integrative levels and involves complex interaction between these levels through multiple neurochemical processes. In attempting to relate a specific neurochemical system to a specific response, the focus of attention must ultimately be the fiber pathways involved in the response. The hypothalamus itself has extremely rich supplies of biologically active substances such as putative neurotransmitters and neurohormones. To understand the role of these substances in behavioral functions, investigations into their impact on local neuronal (or nonneuronal) tissue and, consequently, on physiological and behavioral responses will need to progress towards studies of the cell bodies from which the hypothalamic neurochemical systems originate, as well as studies on the nature of the signals that alter the activity of these cells.

The hypothalamus has been extensively characterized with biochemical and histochemical techniques. The first major studies examined acetylcholinesterase-positive structures and the various monoamine systems. The introduction of immunohistochemical techniques into neurohistochemistry has resulted in a dramatic extension of the possibilities of identifying and mapping neuron systems on the basis of their content of specific macromolecules. This holds true for aminergic systems, where antibodies to the enzymes in the synthesis of catecholamine and 5-hydroxytryptamine are used; for γ-aminobutyric acid-containing neural systems with antibodies to the synthesizing enzyme, glutamate decarboxylase, and for a large number of peptides. The micropunch technique coupled with microenzymatic assays have provided yet another valuable tool for biochemical mapping of neuronal terminations. By considering the distribution pattern of the various systems identified, it is clear that interactions among different neurochemical systems—principally via axodendritic (or axosomatic) or via axoaxonic contacts—are likely to occur in many hypothalamic and extrahypothalamic areas responsible for a complex behavior.

This chapter, completed in January 1979, reviews studies which have attempted to relate brain chemistry to the control of feeding and drinking behavior, as well as water-electrolyte excretion. Each section is organized to provide, first, a brief review of a particular neurochemical system as defined via histochemical and biochemical techniques and, second, the neurobehavioral studies investigating the function of this system and, in some cases, its potential interaction with other neurochemical systems. These studies are clearly only in their very early stages. They provide, however, an important foundation for developing reasonable hypotheses concerning specific functions of discrete hypothalamic areas and particular neurochemical pathways innervating these areas. This is no small achievement, in view of the extreme complexity of the neurocircuitry of the hypothalamus and of the biological control systems under investigation.

II. Noradrenergic and Adrenergic Systems

A. Anatomy of Noradrenergic and Adrenergic Projections

The organization of the hypothalamic catecholaminergic (CA) innervating systems is extremely complex, in that there exist extrinsic afferent and intrinsic intrahypothalamic systems, as well as major CA pathways which course through the hypothalamus on their way to forebrain regions. The CA innervations of the hypothalamus include noradrenergic, dopaminergic, and probably adrenergic neurons. These systems have been studied by means of the Falck-Hillarp formaldehyde histofluorescence method, the glyoxylic acid fluorescence method, and by immunofluorescence of CA-synthesizing enzymes. These methods, in addition to biochemical mapping of neuronal terminations via micropunch technique and microenzymatic assays, have permitted a detailed description of the morphology and organization of the CA-containing neurons, which has then set the stage for a rational approach to experimental investigations into the role of neurotransmitters in behavior and drug actions. Our knowledge of the CA systems which innervate the hypothalamus has recently been reviewed by several investigators (Brownstein, 1977; Hökfelt et al., 1978; Jacobowitz, 1978; Lindvall and Björklund, 1978; Moore and Bloom, 1978, 1979; Swanson and Hartman, 1975). The noradrenergic and adrenergic projection path-

Feeding and Drinking and Water-Electrolyte Excretion

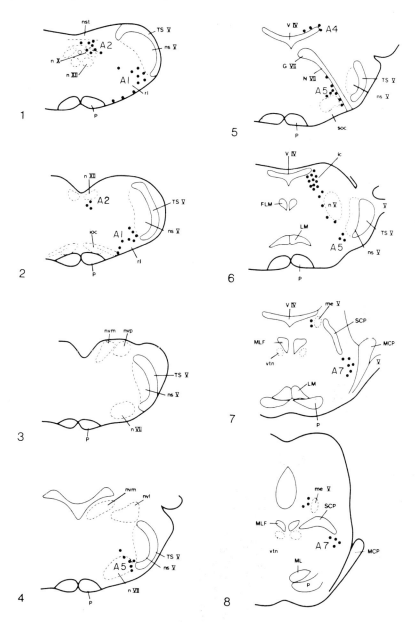

Figure 1 Distribution of NE-containing cell bodies (A1-7) in the lower brain stem of the rat, represented at eight frontal levels. For abbreviations, see page 405. (From Lindvall and Björklund, 1978.)

ways and their preterminal and terminal fields will be briefly described here (see Figs. 1-4).

The evidence suggests that hypothalamic norepinephrine (NE) innervations originate from four cell systems in the lower brain stem. Dahlström and Fuxe (1964) described the distribution of CA-containing perikarya in the lower brain stem and classified them

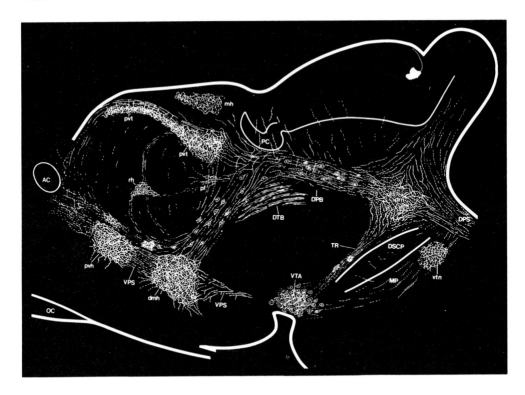
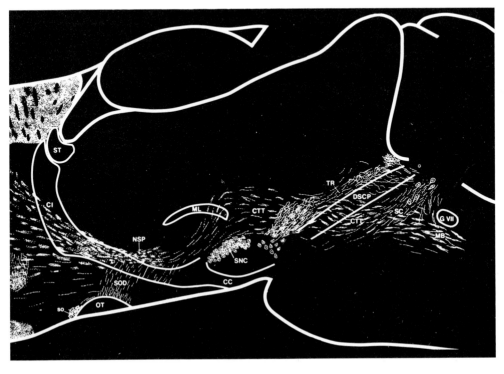

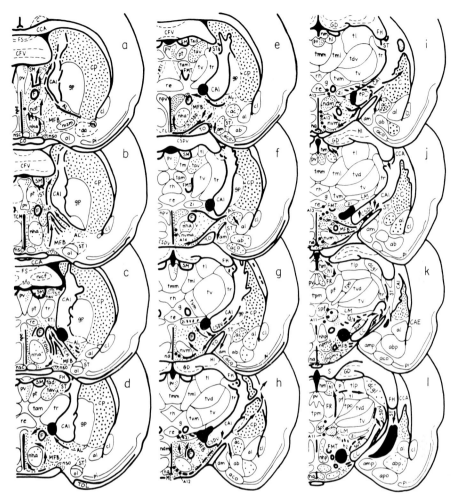

Figure 3 Schematic drawing of preterminal axons showing swellings, distortions, and increased fluorescence after neurotoxic agent or knife lesions. Frontal sections extend from the most rostral aspect of the anterior hypothalamus (a) to the paraventricular nucleus (e), the ventromedial nucleus (h), and the posterior hypothalamus (l). The dorsal NE ascending pathway is shown with fine stripples. The nigrostriatal DA pathway is the large black region medial to the internal capsule. Catecholaminergic varicose fibers are indicated by black dots and cell bodies (groups A11-14) by filled circles. The heavy arrows indicate directionality of proposed noradrenergic axonal pathways. See page 405 for abbreviations. (From Jacobowitz, 1975.)

Figure 2 Semidiagrammatic representation of the ascending CA fiber systems. Top: Periventricular CA fiber system (rostral to the locus ceruleus), the medial fiber flow of the tegmental CA radiations (TR), and the CA fibers of the mamillary peduncle (MP), on a composite drawing of somewhat different paramedian sagittal planes. Bottom: CA fiber systems in the central tegmental tract (CTT) and its caudal extension, the medullary CA bundle (MB). Rostrally, part of the nigrostriatal pathway is represented in its extension through the internal capsule (CI) and the globus pallidus. Composite drawing of different sagittal planes. See page 405 for abbreviations. (From Lindvall and Björklund, 1974b.)

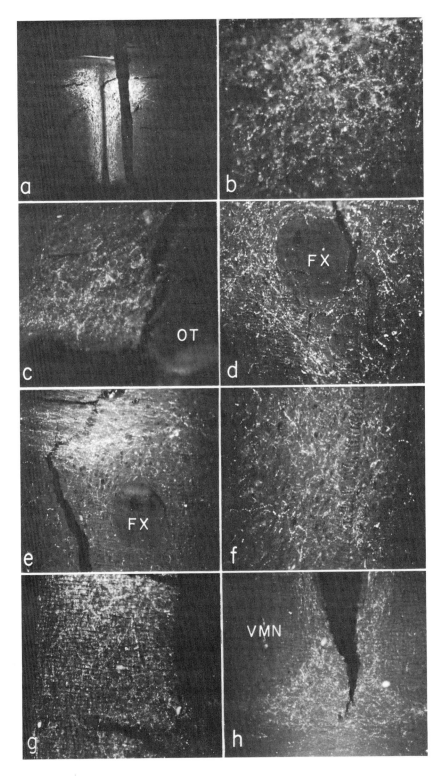

into 12 groups (designated A1-A12). They form the dorsal medullary cell system (A2), the pontine (A5 and A7) and medullary (A1 and A3) parts of the lateral tegmental NE cell system, and the locus ceruleus cell group (A6). The axons of these systems appear to ascend in a position corresponding to the central tegmental tract (CTT) (or ventral NE bundle), as defined by classic neuroanatomical techniques, and also as part of the dorsal tegmental bundle (DTB) (or dorsal NE bundle) located in the dorsomedial aspect of the CTT. The noradrenergic fibers of the CTT originate from the neurons of the lateral tegmental and the dorsal medullary neurons, whereas the DTB fibers originate from the locus ceruleus. The pontine and medullary fibers in the CTT run through and ventral to the decussation of the superior cerebellar peduncles. On its rostral side, the dorsal parts of this fiber system turn rostroventrally to join the tegmental CA radiations, a radially oriented system where the fibers, in going from dorsal to ventral, diverge strongly in the ventrolateral direction in a fan-like arrangement. These CTT fibers, including some axons which have left the DTB in a rostroventral and somewhat lateral direction, follow the radial course of the tegmental radiations for varying distances, and a significant portion leave the radiations after only a short distance to resume their longitudinal course, now in more ventral and lateral positions. Other CTT fibers follow the tegmental radiations down into the ventrolateral tegmentum. At this point, they join the most ventral CTT fibers which have proceeded along a relatively straight course after passing ventral to the decussation of the superior cerebellar peduncles. The axons then assemble in the caudal hypothalamus and ascend along the medial forebrain bundle. They leave this bundle (remaining dorsally or turning ventrally) at various levels medially towards the hypothalamic and preoptic nuclei, sometimes as collaterals of fibers continuing further rostrally.

An alternative route for NE fibers to reach, in particular, the periventricular hypothalamus is along the periventricular fiber system, which in addition to carrying axons (predominantly dopaminergic) originating in the diencephalic periventricular and paramedian cell system, receives fibers originating from pontine and medullary NE cell groups. These NE axons to the hypothalamus join the periventricular fiber system at various levels. In particular, fibers from the CTT and locus ceruleus contribute to this system along its course in the pons and mesencephalon. Furthermore, in the caudal diencephalon, locus fibers from the DTB join the dorsal branch of the hypothalamic periventricular system (the dorsal periventricular bundle) which turns ventrally through the caudal thalamus to innervate the periventricular and medial hypothalamic nuclei. Finally, fibers that have passed through the tegmental radiations run into the ventral branch of

---

Figure 4 Photomicrographs showing CA fluorescence in different hypothalamic areas of intact rats. Brain tissue was treated according to the Falck-Hillarp technique. (a) Paraventricular nucleus at low magnification showing entire nucleus on either side of the third ventricle. (b) Paraventricular nucleus at higher magnification. (c) Supraoptic nucleus lateral to the optic tract (OT). (d) Perifornical area showing CA fluorescence surrounding the fornix (FX) at the level of the ventromedial nucleus. (e) Zona incerta region just dorsal to the fornix (FX) and lateral to the paraventricular nucleus. (f) Lateral hypothalamic region at the level of the ventromedial nucleus. (g) Dorsomedial nucleus. (h) Periventricular area adjacent to the third ventricle, medial and ventral to the ventromedial nucleus (VMN). (Results of collaborative studies conducted by L. L. Brown, Albert Einstein College of Medicine, and S. F. Leibowitz, The Rockefeller University.)

the hypothalamic periventricular system, which courses immediately dorsolateral to the interpeduncular nucleus and then ascends rostrally through the periventricular nucleus and into the dense CA terminal systems of the dorsomedial and paraventricular nuclei.

The contribution of the locus ceruleus to the hypothalamic NE systems is a relatively minor one. The majority of the locus terminals are found in the periventricular and paraventricular nuclei. A greater portion of the hypothalamic NE innervation appears to originate in the lateral tegmental and dorsal medullary cell groups (A1, A2, A5, and A7). The terminals of these NE systems are widely but unevenly distributed in the hypothalamus (Fig. 3 and 4). High terminal densities are found in the paraventricular, periventricular, dorsomedial, and supraoptic nuclei, and in the perifornical area and the median eminence, whereas lower terminal densities occur in the premamillary, mamillary, arcuate, and ventromedial nuclei and the anterior and posterior hypothalamic areas. The lateral hypothalamus and the retrochiasmatic region are distinct in that they are traversed by major CA pathways, the medial forebrain bundle and the supraoptic decussations, respectively. These fiber systems may very possibly make abundant contacts of passage with the neurons in these regions.

Epinephrine (EPI) is also present in the hypothalamus in smaller but significant concentrations. From immunofluorescence studies with antibodies against the EPI-forming enzyme phenylethanolamine-N-methyltransferase, presumed EPI-containing terminals appear to be present in highest concentrations in the paraventricular, dorsomedial, and arcuate nuclei, the perifornical area, and the basal hypothalamus. Few or no EPI fibers are found in the ventromedial, supraoptic, and suprachiasmatic nuclei, the median eminence, and the anterior hypothalamic nucleus. Cell bodies from which these hypothalamic terminals may originate have been identified only in the medulla oblongata and appear to be clustered in two cell groups (referred to as C1 and C2), lying in the region of the NE A1 and A2 cell groups.

B. Feeding Behavior

Early studies using electrical stimulation and electrolytic lesion techniques provided extensive evidence suggesting that the hypothalamus plays an essential role in the regulation of feeding behavior. A major question which these studies could not answer is whether the effects of these hypothalamic manipulations were actually due to changes in hypothalamic cellular function or whether they resulted from activation or destruction of fibers passing through the hypothalamus. An important contribution to this field of research, and a vital impetus to the development of neurochemical constructs of feeding came from the innovative work of Grossman (1962a, b). Using the technique of injecting drugs directly into the brain of rats with chronically implanted brain cannulas, Grossman tested the hypothesis that hypothalamic receptor sites, presumably located on the postsynaptic cellular membranes, may be sensitive to various putative neurotransmitters and may then respond by altering feeding. The outcome of these studies was dramatic and has provided the foundation for extensive research that has been conducted over the past two decades. It was found that, when l-NE was injected into the hypothalamus of fully satiated rats, a vigorous feeding response occurred. Numerous investigators have confirmed this important observation in other species (Baile, 1974; Setler and Smith, 1974; Yaksh and Myers, 1972) as well as in the rat (e.g.,

Booth, 1968; Coury, 1967; Davis and Keesey, 1971; Leibowitz, 1970b, 1973a, 1975a; Slangen and Miller, 1969) (see Fig. 5). In addition, EPI has been found to produce the response, with perhaps even greater potency than NE (Booth, 1968; Grossman, 1962a; Leibowitz, 1975b; Slangen and Miller, 1969), whereas dopamine (DA), serotonin, and the biologically inactive d-isomer of NE were generally ineffective (Booth, 1968; Leibowitz, 1975b; Slangen and Miller, 1969). Dose-response studies with NE and EPI revealed that the elicited eating response was dose dependent (Booth, 1968; Leibowitz, 1975a, b; Miller et al., 1964). Furthermore, in addition to initiating a new feeding response in food-satiated rats, NE and EPI have both been found to potentiate the on-going feeding behavior of already hungry rats (Grossman, 1962a; Leibowitz, 1973a, 1978a; Ritter and Epstein, 1975). There is evidence that this stimulatory effect of NE on feeding may be dependent on the environmental cycle of darkness and light (Margules et al., 1972; Stern and Zurik, 1973). However, the nature of this relationship remains to be delineated, as there is clear evidence to indicate that the phenomenon may reliably occur at both dark and light times in the diurnal cycle (Armstrong and Singer, 1974; Leibowitz, 1978b; Matthews et al., 1978).

In studies with EPI and NE injected into the hypothalamus of hungry rats, it became apparent that adrenergic stimulation of this brain area, in addition to facilitating feeding, may have an inhibitory effect on this response. A suppression of feeding was observed with l-EPI (Leibowitz, 1970a, 1973a) and also with NE (Margules, 1970a). The evidence indicated further that EPI was considerably more potent than NE in producing this effect and that the d-isomer of EPI was totally ineffective (Leibowitz and Rossakis, 1978b). Studies also demonstrated that the selective β-adrenergic receptor agonist isoproterenol could produce a similar suppression of food intake (Goldman et al., 1971; Jackson and Robinson, 1971; Lehr and Goldman, 1973; Leibowitz, 1970a, b; Margules, 1970b).

While these experiments appeared to strengthen the notion that hypothalamic adrenergic receptor mechanisms may be involved in feeding control, they raised many more

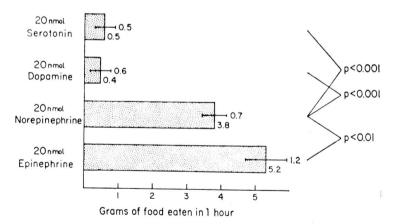

Figure 5 Effects of intrahypothalamic injection of monoaminergic agonists on food intake of satiated rats (N = 12). (From Slangen and Miller, 1969.) Reprinted with permission from *Physiol. Behav. 4*, J. L. Slangen and N. E. Miller, Pharmacological tests for the function of hypothalamic norepinephrine in eating behavior, 1969, Pergamon Press, Ltd.

questions with regard to the direction and nature of the control and the specific receptor types and brain areas involved in this control, as well as basic concerns of specificity and physiological significance of the phenomena. While it is reasonable that hypothalamic adrenergic neurotransmitters may have multiple effects on a particular response, what specifically determines the nature of these effects and how can they be experimentally dissociated? Most important, do the effects of the exogenous substances reflect the action of endogenous neurotransmitters released under natural conditions?

Two major concerns forever present in central injection studies are drug dose and drug spread from the injection site. Most investigations which have examined the effects of drugs in the brain have generally needed to use high concentrations (10^{-2} to 10^{-4} M), which are clearly aphysiological and involve extreme alterations in transmitter levels. Several explanations have been offered, including the possibilities that only a certain portion of the administered drug may actually reach the brain parenchyma, and a considerably smaller portion may then reach and effectively modulate a significant number of the appropriate terminals or receptor sites. While these may be reasonable possibilities, the burden still rests in the hands of investigators who apply chemicals to the brain to show either that lower concentrations are indeed effective or, at least, that different sources of evidence converge to argue in favor of a proposed physiological process. With regard to the question of drug spread from the injection site, this has been examined by a number of investigators (see Myers, 1974; Routtenberg, 1972), and it appears clear that an extensive amount of the injected compound spreads up along the outer surfaces of the cannula shaft and into the ventricles. (Unless specifically angled or aimed at cortical structures, most injection cannulas would be expected to penetrate the ventricular system.) This cannot be avoided but can perhaps be somewhat reduced by using small injection volumes (0.5 μl or less). The importance of ventricular spread for the drug-elicited phenomenon can to some extent be evaluated either by testing cannula implants that do not pass through the ventricles or by shifting the cannula tip to different sites while maintaining a constant position relative to the ventricles. The credibility of results yielded by such mapping procedures ultimately rests on the pattern of spread exhibited by the drug at the site of injection, as revealed through histochemical or autoradiographic techniques. The nature and extent of this spread would undoubtedly depend upon the particular compound examined, as well as the tissue to which it is administered. In one study, crystalline DA placed in the caudate-putamen was shown to produce a spherical spread of 1-2 mm in diameter (Routtenberg et al., 1968). Injection of NE in solution (1 μl) into the perifornical hypothalamus, however, distributed the NE over a somewhat smaller area, approximately 1 mm in diameter (Leibowitz, 1978a). Thus, these findings make clear the limitations of the central-injection technique, which need to be kept clearly in mind when evaluating results of studies reviewed below.

1. Paraventricular and Perifornical Hypothalamus

The evidence summarized above demonstrated that hypothalamic injection of NE or EPI can alter feeding in the rat, as well as in several other species, and that this effect may be either facilitatory or inhibitory in nature. One variable which appears to be important in determining the direction of these adrenergic agonists' effects on feeding is the brain site of injection. While the initial studies with central NE or EPI injection were conducted with hypothalamic cannula implants, there was no evidence for or against the hypothesis that extrahypothalamic structures may be the site of action. In the satiated rat, Coury (1967) examined several forebrain regions and obtained evidence to suggest that the NE-elicited feeding phenomenon may be mediated via the Papez anatomical circuit. Booth (1967), however, pointed out a specific site in the brain that he found to be particularly responsive to NE. This site, which was located just next to the fornix (approximately 1 mm lateral to midline) at the anterior hypothalamic level, has been confirmed by other investigators (Davis and Keesey, 1971; Leibowitz, 1975a; Slangen and Miller, 1969) to be a sensitive area, although further investigations provided some indication that NE may actually be acting at a more medial structure (Davis and Keesey, 1971; Leibowitz, 1970b). In fact, the more lateral hypothalamic sites appeared to be more responsive with regard to the phenomenon of EPI- or NE-induced suppression of feeding (Leibowitz, 1970a, b; Margules, 1970a).

A more systematic study of this problem was recently conducted in over 800 rats which received cannula implants in one of 30 forebrain structures and were tested under satiated or food-deprived conditions (Leibowitz, 1973a, 1978a; Leibowitz and Rossakis, 1979a). Both NE and EPI were examined in these animals, and the pattern of results clearly indicated that the brain sites where the adrenergic agonists facilitated feeding were distinct from those where the agonists inhibited feeding (Fig. 6). Both studies demonstrated that all sites outside the hypothalamus were relatively or totally unresponsive to adrenergic stimulation in terms of having effects on feeding behavior. Within the hypothalamus, both the facilitatory and inhibitory effects were observed and were generally found to be localized to the rostral half of the hypothalamus. Clear dissociation of the antagonistic phenomena occurred within this area, where the medial paraventricular nucleus (PVN) was distinguished as the most effective site for increasing feeding (see also Matthews et al., 1978), whereas the lateral perifornical area at the level of the ventromedial nucleus (VMN) was distinguished as the most effective site for decreasing feeding. Areas greater than 0.5 mm lateral, medial, dorsal, or ventral to these sites yielded significantly smaller effects. With regard to their rostral-caudal extent, both phenomena were relatively weak or absent just rostral to the PVN and caudal to the VMN and were moderate to very strong throughout the level of these two nuclei. With both the facilitatory and inhibitory feeding effects, drug spread via the ventricles did not appear to be important. Dramatic changes in responsiveness to NE or EPI or a reversal of their effect on feeding were observed with cannula shifts that maintained the position of the cannula relative to the third or lateral ventricles but increased its distance relative to the PVN or perifornical area.

When injected into these most sensitive areas, the latency of the drug-induced changes in feeding was quite short. Whereas earlier studies of NE-elicited feeding from structures outside the PVN generally reported latencies of 5-10 min, NE injected into the PVN produced a feeding response frequently within 0.5-2.0 min (Leibowitz, 1978a). The feeding suppression induced by perifornical administration of EPI was similarly observed within 1-2 min after injection (Leibowitz and Rossakis, 1979a), which supports the possibility that the mediating receptor sites are close to the tip of the cannula. Since, in these mapping studies, the animals were handled while being injected, it seems likely that shorter latencies would be observed with less disruptive, remote injection procedures. In addition to this evidence on response latency, results obtained with low doses of the adrenergic agonists argue strongly for PVN and perifornical sites of action. In the case of

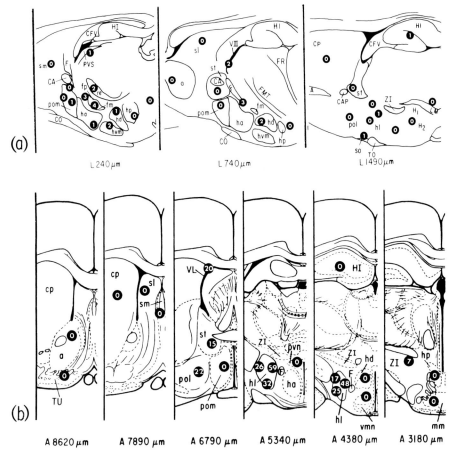

Figure 6 Schematic representation of central injection sites for catecholamines and their effect on food intake. (a) Sagittal sections showing the stimulatory effect of NE on feeding in satiated rats (N = 357). Each site is represented by a circle with a number enclosed. Between 6 and 20 rats had cannulas aimed at a particular area, and the number within each circle indicates the magnitude of the average feeding response (in grams) produced by NE (40 nmol) injected at that site. (b) Frontal sections showing the suppressive effect of EPI (and DA) on feeding in hungry, pargyline-pretreated rats (N = 204). Each site is represented by a circle with a number enclosed. Between 6 and 12 rats had cannulas lowered to within a 0.4-mm radius of the designated areas, and the number within each circle indicates the mean percent suppression of food consumption observed with the CA agonists (EPI and DA scores were similar and therefore averaged) as compared with each rat's own vehicle baseline. See page 405 for abbreviations. (From Leibowitz, 1978a; Leibowitz and Rossakis, 1979a.)

NE injected into the PVN of satiated rats, the threshold dose for eliciting feeding was between 1.0 and 4.2 ng (Leibowitz, 1978b). Ritter and Epstein (1975) observed a reliable potentiation of spontaneously initiated eating with 2.5 ng of NE injected in the rostral hypothalamus. These studies would appear to respond convincingly to the criticism that high doses are necessary to produce effects with central drug injection and further substantiate the particular sensitivity of the PVN to the facilitatory effect of adrenergic stimulation on feeding. Dose-response studies with EPI injected into the perifornical area of rats pretreated with pargyline revealed a reliable inhibition of feeding at a dose as low as 150 ng (Leibowitz and Rossakis, 1978a); however, lower doses were not effective, which, in terms of response sensitivity, clearly distinguishes this phenomenon from the NE facilitatory effect in the PVN. The reasons for this threshold difference are not clear, although there are several possible explanations. One important factor may be that perifornical CA neurons, although identified in distinct clusters around the fornix (Fig. 4d), are clearly more diffuse and less concentrated than those that innervate the PVN (Fig. 4a, b). This characteristic alone may require higher concentrations of the exogenous CA to reach and activate a significant portion of perifornical receptor sites. Another limiting factor may be the close proximity of the PVN feeding-facilitatory receptors which appear to antagonize the function of the perifornical feeding-inhibitory receptors. Since EPI has been shown to activate both populations of receptors (see above), it is not surprising that its effect in the perifornical hypothalamus is significantly potentiated when the PVN receptors are pharmacologically blocked (Leibowitz, 1976; Leibowitz and Rossakis, 1978a, b).

2. Adrenergic Receptor Analyses

The above evidence has distinguished the antagonistic effects of adrenergic neurotransmitters on feeding behavior, in terms of the brain areas where the effects appear to be localized and also in terms of the threshold of sensitivity exhibited by these areas to exogenous stimulation. Pharmacological studies have provided additional evidence for distinction, in terms of the type of adrenergic receptor mediating the behavioral responses. In the periphery, pharmacological criteria have been used to differentiate adrenergic receptors into two categories, namely, α and β (Ahlquist, 1967). β-Adrenergic receptors are characterized by their much higher affinity for the agonist isoproterenol than for NE, which is more closely associated with the α receptor. Epinephrine, in contrast, is found to be active on both types of receptors. Lands et al. (1967) extended this classification to include two distinct types of β-adrenergic receptors, referred to as β_1 and β_2. The β_1 receptor, associated with the heart, exhibited greater affinity for isoproterenol relative to EPI and NE which were equipotent, while the β_2 receptor, in the bronchial smooth muscle for example, displayed essentially equal affinity for isoproterenol and EPI which were considerably more potent than NE. Antagonists such as propranolol and alprenolol have high affinity for boty types of β receptors, in contrast to practolol and butoxamine which are selective antagonists of β_1 and β_2 receptors, respectively.

These pharmacological tools have been applied to the brain in an attempt to characterize the NE- and EPI-receptive sites which facilitate and suppress feeding in the rat. The results are clear in demonstrating that these two phenomena are mediated by distinct types of receptors, namely α adrenergic for the facilitation of feeding (Booth, 1968; Leibowitz, 1975b; Ritter and Epstein, 1975; Slangen and Miller, 1969) and β_2 adrenergic for the suppression of feeding (Goldman et al., 1971; Leibowitz, 1970a; Leibowitz and Rossakis, 1978b). Briefly, the α-adrenergic receptors were sensitive to both NE and EPI, as well as to the α agonists metaraminol and clonidine. They were insensitive to the β agonist isoproterenol and to DA. The effects of centrally injected antagonists on the action of NE or EPI were such that the elicited feeding response was blocked by the α-receptor antagonists phentolamine, phenoxybenzamine, and tolazoline, and unaffected or potentiated by antagonists of β-adrenergic, dopaminergic, and serotonergic receptors. These results contrast dramatically with those obtained with the phenomenon of adrenergic suppression of feeding (Leibowitz and Rossakis, 1978b). In a series of dose-response studies, this effect was exhibited by several adrenergic-receptor stimulants, with the order of potency being l-EPI = l-isoproterenol > dl-salbutamol (β_2) > dl-terbutaline (β_2) > l-NE > l-phenylephrine. The EPI-induced suppression was antagonized by β-adrenergic receptor blockers but not by α-adrenergic blockers. The relative potency of these antagonists was calculated to be l-propranolol > l-alprenolol > dl-butoxamine (β_2) = l-sotalol = dl-pindolol > practolol (β_1), with the α antagonists phentolamine and tolazoline exhibiting no effect. The receptor antagonism produced by propranolol was stereospecific and reversible by EPI. Serotonergic and cholinergic antagonists had little effect. Dopaminergic antagonists, however, were effective in reversing EPI's action on feeding. The order of potency of the neuroleptics and structurally related compounds in producing blockade was haloperidol > fluphenazine > chlorpromazine > pimozide > promazine, with the ineffective neuroleptic promethazine and the tricyclic antidepressants imipramine and desipramine having no such effect. These results indicate that the perifornical EPI-sensitive sites which inhibit feeding have characteristics expected of classic, β-adrenergic receptors, specifically β_2 subtype. Furthermore, the evidence suggests that these sites may be closely associated with functionally similar dopaminergic receptors.

The assumption underlying these pharmacological studies is that the resulting receptor classification reflects the affinities of the adrenergic agents to some hypothetical structure called the α- or β-adrenergic receptor. The recent development of reversible, receptor binding assays permits a direct test of this hypothesis. Using this procedure, several laboratories have studied adrenergic receptors in the brain, as well as in the periphery, and have provided evidence suggesting that α and β receptors do indeed exist in the brain, including the hypothalamus, that these receptors have very similar properties to adrenergic receptors in the periphery, and that brain areas may differ with respect to their relative concentrations of the α-, β_1-, and β_2-receptor subtypes (Alexander et al., 1975; Bylund and Snyder, 1976; Davis et al., 1977; Greenberg et al., 1976; Lefkowitz, 1976; Sporn and Molinoff, 1976; U'Prichard et al., 1977, 1978). In the hypothalamus, the α receptor exists in considerably higher concentration than the β receptor. In the cortex, the β_1 receptor is found to predominate over the β_2 receptor, whereas in the cerebellum, the β_2 receptor predominates. With respect to the subtype of β receptor that may exist in the hypothalamus, detailed ligand-binding studies have not yet explored this question. The above pharmacological studies, however, suggest the existence of β_2 receptors, possibly in the area of the fornix, in contrast to α-adrenergic receptors concentrated in the PVN.

The above evidence also suggests that the β-adrenergic receptors that suppress feeding may in some manner interact with a dopaminergic receptor population which produces a similar effect. Studies to be described below (Section III.B) support this hypothesis but suggest that EPI and DA act directly on distinct, functionally independent β-adrenergic and dopaminergic receptors, respectively, that are located contiguously in the perifornical area of the hypothalamus. It should be noted that, similar to the pharmacological results showing dopaminergic antagonists' inhibition of EPI's suppressive effect on feeding, several biochemical and iontophoretic studies in the brain have found neuroleptics (as well as β-receptor blockers) to antagonize adrenergic effects on the enzyme adenylate cyclase (Blumberg et al., 1976; Bockaert et al., 1977; Horn and Phillipson, 1976; Krueger et al., 1975; Palmer and Manian, 1974), in addition to their effects on neuronal activity (Bloom, 1975; Freedman and Hoffer, 1975). A possible explanation for each of these findings may be obtained from the studies of Weiss and his colleagues (Levin and Weiss, 1975, 1977; Weiss et al., 1974) and Costa and his colleagues (Costa et al., 1977; Gnegy et al., 1976). These studies suggest that neuroleptics, in addition to blocking postsynaptic DA receptors (Creese et al., 1976), may competitively inhibit an "endogenous protein activator," a small-molecular-weight protein located in synaptic membrane and postulated to be a key link in postsynaptic receptor responses to neurotransmitters. If the β-adrenergic, EPI-sensitive neuron in the perifornical hypothalamus is one such neuron that depends on normal functioning of this endogenous protein, neuroleptic antagonism of the EPI response may possible occur via competitive inhibition of this macromolecule. The relationship between the EPI and DA systems suppressing feeding will be discussed further in Section III.B below.

3. Changes in Endogenous Neurotransmitter Function

The PVN and perifornical areas of the hypothalamus, where exogenous NE and EPI are found to alter feeding behavior, are known to receive a particularly dense population of adrenergic neurons, containing either NE or EPI (see Section II.A). A crucial step in determining whether the endogenous neurotransmitter systems have a physiological role in control of naturally motivated behavior involves the demonstration that this behavior can be produced by spontaneous or drug-induced release of the endogenous neurotransmitter. Suggestive evidence linking endogenous NE (or EPI) release to elicited or enhanced eating behavior initially came from two preliminary studies which tested drugs known to increase the synaptic availability of these amines. Slangen and Miller (1969) examined the CA depleter tetrabenazine, which inhibits the storage of NE in the presynaptic granules, in combination with a monoamine oxidase (MAO) inhibitor nialamide, and found that these drugs injected together into the hypothalamus consistently elicited eating in satiated rats. A similar effect was reported for the CA uptake blocker desmethylimipramine, which also enhances synaptic concentration of endogenous NE (Montgomery et al., 1971). This drug, however, appeared effective only in hungry rats, which showed an enhanced eating response relative to their baseline score. A stimulatory effect on feeding has also been observed shortly after hypothalamic injection of the neurotoxin 6-hydroxydopamine (6-OHDA) (Evetts et al., 1972). During the early stages of causing a selective and irreversible degeneration of CA-containing nerve terminals, this compound is believed to cause the release of CA neurotransmitters from the presynaptic terminal, and Evetts postulated that this initial consequence of 6-OHDA's pharmacological action led to the alteration in feeding behavior. In two additional series of studies, the drugs chlorpromazine (Leibowitz and Miller, 1969; Leibowitz, 1976) and amphetamine (Leibowitz, 1970b, 1975c) were examined and were also found to stimulate feeding. Chlorpromazine's predominant pharmacological effect, at least in the peripheral nervous system, is that of α-adrenergic and dopaminergic receptor blockade (Andén et al., 1966; Gokhale et al., 1964; York, 1972). In the central nervous system, this drug has similar receptor antagonist effects, at least in some brain areas, but also seems to have some agonistic effects (Bradley et al., 1966; Schmitt et al., 1973) which may explain its stimulatory action on feeding. Furthermore, chlorpromazine has been shown to increase the neuronal outflow of hypothalamic NE, possibly due to feedback activation of NE neurons as a result of α-adrenergic receptor antagonism (Lloyd and Bartholini, 1975). It is of interest that the latency and duration for NE release are very similar to the time course of the eating response produced by hypothalamic chlorpromazine injection. The stimulatory effect of amphetamine (AMPH) on feeding was not as robust as that of chlorpromazine. This is somewhat surprising, since AMPH is known to be one of the most potent releasers of endogenous CA (Carlsson, 1970; Glowinski, 1970), and it does not appear to antagonize central CA receptors. One possible explanation is that AMPH has a very potent anorectic action which masks its stimulatory effect on feeding (see below).

The above studies provide an initial step in the direction of identifying feeding changes associated with drug-induced release of endogenous CA. The evidence is weak, however, for several reasons. In all studies, the sites of injection were outside the PVN area where NE is most effective in stimulating eating. The areas tested were the preoptic area or lateral ventricle (Evetts et al., 1972), perifornical area (Leibowitz and Miller, 1969; Slangen and Miller, 1969), ventromedial hypothalamus (Leibowitz, 1970b), or lateral hypothalamus (Montgomery et al., 1971). Chlorpromazine's effect in the perifornical area, for example, may have little to do with its effect in activating NE neurons which increase feeding but more to do with its antagonistic effect on the perifornical CA receptors that suppress feeding (Leibowitz, 1978c; see Section III.B). The drugs under investigation are very complex in their action on the central nervous system. Thus, what is clearly needed are more careful pharmacological analyses of the behavioral effects which they produce, to determine whether they are indeed a consequence of functional changes at CA synapses and, if so, what the properties of these changes are. Most of the above studies failed to yield dose-response analyses, as well as studies of the receptors mediating the elicited feeding effect. Furthermore, they also provided little evidence, such as through administration of CA synthesis inhibitors, that endogenous CA release had actually been affected by the drug. The chlorpromazine-induced eating effect was shown to be abolished by the α-adrenergic receptor antagonist phentolamine and actually potentiated by the β-receptor antagonist propranolol (Leibowitz, 1976). (Similar results were obtained with NE-elicited eating [Leibowitz, 1975b]). It was also found to be reversed by central injection of the CA depleter Ro 4-1284 or the synthesis inhibitor diethyldithiocarbamic acid, which in contrast had no effect on the eating response elicited by exogenous NE. While this evidence supports the possibility that chlorpromazine is acting indirectly through endogenous NE release, caution is needed since the doses required to elicit the feeding effect were quite high (Leibowitz and Miller, 1969), and a variety of brain sites have yet to be explored to assess the anatomical specificity of the phenomenon. Evetts et al. (1972) demonstrated that the eating elicited by 6-OHDA was blocked not only by phentolamine but also by the β-receptor antagonist sotalol. It is difficult to evaluate these findings since only one dose of the antagonists was examined. However, it is clear that a more thorough pharmacological and anatomical analysis of this problem is essential.

Neuronal uptake mechanisms are important in the rapid physiological inactivation of CA in the brain (see Iversen, 1973; Horn, 1976). Pharmacological intervention at the sites involved in the uptake process should, therefore, increase the availability of CA at the postsynaptic receptor sites and thus potentiate or prolong the effects produced by the CA. It has been shown that the presynaptic uptake inhibitor desipramine, which is relatively specific in its action on adrenergic neurons (Iversen, 1973; Horn, 1976), can reliably increase the feeding-stimulatory effect of injected NE (Booth, 1968; Slangen and Miller, 1969). This finding emphasizes the importance of neuronal uptake in determining the synaptic concentration of the CA neurotransmitters. In a recent study, several tricyclic antidepressants which affect this process were tested in the PVN to determine whether they alone would cause a sufficient increase in the endogenous NE levels at the receptor to elicit feeding (Leibowitz et al., 1978b). The drugs were desipramine, protriptyline, and amitriptyline, which, in addition to inhibiting presynaptic uptake of NE, are found at high concentrations to cause a release of NE from brain tissue (Hamberger, 1967; Hughes, 1978; Ng et al., 1970; Placheta et al., 1976). When injected into the PVN, each of these drugs reliably elicited eating in fully satiated animals. This effect was positively correlated in magnitude with the same effect observed with exogenous NE in the same animals. As with NE, the antidepressant eating response was abolished by α-adrenergic-receptor blockers in a dose-dependent manner but was unaffected by a variety of β-adrenergic, dopaminergic, serotonergic, histaminergic, and cholinergic blockers. It was also reversed by PVN pretreatment of the CA synthesis inhibitors α-methyltyrosine, Ro 4-4602, Fla-63, at doses that had no effect on the action of exogenous NE and thus left intact the postsynaptic receptor function (Fig. 7). Consistent with these data, which suggest that endogenously released NE may have an effect similar to that of exogenous NE, is the study of Martin et al. (1976), which, via a push-pull cannula, demonstrated an increase in the efflux of ^{14}C-labeled NE after perfusion of desipramine in the region of the PVN and VMN. This study's additional observation, however, of an increased efflux of inert [^{14}C] inulin leads one to question whether the effect of desipramine on [^{14}C] NE reflects a physiological release mechanism restricted to adrenergic neurons, or whether it indicates a nonspecific releasing action on other types of neurons or perhaps merely an extracellular shift of the residual exogenous amine. Thus, on the basis of the available data, one cannot draw any firm conclusions regarding the precise mechanism of antidepressant drug action, except to say that their facilitatory effect on food consumption depends on the integrity of presynaptic adrenergic neurotransmitter stores and postsynaptic α-adrenergic receptors.

Amphetamine is one of the most potent releasers of CA in the brain (Carlsson, 1970; Glowinski, 1970). While this drug, after medial hypothalamic injection, has in two studies been shown to facilitate eating in hungry rats (Leibowitz, 1970b, 1975c), a stimulatory effect in satiated rats has been difficult to reveal or, when apparent, is quite small and variable (Leibowitz, unpublished data). One possible explanation for this inconsistent response lies in the finding that centrally administered AMPH has a potent anorectic effect, which is mediated, at least in part, by CA neurons of the perifornical hypothalamic area (Leibowitz, 1978c; see below). The close proximity of this area to the medial hypothalamus, where adrenergic stimulation increases eating, suggests that AMPH's potential stimulatory effect on eating in the medial area may be camouflaged by its strong inhibitory effect in the lateral region. Tranylcypromine, a structural ana-

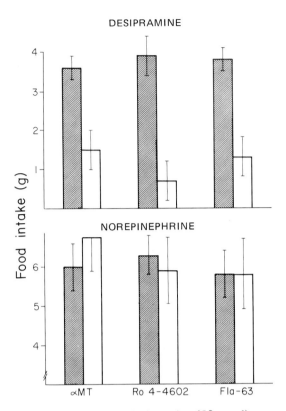

Figure 7 Food intake after injection of desipramine (50 nmol) or norepinephrine (50 nmol) in satiated rats (N = 22). The synthesis inhibitors α-methyltyrosine (αMT) at 160 nmol, Ro 4-4602 at 70 nmol, and Fla-63 at 40 nmol, were injected into the paraventricular nucleus 3 hr before the adrenergic stimulants. ▨, Desipramine or norepinephrine injected alone. ☐, Desipramine or norepinephrine injected in combination with a CA synthesis inhibitor. Each of the three inhibitors was found to inhibit reliably (at least at $p < 0.01$) the eating response induced by desipramine without affecting the norepinephrine response. (From Leibowitz et al., 1978b.) Reprinted with permission from *Progr. Neuro-Psychopharmacol.* 2, S. F. Leibowitz, A. Arcomano, and N. J. Hammer, Potentiation of eating associated with tricyclic antidepressant drug activation of α-adrenergic neurons in the paraventricular hypothalamus, 1978, Pergamon Press, Ltd.

logue of AMPH, does not appear to manifest this potent anorexic effect of AMPH in the perifornical area (Leibowitz and Rossakis, unpublished data), but it does appear to share AMPH's action in releasing NE from adrenergic neurons in the brain and also in blocking the intake of NE into these neurons (Hendley and Snyder, 1968; Schildkraut, 1970; von Euler, 1970; Ziance et al., 1977). In addition, tranylcypromine, used clinically as an antidepressant, is a potent inhibitor of the enzyme MAO. When this compound was injected into the PVN, it too was found to elicit feeding (Leibowitz et al., 1978a). As with the tricyclic antidepressants, the effect was selectively antagonized by α-adrenergic receptor blockers and markedly attenuated by CA synthesis inhibitors. Furthermore, two other MAO inhibitors, namely, pargyline and iproniazid, which are less effective than tranylcypromine in releasing and blocking the uptake of NE, were found to be less effective in eliciting eating.

These findings with different drugs reinforce the concept that endogenously released NE (or EPI), in the region of the PVN, is associated with increased eating behavior. These pharmacological results receive some support from biochemical studies which have correlated changes in eating behavior with changes in the level, turnover, or release of hypothalamic NE. Food deprivation was found to produce a decrease in NE levels (Glick et al., 1973b) and an increase in NE turnover (Friedman et al., 1973) when whole-hypothalamic measurements were taken. Consistent with the mapping study conducted with exogenous NE (Leibowitz, 1978a) and the above reports with PVN injection of antidepressants, biochemical analyses of NE metabolism have distinguished the medial region of the rostral hypothalamus, specifically the PVN and the dorsomedial and ventromedial areas, as exhibiting the most dramatic changes in NE turnover or release as a function of changes in eating behavior (Cruce et al., 1976; Martin and Myers, 1975; van der Gugten, 1977).

When the peripheral route of drug administration is used, the most common effect observed with drugs that stimulate brain CA function is a suppression of feeding. In rats, the antidepressants are found to reduce food intake, while at the same time cause a dramatic increase in activity level. No increase in feeding, like that obtained with PVN injection of these drugs, has yet been observed (Leibowitz, unpublished results). Thus, it appears that the peripheral route of drug administration may be of little help in elucidating the central phenomenon of NE-elicited feeding. One compound that does potentiate feeding when peripherally administered is the glucose analogue 2-deoxy-D-glucose, which inhibits the utilization of glucose in the brain. The results obtained with this drug will be described below in Section II.H. However, it is interesting to note that the effect of this drug on feeding may involve the release of endogenous NE from the medial hypothalamus and its activation of postsynaptic α-adrenergic receptors.

The anorectic property of peripherally injected drugs that affect brain CA mechanisms has been under investigation for many years (see Garattini and Samanin, 1978). The general consensus is that these drugs, specifically AMPH and related compounds, produce their effect through the release of the endogenous neurotransmitters, NE and DA. Until recently, the only studies that have provided information regarding AMPH's locus of action in the brain are electrolytic or neurotoxin lesion studies that have analyzed the potency of peripherally injected AMPH before and after damage to specific brain regions. These studies generally indicate the dependence of AMPH on the integrity of the lateral hypothalamus just lateral to the fornix, in contrast to the medial hypothalamus, anterior hypothalamus, septum, amygdala, area postrema, and caudate-putamen (see Leibowitz, 1978c). Direct evidence in support of the role of the lateral hypothalamus in mediating AMPH's anorectic effect comes from the findings of a central injection study in which AMPH was examined in 20 different sites throughout the mesencephalon, diencephalon, and telencephalon of the rat brain (Leibowitz, 1975c). The results of this study demonstrated that the most effective region for obtaining AMPH's anorectic effect was the perifornical region of the lateral hypothalamus, at the level of the VMN and PVN. This is the same area that was found to be most responsive to the feeding-suppressive effect of EPI as well (Leibowitz and Rossakis, 1979a; Section II.B.1). At this perifornical site, AMPH produced a reliable effect at a dose as low as 0.8 μg. These findings clearly differentiate AMPH's anorectic action from its effect in producing hyperactivity and stereotyped behavior. These latter phenomena have been associated with areas rostral to the hypothalamus, namely, the nucleus accumbens and caudate putamen (Iversen, 1977), structures which are totally unresponsive to the anorectic effect of locally injected AMPH (Leibowitz, 1975c).

There is extensive evidence in the literature to suggest that AMPH produces its behavioral effects through the release of endogenous CA from presynaptic terminals in the brain. Studies that have analyzed the potency of peripherally administered AMPH after depletion of brain CA have found its anorectic effect to be reduced or abolished, either by systemic injection of a CA synthesis inhibitor, α-methyltyrosine (Baez, 1974; Clineschmidt et al., 1974; Dobrzánski and Doggett, 1976; Frey and Schulz, 1973; Weissman et al., 1966) or by central injection of the neurotoxin 6-OHDA (Ahlskog, 1974; Fibiger et al., 1973; Heffner et al., 1977; Hollister et al., 1975; Samanin et al., 1975). Although there is some suggestion that DA may have a more primary role relative to NE (Heffner et al., 1977; Hollister et al., 1975), there is clear evidence to indicate a significant contribution of noradrenergic or adrenergic mechanisms as well (Ahlskog, 1974; Frey and Schulz, 1973; Holtzman and Jewett, 1971; Samanin et al., 1977). The failure of Franklin and Herbert (1977) to reverse AMPH's action with peripheral administration of the dopamine-β-hydroxylase inhibitor Fla-63 (which blocks the synthesis of NE or EPI from their precursor, DA) may be due to the fact that Fla-63 alone produced anorexia and thus precluded an increase of the feeding baseline. Studies in which AMPH as well as the synthesis inhibitors were injected centrally into the perifornical hypothalamic region, where AMPH is most effective, have demonstrated a total reversal of its anorectic action after injection of either α-methyltyrosine or Fla-63 (Leibowitz, 1975d). Neither synthesis inhibitor when injected alone produced a change in feeding, and, when administered in combination with exogenous EPI, a reliable potentiation of this agonist's anorectic action was observed. Until biochemical analyses to confirm local CA depletion are performed, firm conclusions on the significance of these results cannot be made. However, the finding that EPI's receptor action was actually enhanced by the synthesis inhibitors indicates that the postsynaptic adrenergic receptors were not only intact but were functioning at increased levels of sensitivity, possibly due to reduction in CA synthesis (Costentin et al., 1977).

Thus, it appears that AMPH may act presynaptically to release the CA in the area of the perifornical hypothalamus. Studies employing the peripheral route of drug administration have examined the question of the type of postsynaptic receptors which are acted upon by these endogenously released CA to suppress feeding. Dopamine receptor antagonists have consistently been shown to block AMPH anorexia (Barzaghi et al., 1973; Clineschmidt et al., 1974; Frey and Schulz, 1973; Heffner et al., 1977; Zis and Fibiger, 1975), although some question has recently been raised regarding the specificity of this antagonism (Samanin et al., 1978). The evidence obtained with the adrenergic antagonists is not as clear, primarily because of a lack of sufficient experimentation. In one study, the β-adrenergic blocker propranolol was found to abolish AMPH's anorexia effect in a dose-dependent manner (Sanghvi et al., 1975). This reversal, however, was not observed in another study in which a single high dose of propranolol was tested (Lehr and Goldman, 1973). An α-adrenergic blocker, phentolamine, (tested at a single dose) has also been shown to inhibit AMPH's anorectic action (Frey and Schulz, 1973); once again, this was not confirmed in two other investigations (Lehr and Goldman, 1973; Sanghvi et al., 1975). These differing results cannot be fully explained at this time, although it is clear that more extensive analyses of the adrenergic antagonists, with systematic variation of agonist and antagonist dose levels, are needed. With peripheral administration, the adrenergic antagonists, above a certain dose level, can by themselves produce a marked suppression of feeding. Thus, if such dose levels are examined, a reversal of AMPH's own anorexic effect will be difficult to reveal.

Evidence obtained with injection of the antagonists and AMPH directly into the perifornical lateral hypothalamic region indicates that the AMPH suppression of feeding can be blocked by β-adrenergic, as well as dopaminergic, antagonists but not by antagonists of α-adrenergic, serotonergic, or cholinergic receptors (Leibowitz, 1975d). At doses that had no effect on ad libitum food intake, the β-adrenergic blockers propranolol and sotalol totally reversed AMPH's action. The dopaminergic blockers haloperidol and pimozide were similarly effective (see Section III.B). These results, which direct our attention toward two types of receptors (β-adrenergic and dopaminergic) as potential mediators of AMPH anorexia, are further substantiated by the findings of an additional experiment in which the anorexia induced by *peripherally* administered AMPH could be attenuated by bilateral perifornical hypothalamic injection of a β-adrenergic blocker (producing a 42% reduction) and, to a greater extent, a dopaminergic blocker (a 75% reduction) (Leibowitz, 1975d). These results were obtained with low doses of the blockers that had no effect on AMPH anorexia when administered systemically and no effect on ad libitum food intake when centrally administered. These findings support the suggestion that AMPH, at least in part, acts centrally at the perifornical hypothalamic site of injection to produce its suppressive effect on feeding. The greater effectiveness of central dopaminergic blockade as compared with β-adrenergic blockade in antagonizing the anorexia of peripheral AMPH may reflect a more primary mediating role for postsynaptic dopaminergic receptors at this site (see Section III.B). However, the effectiveness of β-adrenergic blockade, particularly on central AMPH, in addition to the reversal obtained by the synthesis inhibitor Fla-63 which leaves DA stores intact, appears to indicate the simultaneous involvement of noradrenergic or perhaps adrenergic receptors in the perifornical hypothalamus. These receptors appear to be particularly sensitive to EPI, in contrast to NE, and to have properties characteristic of the β-adrenergic receptor. The interaction of AMPH with this type of receptor in the brain has been demonstrated via radioactive ligand-binding studies (Banerjee et al., 1978).

In support of the hypothesis that adrenergic-receptor mechanisms are active in the phenomenon of AMPH-induced anorexia is the finding that feeding suppression induced by perifornical hypothalamic EPI injection is reliably potentiated by desipramine, an inhibitor of CA uptake into adrenergic neurons (Leibowitz and Rossakis, 1978a). Benztropine, in contrast, which acts selectively to inhibit DA uptake into dopaminergic neurons, had no effect on the EPI response. Furthermore, *l*-dopa, when injected into the perifornical hypothalamus, has also been shown to suppress feeding, and its effectiveness is positively correlated with that of EPI as well as DA (Leibowitz and Rossakis, 1979c). The anorexia induced by *l*-dopa is found to be attenuated (but not blocked) by β-adrenergic and dopaminergic blockers, as well as by the dopa decarboxylase inhibitors Ro 4-4602 and MK-486. When peripherally administered, *l*-dopa has similarly been reported to induce anorexia (Baez, 1974; Dobrzánski and Doggett, 1976; Heffner et al., 1977; Sanghvi et al., 1975). In one of these studies, this effect was antagonized by the dopaminergic blocker spiroperidol (Heffner et al., 1977) and in another study by the β-adrenergic blocker propranolol (Sanghvi et al., 1975). These results suggest that peripheral *l*-dopa may in part be acting specifically through perifornical CA neurons to inhibit feeding. In support of this proposal, *l*-dopa, after peripheral administration, is known to be taken up within the hypothalamus, most particularly in the CA-rich perifornical and periventricular areas of this structure (Placidi et al., 1976). Furthermore, it has also been demonstrated to increase the synthesis of NE as well as DA in the hypothalamus (Glowinski and Iverson, 1966; Hyppa et al., 1971; Langelier et al., 1973). (Synthesis of EPI has not been examined.)

The above studies have explored the hypothesis that changes in synaptic concentration of brain-synthesized NE (or possibly EPI) are effective in modulating a behavioral response, namely, feeding. Pharmacological studies have revealed that drugs known to act on the presynaptic terminal to release the transmitter produce marked changes in feeding, changes that cannot occur in the absence of endogenous NE stores. Furthermore, biochemical analyses have been successful in correlating hypothalamic NE concentrations or release with changes in spontaneous feeding behavior. While this evidence constitutes an important step in the direction of establishing a neurotransmitter role for NE or EPI in feeding mechanisms, more systematic tests and analyses of more discrete brain areas are required. It is clear that the complexity of the brain would not permit a single neurotransmitter to assume only one function in regulating a behavior, and anatomical as well as pharmacological dissociation constitutes our only hope at present of identifying the various functions it may serve.

C. Drinking Behavior

The evidence reviewed in the previous sections indicates that hypothalamic adrenergic-receptor mechanisms modulate feeding behavior in a variety of species and that, at least in the rat, this modulatory influence occurs in a bidirectional manner, depending upon the brain area involved. In addition to affecting food consumption, however, hypothalamic injection of adrenergic stimulants has also been shown to alter water intake. The relationship between brain CA and drinking behavior has received relatively little attention in the literature. It has been known for some time that peripherally administered AMPH causes a suppression of drinking behavior in the rat (see Soulairac and Soulairac, 1970). However, the mechanism mediating this phenomenon remains obscure. It has been suggested that the drinking suppression is secondary to the suppression of feeding produced by this drug (Dobrzánski and Doggett, 1976; Glick and Greenstein, 1973). However, a drinking suppression can be observed in the absence of food (Dobrzánski and Doggett, 1976; Neilson and Lyon, 1973). The AMPH effect on water intake was inhibited by the CA synthesis inhibitor α-methyltyrosine, which indicates that it may act through the release of endogenous CA (Holtzman and Jewett, 1971). Peripherally injected antidepressant drugs have also been shown to suppress drinking; however, in a test with α-methyltyrosine, which by itself reduced water intake, the effect of the antidepressant drug was not reversed (Zabik et al., 1977).

Peripherally injected α-adrenergic agonists (e.g., metaraminol) also suppress drinking, whereas β-adrenergic agonists (e.g., isoproterenol) are found to stimulate drinking (Lehr et al., 1967). Epinephrine, which acts on both α- and β-adrenergic receptors, can produce either the suppressive or the stimulatory effect, under conditions where one or the other type of receptors is pharmacologically and selectively blocked. The possibility that α- and β-adrenergic receptors might exist in the brain to modulate drinking behavior was first suggested by Lehr et al. (1967) who found that hypothalamically injected isoproterenol produces a small increase in water intake. A similar phenomenon was later described by Leibowitz (1971) and Mountford (1969). Subsequent studies, however, showed that the doses of isoproterenol required to produce drinking with central injection were as high or higher than those used for peripheral administration (Fisher, 1973; Lehr, 1973). This suggests that the centrally injected β agonist may have leaked into the peripheral circulation and acted perhaps on the kidney to release renin and potentiate drinking (Houpt and Epstein, 1971).

Studies with central injection of NE and EPI have provided clearer evidence with regard to the participation of central adrenergic receptors in the control of drinking. Grossman (1962a) first observed an inhibitory effect of lateral hypothalamic NE injection on water consumption in the thirsty rat. This effect was subsequently confirmed by several investigators (Hendler and Blake, 1969; Hutchinson and Renfrew, 1967; Lovett and Singer, 1971). In a series of mapping studies (Leibowitz, 1971, 1973a, 1975e), NE was tested at a variety of sites throughout the forebrain of the thirsty rat and was found to be generally ineffective in suppressing drinking at sites located outside the hypothalamus. Within the hypothalamus, the lateral sites of injection were similarly ineffective, whereas the medial sites along the periventricular region consistently yielded a reliable suppression of drinking. The most responsive anterior-posterior region of the medial hypothalamus was at the level of the PVN. Further analyses of this phenomenon demonstrated that it was dose dependent and that it could occur reliably at a dose of NE as low as 0.9 ng (Leibowitz, 1973a, 1975e). This dose is slightly lower than the threshold dose (4 ng) observed for the phenomenon of feeding stimulation induced by PVN NE injection in satiated rats (Leibowitz, 1978a). Preliminary evidence indicates that the receptors mediating the drinking suppression, like those involved in the feeding stimulation, are α-adrenergic in nature (Leibowitz, 1972a, 1975e). The effect was produced by EPI, as well as by NE, and was antagonized by the α-adrenergic blockers phentolamine and tolazoline but not by the β-adrenergic blockers propranolol and sotalol.

In addition to suppressing the water intake of thirsty rats, NE has been found to attenuate the drinking produced by a variety of dipsogenic stimuli (Leibowitz, 1972b, 1975e; Setler, 1975; Singer and Kelly, 1972). At a dose of 1.5 μg, NE injected into the PVN produced approximately a 50% reduction in water consumption elicited by peripherally administered hypertonic saline or polyethylene glycol and centrally administered carbachol and angiotensin (Leibowitz, 1972b). While there was no evidence that NE may differentially affect the drinking responses elicited by these various dipsogenic stimuli, further dose-response analyses will need to be conducted to test the degree of specificity of this NE phenomenon (Setler, 1975).

In a series of tests in satiated rats, the time course of this NE drinking-suppressive effect was analyzed in conjunction with the stimulatory effects which NE is found to have on ingestive behavior (Leibowitz, 1975a, b). The drinking-suppressive effect appeared to occur at the same time as the feeding-stimulatory effect. This observation raises the possibility that the drinking suppression is simply a consequence of the NE-induced increase in feeding or hunger. Since the drinking suppression occurred whether or not food was present, it appears that this phenomenon does not result from competition at the behavioral level. However, within the central nervous system, it is possible that NE, through stimulation of receptors involved in feeding regulation, may exert an antagonistic effect on other behaviors, in particular drinking, which may interfere with the on-going feeding response. Evidence counter to this possibility is provided by the finding that hypothalamically injected EPI or AMPH can produce a suppression of drinking in the absence of a feeding-enhancement effect (Leibowitz, 1970b, 1973b; Leibowitz and Rossakis, 1978b). Further analysis of AMPH's action revealed a dose-dependent suppression of drinking at doses of 1-100 nmol and a reversal of this effect with hypothalamic injection of an α-adrenergic receptor blocker. It remains to be determined whether this effect of hypothalamic AMPH injection is the same as that observed with peripheral AMPH injection (Soulairac and Soulairac, 1970). In view of AMPH's CA-releasing properties, the above evidence is at least consistent with the hypothesis that hypothalamic NE (or EPI), particularly within the medial hypothalamus, may have a direct inhibitory effect on water ingestion.

While the main impact of NE on drinking in the *thirsty* rat is clearly inhibitory, this agonist has been found to have an additional effect on the drinking pattern of the *satiated* animal (Fig. 8). The above evidence demonstrated that medial hypothalamic injection of NE produced a feeding response associated with a drinking inhibition. Immediately after NE injection, just prior to the onset of eating, satiated rats also exhibited a small but vigorous drinking response (Leibowitz, 1975a, b, 1978a; Slangen and Miller, 1969). This "preprandial" drinking, similar to the feeding, occurred most effectively with stimulation of the PVN, after a latency of less than a minute (Leibowitz, 1978a). The threshold dose for this phenomenon was between 5.6 and 16.9 ng of NE, somewhat higher than the threshold dose (1-4 ng) for the elicited feeding response (Leibowitz, 1978b). The response lasted for 1-3 min, during which the rats drank up to 4 ml of water (5-10% of total daily intake). If food was available, the rat then initiated feeding within a minute after drinking ceased. Pharmacological analyses of this phenomenon indicated that, like the feeding response, the elicited drinking could be antagonized by hypothalamic administration of α-adrenergic receptor blockers, but not by blockers of dopaminergic or cholinergic receptors. Unlike the feeding, however, the drinking response was susceptible to blockade by β-adrenergic receptor antagonists. Further studies will need to be conducted to analyze the significance of this finding and the possible existence of central β-adrenergic receptors that are stimulatory towards drinking.

A particularly important outcome of the above studies on NE injection in the satiated rat was the observation that the sequence of ingestive responses exhibited after NE (Fig. 8) was remarkably similar to the pattern of ingestive behavior exhibited normally by rats under laboratory conditions (Leibowitz, 1975a). Rats normally take between 6 and 10 well-separated meals each day (Kissileff, 1969; Fitzsimons and Le Magnen, 1969). As with the eating response induced by central adrenergic stimulation, the eating exhibited by rats at a normal meal occurs vigorously and continuously over a period of 5-15 min, and the size of the meal in both cases generally varies between 2 and 3 g. Drinking behavior is rarely observed during the meal. However, under normal conditions, rats frequently drink a small amount of water (0.5-3.0 ml) a few minutes before or after the meal. After adrenergic stimulation, the response most often occurs before the meal, and, as with spontaneously elicited preprandial drinking, it lasts 2-3 min, occurring vigorously and continuously throughout this interval, and then stops for a minute or so until eating is initiated. As with the drinking behavior observed under normal conditions, the amount of water ingested by individual rats after NE injection was positively correlated

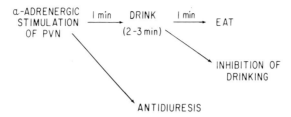

Figure 8 Temporal sequence of α-adrenergic effects on rat ingestive behavior and urine formation observed after paraventricular nucleus (PVN) injection of NE.

with the size of his corresponding meal. Perhaps 70% or more of a rat's normal water intake is closely associated with feeding, and it appears from the above results that this same type of drinking behavior, called "food-associated" drinking (Kissileff, 1969), can also be observed after central adrenergic stimulation.

From the above analysis, it becomes clear that natural ingestive behavior and adrenergically elicited ingestive behavior bear striking similarities. These similarities provide sound support for the hypothesis that the responses observed with exogenous adrenergic stimulation reflect the normal function of an endogenous adrenergic system in controlling ingestive behavior. This hypothesis is further supported by the finding (see above) that quite low doses of NE or EPI, approaching physiological levels, are effective in producing these behaviors.

D. Water Excretion

From the above evidence, it becomes clear that hypothalamic NE or EPI has quite dramatic effects on water consumption. If these neurotransmitters have a more general function in modulating body-water homeostasis, they would be expected to alter water excretion in addition to water ingestion. In a variety of species, the adrenergic stimulants have been examined in the hypothalamus, and their effect on urine formation is found to be predominantly inhibitory in nature.

Studies with peripheral drug administration have generally indicated that α-adrenergic stimulation inhibits release of antidiuretic hormone (ADH) and thus increases urine formation, whereas β-adrenergic stimulation potentiates ADH release and consequently decreases urine formation (see Bisset, 1968; Ginsberg, 1968; Lehr et al., 1967). The pattern observed with central drug administration is quite different. In an early study by Duke et al. (1950) in the dog, EPI injected directly into the supraoptic nucleus (SON) was found to potentiate the antidiuresis induced by cholinergic stimulation. More recent work in the anesthetized dog (Bhargava et al., 1972) and rat (Bridges et al., 1976) revealed an antidiuresis and increased release of ADH after lateral ventricular injection of NE or EPI. This effect was inhibited by the α-adrenergic receptor blocker phentolamine. In the goat (Olsson, 1970; Vandeputte-Van Messom and Peeters, 1973) and rat (Kuhn, 1973, 1974; Morris et al., 1976), a similar pattern of results was observed with NE or EPI injection into the third ventricle. Only a few studies have examined the effects of these CA injected directly into the SON, a likely site for the mediation of ADH release. The consequence of these injections in the rat and cat was, once again, ADH

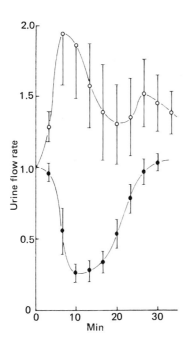

Figure 9 Effects of adrenergic blockers microinjected into the supraoptic nucleus on the antidiuretic response to 5 µg intranuclear norepinephrine. Solid circles represent the mean rate of urine flow in rats pretreated with the β-adrenergic blocker dichlorisoproterenol, which had no impact on the norepinephrine antidiuresis and open circles represent the mean for pretreatment with the α-adrenergic blocker phenoxybenzamine, which reversed the norepinephrine effect. (From Urano and Kobayshi, 1978.)

release, decrease in urine output, and increase in urine osmolality (Garay and Leibowitz, 1974; Milton and Paterson, 1974; Urano and Kobayashi, 1978) (see Fig. 9). This effect of NE and EPI in hydrated rats was found to be dose dependent and could be observed at remarkably low doses, as low as 0.09 ng (Garay and Leibowitz, 1974). It was also mimicked by drugs, namely, desipramine and AMPH, which are known to enhance the release of endogenous NE, and this effect could not occur after local drug-induced inhibition of NE synthesis (Leibowitz and Tom, unpublished data). A mapping study of adrenergically induced antidiuresis revealed that sites outside the hypothalamus (septum, caudate, and nucleus accumbens) were insensitive to NE, as well as sites in the rostral and caudal regions of the lateral hypothalamus (Garay and Leibowitz, 1974). In the medial hypothalamus, the PVN was quite sensitive, although the SON generally yielded a significantly larger antidiuretic response. The ventromedial area of the hypothalamus responded to high doses of NE; this effect, however, may very possibly be due to drug spread along the outside of the cannula shaft and into the area of the PVN.

Pharmacological analysis of NE- or EPI-induced antidiuresis consistently reveal the importance of α-adrenergic receptors in mediation of the response (see Fig. 9) (Garay and Leibowitz, 1974; Milton and Paterson, 1974; Urano and Kobayashi, 1978). In the rat, the α-adrenergic antagonists phentolamine, tolazoline, and phenoxybenzamine were each effective in reversing the antidiuretic action of NE or EPI in the SON, whereas antagonists of dopaminergic, serotonergic, and cholinergic receptors had no effect (Garay and Leibowitz, 1974). Blockade of β-adrenergic receptors generally left the adrenergic antidiuresis intact, although with the β blocker propranolol an inhibition of the effect was occasionally observed. This was sometimes seen with the *d*- as well as the *l*-isomer of propranolol, which indicates that the inconsistent reversal may be attributed to the local anesthetic properties of this antagonist. The β-adrenergic blocker sotalol did not affect the NE-induced antidiuresis. In the cat (Milton and Paterson, 1974), the ADH release produced by SON injection of NE was attenuated by phentolamine and blocked by propranolol. Other β-adrenergic blockers were not tested in this study; thus, it remains unclear whether propranolol's action is due to β-receptor blockade or to its nonspecific depressing effect on neuronal function. If central β-adrenergic receptors do indeed exist to modulate ADH release, there is preliminary evidence to suggest that they may act in an antagonistic fashion to the α-adrenergic receptors. This evidence comes from the study of Bhargava et al. (1972) in the dog, which showed ventricular injection of the β-adrenergic agonist isoproterenol or low doses of EPI to produce diuresis and decreased release of ADH, a propranolol-sensitive phenomenon. In the rat, however, isoproterenol was reported to produce antidiuresis (Garay and Leibowitz, 1974; Morris et al., 1977).

Two exceptions to the accumulated evidence revealing an excitatory effect of central adrenergic stimulation on ADH release are the studies of Blundell and Herberg (1973) and Wolny et al. (1974). Blundell and Herberg demonstrated that NE injected into the "basal diencephalon" of the rat caused an increase in urine output associated with a decrease in osmolality. This effect was positively correlated with the increase of feeding response also induced by NE. On the basis of this evidence, it was hypothesized that NE may be acting on a central mechanism that controls not only food intake but also water intake, urine output, and sodium intake. The evidence bearing on this interesting question will be discussed in sections below. With regard to the stimulatory effect of NE on urine output, further dose-response and pharmacological studies will need to be conducted to evaluate its relation to NE's predominant antidiuretic effect described above. Since the site of injection was not clearly defined in the Blundell and Herberg study, this will also need to be taken into account. It should be noted that the dose of NE used in this study (65 nmol or approximately 21 μg) was very high compared with the doses (0.1-1000 ng) which were found to produce antidiuresis when administered directly into the SON (Garay and Leibowitz, 1974). This dose differential may help to explain the results of Wolny et al. (1974) which also revealed a diuretic effect of NE injected into the lateral ventricles of the rat. These investigators tested 100 μg of NE and found the diuresis produced by this dose to be reversed by α-adrenergic blockade.

There is additional evidence to support the idea that central α-adrenergic receptors mediate the release of ADH. Mills and Wang (1964) demonstrated that electrical stimulation of the SON or PVN caused a release of the hormone, and peripheral injection of the α-adrenergic blocker dibenzyline inhibited this release. Similarly, ADH secretion associated with electrical stimulation of the lower brain stem or of peripheral nerves was also reversed by peripheral injection of α-adrenergic blockers. De Wied and László (1967) and Bridges and Thorn (1970) revealed a similar attenuating effect of α-adrenergic blockade on ADH release caused by peripheral hypertonic saline injections.

This evidence, together with the results of the central injection studies described above, support the existence of central adrenergic neurons, which function through postsynaptic α receptors located in the SON and PVN to produce a release of ADH and a suppression of water excretion. Histochemical, immunocytochemical, and biochemical studies have revealed a relatively dense concentration of NE- and EPI-containing fibers in both nuclei (see Section II.A and Fig. 4), and there is evidence that these fibers make contact with neurosecretory cells in these areas (McNeill and Sladek, 1978). It was found, however, that the major CA innervation to the SON and PVN does not coexist with the major distribution of magnocellular perikarya, which suggests that the CA influences on neurons in these nuclei may be via an axodendritic mechanism or possibly an interneuronal pool. Evidence to suggest that α-adrenergic neurons do not directly excite neurosecretory cells to release ADH is found in the results of microelectrode recording studies. These studies have shown NE and EPI to be predominantly (but not exclusively) inhibitory in their effects on antidromically identified neurosecretory cells of the SON (Barker et al., 1971), and 6-OHDA, which destroys adrenergic innervation of the SON and PVN, caused an increase in firing rate of neurosecretory neurons (Young, 1975). The relationship between the adrenergic synapses and proposed osmoreceptors which promote the release of ADH has not been delineated. Although peripheral administration of α-adrenergic blockers was demonstrated to inhibit the ADH release caused by peripheral hypertonic saline injection (de Wied and László, 1967; Bridges and Thorn, 1970), Milton and Paterson (1974) failed to replicate this effect with SON injection of the adrenergic blockers and hypertonic saline. Injection of 6-OHDA into the lateral ventricles caused a loss of CA terminals in the SON and a decrease in the basal level of ADH release (Milton and Paterson, 1973). In this preparation, ADH release induced by hypertonic saline injected into the SON was abolished. This evidence is consistent with the hypothesis of an adrenergic neuronal link between osmoreception and ADH release, although the adrenergic receptor may simply have a modulatory or permissive role on the complex interactions which occur in the SON.

At least in the rat, adrenergic stimulation of the hypothalamus appears to be associated with water conservation, in terms of ADH-mediated effects on kidney function, but it also appears to signal water excess as reflected by its profound inhibition of water ingesttion. Both of these effects are induced by remarkably low doses of the adrenergic neurotransmitters, which argues in favor of their reflecting a physiological function. The relationship between the two phenomena, however, remains unclear, since they do not lend themselves to any unitary concept of body-water homeostasis. They cannot be dissociated on pharmacological grounds, as they both depend on the α-adrenergic receptor. Mapping studies, however, suggest that they may be anatomically differentiated, with the drinking suppression localized to the periventricular region and the ADH release localized to the SON and PVN. The apparent anatomical contiguity of the receptors involved in drinking suppression and feeding stimulation (see Sections II.B.1 and II.C) suggest that these two particular phenomena may be more closely related and may reflect a unitary process involved in coordinating information relating to the hydrational state. It is interesting to note, however, that with the initiation of eating comes a need for water conservation, as reflected in a rapid antidiuresis (Kakolewski et al., 1968) and a vigorous but brief preprandial drinking response (Kisseleff, 1969; Fitzsimons and Le Magnen, 1969). Both of these effects, as well as the eating itself, are produced by NE injected into the PVN, at doses near physiological levels (Fig. 8).

E. Sodium Ingestion and Excretion

In addition to controlling water balance through ingestion and excretion, the hypothalamus appears to play an important role in electrolyte homeostasis, in particular, sodium homeostasis (see Andersson, 1975, 1977; Covian et al., 1975; Epstein, 1978). The possibility that central adrenergic synapses are involved in the mediation of this process has received limited attention. The available evidence suggests that hypothalamic NE or EPI may promote sodium excretion, through an unknown effector mechanism, and possibly cause an increase in sodium ingestion.

The studies described in Section II.D above clearly demonstrated that hypothalamic adrenergic stimulation causes a release of ADH and a decrease in water excretion. In a few of the investigations, sodium and potassium were measured in the urine, and, after NE injection into the third ventricle (Morris et al., 1976, 1977) or into the SON (Garay and Leibowitz, 1974), an increase in the excretion of both electrolytes was found to accompany the decrease in urine formation. The sodium/potassium ratio was not altered in these studies, which suggests that the observed effects may be attributed to ADH release which causes similar changes in electrolyte as well as water excretion (Chan, 1965). It has been argued, however, that changes in blood levels of ADH are not completely responsible for centrally induced changes in sodium excretion, since the natriuresis produced by central hypertonic saline injection occurred despite the administration of large doses of Pitressin (Zucker et al., 1974). Furthermore, in the study by Morris et al. (1976), the natriuretic and kaliuretic responses induced by NE as well as hypertonic saline injected into the third ventricle were abolished by median-eminence lesions which only partially attenuated the antidiuresis elicited by these substances.

With lateral hypothalamic injection of NE or the α-receptor agonist metaraminol, Pillar et al. (1977) observed an increase in electrolyte excretion plus an increase in the sodium/potassium ratio. In this study, as in the study of Morris et al. (1977) where NE was injected into the third ventricle, the elicited natriuresis was antagonized by α-adrenergic blockade but enhanced by β-adrenergic blockade. In both of these investigations, in contrast, the β-receptor agonist isoproterenol was associated with the opposite effect, namely, a decrease in urinary sodium and potassium. This evidence led these authors to postulate the existence of a hypothalamic adrenergic mechanism consisting of α and β receptors which function antagonistically in the regulation of electrolyte excretion. While central natriuresis has been proposed to involve a natriuretic hormone (Morris et al., 1976, 1977), the possible involvement of alterations in blood pressure, glomerular filtration rate, or renal neural input has yet to be investigated in these studies with central adrenergic stimulation.

The possibility of an adrenergic link in the control of sodium ingestion has been proposed by Chiaraviglio and Taleisnik (1969). These investigators demonstrated that third ventricular injection of NE in the satiated rat produced an increase in NaCl (1%) intake of approximately 5 ml during the 30 min test period (compared with 1-2 ml after injection of the vehicle). Furthermore, the increase in sodium intake exhibited by sodium-depleted animals was found to be blocked by an α-adrenergic blocker. It was also inhibited by injection of the CA depleter reserpine, the synthesis inhibitor α-methyltyrosine, and by midbrain lesions that damage CA fibers ascending to the forebrain (Chiaraviglio, 1972, 1976; Chiaraviglio and Taleisnik, 1969). This evidence, suggesting that sodium ingestion may depend on endogenous NE stores, is consistent with the finding that CA depletion caused by intracisternal injection of 6-OHDA is associated with a reduced preference for saline solution in animals treated with deoxycorticosterone (Breese et al., 1973).

Antunes-Rodrigues and McCann (1970) described a similar effect of ventricular injection of NE on sodium ingestion, although the effect observed (only an increase of 1 ml in saline consumption) was smaller than that described by Chiaraviglio. This may be attributed to the fact that Antunes-Rodrigues and McCann used a more concentrated NaCl solution (1.5%), compared with the more strongly preferred 1% solution used by Chiaraviglio. This would appear to be the case, in light of the additional finding that PVN injection of NE induced a 3- to 5-ml response with a 0.5 to 1% solution but only a 1- to 2-ml response with a 1.5% solution (Leibowitz, unpublished data). Detailed analysis of this response in PVN-injected rats revealed that it generally occurred after a brief latency (1-2 min) and ceased within approximately 10-15 min after injection. Furthermore, it was found to be antagonized by the α-adrenergic blocker phentolamine. While these results are consistent with the proposed role of hypothalamic adrenergic systems in control of sodium intake, it was observed that the ingestive response that occurred with 1% NaCl after PVN NE injection was essentially identical to the previously described "preprandial" drinking response that occurred in the presence of water (Leibowitz, 1975b, 1978a). That is, as discussed in Section II.C above, satiated rats receiving an injection of NE in the PVN exhibited a small (up to 5 ml) but vigorous increase in water ingestion during the first few minutes after injection (Fig. 8). This response was then immediately followed by a feeding response which was also associated specifically with the PVN (Leibowitz, 1978a). As to whether the increase in water intake and NaCl intake are indeed one and the same response, this cannot be determined at this time. In evaluating this question, however, it will be important to distinguish between the ingestion of sodium in response to a homeostatic need versus ingestion in response to the enhanced pallatability of dilute solutions of sodium salts as compared to water. While this distinction may be difficult to make in the case of the small fluid ingestive response elicited by injected NE, it appears that the effects obtained with CA depletion in sodium-depleted rats reflect deficits in sodium balance and occur specifically with saline and not water ingestion (Chiaraviglio and Taleisnik, 1969; Chiaraviglio, 1972).

F. Lesion Studies of Adrenergic Pathways

In the previous sections, it has been shown that hypothalamic adrenergic stimulation has a variety of effects on feeding behavior, water ingestion, water excretion, and possibly sodium ingestion and excretion. The evidence to date reveals the PVN, the SON, and the perifornical area as being the three most responsive sites to the modulatory effects of central NE or EPI injection. As described in Section II.A above, these three areas are found to be densely innervated by CA fibers (Figs. 3 and 4), and the cell bodies from which these fibers originate are located in the pons-medulla region of the brain stem (Fig. 1). Only recently have these central adrenergic projections been investigated with the focus of determining whether they have a physiological function in the regulation or modulation of natural ingestive behavior. This task is made particularly difficult by virtue of the complexity of the ascending CA fiber systems that innervate the hypothalamus. Each area of the hypothalamus may receive input from a number of cell groups in the lower brain stem, and each cell group may project to a number of hypothalamic areas. Furthermore, the axons of the CA systems are highly collateralized, such that the projections to diverse areas of the forebrain may originate from a single cell body. In light of these complexities, it becomes clear that it is essentially impossible to damage selectively a specific CA pathway. Even if direct damage were to occur to only one pathway, indirect, transsynaptic changes would be expected to develop in other CA systems which interact with the damaged pathway. Despite these difficulties, lower brain-stem lesion studies are fruitful under conditions where the above limitations are fully recognized, and the nature and extent of neural damage is carefully determined.

There are a number of studies which have shown midbrain lesions to have profound effects on feeding behavior, drinking behavior, and urine formation and have implicated CA pathways in the mediation of these effects. For example, electrolytic lesions in the area of the locus ceruleus (A6) cell bodies have been shown to produce an increase in water consumption (Osumi et al., 1975; Redmond et al., 1977; D. Roberts et al., 1976). The possibility that this effect may be attributed to damage to the noradrenergic cell bodies has been questioned by D. Roberts et al. (1976) who failed to replicate the effect with locus ceruleus injections of the selective neurotoxin 6-OHDA. It is interesting that knife cuts in the dorsal pontine tegmentum, which undoubtedly damaged an extensive number of the ascending locus ceruleus fibers, were also found to cause hyperdipsia (Grossman, 1978). While these cuts generally reduced hypothalamic NE, the possibility of a causal relationship between these neurochemical and behavioral changes has yet to be explored. The effect of locus ceruleus lesions on food intake is not clear, since either a decrease (Koob et al., 1976; Osumi et al., 1975; Ross et al., 1973), an increase (Halperin et al., 1978; Redmond et al., 1977) or no change (Anlezark et al., 1973) have been reported. It is clearly premature to speculate on the significance of these findings relative to the NE cell body damage and forebrain axon degeneration. What is needed is a histofluorescence analysis of the damage caused by this lesion and discrete modifications of the lesion, to assess whether there exists any correlation between the behavioral consequences and specific fiber loss. In addition to changes in feeding and drinking, these lesions in the area of the locus cell bodies were found to cause a urinary disturbance, characterized by dilatation of the bladder, hematuria, and urine retention for several days after the lesion (Osumi et al., 1975). Tests conducted after this initial effect had ceased revealed no change in urine volume and specific gravity, except under the emergency conditions of water deprivation, where the lesioned animals failed to concentrate their urine as did the control animals. A similar effect on urine excretion was noted in rats with lesions in the dorsal midbrain tegmentum, which would be expected to damage locus fibers as they follow a rostroventral course through the brain stem (Leibowitz and Brown, 1980c). Through fluorescence histochemistry, it was determined that this lesion caused a small reduction in hypothalamic CA innervation, apparently localized to the PVN, and periventricular and dorsomedial nuclei (Leibowitz and Brown, 1980a). The SON was not studied.) This finding is therefore consistent with the evidence that NE stimulation of the PVN (and SON) caused ADH release and a consequent reduction in urine excretion (see Section II.D). The possibility that locus fibers contribute to this hypothalamic function is supported by the biochemical evidence that locus ceruleus lesions produce a decrease in NE content of both the PVN and SON (see Section II.A).

Due to its compact arrangement and relative accessibility, the A6 NE cell group has received considerably more attention in lesion studies compared to that given other pontine or medullary NE cell groups. A5 and A7 lie more ventrally and are more

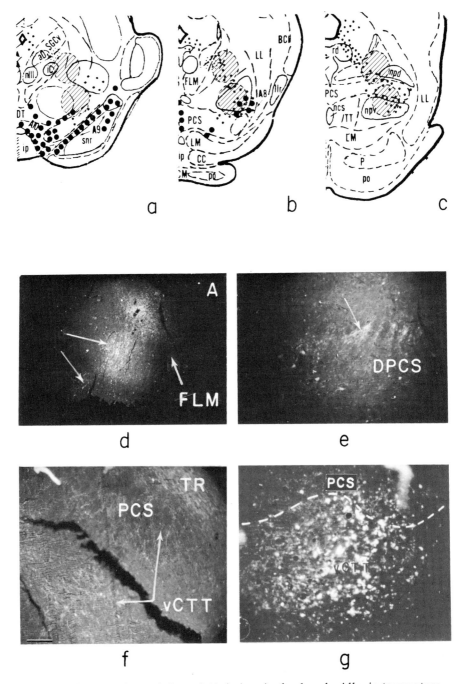

Figure 10 (a-c) Illustrations of electrolytic lesions in the dorsal midbrain tegmentum (ventrolateral to the central gray matter) and the ventral midbrain tegmentum (dorsolateral to the medial lemniscus), superimposed on three frontal sections showing CA varicose fibers (small black dots) and cell bodies (large black dots) as drawn by Palkovits and Jacobowitz (1974). The dorsal lesion overlaps adrenergic fibers coursing dorsal to

Feeding and Drinking and Water-Electrolyte Excretion

scattered, and the medullary cell groups (A1-A4) are difficult to study due to this brain region's role in the control of vital functions (Fig. 1). Most studies have therefore focused on the CA projections at the level of the pons and midbrain, where the fibers unfortunately become more scattered and complex in their arrangement. Consequently, interpretation of the results becomes very difficult. Lesions of the compact DTB, as it courses along the ventrolateral edge of the central gray matter, have generally failed to reveal any dramatic changes in food or water intake (e.g. Ahlskog, 1974; Ungerstedt, 1971b). However, electrolytic midbrain lesions ventrolateral to the central gray matter and dorsolateral to the red nucleus have been shown to result in severe aphagia and adipsia in the rat (Gold, 1967; Lyon et al., 1968; Parker and Feldman, 1967). (See dorsal lesion in Fig. 10.) Further analyses of this phenomenon revealed that it was associated with hyperactivity and hyperexploration, as well as a failure of the rats to attend to and localize sensory stimuli and to utilize relevant sensory information in making appropriate adaptive responses to the environment (Leibowitz and Brown, 1980c). These functional deficits, which may be attributed to the destruction of multiple, long, ascending and descending pathways that provide the basis for normal sensory integration (Sprague et al., 1961), very possibly contributed to the decrease in food or water intake but would appear not to be the primary cause in view of the differences observed in the time course of the recovery of these deficits. Damage to visceral and gustatory afferents which course through the dorsal midbrain tegmentum (Norgren, 1976) may also have contributed to the feeding and drinking deficit. However, what is most important for the present discussion is the finding that 6-OHDA lesions in the same area of the dorsal midbrain tegmentum did not produce the aphagia and adipsia observed with the electrolytic lesion but instead produced a very small reduction (10 to 15%) of food intake associated with an increase in water intake (Leibowitz and Brown, 1980a). This clearly establishes, therefore, that the aphagia and adipsia were not a consequence of damage to specific CA projections.

Figure 10 *(continued)*
the superior cerebellar peduncle (PCS) in (c), whereas the ventral lesion overlaps fibers coursing ventral to the PCS. See page 405 for abbreviations. (d-e) Damaged dorsal central tegmental tract (CTT) fibers and tegmental radiation (TR) fibers shown in fluorescence photomicrographs caudal to the dorsal tegmental lesion illustrated in (a-c). In (d), the aqueduct (A) and medial longitudinal fasciculus (FLM) appear to the right, and the top arrow shows the build-up of CA fluorescence in the dorsal CTT fibers and the bottom arrow shows the CA fluorescence in a few TR fibers. Additional fluorescence can be seen at the top of the picture within the central gray matter (to the right) and is a normal, undamaged preterminal area. In (e), the arrow indicates appearance of damaged dorsomedial TR fibers just anterior to (d) and ventrolateral to the central gray matter. Fibers radiate along the rostral-dorsal edge of the decussation of the superior cerebellar peduncles (DPCS) in the dorsal tegmentum. (f-g) Ventral CTT fibers (vCTT) coursing ventral to the PCS just caudal to the level of the ventral tegmental lesion shown in (a-c). In (f), the fluorescence photomicrograph of brain tissue from an intact rat shows some vCTT fibers following a longitudinal course ventral to and within the PCS and TR fibers radiating along the dorsal surface of the PCS. In (g), the photomicrograph shows the autofluorescent reaction that is apparent immediately behind the ventral tegmental lesion, in an animal killed two months after lesion. This reaction falls directly in line with the adrenergic vCTT projection shown in (f). (Based on data from Leibowitz and Brown, 1980a,b.)

There is some evidence to suggest, however, that more subtle changes in food or water ingestion may have resulted from CA loss (Leibowitz and Brown, 1980a, c; Leibowitz, unpublished data). Electrolytic lesions just ventrolateral to the central gray matter were found to produce persisting deficits in food-intake regulation under specific circumstances. These deficits were: (1) a loss of eating in response to glucoprivation, (2) a failure to increase eating in compensation for food deprivation, (3) a decrease of sucrose consumption, and (4) a decreased ability to sustain a nocturnal pattern of eating. So far, tests in dorsal midbrain 6-OHDA-lesioned animals have revealed two of these deficits, namely, loss of glucoprivic eating and decrease of sucrose consumption. This constellation of functional deficits, which interestingly has also been discovered in the intact hamster (Ritter and Balch, 1978; Silverman and Zucker, 1976), was accompanied by the small suppression of food intake and increase in water intake mentioned above. These deficits were associated with damage to specific noradrenergic fibers coursing through the dorsal tegmentum (Fig. 10d, e) and a detectable loss of CA fluoressence in the PVN, dorsomedial and periventricular nuclei but little or no change in the perifornical and lateral hypothalamic areas. As with all such lesions, the task of demonstrating a causal relationship between such neurochemical and behavioral changes is a very difficult one indeed and requires that one bring to bear on this issue a variety of evidence, obtained using many different techniques. Such evidence will be discussed in Section II.H below, which evaluated the possibility that noradrenergic projections to the medial hypothalamus (in particular the PVN and periventricular nucleus) that are damaged by these dorsal midbrain lesions participate in the integration of signals relating behavior to brain glucose metabolism.

In contrast to these lesions in the dorsal midbrain tegmentum, electrolytic lesions in the ventral region of the pons and midbrain in the area of and just dorsal to the medial lemniscus were found to produce hyperphagia and obesity, associated with a dramatic loss of CA fluorescence in the hypothalamus and specific damage to ventral adrenergic fibers (Ahlskog and Hoebel, 1973; Ahlskog, 1974) (see Fig. 10.) On the basis of evidence showing a similar effect with 6-OHDA, these authors proposed that the behavioral change was a consequence of damage to NE- or EPI-containing fibers which course through the ventral tegmentum and innervate the hypothalamus. A similar suggestion was made by Kapatos and Gold (1973) on the basis of their results demonstrating that knife cuts along the midlateral portion of the hypothalamus, which would be expected to sever a significant portion of these CA fibers ascending from the ventral tegmentum, also produce hyperphagia. This association realized by these studies was questioned by Grossman (1978) who employed the knife-cut procedure and by Lorden et al. (1976) and Oltmans et al. (1977) who examined a variety of lesions, namely, electrolytic, 6-OHDA, copper sulfate, and 5, 6-dihydroxytryptamine. With these procedures, the investigators failed to demonstrate a correlation between the extent of hyperphagia and extent of hypothalamic NE loss. Although this evidence necessitates a reevaluation of the hypothesized causal relationship between hyperphagia and noradrenergic (or adrenergic) fiber damage, it cannot be taken as contradictory evidence, since histofluorescence analyses were not performed to confirm the nature and extent of midbrain CA fiber damage and measurements of CA in the whole hypothalamus, as opposed to discrete areas, were taken.

While these studies have taken innovative steps in an attempt to ascribe a functional role to specific CA pathways, their interpretation remains difficult at best, due in part

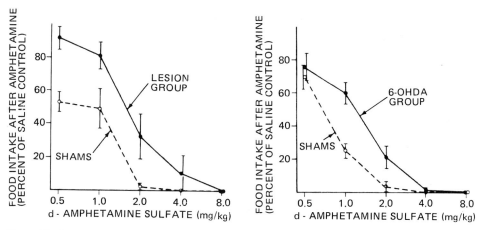

Figure 11 Lesions or 6-OHDA injection in the vicinity of the ventral adrenergic fibers of the central tegmental tract, caudal to the midbrain DA cell bodies, attenuated AMPH-induced anorexia by approximately 50%. (From Ahlskog, 1974.)

to the complexity and diversity of the systems which have impact on long-term, food-intake regulation and body-weight gain. Another, perhaps more direct approach to the problem of relating neurochemistry to behavior has been that of combining brain-stem lesions with peripheral or central injections of drugs. If it is known where and how a drug is acting to produce a specific behavioral change, the central injection approach in particular permits one to probe a discrete brain area and then associate this area, and the behavior it affects, with specific consequences of CA pathway destruction. With peripheral drug administration, this procedure has been helpful in identifying the CA pathways which may be involved in mediating the anorectic effect of AMPH. The results indicate that 6-OHDA or electrolytic lesions in the ventral pontine and midbrain tegmentum, which produced extensive damage to ventral noradrenergic and probably adrenergic fibers of the central tegmental tract but had little effect on DA projections (Fig. 10c, f, and g), caused approximately a 50% reduction in the anorectic potency of peripherally injected AMPH (Fig. 11) (Ahlskog, 1974; Carey, 1976; Leibowitz and Brown, 1980c; Samanin et al., 1977). This loss of potency was associated with a marked reduction of CA innervation throughout the hypothalamus, including the perifornical area where AMPH is believed to exert its action (Leibowitz, 1975d). In a similar study where AMPH was injected directly into the perifornical hypothalamus, a 40-45% loss of responsiveness was obtained with such a lesion (Leibowitz, 1979b; Leibowitz and Brown, 1980b). This attenuation of AMPH's action was associated with a potentiation of the anorectic action of EPI injected into the perifornical area. This latter finding suggests that receptor supersensitivity may have developed in these animals as a consequence of the damage caused specifically to adrenergic fibers innervating the perifornical region (Fig. 4d). Taken together, these lesion results support the existence of adrenergic synapses in this area which when activated produce a suppression of feeding behavior. The evidence further suggests that the adrenergic receptors involved in this phenomenon are located on the postsynaptic cell membrane; that the afferent fibers to this system project through the ventral portion of the pontine and midbrain tegmentum (Fig. 10); and that these fibers originate from one of the lateral tegmental or medullary cell groups (Fig. 1).

In these studies, lesions in the dorsal tegmentum (Fig. 10) were not found to have any impact on the anorectic effect of AMPH injected either peripherally or directly into the perifornical area. These lesions, however, did have a profound effect on the feeding changes evoked by drug injection into the PVN (Leibowitz, 1979b; Leibowitz and Brown, 1980a). As described in Section II.B, injection of NE or EPI into this nucleus causes an increase of feeding, which is mediated via α-adrenergic receptors. This response is mimicked by antidepressant drugs which potentiate and depend upon the release of the endogenous amines into the synaptic cleft. Through a combination of lower brain-stem lesions and PVN drug injection, evidence was obtained to suggest that the adrenergic fibers involved in this response course through the dorsal region of the pons and midbrain tegmentum and possibly originate from either the locus ceruleus or the lateral tegmental cell groups (Fig. 10) (Leibowitz and Brown, 1980a). Electrolytic and 6-OHDA lesions in the dorsal tegmental area, ventrolateral to the central gray matter, abolished the feeding response elicited by PVN injection of the antidepressant drugs. The CA fibers destroyed by the lesion showed build-up of the neurotransmitter immediately caudal to it and exhibited a radiating pattern as they turned rostroventral along the dorsal surface of the superior cerebellar peduncles (Fig. 10d and e). The reduction of CA fluorescence in the hypothalamus was small but detectable, specifically within the PVN (the area of drug injection), as well as the periventricular and dorsomedial nuclei. This loss of CA innervation was associated with a dramatic potentiation of the feeding response elicited by PVN injection of NE, suggesting the development of denervation supersensitivity that has functional significance. No change of fluorescence was seen in the perifornical or more lateral areas. Furthermore, ventral tegmental lesions, which were shown to abolish AMPH action in the perifornical region (see above), generally potentiated the antidepressant feeding response produced in the PVN. These lesion studies clearly dissociate the antagonistic effects of hypothalamic adrenergic stimulation on feeding and provide some insight into the projection route taken by the adrenergic fibers that mediate these effects.

G. Hypothalamic Lesion Studies

In the above studies, damage to lower brain-stem adrenergic projections has been associated with changes in feeding, drinking, and urine formation, as well as specific changes in CA innervation to the hypothalamus. While there is some evidence to suggest a causal link between these neurochemical and behavioral changes, further research is clearly necessary to firmly establish and characterize this relationship. Thus, despite the profound changes that have been observed with hypothalamic adrenergic stimulation, the precise function of the proposed adrenergic systems in the control of natural behavior remains to be discovered. Discrete hypothalamic lesions have been known for some time to affect ingestive behavior and water-electrolyte balance. The question is whether these lesion effects might be related to and a consequence of damage to the adrenergic receptor systems located in the hypothalamus.

Electrolytic lesions in the ventromedial hypothalamus produce hyperphagia and obesity (Brobeck, 1945; Graff and Stellar, 1962). This effect does not appear to be attributed to damage to the VMN itself but rather to damage lateral or rostral to the nucleus (Gold, 1973; Joseph and Knigge, 1968). Furthermore, it has been proposed that the hyperphagia might be due to the destruction of projections to or from the lower brain stem (Brobeck, 1946; Gold, 1973), possibly fibers of the noradrenergic or adrenergic systems which course just lateral to the VMN (Kapatos and Gold, 1973). Indeed, Ahlskog and Hoebel (1973) have demonstrated hyperphagia after midbrain lesions that damage adrenergic fibers coursing through the ventral tegmentum. However, further analyses have revealed distinct differences between specific characteristics of the behavioral deficits resulting from the midbrain and ventromedial hypothalamic lesions (Hoebel, 1977). This suggests that they may involve different neural substrates, although they may still have in common damage to a specific subgroup of fibers.

Coscina et al. (1976) and Glick et al. (1973a) have demonstrated that ventromedial hypothalamic lesions cause a decrease in diencephalic NE, which is positively correlated with the increase in food intake. Since NE injected into the medial hypothalamus elicits feeding (Leibowitz, 1970b; 1978a), this evidence, together with the lesion studies, might suggest that injected NE is acting to inhibit a medial hypothalamic "satiety" system and that destruction of this system causes hyperphagia (Leibowitz, 1970b; Herberg and Franklin, 1972; Ritter and Epstein, 1975). Signs of motivational deficits observed with NE-elicited feeding, similar to those associated with lesion-induced hyperphagia, support this conclusion (Booth and Quartermain, 1965; Coons and Quartermain, 1970). Furthermore, iontophoretic studies reveal a predominantly inhibitory effect of NE on neuronal firing in the medial hypothalamus, although some excitatory effects have been described particularly in the case of nonneurosecretory cells in the PVN (Bloom et al., 1963; Cross et al., 1975; Moss et al., 1972). Herberg and Franklin (1972) demonstrated that ventromedial hypothalamic lesions that produce obesity disrupt the feeding response elicited by NE injections. However, some lesions were ineffective, and there was no clear relationship between the magnitude of the obesity and the loss of NE responsiveness. Since the most sensitive site for NE-elicited feeding has been found to be the PVN and not the ventromedial hypothalamus (Leibowitz, 1978a), the disruption of this response after ventromedial lesions most probably does not reflect direct damage to the NE receptors. Instead, it may be due to the destruction of efferent fibers which mediate the NE response or perhaps regulate the release of hormones upon which NE feeding depends (Leibowitz, 1977). Thus, the failure to establish a relationship between the obesity and NE eating may be due to the difficulty in lesioning a significant and consistent portion of the mediating fibers. A further point of consideration is the fact that the effective locus of the obesity-producing medial hypothalamic lesion is not within the VMN itself but is actually rostral to the nucleus, possibly ventral to the PVN (Gold, 1973). It may actually involve the PVN itself, since lesions restricted to this nucleus are also effective in producing hyperphagia and obesity (Eng et al., 1979; Heinbecker et al., 1944; Leibowitz, 1979c) (see Fig. 12b.)

These lesion and chemical stimulation studies, therefore, focus on the medial hypothalamus rostral to the VMN as providing an important link in the process of feeding regulation. Studies employing knife cuts similarly converge on this area as yielding the most robust hyperphagia after damage. Parasagittal knife cuts aimed between the medial and lateral hypothalamic regions produced overeating and obesity (Grossman, 1975), and the most effective site for revealing this phenomenon appeared to be at the level of the PVN (Gold, 1970; Grossman, 1975; Paxinos and Bindra, 1973; Sclafani, 1971).

Figure 12 (a) Montage (fluorescence photomicrograph) through the paraventricular region (PVN borders marked with dotted black line) seven days after parasagittal knife cut (arrow) just lateral to the fornix. Decreased CA fluorescence can be seen ipsilateral to the transection compared to the contralateral side. Build-up of CA fluorescence can be seen lateral to the knife cut. V, third ventricle. (From O'Donohue et al., 1978.)

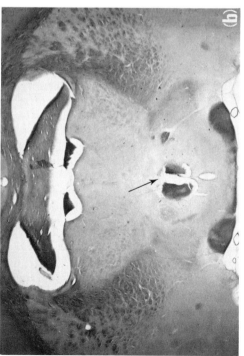
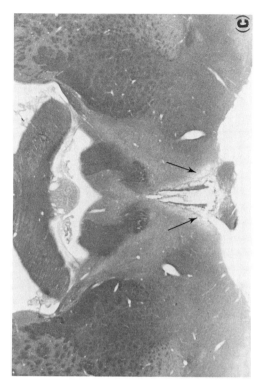

Figure 12 (b,c) Frontal cresyl violet-stained sections of rat brain showing electrolytic lesion of the : (b) PVN) (arrow), located rostral and dorsal to the VMN, which caused a twofold increase in food intake and a fourfold increase in body-weight gain. (From Leibowitz, 1979c). (c) Anteroventral third ventricle area (arrows), just rostral to the PVN, which caused hypodipsia and reduced drinking responses to angiotensin II and hypertonic saline. (From Buggy and Johnson, 1977.)

Coronal knife cuts also caused hyperphagia and increased weight gain (Albert et al., 1971; Grossman, 1975; Sclafani and Berner, 1977), and this effect, which could be observed with cuts in the midbrain as well as throughout the hypothalamus, disappeared immediately rostral to the PVN (Gold et al., 1977).

The question of the relationship between this phenomenon and the NE-elicited feeding response has recently been investigated by Aravich et al. (1978) in rats with combined knife cuts and PVN cannulas. This study demonstrated that animals made obese with parasagittal cuts lateral to the PVN and VMN exhibited at least a normal but in many cases increased responsiveness to NE. This finding would appear to indicate that the obesity does not primarily result from destruction of *efferent* fibers that mediate the NE feeding response. However, direct damage to the neurons intrinsic to the medial hypothalamus, upon which NE axons terminate, may to some extent contribute to the hyperphagia syndrome, since lesions within the PVN (Fig. 12b), that have no impact on tissue immediately ventral or caudal to the nucleus, result in a marked increase in food intake and body weight (Leibowitz, 1979c).

With regard to the adrenergic *afferents* to the PVN, the concept of NE acting in an inhibitory fashion to suppress neurons of satiety would actually lead to the prediction that destruction of the ascending fibers yield the opposite behavioral effect, namely, increased satiety, decreased food intake, and loss of body weight. In this context, it is interesting to note the result of Richardson and Jacobowitz (1973) and Richardson et al. (1974) which demonstrated that intraventricular injection of the neurotoxin 6-OH-DOPA temporarily reduced feeding (for approximately 5 days), an effect that was associated with a relatively selective loss of CA fluorescence within the PVN, periventricular and dorsomedial nuclei adjacent to the third ventricle. Furthermore, 6-OHDA damage to noradrenergic fibers of the dorsal tegmentum, which appear to project to the PVN (Fig. 10d,e) as well as mediate the feeding elicited through presynaptic NE release, also produced a small but consistent decrement in daily food intake (10-20%) and body weight (3-4%) over at least a two-week period (Leibowitz and Brown, 1980a). These neurotoxin effects on 24-hr food intake require further investigation to understand their relationship to the different hypothalamic-lesion syndromes, as well as to the stimulatory effect of NE on a single meal.

While it is likely that multiple systems are involved in the hyperphagia- and obesity-producing effects of hypothalamic damage, another way in which the medial hypothalamic NE system may contribute to this phenomenon, at least with the knife cuts, is through the development of adrenergic receptor supersensitivity and axon regeneration within the medial hypothalamus, as a result of the destruction of innervating fibers. The *initial* consequence of damage to these fibers would be a reduction in feeding associated with a reduction specifically in paraventricular and periventricular concentrations of NE. This pattern of change has been described for selective neurotoxin lesions of ascending NE fibers (see above), and it has also recently been observed with parasagittal knife cuts just lateral to the fornix at the level of the PVN (Fig. 12a) (O'Donohue et al., 1978). After this initial deficit, however, compensatory changes are likely to occur, which may not only reverse the hypophagia but may actually lead to hyperphagia. There are three independent studies which are consistent with this proposal. Direct evidence for the development of NE receptor supersensitivity in the medial hypothalamus has been obtained in animals with selective neurotoxin damage to afferent noradrenergic fibers

(Leibowitz and Brown, 1980a). Furthermore, CA axon sprouting (Richardson and Jacobowitz, 1973) and a rebound of NE content (O'Donohue et al., 1978) have also been detected in this region after damage to the noradrenergic input. Finally, animals with parasagittal knife cuts lateral to the PVN may exhibit increased sensitivity to NE injection in the PVN (Aravich et al., 1978). Thus, it appears that an abnormal condition of enhanced noradrenergic-receptor function can occur in these animlas, which may at least have an aggravating influence on the development of long-term hyperphagia and obesity.

Along with this evidence that hyperphagia may result in part from denervation-induced, NE-receptor supersensitivity or from the destruction of intrinsic medial hypothalamic neurons upon which the NE receptors are located, there is the additional possibility that CA fibers coursing through or terminating in the perifornical region of the hypothalamus, which may be damaged by the medial hypothalamic lesions as well as the knife cuts, are also involved in the development of overeating. This proposal is made in light of the evidence that CA receptors (dopaminergic and β-adrenergic) within the lateral perifornical hypothalamus at the level of the VMN and PVN exert an inhibitory influence over food intake (see above and Section III.B on DA). There is evidence that medial hypothalamic lesions produce the strongest hyperphagia when they penetrate tissue lateral to the VMN (see Grossman, 1975). In mapping studies on the obesity caused by coronal knife cuts, the perifornical area is found to be the most effective region. Damage limited to the midline had little effect on feeding, and midlateral cuts, precisely overlapping the perifornical area, did not need to extend medially to produce a reliable effect (Gold et al., 1977; Grossman and Hennessy, 1976; Paxinos and Bindra, 1972, 1973; Sclafani, 1971). Furthermore, the hyperphagia produced by coronal cuts disappeared just rostral to the PVN, where CA-induced anorexia also ceased.

The coronal perifornical cuts would be expected to sever the ascending CA fibers which terminate in the perifornical area, although they could also damage axons that ultimately course medially toward the PVN (see Section II.A). The parasagittal cuts, depending upon their position relative to the fornix, might sever some perifornical fibers but would appear to have their most dramatic effects on medially directed axons. To assess the accuracy of these predictions and to examine the relationship between perifornical hypothalamic CA neurons and knife-cut induced feeding disorders, histofluorescence analyses of hypothalamic CA innervation need to be conducted on brains subjected to perifornical cuts. In a recent study of Sclafani and Berner (1977), the feeding suppressive effect of peripheral AMPH injection remained essentially intact after hyperphagia-producing coronal or parasagittal knife cuts. This finding would appear to dissociate perifornical hyperphagia from AMPH anorexia; in the absence of histofluorescence analyses, however, it unfortunately does not permit one to draw any conclusions regarding the relationship of these behavioral phenomena to perifornical CA innervation. Since discrete midlateral lesions, along the lateral border of the fornix at the level of the VMN, are effective in attenuating the anorexia induced by peripheral AMPH (Blundell and Leshem, 1974), it is possible that knife-cut damage in the perifornical region may not be severing an adequate portion of the innervating adrenergic fibers or efferent neural elements to significantly disrupt AMPH's action in this area. The histofluorescence analysis of O'Donohue et al. (1978) in animals with parasagittal cuts just lateral to the fornix revealed some CA depletion in the perifornical area, along with a 30-40% loss of fluorescence in the PVN (Fig. 12a). However, the interpretation of these results relative to perifornical function is somewhat complicated by the additional observation that CA build-up, within severed neurons proximal to the cut, occurred just lateral to the fornix. The impact of this build-up (which lasts one to two weeks) on the function of perifornical neurons and consequently on feeding behavior needs to be taken into consideration.

In addition to their effect on food intake, hypothalamic NE and EPI have also been shown to have impact on water and electrolyte balance (see Sections II.C-E). While, through lesion and electrophysiological studies, extensive literature has accumulated to suggest a vital role of the hypothalamus in these functions (see chapter by Hatton and Armstrong in this volume), there are few studies that have attempted to relate these results with the effects observed with hypothalamic adrenergic stimulation. As described in Section II.C, the periventricular region and the PVN have been distinguished as the most sensitive sites with respect to two NE-elicited phenomena, namely, a transient preprandial drinking response, followed by a more potent and long-lasting inhibition of drinking. In light of this evidence, it is noteworthy that lesions of the periventricular region just rostral to the PVN (Fig. 12c) produce reversible adipsia associated with persisting deficits in the regulation of food-associated drinking and in the drinking response to angiotensin and hypertonic saline (Buggy and Johnson, 1977). At present, the relationship between these phenomena is not understood, although one may speculate (as above) that NE is primarily inhibitory in its action on periventricular neurons. Thus, if one assumes that these neurons are involved in activating thirst (see Buggy and Johnson, 1977; Andersson, 1975), their inhibition via NE input, as well as their destruction by lesion, would be expected to cause a reduction in water consumption. While a variety of neural substrates are undoubtedly involved, a further prediction of the above model would be that damage to the noradrenergic afferents would cause a release of the NE inhibition, leading to an increase in water ingestion. In the study of O'Donohue et al. (1978), this prediction was borne out. These investigators found that parasagittal knife cuts just lateral to the PVN produced hyperdipsia, associated with a partial loss of NE in the PVN and periventricular nucleus (Fig. 12a). These authors suggested that this change in drinking might reflect damage to preterminal noradrenergic fibers of the tractus filiformis lateralis, which pass dorsal to the fornix to innervate these medial hypothalamic nuclei and which originate from the noradrenergic cell groups A1, A2, or A6. This suggestion is consistent with the evidence that lesions of the locus ceruleus nucleus cause hyperdipsia (Osumi et al., 1975; D. Roberts et al., 1976), although there is a question as to whether this phenomenon is specifically dependent upon the noradrenergic cell bodies of this nucleus (D. Roberts et al., 1976).

An additional hypothalamic area that may be involved in CA-induced drinking suppression is the zona incerta, which is known to have a dense CA innervation (Fig. 4e). Walsh and Grossman (1976) have shown lesions in this area to produce changes in water consumption, which include hypodipsia under ad libitum conditions (see also Evered and Mogenson, 1976) and a marked decrement of food-associated drinking and of drinking in response to hypertonic saline and angiotensin. The knife cuts of O'Donohue et al. (1978) clearly damaged CA input to the zona incerta, as well as to the PVN and periventricular area. It is surprising that biochemical assays of the whole hypothalamus failed to reveal a change in NE or DA concentrations after zona incerta lesions

(Walsh et al., 1977), since numerous CA fibers which innervate the hypothalamus are known to course through this area. It would appear that the measurements taken of the whole hypothalamus may have masked more subtle changes which occurred in discrete hypothalamic nuclei. Another possible explanation of the hyperdipsia produced by parasagittal knife cuts is that the paraventricular-supraopticohypophyseal tract had been cut (Gold et al., 1977). Records of urine output were not taken; thus, it is not known whether the hyperdipsia is secondary to polyuria.

The evidence summarized in Section II.D supports the existence of an adrenergic synaptic link in the process of ADH release and water excretion. The role of the SON and PVN in the synthesis and release of this hormone has been well established (e.g., Stutinsky, 1974; Defendini and Zimmerman, 1978), and it appears that these nuclei are most sensitive to the antidiuretic effect of NE or EPI stimulation, although extensive mapping studies have not yet been conducted. Iontophoretic studies have revealed a predominantly inhibitory effect of the agonists on antidromically identified neurosecretory cells of the SON (Barker et al., 1971), although excitatory effects on cells of the PVN and SON have been revealed (Cross et al., 1975). If the adrenergic innervation of these nuclei acts through neuronal inhibition, it may possibly occur via an interneuron which translates the inhibition into an excitation of the neurosecretory cells. This possibility receives support from the effects observed with 6-OHDA injections, which destroyed adrenergic input to the SON and caused an increase in firing rate of neurosecretory neurons (Young, 1975), as well as a decrease in ADH release in response to hyperosmotic stimulation (Milton and Paterson, 1973).

On the basis of lesion and electrical stimulation studies, several hypothalamic, as well as forebrain and brain-stem, structures have been implicated in the control of sodium intake and excretion. From these studies, it is very difficult to determine those areas which may have receptors sensitive to Na^+ concentration or may be a part of the effector pathway and those which, when lesioned or stimulated, alter other central-peripheral functions and then indirectly change sodium balance. Based on perfusion studies of the cerebrospinal fluid, Andersson (1977) has proposed that cerebral receptors regulating Na^+ excretion may be located along the third and fourth ventricles. Lesions of the medial preoptic area and PVN, as well as the posterior hypothalamus, have been shown to cause natriuresis (Covian et al., 1975; Keeler, 1959, 1972). Furthermore, damage to the PVN produced a marked and specific decrease in aldosterone production in sodium-deficient rats (de Wied et al., 1972). There are various neural pathways whereby the brain influences the kidney to change Na^+ excretion, and the most feasible possibilities appear to be changes in renal hemodynamics, in renal nerve activity, and/or via hormones. The existence of a "natriuretic factor" has been proposed to account for experimentally induced changes which are too rapid to be accounted for by the release of aldosterone. The possibility that any or all of these mechanisms are involved in adrenergically induced natriuresis remains to be explored (see Section II.E).

With regard to sodium-intake regulation, hypothalamic lesions have been found to impair sodium appetite in response to a variety of stimuli (e.g., Wolf et al., 1974). Studies on spontaneous sodium intake have revealed a suppression of appetite after rostral medial hypothalamic lesions (Antunes-Rodrigues et al., 1970) and an increase in intake after lesions in the lateral hypothalamus (Antunes-Rodrigues et al., 1970) as well as the SON (Covian and Antunes-Rodrigues, 1963; Grace, 1968). While the phenomenon of adrenergic stimulation of sodium intake has not yet been localized to a specific hypothalamic area, the medial hypothalamus adjacent to the third ventricle appears to be a likely candidate (see Section II.E). The relationship between this adrenergic response and the effects of rostral medial hypothalamic lesions on sodium appetite (Antunes-Rodrigues et al., 1970), as well as on drinking in response to NaCl injections (see Andersson, 1975; Buggy and Johnson, 1977), has yet to be examined.

H. Functional Analysis of Adrenergic Systems

The studies described in preceding sections have provided anatomical and pharmacological evidence that hypothalamic adrenergic projections have a strong influence over ingestive behavior. While the direction of this influence, either stimulatory or inhibitory, has been defined at specific terminal areas, the precise function of the adrenergic systems in their control over normal ingestion is still unknown. The variables which come to bear on normal regulation of food and water intake are innumerable. However, as described below, there is some evidence suggesting that hypothalamic NE or EPI may have distinct modulatory influences over the process of initiating or terminating a meal; that they may alter preferences for specific dietary constituents and possibly respond to internal signals of deficiency in these constituents; and that they may interact with and/or depend upon peripheral autonomic and endocrine systems regulating feeding.

1. Meal Patterns

It has become increasingly apparent that in order to dissect out the multiple controls over feeding activities, more sensitive behavioral techniques for detecting subtle changes in food intake are needed. Most pharmacological studies described earlier, which related CA to food-intake regulation, have taken gross measurements of food intake, under unnatural conditions during which the animal was disturbed by handling and experimenter-imposed changes in hunger. To understand the CA mechanisms in terms of their specific function in feeding control, more precise measurements of their effects on various food-intake patterns are needed, in undisturbed conditions where the animal is allowed to exhibit natural, spontaneously initiated behaviors. Detailed analyses of meal patterns in rats have shown that, under normal circumstances, these animals consume 8-10 discrete meals a day, primarily at night (Fitzsimons and Le Magnen, 1969; Kissileff, 1969). Any alterations in amount of food intake over a 24-hr period are found to occur either through changes in meal frequency, that is, the rate at which new meals are initiated (a possible measure of "hunger"), or in meal size, that is, the point at which the meal is terminated (a possible measure of "satiety") (Le Magnen, 1971). Experimental manipulations of these two parameters have indicated that they may be independently responsive to a variety of physiological events. For example, the process of initiating feeding (meal frequency) is influenced by cold stress (Kissileff, 1968) and insulin (Larue, 1977), whereas the process of terminating feeding (which determines meal size) is influenced to meet metabolic demands, such as prior food deprivation (Levitsky, 1970), caloric dilution of liquid diets (Snowdon, 1969), and lactation (Kissileff and Becker, 1974).

With this information at hand, a more sensitive microanalysis of brain CA effects on behavior should significantly increase our understanding of how the CA neurotransmitters function to influence feeding. Such procedures have been employed only recently in studies of drug action. Using the central route of drug administration, Ritter and Epstein (1975) examined the stimulatory effects of NE injection on feeding. These investigators found that this agonist, injected into the anteromedial hypothalamus, sustained a normal, spontaneously initiated meal (that is, increased the meal size) at quite low doses (2.5 ng). However, no effects of NE on meal frequency were detected; that is, when NE was injected during the intermeal interval at low doses, no change in the time of meal initiation was observed. As pointed out by the authors, only a single intermeal injection time was tested (60 min before the next meal), which leaves open the possibility that NE injected somewhat closer to that meal might actually have shifted its time of initiation. This evidence provides support, however, for the possibility that NE induces feeding through the inhibition of satiety, rather than through the stimulation of hunger. This hypothesis receives further support from lesion studies which have shown that destruction of the medial hypothalamic area where NE is most effective in potentiating eating results in hyperphagia and obesity (see Section II.G).

In a series of studies on meal patterns conducted with the anorectic drug AMPH, it was demonstrated that the main effect of this compound, when peripherally injected, was on meal frequency rather than on meal size (Blundell and Latham, 1978). That is, the feeding suppression induced by AMPH appeared to be predominantly a consequence of the reduction in the total number of meals, as opposed to a decrease in the size of the meals. This pattern of effect was observed with the anorectic drug mazindol, which is also believed to act via endogenous CA. Interestingly, this feeding suppression produced by these compounds was also associated with an increase in the rate of eating, which appeared to be independently regulated from meal size. Furthermore, at low doses, AMPH's anorectic action disappeared, and a stimulatory effect was actually revealed, associated with an increase in meal frequency. This stimulatory action is consistent with the small increase in feeding observed with AMPH injection into the medial hypothalamus in contrast to the perifornical hypothalamus where AMPH reveals its more potent anorectic action (Leibowitz, 1970b; Leibowitz, 1975c). While it is clearly premature to speculate on these patterns of changes in terms of the type of signals (e.g., metabolic versus nonmetabolic) to which the central adrenergic systems may be responsive, these results provide an important first step in defining the specific characteristics of the control mechanisms under study and in establishing a profile for comparing the action of a variety of drugs.

2. Dietary Parameters

Studies of the effects of drugs on feeding behavior have traditionally examined total food consumption of a single, standard diet. This procedure allows the investigator to detect changes in the total number of calories that the animal consumes. It does not, however, permit detection of any alterations that the drug might cause in food preference; thus, it may actually produce misleading results in cases where dietary constituents are differentially affected. When animals under ad libitum conditions are given an opportunity to select among a variety of foods (a self-selection feeding paradigm), they are found to maintain an appropriate balance between the different components of the diet. There is a good deal of evidence suggesting that their specific selections and the amounts consumed are geared toward energy homeostasis. Collier and Anderson have advanced this idea to suggest that, in addition to regulating calorie intake, animals adjust their food intake to achieve dietary balance, specifically a constant ratio between protein and carbohydrate consumption (Anderson, 1979; Collier et al., 1969; Leshner et al., 1971). Anderson and Ashley have recently proposed that protein intake may be independently regulated and that this regulation may be reflected in plasma amino-acid ratios (Anderson, 1979; Anderson and Ashley, 1978; Ashley et al., 1979). Furthermore, these investigators have suggested that brain monamines may be involved in this regulation.

This focus on dietary parameters provides an important dimension to our attempt to identify the specific function of hypothalamic CA neurotransmitters in control of feeding. There appear to be no studies which have addressed the possibility that adrenergic stimulation may differentially affect intake of specific dietary constituents, although two earlier studies with hypothalamic NE injection have indicated that diet-related parameters are important in determining the magnitude of the eating response (Booth and Quartermain, 1965; Broekkamp et al., 1974). It has been found, however, that the antidepressant amitriptyline, which in humans produces carbohydrate craving without concomitant hypoglycemia (Needleman and Waber, 1976; Paykel et al., 1973), elicits eating when injected into the PVN of rats, particularly when the diet is sweet and high in carbohydrate content (Leibowitz et al., 1978b). This effect was abolished by α-adrenergic blockers and NE synthesis inhibitors, indicating that endogenous NE (or EPI) may be mediating the response. Further studies have revealed that hypothalamic NE injection may differentially increase the animals' preference for sweet or nonsweet carbohydrate (S. F. Leibowitz, J. R. Tretter and A. Kirchgessner, in preparation). For example, rats receiving PVN injections of NE eat pure sugar and drink sucrose or sweet-milk solutions (but not saccharin solutions). When they are permitted to select from two highly palatable diets simultaneously presented, one containing sugar and one containing fat, the proportion of sugar-diet eaten after NE administration is considerably greater than that selected under nondrug conditions. Furthermore, when given a choice between pure protein (casein), carbohydrate (dextrin), and fat (corn oil), the NE-injected rats show a proportional and frequently selective increase in their consumption of the dextrin.

These results direct our attention toward carbohydrate as possibly being the nutrient specifically preferred by rats induced to eat by PVN adrenergic stimulation. This finding is interesting in light of the evidence that genetically obese Zucker rats, which have reduced NE levels specifically within the PVN (Cruce et al., 1976), also exhibit a selective decrement in sweet preference (Grinker et al., 1977). Furthermore, a chronic reduction in hypothalamic CA caused by ventricular 6-OHDA injection in nonobese rats is associated with a specific decrease in glucose-solution intake (Breese et al., 1973; Coscina et al., 1973; Ellison and Bresler, 1974). This effect did not appear to be attributed to general alteration in gustatory acuity, since the rats showed normal responsiveness to bitter-tasting solutions. The possibility remains, however, that the CA damage produced a change in gustatory responsiveness specific to sweet-tasting foods.

From the 6-OHDA studies, the specific CA projections which may be involved in the response to glucose is not made clear. Adrenergic (Coscina et al., 1973; Ellison and Bresler, 1974), as well as dopaminergic (Breese et al., 1973), projections contribute to the behavioral deficit. The evidence that PVN NE stimulation may evoke a specific carbohydrate preference (see above) suggests that at least one set of fibers whose damage may account for the 6-OHDA-induced decrease in glucose intake are those which project to the PVN and mediate noradrenergic stimulation of feeding (Leibowitz and Brown, 1980a) (see Fig. 10d,e). 6-Hydroxydopamine injections into the path of these fibers (as they course in the dorsal pontine and midbrain tegmentum) have been shown to produce a small suppression of daily food intake (10-20%), associated with a small but consistent loss of CA fluorescence specifically in the PVN and periventricular nucleus (Leibowitz and Brown, 1980a). Preliminary studies with lesions to this adrenergic projection have revealed, in addition, a decreased preference for diets containing sucrose, along with generally normal responsiveness to diets adulterated with quinine (Leibowitz and Brown, 1980c; Leibowitz, unpublished data).

This evidence for a role of hypothalamic adrenergic neurons in the control of a specific appetite for sugar or carbohydrate in general takes on broader significance in light of the additional finding that damage to these neurons also decreases the animals' responsiveness to the nonmetabolizable glucose analogue 2-deoxy-D-glucose (2-DG). This drug is known to increase food intake in intact rats (Smith and Epstein, 1969) as well as in humans (Thompson and Campbell, 1977), presumably by causing a profound and reversible intracellular blockade of glycolysis. In the electrolytic or 6-OHDA midbrain-lesioned animals, however, peripherally administered 2-DG was relatively ineffective in eliciting feeding, just as was PVN injection of antidepressant drugs which, in intact animals, are known to cause eating through the release of endogenous NE (Leibowitz and Brown, 1980a, c; Leibowitz, unpublished data). Other studies have also demonstrated an attenuation of 2-DG-elicited feeding, after lesions in the midbrain (see Grossman, 1978) or the anterior zona incerta just lateral or caudal to the PVN (Walsh and Grossman, 1975).

A relationship between hypothalamic NE-induced eating and the 2-DG feeding response was originally suggested by Müller et al. (1972) who found the 2-DG eating to be blocked by peripheral injection of the CA synthesis inhibitor α-methyltyrosine and ventricular injection of 6-OHDA or the α-adrenergic blocker phentolamine. Since then, a number of studies have tended to build on this hypothesis from a variety of directions. 2-Deoxy-D-glucose has been shown to cause an increase in turnover of hypothalamic (as opposed to telencephalic) NE (Ritter and Neville, 1977) and an increase in the efflux of [^{3}H] NE within the hypothalamus (McCaleb et al., 1978). Furthermore, stress-induced depletion of hypothalamic NE leads to a decrease in responsiveness to 2-DG, which is positively correlated with the extent of NE loss (Ritter et al., 1978). Ventricular injections of 2-DG induce feeding (Fig. 13), although the responsiveness of animals to intrahypothalamic injections remains in dispute (see Berthoud and Mogenson, 1977). These investigators compared the behavioral responses elicited by ventricular 2-DG and NE injections and found that these drugs produced a similar sequence and time course of ingestive responses, consisting of initial drinking followed by eating. As demonstrated by Leibowitz (1975a) with NE, these responses induced by 2-DG were selectively

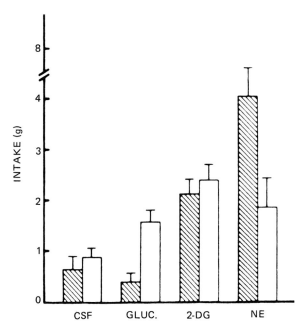

Figure 13 Effect of 3-µl injections of various solutions into cerebral ventricles on food ▨ and water ☐ intake of the rat. CSF, cerebrospinal fluid; glucose, 3.7 µmol; 2-DG 2-deoxy-D-glucose, 3.7 µmol; NE, norepinephrine bitartrate, 55 nmol. (From Berthoud and Mogenson, 1977.)

antagonized by α-adrenergic blockade. Furthermore, the 2-DG feeding response, like the NE response, was disrupted by quinine-adulteration of the diet (Booth and Quartermain, 1965; Kanarek and Mayer, 1978) and was associated with a specific preference for sugar (Thompson and Campbell, 1977).

Taken together, this evidence supports the idea that specific midbrain-medial hypothalamic adrenergic neurons provide a link in the process of receiving information regarding glucose utilization in the brain and transducing this information into an effective eating response for restoring a deficit in glucose utilization.

3. Hormone Interactions

In the context of the research just reviewed, in which noradrenergic systems have been linked to a metabolic function, it is particularly noteworthy that feeding elicited by PVN injection of NE is dependent upon circulating adrenal hormones (Leibowitz et al., 1976; Leibowitz, 1977) (see Fig. 14). Hypophysectomy was found to abolish the feeding response to NE, while ingestive responses elicited by other drugs remained intact. Furthermore, adrenalectomy, as opposed to thyroidectomy or gonadectomy, caused approximately a 75% loss of the response, and this change was correlated with plasma corticosterone levels. Hormone replacement experiments showed the behavioral deficit to be restored specifically by corticosterone injections or implants.

The primary physiological effects of glucocorticoids are believed to involve enzyme induction and protein synthesis, which work in concert toward the goal of maintaining or restoring normal glucose levels, while specifically favoring the brain. In the periphery, glucocorticoids generally act in a "permissive" manner, particularly in a fasting state. They are required for normal activation of gluconeogenesis, glycogenolysis, and lipolysis production by the CA. The mechanism of this interaction is not clearly understood, although it has been proposed to involve changes beyond the point of cyclic nucleotide formation, perhaps having effects on intracellular ion metabolism (e.g., Guidieri et al., 1975; Lehr, 1972; Wolfe et al., 1976). Steroids, furthermore, have been shown to block CA uptake mechanisms, leading to a marked potentiation of CA effects at the receptor site (Iverson and Salt, 1970). Hypothalamic monamines are believed to modulate the secretion of glucocorticoids from the adrenals, and in reverse, circulating glucocorticoids modulate brain monamine synthesis and release (see Krieger and Ganong, 1977). The corticosterone-releasing factor has been identified in the hypothalamic PVN, and a stimulatory pathway for the release of adrenocorticotropic hormone (ACTH), possibly involving vasopressin release, has recently been proposed to course from the medulla to the region of the locus ceruleus and then further rostral to the magnocellular perikarya in the PVN (see Krieger and Ganong, 1977).

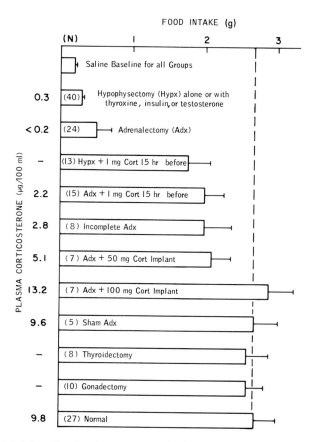

Figure 14 Food intake 60 min after paraventricular nucleus injection of NE in rats receiving a variety of hormone manipulations, including corticosterone (Cort) replacement. (From Leibowitz, 1977.)

The adrenal cortex has long been known to influence feeding behavior. Addison's disease involves an increase in appetite for salt but eventually a decrease in general appetite, anorexia, and body-weight loss. Cushing's syndrome, in contrast, is often characterized by the development of a rather specific obesity, enhanced by retention of salt and water. The specific influence of ACTH and adrenocorticosteroids on food intake remains unexplained. Clinical studies have often reported an increase in food intake of patients treated with glucocorticoids, but here the issue is clouded by the fact that the patients are frequently underweight at the start of treatment and the hormone effect on food intake is possibly secondary to its effect on general health and well-being.

Animal studies have provided some evidence which links glucocorticoids to feeding. For example, in rats and monkeys maintained under restricted feeding conditions, peaks of glucocorticoid secretion are found to precede the onset of the feeding response (Moberg et al., 1975; Holaday et al., 1977; Krieger, 1974). A common effect observed with adrenalectomy is a reduction of food intake and body weight, although these changes may be small if the animals are maintained under nonstressful conditions with palatable food and salt replacement continuously available. Particularly when the adrenalectomized animal is deprived of food even for just a few hours do the effects of glucocorticoid deficiency (e.g., hypoglycemia) become manifest. It has been reported that adrenalectomized animals, when given a choice of diets, may decrease their intake specifically of dextrose and protein, while increasing their intake of fat (Clark and Clausen, 1943; Richter, 1941). Peripheral injections of relatively low doses of glucocorticoids may increase food intake and weight gain in both adrenalectomized and intact animals, whereas high doses may either have no effect or actually cause a decrease in food intake (see Lytle, 1977; Simpson et al., 1974). In humans, 2-DG, which produces a fast-like state, causes an increase of feeding and enhancement of sucrose preference (Thompson and Campbell, 1977). It is interesting that these effects, similar to those produced by hypothalamic NE injection in the rat (see above), are correlated in magnitude with glucocorticoid (cortisol) secretion.

It is known that metabolic processes often have behavioral expression. That is, satisfaction of certain metabolic requirements is sometimes achieved by alterations in behavior. It is possible that hypothalamic adrenergic feeding systems participate in this interphase between metabolism and behavior. To maintain homeostasis, under normal conditions as well as in response to food deprivation or, specifically, glucoprivation, an organism is known to react in numerous ways. One such way may include the release of glucocorticoids which work in concert not only with the catecholamines peripherally, to produce various metabolic changes, but also with NE centrally, to evoke food consumption and, in particular, carbohydrate preference.

4. Autonomic Variables

In addition to its role in the control of neuroendocrine functions, the hypothalamus is known to have a primary influence over the autonomic nervous system. In general, it appears that the medial hypothalamus exerts a tonic sympathetic control over visceral function, whereas the lateral hypothalamus exerts a parasympathetic effect (Ban, 1966). Consistent with this suggestion are the results obtained with medial hypothalamic lesions, which, presumably by altering the sympathetic/parasympathetic balance in favor of lateral hypothalamic parasympathetic outflow, produce such effects as an increase in insulin release and gastric acid secretion. These parasympathetic effects have been shown to be blocked by transection of the vagus nerve, which transmits the information to the viscera, as well as by atropine, which blocks the mediating cholinergic (muscarinic) receptors (Lanciault et al., 1973; Stephens and Morrissey, 1975; Woods and Porte, 1974).

The possibility that changes in peripheral autonomic function may participate in the process of central adrenergic control of feeding behavior has received little attention. This possibility becomes reasonable in light of the recent findings that the PVN is the most effective site for producing NE-elicited feeding (Leibowitz, 1978a), and that there exist long ascending and descending fibers connecting the PVN and the vagal nuclei of the brain stem (Conrad and Pfaff, 1976; Ricardo and Koh, 1977; Saper et al., 1976; Swanson, 1977). In fact, NE injection into the hypothalamus, either just lateral to the PVN or into the medial anterior hypothalamus, has been shown to cause an increase in gastric-acid secretion and decrease in heart rate (Carmona and Slangen, 1973, 1974), as well as an increase in insulin release (de Jong et al., 1977).

The question of whether the feeding changes induced by hypothalamic adrenergic stimulation are in any way related to or dependent upon peripheral autonomic function has recently been evaluated in animals that had sustained pharmacological or neural blockade of parasympathetic information to the viscera. The pharmacological block was produced by systemic administration of the cholinergic antagonists atropine and scopolamine, and the results with these agents indicated a strong dependence of the NE-elicited feeding response on intact peripheral cholinergic receptors (Fig. 15) (Leibowitz,

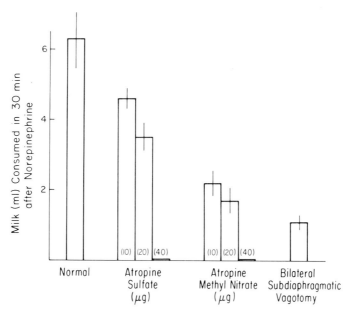

Figure 15 Ingestion of milk after paraventricular nucleus injection of NE in animals receiving intraperitoneal injections of the cholinergic antagonist atropine or after bilateral subdiaphragmatic vagotomy.

unpublished data). At remarkably low doses (as low as 5 μg), both blockers, as well as their quarternary ammonium derivatives which have little effect on the brain, were effective in attenuating or abolishing the initiation of feeding (of either sweet milk or mash) observed with NE injection into the PVN. The lower doses effective in blocking this feeding response had no effect on feeding elicited by mild food deprivation.

Neural block of visceral parasympathetic function, via subdiaphragmatic vagotomy, was also found to disrupt eating elicited by NE at a wide range of doses (Sawchenko and Deluca, 1978; Leibowitz, 1979c) (see Fig. 15). The specificity of this blockade to NE feeding elicitation, as opposed to the process of meal initiation in general, needs to be seriously questioned. The vagus has been directly linked with meal initiation (Penaloza-Rojas and Russek, 1963; Penaloza-Rojas et al., 1969; Snowdon, 1970), and its removal has been shown to alter the animals' meal patterns, causing a decrease in meal size and an increase in meal frequency (Davis and Booth, 1974; Snowdon and Epstein, 1970). Further tests with the vagotomized rats, however, revealed normal, short-term feeding in response to mild food deprivation and other stimulating drugs such as insulin. This suggests that a general limitation on alimentary reflexes was not a primary factor in the disruption of NE-elicited feeding. Vagal afferents to the brain may contribute to the response, although the data obtained with peripheral cholinergic blockade focus our attention on the efferent, parasympathetic fibers of the vagus nerve.

While these results reveal a link between central adrenergic elicitation of feeding and peripheral autonomic systems, the nature of this relationship and the neural pathways involved need to be elucidated. Vagotomy is also found to disrupt drinking behavior in response to a variety of stimuli (Kraly et al., 1975); thus, the specificity of the effect on NE feeding remains an open question. Since NE in intact animals appears to produce an increase in carbohydrate (or sugar) preference (see Section II.H.2), and vagotomy produces a decrement in sucrose ingestion (Fox et al., 1976), the decrease in NE feeding may be attributed to this shift in preference. Although the vagus may have a function in relaying information (to the brain) concerning absorption and/or storage of glucose, the effects of vagotomy on sucrose ingestion may simply be a consequence of the dumping syndrome or the hypovolemia resulting from ingestion of hypertonic sucrose solution.

While the functions of the vagus nerve are numerous and varied, one function in particular, namely pancreatic β-cell release of insulin, would appear to provide a potential source of interaction between hypothalamic adrenergic neurons and peripheral autonomic systems. The evidence includes the finding that immediately at the start of a spontaneously elicited meal, as well as after NE injection, a dramatic and rapid increase in the release of insulin occurs (de Jong et al., 1977; Steffens, 1969). Although this effect of NE is not blocked by peripheral atropine injection, the importance of insulin release for NE-induced eating is suggested by two studies which have demonstrated an attenuation or abolition of the NE feeding response after selective dissection of the coeliac vagal branch (P. E. Sawchenko, R. M. Gold and S. F. Leibowitz, in preparation) and after the establishment of diabetes through alloxan treatment (Kanner, 1974). Davis and Keesey (1971), furthermore, have provided preliminary evidence that NE and insulin may interact synergistically to facilitate feeding behavior, and additional studies of eating elicited by 2-DG have suggested a possible relationship between intracellular glucose utilization and noradrenergic function (Berthoud and Mogenson, 1977; Müller et al., 1972).

To understand the contribution made by insulin to noradrenergic stimulation of feeding, further research will need to be conducted to determine whether "insulin-sensitive" elements actually exist in brain neural tissue or whether they exist in the brain vasculature (van Houten and Posner, 1979) which then influences the tone of neural tissue. Furthermore, if the NE feeding response does specifically require the transmission of information from the medial hypothalamus to the peripheral autonomic system, the central neural pathway involved in this process will need to be established. Direct (monosynaptic) connections between the perikarya of the PVN and vagal nuclei of the hindbrain have been identified (see above). However, studies involving animals with midbrain lesions or knife cuts in the path of these projections have so far failed to reveal any disruption of the feeding response elicited by PVN injection of NE (Aravich et al., 1978; Leibowitz and Brown, 1980a). More specific tests, such as with lesions directly to the vagal nuclei, will need to be conducted to evaluate the neuroanatomical link between the PVN noradrenergic system and peripheral autonomic function.

I. Catecholaminergic Systems and Cerebral Circulation and Metabolism

When interpreting the behavioral effects produced by changes in central neurochemical stimulation, it is generally assumed that the endogenous substance under investigation has a neurotransmitter role and is acting at a receptor site within a synpase to alter the ion permeability characteristics of the neuronal membrane. There is extensive evidence to suggest that NE, DA, and possibly EPI mediate the synaptic transmission of neural impulses in the brain and that this function may involve a sequence of events, including the activation of the enzyme adenylate cyclase; a consequent increase in cyclic AMP; the activation of protein kinases; and the subsequent phosphorylation of specific substrate proteins, and the consequent alteration in membrane ion permeability (see Bloom, 1975; Blumberg et al., 1976; Bockaert et al., 1977; Greengard, 1976; Iversen, 1975; Krueger et al., 1975). While the CA agonists may indeed modulate synaptic activity in this manner, it is important to consider the possibility that they may affect nonneuronal as well as neuronal tissue in the brain and that they may also evoke changes in cerebral metabolism or circulation. The catecholamines EPI, NE, and isoproterenol, as well as AMPH and *l*-dopa, have each been found to cause an increased breakdown of glycogen in brain tissue, and this effect appears to be accompanied by an increase in oxygen consumption and glucose metabolism (e.g. Gilman and Schrier, 1972; Kakimoto et al., 1964; MacKenzie et al., 1976a, b; Nahorski and Rogers, 1975; Nahorski et al., 1975; Panksepp and Reilly, 1975; Schwartz, 1978). There is evidence to suggest that cyclic AMP may be involved in these CA-induced metabolic changes, since the stimulatory effect of the CA appears to be accompanied by increased conversion of phosphorylase *b* to phosphorylase *a*, as well as the consequent increase in glycogenolysis (Breckenridge, 1964; Drummond and Bellward, 1970; Edwards et al., 1974; Nahorski and Rogers, 1974; Nahorski et al., 1975). Furthermore, brain-stem lesions that damage ascending CA projections have been shown to cause a decrease in glucose metabolism within the forebrain (Hass et al., 1977; Heller and Hoffman, 1975; Schwartz, 1978; Schwartz et al., 1976). Consistent with the association found to exist in the periphery between local blood flow and metabolism, a number of these studies in the brain have found that alterations in CA activity and metabolic demand are accompanied by changes in cerebral blood flow (see Bates et al., 1977; de la Torre et al., 1977; Hass et al., 1977; Kuschinsky and Wahl, 1978; Lavyne et al., 1977; McCulloch and Harper, 1977; MacKenzie, 1976). A clear association between adrenergic varicosities and small blood vessels in the hypothalamus has been demonstrated (see de la Torre et al., 1977; Swanson et al., 1977), and this association has been proposed to result in changes in capillary permeability as well as in blood flow (Preskorn et al., 1978; Raichle et al., 1975).

The significance of the above findings for the effects of hypothalamic adrenergic stimulation on ingestive behavior or water balance is not known at this time. The energy requirement of the brain and therefore that for O_2 and its substrate, is substantial and continuous, and a large number of adaptive mechanisms have evolved for maintaining a constancy in the supply of these materials. In light of the evidence reviewed above, the hypothalamic adrenergic systems appear to be particularly well suited to enhance this process, not only via activation of neural mechanisms modulating the ingestion of specific nutrients (a glucocorticoid-dependent process) but also via their effects on cerebral carbohydrate metabolism and blood flow to meet local metabolic needs. As described above, a deficit in each of these effects is found to occur with damage to specific brainstem CA cells that project to the hypothalamus.

The relationship between the various changes in central neural and metabolic function associated with adrenergic stimulation of the brain has not been investigated. It is not known, for example, whether both changes must necessarily occur to evoke a response and, if so, whether they occur in parallel or in sequence. Studies of the fine structure of adrenergic innervation of the PVN have demonstrated that varicosities containing small granular vesicles (presumed to be noradrenergic) are associated with other structures in two distinct ways (Swanson et al., 1978). Approximately 20% exhibited distinct synaptic membrane specializations with the dendrites of other neurons, whereas a smaller percentage (5%) were directly apposed to the basal lamina of microvessels in the nucleus. The remainder of the varicosities, however, had no clear association with particular postsynaptic structures, although they were close to many different elements including neurons, glial cells, and blood vessels. On the basis of these findings, it appears that adrenergic systems may influence the function of the PVN via conventional synaptic interaction with its neurons, as well as via changes in the flow and permeability of its vascular supply. Furthermore, in light of the large number of labeled varicosities without synaptic membrane specializations, Swanson and his colleagues proposed that NE or EPI released into the extracellular fluid from these varicosities may also exert a "neurohumoral" effect over a broad area, on any cellular element (including glia) with appropriate receptor sites. Their function here might be to prepare the postsynaptic neurons metabolically to receive input through conventional adrenergic synapses. This evidence for combined synaptic and neurohumoral effects (similar to what occurs in the peripheral sympathetic system) emphasizes the need to broaden our perspective in evaluating central neurochemical effects on behavior and taking into consideration a variety of effects that may contribute to the final outcome.

III. Dopaminergic Systems

A. Anatomy of Dopaminergic Projections

The principal morphologic difference between the DA and NE systems of the brain is that the DA systems appear to be "local" systems with highly specified, topographically organized projections. From earlier work, the general view of the organization of the DA systems had been that there exist three major forebrain systems, namely, a nigrostriatal system arising from the Dahlström and Fuxe (1964) midbrain cell groups A8 and A9, a mesolimbic system arising from midbrain cell group A10, and a tuberohypophyseal system arising from hypothalamic cell group A12 (see Lindvall and Björklund, 1978; Moore and Bloom, 1978; and Ungerstedt, 1971a for reviews). With the introduction of the glyoxylic acid fluorescence histochemical method (Lindvall and Björklund, 1974) and a variety of new powerful neuroanatomical methods, however, there has been a marked increase in our understanding of the extent and organization of DA neuron systems, which, in addition to the mesotelencephalic and tuberohypophyseal projections, include an incertohypothalamic DA system and a periventricular DA system (see Hökfelt et al., 1978; Jacobowitz, 1978; Lindvall and Björklund, 1978, Moore and Bloom, 1978) (see Fig. 16).

The mesencephalic DA cell system lies in the ventral mesencephalon, specifically within the substantia nigra pars compacta and pars lateralis (group A9), the ventromedial tegmentum (A10), and the ventrolateral tegmentum extending caudal and dorsal to the substantia nigra and caudal and ventrolateral to the red nucleus (A8). Since no sharp boundaries can be drawn between these three parts of the system, Moore and Bloom (1978) have not used the nuclear designations A8, A9, and A10 of Dahlström and Fuxe (1964) but, instead, the term mesotelencephalic to apply to the entire mesencephalic DA system which projects largely upon the telencephalon (neostriatum, isocortex, and allocortex). The cells of the tuberohypophyseal system (A12) are located in the hypothalamic arcuate nucleus and the immediately adjacent periventricular nucleus. These cells project exclusively to the neurohypophyseal complex and to the pars intermedia of the adenohypophysis. The DA cell bodies of the incertohypothalamic system are found in the caudal thalamus and the posterior hypothalamic area (A11); the medial zona incerta (A13) just dorsal to the dorsomedial hypothalamic nucleus, and the periventricular nucleus of the anterior hypothalamus and preoptic region (A14). The incertohypothalamic system is essentially an intradiencephalic system; it consists of short, delicate axons which project locally, predominantly to the medial zona incerta and the dorsal and anterior hypothalamic areas (from A11 and A13 cells) and the anterior periventricular nucleus and medial preoptic area (from A14 cells). The periventricular CA system contains both DA and NE axons as well as DA cell bodies. The major source of NE axons to this system is the locus ceruleus and the dorsal medullary cell group (see Section II.A). The DA cell bodies that give rise to the periventricular system are located principally in the periventricular and periaqueductal gray matter, distributed diffusely from the medulla to the rostral third ventricle. The cells along the mesencephalic periaqueductal gray matter and the periventricular gray matter of the caudal thalamus are referred to as group A11. Those along the mesencephalic raphe, extending from the dorsal raphe nucleus ventrorostrally toward the interpeduncular nucleus, represent the

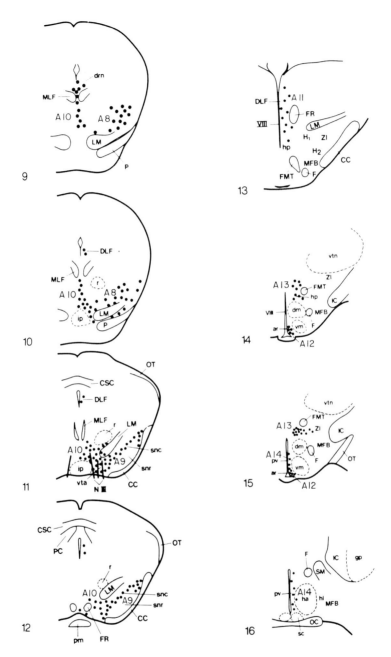

Figure 16 Distribution of CA-containing cell bodies (A8-14) in the pons, mesencephalon, and diencephalon of the rat brain, represented at eight frontal levels continuing rostral to Figure 1. For abbreviations, see page 405. (From Lindvall and Björklund, 1978.)

dorsal, midline part of the A10 system. The periventricular system has been divided into two components: a dorsal periventricular system, corresponding in large part to at least one component of the dorsal longitudinal fasciculus, and a ventral periventricular system, extending along the periventricular region of the hypothalamus. The dorsal system appears to be composed of neurons with relatively short axons that project predominantly into the periaqueductal and periventricular regions. The ventral periventricular bundle is formed in the supramammillary region and runs rostrally into the dorsomedial hypothalamic nucleus. At this level, fibers from the dorsal periventricular bundle join it, and the ventral periventricular bundle ascends in the periventricular nucleus, innervating that nucleus, as well as the PVN. The SON may also receive axons from the periventricular DA system.

The distribution of DA nerve terminals in the hypothalamus is so far only incompletely understood. It was originally assumed that virtually all hypothalamic DA nerve terminals have their cell bodies in the hypothalamus, belonging to cell group A12, and therefore, little was known about DA projections to hypothalamic areas other than the median eminence (to which A12 projects). However, it has been demonstrated that only 20% of hypothalamic DA can be attributed to the arcuate-periventricular projection (Björklund et al., 1973). Part is contributed by the incertohypothalamic system (Björklund et al., 1975), but some is very possibly contributed by components of the mesotelencephalic system. It has recently been shown that caudal hypothalamic deafferentation, as well as lesions in the area of the midbrain A8 cells, or the A9-A10 cells, result in a significant (40%) decrease in DA concentration of specific hypothalamic nuclei (Kizer et al., 1976; Palkovits, 1977). Furthermore, autoradiographic analysis of the projection of ventral midbrain neurons in the area of the DA cell groups has revealed terminal areas within the hypothalamus, including the periventricular and perifornical regions (Fallon and Moore, 1978). While further studies are clearly required to understand the hypothalamic DA systems, with respect to their origin as well as their terminal fields, the evidence indicates so far that midbrain, as well as diencephalic, DA cell groups project to the hypothalamus. Biochemical assays of hypothalamic DA content have revealed DA in all hypothalamic nuclei. While the arcuate nucleus showed the highest concentration, the PVN and dorsomedial nucleus were also quite high, and somewhat lower concentrations were found in the periventricular and supraoptic nuclei and the perifornical area (Palkovits et al., 1974; Versteeg et al., 1976).

B. Feeding Behavior

1. Hypothalamic and Extrahypothalamic Systems

The first studies to implicate central DA in the control of feeding started with the innovative work of Ungerstedt (1971b) who suggested that the disturbances in ingestive behavior associated with far-lateral hypothalamic lesions were due to the transsection of the nigrostriatal DA pathways which originate in the substantia nigra and ascend through the lateral hypothalamus to innervate the neostriatum. Ungerstedt's observations, that electrolytic or 6-OHDA lesions of the substantia nigra produce aphagia and adipsia, have subsequently been replicated with intraventricular as well as intracerebral 6-OHDA injections and related not only to the transient aphagia and adipsia but also to persisting regulatory deficits in food and water intake (see Grossman, 1975; Marshall, 1978; Stricker and Zigmond, 1976; Teitelbaum and Wolgin, 1975; Zigmond and Stricker, 1974, for reviews). These studies demonstrated that the feeding disorder produced by 6-OHDA was quite similar, although not identical, to that observed after lateral hypothalamic lesions; that the magnitude of the specific deficits in feeding regulation was correlated with the los of striatal DA innervation; and that the feeding capacities of the lesioned rats could be restored with peripheral injection of a DA agonist. In addition to the changes in food intake, however, destruction of the nigrostriatal system also caused profound disturbances in sensorimotor integration, as well as a variety of other behavior dysfunctions (such as in learning and sexual and aggressive responses). From these observations, therefore, it has been proposed that the nigrostriatal DA neurons are not involved in the specific mediation of individual activities (such as feeding or drinking) but instead provide some common activational component of all motivated behavior, thereby determining the animal's responsiveness to external stimuli (Marshall, 1978; Stricker and Zigmond, 1976).

The relationship between this telencephalic DA system and hypothalamic CA systems involved in feeding regulation remains to be elucidated. It has been proposed that the nigrostriatal DA pathway mediates the arousal and motor components of feeding, whereas the hypothalamic CA mechanisms provide the direct stimuli for initiating and terminating this behavior (Ungerstedt, 1974). While there appear to be no studies that have directly tested this hypothesis, there is evidence that hyperphagia and obesity induced by ventromedial hypothalamic lesions are associated not only with a reduction in hypothalamic NE concentration (Coscina et al., 1976) but also with a transient increase in striatal DA turnover (Marshall and Teitelbaum, 1977) and an increase in the ratio of striatal DA to hypothalamic NE (Glick and Stanley, 1975). Marshall (1978) found that intranigral 6-OHDA, which depleted striatal DA by 90%, blocked the hyperphagia induced by hypothalamic lesions, suggesting a permissive role for the nigrostriatal DA system in this feeding disorder. Coscina et al. (1973), however, failed to observe this relationship, possibly due to the fact that striatal DA was reduced only by 65-80% (see Marshall, 1978). It is interesting to note in this context that Schwartz et al. (1976) observed a significant decrease in glucose uptake in the hypothalamus after lesions in the substantia nigra. This may reflect a possible source of interaction between nigrostriatal and hypothalamic mechanisms.

There is one study which directly tested the NE feeding stimulation phenomenon in animals with altered nigrostriatal function (Oades, 1977). In this investigation, the animals received a unilateral 6-OHDA lesion of the substantia nigra and were found to exhibit a normal feeding response to hippocampal injection of NE. This result, suggesting that there may not be an interaction between these two neurochemical systems, is difficult to interpret, since the NE injected into the hippocampus may very possibly have been acting on both sides of the brain, due to its spread (via the ventricles) to the most sensitive receptor sites in the hypothalamic PVN (Leibowitz, 1978a).

Possibly related to the finding that nigrostriatal DA lesions produce aphagia and adipsia are the results obtained with peripherally injected neuroleptics. The butyrophenones in particular, as well as the diphenylbutyl-piperidine pimozide, are found to cause a suppression of food and water intake in response to deprivation, electrical brain stimulation, and acute homeostatic challenges (Block and Fisher, 1975; Phillips and Nikaido, 1975; Rolls et al., 1974; Rowland and Engle, 1977; Zis and Fibiger, 1975). As with the effects of striatal DA lesions, questions regarding the specificity of the neuroleptic effects makes the interpretation of these findings very difficult. The role of motor disturbances (associated with neuroleptic treatment) in the attenuation of feeding and drinking behavior remains unclear, although the effects do not appear to be attributed to a simple motor incapacitation. There is evidence to suggest that water intake may be more susceptible to the suppressive action of the neuroleptics (Block and Fisher, 1975; Rowland and Engle, 1977), although there is conflicting evidence on this point (Rolls et al., 1974). Finally, Zis and Fibiger (1975) indicated that, as with DA lesions (Stricker and Zigmond, 1976), the neuroleptics may be more effective in attenuating responses to acute metabolic challenges than to naturally occurring drive states. However, this was not found to be the case in the studies by Block and Fisher (1975) and Rowland and Engle (1977).

Thus, whether these effects produced by blockade of the DA receptor are due to interruption of the nigrostriatal DA system, as opposed to DA receptors in other brain areas or possibly in the periphery, remains to be determined. In contrast to the butyrophenones, the phenothiazine neuroleptics have actually been demonstrated to stimulate feeding behavior. There are numerous reports of an increase in appetite and obesity during chlorpromazine treatment in humans (e.g. Haase and Janssen, 1965; Hollister, 1965), and, in animals, a stimulation of feeding with peripheral (Reynolds and Carlisle, 1961; Robinson et al., 1975; Stolerman, 1970) and central (Leibowitz, 1976; Leibowitz and Miller, 1969) injections of chlorpromazine has been observed. In these latter studies, chlorpromazine was injected into the perifornical region of the hypothalamus, just lateral to the PVN. At this site, EPI has been found to suppress feeding in hungry animals, an effect which is mediated by β-adrenergic receptors but is also antagonized by neuroleptics (Leibowitz, 1970a; Leibowitz and Rossakis, 1978b). In light of this evidence, it is possible that chlorpromazine is increasing food intake via antagonism of perifornical CA receptors which are inhibitory towards feeding. This possibility is strengthened by the additional finding that DA activation of the perifornical area, like β-adrenergic stimulation, causes a suppression of feeding, and DA blockade, produced by haloperidol, leads to an increase in feeding (see below).

It has also been suggested, however, that chlorpromazine may facilitate feeding through activation of an α-adrenergic, feeding-stimulatory system (see Section II.B.3 above). This receptor system, concentrated in the PVN, has been shown to be activated by perifornical NE injections (Booth, 1968; Leibowitz, 1975a; Slangen and Miller, 1969), presumably via spread of the agonist to the nucleus which lies just 0.5 mm medial to the perifornical injection site (Leibowitz, 1978a). Perhaps the chlorpromazine may similarly be spreading to the PVN and activating the α-adrenergic receptors. In support of this suggestion, the feeding effect of chlorpromazine can be antagonized by the α-adrenergic blockers phentolamine and tolazoline (Leibowitz, 1976), and medial hypothalamic infusions of chlorpromazine have been shown to be effective in releasing endogenous NE (Lloyd and Bartholini, 1975). Furthermore, there is evidence that chlorpromazine, which in the periphery is known to antagonize α-adrenergic receptors, may at least in some brain areas actually stimulate α-adrenergic receptors (e.g., Bradley et al., 1966; Schmitt et al., 1973). Until further studies are conducted with centrally injected chlorpromazine, these questions regarding its mechanism of action cannot be resolved. From the above evidence, however, both α-adrenergic stimulation in the PVN and dopaminergic blockade in the perifornical area seem reasonable possibilities. Further insight into this problem may come from the recent observation that peripheral injection of clozapine, another neuroleptic, also increased feeding (Antelman et al., 1977). The pharmacological profile of this drug is quite different from that of the more classic neuroleptic agents, although it shares with chlorpromazine its actions on adrenergic receptor systems.

These studies with peripherally injected neuroleptics are further complicated in their interpretation by the finding that peripheral injection of DA stimulants can also suppress feeding behavior. It has been known for some time that AMPH, which releases brain CA, is a potent anorectic, and accumulating evidence has suggested that this drug's action may be mediated via endogenous DA, as well as NE or EPI (see below). Few studies, however, have actually tested selective DA agonists that act directly on the receptor site. Using the peripheral route of drug administration, Barzaghi et al. (1973) and subsequently Heffner et al. (1977) demonstrated a dose-dependent anorectic effect with the DA agonist apomorphine. This effect was blocked by the DA antagonists pimozide or spiroperidol. Two studies in which drugs were administered into the ventricles of the brain revealed a suppression of feeding with DA itself, at high doses ranging from 100 (Kruk, 1973) to 1000 μg (Hansen and Whishaw, 1973). A similar effect was observed with apomorphine, and both the DA and apomorphine feeding suppression was antagonized by peripherally administered pimozide (Kruk, 1973).

This evidence, revealing an anorectic effect with both a decrease and an increase in central DA activation, was interpreted by Heffner et al. (1977) as indicating that DA-containing neurons controlled feeding through an inverted U-shaped function, whereby deviations in either direction from the optimal point resulted in behavioral disturbances. This principle may, indeed, apply to most biological systems which have optimal conditions under which they function. Another possible explanation is that there exists more than one population of central DA receptors controlling feeding, perhaps in different brain areas, that are being affected by the drug manipulations used in the above studies. It was suggested earlier that DA innervation of the neostriatum may alter feeding behavior as a consequence of its more general effects on arousal and motor function. Thus, a decrease in DA activation within this system (through 6-OHDA injections or neuroleptic injections) would be expected to cause a deficit in feeding behavior (as well as other motivated behaviors), and DA stimulation would be predicted to restore these behaviors. In addition to this effect and in a quite separate system, however, it is possible that DA receptors might have a more direct function in controlling feeding, a function which is inhibitory in nature. While the above studies with peripheral or ventricular injections of AMPH, apomorphine, or DA would suggest this, they give no indications as to where these receptors may exist.

With administration of DA directly into different areas of the brain, evidence has recently been obtained which demonstrates that this agonist may produce anorexia through stimulation of receptors located specifically in the lateral perifornical region of the hypothalamus (Leibowitz, 1978c; Leibowitz and Rossakis, 1979a, b, c). Sites outside the hypothalamus (including the DA-rich neostriatum and nucleus accumbens) or within the medial or far-lateral hypothalamus were generally unresponsive to DA. The midlateral perifornical region, in contrast, from the caudal aspect of the PVN to the caudal aspect of the VMN, yielded a 50-70% suppression of feeding after injection of DA into pargyline-pretreated animals (Fig. 17). This brain area, which through histofluorescence has been shown to contain a fairly dense cluster of CA (DA- as well as NE- and EPI-containing) varicosities (Fig. 4d), was also found to be most responsive to the anorectic effect of EPI (see Section II.B.1), and the magnitudes of the effects produced by EPI and DA in the same animals were significantly and positively correlated. The threshold dose for DA's anorectic action in the perifornical area was less than 31 ng (Leibowitz and Rossakis, 1978a). Furthermore, its effect was potentiated by the DA uptake inhibitor benztropine but not by the NE uptake inhibitor desipramine. The reverse was observed in the case of EPI, which indicates that there may exist distinct presynaptic nerve endings for the release of DA and EPI.

The receptor sites for these two CA agonists also appeared to be distinct, although contiguous. Consistent with the literature obtained with peripheral drug injection, the anorectic effect of perifornical DA injection was antagonized by locally administered neuroleptics (Leibowitz and Rossakis, 1979b). The relative potency of these and structurally related compounds was haloperidol > fluphenazine > chlorpromazine > pimozide

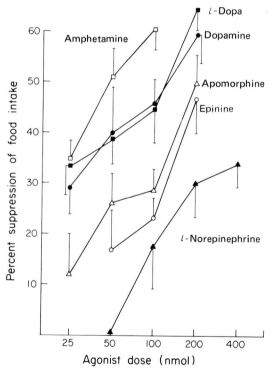

Figure 17 Percent suppression of food intake (relative to vehicle baseline) after perifornical hypothalamic injection of CA stimulants in pargyline-pretreated, hungry rats. (Based on data from Leibowitz and Rossakis, 1979c.)

> promazine. The stereospecificity of the hypothalamic DA blockade was demonstrated with (+)butaclamol versus the inactive (−)butaclamol. The relatively ineffective neuroleptic promethazine, the tricyclic antidepressants, imipramine and desipramine, and the antagonists of α-adrenergic, β-adrenergic, cholinergic, and serotonergic receptors did not manifest the ability to reverse DA's action. The inhibitory effect of DA in the perifornical area was dose related and mimicked by other CA stimulants, with the order of potency being DA > apomorphine = epinine > NE (Fig. 17). Additional tests with the DA-receptor antagonists injected alone indicated that haloperidol and chlorpromazine, at doses higher than those required to inhibit DA's action, were effective in causing a facilitation of feeding.

This evidence, therefore, obtained with the various agonists and antagonists, has revealed properties of the hypothalamic DA-sensitive sites which closely resemble those expected of DA receptors, in the periphery and in other brain areas, that have been characterized indirectly through adenylate cyclase or, more directly, via radioactive ligand binding (e.g. Bockaert et al., 1977; Burt et al., 1976; Creese et al., 1976; Iversen, 1975; Krueger et al., 1975). As is typical of the DA receptor, the DA-receptive sites in the hypothalamus exhibited little affinity for the adrenergic agonist NE. It was mentioned above, however, that EPI injection into the perifornical area has a similar effect to that of DA on feeding, with comparable potency, and while EPI's action appears to be specifically mediated by β_2-adrenergic receptors, it is also found to be antagonized by DA blockers (see Section II.B.2 above). While this finding suggests a close relationship between the adrenergic and dopaminergic receptors, the observation that β-adrenergic antagonists had no effect on DA's receptor action even at high doses demonstrates that EPI and DA are acting on two distinct and functionally independent populations of hypothalamic receptors (see Section II.B.3 for further discussion of this point).

In an earlier study, Friedman et al. (1973) revealed a facilitatory effect on feeding of lateral hypothalamic injection of *l*-dopa. This phenomenon, which occurred in hungry but not satiated rats, has been examined in greater detail by Starr and Coons (Starr, 1975; Starr and Coons, 1975), who found that the brain site most responsive to this effect was in the dorsocaudal far-lateral hypothalamus, in the path of the nigrostriatal projection as it courses along the medial edge of the internal capsule. The facilitatory effect in this region was observed with DA as well as *l*-dopa; the effective doses ranged from 2 to 10 or 20 μg, which each yielded the same magnitude of response, approximately a 30% increase in feeding. While a pharmacological study of the receptors mediating this phenomenon has not yet been conducted, Starr and Coons observed the opposite change in feeding, namely, a feeding suppression after far-lateral injection of the DA blockers haloperidol and pimozide, or the β-adrenergic blockers propranolol and LB-46. The significance of these feeding changes and their relationship to the phenomenon of CA-induced anorexia associated with the perifornical area (which lies more medial and ventral) will need to be explored further. The lack of a dose response with the DA facilitation of feeding in the far-lateral hypothalamus may be indicative of an all-or-none effect, perhaps on the nigrostriatal projection system coursing through this area. It is interesting to note that, just as DA injection facilitated feeding in food-deprived rats, it also facilitated the drinking response of water-deprived animals (see Section III.C). These effects, therefore, may reflect a more general activation state concordant with a general function of the DA nigrostriatal system in arousal and sensorimotor integration, as proposed above. An elaboration of this hypothesis, however, may be necessary in light of the evidence that DA injection facilitated only the specific consummatory response that was appropriate to the animals' motivational state; feeding was not potentiated in thirsty rats, nor was drinking potentiated in hungry animals.

2. Changes in Endogenous Transmitter Function

The above evidence, consistent with peripheral injection studies, makes it clear that hypothalamic DA stimulation, specifically in the perifornical area, can produce a decrease in food intake. The question now is whether this effect of exogenous DA actually reflects a true function of hypothalamic CA mechanisms in control of spontaneous feeding. One source of evidence in support of this proposal is the finding that AMPH, which is known to release endogenous DA from presynaptic endings (Carlsson, 1970; Glowinski, 1970; Lloyd and Bartholini, 1975), is effective in suppressing feeding (Fig. 17) after injection into the perifornical hypothalamus (as opposed to other brain areas) in hungry animals (Leibowitz, 1975c). It is also effective, when peripherally injected, in inhibiting feeding elicited by lateral hypothalamic electrical stimulation (see Wisehart and Walls, 1974). Furthermore, the anorectic effect of AMPH in hungry animals, whether produced via peripheral or hypothalamic drug injection, can be antagonized by CA synthesis inhibitors as well as by blockers of dopaminergic (and adrenergic) receptors, and perifornical injection of the DA antagonists haloperidol and pimozide significantly attenuated the anorexia caused by peripheral AMPH injection (see Leibowitz, 1975d, 1978c).

Additional studies with presynaptically acting drugs have demonstrated that mazindol, as well as methamphetamine and phenmetrazine, are similarly effective as anorectic agents after injection into the perifornical hypothalamic region (Leibowitz and Rossakis, 1979a). Mazindol, although it possesses a different chemical structure from phenylethylamine derivatives such as AMPH, is believed, like AMPH, to act indirectly through endogenous CA neurons (dopaminergic as well as adrenergic), although there appear to be some differences between these two drugs' over-all pharmacological profile (see Garattini and Samanin, 1978). When injected peripherally or into the perifornical hypothalamus, mazindol's anorectic action was attenuated or blocked by DA antagonists (Garattini and Samanin, 1978; Leibowitz and Rossakis, 1978a).

The precursor to DA as well as to NE and EPI, l-dopa, has also been found to suppress feeding. This occurred when the drug was administered peripherally (Baez, 1974; Dobrzánski and Doggett, 1976; Heffner et al., 1977; Sanghvi et al., 1975) as well as directly into the perifornical hypothalamus (Fig. 17) (Leibowitz and Rossakis, 1979c). In the perifornical area, this effect was positively correlated in magnitude with the same effect produced by the CA agonists DA and EPI and by AMPH. The action of l-dopa was antagonized by local administration of the dopa decarboxylase inhibitors Ro 4-4602 and MK-486; it was only partially reversed by separate injections of the DA blocker haloperidol and β-adrenergic blocker propranolol but totally abolished by combined injections of these two antagonists. Thus, it appears that l-dopa may be exerting its action at least in part through increased CA synthesis, specifically within dopaminergic and adrenergic neurons of the perifornical hypothalamic region. In support of this hypothesis, an increase in synthesis of both DA and NE (EPI was not assayed) has been demonstrated in the hypothalamus after peripheral l-dopa injection (Glowinski and Iversen, 1966; Hyppa et al., 1971; Langelier et al., 1973).

These studies, therefore, provide pharmacological evidence for the hypothesis that changes in activity of endogenous DA may be associated with predictable changes in feeding. A few studies have employed biochemical methods to analyze changes in hypothalamic DA levels or turnover as a function of feeding state; however, inconsistent results have been obtained. Friedman et al. (1973) found food deprivation to result in an increase in whole hypothalamic DA levels, as well as an increase in the loss of DA after injection of α-methyltyrosine. Measurements of concentration and turnover in discrete hypothalamic areas, however, failed to reveal any changes in DA (van der Gugten, 1977). Biggio et al. (1977) observed a decrease in brain DA metabolites in hungry animals before eating and a rise in metabolites, possibly reflecting increased turnover, during the next 2 hr of feeding. In measuring the efflux of [^{14}C] DA via a push-pull cannula, Martin and Myers (1976) observed an increase during the feeding process in circumscribed sites, namely, the nucleus reuniens and zona incerta. With measurements of endogenous DA release, however, van der Gugten and Slangen (1977) failed to observe reliable changes in any of the hypothalamic areas analyzed. Further studies are clearly necessary to resolve this difficult question concerning simultaneous changes in brain neurochemistry and behavior. The techniques are very complex, as are the neural circuitry and behaviors under investigation. Thus, variable results would not be surprising with only small changes in procedures, placements, or definitions of behavior.

With gross measurements of food intake, the studies reviewed in the previous section clearly revealed that total food intake could be inhibited by dopaminergic stimulation of the hypothalamus. In order to achieve a better understanding of the function subserved by DA in producing this effect, Blundell and Latham (1978) analyzed various parameters of feeding, including latency to eat, rate of eating, and meal frequency, size, and duration, in animals receiving peripheral injection of various drugs. The general pattern of results indicated that DA may be specifically involved in adjusting the rate of eating and that its anorectic effect may be attributed primarily to an effect on meal frequency as opposed to meal size or latency to eat. Both amphetamine and mazindol reduced the number of meals taken by the animals but had little effect on the size of the meal. Amphetamine appeared to delay the initiation of feeding, but this effect could not be replicated with mazindol. With regard to the rate of eating, DA-receptor activation affected this parameter but in the opposite direction one might expect. The DA stimulants AMPH and mazindol increased the rate of eating, which had no consequent impact on the total food consumed during a meal. Dopaminergic antagonists, in contrast, caused the opposite effect, namely, a decrease in rate of eating. The relationship of this phenomenon to the anorectic action of AMPH or mazindol remains to be explored. Blundell and Latham proposed, however, that meal size (perhaps mediated by serotonergic systems as described in Section IV.B) and rate of eating (a dopaminergic function) can be independently regulated, and that meal frequency may be an important parameter in translating hypothalamic DA stimulation into a suppression of feeding.

3. Lesion Studies of Dopaminergic Pathways

Attempts to relate specific dopaminergic pathways to the control of feeding and drinking behavior have focused almost exclusively on the DA projection from the substantia nigra (A9 cell group) which innervates the neostriatum (Figs. 2 and 16). As described in Section III.B.1, damage to this nigrostriatal pathway has been found to produce aphagia and adipsia, as well as a variety of persisting regulatory deficits. There is evidence to suggest, however, that these effects result from a more general disturbance in arousal and motor function, rather than an impairment in the control of a specific behavior such as feeding or drinking (Marshall, 1978; Stricker and Zigmond, 1976). Lesions in the medial region of the ventral tegmentum, where the A10 DA cell bodies are concentrated, have been associated with a hyperactivity syndrome, with apparently no specific disturbances in feeding or drinking (Le Moal et al., 1977). The midbrain DA A8 cell group, which is located just dorsal and caudal to the substantia nigra (Fig. 16), has not specifically been investigated with respect to its role in the control of feeding and drinking behavior. However, ventral tegmental lesions (electrolytic or 6-OHDA) which produced significant damage to these cells, or at least impinged on their caudal aspect, have been reported to produce hyperphagia and obesity (Ahlskog, 1974; Leibowitz and Brown, 1980c). Ahlskog attributed this behavioral effect to the destruction of ascending noradrenergic or adrenergic fibers, which course through the ventral midbrain tegmentum in the same region as the DA A8 cells (see Section II.F). While biochemical assays in Ahlskog's study demonstrated a large reduction of forebrain NE, they also revealed a small decrease in DA as well. Thus, the possibility exists that damage to the midbrain DA cells may have contributed to the resulting hyperphagia syndrome. While this suggestion is speculative at this point, it has merit in pointing out the problems that may arise from lower brain-stem lesions that sever multiple CA as well as non-CA systems. It will be very difficult with our present techniques to identify the function of the DA A8 cells, since they lie scattered throughout a significant portion of the ventral tegmentum.

Through the combination of midbrain lesions and hypothalamic cannulas, however, some evidence has been obtained to suggest that DA cells (A8 and A9) of the ventral midbrain (Fig. 18), along with a ventral adrenergic projection (Fig. 10), may have a function in mediating the inhibition of feeding produced by perifornical hypothalamic CA stimulation (Leibowitz, 1979b; Leibowitz and Brown, 1980b). The primary evidence for this conclusion is that far-ventral tegmental electrolytic or 6-OHDA lesions which damaged specifically these DA cell groups (Fig. 18) invariably caused: (1) a marked reduction of CA varicosities in the perifornical area; (2) a strong reduction or loss of the anorectic response produced by perifornical injection of the presynaptically acting drugs AMPH and mazindol; and (3) a potentiation of the same response produced by the CA agonist DA. Lesions in the dorsal midbrain tegmentum, which left the ventral adrenergic and dopaminergic fibers intact but damaged several other CA projections, had no apparent effect on the responsiveness of the perifornical hypothalamus to CA drug stimulation as well as on the CA fluorescence in that region. Lesions in the area of the DA A10 cells potentiated the effectiveness of the anorexigenic drugs in the perifornical hypothalamus.

Electrolytic (Carey and Goodall, 1975) and 6-OHDA (Fibiger et al., 1973) lesions in the vicinity of the substantia nigra A9 cells have previously been found to attenuate (by approximately 50%) the anorexic effect of peripheral AMPH injection. This evidence was taken to support the hypothesis that the nigrostriatal DA system plays a specific role in CA-induced anorexia. It is likely, however, that both of these lesions were also associated with damage to other CA (dopaminergic or adrenergic) projections that innervate other forebrain areas (Fig. 18). Furthermore, 6-OHDA injection into the caudate nucleus

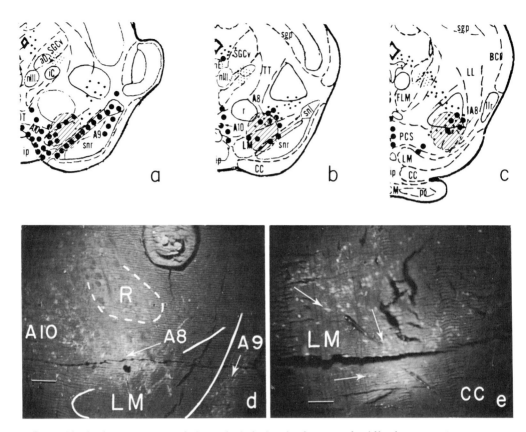

Figure 18 (a-c) Illustrations of electrolytic lesion in the ventral midbrain tegmentum (in the area of the medial lemniscus), superimposed on three frontal sections showing CA varicose fibers (small black dots) and cell bodies (large black dots) as drawn by Palkovits and Jacobowitz (1974). The lesion overlaps the DA cell bodies of A8 and some of the A9 cells within the substantia nigra. See page 405 for abbreviations. (d) Fluorescence photomicrograph showing the midbrain DA cell groups (A8, A9, and A10) at the level of the red nucleus (R). A8 and A9 cells are located within and along the dorsomedial surface of the medial lemniscus (LM), whereas A10 cells are located medially and substantia nigra A9 cells ventral to LM. (e) Caudal to red nucleus, this photomicrograph shows scattered A8 cells (arrows) dorsal and lateral to the LM. The cells immediately dorsal to the crus cerebri (CC) fall within the recitulata zone of the substantia nigra and merge with the A8 cells. (Based on data from Leibowitz and Brown, 1980b.)

had no effect on AMPH's anorexigenic action (Samanin et al., 1977), and injections of CA stimulants directly into the caudate nucleus produced no change in baseline food intake (Leibowitz, 1975c; Leibowitz and Rossakis, 1979a). These central mapping studies and the above midbrain lesion results argue in favor of a midbrain-hypothalamic DA projection which originates from cell bodies in the ventral midbrain (ventral A8 and A9) and terminates in the perifornical hypothalamic area. Evidence to support the existence of such a projection and its proposed function in suppressing feeding behavior is provided by the finding that lesions in the vincinity of the DA A8 and A9 cells (Fig. 18) potentiate the anorexic effect of DA receptor stimulation of the perifornical hypothalamus, while causing a total loss of AMPH anorexia and a marked reduction of CA varicosities in that area (Leibowitz and Brown, 1980a,b). Biochemical assays, furthermore, have found similar ventral tegmental lesions to cause a significant reduction of hypothalamic DA levels (Kizer et al., 1976), and autoradiographic studies have shown ventral midbrain cells projecting to the perifornical hypothalamus (Fallon and Moore, 1978).

It is possible that diencephalic cells belonging to the incertohypothalamic DA system described by Bjorklund et al (1975) (Fig. 16) contribute to the perifornical DA innervation involved in feeding inhibition. These cells project short axons to the rostral half of the hypothalamus and may possibly send collaterals to the perifornical region. The strongest argument against this proposal, however, is that ventral midbrain lesions, which had no apparent impact on these cells, totally abolished the animals' responsiveness to perifornical (Leibowitz and Brown, 1980b) as well as to peripheral (Ahlskog, 1974; Leibowitz and Brown, 1980c) AMPH injection. Thus, it would appear that the dopaminergic cell bodies which are important for hypothalamic CA-induced suppression of feeding are located in the ventral midbrain region. The adrenergic cell bodies also involved in the response appear to be located in the medulla (cell groups A1, C1, and possibly A2 and C2) or the pons (A5 and A7) (see Section II.F). The exact rostral course of these projections from the midbrain to the hypothalamus is not known, although they very likely join the medial forebrain bundle (Fig. 2). Damage to this bundle, either just lateral to the fornix (Blundell and Leshem, 1974) or medial to the internal capsule (Carlisle, 1964), has been shown to attenuate the anorexigenic action of peripherally injected AMPH. In contrast, medial hypothalamic lesions (Epstein, 1959), parasagittal knife cuts just medial to the fornix (Sclafani and Berner, 1977), and medial coronal cuts (caudal to the VMN) extending just to the medial edge of the lateral hypothalamus (Sclafani and Berner, 1977), produced no change or actually potentiated AMPH's action. From this evidence, it appears that the fibers mediating the CA feeding suppression remain within the medial forebrain bundle of the lateral hypothalamus before turning medially towards the fornix at the level of the VMN.

The finding that midbrain lesions of specific CA projections selectively disrupt the presynaptic action of AMPH, while leaving intact the effect of DA and EPI, indicates that the perifornical dopaminergic and adrenergic receptors upon which these CA agonists act are located on the postsynaptic cell membrane (Leibowitz and Brown, 1980b). While these CA receptors appear to be anatomically contiguous (Leibowitz and Rossakis, 1979a), the evidence that two quite different lesions, one aimed at the A8-A9 cells (Fig. 18) and one at the ventral adrenergic fibers caudal to the DA cell bodies (Fig. 10), selectively potentiate the receptor action of DA and EPI, respectively, supports the proposal (Leibowitz, 1978c) that the two receptor systems are distinct and functionally independent. This evidence for denervation supersensitivity indicates further that the dopaminergic and adrenergic receptor sites are neuronally linked to specific CA cell bodies in the lower brain stem.

In addition to this lateral perifornical CA system which is involved in the suppression of feeding, there is evidence, through pharmacological and histochemical studies, that the medial paraventricular and periventricular hypothalamic nuclei, and dorsal midbrain noradrenergic projections to these nuclei, play a role in the mediation of CA-induced stimulation of feeding (see Section II.F). α-Noradrenergic activation of the PVN, via exogenous or endogenously released NE, was found to induce eating in satiated rats. Furthermore, lesions in the dorsal midbrain, which damage specifically the dorsal fibers of the central tegmental tract, disrupted the action of drugs (antidepressant agents) which act presynaptically in the PVN to elicit feeding and potentiated the postsynaptic effect of exogenous NE. In the study of Leibowitz and Brown (1980b), these same dorsal lesions were found to have no effect on the feeding-suppressive effect induced by perifornical CA stimulation. This supports the suggestion that these antagonistic hypothalamic systems are mediated via different CA projections. These CA projections, however, may in some manner interact, as evidenced by the finding that ventral midbrain lesions which disrupted the anorexic effect of AMPH in the perifornical area significantly *potentiated* the feeding stimulation effect of presynaptically acting drugs (antidepressants) in the paraventricular area (Leibowitz and Brown, 1980a). The level at which this interaction takes place is not known, although it would appear to be at some presynaptic point since the postsynaptic feeding stimulation effect of exogenous NE in the PVN was only minimally potentiated by the ventral lesions.

C. Drinking Behavior and Water Excretion

Very few studies have attempted to relate brain DA mechanisms to the control of water ingestion. Peripherally injected amphetamine has been found to suppress drinking behavior, an effect which is reversed by inhibition of brain CA synthesis (see Section II.C). Neilson and Lyon (1973) observed a similar effect with the DA agonist apomorphine and reported that the drinking suppression produced by AMPH or apomorphine could be partially blocked by DA antagonists. This is consistent with the finding that perifornical injection of DA or AMPH suppressed drinking in addition to feeding (Leibowitz, 1973b; Leibowitz and Rossakis, 1979b). However, in these central injection studies (conducted in hungry animals), the drinking suppression induced by DA was attenuated by α-adrenergic blockers and unaffected by antagonists of dopaminergic, β-adrenergic, or serotonergic receptors. This finding is consistent with the evidence that α-adrenergic stimulation of the hypothalamus is generally inhibitory in its effect on water consumption (see Section II.C). It also indicates that, while hypothalamic DA can suppress drinking as well as feeding behavior, these effects involve different populations of receptors and therefore can be pharmacologically and behaviorally dissociated.

Peripheral injections of drugs which antagonize DA receptors have also been associated with a suppression of water intake. This has been observed with drinking responses induced by deprivation and by acute homeostatic challenges (Block and Fisher, 1975; Phillips and Nikaido, 1975; Rolls et al., 1974; Rowland and Engle, 1977; Zis and Fibiger, 1975). A suppression of feeding, as well as other behaviors, has also been observed with the DA antagonists, and the possibility was suggested that instead of acting on a dopaminergic mechanism specifically involved in regulation of feeding and drinking behavior, these neuroleptics were having a more general affect on arousal or motor function, perhaps via the nigrostriatal DA system (see Section III.B.1 above).

With regard to a possible stimulatory role of central DA on water intake, this was proposed by Setler and Fitzsimons (see Setler, 1977, for review) who revealed that drinking induced by central angiotensin injection or peripheral isoproterenol injection was antagonized by central injection of the DA antagonist haloperidol. Attempts to demonstrate a stimulatory effect with central injections of DA or the dopaminergic agonist apomorphine, however, have not been very successful, revealing only a weak dipsogenic effect with either of these agonists (Fisher, 1973; Setler, 1977). In tests with DA injections into various hypothalamic areas of thirsty or hungry animals, the predominant effect of this agonist was found to be inhibitory, and in food- and water-satiated animals, no increase in spontaneous water intake was observed (Leibowitz, 1975b; Leibowitz and Rossakis, 1979a, b). However, there is a report that DA injected into the far-lateral hypothalamus, in the path of the nigrostriatal DA projection, may facilitate an on-going drinking response produced by water deprivation (Starr, 1975). Further research is required to understand the relationship of this phenomenon to a proposed function of the nigrostriatal system in the control of feeding and drinking behavior (see above).

Earlier studies on the effect of central DA on water excretion or ADH release failed to reveal any significant change with this agonist (e.g., Bhargava et al., 1972; Olsson, 1970). A few more recent studies, however, have given some indication that hypothalamic DA may exert an effect similar to that of α-adrenergic stimulation (see Section II.D above), namely, antidiuresis as a consequence of ADH release. This effect was observed with ventricular injections of DA in the anesthetized rat and found to be antagonized by the DA blockers, haloperidol and pimozide, and also by the α-adrenergic blocker phentolamine (Bridges et al., 1976). In an in vitro experiment with hypothalamic slices containing the SON, PVN, median eminence, and proximal pituitary stalk, Bridges and his colleagues found apomorphine as well as DA to be effective in releasing ADH. Wolny et al. (1974) observed a diuretic effect of DA after ventricular injection; however, this was obtained at very high doses (100 μg) of the agonist.

There appear to be only two studies which have examined the effects of DA injected directly into the SON. Urano and Kobayashi (1978) tested DA's effect on urine flow in anesthetized rats and observed a decrease in urine output at doses of 10-20 μg and a blockage of this effect with peripheral injection of pimozide. In awake and hydrated animals, an antidiuretic effect (reduced urine output and increased urine osmolarity) of DA injected into the SON was revealed at considerably lower doses (Garay and Leibowitz, 1974). This effect was reliable at a dose as low as 0.8 ng, 10-fold higher than the threshold dose for NE (see Section II.D) and showed a monotonic increase in magnitude with doses up to 0.1 μg. (Higher doses were sometimes less effective.) Tests with receptor antagonists injected directly into the SON in combination with DA revealed an inhibition of DA-induced antidiuresis with the α-adrenergic blocker phentolamine and a reversal of the effect (towards diuresis) with the dopaminergic antagonist haloperidol. The β-adrenergic blocker LB-46 had no effect.

Thus, it appears that DA receptors in the SON, responsive to haloperidol blockade, may be involved in the control of ADH release. In view of the evidence that α-adrenergic receptor blockade also disrupts DA's action, it would appear either that the DA receptors work in conjunction with α-adrenergic receptors to produce antidiuresis or possibly that the injected DA works, in part, through the synthesis and release of endogenous NE or EPI. The possibility that endogenous CA neurotransmitter release can mimic the effect of the exogenous CA agonists on ADH release is suggested by the finding that AMPH injected into the SON of paragyline-pretreated rats similarly produced antidiuresis, and this effect was inhibited by local administration of the CA synthesis inhibitor α-methyltyrosine (Leibowitz and Tom, unpublished data).

In addition to producing an antidiuresis, DA injected into the SON also caused an increase in the excretion of NA^+ and K^+ (Garay and Leibowitz, 1974). This effect, however, would appear to be due to ADH release, since the ratio of these two electrolytes remained stable. Two additional studies in which DA was injected into the ventricles also failed to reveal a specific effect of DA on electrolyte excretion (Bhargava et al., 1972; Morris et al., 1977). This would appear to distinguish DA from the adrenergic agonists, which caused an increase in urinary sodium in addition to a decrease in water excretion (see Section II.E above).

IV. Serotonergic Systems

A. Anatomy of Serotonergic Projections

The fluorescence histochemical technique is considerably less sensitive for neurons containing serotonin (5-HT) as compared to the catecholamines, and, consequently, our knowledge of the 5-HT pathways is incomplete. However, the additional use of the autoradiographic tracing technique and enzymatic-isotopic microassays, along with various drug regimens including the neurotoxic dihydroxytryptamines selective for 5-HT neurons, has opened up new possibilities in this field and been particularly invaluable in the mapping of 5-HT fibers (see Azmitia, 1978; Fuxe and Jonsson, 1974; Hökfelt et al., 1978; Jacobowitz, 1978; Kent and Sladek, 1978; Moore et al., 1978; and Morgane and Stern, 1974 for reviews). From fluorescence histochemical evidence, it is clear that most of the 5-HT cell bodies are localized to the raphe region of the lower brain stem (designated cell groups B1-B9 by Dahlström and Fuxe [1964]). The fibers ascending to the forebrain, including the hypothalamus, appear to originate perdominantly from the midbrain raphe nuclei, specifically, the dorsal raphe (B7) and median raphe (B8) nuclei (Fig. 19a). However, several investigators have performed complete deafferentation of the hypothalamus and found 5-HT levels in the hypothalamic island to be 30-40% of normal. This finding has led investigators to propose the existence of 5-HT cell bodies within the hypothalamus (see above references), and, in support of this proposal, there is recent evidence to suggest that such cells may be located in the region of the arcuate nucleus (Kent and Sladek, 1978). The dorsal raphe nucleus appears to be the largest single contributor to the hypothalamic 5-HT projections, as lesions of this nucleus produce a 50-65% loss of 5-HT content in various hypothalamic nuclei (Palkovits et al., 1977).

Histochemically, the highest numbers of 5-HT nerve terminals in the hypothalamus are found in the suprachiasmatic nucleus, but 5-HT terminals are spread all over the hypothalamus. Analysis of the regional distribution of 5-HT, via the enzymatic radioisotopic method, has revealed the highest concentrations in the basal and posterior hypothalamus (including the arcuate, suprachiasmatic, and posterior nuclei as well as the perifornical area), somewhat lower concentrations in the PVN and dorsomedial nucleus and still lower concentrations in the SON and VMN (Palkovits et al., 1977; Saavedra et al., 1974). Results of autoradiographic studies (Fig. 19b-f) are generally consistent with this pattern of distribution (see Azmitia, 1978; Moore et al., 1978). Furthermore, they demonstrate that the 5-HT axons projecting to the hypothalamus, originating predomi-

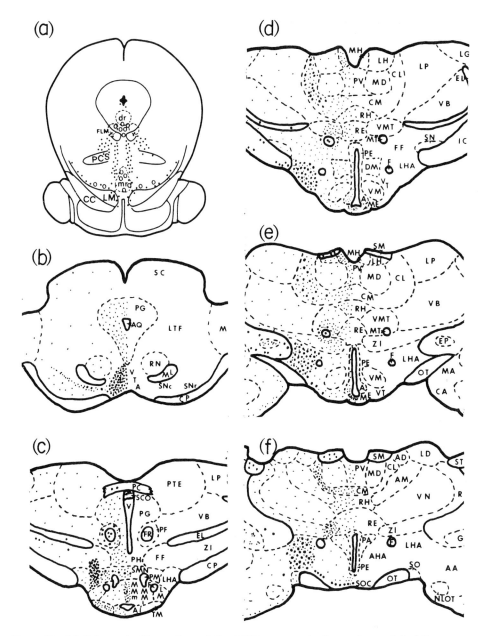

Figure 19 (a) Schematic illustration of the 5-HT neurons in the raphe nuclei of the mesencephalon, based on evidence obtained with fluorescence microscopy in animals treated with 5-HT neurotoxins. Open circles: cell bodies; black dots: 5-HT axons. (From Fuxe and Jonsson, 1974.) (b-f) Drawings of coronal sections showing autoradiographic labeling on the left side of the brain produced by injection of tritiated proline into the raphe nuclei of the midbrain. Large dots represent labeling of fibers of passage; small dots represent terminal field labeling. See page 405 for abbreviations. (From Moore et al., 1978.)

nantly from the midbrain raphe cell groups, course mainly along the periventricular region of the third ventricle; ventrolaterally to the medial forebrain bundle in the supraoptic commissure of Meynert; and ventromedially within the medial forebrain bundle. The periventricular tract, originating in dorsal cell group B7, constitutes the main projection to the medial hypothalamus, including the periventricular nuclei, the arcuate nucleus, and the internal layer of the median eminence. The tract which travels ventrolateral to the medial forebrain bundle (between the crus cerebri and optic tract) also originates from B7 and projects predominantly to the suprachiasmatic nuclei. The fibers from the median cell group B8 to the lateral hypothalamus and the preoptic area travel in the ventromedial aspect of the medial forebrain bundle. These fibers remain confined mainly to the lateral hypothalamic nuclei until the level of the suprachiasmatic nuclei, where they move ventromedially to innervate both the PVN and suprachiasmatic nuclei.

B. Feeding Behavior

1. Serotonergic Receptor Stimulation

In contrast to the extensive evidence (see above) in support of a role for hypothalamic CA systems in control of feeding, there are considerably fewer studies suggesting a similar function for hypothalamic 5-HT (see Blundell, 1977; and Garattini and Samanin, 1978 for review). Perhaps the strongest evidence has come from peripheral injection studies which have examined the mechanism of action of the anorectic drug fenfluramine. This drug, in contrast to AMPH, appears to act predominantly on 5-HT systems and minimally on CA systems (see Section IV.B.2 below). If the 5-HT systems involved in this effect are located centrally, one should find central injection of 5-HT to cause a suppression of feeding. However, there have been several reports of negative effects (in satiated or hungry animals) with 5-HT injection into the medial or lateral hypothalamus or lateral ventricles (see Baile, 1974; Blundell, 1977). The doses used in these studies were between 5 and 20 μg. At a higher dose of 100 μg, Kruk (1973) observed a strong suppression of feeding after ventricular injection, and this effect was blocked by the 5-HT antagonist cyproheptadine but not by the DA antagonist pimozide. Lehr and Goldman (1973) demonstrated a similar effect with perifornical hypothalamic injection of 5-HT (at 18 μg) but found this to be counteracted by the β-adrenergic blocker propranolol. The possibility that this phenomenon may be related to the finding that β-adrenergic stimulation of the same perifornical area (with EPI or isoproterenol) also suppresses feeding (Leibowitz, 1978c) has not been explored.

The main question which needs to be addressed in these studies of central 5-HT injection is whether the inhibitory effects observed are specific to feeding behavior or whether they reflect a general behavioral depression. Brain 5-HT has been implicated in the control of such behaviors as sleep (Morgane and Stern, 1974), aggression (Reis, 1974), and mating (Soulairac and Soulairac, 1975), and the effect it exerts is generally believed to be inhibitory in nature. Hypothalamic injection of 5-HT, particularly at doses above 10 μg, is found to cause sleep in the rat (Leibowitz and Papadakos, unpublished data), and this effect would obviously be expected to compete with a feeding response. In fact, when injected into the perifornical hypothalamus at lower doses (1-5 μg) which do not cause sleepiness, feeding in hungry rats may actually be enhanced (Leibowitz and Papadakos, 1978), suggesting that the inhibition observed at higher doses (Lehr and Goldman, 1973) may be a consequence of the sedative effect.

In a series of studies with 5-HT injection into the medial PVN of the rat, a distinct inhibition of feeding, without effects on general arousal, has been reported (Leibowitz and Papadakos, 1978). In hungry rats, doses of 1-10 μg produced a reliable, dose-dependent suppression of up to 50% during a 60-min period after injection. A more potent inhibitory effect was revealed in satiated rats that were eating in response to PVN injection of NE (Fig. 20). That is, 5-HT, when injected into the PVN immediately prior to NE, antagonized by up to 90% the eating response induced by NE. This action of 5-HT was dose dependent and occurred reliably at a dose as low as 100 ng. Furthermore, this inhibition appeared to be relatively specific to 5-HT, since the catecholamines, namely, DA and isoproterenol, failed to interact with NE feeding in this manner, except at doses above 10 μg.

Tests with receptor antagonists injected into the PVN immediately prior to 5-HT (and NE) revealed a dose-dependent antagonism of 5-HT's action with the 5-HT blockers, methysergide and cinanserin. This evidence argues in favor of the existence of 5-HT receptors in the PVN for inhibition of feeding. It was also found, however, that the effect of 5-HT could be counteracted by injection of the β-adrenergic blockers propranolol and alprenolol, as well as by the phenothiazine dopaminergic antagonists fluphenazine and chlorpromazine. The butyrophenone neuroleptic haloperidol, and the diphenylbutylpiperidine pimozide, in contrast, were without effect, as were the cholinergic antagonist atropine and the histaminergic antagonist dexbrompheniramine. This finding reveals, in a specific brain area, the interaction that a number of biochemical studies have shown to occur between 5-HT receptors and CA receptor antagonists. That is, radioactive ligand-binding studies have demonstrated a strong affinity of β-adrenergic and dopaminergic antagonists for 5-HT receptors (e.g. Creese and Snyder, 1978; Middlemiss et al., 1977), and central neurochemical or physiological effects of 5-HT have been shown to be abolished by these receptor antagonists (e.g., Ahn and Makman, 1978; Weinstock et al., 1977).

The pharmacological characteristics of 5-HT and CA receptors appear to be similar in many respects, and thus, it will be difficult to differentiate these receptors with the use of the antagonists. However, with respect to the proposed PVN 5-HT receptors for feeding inhibition (Leibowitz and Papadakos, 1978), these appear to be distinct from the dopaminergic receptor, which should be antagonized by all potent neuroleptics (including haloperidol and pimozide which had no effect on 5-HT's action) and more effectively stimulated by DA (which in the PVN suppressed feeding only at considerably higher doses than 5-HT). The 5-HT receptors also appear to be distinct from the β-adrenergic receptor, since only at very high doses did the potent β agonist isoproterenol mimic 5-HT's inhibitory effect. While this evidence argues in favor of a specific 5-HT receptor-mediated response, it by no means rules out a possible interaction of serotonergic, dopaminergic, and β-adrenergic receptors in this area of the brain. They may work synergistically to inhibit the potent feeding-stimulatory effect of α-adrenergic activation in the PVN. However, the question of whether they actually exist and are related in any way to dopaminergic and β-adrenergic receptors localized to the perifornical hypothalamic area (see Sections II.B and III.B) remains to be answered by future studies.

2. Pharmacological Manipulations of Endogenous Serotonin

The above evidence supports the concept that hypothalamic 5-HT receptor systems, in particular within the medial hypothalamus, are inhibitory towards feeding. Studies with peripheral injection of 5-HT precursors have yielded results that are consistent with this hypothesis, although the specificity of the effects observed remains open to question (Blundell, 1977; Garattini and Samanin, 1978). Intraperitoneal injection of 5-hydroxytryptophan (5-HTP), which increases brain 5-HT levels, has been shown to produce a dose-dependent suppression of feeding (e.g. Blundell and Leshem, 1975; Joyce and Mrosovsky, 1964). Since these experiments did not use peripheral decarboxylase inhibitors, which help to reverse the general depressant effect of 5-HTP alone, the significance of the feeding suppression remains unclear. A similar effect to that of 5-HTP has been observed with the 5-HT precursor tryptophan. Peripheral injection of this amino acid was shown to suppress feeding (Fernstrom and Wurtman, 1972), in addition to causing an increase in 5-HT levels and synthesis (Schubert and Sedvall, 1972).

There appears to be only one study which investigated the effects produced by central injection of a 5-HT precursor (Fig. 20). In this study, 5-HTP was injected into the PVN of the rat and found to inhibit the feeding response elicited by NE (Leibowitz and

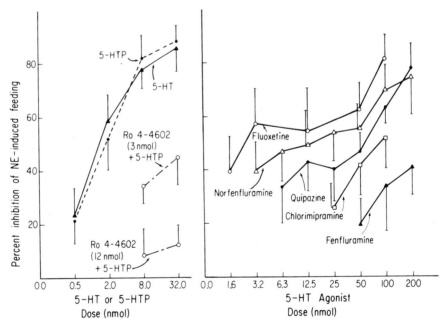

Figure 20 Feeding suppressive effect induced by paraventricular nucleus injection of 5-HT stimulants at a wide range of doses. Rats were injected with these stimulants a few minutes prior to NE; the feeding response normally elicited by NE was inhibited by 5-HT and its precursor 5-HTP, as well as by the 5-HT uptake inhibitors fluoxetine and chlorimipramine, the 5-HT-releasing drugs norfenfluramine and fenfluramine, and the 5-HT-receptor agonist quipazine. The feeding inhibition produced by 5-HTP (but not 5-HT) was attenuated by central injection of the decarboxylase inhibitor Ro 4-4602. (From Leibowitz and Papadakos, 1978.)

Papadakos, 1978). The effect was dose dependent, occurring reliably at a dose as low as 100 ng. This dose is approximately four to six orders of magnitude lower than peripherally effective doses. The 5-HTP-induced feeding suppression, furthermore, was reversed by central injection of the decarboxylase inhibitors Ro 4-4602 and MK-486, which had no effect on the response to 5-HT (Fig. 20). These findings suggest that 5-HTP is acting through endogenous 5-HT synthesis rather than through direct receptor activation. On the basis of this evidence alone, however, it cannot be concluded that the increased synthesis is occurring specifically in serotonergic neurons, as opposed to catecholaminergic neurons. The enzyme which decarboxylates 5-HTP is capable of decarboxylating similar, naturally found, aromatic amino acids (Lovenberg et al., 1962); thus, 5-HTP may be taken up and decarboxylated in catecholaminergic neurons and brain capillary walls, as well as in serotonergic neurons. Evidence that argues against this possibility is the finding that 5-HTP-induced inhibition of feeding is not reversed by PVN injection of the CA synthesis inhibitor α-methyltyrosine (Leibowitz and Papadakos, 1978).

In addition to this evidence obtained with the 5-HT precursors, there is an extensive literature describing an inhibitory effect on feeding observed after peripheral injection of drugs which enhance the release or block the neuronal uptake of endogenous 5-HT. A great deal of work has been done on the well-known anorectic drug fenfluramine (FENF) and its major de-ethylated metabolite norfenfluramine. Peripheral injections of these drugs produce a suppressant effect on food intake which appears to be mediated by brain 5-HT systems (see Blundell, 1977; and Garattini and Samanin, 1978, for review). This effect of FENF is antagonized by 5-HT antagonists. However, there are conflicting reports as to whether it is reversed by neurotoxin or electrolytic lesions which destroy central 5-HT neurons. For example, intraventricular injections of the 5-HT neurotoxin 5,6-dihydroxytryptamine attenuated the anorexic effect of FENF (Clineschmidt, 1973), as did lesions of the median raphe nuclei (Clineschmidt et al., 1978; Samanin et al., 1972). Other studies, however, have failed to replicate these results using somewhat different procedures (Hoebel et al., 1978; Hollister et al., 1975; Sugrue et al., 1975). It is possible that these inconsistencies are due to differences in the pattern of 5-HT depletion in discrete brain areas, although this is difficult to test since the site of FENF's action is not known.

It is important to give serious consideration to the possibility that serotonergic stimulants, including FENF, may be inhibiting feeding through their action on peripheral mechanisms, specifically, the gut, where 5-HT is concentrated (see Blundell, 1977). It has been shown, however, that norfenfluramine injected into the ventricles at a lower dose (100 μg) than is effective peripherally can strongly suppress deprivation-induced feeding, suggesting that central 5-HT mechanisms participate in the response (Kruk, 1973). A lower dose (20 μg) of norfenfluramine was effective after injection into the interstitial nucleus of the stria terminalis or the neostriatum (Broekkamp et al., 1975), and a similar dose of FENF was reported to be effective in the lateral hypothalamus (Blundell and Leshem, 1973). There is evidence, however, that a more effective site may be in the medial PVN, where feeding elicited by NE was suppressed by norfenfluramine injection at doses as low as 0.5 μg (Fig. 20) (Leibowitz and Papadakos, 1978). More extensive mapping studies are clearly required to determine whether the PVN, or any other brain area, may be important in the mediation of FENF anorexia. Lesion studies, in animals tested with peripheral injection of FENF as well as the CA-releasing drug AMPH, have so far failed to distinguish a particular hypothalamic (or extrahypothalamic) site. The anorectic effect of FENF appears to be unaffected by lesions in the lateral hypothalamus, anterior hypothalamus, and ventromedial area (see Blundell, 1977), in contrast to AMPH anorexia which is abolished specifically by lateral hypothalamic damage (see Leibowitz, 1978c). As to whether the dorsomedial paraventricular area is of particular importance will need to be examined in future lesion experiments.

In addition to FENF or norfenfluramine, there are other drugs which activate endogenous 5-HT systems and which have been shown to suppress feeding. Chlorimipramine, which inhibits 5-HT uptake and thus increases synaptic concentration of endogenous 5-HT, reduced food intake after peripheral injection into food-deprived animals (Blundell, 1977), as well as after central injection into animals induced to eat through PVN α-adrenergic stimulation (Fig. 20) (Leibowitz and Papadakos, 1978). The drug Lilly 110140 (fluoxetine) appears to be a more specific inhibitor of the 5-HT membrane pump than chlorimipramine (Fuller et al., 1974b) and has been reported to increase hypothalamic levels of 5-HT without affecting concentrations in other brain areas (Fuller et al., 1974a). Consistent with this biochemical evidence, peripheral injection of fluoxetine caused a dramatic suppression of food intake (Goudie et al., 1976), and, when injected directly into the medial hypothalamus, fluoxetine was more effective than chlorimipramine in suppressing α-adrenergically elicited feeding (Fig. 20), causing a reliable inhibition at a dose at least as low as 100 ng (Leibowitz and Papadakos, 1978). A selective 5-HT stimulant, quipazine, which appears to act directly on the receptor site rather than through endogenous 5-HT stores, also suppressed feeding when peripherally (see Garattini and Samanin, 1978) and centrally (Leibowitz and Papadakos, 1978) injected (Fig. 20). The action of this compound was blocked by the 5-HT antagonist methergoline but not by median raphe lesions or by CA receptor antagonists and 6-OHDA injections (see Samanin et al., 1978).

In support of the above evidence that serotonergic-receptor stimulation, either via exogenous or endogenous 5-HT, exerts an inhibitory effect on feeding, there are a number of reports that serotonergic-receptor blockade has a stimulatory effect on feeding (see Blundell, 1977, for review). This was initially observed with the serotonergic (and histaminergic) antagonist cyproheptadine, which increased appetite in asthmatic children as well as in underweight or anorexic adults. Peripheral injection of cyproheptadine in animals has similarly been observed to increase food consumption and body weight. However, there appears to be no evidence that central cyproheptadine injections may produce this response, although changes in firing rate of hypothalamic neurons have been observed (Oomura et al., 1973). The serotonergic antagonists methysergide and BC-105 have also been associated with increased appetite and weight gain in humans, although there appears to be only one study in animals which revealed such an effect with peripheral injection of methysergide (see Blundell, 1977).

Another class of drugs which may be affecting feeding, at least in part, through central serotonergic systems is the benzodiazepines. These minor tranquilizers have been reported to decrease the turnover rate of brain 5-HT (see Costa and Greengard, 1975), and a behavioral consequence of their injection in animals is an increase in eating behavior (see Poschel, 1971; Soper and Wise, 1971) as well as a reversal of the anorectic action of AMPH (Abdallah et al., 1974). A direct association between these neurochemical and behavioral effects of the benzodiazepines has yet to be established. It has been argued that their stimulatory effect on feeding is more a nonspecific consequence of their antianxiety properties rather than a direct effect on food-specific mechanisms (Poschel, 1971; Tye et al., 1976). Moreover, the benzodiazepines have been shown to alter the metabolism of brain CA as well as 5-HT (Costa and Greengard, 1975). Additional pharmacological tests are clearly needed to evaluate the role of these different monoamines in the feeding response.

As to the question of whether the benzodiazepines are acting on a central feeding mechanism, Wise and Dawson (1974) have argued in favor of this point, in light of their finding that: (1) the benzodiazepine diazepam was effective in producing specifically feeding (and not drinking) under familiar conditions where there was minimal source of anxiety; (2) diazepam induced animals to perform or learn to perform a task in order to obtain food; and (3) diazepam-induced feeding, like deprivation-induced feeding, appeared to be terminated by the postingestional, satiating consequences of eating. These authors did not propose that altered hunger levels were exclusively involved in the benzodiazepine feeding phenomenon but suggested that the relative importance of the hunger-inducing and response-releasing properties of the drugs may depend on the various testing conditions in which food-related behaviors occur.

3. Serotonergic Versus Catecholaminergic Systems

It has been demonstrated that the catecholamines, DA and EPI, as well as the CA-releasing drug AMPH, are inhibitory towards feeding and exert this action, at least in part, through dopaminergic and β-adrenergic receptors in the lateral perifornical region of the hypothalamus (see Sections II.B and III.B). From the data summarized above, it becomes clear that 5-HT produces a similar effect, and although the location of the serotonergic receptors mediating this phenomenon has not been established, a brain area particularly sensitive to serotonergic stimulation is the medial paraventricular hypothalamus, in contrast to the lateral perifornical region which is generally insensitive. Lesion studies have provided clear evidence that the neuroanatomical pathways mediating the CA and 5-HT suppressive effects on feeding are different (e.g., Blundell and Leshem, 1974; Carey and Goodall, 1975; Fibiger et al., 1973; Hoebel, 1977; Hoebel et al., 1978). There are two additional pieces of evidence described below which indicate that the specific behavioral mechanisms through which these monamines act are also different.

Microanalysis of the animals' feeding patterns after injection of 5-HT stimulants (including FENF) demonstrated a distinctive behavioral profile characterized by a decrease in meal size associated with a decrease in the rate of eating (Blundell and Latham, 1978). Since serotonergic stimulation appeared to have little or no effect on the initiation of eating or the frequency of the meals, this effect on meal size was interpreted as an action of 5-HT mechanisms on a satiety process, whereby food intake is suppressed specifically through an early termination of eating. In support of this proposal is the finding that medial hypothalamic injection of NE (through α-adrenergic receptors) stimulates eating apparently via an increase in meal size (rather than meal frequency) (Ritter and Epstein, 1975), and injection of 5-HT into the same hypothalamic site antagonizes the action of NE (Leibowitz and Papadakos, 1978). Furthermore, there is extensive evidence for a function of the medial hypothalamus in the mediation of satiety (see Section II.G above), and whereas NE may be acting to inhibit this satiety area (disinhibit eating through increasing meal size), 5-HT may be activating this area (decreasing meal size) through antagonism of noradrenergic function.

This inhibitory effect of 5-HT stimulants on feeding, produced via a reduction in meal size, contrasts with the mechanism through which the drugs AMPH and mazindol appear to suppress feeding (Blundell and Latham, 1978). These drugs, which are believed to act through central dopaminergic and β-adrenergic mechanisms, have little effect on meal size and on the latency to eating. Their main effects appear to be on meal frequency, which was reduced, and also on rate of eating, which surprisingly was increased. Thus, the main parameter which accounts for the suppression of total food intake with these CA stimulants seems to be a decrease in number of meals, rather than a decrease in the meal size as with 5-HT stimulants. The finding that AMPH and mazindol increase rate of eating, whereas dopaminergic antagonists decreased rate of eating, led Blundell and Latham (1978) to propose a role for DA in adjusting the pace of eating. The dopaminergic and serotonergic systems may interact at this level, since the 5-HT stimulants were also found to alter eating rate.

A further dissociation between these monoamine systems, in terms of their mechanism for suppressing feeding, has been revealed through experiments in which different diets were tested. Anderson and Ashley have recently proposed that protein and energy intake may be independently regulated and that this regulation may be reflected in plasma amino-acid ratios (Anderson, 1979; Anderson and Ashley, 1978). Furthermore, these investigators have suggested that brain monoamines may be involved in this regulation, and that the CA may specifically affect energy intake, whereas 5-HT may affect protein intake. A considerable amount of work will be required to test this hypothesis. At present, there are two studies, bearing directly on this issue, which have taken advantage of the self-selection feeding paradigm to examine differential effects of serotonergic and catecholaminergic drugs on dietary choice. In the study by Ashley et al., (1979), depletion of brain 5-HT, through injection of 5-HT neurotoxins, selectively decreased protein intake while leaving energy uptake unaffected. However, Wurtman and Wurtman (1977), in short-term studies, examined the effects of peripherally injected serotonergic stimulants, namely, FENF and fluoxetine, and found these anorectic agents to preserve (and possibly even potentiate) protein intake, while causing a selective decrease in calorie intake. The suggestion of this latter finding, that 5-HT stimulation may work specifically through calorie reduction to suppress total food intake, is particularly interesting in light of the evidence that medial hypothalamic NE, which is specifically antagonized by 5-HT, produces a preferential increase in carbohydrate consumption (see Section II.H.2 above). Further, in contrast to 5-HT stimulation, the anorectic effect of AMPH was found by Wurtman and Wurtman to be associated with a decrease in protein intake as well as a decrease in calorie intake. These studies, therefore, provide another avenue for understanding and characterizing the mechanisms through which brain monoamines may suppress or stimulate feeding.

4. Lesion Studies of Monoaminergic Pathways

The hypothesis that brain serotonergic systems act predominantly in an inhibitory fashion in their effect on feeding control mechanisms leads to the prediction that depletion or destruction of these systems might result in an increase in feeding and body weight. While there is some evidence in support of this prediction, the effects observed with 5-HT depleters, neurotoxins, and raphe lesions are sometimes small or variable and are questionable with respect to their behavioral or neurochemical specificity (see Blundell, 1977; Coscina, 1978; Grossman, 1978; and Hoebel et al., 1978, for reviews). Parachlorophenylalanine (PCPA), which inhibits the synthesis of 5-HT, has generally been found to decrease food intake when peripherally injected, despite a marked reduction of brain 5-HT. It has been suggested that this feeding suppression may be a consequence of gut irritation and nausea associated with this drug after peripheral administration (see Blundell, 1977). These effects may be overcome with central injection of this synthesis inhibitor, and, indeed, it was demonstrated that intraventricular injection of PCPA in rats produced a transient hyperphagia and increased weight gain, which was accompanied by a selective depletion of brain 5-HT (Hoebel et al., 1978). This alteration in feeding diminished with the recovery of 5-HT levels towards normal and was reversed by peripheral injection of the 5-HT precursor 5-HTP. This evidence supports the concept of an inhibitory serotonergic mechanism in the control of feeding. The specificity of the PCPA effects, however, has recently been brought into question by Coscina et al. (1978), who compared the efficacy of PCPA to that of its parent amino acid, phenylalanine, in modifying food intake. These investigators found phenylalanine and PCPA to be equally potent in producing hyperphagia, but neither compound produced a reliable change in forebrain concentrations of 5-HT, DA, or NE. These findings question the importance of 5-HT in the behavioral effects of PCPA, although it is not clear why PCPA was ineffective in altering 5-HT levels, and, furthermore, it is possible that the measurements of forebrain monoamines may have failed to detect significant depletions in specific, discrete regions.

Consistent with the PCPA-induced hyperphagia is the finding that the 5-HT derivatives, 5,6-dihydroxytryptamine (5,6-DHT), or 5,7-DHT, which cause selective degeneration of 5-HT neurons, produce an increase in food consumption and body weight after injection into the ventricles (Saller and Stricker, 1976). This effect, however, appeared to be associated with increased growth as well as changes in gastrointestinal function and thus may not necessarily be attributed directly to alterations in feeding mechanisms. Moreover, the changes in food intake did not occur in all animals, despite reductions in brain 5-HT, and other investigators have had difficulty in replicating the phenomenon (Coscina, 1978; Hoebel et al., 1978). In contrast to these results with intraventricular injection, Myers (1978) demonstrated that 5,6-DHT injected into the rostral hypothalamic region caused a transient decrease in food intake. This effect was attributed to the agonistic properties of the neurotoxin which occur as a result of 5-HT release from the degenerating neurons.

Serotonergic projections to the forebrain appear to originate predominantly from cell bodies (B7 and B8) located in the midbrain raphe nuclei (see Section IV.A). If these projections mediate an inhibitory effect on feeding behavior, lesions of the raphe nuclei might be expected to cause hyperphagia and possibly obesity. In animals receiving knife cuts in the midbrain tegmentum, Grossman (1978) observed a positive correlation between the development of hyperphagia and the depletion of 5-HT in the telencephalon. In several studies, however, which examined the effects of localized destruction of the dorsal (B7) and median (B8) raphe nuclei, no change or actually a reduction in food intake and body weight was obtained (see Blundell, 1977). This result by no means negates the possibility that forebrain 5-HT inhibits feeding, and, in fact, a positive result obtained with such widespread destruction of many complex and interacting serotonergic projections might actually have been more surprising. Only the simplest model predicts opposite effects to be obtained with 5-HT stimulation and depletion, and even if the precise serotonergic pathway mediating the effects on feeding were known, the expected complex interaction of this system with other neurochemical systems could reasonably preclude specific changes in feeding even after selective damage to the appropriate serotonergic fibers. There is as yet no evidence that brain 5-HT plays any essential role, as opposed to a permissive or anticipatory role, in the control of food intake, and for this reason alone, destruction of the serotonergic system may have inconsistent or minimal effects on ad libitum food intake. Furthermore, although drugs which stimulate 5-HT receptor mechanisms produce immediate and dramatic changes in feeding, the function actually served by the serotonergic fibers may not be one of discrete information processing. Rather, the wide distribution and extensive collateralization of the serotonergic system might suggest that the function of the raphe neurons is to provide some modulating effects which would bias and direct the responses of innervated neurons to more specific input. A consequence of damage to the serotonergic neurons would therefore depend on the pattern of input, from a variety of neurochemical systems, to each particular area innervated by the serotonergic neuron system.

The inevitable interactions which occur between neurochemical systems in the control of specific behavioral processes have been examined by a number of investigators (e.g., Butcher, 1977; Garattini et al., 1978; Samanin and Garattini, 1975). Relevant to studies which have manipulated brain 5-HT and observed alterations in feeding behavior is the finding that median raphe lesions cause an increase in NE turnover rate in forebrain structures, as well as a depletion of 5-HT (see Samanin and Garattini, 1975). Furthermore, peripheral injections of 5-HTP, which suppress feeding, produce a decrease in forebrain DA concentrations, as well as an increase in 5-HT synthesis. These results are but two examples of the complex neurochemical interactions that may occur in the brain, and they lead us to view critically the outcome of studies employing such brain manipulations and their interpretation with respect to the primary cause of behavioral change.

The reverse situation, where direct manipulations of brain CA alter brain 5-HT, has also been demonstrated (Garattini and Samanin, 1975). Locus ceruleus lesions or injections of 6-OHDA or CA synthesis inhibitors consistently produce an increase in the turnover rate of forebrain 5-HT. Thus, the changes in feeding induced by these manipulations of brain CA (see Section II.F) may actually be attributed to the changes in serotonergic function. It is interesting that lesions of the adrenergic fibers coursing through the ventral pons and midbrain tegmentum have been shown to increase 5-HT levels in the region of the cells of the median raphe nucleus (Massari et al., 1977). Perhaps the resulting hyperphagia (Ahlskog, 1974), which has been questioned with regard to its relationship to NE systems (Oltmans et al., 1977), may in part be attributed to an insufficient release of 5-HT from forebrain terminals. Consistent with this suggestion is the finding that ventral tegmental lesions, which destroy the ascending adrenergic projection and produce hyperphagia, enhance the anorectic effect of peripherally injected FENF, while attenuating the action of AMPH (Hoebel et al., 1978).

The neurochemical basis of the hyperphagia induced by medial hypothalamic lesions has also been studied, and it has been demonstrated, not surprisingly, that a number of changes in neurotransmitter function occur as a result of this lesion (see Coscina, 1978, and Section II.G above). The hyperphagia produced by medial hypothalamic lesions, as well as by gold thioglucose injections, was associated with a decrease in 5-HT levels, which was inversely correlated with body-weight gain. While dorsal and medial raphe lesions, which led to a 75% reduction in forebrain 5-HT, had no effect of their own on feeding, they were found to block the hyperphagia and obesity induced by the ventromedial hypothalamic lesion (Coscina, 1978). This result would suggest that the integrity of the serotonergic neurons may be important for interacting with other neurochemical systems in the control of feeding and body-weight gain. Unfortunately for this hypothesis, injections of the neurotoxin 5, 7-DHT, which also depressed forebrain 5-HT, failed to disrupt the hypothalamic hyperphagia effect. While there are many possible explanations for this inconsistency (see Coscina, 1978), these results exemplify the many complexities that one must inevitably encounter in an attempt to relate brain neurochemistry to behavior.

C. Drinking Behavior and Water Excretion

In contrast to the studies indicating a possible inhibitory role of brain 5-HT in the control of feeding behavior, there appears to be little evidence to suggest that this indoleamine may have impact on water homeostasis, either through ingestion or excretion of water. Peripheral injections of 5-HT have been shown to produce a strong increase in water ingestion; this phenomenon, however, appears to be mediated predominantly through peripheral mechanisms, possibly the renin-angiotensin system of the kidney (Lehr and Goldman, 1973; Meyer et al., 1974). With respect to central injection of the indoleamine, there is one report that describes a stimulatory effect, in the monkey, of hypothalamic 5-HT administration on drinking (Sharpe and Myers, 1969). In the rat, tests in numerous diencephalic or telencephalic sites have generally failed to reveal any significant alterations in water ingestion. The PVN, however, where 5-HT acts to inhibit feeding (see Section IV.B), appeared to be an exception; injection of 5-HT into this nucleus induced a small to moderate drinking response (lasting 5-10 min) in fully satiated animals (Leibowitz, Papadakos, and Kirchgessner, unpublished observations). Midbrain raphe lesions or injections of serotonergic neurotoxins are generally associated with either no change or a transient decrease in water intake (see Section IV.B.4 above for references), although Coscina et al. (1972) and Lorens and Yunger (1974) have reported a transient increase in this behavior. More localized brain manipulations and analyses are clearly needed to evaluate the relationship between these lesion results and the increased water ingestion observed with centrally injected 5-HT. If further work, however, substantiates this 5-HT drinking phenomenon, this evidence will provide a basis for an extension of the hypothesis (see above) that 5-HT and NE, perhaps within the PVN, are antagonistic in their action, with 5-HT inhibiting feeding and stimulating drinking and NE causing the opposite effects (see Section II.B and C).

In contrast to the CA neurotransmitters which are also effective in releasing ADH and thereby reducing water excretion (Section II.D), there is little evidence to suggest that 5-HT may exert a similar or opposite effect on these parameters. Bharghava et al. (1972) and Olsson (1970) observed no change in diuresis or ADH release after injection of 5-HT into the ventricles. There was also no change in electrolyte excretion after ventricular or septal administration of 5-HT (Bhargava et al., 1972; Camargo et al., 1976). There appears to be only one report of a positive effect of 5-HT on urinary parameters. This was observed by Urano and Kobayashi (1978) who revealed a decrease in urine flow after SON injection of 5-HT in the anesthetized rat. The mechanism and significance of this effect and the reasons for the negative results obtained in the other studies will need to be explored in future investigations.

V. Cholinergic Systems

A. Anatomy of Cholinergic Projections

Studies of the cholinergic neurons in the brain have lagged substantially behind those of the monoaminergic neurons. The major problem has been the lack of sensitive and specific chemical methods for the measurement of acetylcholine (ACh) and the enzymes associated with its synthesis or degradation. Since bioassays are difficult and lack specificity, the single most important stimulus to research into cholinergic mechanisms has been the development of various chemical methods for ACh determination. The lack of reliable histochemical methods for localizing a biochemical marker that is unique to cholinergic neurons has been a serious problem. There have been extensive studies on the localization of the catabolic enzyme acetylcholinesterase (AChE) in the brain. However, this enzyme marks some neurons that are not cholinergic. Recent developments in this field include a precipitation technique and immunohistochemical approach to visualize the synthesizing enzyme choline acetyltransferase, which may make a more detailed mapping of cholinergic systems possible.

What appears to emerge from the evidence obtained with the available techniques is the concept of several cholinergic pathways throughout the forebrain that may be interconnected via the ramifications of the major route, namely, the diffuse ascending tegmental-mesencephalic-cortical system (Fig. 21) (see Hoover et al., 1978; Karczman and Dun, 1978; Kobayashi et al., 1978; Lewis and Shute, 1978; Parent and Butcher, 1976). This system abuts on the ventral and dorsal pathways; the ventral pathway originates from the substantia nigra and ventral tegmental area and the dorsal pathway from the dorsal tegmental area. This dorsal tegmental pathway supplies the colliculi, geniculate bodies, pretectal area, and thalamus, whereas the ventral tegmental pathway ascends through the subthalamus and lateral hypothalamus to basal diencephalic and telencephalic areas. Evidence obtained from lesions in the forebrain indicates the existence of descending cholinergic pathways, possibly originating in the pallidum, that tend to innervate the lateral parts of the hypothalamus and the thalamus predominantly, in contrast to the ascending pathways which project to the medial parts of the diencephalon.

Analyses of the hypothalamus have revealed staining for AChE as well as binding of radiolabeled cholinergic antagonists, suggesting the existence of cholinoceptive sites. These sites, however, appear to be considerably fewer in number relative to that of various telencephic regions. Neurons containing AChE have been shown to exist in the dorsal and posterior hypothalamus. They have also been located in the paraventricular and supraoptic nuclei. Radiolabeled cholinergic antagonists have been shown to bind in significant quantities to essentially all hypothalamic areas, showing a density of approxi-

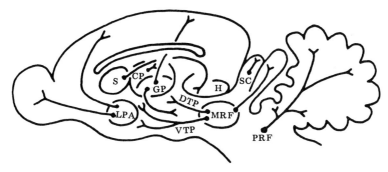

Figure 21 Some AChE-containing presumed cholinergic pathways of the brain. CP, caudate-putamen (striatum); DTP, dorsal tegmental pathway; GP, globus pallidus (pallidum); H, hippocampal formation; LPA, lateral preoptic area (substantia innominata); MRF, midbrain reticular formation; PRF, pontine reticular formation; S, medial septal and diagonal band nuclei; SC, superior colliculus; VTP, ventral tegmental pathway. (From Lewis and Shute, 1978.)

mately one-third that exhibited by the densest telencephalic area, the caudate-putamen. Biochemical studies have revealed significant concentrations of ACh and choline actyltransferase, as well as AChE, throughout the hypothalamus.

B. Feeding Behavior

Compared with the research on monoaminergic systems, there are relatively few studies which have examined a possible role of cholinergic systems in the control of feeding. The available evidence is derived primarily from studies with the synthetic cholinergic (muscarinic) stimulant carbamylcholine (carbachol or CARB), which is relatively resistant to hydrolysis by endogenous enzymes and exerts longer-lasting depolarization of the cholinoceptive membrane than does the neurotransmitter ACh. An early report of Grossman (1962a) revealed an inhibitory effect of lateral hypothalamic injection of CARB on feeding in hungry rats. Singer and Kelly (1972) replicated this finding and noted that CARB was more effective in suppressing feeding than was pharmacological blockade of the α-adrenergic receptors which are believed to mediate feeding stimulation (see Section II.B above). These authors also found NE injection to be more effective in suppressing drinking than was pharmacological blockade of the cholinergic receptors that are believed to mediate drinking stimulation (see below). On the basis of this evidence, they proposed that adrenergic and cholinergic neuronal systems have opposite effects on feeding and drinking behavior and that they have direct inhibitory influence on each other's functional activity. In the monkey, a similar suppression of feeding was observed with hypothalamic injection of CARB, as well as ACh and the cholinesterase inhibitor physostigmine which potentiates the action of endogenous ACh (Sharpe and Myers, 1969). This effect was blocked by the cholinergic (muscarinic) antagonist atropine. Furthermore, peripheral injection of the irreversible cholinesterase inhibitor diisopropylflurophosphate produced a long-term (10-day) feeding suppression (Russell et al., 1971).

The area of the brain and the type of mechanism mediating this inhibitory effect is not known. Hoebel (1977) proposed that ACh, which has been associated with drinking behavior and antidiuresis (see below), may be exerting an effect on satiety as part of its more general function in controlling body-fluid homeostasis. It is known that osmotic balance may be achieved through alterations in the ingestion or excretion of food, as well as water, and furthermore, that blood osmolality may have impact on food as well as water ingestion (see Russek, 1975). To test this hypothesis and to determine whether ACh acts directly on a feeding mechanism, as opposed to secondarily via a drinking mechanism, it will be necessary to locate the cholinergic receptors which mediate the different effects of this neurotransmitter. Little progress has been made in this regard, since CARB is generally used and found to be effective in many brain areas. This lack of anatomical specificity may be attributed to CARB's particular potency, its potential effects on axons of passage, and its spread via the ventricles to distant brain sites.

In addition to suppressing feeding, central CARB injections are also shown to have a stimulatory effect on feeding. This phenomenon has been demonstrated in several species, including the rat, cat, sheep, and rabbit (see Baile, 1974; Hoebel, 1977). This effect has not been shown to be anatomically specific; it occurs after CARB injection into diverse brain areas including the rostral and lateral hypothalamus, mamillary bodies, preoptic area, septum, hippocampus, and nucleus reuniens (e.g., Sommer et al., 1967; Wise, 1972). Moreover, it is not a very robust phenomenon and may take several days of repeated injections to be observed (Chance and Lints, 1977; Wise, 1972). Wise (1972) argued that the phenomenon may simply represent a rebound feeding response as a consequence of CARB's initial and primary stimulatory effect on drinking associated with feeding inhibition (see above). The time course of these effects appeared to support this view. However, Chance and Lints (1977) demonstrated that, while this sequence of drinking first followed by feeding may be typical of the first few tests with CARB, subsequent tests in the same animals showed that these responses may overlap during the initial periods of the test.

Some evidence for site specificity was revealed by the work of Sciorelli et al. (1972), who found CARB-elicited feeding to occur after drug injection into the perifornical hypothalamus at the level of the VMN but not more rostrally at the level of the PVN. These investigators also observed a similar eating response after injection of dibutyryl cyclic AMP into the perifornical area and found this response to be potentiated by physostigmine and antagonized by atropine and hemicholinium, a depleter of endogenous ACh. It was therefore proposed that cyclic AMP may be eliciting eating through the release of ACh. There is additional evidence that feeding elicited by electrical stimulation of the lateral hypothalamus may also involve a cholinergic link, since peripheral physostigmine injection stimulated this response, an effect reversed by atropine (Stark et al., 1971). It is interesting that these investigators found AMPH to produce the opposite effect to CARB, namely, a suppression of the feeding elicited by hypothalamic stimulation. In the perifornical hypothalamus, where CARB and cyclic AMP were particularly effective in stimulating feeding (Sciorelli et al., 1972), AMPH and the CA agonists were also shown to be most effective in suppressing feeding (Leibowitz, 1978c). This raises the interesting possibility that CA and ACh mechanisms may interact in this area of the brain, as they have been demonstrated to interact in other brain areas (Butcher, 1977). Unfortunately, this hypothesis will be difficult to test until the problems surrounding the CARB eating phenomenon are resolved.

It has been demonstrated that peripheral injection of cholinergic antagonists can produce an inhibition of feeding (Pradhan and Roth, 1968). While it was suggested that this effect may be mediated peripherally, perhaps via the drugs' effects on salivation, Glick and Greenstein (1973) argued in favor of a central site of action, since they found the quarternary form of the cholinergic antagonist (which passes minimally into the brain) to be considerably less effective than the tertiary form in suppressing feeding. There is some support for this idea, since central injections of the cholinergic antagonists have also been observed to produce this effect (e.g., Neil and Grossman, 1973). However, this central response appears to be small (Singer and Kelly, 1972), considerably smaller than the suppression observed with peripheral injections of the quarternary cholinergic antagonists. Since peripheral administration of atropine methyl nitrate reduced ingestion of a liquid diet as well as a solid diet, but not water intake, Lorenz et al. (1978) proposed that this effect may be a consequence of drug interaction with peripheral satiety receptors, rather than with receptors involved in specific ingestive responses such as licking, swallowing, and salivation.

C. Drinking Behavior and Water-Electrolyte Excretion

Extensive literature has accumulated over the past decade and a half, based on the original finding of Grossman (1962a) that CARB injected into the rat hypothalamus caused an increase in drinking behavior. This phenomenon, in contrast to the feeding-stimulation effect (see above), was very robust and could be observed with injections of ACh itself (Grossman, 1962a; Simpson and Routtenberg, 1974) as well as with inhibitors of cholinesterase (Winson and Miller, 1970). The response produced by CARB was dose dependent and could be blocked by anticholinergic drugs given centrally or systemically (see Setler, 1977) as well as by central injection of NE (see Section II.C). It was not affected by adrenolytic agents or by 6-OHDA-induced depletion of brain CA (Setler, 1977). Carbachol-elicited drinking appeared to have several similarities, in terms of its motivational properties, to drinking induced by water deprivation. It caused an animal to perform operant tasks to obtain water, and the preferences for specific solutions (e.g., water, sucrose, quinine) exhibited by CARB-injected rats were similar to those of water-deprived animals (see Fisher, 1973; Setler, 1977).

This evidence has led investigators to propose that brain cholinergic mechanisms provide a link in the process of regulating water ingestion. One source of difficulty for this hypothesis is the fact that drinking elicited by central cholinergic stimulation is almost exclusively associated with the rat. Tests in other species have generally yielded negative results (Levitt, 1971), although there is one report in dog where CARB injected into the third ventricle produced a small increase in water ingestion (Ramsay and Reid, 1975). In the monkey, nicotine also elicited drinking (Myers et al., 1973), although the relationship of this phenomenon to the muscarinic response of CARB remains to be established.

An additional problem has been that of determining the site in the brain where cholinergic receptors exist to stimulate drinking behavior. Early reports demonstrated that CARB was effective after injection into numerous telencephalic as well as diencephalic areas, and it was proposed that there existed a cholinergic drinking circuit which corresponded approximately to the Papez anatomical circuit (of emotion) (Fisher and

Coury, 1962). This suggestion of a cholinergic limbic-system pathway received support from the finding that atropine injected at various points along this circuit could suppress the drinking response elicited by CARB injection at a different point within the circuit (Levitt, 1971). Furthermore, cholinergic stimulation at one site was found to alter neuronal firing at an ipsilateral or contralateral site within the circuit, the time course of which appeared similar to that of the drinking response (Buerger et al., 1973; Snyder and Levitt, 1975). However, changes in neuronal firing were not observed at all sites of the circuit where atropine was effective in blocking CARB-elicited water intake, and, moreover, electrolytic lesions at various points in the circuit (e.g., the preoptic area, septum, and posterior hypothalamus), failed to affect the CARB response (see Levitt, 1971; Setler, 1977). A notable exception is the lateral hypothalamus, where a lesion was effective in abolishing CARB drinking. This effect must be questioned in light of the evidence that lateral hypothalamic lesions produce a wide variety of deficits in feeding and drinking behavior.

The concept of a limbic cholinergic circuit, based on the finding that numerous brain sites are responsive to CARB, was challenged by Routtenberg (1967) who questioned that CARB may not be acting at the site of application but rather may be spreading up the shaft of the injection cannula into the ventricles (through which the cannula generally penetrates) and exerting its action at a more distant site (see Routtenberg, 1972). Levitt (1971) argued against this suggestion and cited the evidence indicating that the position of the cannula relative to the ventricles did not appear to be important in determining the sensitivity of a particular site to CARB, and direct injection of CARB into the ventricles was relatively ineffective in eliciting drinking. Peck (1976), however, suggested that the ineffectiveness of the ventricular injections may have resulted from the use of high doses of the drug, which are reported to produce side effects that may interfere with drinking. In his studies, Peck found CARB to be equally potent in producing drinking, whether injected into the lateral ventricles or preoptic-anterior hypothalamic area. The threshold dose for these two sites was the same, that is, 2 ng.

In a more extensive mapping study with CARB (Fig. 22) Swanson and Sharpe (1973) (using a $0.1\text{-}\mu$l injection) obtained no evidence to support the idea of a cholinergic limbic-system circuit but, rather, found medial diencephalic sites, adjacent to the third ventricle and at the level of or rostral to the VMN, to be most responsive to the CARB-elicited drinking effect in contrast to lateral diencephalic sites which were generally ineffective. The threshold dose of CARB along the midline was 2.5 ng (Swanson et al., 1973). While Swanson and Sharpe (1973) suggested that the ventricles were not involved, Simpson and Routtenberg (1972) have presented evidence that the subfornical organ (SFO), a circumventricular organ of the dorsal third ventricle, may contain the cholinergic receptors upon which CARB is acting to elicit drinking. This structure yielded a larger response, with a shorter latency, than did the lateral hypothalamus or a more caudal ventricular site, and SFO lesions were found to attenuate (but not abolish) CARB-elicited drinking. Acetylcholine was effective after injection into the SFO, apparently more effective at this site than after injection into the third ventricle (Mangiapane and Simpson, 1978; Simpson and Routtenberg, 1974). The possibility that the SFO is CARB's site of action is supported by Swanson and Sharpe's results (1973) showing greatest sensitivity along midline or medial areas, at the level of or rostral to the VMN, where CARB elicits a reliable response at a dose of 2.5 ng.

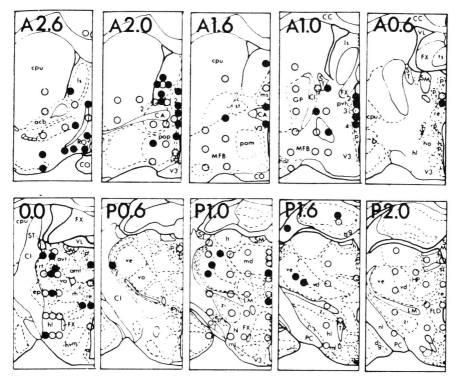

Figure 22 Brain sites from which a maximally effective dose of carbachol (2.2 nmol) did or did not elicit drinking in the rat. Black circle indicates a drinking response greater than or equal to 5 ml of water; open circle indicates no response. Tissue responsive to carbachol was generally confined to a medial zone around the anterior aspect of the third ventricle. See page 405 for abbreviations. (From Swanson and Sharpe, 1973.)

Also consistent with this evidence is the suggestion that CARB may be acting at a site located in the anteroventral portion of the third ventricle (at the level of the preoptic area) perhaps in addition to the SFO (Buggy and Fisher, 1976; Buggy, 1978). Carbachol was effective after injection into this area, at least at the higher dose levels (dose-response studies have not yet been conducted), and cold-cream plugs designed to obstruct this portion of the ventricles were found to abolish the CARB drinking response. Moreover, SFO lesions combined with CARB injections into this ventral region had no effect on the response, in contrast to the study of Simpson and Routtenberg (1972) where SFO lesions suppressed the effect of dorsal ventricular injection of CARB. Buggy and Fisher (1976) suggested that the effectiveness of SFO lesions in Simpson and Routtenberg's study was due to the fact that they obstructed circulation of the cerebrospinal fluid and thereby prevented CARB from reaching the cholinergic receptors within the anteroventral region of the third ventricle. The actual location of these receptors in this ventral area is not known, although the authors suggested two possibilities, namely, the organum vasculosum of the lamina terminalis (OVLT) (which like the SFO receives cholinergic input) and the periventricular nuclear region at the level of the preoptic area. A discrete lesion of the periventricular area was found to abolish CARB-induced drinking, although the possibility that this effect may be due to the destruction of fibers of passage, rather than the cholinergic terminal field, cannot be ruled out.

These studies clearly focus our attention on the third ventricular region, in the rostral forebrain, as being the site where cholinergic stimulation is most effective in promoting water ingestion. The evidence suggests that circumventricular organs (SFO and OVLT) may be involved, although further studies are required to determine the relative importance of these two structures. The periventricular area of the rostral diencephalon may also contain cholinoceptive sites. Each of these areas shows evidence of receiving a cholinergic input (see references in Section V.A), and also of having a function in body-fluid homeostasis (see Buggy and Johnson, 1977; Simpson et al., 1978a, b).

The process of maintaining water balance requires not only adjustments in water ingestion but also in the excretion of water and electrolytes. Cholinergic stimulation of the brain has been shown by a number of studies to affect these parameters, and the nature of the change has invariably been in the direction of ADH release, decreased urine output, and increased electrolyte excretion, in some cases preferential towards sodium.

Early indications that cholinergic mechanisms may be involved in this function came from the study of Pickford (1947) showing that SON injection of ACh (plus physostigmine) in the dog produced antidiuresis, an effect not correlated with systemic blood pressure changes, and from the studies of de Wied and László (1967) and Bridges and Thorn (1970) which demonstrated that peripheral injections of the anticholinergic atropine could block the ADH release caused by infusions of hypertonic saline. The brain sites most sensitive to the ADH-releasing property of cholinergic stimulation have not been extensively explored. A majority of the studies in this area of research have injected either CARB or ACh into the ventral regions of the third ventricle (e.g., Dorn et al., 1970; Kuhn, 1973; Kuhn and McCann, 1971; Morris et al., 1977; Olsson, 1970; Vandeputte-Van Messom and Peeters, 1973) and observed either antidiuresis or ADH release, frequently associated with natriuresis, and a block of these effects with atropine. In vitro studies on tissue from the area of the SON, PVN, and median eminence, have also revealed a similar increase in ADH with cholinergic stimulation (Bridges et al., 1976; Sladek and Robert, 1978).

In terms of localizing this effect, there are apparently only two studies, in addition to that of Pickford (1947), which have injected drugs directly into the region of the SON, and the outcome was the same as that observed with ventricular injections. Urano and Kobayashi (1978) described a decrease in urine flow after injection of ACh in the rat and an inhbition of this effect with cholinergic (muscarinic) receptor blockade. Milton and Paterson (1974), in contrast, observed an increase in ADH release in the cat after SON injection of nicotine and a reversal of this effect after nicotine receptor blockade produced by hexamethonium and pempidine. This result is identical to that observed by Sladek and Robert (1978) in an in vitro study of rat hypothalamus and substantiates the original finding of Bridges and Thorn (1970) that peripheral pempidine, as well as atropine, is effective in attenuating hyperosmotic-induced ADH release. Thus, it appears that nicotinic, as well as muscarinic, receptors may participate in cholinergic-induced antidiuresis. This evidence is consistent with the results of microiontophoretic studies (Barker et al., 1971) which have revealed changes in the firing of SON neurons (20% antidromically identified as neurosecretory cells) after both nicotinic and muscarinic stimulation.

While this evidence would support the existence of cholinoceptive sites in the SON, there is no assurance that the cannula-injected cholinergic drugs, which release ADH, are actually acting on these sites, as opposed to or in addition to sites in other brain regions. Adequate mapping studies have not been performed, and the problems with ventricular spread of the drugs from the injection site, discussed above in relation to cholinergic-induced drinking, also apply to these studies of cholinergic antidiuresis and electrolyte excretion. Peck (1976) has demonstrated that CARB injected into the lateral ventricles is effective in producing antidiuresis and natriuresis at a dose as low as 20 ng. When injected into the hypothalamus just rostral to the PVN (which appeared more responsive than the SON), the threshold dose for these effects was somewhat lower, namely, 2 ng. While these results may indicate some localization of the cholinoceptive sites, Peck also found that cannulas which reached apparently sensitive hypothalamic tissue without penetrating through a ventricle failed to yield a response after CARB injection. Mangiapane and Simpson (1979) have actually proposed that SFO cholinergic receptors may be involved in ADH release, perhaps via projections directly to the SON (Miselis et al., 1978), in addition to the drinking and pressor responses associated with this organ.

At this time, the evidence is clear that cholinergic stimulation of the brain causes increased water ingestion and decreased water excretion (ADH release) associated with natriuresis, and the receptors involved appear to be muscarinic and possibly nicotinic in nature. While important steps have been made in the direction of localizing these receptors, apparently to rostral ventricular and periventricular regions and also to the SON and PVN, further work is needed to determine whether they are concentrated in one specific structure or whether several areas are involved. An important issue which requires attention at this point concerns the question of what specific mechanism the neurotransmitter ACh relates to in producing its effects on physiological and behavioral responses of body-fluid homeostasis.

Buggy and Fisher (1974) observed that CARB would elicit ingestion of water but not hypertonic saline, in contrast to another dipsogen angiotensin which increased consumption of both solutions. This pattern of effect was even observed in animals that had been deprived of sodium. These authors suggested, therefore, that cholinergic mechanisms were involved in the mediation of drinking induced by intracellular dehydration rather than alterations in the extracellular fluid compartment. This proposal is supported by the evidence revealing a function of the anteroventral third ventricle—where cholinoceptive sites are believed to exist (see above)—in the mediation of water ingestion evoked by hypertonicity (e.g., Andersson, 1975; Buggy and Johnson, 1977).

While anticholinergic agents have been found to reduce drinking induced by hypertonic saline, this suppressive effect was not specific to this challenge but occurred with all dipsogenic agents (Block and Fisher, 1975). Mangiapane and Simpson (1979) proposed that cholinergic input to the SFO may function in parallel with angiotensin mechanisms to mediate synergistic regulatory responses (drinking as well as pressor effects and antidiuresis) to changes in vascular fluid volume. This suggestion was based in part on the evidence that SFO injections of hypertonic saline did not elicit drinking (Simpson and Routtenberg, 1974), and that ablation of this organ failed to affect drinking elicited by systemic hypertonic saline injection (Simpson et al., 1978a). Rather, cholinergic stimulation of the SFO, like angiotensin stimulation, caused a pressor as well as a drinking response (Mangiapane and Simpson, 1978; Simpson et al., 1978a), and it has been demonstrated that the SFO projects to the SON and, thereby, may affect ADH release (Miselis et al., 1978). These findings have led to the proposal that cholinoceptive sites in the SFO are more involved in vascular homeostasis than in regulating intracellular fluid balance.

The possibility that central cholinergic systems may affect fluid balance through multiple routes, including mechanisms responsive to extracellular changes in the SFO and those responsive to intracellular changes in the anteroventral regions of the third ventricle, is not unreasonable and is actually consistent with the need for integrating signals of body-fluid composition with signals for pressure and volume of the intravascular circulation and, ultimately, for maintaining body-water constancy through control mechanisms of ingestion as well as excretion. The main problem lies in establishing that these different brain areas and specific neurochemical receptors therein are part of the physiological apparatus essential for body-water homeostasis. Ablation of the SFO or anteroventral third-ventricle region has only temporary effects on an animal's ability to adjust body fluids (see above references). While this suggests that these regions are not essential for life preservation, the enormous redundancy in the brain could easily permit other mechanisms to assume responsibility for functions normally served by the ablated areas. It is not necessarily true that cholinergic mechanisms are called upon for every occasion where adjustments are required, and only further research will permit us to identify the precise conditions under which these mechanisms operate. The manner in which cholinergic systems interact with adrenergic systems to achieve their goal (Bhargava et al., 1972; Morris et al., 1977) will also need to be explored.

VI. Histaminergic Systems

A. Anatomy of Histaminergic Projections

Histamine (HA) is classically associated with various pathological processes, such as allergy, anaphylaxis, injury, and stress; in these processes, HA is liberated from mast cells in the body to produce characteristic physiological responses, such as vasodilitation, itching, and edema. Several lines of evidence, however, suggest that HA may have a neurotransmitter function in the brain, as reviewed by Calcutt (1976), Green et al. (1978), Schwartz (1977) and Taylor (1975). Histamine is present in the brain and, together with its synthesizing enzyme, l-histidine decarboxylase, is localized in synaptosomes (pinched-off nerve terminals) prepared from brain homogenates. Endogenous HA is selectively released from slices of brain tissue by potassium depolarization, accompanied by an increase in HA synthesis. Its turnover in the brain is very rapid (although not in all cellular compartments), and its formation is found to be accelerated by physiological stress and decreased by barbiturates, hypnotics, and histidine decarboxylase inhibitors. Specific neuronal receptors to HA (referred to as H_1 and H_2) are present in the brain, as demonstrated by various electrophysiological and iontophoretic studies, as well as by investigations of HA's stimulatory effects on cyclic AMP.

Histamine concentrations in the brain are approximately one-tenth the concentrations of NE or 5-HT. Its relative distribution, however, appears to parallel that of these monoamines, with the highest concentration in the hypothalamus, intermediate levels in the mesencephalon and telencephalon, and lowest levels in the cerebellum or medulla-pons. Radiolabeled HA antagonists have been shown to bind to brain tissue, and the regional distribution of [^{3}H] mepyramine (H_1-antagonist) binding in the rat resembled that of endogenous HA (Chang et al., 1978), whereas the [^{3}H] cimetidine (H_2-antagonist) binding in the guinea pig showed a somewhat different pattern (Burkard, 1978). Within the hypothalamus, the HA content is unevenly distributed among various nuclei. The ventral and posterior parts of the hypothalamus have the highest HA levels, as well as l-histidine decarboxylase activity, and intermediate levels are found to exist in a number of other hypothalamic nuclei, including the SON and PVN.

The lack of adequate histochemical methods has required that neurochemical and lesion techniques be combined to investigate the anatomical disposition of HA neurons in the brain (see Schwartz, 1977). It has been found, for example, that lesions of the medial forebrain bundle as it courses through the lateral hypothalamus significantly diminished (50%) HA content and synthesis and l-histidine decarboxylase activity in areas rostral to the lesion (including anterior hypothalamus, cortex, and hippocampus) but not caudal, suggesting that the HA fibers comprise an ascending system. While the precise localization of HA cell bodies is not known, there is evidence from lesion studies that suggests that ascending fibers emanate from the posterior hypothalamus or rostral midbrain. There also appears to be a descending system which innervates the brain stem, notably the periventricular gray matter. Some brain HA is also believed to be located in nonneuronal cells, apparently mast cells, and structures with much of their rich HA concentration located in these cells appear to include the median eminence, pituitary stalk, infundibulum, and the circumventricular organs. Histamine may also exist in the vascular wall in cells distinct from mast cells.

B. Feeding and Drinking Behavior

There is little evidence to suggest that brain HA may affect food intake. In the cat (Clineschmidt and Lotti, 1973), intraventricular HA injections produced a small (25%) suppression of food intake. In the rat, a similar, although considerably more robust, effect was observed with systemic injection of HA, while injection into the lateral hypothalamus caused a small increase of food intake (Leibowitz, 1973c). The significance of these effects on feeding is not known, although it is very likely that they are related to, and perhaps secondary to, HA's potent stimulatory effect on water ingestion.

The first indication that HA might alter drinking behavior was provided by a study of Gerald and Maickel (1972), in which peripherally administered antihistamines were found to markedly suppress the water intake of thirsty rats, and lateral hypothalamic injections of HA (at relatively large doses of 40-160μg) potentiated water intake. Shortly thereafter, two independent reports simultaneously appeared in which a potent stimulatory effect of peripherally administered HA on drinking was described (Gutman and Krausz, 1973; Leibowitz, 1973d). This effect occurred in a dose-dependent manner, at doses ranging from 0.38 to 24 mg/kg (see Leibowitz, 1979a). The response developed within a few minutes after injection and proved to be extremely reliable and robust.

In the periphery, two types of HA receptors have been pharmacologically characterized in the mammalian species (Ash and Schild, 1966; Black et al., 1972). The first type, referred to as H_1, is found to be selectively stimulated by the compounds 2-methylhistamine and pyridyl etmylamine and selectively blocked by the classic antihistamine agents. The second type, H_2, in contrast, is selectively stimulated by the compounds 4-methylhistamine and dimaprit and selectively blocked by recently developed H_2-receptor blockers. Using these drugs, the HA receptors mediating the drinking response were characterized and found to have properties of both the H_1 and H_2 receptors (Leibowitz, 1979a). The agonists produced a reliable increase in water intake, with a relative potency in the order of HA > pyridyl etmylamine = 4-methylhistamine = dimaprit > 2-methylhistamine. Furthermore, the response elicited by HA was partially blocked (approximately 50%) by each of a variety of H_1 and H_2 antagonists tested, at doses which had no effect on water ingestion evoked by deprivation, isoproterenol, or hypertonic saline. When injected in combination, the H_1 and H_2 antagonists showed an additive blockade of the HA response of up to 82%. Antagonists of α-adrenergic, serotonergic, dopaminergic, and cholinergic receptors had no effect on the HA response, except perhaps at high doses which exerted a nonspecific suppression of all dipsogenic stimuli.

With regard to the β-adrenergic blocker propranolol, this was found in a study of Gutman and Krausz (1973) to cause a 50% inhibition of the drinking elicited by peripheral HA. This finding was confirmed (Leibowitz, 1979a), but the partial block appeared to be independent of dose, occurring with doses ranging from 0.08 to over 6 mg/kg; furthermore, a similar effect of propranolol was also observed in the case of hyperosmotic and deprivation-induced drinking.

Studies with central drug injections in the rat have revealed similar effects of HA on water ingestion, at doses approximately three to four orders of magnitude lower than those used peripherally (Fig. 23) (Leibowitz, 1973d; 1979a). Sites outside the hypothalamus, in the caudal hypothalamus or in the SON, were generally unresponsive to HA. The most effective sites were in the rostral hypothalamus, in the paraventricular and perifornical areas, where HA elicited a small to moderate drinking response (4-6 ml) after a latency of 2-10 min. The threshold dose for this effect in the PVN (the most sensitive area) was 60 ng. Similar to the results obtained with peripheral drug injection, tests with centrally administered HA antagonists revealed a partial block of this phenomenon with both the H_1- and H_2-receptor antagonists, and no effect with catecholaminergic, cholinergic, and serotonergic antagonists.

Systemic injection of HA is known to have dramatic effects on the cardiovascular system. It causes hypotension, as a consequence of its potent vasodilatory action on minute blood vessels and capillaries and its ability to increase capillary permeability. Other hypotensive drugs have been found to increase water consumption, and this effect is generally associated with an increase in renin release and is abolished by nephrectomy as well as the β-adrenergic blocker propranolol (see Fitzsimons, 1972). Histamine appears to be distinguishable from these drugs, since its drinking effect was only partially (50%) inhibited by propranolol (perhaps a nonspecific block) and only slightly (20 to 30%) inhibited by nephrectomy (Gutman and Krausz, 1973; Leibowitz, 1979a). The results would appear to argue against an essential role for the renin-angiotensin system in HA-induced drinking. The primary difference between HA and other hypotensive agents seems to lie in HA's additional action on vascular permeability. Similar to the effects

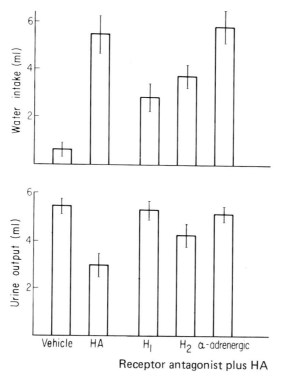

Figure 23 Effects on water intake and urine excretion observed after hypothalamic injection of HA (25 nmol) in the rat. Histamine injection into the paraventricular nucleus (upper chart) stimulated drinking, a response partially attenuated by local administration of H_1- and H_2-receptor antagonists but unaffected by an α-adrenergic antagonist. In the supraoptic nucleus (lower chart), HA caused antidiuresis, and this effect was abolished by the H_1 and α-adrenergic antagonists and attenuated by an H_2 antagonist. (Based on results from Leibowitz, 1979a.)

observed with polyethylene glycol and formalin (Stricker, 1973), HA, by causing an increase in capillary permeability, might be expected to produce a decrease in blood volume which, through baroreceptor mediation, would lead to the activation of thirst-stimulating mechanisms in the brain. Further work will need to be done to evaluate the contribution of this mechanism to HA-induced drinking. It appears consistent with the available evidence, however, to postulate at least a partial mediating role for this series of events, perhaps somewhat aggravated by HA's potent vasodilatory effect.

The possibility that peripheral HA is acting directly on the brain to stimulate drinking needs to be evaluated in light of the finding that relatively little of systemically administered [^{14}C] HA penetrates into brain tissue, although a minute amount was detected over and above that expected to be within the vasculature (Snyder et al., 1964). This would appear to suggest that the site of action of peripheral HA, if it is indeed acting centrally, might be either on the cerebral vasculature or on brain areas, such as circumventricular structures, which lie outside the blood-brain barrier. In support of this is the finding that peripheral HA-induced water intake was inhibited by H_1- and H_2-receptor antagonists which are essentially incapable of penetrating into brain tissue with a barrier (Leibowitz, 1979a). Another dipsogenic substance, angiotensin, is believed to produce its effects by acting on a circumventricular organ (Epstein, 1978), and it is possible that HA might be operating in a similar fashion. In a study by Nicolaidis and Fitzsimons (1975), evidence was provided to suggest that angiotensin's thirst-provoking action may be mediated by its effects on mechanoreceptors which respond to alterations in the volume of the cerebral blood vessels. The circumventricular structures are highly vascularized, and the effect of angiotensin in this brain region was found to be antagonized by drugs which decrease vascular motility. The importance of the vasculature in determining brain function has been seriously ignored, and, in light of HA's potent effects on the cerebral as well as peripheral vascular system, this hypothesis relative to HA's dipsogenic action deserves careful and thorough consideration (see Leibowitz, 1979a).

With regard to the relationship between the drinking response induced by centrally and peripherally injected HA, the differential in threshold dose (approximately three orders of magnitude) would argue against the possibility that central HA is leaking from the brain to activate peripheral HA receptors. The possibility that peripheral and central HA are acting on similar receptors, perhaps located in the cerebral vasculature, needs to be seriously considered. However, in light of the evidence for a neurotransmitter role of HA in the brain (see Section VI.A above), it is also reasonable to propose a direct action of HA on central neural tissue. The significance of such a receptor mechanism for maintaining body-fluid homeostasis is not known at this time. Further investigations to examine the validity of this working hypothesis will need to employ more extensive pharmacological manipulations, such as drugs that release endogenous HA, as well as biochemical analyses correlated with behavior. The PVN and other hypothalamic structures contain high concentrations of endogenous HA, and when applied iontophoretically to the hypothalamus, HA has a predominantly excitatory effect on neuronal activity (Haas and Wolf, 1977). These findings are consistent with the idea that endogenous HA has a neurotransmitter role in the control of drinking behavior, and that this control involves specific H_1- and H_2-neuronal receptors. It should be pointed out, however, that in the iontophoretic studies of Haas and Wolf, the neuronal response elicited by HA was sometimes slow and variable. Furthermore, in the behavioral studies (see above), the drinking response induced by microinjected HA had a relatively long latency, varying between 2 and 10 min.

There are numerous reasons why HA might be slow in acting, and three possibilities are that: (1) HA may produce its effect after spread to another brain site, although its effect appeared to have anatomical specificity (see above); (2) HA, rather than causing direct activation of thirst-stimulating neurons, is acting indirectly through some other neural mechanism (e.g., an efferent heat-loss mechanism); and (3) HA may be acting through another neurotransmitter substance in the brain (see Taylor, 1975).

There is accumulating evidence to suggest that, in addition to its neuronal localization, HA in the brain may also exist in nonneuronal cells, probably mast cells (see Green et al., 1978; Schwartz, 1977). While the function of these cells in the hypothalamus is not known, it is interesting to note that they are especially numerous along blood vessels and meninges, although they are also found in the parenchyma. The close proximity of these HA-containing cells to the vasculature returns us to the first suggestion that HA might elicit drinking behavior through its actions on the cerebral blood vessels. This

Feeding and Drinking and Water-Electrolyte Excretion

mode of action, which might of course occur in addition to a direct HA effect on brain neurons, may be a component of the immune or inflammatory processes characteristically associated with mast-cell HA. Goldstein and Halperin (1977), through studies with peripheral drug injections, concluded that mast cells may have a role in the detection of changes in tonicity of body fluids and that HA released by the mast cells participates in the afferent link of drinking behavior elicited by hypertonic stimulation. Several characteristics of the mast cells, which might be expected to be present in an osmotic receptor, include their perivascular distribution, their high sensitivity to changes in tonicity, and their ability to synthesize and release amines (namely HA and 5-HT) which activate water ingestion.

C. Water Excretion

Histamine, when peripherally administered, has been known for many years to cause a reduction in urine output (see Leibowitz, 1979a, for review). While direct effects on renal function may contribute to this response, there is additional evidence that the effect is hormonally mediated, via the release of ADH (Dogterom et al., 1976; Leibowitz, 1979a). With peripheral drug injection, HA's antidiuretic effect occurred in a dose-dependent fashion, at doses ranging from 0.1 to 6.0 mg/kg (Leibowitz, 1979a) or up to 50 mg/kg (Dogterom et al., 1976). This effect was abolished by H_1-receptor antagonists and attenuated (up to 50%) by an H_2-receptor antagonist (Leibowitz, 1979a). Blockers of adrenergic, cholinergic, dopaminergic, and serotonergic receptors had no effect.

As observed with HA-induced drinking, central drug effects on antidiuresis were similar to those obtained with peripheral drug injections, although at considerably lower doses. In the dog, ADH release was manifested with ventricular HA injection, although in this particular study relatively high doses (25 to 200 μg) were employed (Bhargava et al., 1973). Somewhat lower doses (15-60 μg) were found to be effective in producing antidiuresis in the cat after direct injection into the SON (Bennett and Pert, 1974). This effect was abolished by destruction of the supraopticohypophyseal tract. In the rat, both the SON and PVN were responsive to central HA injection, while the perifornical lateral hypothalamus (1 mm lateral to the PVN) was generally insensitive (Fig. 23) (Leibowitz, 1979a). Moderate to strong antidiuresis occurred at doses of 3-12 μg, and the threshold dose for the SON, the most effective area, was at least as low as 40 ng.

Tests with centrally injected HA antagonists demonstrated that H_1-receptor blockers essentially abolished the HA-induced antidiuresis (Bennett and Pert, 1974; Bhargava et al., 1973; Leibowitz, 1979a). An H_2-receptor antagonist yielded a partial blockade of HA's action (Leibowitz, 1979a), similar to that observed with peripheral drug injection (see above). This pattern of results contrasts with that obtained with the HA drinking response, which was partially antagonized by the H_1- as well as the H_2-receptor blockers. Additional tests with dopaminergic, β-adrenergic, and cholinergic receptor blockers failed to reveal any effects of these drugs on HA's antidiuretic response, emphasizing the specificity of the HA-receptor action. With the α-adrenergic receptor antagonist, however, a nearly total block of HA's effect was observed (Fig. 23). This result, which clearly distinguishes the antidiuretic response to HA from its elicited drinking response that remained unaffected by α-adrenergic blockade, is consistent with the finding of Bhargava et al. (1973) that phenoxybenzamine, another α-adrenergic antagonist, could block the

ADH-releasing effect of cerebroventricular HA in the dog. Bennett and Pert (1974), however, did not observe this blockade effect with injection into the cat SON.

Studies of HA's action on the kidney, or of reflexive changes that occur as a consequence of HA-induced change in systemic blood pressure, have indicated that HA's antidiuretic effect occurs independently of changes in renal hemodynamics (see Leibowitz, 1979a). The evidence cited above supports the proposal that HA, peripherally or centrally administered, affects the tubular reabsorption of water presumably via the release of ADH. Whether this reflects the consequence of HA's potent effects on the vasculature (decreased blood pressure due to vasodilatation, increased capillary permeability, and thus decreased blood volume) remains to be determined. These intravascular changes would be expected to initiate a variety of cardiovascular adjustments which contribute to blood-volume regulation, as well as activate complementary behavioral (water-ingestion) and physiological (ADH-release) adjustments to help restore normal body-fluid volume. The importance of HA's effect on capillary permeability and blood volume and the consequent changes in vasoreceptor activity will need to be carefully examined in terms of the pharmacological properties of this phenomenon and its predictive ability for HA-induced antidiuresis. At the moment, this mechanism, and perhaps to a lesser extent the renin-angiotensin system, would appear to provide a reasonable explanation for the effect of peripherally administered HA on ADH release.

Histamines' actions, however, may not end here, and, as discussed for the elicited drinking response, the phenomenon of antidiuresis may involve an additional set of HA receptors, perhaps located in a region of the brain. In view of the results of Snyder et al. (1964) indicating that systemic HA fails to pass the blood-brain barrier in any significant amounts, HA's central activity would most likely be confined to structures outside the barrier or perhaps in the cerebral vasculature. Evidence to support this suggestion is provided by the finding that antihistamines, which are essentially incapable of penetrating the blood-brain barrier, were effective in attenuating or abolishing the antidiuresis produced by systemic HA (Leibowitz, 1979a).

Based on this argument, it seems that the H_1 and H_2 receptors identified by central injection studies in the SON and PVN are inaccessible to peripheral HA, unless, of course, these receptors exist in or near to the vasculature or in some unique location lying outside the barrier. In view of HA's potent actions on cerebral vessels, the suggestion of a nonneuronal site of action for central as well as peripheral histamine remains a real possibility. With regard to HA's actions on neural structures lying outside the blood-brain barrier, the most likely candidates are the circumventricular structures which have been postulated to have a role in the mediation of angiotensin-induced thirst (Epstein, 1978) and the median eminence. The median eminence and the pituitary are found to have the highest concentrations of histamine of any brain regions assayed. The pituitary, however, does not seem to be essential for the antidiuretic effect (nor presumably the ADH release) elicited by peripheral HA, since this response was found to occur equally in hypophysectomized and normal animals (Leibowitz, 1979a). The median eminence, in contrast, may be a reasonable site for peripheral HA's receptor action since, in addition to its dense concentration of endogenous HA and its location outside the blood-brain barrier, the median eminence (zona externa) is believed to play a role in ADH release (Defendini and Zimmerman, 1978). Blackmore and Cherry (1955) found the destruction of this area to abolish the antidiuresis induced by systemic HA. This experiment, however, does not allow us to distinguish the possibility that peripheral HA acts directly on median-eminence histaminergic receptors, from the possibility of its acting at some other site which then requires an intact supraopticohypophyseal pathway for release of ADH.

The question of whether peripheral HA acts on H_1 and H_2 receptors located in the SON and PVN cannot be resolved at the present time. This remains a distinct possibility, the above considerations notwithstanding. The pharmacological analyses conducted with peripheral and central drug injections revealed similar results with the H_1- and H_1-receptor blockers. A clear distinction, however, between the antidiuretic effects produced by these two routes of administration was observed with an α-adrenergic blocker which essentially abolished central HA's actions while leaving intact the effect of subcutaneous HA (Bhargava et al., 1973; Leibowitz, 1979a). This finding suggests that peripheral and central HA may act at different sites, or at least involve different neurochemical mechanisms. It is possible, of course, that in the SON, where phentolamine blocked HA's antidiuretic effect, this α-adrenergic antagonist is exhibiting antihistamine properties. This action, however, was not observed with phentolamine in the PVN in the studies of HA-elicited drinking (see above). Furthermore, Bhargava et al. (1973) also found tetrabenzine, a NE-depleting agent, to block HA's actions, suggesting that HA was operating indirectly through the release of NE. This is certainly a reasonable possibility, in light of the extensive evidence demonstrating an antidiuretic effect of NE in the SON and PVN (see Section II.D above) and the finding that HA increases the efflux of radiolabeled CA in the hypothalamus (Subramanian and Mulder, 1977).

VII. GABAergic Systems

A. Biochemical Studies

Research over the past several years has provided evidence that amino acids may be the primary candidates for mediating neurotransmission in the mammalian brain. By neurophysiological techniques, two classes of amino acids have been delineated, namely, excitatory amino acids (glutamic and aspartic acids, and cysteic and homocysteic acids), which depolarize most central neurons, and inhibitory amino acids (γ-aminobutyric acid [GABA], glycine, β-alanine, and taurine), which hyperpolarize the neuronal membrane (for review, see Aprison, 1978; Fonnum and Storm-Mathison, 1978; Garattini et al., 1978; Hökfelt et al., 1978; Iversen, 1978; Snyder, 1975; Werman, 1972). In quantitative terms, amino acids seem to be the major neurotransmitters, with the better-known neurotransmitters—NE, DA, 5-HT, and ACh—accounting for transmission at only a small percentage of central nervous synapses. Neurophysiological studies on the important inhibitory amino acids, GABA and glycine, indicate that GABA may be active throughout the neuroaxis, whereas glycine hyperpolarizes neurons only in the spinal cord, brainstem, and diencephalon. This is consistent with the biochemical evidence showing that GABA is present in high concentrations throughout the brain, with highest levels in the hypothalamus, basal ganglia, colliculi, and dendate nucleus, whereas glycine levels are highest in the spinal cord and brain stem, intermediate in the midbrain, and lowest in forebrain structures.

The focus of this section will be on GABA, which has been shown to have effects on ingestive behavior. Studies with the other amino acids are either nonexistent or have failed to produce significant behavioral changes. The role of GABA as a major inhibitory transmitter is now generally accepted (E. Roberts et al., 1976). This amino acid is ubiquitously distributed in the brain and virtually all central neurons are powerfully inhibited by GABA, which may represent the transmitter used by up to 50% of all synaptic terminals in certain regions of the brain (see Fonnum and Storm-Mathisen, 1978; Iversen, 1978). There is good correlation between the levels of GABA and its synthesizing enzyme, glutamic acid decarboxylase (GAD), in the different brain regions, and the hypothalamus is one of the structures that contains large amounts. Within the hypothalamus, the areas with the highest concentrations of GAD were found to be in the preoptic, anterior, and dorsomedial nuclei; intermediate levels in the mamillary bodies, PVN, and VMN; and low levels in the arcuate and supraoptic nuclei and median eminence. Evidence obtained in lesion studies indicates that most of the GAD-containing cells have their origin inside the hypothalamus. This suggests that most of the GABAergic neurons are likely to be short interneurons providing intrahypothalamic connections.

B. Feeding Behavior

Recent studies of the effects of central GABAergic stimulation on ingestive behavior have provided evidence that this amino-acid neurotransmitter may affect food ingestion and that it may mediate changes induced by the monoaminergic neurotransmitters as well as by certain metabolic signals. Grandison and Guidotti (1977) reported that ventromedial hypothalamic injection of muscimol, a GABAergic agonist, in the satiated rat evoked a feeding response which resembled that observed after injection of NE (Fig. 24). It was demonstrated that the action of muscimol could be blocked by the GABAergic receptor antagonist bicuculline but not by the α-adrenergic antagonist phentolamine. The NE-elicited feeding response, in contrast, was antagonized by α-adrenergic blockade (consistent with findings of other investigators—see Section II.B), as well as by the GABAergic blocker. This finding established a possible GABAergic link in the process of noradrenergic control of feeding.

Kelly et al. (1977) reported a similar increase in feeding in the rat after GABA injection into the ventromedial hypothalamus. This finding has been extended in more recent studies, which revealed antagonistic feeding effects of GABAergic receptor stimulation of the medial and lateral hypothalamus (Kelly et al., 1977; Kelly, 1978). Both GABA and muscimol produced an increase in feeding after injection into the medial hypothalamus, and the GABAergic antagonists bicuculline and picrotoxin suppressed food intake. Mapping studies indicated that the most effective site for revealing enhanced feeding with GABAergic stimulation was actually rostral to the VMN in the area of the PVN. This nucleus has also been demonstrated to be the most sensitive site for eliciting eating with NE injections or with drugs that release endogenous NE (see Section II.B.1). A contrasting effect of the GABAergic drugs in the lateral hypothalamus was also described. Injection of muscimol into this area inhibited feeding, whereas bicuculline stimulated feeding. These effects were associated with the perifornical (midlateral) region of the lateral hypothalamus and diminished or disappeared as the injection site was moved towards the far-lateral region or outside the lateral hypothalamus. Once again, the site of sensitivity to GABAergic stimulation appeared to correspond to the perifornical area demonstrated to be most responsive to the anorectic action of the CA agonists and of AMPH (see Section II.B.1 and 3).

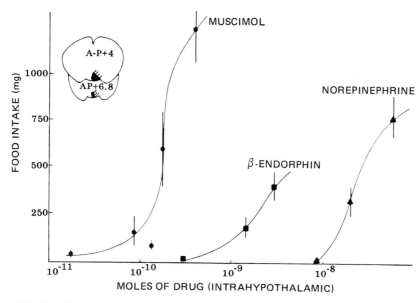

Figure 24 Food intake after ventromedial hypothalamic injection (see insert) of the GABA agonist muscimol, the adrenergic agonist norepinephrine, and the opiate peptide β-endorphin in the satiated rat. (From Grandison and Guidotti, 1977.) Reprinted with permission from *Neuropharmacology 16*, L. Grandison and A. Guidotti, Stimulation of food intake by muscimol and beta endorphin, 1977, Pergamon Press, Ltd.

Further tests conducted by Kelly and his colleagues revealed that the feeding stimulation effect of GABA in the PVN was specific to this inhibitory amino acid, as little or no change in feeding occurred with injection of glycine or its antagonist strychnine. The muscimol effect was dependent upon dose and was inhibited by prior injection of the antagonist bicuculline. Furthermore, PVN injection of the water-soluble benzodiazepine flurazepam also showed a small increase in feeding, suggesting that this class of drugs, which produce feeding after peripheral administration, may be acting through central GABAergic mechanisms, in addition to serotonergic mechanisms (see Section IV.B.2). There is considerable evidence that the effects of the benzodiazepines may be mediated by the direct or indirect activation of GABAergic receptors (see Garattini et al., 1978; E. Roberts et al., 1976).

In the perifornical area, where GABAergic stimulation and CA stimulation (through AMPH and CA agonists) cause inhibition of feeding, Kelly examined the effect of the enzyme inhibitors allylglycine and ethanolamine-O-sulphate on peripherally administered AMPH. Allylglycine is a potent inhibitor of GAD, the synthesizing enzyme of GABA, and thus leads to a depletion of the amino acid. Ethanolamine-O-sulphate, in contrast, is an inhibitor of GABA-transaminase, the degradative enzyme of GABA, and thus leads to an increase in the availability of GABA. The question addressed here was, if the anorectic drug AMPH is acting through CA mechanisms in the perifornical area and if these CA mechanisms are dependent upon subsequent mediation of GABAergic mechanisms, perifornical injection of allylglycine, through GABA depletion, should be expected to reduce the potency of peripheral AMPH injection, whereas ethanolamine-O-sulphate, through increased GABA availability, should potentiate the AMPH effect. The results obtained were essentially consistent with this prediction. The most dramatic observation was made with allylglycine, which after bilateral perifornical administration was effective in abolishing AMPH anorexia. Tests with the anorectic drug FENF, which is believed to act through serotonergic systems (see Section IV.B.2), showed this drug's action to remain intact. The results with ethanolamine-O-sulphate were somewhat ambiguous, since this drug alone, presumably through increased GABA stimulation, produced an inhibitory effect on feeding after injection into the perifornical area. The outcome of the tests with AMPH revealed a very strong anorectic effect of these drugs in combination, clearly an enhancement over the effect of AMPH alone. Further tests are required to determine at what level, neurochemical or behavioral, this interaction may have occurred.

Kelly and his colleagues have offered these results as evidence that afferent CA systems in the medial (PVN) and lateral (perifornical) regions of the hypothalamus alter feeding through hypothalamic GABAergic interneurons which then modulate the activity of primary efferents of the neural circuitry of feeding. This proposal is based on the suggestion of Roberts (1977) that GABA has a primary role in the mediation of afferent influences on control or command neurons within a particular circuit and also on the evidence that a catecholaminergic-GABAergic interaction of this nature occurs in the extrapyramidal system and the cerebellum (see Bloom et al., 1971; Gale et al., 1977; Grzanna et al., 1977; Landis et al., 1975; Roberts and Hammerschlag, 1976). Reubi et al. (1977) have recently demonstrated that DA and its agonists selectively stimulate the release of [^{3}H] GABA in the substantia nigra in vitro. It is suggested that the CA in the hypothalamus lead to the same effect in producing their effects on feeding.

In light of the high density of GABA in the hypothalamus and the fact that GABAergic synapses are so prevalent, the hypothesis certainly deserves further attention. More extensive biochemical as well as pharmacological studies of this neurochemical interaction may help us to understand the relationship between these neurotransmitter systems and the profound changes in feeding behavior that occur after hypothalamic lesions or knife cuts. Such manipulations would clearly have significant impact on the control exerted by GABAergic neurons on efferent activity.

A possible link between hypothalamic GABAergic mechanisms and metabolic signals modulating food intake has been suggested by the evidence of Kimura and Kuryama (1975) that peripheral injection of insulin (which produces hypoglycemia and increased feeding) increased GABA levels in the ventromedial hypothalamus and decreased GABA levels in the lateral hypothalamus. In contrast, alloxan-induced hyperglycemia was associated with an increase in GABA levels of the lateral hypothalamus. Meeker and his colleagues have recently advanced the idea that hypothalamic cells sensitive to glucose metabolism may encode metabolic information through the formation of GABA (via substrate entry into the GABA shunt) and that the flow of information through the GABA shunt, which parallels energy flow through the tricarboxylic acid cycle, may provide the link between energy metabolism and long-term changes in functional activity (Meeker and Panksepp, 1977; Meeker et al., 1977; Meeker and Myers, 1979). In support of this suggestion, these investigators have demonstrated that [U-^{14}C] glucose becomes incorporated into amino acids within the hypothalamus, predominantly GABA and glutamate (as opposed to glycine, aspartate, alanine, and taurine). Food deprivation enhanced this process, and, during feeding (after deprivation), an increase in the efflux of [^{3}H] GABA and [^{14}C] glutamate was observed.

This work adds an important dimension to the hypothesis that hypothalamic CA neurons may be involved in receiving information of glucose utilization and transducing this information into an effective eating response for rapidly restoring glucose deficits (see Sections II.H.2 and 3). Interneurons that are GABAergic in nature may respond to CA afferents (see Grandison and Guidotti, 1977; Kelly, 1978; Kelly et al., 1977), and, depending upon the intrinsic activity of these interneurons, which reflects the metabolic state of the animal, efferent systems may be activated to produce a change (increase or decrease) in feeding behavior. The area of the hypothalamus which Meeker and his colleagues proposed as the site of interaction between glucose metabolism and GABA synthesis was along the lateral border of the VMN. The biochemical changes cited above appeared to occur most strongly at this site. Furthermore, drugs which modulate GABA-ergic activity were most effective in altering long-term feeding at this site (Meeker and Panksepp, 1977). That is, injections of D-glucose, GABA, and aminooxyacetic acid (an inhibitor of GABA-transaminase) caused a suppression of 24-hr food intake, whereas bicuculline and γ-hydroxybutyrate (which inhibits glucose incorporation into GABA or glutamate) injections increased 24-hr feeding. (Glutamate and glutamine had no effect.)

While the direction of these effects is consistent with the feeding changes observed by Kelly (1978) in the perifornical region directly dorsal to Meeker's most effective site (just lateral to the VMN), it is difficult to compare these studies in light of the extensive differences in the procedures used. In particular, Kelly examined the immediate drug effects on feeding, whereas Meeker and Panksepp focused on long-term changes. The latter investigators, furthermore, used somewhat higher doses of the drugs. The anatomical characteristics of these systems clearly need to be analyzed further. The specificity of the drug effects also requires careful evaluation, in light of evidence that other amino acids may affect feeding (Panksepp, 1972). It has been demonstrated, furthermore, that injections of amino acids into the medial hypothalamus may cause permanent cellular damage sufficient to produce long-term overeating and obesity (Simson et al., 1977). All these factors notwithstanding, this research clearly has great potential in helping us to understand how the hypothalamus integrates information from metabolic and neurochemical sources to modulate food consumption.

VIII. Peptidergic Systems

A. Peptides as Neuromodulators

The rapid development in recent years in our knowledge of neuroregulation in the brain has focused overwhelmingly on the biogenic amines. However, substances other than amines have long been known to act on the brain. The introduction of peptide hormone treatment in human therapy revealed that these hormones can drastically change the excitatory state of the brain. It was therefore proposed that peptides may act as modulators of nervous activity, altering synaptic transmission by facilitating transmission through normally active pathways. However, the evidence obtained has been considered to be mostly of pharmacological interest, since the peptides are not normally expected to cross the blood-brain barrier to any significant extent. It is only recently that certain lines of investigation have led to an increased appreciation that highly active peptide substances, including those traditionally considered to be peripheral hormones, are synthesized and distributed fairly widely in the brain and are highly concentrated in the hypothalamus (see Hökfelt et al., 1978).

The role of these peptides in the brain may be as varied as the peptides themselves (see Lipton et al., 1978; and Usdin et al., 1977, for reviews). They can affect neurotransmitter metabolism, or they may be transmitters themselves. They may alter membrane permeability and cyclic nucleotide activity, and they may affect cell growth and RNA, DNA, and protein synthesis. Subcellular localization studies have indicated that these peptides or their synthesizing enzymes are not confined to such nonneural elements as vascular tissue or glial cells but are present in neural cells and in synaptosomes.

Whether the peptides are involved in transmission at chemically mediated synapses remains an open question. In general, putative neurotransmitter agents considered to mediate rapid communication between neurons at synaptic sites are low-molecular-weight substances (e.g., ACh, CA, and amino acids). Actually, few of these agents have clearly been demonstrated to fulfill all criteria established as requirements for identification of a substance as a neurotransmitter, and a reluctance to accept larger molecules (e.g., peptides) as neurotransmitters and a lack of knowledge regarding their precise action on neural tissue have led to the emergence of the term neuromodulator. While the definition of this term may vary from one investigator to the next, it generally appears to imply the ability to influence neuronal excitability in a less direct manner than the traditionally accepted neurotransmitters, perhaps by interacting with or modifying the effects of these neurotransmitters and having a time course ranging from seconds to days.

Peptide neuromodulators may exert their effects presynaptically by modifying either synthesis or release of transmitters from the same or separate nerve terminals. They may also act postsynaptically, possibly at the receptor level. There are a large number of peptides that have been found to exist in the brain, as well as the hypothalamus, and these include luteinizing-hormone-releasing hormone, thyrotropin-releasing hormone, somatostatin, oxytocin, and vasopressin (plus their carrier substance neurophysin), substance P, enkephalins, angiotensin II, neurotensin, gastrin, cholecystokinin, vasoactive intestinal peptide, and prolactin (see Hökfelt et al., 1978; Defendini and Zimmerman, 1978; Snyder, 1978). There is extensive evidence to suggest that many of these peptides are active in neuronal function, and a review of this evidence may be found in books edited by Lipton et al. (1978) and Usdin et al. (1977).

B. Feeding and Drinking Behavior

There are a number of studies which have tested the effects of peripheral and central peptide administration on ingestive behavior. While a variety of effects have been reported, there are only a very few peptides which have been sufficiently explored to raise any questions concerning their potential regulatory function in energy or fluid homeostasis. A comprehensive review of this evidence is beyond the scope of this chapter and in some cases can be found in recently published works (see references below). A brief summary, however, would be helpful in illustrating the dramatic behavioral changes that can occur with peptide administration.

A great deal of attention has been given to the octapeptide angiotensin II. The renin-angiotensin system in the periphery is well known for its physiological role in the control of the secretion of the mineralocorticoid hormone aldosterone and in the regulation of

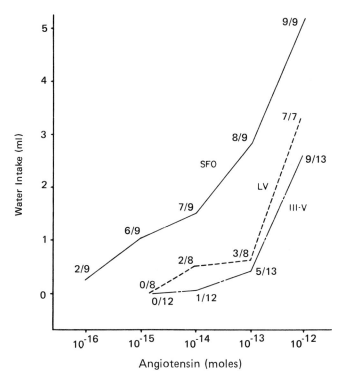

Figure 25 Total water intake 15 min postinjection as a function of moles of Ile^5-angiotension II per 1.0 μl applied topically to the subfornical organ (SFO), dorsal third ventricle (III-V), or lateral ventricle at the level of the interventricular foramen (LV). At each data point, the numerator indicates the number of animals ingesting water, and the denominator indicates the number of animals injected at that dose. (From Simpson et al., 1978a.) Copyright 1978 by the American Psychological Association. Reprinted by permission.

blood pressure. There is accumulating evidence to suggest that this sytem, through synthesis of angiotensin II in the periphery (and possibly in the brain), may play an important role in a variety of species in the mediation of drinking behavior and ADH secretion, and, possibly, in the ingestion of sodium (Epstein, 1978; Fregly, 1978; Ganten et al., 1976, 1978; Hoffman et al., 1977; Severs and Daniels-Severs, 1973). When centrally injected, angiotensin has been found to be effective, at very low doses, in eliciting water ingestion and evoking ADH release (Fig. 25). At least in the case of drinking, this peptide is believed to act on circumventricular organs which lie outside the blood-brain barrier and are, therefore, accessible to circulating angiotensin. There is evidence that catecholaminergic mechanisms may be involved in the mediation of angiotensin-induced water and sodium intake and antidiuresis, although further critical studies investigating this relationship are required.

Additional studies of peptide effects on drinking behavior have revealed a potent dipsogenic effect in the pigeon with the structurally similar substances, eledoisin and substance P (Evered et al., 1977). These peptides were effective at brain areas responsive to angiotensin, and the drinking behavior elicited by these agents had a similar time course and dose-response relationship. Other peptides, namely, vasopressin, vasotocin, and bradykinin, as well as the potent excitatory substance glutamate, were without effect. In rats, central administration of vasopressin or neurotensin have yielded negative results, although generally only one dose or one injection site was examined (Grandison and Guidotti, 1977; Kenny et al., 1978; Severs et al., 1978). In one study, ventricular injection of substance P has been associated with an increase in food ingestion (Childress, 1978), and the releasing hormones (luteinizing-hormone-releasing hormone and thyrotropin-releasing hormone) and somatostatin have been reported to inhibit water ingestion (Vijayan and McCann, 1977).

A particularly potent antidipsogenic substance appears to be the prostaglandin Es (PGE) which are synthesized within the brain (see Epstein, 1978). Baile (1974) first reported that hypothalamic injection of PGE inhibited food as well as water ingestion in the rat; that prostaglandin antagonist elicited feeding at sites responsive to NE; and that the feeding effect induced by this antagonist was selectively abolished by an α-adrenergic-receptor blocker. The implication of these findings, that hypothalamic prostaglandins may modulate adrenergic-receptor function to regulate energy intake or storage, has yet to be examined. However, Kenny and Epstein have examined further the inhibitory effect of PGE on water ingestion and found it to occur independent of the feeding-inhibitory effect at considerably lower doses (as low as 10 ng) (see Epstein, 1978). At these low doses, PGE was most effective in inhibiting angiotensin-induced water ingestion, although at higher doses the antidipsogenic effect expanded to include drinking after central carbachol, systemic hypovolemia, and water deprivation. Based on this evidence, it was proposed that angiotensin, while stimulating drinking, may simultaneously activate the synthesis of the prostaglandins, which then through their vasodilator action lead to the inhibition of the angiotensin response. This proposal is consistent with other evidence linking angiotensin drinking to this peptide's constrictor action on the cerebral vasculature (Epstein, 1978; Nicolaidis and Fitzsimons, 1975).

Another peptide which has been the focus of attention is the intestinal hormone cholecystokinin. When peripherally administered, this substance inhibited feeding and elicited behaviors typically associated with satiety (Antin et al., 1975; Gibbs et al., 1973). Studies with central administration of cholecystokinin have yielded conflicting results (e.g. Maddison et al., 1978; Nemeroff et al., 1978; Stern et al., 1976), suggesting that this peptide's primary site of action may be in the periphery, possibly involving an abdominal site innervated by the vagus nerve (Smith and Cushin, 1978). The additional possibility of a central mechanism of action has been reinforced by the recent evidence that genetically obese mice may contain an abnormally low level of cholecystokinin in the brain (Straus and Yalow, 1979). This change was not revealed in studies from another laboratory in which similar procedures were used to examine a variety of genetically obese animals, as well as different nutritional states (Schneider et al., 1979).

Caerulein, a decapeptide chemically and physiologically similar to cholecystokinin, has been tested after central injection in the rat and been found to inhibit feeding after administration into the ventromedial but not the lateral hypothalamus (Stern et al., 1976). Ventromedial hypothalamic lesions, furthermore, attenuated the effectiveness of peripheral cerulin in suppressing feeding, although in a separate study such lesions failed to inhibit the anorexia induced by cholecystokinin (Kulkosky et al., 1976). There is evidence that calcitonin, which increases after a meal and is released by cholecystokinin, may also be involved in satiety (Freed et al., 1978). Intraventricular injection of this peptide reduced food intake at doses 75 times lower than those found to be effective upon peripheral administration. Another gastrointestinal hormone, somatostatin, also suppressed feeding upon peripheral administration but had no observable effect in the brain, at least after ventricular or medial hypothalamic injection (Grandison and Guidotti, 1977; Kenney et al., 1978).

The peptide β-endorphin, in contrast, has actually been shown to stimulate eating, either with ventricular injection (Kenney et al., 1978) or ventromedial hypothalamic injection (Fig. 24) (Grandison and Guidotti, 1977). Interestingly, the eating effect in the ventromedial hypothalamus, like that produced by GABA or NE (see Sections II.B and VII.B), was found to be antagonized by the GABAergic antagonist, bicuculline, suggesting a potential relationship between these three neurochemical systems in modulating feeding. Consistent with this evidence is the finding of Margules et al. (1978) that peripheral administration of the opiate antagonist naloxone abolished overeating in genetically obese mice or rats, and that the obese animals showed an elevated β-endorphin concentration in the pituitary (but not the brain). Genetically obese rats have also been distinguished by their altered amine levels in the brain, specifically within the PVN and median eminence where NE content is abnormally low (Cruce et al., 1976). In light of this evidence, it would be of interest to examine the responsiveness of these particular rats to medial hypothalamic injection of NE, GABA, or β-endorphin.

The effects observed in the above studies clearly show the dramatic behavioral changes that may be associated with central, as well as peripheral, peptide stimulation. The mechanism through which these gastrointestinal hormones act, however, has yet to be established. Several important questions which need to be answered are: (1) whether the hormone released peripherally affects and acts through the brain; (2) whether the central system synthesizing the same hormone acts in conjunction with or independently of the peripheral system; and (3) whether these hormones are released under physiological conditions to affect naturally motivated feeding behavior. All studies examining the central effects of the various peptides must ultimately address the issues concerning their mode of action at the cellular level; this includes the question as to their role as neurotransmitters, neuromodulators, or neurohormones and also the possibility that they act directly on nonneural tissue. There is the additional problem of determining whether the peptide itself is the active substance, as opposed to its constituent amino acids. These difficulties notwithstanding, the evidence is clear that peptides have profound effects on feeding and drinking behavior, and thus, these substances, along with the amines and amino acids, must be considered as participants in the complex process of mediating energy and fluid homeostasis.

IX. Summary and Conclusions

The evidence reviewed in this chapter clearly indicates that responses involved in energy and fluid homeostasis are affected by numerous neurochemical substances endogenous to the brain, including the biogenic amines, amino acids, and peptides. The findings indicate that monoaminergic neurons, containing NE, DA, EPI, or 5-HT, may be involved in modulating or controlling feeding behavior, particularly within areas of the hypothalamus at the level of or rostral to the VMN. In contrast, ACh and possibly HA, acting specifically on circumventricular structures or areas adjacent to the third ventricle, appear to be more active in the control of fluid homeostasis. Adrenergic neurons may also help to regulate body fluids, possibly as part of a more complex series of events focused on integrating nutritional and hydrational needs. The possibility that adrenergic neurons, perhaps through GABAergic interneurons, provide an avenue of communication between metabolism and neuronal processes, may prove to be an important concept in defining the specific functions served by these hypothalamic neurochemical systems. Peptides present in the hypothalamus may also participate as neuroregulators in behavioral or physiological processes, although their relationship to the amines has yet to be determined.

It is evident that the idea of a single substance controlling a single behavior, a concept which dominated earlier research, has little value in explaining the operations of brain-behavioral mechanisms. Although at the experimental level, such focus on a single neuroregulator or neural system may be necessary and revealing of important insights, one must remember that neurons of the central nervous system are not isolated entities but are parts of complex and dynamic neuronal circuitries in which each component receives, elaborates, and exchanges information with a select number of other components involving several neurochemical substances even perhaps within the same neuron. In a given circuit, furthermore, the classic concept of synaptic contacts between presynaptic axon terminals and postsynaptic dendrites or soma can be considered only as a special aspect of a much more complex general picture which includes a vast neuronal network of dendrodendritic, somatodendritic, axosomatic, and axoaxonic chemical synapses. For example, a variety of histochemical, pharmacological, and electrophysiological evidence supports the view that dendrites, more than merely providing a passive postsynaptic receptor surface, may actively release excitatory and inhibitory transmitter substances. Through an intricate system of serial or reciprocal synapses, such dendrites can influence the very terminals which synapse on them, allowing for a fine level of local neural information processing. Presynaptic receptors, which are concerned with feedback inhibition of transmitter synthesis and/or release, are yet another aspect of the local neuronal circuitry which, through interconnecting projections, may have profound impact on distant areas of the brain.

These complexities in local synaptic function become magnified as we broaden our perspective to focus on the interactions between multiple neurotransmitter systems and brain areas. The coordinated activity of neuronal populations is clearly more significant to the behavioral functions of central networks than is the activity of any individual constituent neuron. Importantly, various mechanisms of interaction which are intrinsic to neuronal populations exert profound influences on the dynamic behavior of individual elements within the population and may markedly affect the behavior of such elements in response to pharmacological agents. Unfortunately, lack of specific knowledge on how multiple neurotransmitter systems are integrated in any given circuitry poses at once an obstacle and a challenge to the study of how neuroactive drugs modulate the synaptic network. To obtain such information, the neuropharmacologist is faced with the need to develop experimental approaches, since classic techniques are often inadequate. Such novel approaches must increasingly involve the combination of techniques from a variety of disciplines.

From the overview presented in this chapter, it becomes apparent that this area of research, on brain neurochemistry in relation to feeding, drinking and water-electrolyte excretion, has barely reached the point of examining and understanding the nature of interaction between the various neuroactive substances affecting these physiological and behavioral responses. However, with the recent development of sensitive biochemical and histochemical techniques that permit more precise characterization of discrete brain areas and neural projections, one is hopeful that the challenge of relating multiple neurochemical systems in control of behavior will be met. It is particularly encouraging that recent studies have revealed anatomical specificity for some neurochemically receptive units within the hypothalamus. Of equal importance is the initial progress that has been made towards characterizing a potential function for these neural processes and in defining the circuitry, involving extrahypothalamic and hypothalamic structures, which may underly this function. The tools and procedures used in these studies are generally crude and associated with several weaknesses and a potential for misleading results. Fortunately, the need for refining our techniques, as well as combining a variety of approaches within a given study, is becoming realized more and more each day. Advances in this realm, associated with the greater appreciation which appears to be emerging for an interaction between neural systems, should help us to progress towards a more accurate description of the precise role served by neuroregulatory substances in control of feeding and drinking behavior and related responses. A broader analysis along these lines, involving a variety of species, will also be important for bridging the gap between laboratory and clinical research. There are remarkable similarities which have been noted between the responses of human and nonhuman subjects to pharmacological agents, which gives us hope that our knowledge of central neurochemical systems in animals may be fruitfully applied towards the ultimate goal of understanding brain-behavioral functions in man.

Acknowledgments

The preparation of this review was supported by National Institute of Mental Health grant MH 22879, by an Alfred P. Sloan Fellowship, and by funds from the Whitehall Foundation.

Abbreviations

Selected abbreviations are listed here. See cited references for additional abbreviations.

A1-14	CA-containing cell groups according to Dahlstrom and Fuxe (1964)	CAI	capsula interna
		CC	crus cerebri
		CI	internal capsule
A	arcuate nucleus	CO	chiasma opticum
ah	anterior hypothalamic nucleus	cp	caudatus putamen
AHA	anterior hypothalamic area	CSDV	commissura supraoptica dorsalis, pars ventralis (Meynert)
AL	ansa lenticularis		
AQ	cerebral aqueduct		
ar	arcuate nucleus	DLF	dorsal longitudinal fasciculus

dmh	dorsomedial hypothalamic nucleus	nhp	nucleus hypothalamicus posterior
DPB	dorsal periventricular bundle	npe	nucleus periventricularis hypothalami
DPS	dorsal periventricular system	npv	nucleus paraventricularis
drn	dorsal raphe nucleus	ns V	spinal trigeminal nucleus
DSCP, DPCS	decussation of the superior cerebellar peduncles	nsc	nucleus suprachiasmaticus
DTB	dorsal tegmental bundle	nso	nucleus supraopticus
F, FX	fornix	NSP	nigrostriatal pathway
FLM	fasciculus longitudinalis medialis	nst	nucleus of the solitary tract
		nvm	nucleus ventromedialis
fm	nucleus paraventricularis, pars magnocellularis	OC	optic chiasm
		OT	optic tract
FMT	fasciculus mamillothalamicus	P	pyramidal tract
fp	nucleus paraventricularis, pars parvocellularis	PA	paraventricular hypothalamic nucleus
FR	fasciculus retroflexus	PCS	superior cerebellar peduncle
ha	anterior hypothalamic area	PE	periventricular hypothalamic nucleus
hd	dorsomedial hypothalamic nucleus	PG	periaqueductal gray matter
hl	lateral hypothalamic area	pv	periventricular hypothalamic nucleus
hp	posterior hypothalamic area		
hvm	nucleus ventromedialis	pvh, pvn	paraventricular hypothalamic nucleus
IC	internal capsule		
ip	interpeduncular nucleus	PVS	periventricular CA system
lc	locus ceruleus	r, RN	red nucleus
LM	lemniscus medialis	re	nucleus reuniens
MCP	medial cerebellar peduncle	SC	nucleus subceruleus
ME	median eminence	sc, SCN	suprachiasmatic nucleus
me V	mesencephalic trigeminal nucleus	SCP	superior cerebellar peduncle
		sfo	subfornical organ
MFB	medial forebrain bundle	SM	stria medullaris
ML	medial lemniscus	snc, SNp	substantia nigra, pars compacta
MLF	medial longitudinal fasciculus	snr, SNr	substantia nigra, pars reticulata
MP	mamillary peduncle	so	supraoptic nucleus
MT	mamillothalamic tract	soc	superior olivary complex
N III	oculomotor nerve	SOD	supraoptic decussations
N VII	facial nerve	TO	tractus opticus
n III	oculomotor nucleus	TR	tegmental CA radiations
n V	principal trigeminal nucleus	V III	third ventricle
n VII	facial nucleus	V IV	fourth ventricle
n X	dorsal motor nucleus of the vagus	VM, vmn	ventromedial hypothalamic nucleus
n XII	hypoglossal nucleus	VPS	ventral periventricular CA system
na	nucleus arcuatus		
ndm	nucleus dorsomedialis	VTA, vta	ventral tegmental area
nha	nucleus hypothalamicus anterior	vtn	ventral tegmental nucleus
		ZI, zi	zona incerta

References

Abdallah, A. H., White, H. D., and Kulkaini, A. S. (1974). Interaction of d-amphetamine with central nervous system depressants on food intake and spontaneous motor activity of mice. *Eur. J. Pharmacol. 26,* 119-121.

Ahlquist, R. P. (1967). Development of the concept of alpha and beta adrenotropic receptors. *Ann. N.Y. Acad. Sci. 139,* 549-552.

Ahlskog, J. E. (1974). Food intake and amphetamine anorexia after selective forebrain norepinephrine loss. *Brain Res. 82,* 211-240.

Ahlskog, J. E., and Hoebel, B. G. (1973). Overeating and obesity from damage to a noradrenergic system of the brain. *Science 182,* 166-169.

Ahn, H. S., and Makman, M. H. (1978). Serotonin sensitive adenylate cyclase activity in monkey anterior limbic cortex — antagonism by molindone and other antipsychotic drugs. *Life Sci. 23,* 507-512.

Albert, D. J., Storlien, L. H., Albert, J. G., and Mah, C. J. (1971). Obesity following disturbance of the ventromedial hypothalamus: a comparison of lesions, lateral cuts, and anterior cuts. *Physiol. Behav. 7,* 135-141.

Alexander, R. W., Davis, J. N., and Lefkowitz, R. J. (1975). Direct identification and characterization of β-adrenergic receptors in rat brain. *Nature (London) 258,* 437-440.

Andén, N.-E., Dhalström, A., Fuxe, K., and Hökfelt, T. (1966). The effect of haloperidol and chlorpromazine on the amine levels of central monoamine neurons. *Acta Physiol. Scand. 68,* 419-420.

Anderson, G. H. (1979). Control of protein and energy intake: Role of plasma amino acids and brain neurotransmitters. *Canad. J. Physiol. Pharm. 57,* 1043-1057.

Anderson, G. H., and Ashley, D. V. M. (1978). Plasma amino acids, brain mechanisms and the control of protein intake. In *Nutrition in Transition.* Proceedings of 5th Western Hemisphere Nutrition Congress, P. L. White and N. Selvey (Eds.), American Medical Assoc., Munroe, Wisconsin, pp. 237-246.

Andersson, B. (1975). The central control of water and salt balance. In *Neural Integration of Physiological Mechanisms and Behavior,* G. J. Mogenson and F. R. Calaresu (Eds.). University of Toronto Press, Toronto, pp. 213-225.

Andersson, B. (1977). Regulation of body fluids. *Ann. Rev. Physiol. 39,* 185-200.

Anlezark, G. M., Crow, T. J., and Greenway, A. P. (1973). Impaired learning and decreased cortical norepinephrine after bilateral locus coeruleus lesions. *Science 181,* 682-694.

Antelman, S. M., Black, C. A., and Rowland, N. E. (1977). Clozapine induces hyperphagia in undeprived rats. *Life Sci. 21,* 1747-1750.

Antin, J., Gibbs, J., Holt, J., Young, R. C., and Smith, G. P. (1975). Cholecystokinin elicits the complete behavioral sequence of satiety in rats. *J. Comp. Physiol. Psychol. 89,* 784-790.

Antunes-Rodrigues, J., and McCann, S. M. (1970). Water, sodium chloride, and food intake induced by injections of cholinergic and adrenergic drugs into the third ventricle of the rat brain. *Proc. Soc. Exp. Biol. Med. 133,* 1464-1470.

Antunes-Rodrigues, J., Gentil, C. G., Negro-Vilar, A., and Covian, M. R. (1970). Role of adrenals in the changes of sodium chloride intake following lesions in the central nervous system. *Physiol. Behav. 5,* 89-93.

Aprison, M. H. (1978). Glycine as a neurotransmitter. In *Psychopharmacology: A*

Generation of Progress, M. A. Lipton, A. DiMascio, and K. F. Killam (Eds.). Raven, New York, pp. 333-346.

Aravich, P. F., Sclafani, A., and Leibowitz, S. F. (1978). The effects of knife cuts on eating elicited by noradrenergic stimulation of the hypothalamus. *Neurosci. Abstr. 4,* 171.

Armstrong, S., and Singer, G. (1974). Effects of intrahypothalamic administration of norepinephrine on the feeding response of the rat under conditions of light and darkness. *Pharmacol. Biochem. Behav. 2,* 811-815.

Ash, A. S. F., and Schild, H. O. (1966). Receptors mediating some actions of histamine. *Br. J. Pharmacol. 27,* 427-439.

Ashley, D. V. M., Coscina, D. V., and Anderson, G. H. (1979). Selective decrease in protein intake following brain serotonin depletion. *Life Sci. 24,* 973-984.

Azmitia, E. C., Jr. (1978). The serotonin-producing neurons of the midbrain median and dorsal raphe nuclei. In *Handbook of Psychopharmacology,* Vol. 9, L. L. Iversen, S. D. Iversen, and S. H. Snyder (Eds.). Plenum, New York, pp. 233-314.

Baez, L. A. (1974). Role of catecholamines in the anorectic effects of amphetamine in rats. *Psychopharmacologia 35,* 91-98.

Baile, C. A. (1974). Putative neurotransmitters in the hypothalamus and feeding. *Fed. Proc. 33,* 1166-1175.

Ban, T. (1966). The septo-preoptico-hypothalamic system and its autonomic function. In. *Progress in Brain Research,* Vol. 21A, T. Tokizane and J. P. Schacte (Eds.). Elsevier, Amsterdam, pp. 1-43.

Banerjee, S. P., Sharma, V. K., Kung, L. S., and Chanda, S. K. (1978). Amphetamine induces β-adrenergic receptor supersensitivity. *Nature (London) 271,* 380-381.

Barker, J. L., Crayton, J. W., and Nicoll, R. A. (1971). Supraoptic neurosecretory cells – adrenergic and cholinergic sensitivity. *Science 171,* 208-210.

Barzaghi, R., Gropetti, A., Mantegazza, P., and Muller, E. E. (1973). Reduction of food intake by apomorphine: A pimozide-sensitive effect. *J. Pharm. Pharmacol. 25,* 909-911.

Bates, D., Weinshilboum, R. M., Campbell, R. J., and Sundt, T. M., Jr. (1977). The effect of lesions in the locus coeruleus on the physiological responses of the cerebral blood vessels in cats. *Brain Res. 136,* 431-443.

Bennett, C. T., and Pert, A. (1974). Antidiuresis produced by injections of histamine into the cat supraoptic nucleus. *Brain Res. 78,* 151-156.

Berthoud, H. R., and Mogenson, G. J. (1977). Ingestive behavior after intracerebral and intracerebroventricular infusions of glucose and 2-deoxy-D-glucose. *Am. J. Physiol. 233,* R127-R133.

Bhargava, K. P., Kulshrestha, V. K., and Srivastava, Y. P. (1972). Central cholinergic and adrenergic mechanisms in the release of ADH. *Br. J. Pharmacol. 44,* 617-627.

Bhargava, K. P., Kulshrestha, V. K., Santhakumari, G., and Srivastava, Y. P. (1973). Mechanism of histamine-induced antidiuretic response. *Br. J. Pharmacol. 47,* 700-706.

Biggio, G., Porceddu, M. L., Fratta, W., and Gessa, G. L. (1977). Changes in dopamine metabolism associated with fasting and satiation. In *Advances in Biochemical Psychopharmacology,* Vol. 16, *Nonstriatal Dopaminergic Neurons,* E. Costa and G. L. Gessa (Eds.). Raven, New York, pp. 377-380.

Bisset, G. -W. (1968). The milk ejection reflex and the actions of oxytocin, vasopressin and synthetic analogues on the mammary gland, neurohypophysial hormones and similar peptides. In *Handbook of Experimental Pharmacology,* B. Berde (Ed.). Springer-Verlag, New York, pp. 483-492.

Björklund, A., Falck, B., Nobin, A., and Stenevi, U. (1973). Organization of the dopamine noradrenaline innervations of the median eminence-pituitary region in the rat. In *Neurosecretion – The Final Neuroendocrine Pathway,* F. Knowles and L. Vollratn (Eds.). Springer, New York, pp. 209-222.

Björklund, A., Lindvall, O., and Nobin, A. (1975). Evidence of an incerto-hypothalamic dopamine neurone system in the rat. *Brain Res. 89,* 29-42.

Black, J. W., Duncan, W. A. M., Durant, D. J., Ganellin, C. R., and Parsons, E. M. (1972). Definition and antagonism of histamine H_2 receptors. *Nature (London) 236,* 385-390.

Blackmore, W. P., and Cherry, G. R. (1955). Antidiuretic action of histamine in the dog. *Am. J. Physiol. 180,* 596-598.

Block, M. L., and Fisher, A. E. (1975). Cholinergic and dopaminergic blocking agents modulate water intake elicited by deprivation, hypovolemia, hypertonicity, and isoproterenol. *Pharmacol. Biochem. Behav. 3,* 251-262.

Bloom, F. E. (1975). Amine receptors in CNS. I. Norepinephrine. In *Handbook of Psychopharmacology.* Vol. 6, *Biogenic Amine Receptors,* L. L. Iversen, S. D. Iversen, and S. H. Snyder (Eds.). Plenum, New York, pp. 1-22.

Bloom, F. E., Oliver, A. P., and Salmoiraghi, G. C. (1963). The responsiveness of individual hypothalamic neurons to microelectrophoretically administered endogenous amines. *Internatl. J. Neuropharmacol. 2,* 181-193.

Bloom, F. E., Hoffer, B. J., and Siggens, G. R. (1971). Studies on norepinephrine containing afferents to Purkinje cells of rat cerebellum. I. Localization of the fibers and their synapses. *Brain Res. 25,* 501-521.

Blumberg, J. B., Vetulani, J., Stawarz, R. J., and Sulser, F. (1976). The noradrenergic cyclic AMP generating system in the limbic forebrain: Pharmacological characterization in vitro and possible role of limbic noradrenergic mechanisms in the mode of action of antipsychotics. *Eur. J. Pharmacol. 37,* 357-366.

Blundell, J. E. (1977). Is there a role for serotonin (5-hyrdroxytryptamine) in feeding? *Int. J. Obesity 1,* 15-42.

Blundell, J. E., and Herberg, L. J. (1973). Primary polyuria accompanies "hunger" elicited by intrahypothalamic injection of noradrenaline. *Neuropharmacology 12,* 597-599.

Blundell, J. E., and Latham, C. J. (1978). Pharmacological manipulation of feeding behavior: possible influences of serotonin and dopamine on food intake. In *Central Mechanisms of Anorectic Drugs,* S. Garattini and R. Samanin (Eds.). Raven, New York, pp. 83-109.

Blundell, J. E., and Leshem, M. B. (1973). Dissociation of the anorexic effects of amphetamine and fenfluramine following intrahypothalamic injection. *Br. J. Pharmacol. 47,* 183-185.

Blundell, J. E., and Leshem, M. B. (1974). Central action of anorexic agents: effects of amphetamine and fenfluramine in rats with lateral hypothalamic lesions. *Eur. J. Pharmacol. 28,* 81-88.

Blundell, J. E., and Leshem, M. B. (1975). The effect of 5-hydroxytryptophan on food intake and on anorexic action of amphetamine and fenfluramine. *J. Pharm. Pharmacol.* 27, 31-37.

Bockaert, J., Premont, J., Glowinski, J., Tassin, J. P., and Thierry, A. M. (1977). Topographic distribution and characteristics of dopamine and β-adrenergic sensitive adenylate cyclases in the rat frontal cerebral cortex, striatum and substantia nigra. In *Nonstriatal, Dopaminergic Neurons,* E. Costa and G. L. Gessa (Eds.). Raven, New York, pp. 29-37.

Booth, D. A. (1967). Localization of the adrenergic feeding system in the rat diencephalon. *Science 158,* 515-517.

Booth, D. A. (1968). Mechanism of action of norepinephrine in eliciting an eating response on injection into the rat hypothalamus. *J. Pharmacol. Exp. Ther. 160,* 336-348.

Booth, D. A., and Quartermain, D. (1965). Taste sensitivity of eating elicited by chemical stimulation of the rat hypothalamus. *Psychon. Sci. 3,* 525-526.

Bradley, P. B., Wolstencroft, J. H., Hösli, L., and Avanzino, G. L. (1966). Neuronal basis for the central action of chlorpromazine. *Nature (London) 212,* 1425-1427.

Bray, G. A. (1978). *Recent Advances in Obesity Research: II.* Newman, London.

Breckenridge, B. McL. (1964). The measurement of cyclic adenylate in tissues. *Proc. Natl. Acad. Sci. U.S.A. 52,* 1580-1586.

Breese, G. R., Smith, R. D., Cooper, B. R., and Grant, L. D. (1973). Alterations in consummatory behavior following intracisternal injection of 6-hydroxydopamine. *Pharmacol. Biochem. Behav. 1,* 319-328.

Bridges, T. E., and Thorn, N. A. (1970). The effect of autonomic blocking agents on vasopressin release in vivo induced by osmoreceptor stimulation. *J. Endocrinol. 48,* 265-276.

Bridges, T. E., Hillhouse, E. W., and Jones, M. J. (1976). The effect of dopamine on neurohypophysial hormone release in vivo and from the rat neural lobe and hypothalamus in vitro. *J. Physiol. 260,* 647-666.

Brobeck, J. R. (1946). Mechanisms of the development of obesity in animals with hypothalamic lesions. *Psychol. Rev. 26,* 541-559.

Broekkamp, C. L. E., Honig, W. M. M., Pauli, A. I., and Van Rossum, J. M. (1974). Pharmacological suppression of eating behavior in relation to diencephalic noradrenergic receptors. *Life Sci. 14,* 473-481.

Broekkamp. C. L. E., Weemaes, A. J. M., and Van Rossum, J. M. (1975). Does fenfluramine act via nonfenfluramine? *J. Pharm. Pharmacol. 27,* 129-130.

Brownstein, M. (1977). Neurotransmitters and hypothalamic hormones in the central nervous system. *Fed. Proc. 36,* 1960-1963.

Buerger, P. B., Levitt, R. A., and Irwin, D. A. (1973). Chemical stimulation of the brain: Relationship between neural activity and water ingestion in the rat. *J. Comp. Physiol. Psychol. 82,* 278-285.

Buggy, J. (1978). Block of cholinergic-induced thirst after obstruction of anterior ventral third ventricle or periventricular preoptic ablation. *Neurosci. Abstr. 4,* 172.

Buggy, J., and Fisher, A. E. (1974). Evidence for a dual central role for angiotensin in water and sodium intake. *Nature (London) 250,* 733-735.

Buggy, J., and Fisher, A. E. (1976). Anteroventral third ventricle site of action for angiotensin-induced thirst. *Pharmacol. Biochem. Behav. 4,* 651-660.

Buggy, J., and Johnson, A. K. (1977). Preoptic-hypothalamic periventricular lesions: thirst deficits and hypernatremia. *Am. J. Physiol. 233,* R44-R52.

Burkard, W. P. (1978). Histamine H_2-receptor binding with ^{3}H-cimetidine in brain. *Eur. J. Pharmacol. 50,* 449-450.

Burt, D. R., Creese, I., and Snyder, S. H. (1976). Properties of [^{3}H]haloperidol and [^{3}H]dopamine binding associated with dopamine receptors in calf brain membranes. *Mol. Pharmacol. 12,* 800-812.

Butcher, L. L. (1977). Minireview — nature and mechanisms of cholinergic-monoaminergic interactions in brain. *Life Sci. 21,* 1207-1226.

Bylund, D. B., and Snyder, S. H. (1976). Beta-adrenergic receptor binding in membrane preparations from mammalian brain. *Mol. Pharmacol. 12,* 568-580.

Calcutt, C. R. (1976). The role of histamine in the brain. *Gen. Pharmacol. 7,* 15-25.

Camargo, L. A. A., Saad, W. A., Silva Netto, C. R., Gentil, C. G., Antunes-Rodrigues, J., and Covian, M. R. (1976). Effects of catecholamines injected into the septal area of the rat brain on natriuresis, kaliuresis and diuresis. *Can. J. Physiol. Pharmacol. 54,* 219-228.

Carey, R. J. (1976). Effects of selective forebrain depletions of norepinephrine and serotonin on the activity and food intake effects of amphetamine and fenfluramine. *Pharmacol. Biochem. Behav. 5,* 519-523.

Carey, R. J., and Goodall, E. B. (1975). Attenuation of amphetamine anorexia by unilateral nigra striatal lesions. *Neuropharmacology 14,* 827-834.

Carlisle, H. J. (1964). Differential effects of amphetamine on food and water intake in rats with lateral hypothalamic lesions. *J. Comp. Physiol. Psychol. 58,* 47-54.

Carlsson, A. (1970). Amphetamine and brain catecholamines. In *International Symposium on Amphetamines and Related Compounds,* E. Costa and S. Garattini (Eds.). Raven, New York, pp. 289-300.

Carmona, A., and Slangen, J. L. (1973). Effects of chemical stimulation of the hypothalamus upon gastric secretion. *Physiol. Behav. 10,* 657-661.

Carmona, A., and Slangen, J. L. (1974). EEG and heart rate changes elicited by chemical stimulation of the lateral hypothalamus. *Pharmacol. Biochem. Behav. 2,* 531-536.

Chan, W. Y. (1965). Effects of neurohypophysial hormones and their deamino analogues on renal excretion of Na, K, and water in rats. *Endocrinology 77,* 1097-1104.

Chance, W. T., and Lints, C. E. (1977). Eating following cholinergic stimulation of the hypothalamus. *Physiol. Psychol. 5,* 440-444.

Chang, R. S. L., Tran, V. T., and Snyder, S. H. (1978). Histamine H_1-receptors in brain labeled with ^{3}H-mepyramine. *Eur. J. Pharmacol. 48,* 463-464.

Chiaraviglio, E. (1972). Mesencephalic influences on the intake of sodium chloride and water in the rat. *Brain Res. 44,* 73-82.

Chiaraviglio, E. (1976). Angiotensin-norepinephrine interaction on sodium intake. *Behav. Biol. 17,* 411-416.

Chiaraviglio, E., and Taleisnik, W. (1969). Water and salt intake induced by hypothalamic implants of cholinergic and adrenergic agents. *Am. J. Physiol. 216,* 1418-1422.

Childress, A. R. (1978). Substance P increase feeding and food-reinforced behavior in satiated rats. *Eastern Psychological Association, 49th Annual Meeting,* p. 186.

Clark, W. G., and Clausen, D. F. (1943). Dietary "self-selection" and appetites of untreated and treated adrenalectomized rats. *Am. J. Physiol. 139,* 70-79.

Clineschmidt, B. V. (1973). 5, 6-Dihydroxytryptamine: suppression of anorexigenic effect of fenfluramine. *Eur. J. Pharmacol. 24,* 405-409.

Clineschmidt, B. V., and Lotti, V. J. (1973). Histamine: intraventricular injection suppresses ingestive behavior of the cat. *Arch. Int. Pharmacodyn. 206,* 288-298.

Clineschmidt, B. V., McGuffin, J. C., and Werner, A. B. (1974). Role of monoamines in anorexigenic actions of fenfluramine, amphetamine, and parachloromethamphetamine. *Eur. J. Pharmacol. 27,* 313-332.

Clineschmidt, B. V., Zacchei, A. G., Totaro, J. A., Pflueger, A. B., McGuffin, J. C., and Wishovsky, T. I. (1978). Fenfluramine and brain serotonin. *Ann. N. Y. Acad. Sci. 305,* 222-241.

Collier, G., Leshner, A. I., and Squibb, R. L. (1969). Self-selection of natural and purified dietary protein. *Physiol. Behav. 4,* 83-86.

Conrad, L. C. A., and Pfaff, D. W. (1976). Efferents from medial basal forebrain and hypothalamus in the rat. II. An autoradiographic study of the anterior hypothalamus. *J. Comp. Neurol. 169,* 221-261.

Coons, E. E., and Quartermain, D. (1970). Motivational depression associated with norepinephrine-induced eating from hypothalamus: resemblance to the ventromedial hyperphagic syndrome. *Physiol. Behav. 5,* 687-692.

Coscina, D. V. (1978). Effects of central 5, 7-dihydroxytryptamine on the medial hypothalamic syndrome in rats. *Ann. N. Y. Acad. Sci. 305,* 727-644.

Coscina, D. V., Daniel, J. V., Li, P., and Warsh, J. J. (1978). Equivalence of intraventricular p-chlorophenylalanine and phenylalanine in producing hyperphagia and obesity in rats. *Neurosci. Abstr. 4,* 173.

Coscina, D. V., Godse, D. D., and Stancer, H. C. (1976). Neurochemical correlates of hypothalamic obesity in rats. *Behav. Biol. 16,* 365-372.

Coscina, D. V., Grant, L. D., Balagura, S., and Grossman, S. P. (1972). Hyperdipsia after serotonin-depleting midbrain lesions. *Nat. New Biol. 235,* 63-64.

Coscina, D. V., Rosenblum-Blinick, C., Godse, D. D., and Stancer, H. C. (1973). Consummatory behaviors of hypothalamic hyperphagic rats after control injection of 6-hydroxydopamine. *Pharmacol. Biochem. Behav. 1,* 629-642.

Costa, E., and Greengard, P. (Eds.) (1975). *Advances in Biochemistry and Pharmacology,* Vol. 14, *Mechanism of Action of Benzodiazepines.* Raven, New York.

Costa, E., Gnegy, M. E., and Uzunov, P. (1977). Regulation of dopamine receptor sensitivity by an endogenous protein activator of adenylate cyclase. *Arch. Pharmacol. 297* (Suppl. 1), S47-S48.

Costentin, J., Marcais, H., Protais, P., Maudry, M., Delabaume, S., Martres, M. P., and Schwartz, J. C. (1977). Rapid development of hypersensitivity of striatal dopamine receptors induced by alpha-methylparatyrosine and its prevention by protein synthesis inhibitors. *Life Sci. 21,* 307-314.

Coury, J. N. (1967). Neural correlates of food and water intake in the rat. *Science 156,* 1763-1765.

Covian, M. R., and Antunes-Rodrigues, J. (1963). Specific alterations in sodium chloride intake after hypothalamic lesions in the rat. *Am. J. Physiol. 205,* 922-926.

Covian, M. R., Antunes-Rodrigues, J., Gentil, C. G., Saad, W. A., Camargo, L. A., and Silva Netto, C. R. (1975). Central control of salt balance. In *Neural Integration of Physiological Mechanisms and Behavior,* G. J. Mogenson and F. R. Calaresu (Eds.). University of Toronto Press, pp. 267-282.

Creese, I., and Snyder, S. H. (1978). ^{3}H-Spiroperidol labels serotonin receptors in rat cerebral cortex and hippocampus. *Eur. J. Pharmacol. 49,* 201-202.

Creese, I., Brut, D. R., and Snyder, S. H. (1976). Dopamine receptor binding predicts clinical and pharmacological potencies of antischizophrenic drugs. *Science 192,* 481-483.

Cross, B. A., Dyball, R. E. J., Dyer, R. G., Jones, C. W., Lincoln, D. W., Morris, J. F., and Pickering, B. T. (1975). Endocrine neurons. *Recent Prog. Hormone Res. 31,* 243-294.

Cruce, J. A. F., Thoa, N. B., and Jacobowitz, D. M. (1976). Catecholamines in the brains of genetically obese rats. *Brain Res. 101,* 165-170.

Dahlström, A., and Fuxe, K. (1964). Evidence for the existence of monoamine-containing neurons in the central nervous system. I. Demonstration of monoamines in the cell bodies of brainstem neurons. *Acta Physiol. Scand. 61* (Suppl. 232), 1-55.

Davis, J. D., and Booth, D. A. (1974). Vagotomy in the rat reduces meal size of diets containing fat. *Physiol. Behav. 12,* 685-688.

Davis, J. R., and Keesey, R. E. (1971). Norepinephrine-induced eating − its hypothalamic locus and an alternative interpretation of action. *J. Comp. Physiol. Psychol. 77,* 394-402.

Davis, J. N., Strittmatter, W. J., Hoyler, E., and Lefkowitz, R. J. (1977). [^{3}H] Dihydroergocryptine binding in rat brain. *Brain Res. 132,* 327-336.

Defendini, R., and Zimmerman, E. A. (1978). The magnocellular neurosecretory system of the mammalian hypothalamus. In *The Hypothalamus,* S. Reichlin, R. J. Baldessarini, and J. B. Martin (Eds.). Raven, New York, pp. 137-152.

de Jong, A., Strubbe, J. H., and Steffens, A. B. (1977). Hypothalamic influence on insulin and glucagon release in the rat. *Amer. J. Physiol. 233,* E380-E388.

de la Torre, J. C., Surgeon, J. W., and Walker, R. R. (1977). Effects of locus coeruleus stimulation on cerebral blood flow in selected brain regions. *Acta Neurol. Scand. 56* (Suppl. 64), 104-105.

de Wied, D., and László, F. A. (1967). Effect of autonomic blocking agents on ADH-release induced by hyperosmoticity. *J. Endocrinol. 38,* xvi.

de Wied, D., Palkovits, M., Lee, T. C., VanderWal, B., and de Jong, W. (1972). Central nervous and pituitary control of aldosterone secretion in the rat. *Progr. Brain Res. 38,* 279-292.

Dobrzánski, S., and Doggett, N. S. (1976). On the relation between hypodipsia and anorexia induced by (+)-amphetamine in the mouse. *J. Pharm. Pharmacol. 28,* 922-924.

Dogterom, J., Van Wimersma, Tj. B., and de Wied, D. (1976). Histamine as an extremely potent releaser of vasopressin in the rat. *Experientia 32,* 659-660.

Dorn, J., Antunes-Rodrigues, J., and McCann, S. M. (1970). Natriuresis in the rat following intraventricular carbachol. *Am. J. Physiol. 219,* 1292-1298.

Drummond, G. I., and Bellward, G. (1970). Studies on phosphorylase b kinase from neural tissue. *J. Neurochem. 17,* 475-482.

Duke, H. N., Pickford, M., and Watt, J. A. (1950). The immediate and delayed effects of disopropyl-fluorophosphate injected into the supraoptic nuclei of dogs. *J. Physiol. 111,* 81-88.

Edwards, C., Nahorski, S. R., and Rogers, K. J. (1974). In vivo changes of cerebral cyclic adenosine 3',5'-monophosphate induced by biogenic amines: association with phosphorylase activation. *J. Neurochem. 22,* 565-572.

Ellison, G. D., and Bresler, D. E. (1974). Tests of emotional behavior in rats following depletion of norepinephrine, of serotonin, or of both. *Psychopharmacologia 34*, 275-288.

Eng, R., Gold, R. M., and Nunez, A. (1979). Elemental hypothalamic obesity after discrete lesions of the paraventricular nucleus. *Neurosci. Abstr. 5*, 216.

Epstein, A. (1959). Suppression of eating and drinking by amphetamine and other drugs in normal and hyperphagic rats. *J. Comp. Physiol. Psychol. 52*, 37-45.

Epstein, A. N. (1978). The neuroendocrinology of thirst and salt appetite. In *Frontiers in Neuroendocrinology*, Vol. 5, W. F. Ganong and L. Martini (Eds.). Raven, New York, pp. 101-134.

Evered, M. D., and Mogenson, G. J. (1976). Regulatory and secondary water intake in rats with lesions of the zona incerta. *Am. J. Physiol. 230*, 1049-1057.

Evered, M. D., Fitzsimons, J. T., and de Caro, G. (1977). Drinking behaviour induced by intracranial injections of eledoisin and substance P in the pigeon. *Nature (London) 268*, 332-333.

Evetts, K. D., Fitzsimons, J. T., and Setler, P. E. (1972). Eating caused by 6-hydroxy-dopamine-induced release of noradrenaline in the diencephalon of the rat. *J. Physiol. 223*, 35-47.

Fallon, J. H., and Moore, R. Y. (1978). Catecholamine innervation of the basal forebrain. IV. Topography of the dopamine projection to the basal forebrain and neostriatum. *J. Comp. Neurol. 180*, 545-580.

Fernstrom, J. E., and Wurtman, R. J. (1972). Brain serotonin content: Physiological regulation by plasma neutral amino acids. *Science 178*, 414-416.

Fibiger, H. C., Zis, A. P., and McGeer, E. G. (1973). Feeding and drinking deficits after 6-hydroxydopamine administration in the rat: similarities to the lateral hypothalamic syndrome. *Brain Res. 55*, 135-148.

Fisher, A. E. (1973). Relationships between cholinergic and other dipsogens in the central mediation of thirst. In *The Neuropsychology of Thirst: New Findings and Advances in Concepts*, A. N. Epstein, H. R. Kissileff, and E. Stellar (Eds.). Winston & Sons, Washington, D. C., pp. 243-278.

Fisher, A. E., and Coury, J. N. (1962). Cholinergic tracing of a central neural circuit underlying the thirst drive. *Science 138*, 691-693.

Fitzsimons, J. T. (1972). Thirst. *Physiol. Rev. 52*, 468-561.

Fitzsimons, J. T., and Le Magnen, J. (1969). Eating as a regulatory control of drinking in the rat. *J. Comp. Physiol. Psychol. 67*, 273-283.

Fonnum, F., and Storm-Mathisen, J. (1978). Localization of GABA-ergic neurons in the CNS. In *Handbook of Psychopharmacology*, Vol. 9, *Chemical Pathways in the Brain*, L. L. Iversen, S. D. Iversen, and S. H. Snyder (Eds.). Plenum, New York, pp. 357-401.

Fox, K. A., Kipp, S. C., and VanderWeele, D. A. (1976). Dietary self-selection following subdiaphragmatic vagotomy in the white rat. *Am. J. Physiol. 231*, 1790-1793.

Franklin, K. B., and Herberg, L. J. (1977). Amphetamine induces anorexia even after inhibition of noradrenaline synthesis. *Neuropharmacology 16*, 45-46.

Freed, W. J., Perlow, M. J., Carman, J. S., and Hyatt, R. J. (1978). Calcitonin and feeding. *Neurosci. Abstr. 4*, 174.

Freedman, R., and Hoffer, B. J. (1975). Phenothiazine antagonism of the noradrenergic inhibition of cerebellar Purkinje neurons. *J. Neurobiol. 6*, 277-278.

Fregly, M. J. (1978). Angiotensin-induced thirst: peripheral and central mechanisms. *Fed. Proc. 37*, 2667-2716.

Frey, H. H., and Schulz, R. (1973). On the central mediation of anorexigenic drug effects. *Biochem. Pharmacol. 22*, 3041-3050.

Friedman, E., Staff, N., and Gershon, S. (1973). Catecholamine synthesis and the regulation of food intake in the rat. *Life Sci. 12*, 317-326.

Fuller, R. W., Perry, K. W., and Molloy, B. B. (1974a). Effect of an uptake inhibitor of serotonin metabolism in rat brain. *Life Sci. 15*, 1161-1171.

Fuller, R. W., Perry, K. W., Snoddy, H. D., and Molloy, B. B. (1974b). Comparison of the specificity of 3(p-trifluoromethylphenoxy)-N-methyl-d-phenylpropylamine and chlorimipramine as amine uptake inhibitors in mice. *Eur. J. Pharmacol. 28*, 233-236.

Fuxe, K., and Jonsson, G. (1974). Further mapping of central 5-hydroxytryptamine neurons: studies with the neurotoxic dihydroxytryptamines. In *Advances in Biochemical Psychopharmacology*, Vol. 10, *Serotonin: New Vistas of Histochemistry and Pharmacology*, E. Costa, G. L. Gessa, and S. Merton (Eds.). Raven, New York, pp. 1-12.

Gale, K., Guidotti, A., and Costa, E. (1977). Dopamine sensitive adenylate cyclase locations in substantia nigra. *Science 195*, 503-505.

Ganten, D., Hutchinson, J. S., Schelling, P., Ganten, U., and Fischer, R. (1976). The isorenin angiotensin systems in extrarenal tissue. *Clin. Exp. Pharmacol. Physiol. 3*, 103-126.

Ganten, D., Fuxe, K., Phillips, M. I., Mann, J. F. E., and Ganten, U. (1978). The brain isorenin-angiotensin system: Biochemistry, localization, and possible role in drinking and blood pressure regulation. In *Frontiers in Neuroendocrinology*, Vol. 5, W. F. Ganong, and L. Martini (Eds.). Raven, New York, pp. 61-69.

Garattini, S., and Samanin, R. (Eds.) (1978). *Central Mechanisms of Anorectic Drugs*. Raven, New York.

Garattini, S., Pujol, J. F., and Samanin, R. (Eds.). (1978). *Interactions of Putative Neurotransmitters in the Brain*. Raven, New York.

Garay, K. F., and Leibowitz, S. F. (1974). Antidiuresis is produced by adrenergic receptor stimulation of the rat supraoptic nucleus. *Fed. Proc. 33*, 563.

Gerald, M. C., and Maickel, R. P. (1972). Studies on the possible role of brain histamine in behaviour. *Br. J. Pharmacol. 44*, 462-471.

Gibbs, J., Young, R. C., and Smith, G. P. (1973). Cholecystokinin decreases food intake in rats. *J. Comp. Physiol. Psychol. 84*, 488-495.

Gilman, A. G., and Schrier, B. K. (1972). Adenosine cyclic 3', -5'-monophosphate in fetal rat brain cell cultures. *Mol. Pharmacol. 8*, 410-416.

Ginsberg, M. (1968). Production, release, transportation and elimination of the neurohypophysial hormones. In *Handbook of Experimental Pharmacology*, Vol. 23, *Neurohypophysial Hormones and Similar Polypeptides*, B. Berde (Ed.). Springer, Berlin, pp. 286-371.

Glick, S. D., and Greenstein, S. (1973). Pharmacological inhibition of eating, drinking, and prandial drinking. *Behav. Biol. 8*, 55-61.

Glick, S. D., and Stanley, M. E. (1975). Neurochemical correlate of body weight in rats. *Brain Res. 96*, 153-155.

Glick, S. D., Greenstein, S., and Waters, D. H. (1973a). Ventromedial hypothalamic lesions and brain catecholamines. *Pharmacol. Biochem. Behav. 1,* 591-592.

Glick, S. D., Waters, D. H., and Milloy, S. (1973b). Depletion of hypothalamic norepinephrine by food deprivation and interaction with d-amphetamine. *Res. Commun. Chem. Pathol. Pharmacol. 6,* 775-778.

Glowinski, J. (1970). Effects of amphetamine on various aspects of catecholamine metabolism in the central nervous system of the rat. In *Amphetamines and Related Compounds,* E. Costa and S. Garattini (Eds.). Raven, New York, pp. 301-316.

Glowinski, J., and Iversen, L. L. (1966). Regional studies of catecholamines in the rat brain. I. The disposition of [^{3}H] norepinephrine, [^{3}H] dopamine and [^{3}H] dopa in various regions of the brain. *J. Neurochem. 13,* 655-669.

Gnegy, M. E., Costa, E., and Uzunov, P. (1976). The regulation of second messenger response elicited trans-synaptically: participation of the endogenous phosphodiesterase activator and protein kinase. *Proc. Natl. Acad. Sci. U.S.A. 73,* 352-353.

Gokhale, S. D., Gulati, O. D., and Parikh, H. M. (1964). An investigation of the adrenergic blocking action of chlorpromazine. *Br. J. Pharmacol. 23,* 508-520.

Gold, R. M. (1967). Aphagia and adipsia following unilateral and bilaterally asymmetrical lesions in rats. *Physiol. Behav. 2,* 211-220.

Gold, R. M. (1970). Hypothalamic hyperphagia produced by parasagittal knife cuts. *Physiol. Behav. 5,* 23-25.

Gold, R. M. (1973). Hypothalamic obesity: the myth of the ventromedial nucleus. *Science 182,* 488-490.

Gold, R. M., Jones, A. P., Sawchenko, P. E., and Kapatos, G. (1977). Paraventricular area: critical focus of a longitudinal neurocircuitry mediating food intake. *Physiol. Behav. 18,* 1111-1119.

Goldman, H. W., Lehr, D., and Friedman, E. (1971). Antagonistic effects of alpha- and beta-adrenergically coded hypothalamic neurons on consummatory behaviour in the rat. *Nature (London) 231,* 453-455.

Goldstein, D. J., and Halperin, J. A. (1977). Mast cell histamine and cell dehydration thirst. *Nature (London) 267,* 250-252.

Goudie, A. J., Thornton, E. W., and Wheeler, T. J. (1976). Effects of Lilly 110140, a specific inhibitor of 5-hydroxytryptamine uptake, on food intake and on 5-hydroxytryptophan-induced anorexia. Evidence for serotoninergic inhibition of feeding. *J. Pharm. Pharmacol. 28,* 318-320.

Grace, J. E. (1968). Central nervous system lesions and saline intake in the rat. *Physiol. Behav. 3,* 387-393.

Graff, H., and Stellar, E. (1962). Hyperphagia, obesity and finickiness. *J. Comp. Physiol. Psychol. 55,* 418-428.

Grandison, L., and Guidotti, A. (1977). Stimulation of food intake by muscimol and beta endorphin. *Neuropharmacology 16,* 533-536.

Green, J. P., Johnson, C. L., and Weinstein, H. (1978). Histamine as a neurotransmitter. In *Psychopharmacology: A Generation of Progress,* M. A. Lipton, A. DiMascio, and K. F. Killam (Eds.). Raven, New York, pp. 319-332.

Greenberg, D. A., U'Prichard, D. C., and Snyder, S. H. (1976). Alpha-noradrenergic receptor binding in mammalian brain: differential labeling of agonist and antagonist states. *Life Sci. 19,* 69-76.

Greengard, P. (1976). Possible role for cyclic nucleotides and phosphorylated membrane proteins in postsynaptic actions of neurotransmitters. *Nature (London) 260,* 101-108.

Grinker, J., Enns, M., and Nowlis, G. (1977). Two experimental examinations of sweet taste in genetic obesity. *Second International Congress on Obesity,* p. 13.

Grossman, S. P. (1962a). Direct adrenergic and cholinergic stimulation of hypothalamic mechanisms. *Am. J. Physiol. 202,* 872-882.

Grossman, S. P. (1962b). Effects of adrenergic and cholinergic blocking agents on hypothalamic mechanisms. *Am. J. Physiol. 202,* 1230-1236.

Grossman, S. P. (1975). Role of the hypothalamus in the regulation of food and water intake. *Psychol. Rev. 82,* 200-224.

Grossman, S. P. (1978). Correlative analyses of ingestive behavior and regional amine depletions after surgical transections of neural pathways in the mesencephalon, diencephalon, and striatum. In *Central Mechanisms of Anorectic Drugs,* S. Garattini and R. Samanin (Eds.). Raven, New York, pp. 1-37.

Grossman, S. P., and Hennessy, J.W. (1976). Differential effects of cuts through the posterior hypothalamus on food intake and body weight in male and female rats. *Physiol. Behav. 17,* 89-102.

Grzanna, R., Morrison, J. H., Coyle, J. T., and Mollier, M. E. (1977). The immunocytochemical demonstration of noradrenergic neurons in the rat brain: the use of homologous antiserum to dopamine-beta-hydroxylase. *Neurosci. Lett. 4,* 127-134.

Guideri, G., Barletta, M., Chau, R., Green, M., and Lehr, D. (1975). Method for the production of severe ventricular dysrhythmias in small laboratory animals. In *Recent Advances in Studies on Cardiac Structure and Metabolism,* Vol. 10, *The Metabolism of Contraction,* P. E. Roy and G. Rona (Eds.). University Park Press, Baltimore, pp. 661-679.

Gurman, Y., and Krausz, M. (1973). Drinking induced by dextran and histamine: relation to kidneys and renin. *Eur. J. Pharmacol. 23,* 256-263.

Haas, H. L., and Wolf, P. (1977). Central actions of histamine: microelectrophoretic studies. *Brain Res. 122,* 269-279.

Haase, H. J., and Janssen, P. A. J. (1965). *The Action of Neuroleptic Drugs.* North-Holland, Amsterdam.

Halperin, J. M., Stern, Y., Burt, B., Farrell, R., Pavlides, C., Steiner, S. S., and Ellman, S. J. (1978). Alterations in feeding behavior following dorsal brainstem lesions in rats. *Neurosci. Abstr. 4,* 174.

Hamberger, B. (1967). Reserpine-resistant uptake of catecholamines in isolated tissues of the rat. *Acta Physiol. Scand. Suppl. 295,* 1-56.

Hansen, M. G., and Whishaw, I. Q. (1973). The effects of 6-hydroxydopamine, dopamine, and *dl*-norepinephrine on food intake and water consumption, self-stimulation, temperature, and electroencephalographic activity in the rat. *Psychopharmacologia 29,* 33-44.

Hass, W. K., Hawkins, R. A., and Ransohoff, J. (1977). Cerebral blood flow, glucose utilization and oxidative metabolism after bilateral reticular formation lesions. *Acta Neurol. Scand. 56* (Suppl. 64), 240-241.

Hatton, G. I., and Armstrong, W. E. (1980). Hypothalamic function in the behavioral and physiological control of body fluids. In *Handbook of the Hypothalamus,* Vol. 3 (Part B), P. J. Morgane and J. Panksepp (Eds.). Marcel Dekker, New York, pp. 1-105.

Heffner, T. G., Zigmond, M. J., and Stricker, E. M. (1977). Effects of dopaminergic agonists and antagonists on feeding in intact and 6-hydroxydopamine-treated rats. *J. Pharmacol. Exp. Ther. 201,* 386-399.

Heinbecker, P., White, H. L., and Rolf, D. (1944). Experimental obesity in the dog. *Am. J. Physiol. 141,* 549-565.

Heller, A., and Hoffman, P. C. (1975). Neuronal control of neurochemical processes in the basal ganglia. In *Brain Mechanisms in Mental Retardation,* N. A. Buchwald and M. A. B. Brazier (Eds.). Academic, New York, pp. 205-218.

Hendler, N. H., and Blake, W. D. (1969). Hypothalamic implants of angiotensin II, carbachol, and norepinephrine on water and NaCl solution intake in rats. *Commun. Behav. Biol. 4,* 41-48.

Hendley, E. D., and Snyder, S. H. (1968). Relationship between the action of monoamine oxidase inhibitors on the noradrenaline uptake system and their antidepressant efficacy. *Nature (London) 220,* 1330-1331.

Herberg, L. J., and Franklin, K. B. J. (1972). Adrenergic feeding: its blockade or reversal by posterior VMH lesions: and a new hypothesis. *Physiol. Behav. 8,* 1029-1034.

Hoebel, B. G. (1977). The psychopharmacology of feeding. In *Handbook of Psychopharmacology,* Vol. 8, *Drugs, Neurotransmitters, and Behavior,* L. L. Iversen, S. D. Iversen, and S. H. Snyder (Eds.). Plenum, New York, pp. 55-129.

Hoebel, B. G., Zemlam, F. P., Trulson, M. E., MacKenzie, R. G., Ducret, R. P., and Norelli, C. (1978). Differential effects of p-chlorophenylalanine and 5, 7-dihydroxytryptamine on feeding rats. *Ann. N. Y. Acad. Sci. 305,* 590-594.

Hoffman, W. E., Phillips, M. I., and Schmid, P. (1977). The role of catecholamines in central antidiuretic and pressor mechanisms. *Neuropharmacology 16,* 563-569.

Hökfelt, T., Elde, R., Fuxe, K., Johansson, O., Ljungdahl, Å., Goldstein, M., Luft, R., Efendic, S., Nilsson, G., Terenius, L., Ganten, D., Jeffcoate, S. L., Rehfeld, J., Said, S., Perez de la Mora, M., Possani, L., Tapia, R., Teran, L., and Palacios, R. (1978). Aminergic and peptidergic pathways in the nervous system with special reference to the hypothalamus. In *The Hypothalamus,* S. Reichlin, R. J. Baldessarini, and J. B. Martin (Eds.). Raven, New York, pp. 69-135.

Holaday, J. W., Meyerhoff, J. L., and Natelson, B. H. (1977). Cortisol secretion and clearance in the rhesus monkey. *Endocrinology 100,* 1178-1185.

Hollister, A. S., Ervin, G. N., Copper, B. R., and Breese, G. R. (1975). The roles of monoamine neural systems in the anorexia induced by (+)-amphetamine and related compounds. *Neuropharmacology 14,* 715-723.

Hollister, L. E. (1965). Toxicity of psychotherapeutic drugs. *Practitioner 194,* 72-84.

Holtzman, S. G., and Jewett, R. E. (1971). The role of brain norepinephrine in the anorexic effects of dextroamphetamine and monoamine oxidase inhibitors in the rat. *Psychopharmacologia 22,* 151-161.

Hoover, D. B., Muth, E. A., and Jacobowitz, D. M. (1978). A mapping of the distribution of acetylcholine, choline acetyltransferase and acetylcholinesterase in discrete areas of rat brain. *Brain Res. 153,* 295-306.

Horn, A. S. (1976). Characteristics of transport in dopaminergic neurons. In *The Mechanism of Neuronal and Extraneuronal Transport of Catecholamines,* D. M. Paton (Ed.). Raven, New York, pp. 195-214.

Horn, A. S., and Phillipson, O. T. (1976). A noradrenaline sensitive adenylate cyclase in the rat limbic forebrain: Preparation properties and the effects of agonists, adrenolytics and neuroleptic drugs. *Eur. J. Pharmacol. 37,* 1-11.

Houpt, K. A., and Epstein, A. N. (1971). The complete dependence of beta-adrenergic drinking on the renal dipsogen. *Physiol. Behav. 7,* 897-902.

Hughes, I. E. (1978). The effect of amitriptyline on presynaptic mechanisms in noradrenergic nerves. *Br. J. Pharmacol. 63,* 315-321.

Hutchinson, R. R., and Renfrew, J. W. (1967). Modification of eating and drinking: interactions between chemical agent, deprivation state, and site of stimulation. *J. Comp. Physiol. Psychol. 63,* 408-416.

Hyppa, M., Lentinen, P., and Rinne, U. K. (1971). Effect of l-dopa on the hypothalamic, pineal and strial monoamines and on the sexual behavior of the rat. *Brain Res. 30,* 265-272.

Iversen, L. L. (1973). Catecholamine uptake processes. *Br. Med. Bull. 29,* 130-135.

Iversen, L. L. (1975). Dopamine receptors in the brain. *Science 188,* 1084-1089.

Iversen, L. L. (1978). Biochemical psychopharmacology of GABA. In *Psychopharmacology: A Generation of Progress,* M. A. Lipton, A. DiMascio, and R. F. Killam (Eds.). Raven, New York, pp. 25-38.

Iversen, L. L., and Salt, P. J. (1970). Inhibitions of catecholamine uptake by steroids in the isolated rat heart. *Br. J. Pharmacol. 40,* 528-530.

Iversen, S. D. (1977). Brain dopamine systems and behavior. In *Handbook of Psychopharmacology,* Vol. 8, *Drugs, Neurotransmitters, and Behavior,* L. L. Iversen, S. D. Iversen, and S. H. Snyder (Eds.). Plenum, New York, pp. 333-384.

Jackson, H. M., and Robinson, D. W. (1971). Evidence for hypothalamic adrenergic receptors involved in the control of food intake of the pig. *Br. Vet. J. 127,* 51-53.

Jacobowitz, D. M. (1975). Fluorescence microscopic mapping of CNS norepinephrine systems in the rat forebrain. In *Anatomical Neuroendocrinology,* W. E. Stumpf and L. D. Grant (Eds.). S. Karger, Basel, pp. 368-380.

Jacobowitz, D. M. (1978). Monoaminergic pathways in the central nervous system. In *Psychopharmacology: A Generation of Progress,* M. A. Lipton, A. DiMascio, and K. F. Killam (Eds.). Raven, New York, pp. 119-129.

Joseph, S. A., and Knigge, K. M. (1968). Effects of VMH lesions in adult and newborn guinea pigs. *Neuroendocrinology 3,* 309-331.

Joyce, D., and Mrosovsky, N. (1964). Eating, drinking and activity in rats following 5-hydroxytryptophan (5-HTP) administration. *Psychopharmacologia 5,* 417-423.

Kakimoto, Y., Nakajima, T., Tadesada, M., and Sano, I. (1964). Changes in carbohydrate metabolism of the mouse brain following the administration of 3, 4-dihydroxy-2-phenylalanine. *J. Neurochem. 11,* 431-437.

Kakolewski, J. W., Cox, V. C., and Valenstein, E. S. (1968). Short-latency antidiuresis following the initiation of food ingestion. *Science 16,* 458-460.

Kanarek, R. B., and Mayer, J. (1978). 2-Deoxy-D-glucose induced feeding: Relation to diet palatability. *Pharmacol. Biochem. Behav. 8,* 615-617.

Kanner, M. (1974). Neurochemistry of feeding in the diabetic rat. *XXVI International Congress of Physiological Sciences, Jerusalem,* p. 80.

Kapatos, G., and Gold, R. M. (1973). Evidence for ascending noradrenergic mediation of hypothalamic hyperphagia. *Pharmacol. Biochem. Behav. 1,* 81-87.

Karczmar, A. G., and Dun, N. J. (1978). Cholinergic synapses: physiological, pharmacological, and behavioral considerations. In *Psychopharmacology: A Generation of Progress*, M. A. Lipton, A. DiMascio, and K. F. Killam (Eds.). Raven, New York, pp. 293-305.

Keeler, R. (1959). Effect of hypothalamic lesions on renal excretion of sodium. *Am. J. Physiol. 197*, 847-849.

Keeler, R. (1972). Natriuresis after preoptic lesions in rats. *Can. J. Physiol. Pharmacol. 50*, 561-567.

Kelly, J. (1978). GABA: a possible neurochemical substrate for hypothalamic feeding mechanisms. Ph.D. thesis, University of Chicago.

Kelly, J., Alheid, G. F., Newberg, A., and Grossman, S. P. (1977). GABA stimulation and blockage in the hypothalamus and midbrain: effects on feeding and locomotor activity. *Pharmacol. Biochem. Behav. 7*, 537-541.

Kenney, N. J., McKay, L. D., and Woods, S. C. (1978). Effect of intraventricular β-endorphin on food intake. *Neurosci. Abstr. 4*, 176.

Kent, D. L., and Sladek, J. R., Jr. (1978). Histochemical, pharmacological and microspectrofluorometric analysis of new sites of serotonin localization in the rat hypothalamus. *J. Comp. Neurol. 180*, 221-236.

Kimura, H., and Kuryama, K. (1975). Distribution of gamma-aminobutyric acid (GABA) in the rat hypothalamus: Functional correlates with activities of appetite-controlling mechanisms. *J. Neurochem. 24*, 903-907.

Kissileff, H. R. (1968). Effect of water loads and ambient temperature change on temporal patterns of food and water intake in the rat. *Proc. III International Conference on Regulation of Food and Water Intake, Haverford, Pa.*

Kissileff, H. F. (1969). Food-associated drinking in the rat. *J. Comp. Physiol. Psychol. 67*, 284-300.

Kissileff, H. R., and Becker, E. (1974). Influence of lactation on meal taking patterns in rats. *Proc. XXVI International Congress of Physiological Sciences, Jerusalem, Satellite Symposia*, p. 84.

Kizer, J. S., Palkovits, M., and Brownstein, M. J. (1976). The projections of the A8, A9, and A10 dopaminergic cell bodies: evidence for a nigral-hypothalamic-median eminence dopaminergic pathway. *Brain Res. 108*, 363-370.

Kobayashi, R. M., Palkovits, M., Hruska, R. E., Rothschild, R., and Yamamura, H. I. (1978). Regional distribution of muscarinic cholinergic receptors in brain. *Brain Res. 154*, 13-23.

Koob, G. F., Sessions, G. R., Kant, G. J., and Meyerhoff, J. L. (1976). Dissociation of hyperdipsia from the destruction of the locus coeruleus in rats. *Brain Res. 116*, 339-345.

Kraly, F. S., Gibbs, J., and Smith, G. P. (1975). Disordered drinking after abdominal vagotomy in rats. *Nature (London) 258*, 226-228.

Krieger, D. T. (1974). Food and water restriction shifts corticosterone, temperature, activity and brain amine periodicity. *Endocrinology 95*, 1195-1201.

Krieger, D. T., and Ganong, W. F. (Eds.) (1977). ACTH and related peptides: Structure, regulation, and action. *Ann. N. Y. Acad. Sci. 297*, 1-664.

Krueger, B. K., Forn, J., and Greengard, P. (1975). Dopamine-sensitive adenylate cyclase and protein phosphorylation in the rat caudate nucleus. In *Pre- and Postsynaptic*

Receptors, E. Usdin and W. E. Bunney, Jr., (Eds.). Marcel Dekker, New York, pp. 123-147.

Kruk, Z. L. (1973). Dopamine and 5-hydroxytryptamine inhibit feeding in rats. *Nature (London) 246,* 52-53.

Kühn, E. R. (1973). The release of antidiuretic hormone and electrolyte excretion after injection of cholinergic and adrenergic agents into the third ventricle of the male rat. *7th Conference of European Comparative Endocrinologists,* P. 139.

Kühn, E. R. (1974). Cholinergic and adrenergic release mechanism for vasopressin in the male rat: a study with injections of neurotransmitters and blocking agents into the third ventricle. *Neuroendocrinology 16,* 255-264.

Kühn, E. R., and McCann, S. M. (1971). Release of oxytocin and vasopressin in lactating rats after injection of carbachol into the third ventricle. *Neuroendocrinology 8,* 48-58.

Kulkosky, P. J., Breckenridge, C., Krunsky, R., and Woods, S. C. (1976). Satiety elicited by the C-terminal octapeptide of cholecystokinin-pancreozymin in normal and VMH-lesioned rats. *Behav. Biol. 18,* 227-334.

Kuschinsky, W., and Wahl, M. (1978). Local chemical and neurogenic regulation of cerebral vascular resistance. *Physiol. Rev. 58,* 656-689.

Lanciault, G., Bonoma, C., and Brooks, F. P. (1973). Vagal stimulation, gastrin release, and acid secretion in anesthetized dogs. *Am. J. Physiol. 225,* 546-552.

Landis, A. M., Luduena, F. P., and Buzzo, H. J. (1967). Differentiation of receptors responsive to isoproterenol. *Life Sci. 6,* 2241-2249.

Landis, S. C., Shoemaker, W. J., Schlumpf, M., and Bloom, F. E. (1975). Catecholamines in mutant mouse cerebellum. Fluorescence microscopic and chemical studies. *Brain Res. 93,* 253-266.

Langelier, P., Roberge, A. G., Boucher, R., and Poirier, L. J. (1973). Effects of chronically administered *l*-DOPA in normal and lesioned cats. *J. Pharmacol. Exp. Ther. 187,* 15-26.

Larue, C. (1977). Different effects of continuously infused insulin on night and daytime meal patterns in rats. *6th International Conference on the Physiology of Food and Fluid Intake, Paris.*

Lavyne, M. H., Koltun, W. A., Clement, J. A., Rosene, D. L., Pickren, K. S., Zervas, N. T., and Wurtman, R. J. (1977). Decrease in neostriatal blood flow after D-amphetamine administration or electrical stimulation of the substantia nigra. *Brain Res. 135,* 77-86.

Lefkowitz, R. J. (1976). The β-adrenergic receptor. *Life Sci. 18,* 461-472.

Lehr, D. (1972). Isoproterenol and sudden death of asthmatic patients in ventricular fibrillation. *N. Engl. J. Med. 287,* 987-988.

Lehr, D. (1973). Invited comment: Comments to papers on "thirst" by Drs. Fisher, Harvey, and Setler. In *The Neuropsychology of Thirst: New Findings and Advances in Concepts,* A. N. Epstein, H. R. Kissileff, and E. Stellar (Eds.). Winston, Washington, D.C., pp. 307-315.

Lehr, D., and Goldman, W. (1973). Continued pharmacologic analysis of consummatory behavior in the albino rat. *Eur. J. Pharmacol. 23,* 197-210.

Lehr, D., Mallow, J., and Krukowski, M. (1967). Copious drinking and simultaneous inhibition of urine flow elicited by beta-adrenergic stimulation and contrary effect of alpha-adrenergic stimulation. *J. Pharmacol. Exp. Ther. 158,* 150-163.

Leibowitz, S. F. (1970a). Hypothalamic β-adrenergic "satiety" system antagonizes an α-adrenergic "hunger" system in the rat. *Nature (London) 226,* 963-964.

Leibowitz, S. F. (1970b). Reciprocal hunger-regulating circuits involving alpha- and beta-adrenergic receptors located, respectively, in the ventromedial and lateral hypothalamus. *Proc. Natl. Acad. Sci. U.S.A. 67,* 1063-1070.

Leibowitz, S. F. (1971). Hypothalamic alpha- and beta-adrenergic systems regulate both thirst and hunger in the rat. *Proc. Natl. Acad. Sci. U.S.A. 68,* 332-334.

Leibowitz, S. F. (1972a). Central adrenergic receptors and the regulation of hunger and thirst. In *Neurotransmitters,* I. J. Kopin (Ed.). *Res. Publ. ARNMD 50,* 327-358.

Leibowitz, S. F. (1972b). Hypothalamic alpha-adrenergic suppression of drinking: effects on several types of thirst. *Proc. 80th APA Annu. Convention,* pp. 845-846.

Leibowitz, S. F. (1973a). Brain norepinephrine and ingestive behaviour. In *Frontiers in Catecholamine Research,* E. Usdin and S. Snyder (Eds.). Pergamon, Oxford, pp. 711-713.

Leibowitz, S. F. (1973b). Alpha-adrenergic receptors mediate suppression of drinking induced by hypothalamic amphetamine injection. *Fed. Proc. 32,* 754.

Leibowitz, S. F. (1973c). Central histaminergic control of ingestive behavior in the rat. *Proc. APA 81st Annu. Convention,* pp. 1049-1050.

Leibowitz, S. F. (1973d). Histamine: A stimulatory effect on drinking behavior in the rat. *Brain Res. 63,* 440-444.

Leibowitz, S. F. (1975a). Pattern of drinking and feeding produced by hypothalamic norepinephrine injection in the satiated rat. *Physiol. Behav. 14,* 731-742.

Leibowitz, S. F. (1975b). Ingestion in the satiated rat: role of alpha and beta receptors in mediating effects of hypothalamic adrenergic stimulation. *Physiol. Behav. 14,* 743-754.

Leibowitz, S. F. (1975c). Amphetamine: Possible site and mode of action for producing anorexia in the rat. *Brain Res. 84,* 160-167.

Leibowitz, S. F. (1975d). Catecholaminergic mechanisms of the lateral hypothalamus: their role in the mediation of amphetamine anorexia. *Brain Res. 98,* 529-545.

Leibowitz, S. F. (1975e). Central neurochemical-neuroendocrine aspects of adrenergic suppression of drinking behavior in the rat. Paper presented at Eastern Psychological Association, New York.

Leibowitz, S. F. (1976). Brain catecholaminergic mechanisms for control of hunger. In *Hunger: Basic Mechanisms and Clinical Implications,* D. Novin, W. Wyrwicka, and G. Bray (Eds.). Raven, New York, pp. 1-18.

Leibowitz, S. F. (1977). Feeding elicited by noradrenergic stimulation of the paraventricular nucleus: meal size as related to varying plasma corticosterone levels. Paper presented at the VIth International Conference on the Physiology of Food and Fluid Intake, Paris, France.

Leibowitz, S. F. (1978a). Paraventricular nucleus: a primary site mediating adrenergic stimulation of feeding and drinking. *Pharmacol. Biochem. Behav. 8,* 163-175.

Leibowitz, S. F. (1978b). Adrenergic stimulation of the paraventricular nucleus and its effects on ingestive behavior as a function of drug dose and time of injection in the light-dark cycle. *Brain Res. Bull. 3,* 357-363.

Leibowitz, S. F. (1978c). Identification of catecholamine receptor mechanisms in the perifornical lateral hypothalamus and their role in mediating amphetamine and L-DOPA anorexia. In *Central Mechanisms of Anorectic Drugs,* S. Garattini and R. Samanin (Eds.). Raven, New York, pp. 39-82.

Leibowitz, S. F. (1979a). Histamine: Modification of behavioral and physiological components of body fluid homeostasis. In *Histamine Receptors,* T. O. Yellin (Ed.). Spectrum, New York, pp. 219-253.

Leibowitz, S. F. (1979b). Midbrain-hypothalamic catecholamine projection systems mediating feeding stimulation and inhibition in the rat. In *Catecholamines: Basic and Clinical Frontiers,* Vol. 2, E. Usdin, I. J. Kopin, and J. D. Barchas (Eds.). Pergamon, New York, pp. 1675-1677.

Leibowitz, S. F. (1979c). Functional and anatomical studies of noradrenergic system of the paraventricular hypothalamus that controls feeding behavior. *Neurosci. Abstr. 5,* 220.

Leibowitz, S. F., and Brown, L. L. (1980a). Histochemical and pharmacological analysis of noradrenergic projections to the paraventricular hypothalamus in relation to feeding stimulation. *Brain Res.,* in press.

Leibowitz, S. F., and Brown, L. L. (1980b). Histochemical and pharmacological analysis of catecholaminergic projections to the perifornical hypothalamus in relation to feeding inhibition. *Brain Res.,* in press.

Leibowitz, S. F., and Brown, L. L. (1980c). Analysis of behavioral deficits produced by lesions in the dorsal and ventral midbrain tegmentum. *Physiol. Behav.,* in press.

Leibowitz, S. F., and Miller, N. E. (1969). Unexpected adrenergic effect of chlorpromazine: Eating elicited by injection into rat hypothalamus. *Science 165,* 609-611.

Leibowitz, S. F., and Papadakos, P. J. (1978). Serotonin-norepinephrine interaction in the paraventricular nucleus: antagonistic effects on feeding behavior in the rat. *Neurosci. Abstr. 4,* 542.

Leibowitz, S. F., and Rossakis, C. (1978a). Analysis of feeding suppression produced by perifornical hypothalamic injection of catecholamines, amphetamines, and mazindol. *Eur. J. Pharmacol. 53,* 69-81.

Leibowitz, S. F., and Rossakis, C. (1978b). Pharmacological characterization of perifornical hypothalamic β-adrenergic receptors mediating feeding inhibition in the rat. *Neuropharmacology 17,* 691-702.

Leibowitz, S. F., and Rossakis, C. (1979a). Mapping study of brain dopamine- and epinephrine-sensitive sites which cause feeding suppression in the rat. *Brain Res. 172,* 101-113.

Leibowitz, S. F., and Rossakis, C. (1979b). Pharmacological characterization of perifornical hypothalamic dopamine receptors mediating feeding inhibition in the rat. *Brain Res. 172,* 115-130.

Leibowitz, S. F., and Rossakis, C. (1979c). L-DOPA feeding suppression: effect on catecholamine neurons of the perifornical lateral hypothalamus. *Psychopharmacology 61,* 273-280.

Leibowitz, S. F., Chang, K., and Oppenheimer, R. L. (1976). Feeding elicited by noradrenergic stimulation of the paraventricular nucleus: effects of corticosterone and other hormone manipulations. *Neurosci. Abstr. 2,* 292.

Leibowitz, S. F., Arcomano, A., and Hammer, N. J. (1978a). Tranylcypromine: stimulation of eating through α-adrenergic neuronal system in the paraventricular nucleus. *Life Sci. 23,* 749-758.

Leibowitz, S. F., Arcomano, A., and Hammer, N. J. (1978b). Potentiation of eating associated with tricyclic antidepressant drug activation of α-adrenergic neurons in the paraventricular hypothalamus. *Progr. Neuro-Psychopharmacol. 2,* 349-358.

Le Magnen, J. (1971). Advances in studies on the physiological control and regulation of food intake. In *Progress in Physiological Psychology,* Vol. 4, E. Stellar and J. Sprague (Eds.). Academic, New York, pp. 203-261.

Le Moal, M., Stinus, L., Simon, H., Tassin, J. P., Thierry, A. M., Blanc, G., Glowinski, J., and Cardo, B. (1977). Behavioral effects of a lesion in the ventral mesencephalic tegmentum: evidence for involvement of A10 dopaminergic neurons. In *Advances in Biochemical Psychopharmacology,* Vol. 16, *Nonstriatal Dopaminergic Neurons,* E. Costa and G. L. Gessa (Eds.). Raven, New York, pp. 237-245.

Leshner, A. I., Collier, G. H., and Squibb, R. L. (1971). Dietary self-selection at cold temperatures. *Physiol. Behav. 6,* 1-3.

Leverenz, K., Redmond, D. E., Jr., and Huang, Y. H. (1978). Suppression of feeding behavior in food-deprived monkeys by locus coeruleus stimulation. *Neurosci. Abstr. 4,* 177.

Levin, R. M., and Weiss, B. (1975). Mechanisms by which psychotropic drugs inhibit adenosine cyclic $3',5'$-monophosphate phosphodiesterase of brain. *Mol. Pharmacol. 12,* 581-589.

Levin, R. M., and Weiss, B. (1977). Binding of trifluoperazine to the calcium-dependent activator of cyclic nucleotide phosphodiesterase. *Mol. Pharmacol. 13,* 690-697.

Levitsky, S. A. (1970). Feeding patterns of rats in response to fasts and changes in environmental conditions. *Physiol. Behav. 5,* 291-300.

Levitt, R. A. (1971). Cholinergic substrate for drinking in the rat. *Psychol. Rep. 29,* 431-448.

Lewis, P. R., and Shute, C. C. D. (1978). Cholinergic pathways in CNS. In *Handbook of Psychopharmacology,* Vol. 9, *Chemical Pathways in the Brain,* L. L. Iversen, S. D. Iversen, and S. H. Snyder (Eds.). Plenum, New York, pp. 315-355.

Lindvall, O., and Björklund, A. (1974a). The glyoxylic acid fluorescence histochemical method: A detailed account of the methodology for the visualization of central catecholamine neurons. *Histochemistry 39,* 97-127.

Lindvall, O., and Björklund, A. (1974b). The organization of the ascending catecholamine neuron systems in the rat brain as revealed by the glyoxylic acid fluorescence method. *Acta Physiol. Scand. Suppl. 412,* 1-48.

Lindvall, O., and Björklund, A. (1978). Organization of catecholamine neurons in the rat central nervous system. In *Handbook of Psychopharmacology,* Vol. 9, *Chemical Pathways in the Brain,* L. L. Iversen, S. D. Iversen, and S. H. Snyder (Eds.). Plenum, New York, pp. 139-231.

Lipton, M. A., Di Mascio, A., and Killam, K. F. (1978). *Psychopharmacology: A Generation of Progress,* Raven, New York.

Lloyd, K. G., and Bartholini, G. (1975). The effect of drugs on the release of endogenous catecholamines into the perfusate of discrete brain areas of the cat in vivo. *Experientia 31,* 560-562.

Lorden, J., Oltmans, G. A., and Margules, D. L. (1976). Central noradrenergic neurons: Differential effects on body weight of electrolytic and 6-hydroxydopamine lesions in rats. *J. Comp. Physiol. Psychol. 90,* 144-155.

Lorens, S. A., and Yunger, L. M. (1974). Morphine analgesia, two-way avoidance, and consummatory behavior following lesions in the midbrain raphe nuclei of the rat. *Pharmacol. Biochem. Behav. 2,* 215-221.

Lorenz, D., Nardi, P., and Smith, G. P. (1978). Atropine methyl nitrate inhibits sham feeding in the rat. *Pharmacol. Biochem. Behav. 8,* 405-407.

Lovenberg, W., Weissbach, H., and Udenfriend, S. (1962). Aromatic l-amino acid decarboxylase. *J. Biol. Chem. 237,* 89-93.

Lovett, D., and Singer, G. (1971). Ventricular modification of drinking and eating behavior. *Physiol. Behav. 6,* 23-26.

Lyon, M., Halpern, M., and Mintz, E. (1968). The significance of the mesencephalon for coordinated feeding behavior. *Acta Neurol. Scand. 44,* 323-346.

Lytle, L. D. (1977). Control of eating behavior. In *Nutrition and the Brain,* Vol. 2, R. J. Wurtman and J. J. Wurtman (Eds.). Raven, New York, pp. 1-145.

McCaleb, M. L., Myers, R. D., Singer, G., and Willis, G. (1978). Hypothalamic norepinephrine in the rat during feeding and push-pull perfusion with glucose, 2-DG, or insulin. *Am. J. Physiol. 236,* 312-321.

McCulloch, J., and Harper, A. M. (1977). Cerebral circulatory and metabolic changes following amphetamine administration. *Brain Res. 121,* 196-199.

MacKenzie, E. T. (1976). Amine mechanisms in the cerebral circulation. *Pharmacol. Rev. 28,* 275-353.

MacKenzie, E. T., McCulloch, J., O'Keane, M., Pickard, J. D., and Harper, A. M. (1976a). Cerebral circulation and norepinephrine: relevance of the blood-brain barrier. *Am. J. Physiol. 231,* 483-488.

MacKenzie, E. T., McCulloch, J., and Harper, A. M. (1976b). Influence of endogenous norepinephrine on cerebral blood flow and metabolism. *Am. J. Physiol. 231,* 489-494.

McNeill, T. H., and Sladek, J. R., Jr. (1978). Comparative catecholamine-neurophysin morphology in the rat supraoptic and paraventricular nuclei. *Neurosci. Abstr. 4,* 279.

Maddison, S. (1977). Intraperitoneal and intracranial cholecystokinin depress operant responding for food. *Physiol. Behav. 19,* 819-824.

Mangiapane, M. L., and Simpson, J. B. (1978). Subfornical organ: site of pressor and drinking actions of acetylcholine. *Neurosci. Abstr. 4,* 547.

Mangiapane, M. L., and Simpson, J. B. (1979). Pharmacologic independence of subfornical organ receptors mediating drinking. *Brain Res. 178,* 507-517.

Margules, D. L. (1970a). Alpha-adrenergic receptors in hypothalamus for the suppression of feeding behavior by satiety. *J. Comp. Physiol. Psychol. 73,* 1-12.

Margules, D. L. (1970b). Beta-adrenergic receptors in the hypothalamus for learned and unlearned taste aversions. *J. Comp. Physiol. Psychol. 73,* 13-21.

Margules, D. L., Lewis, M. J., Dragonica, J. A., and Margules, A. S. (1972). Hypothalamic norepinephrine: Circadian rhythms and the control of feeding behavior. *Science 178,* 640-642.

Margules, D. L., Moisset, B., Lewis, M. J., Shibuya, H., and Pert, C. B. (1978). β-Endorphin is associated with overeating in genetically obese mice (*ob/ob*) and rats (*fa/fa*). *Science 202,* 988-991.

Marshall, J. F. (1978). The role of central catecholamine-containing neurons in food intake. *Recent Advances in Obesity Research: II,* G. Bray (Ed.). Newman, London, pp. 6-16.

Marshall, J. F., and Teitelbaum, P. (1977). New considerations in the neuropsychology of motivated behavior. In *Handbook of Psychopharmacology,* Vol. 7, L. L. Iversen, S. D. Iversen, and S. H. Snyder (Eds.). Plenum, New York, pp. 201-229.

Martin, G. E., and Myers, R. D. (1975). Evoked release of [^{14}C]norepinephrine from the rat hypothalamus during feeding. *Am. J. Physiol. 229,* 1547-1555.

Martin, G. E., and Myers, R. D. (1976). Dopamine efflux from the brain stem of the rat during feeding, drinking, and lever-pressing for food. *Pharmacol. Biochem. Behav. 4,* 551-560.

Martin, G. E., Myers, R. D., and Newberg, D. C. (1976). Catecholamine release by intracerebral perfusion of 6-hydroxydopamine and desipramine. *Eur. J. Pharmacol. 36,* 299-311.

Massari, W. J., Tizabi, Y., and Jacobowitz, D. M. (1977). The effect of lesions of the ventral noradrenergic bundle on serotonin and norepinephrine concentration in discrete nuclei. *Neurosci. Abstr. 3,* 255.

Matthews, J. W., Booth, D. A., and Stolerman, I. P. (1978). Factors influencing feeding elicited by intracranial noradrenaline in rats. *Brain Res. 141,* 119-128.

Meeker, R., and Myers, R. D. (1979). In vivo ^{14}C-amino acid profiles in discrete hypothalamic regions during push-pull perfusion in the unrestrained rat. *Neuroscience 4,* 495-506.

Meeker, R., and Panksepp, J. (1977). The role of GABA in ventromedial hypothalamic regulation of food intake. Paper presented at 6th International Conference on the Physiology of Food and Fluid Intake, Paris, France.

Meeker, R. B., McCaleb, M. L., and Myers, R. D. (1977). Changes in catecholaminergic and amino acid activity in the hypothalamus during feeding. Paper presented at the 6th International Conference on Physiology of Food and Fluid Intake, Paris, France.

Meyer, D. R., Abele, M., and Hertting, G. (1974). Influence of serotonin on water intake and the renin-angiotensin system in the rat. *Arch. Int. Pharmacodyn. 212,* 130-140.

Middlemiss, D. N., Blakeborough, L., and Leather, S. R. (1977). Direct evidence for an interaction of β-adrenergic blockers with the 5-HT receptor. *Nature (London) 267,* 289-290, 1977.

Miller, N. E., Gottesman, K. S., and Emery, N. (1964). Dose response to carbachol and norepinephrine in rat hypothalamus. *Am. J. Physiol. 206,* 1384-1388.

Mills, E., and Wang, S. C. (1964). Liberation of antidiuretic hormone: pharmacologic blockade of ascending pathways. *Am. J. Physiol. 207,* 1405-1410.

Milton, A. S., and Paterson, A. T. (1973). Intracranial injections of 6-hydroxydopamine (6-OH-DA) in cats: Effects on the release of antidiuretic hormone. *Brain Res. 61,* 423-427.

Milton, A. S., and Paterson, A. T. (1974). A microinjection study of the control of antidiuretic hormone release by the supraoptic nucleus of the hypothalamus in the cat. *J. Physiol. 241,* 607-628.

Miselis, R. R., Shapiro, R. E., and Hand, P. J. (1978). Subfornical organ efferent and supraoptic nucleus afferent connections and their implications in the control of water balance. *Neurosci. Abstr. 4,* 179.

Moberg, G. P., Bellinger, L. L., and Mendel, V. E. (1975). Effect of meal feeding on daily rhythms of plasma corticosterone and growth hormone in the rat. *Neuroendocrinology 19,* 160-169.

Montgomery, R. B., Singer, G., and Purcell, A. T. (1971). The effects of intrahypothalamic injections of desmethylimipramine on food and water intake of the rat. *Psychopharmacologia 19,* 81-86.

Moore, R. Y., and Bloom, F. E. (1978). Central catecholamine neuron systems: Anatomy and physiology of the dopamine systems. *Ann. Rev. Neurosci. 1,* 129-169.

Moore, R. Y., and Bloom, F. E. (1979). Central catecholamine neuron systems: Anatomy and physiology of the norepinephrine and epinephrine systems. *Ann Rev. Neurosci. 2,* 113-168.

Moore, R. Y., Halaris, A. E., and Jones, B. E. (1978). Serotonin neurons of the midbrain raphe. Ascending projections. *J. Comp. Neurol. 180*, 417-437.

Morgane, P. J., and Stern, W. C. (1974). Chemical anatomy of brain circuits in relation to sleep and wakefulness. *Adv. Sleep Res. 1*, 1-131.

Morris, M., McCann, S. M., and Orias, R. (1976). Evidence for hormonal participation in the natriuretic and kaliuretic responses to intraventricular hypertonic saline and norepinephrine. *Proc. Soc. Exp. Biol. Med. 152*, 95-98.

Morris, M., McCann, S. M., and Orias, R. (1977). Role of transmitters in mediating hypothalamic control of electrolyte excretion. *Can. J. Physiol. Pharmacol. 55*, 1143-1154.

Moss, R. L., Urban, I., and Cross, B. A. (1972). Microelectrophoresis of cholinergic and aminergic drugs on paraventricular neurons. *Am. J. Physiol. 223*, 310-318.

Mountford, D. (1969). Alterations in drinking following isoproterenol stimulation of hippocampus. *Physiologist 12*, 309.

Müller, E. E., Cocchi, D., and Mantegazza, P. (1972). Brain adrenergic system in the feeding response induced by 2-deoxy-D-glucose. *Am. J. Physiol. 223*, 945-950.

Myers, R. D. (1974). *Handbook of Drug and Chemical Stimulation of the Brain.* Van Nostrand-Reinhold, New York.

Myers, R. D. (1978). Hypothalamic actions of 5-hydroxytryptamine neurotoxins: feeding, drinking, and body temperature. *Ann. N. Y. Acad. Sci. 305*, 556-575.

Myers, R. D., Hall, G. D., and Rudy, T. A. (1973). Drinking in the monkey evoked by nicotine or angiotensin II microinjected in hypothalamic and mesencephalic sites. *Pharmacol. Biochem. Behav. 1*, 15-22.

Nahorski, S. R., and Rogers, K. J. (1974). The incorporation of glucose into brain glycogen and the activities of cerebral glycogen phosphorylase and synthetase: some effects of amphetamine. *J. Neurochem. 22*, 579-587.

Nahorski, S. R., and Rogers, K. J. (1975). The role of catecholamines in the action of amphetamine and L-DOPA on cerebral energy metabolism. *Neuropharmacology 14*, 283-290.

Nahorski, S. R., Rogers, K. J., and Edwards, C. (1975). Cerebral glycogenolysis and stimulation of β-adrenoreceptors and histamine H_2 receptors. *Brain Res. 92*, 529-533.

Needleman, H. L., and Waber, D. (1976). Amitriptyline therapy in patients with anorexia nervosa. *Lancet 2*, 580.

Neil, D. B., and Grossman, S. P. (1973). Effects of intrastriatal injections of scopolamine on appetitive behavior. *Pharmacol. Biochem. Behav. 1*, 313-318.

Neilson, E. B., and Lyon, M. (1973). Drinking behaviour and brain dopamine: antagonistic effect of two neuroleptic drugs (pimozide and spiraminde) upon amphetamine- or apomorphine-induced hypodipsia. *Psychopharmacologia 33*, 299-308.

Nemeroff, C. B., Osbahr, A. J., III, Bissette, G., Jahnke, G., Lipton, M. A., and Prange, A. J., Jr. (1978). Cholecystokinin inhibits tail pinch-induced eating in rats. *Science 200*, 793-794.

Ng, K. Y., Chase, T. N., and Kopin, I. J. (1970). Drug-induced release of ^{3}H-norepinephrine and ^{3}H-serotonin from brain slices. *Nature (London) 228*, 468-469.

Nicolaidis, S., and Fitzsimons, J. T. (1975). La dépendance de la prise d'eau induite par l'angiotensine II envers la fonction vasomotrice cérébrale local chez le rat. *C. R. Acad. Sci. Paris 281*, 1417-1420.

Norgren, R. (1976). Taste pathways to the hypothalamus and amygdala. *J. Comp. Neurol. 166*, 17-30.

Oades, R. D. (1977). The effects of unilateral 6-hydroxydopamine lesions in the substantia nigra on hippocampal noradrenaline-induced feeding and other behaviour in the rat. *Neurosci. Lett. 4*, 287-291.

O'Donohue, T. L., Browley, W. R., and Jacobowitz, D. M. (1978). Changes in ingestive behavior following interruption of a noradrenergic projection to the paraventricular nucleus: Histochemical and neurochemical analyses. *Pharmacol. Biochem. Behav. 9*, 99-105.

Olsson, K. (1970). Effects on water diuresis of infusions of transmitter substances into the third ventricle. *Acta Physiol. Scand. 79*, 133-135.

Oltmans, G. A., Lorden, J. F., and Margules, D. L. (1977). Food intake and body weight: effects of specific and nonspecific lesions in the midbrain path of the ascending noradrenergic neurons of the rat. *Brain Res. 128*, 293-308.

Oomura, Y., Ono, T., Sugimari, M., and Nakamura, T. (1973). Effects of cyproheptadine on the feeding and satiety centres in the rat. *Pharmacol. Biochem. Behav. 1*, 449-459.

Osumi, Y., Oishi, R., Fujiwara, H., and Takaori, S. (1975). Hyperdipsia induced by bilateral destruction of the locus coeruleus in rats. *Brain Res. 86*, 419-427.

Palkovits, M. (1977). Transmitters of the surpainfundibular system. In *Advances in Biochemical Psychopharmacology*, Vol. 16, E. Costa and G. L. Gessa (Eds.). Raven, New York, pp. 71-78.

Palkovits, M., and Jacobowitz, D. M. (1974). Topographic atlas of catecholamine and acetylcholinesterase-containing neurons in the rat brain. II. Hindbrain (mesencephalon, rhombencephalon). *J. Comp. Neurol. 157*, 29-42.

Palkovits, M., Brownstein, M., Saavedra, J. M., and Axelrod, J. (1974). Norepinephrine and dopamine content of hypothalamic nuclei of the rat. *Brain Res. 77*, 137-149.

Palkovits, M., Saavedra, J. M., Jacobowitz, D. M., Kizer, J. S., Záborsky, L., and Brownstein, M. J. (1977). Serotonergic innervation of the forebrain: Effect of lesions on serotonin and tryptophan hydroxylase levels. *Brain Res. 130*, 121-134.

Palmer, G. C., and Manian, A. A. (1974). Effects of phenothiazines and phenothiazine metabolites on adenyl cyclase and the cyclic AMP response in rat brain. In *Phenothiazines and Structurally Related Drugs*, I. S. Forest, C. J. Carr, and E. Usdin (Eds.). Raven, New York, pp. 749-767.

Panksepp, J. (1972). Hypothalamic radioactivity after intragastric glucose-^{14}C in rats. *Am. J. Physiol. 223*, 396-401.

Panksepp, J., and Reilly, P. (1975). Medial and lateral hypothalamic oxygen consumption as a function of age, starvation, and glucose administration. *Brain Res. 94*, 133-140.

Parent, A., and Butcher, L. L. (1976). Organization and morphologies of acetylcholinesterase-containing neurons in the thalamus and hypothalamus of the rat. *J. Comp. Neurol. 170*, 205-226.

Parker, S. W., and Feldman, S. M. (1967). Effect of mesencephalic lesions on feeding behavior in rats. *Exp. Neurol. 17*, 313-326.

Paxinos, G., and Bindra, D. (1972). Hypothalamic knife cuts: effects on eating, drinking, irritability, aggression, and copulation in the male rat. *J. Comp. Physiol. Psychol. 79*, 219-229.

Paxinos, G., and Bindra, D. (1973). Hypothalamic and midbrain neural pathways involved in eating, drinking, irritability, aggression, and copulation in rats. *J. Comp. Physiol. Psychol. 82*, 1-14.

Paykel, E. S., Muelter, P. S., and de la Vergue, P. M. (1973). Amitriptyline, weight gain and carbohydrate craving: a side effect. *Br. J. Psychiat. 123*, 501-507.

Peck, J. W. (1976). Carbachol, angiotensin-II, ventricular spread, and water balance in rats. *Pharmacol. Biochem. Behav. 5*, 591-595.

Penaloza-Rojas, J. H., and Russek, M. (1963). Anorexia produced by direct current blockade of the vagus nerve. *Nature (London) 200*, 176.

Penaloza-Rojas, J. H., Barrera-Mera, B., and Kubli-Garfias, C. (1969). Behavioral and brain electrical changes after vagal stimulation. *Exp. Neurol. 23*, 378-383.

Phillips, A. G., and Nikaido, R. S. (1975). Disruption of brain stimulation-induced feeding by dopamine receptor blockade. *Nature (London) 258*, 750-751.

Pickford, M. (1947). The action of acetylcholine in the supraoptic nucleus of the chloralosed dog. *J. Physiol. 106*, 264-270.

Pillar, A. X., Silva-Netto, C. R., Camargo, L. A. A., Saad, W. A., Antunes-Rodrigues, J., and Covian, M. R. (1977). Adrenergic stimulation of the lateral hypothalamic area on sodium and potassium excretion. *Pharmacol. Biochem. Behav. 6*, 145-149.

Placheta, P., Singer, E., Kriwanek, W., and Hertting, G. (1976). Mepiprazole, a new psychotropic drug: effects on uptake and retention of monoamines in rat brain synaptosomes. *Psychopharmacology 48*, 295-301.

Placidi, G. F., Fornaro, P., Papeschi, R., and Cassano, G. B. (1976). Autoradiographic distribution study of ^{14}C-DOPA in cat brain. *Arch. Int. Pharmacodyn. 220*, 287-294.

Poschel, B. P. H. (1971). A simple and specific screen for benzodiazepine-like drugs. *Psychopharmacologia 19*, 193-198.

Pradhan, S. N., and Roth, T. (1968). Comparative behavioral effects of several anticholinergic agents in rats. *Psychopharmacologia 12*, 358-366.

Preskorn, S. H., Clark, H. B., Hartman, B. K., and Raichle, M. (1978). Amitriptyline induced alterations in cerebral capillary permeability. *Neurosci. Abstr. 4*, 500.

Raichle, M. E., Hartman, B. K., Eichling, J. O., and Sharpe, L. G. (1975). Central noradrenergic regulation of cerebral blood flow and vascular permeability. *Proc. Natl. Acad. Sci. U.S.A. 72*, 3726-3730.

Ramsay, D. J., and Reid, I. A. (1975). Some central mechanisms of thirst in the dog. *J. Physiol. 253*, 517-525.

Redmond, E. D., Jr., Huang, H. Y., Baulu, J., Snyder, R. D., and Maas, W. J. (1977). Norepinephrine and satiety in monkeys. In *Anorexia Nervosa*, R. Vigersky (Ed.). Raven, New York, pp. 81-96.

Reis, D. J. (1974). Central neurotransmitters in aggression. *Res. Publ. ARNMD 52*, 119-146.

Reubi, J. C., Iversen, L. L., and Jessel, T. M. (1977). Dopamine selectively increases ^{3}H-GABA release from slices of rat substantia nigra in vitro. *Nature (London) 268*, 652-654.

Reynolds, R. W., and Carlisle, H. S. (1961). The effect of chlorpromazine on food intake in the albino rat. *J. Comp. Physiol. Psychol. 54*, 354-356.

Ricardo, J. A., and Koh, E. T. (1977). Direct projection from the nucleus of the solitary tract to the hypothalamus, amygdala, and other forebrain structures in the rat. *Anat. Rec. 187*, 693.

Richardson, J. S., and Jacobowitz, D. M. (1973). Depletion of brain norepinephrine by intraventricular injection of 6-hydroxydopa: a biochemical, histochemical and behavioral study in rats. *Brain Res. 58*, 117-133.

Richardson, J. S., Coran, N., Hartman, R., and Jacobowitz, D. (1974). On the behavioral and neurochemical actions of 6-hydroxydopa and 5,6-dihydroxytryptamine in rats. *Res. Commun. Chem. Pathol. Pharmacol. 8*, 29-44.

Richter, C. P. (1941). Sodium chloride and dextrose appetite of untreated and treated adrenalectomized rats. *Endocrinology 29*, 115-125.

Ritter, R. C., and Balch, O. K. (1978). Feeding in response to insulin but not to 2-deoxy-D-glucose in the hamster. *Am. J. Physiol. 234*, E20-E24.

Ritter, R. C., and Epstein, A. N. (1975). Control of meal size by central noradrenergic action. *Proc. Natl. Acad. Sci. U.S.A. 72*, 3740-3743.

Ritter, R. C., and Neville, M. (1977). Hypothalamic noradrenaline turnover is increased during glucoprivic feeding. *Fed. Proc. 35*, 642.

Ritter, S., Wise, C. D., and Stein, L. (1975). Neurochemical regulation of feeding in the rat: facilitation by α-noradrenergic, but not dopaminergic, receptor stimulants. *J. Comp. Physiol. Psychol. 88*, 778-784.

Ritter, S., Plezer, N. L., and Ritter, R. C. (1978). Absence of glucoprivic feeding after stress suggests impairment of noradrenergic neuron function. *Brain Res. 149*, 399-411.

Roberts, D. C. S., Price, M. T. C., and Fibiger, W. C. (1976). The dorsal tegmental noradrenergic projection: An analysis of its role in maze learning. *J. Comp. Physiol. Psychol. 90*, 363-372.

Roberts, E. (1977). The γ-aminobutyric acid system and schizophrenia. In *Neuroregulators and Psychiatric Disorders*, E. Usdin, D. A. Hamburg, and J. D. Barchas (Eds.). Oxford University Press, New York, pp. 347-357.

Roberts, E., and Hammerschlag, R. (1976). Amino acid transmitters. In *Basic Neurochemistry*, G. J. Siegel, R. W. Albins, R. Katzman, and B. W. Agranoff (Eds.). Little, Brown, Boston, pp. 218-245.

Roberts, E., Chase, T. N., and Tower, D. B. (Eds.) (1976). *GABA in Nervous System Function*. Raven, New York.

Robinson, R. G., McHugh, P. R., and Bloom, F. E. (1975). Chlorpromazine induced hyperphagia in the rat. *Psychopharmacol. Commun. 1*, 37-50.

Rolls, E. T., Rolls, B. J., Kelly, P. H., Shaw, S. G., Wood, R. J., and Dale, R. (1974). The relative attenuation of self stimulation eating and drinking produced by dopamine-receptor blockade. *Psychopharmacologia 38*, 219-230.

Ross, R. A., Smith, G. P., and Reis, D. J. (1973). Effects of lesions of locus coeruleus on regional distribution of dopamine-β-hydroxylase (DBH) in brain and feeding and drinking behavior in rat. *Fed. Proc. 32*, 708.

Routtenberg, A. (1967). Drinking induced by carbachol: thirst circuit or ventricular modification? *Science 157*, 838-839.

Routtenberg, A. (1972). Intracranial chemical injection and behavior: a critical review. *Behav. Biol. 7*, 601-641.

Routtenberg, A., Sladek, J., and Bondareff, W. (1968). Histochemical fluorescence after application of neurochemicals to caudate nucleus and septal area in vivo. *Science 161*, 272-274.

Rowland, N., and Engle, D. J. (1977). Feeding and drinking interactions after acute butyrophenone administration. *Pharmacol. Biochem. Behav. 7*, 295-301.

Russek, M. (1975). Current hypotheses in the control of feeding behaviour. In *Neural Integration of Physiological Mechanisms and Behaviour*, G. T. Mogenson and F. R. Calaresu (Eds.). University of Toronto Press, pp. 128-147.

Russell, R. W., Vasquez, G. J., Overstreet, D. H., and Dalglish, F. W. (1971). Consummatory behavior during tolerance to and withdrawal from chronic depression of cholinesterase activity. *Physiol. Behav. 7*, 523-528.

Saavedra, J. M., Palkovits, M., Brownstein, M. J., and Axelrod, J. (1974). Serotonin distribution in the nuclei of the rat hypothalamus and preoptic region. *Brain Res. 77*, 157-165.

Saller, C. F., and Stricker, E. M. (1976). Hyperphagia and increased growth in rats after intraventricular injections of 5,7-dihydroxytryptamine. *Science 192*, 385-387.

Samanin, R., and Garattini, S. (1975). Serotonergic system in the brain and its possible functional connections with other aminergic systems. *Life Sci. 17*, 1201-1210.

Samanin, R., Ghezzi, D., Valgelli, L., and Garattini, S. (1972). The effects of selective lesioning of brain serotonin or catecholamine containing neurons on the anorectic activity of fenfluramine and amphetamine. *Eur. J. Pharmacol. 19*, 318-322.

Samanin, R., Bernasconi, S., and Garattini, S. (1975). The effect of selective lesioning of brain catecholamine-containing neurons on the activity of various anorectics in the rat. *Eur. J. Pharmacol. 34*, 373-375.

Samanin, R., Bendotti, C., Bernasconi, S., Borroni, E., and Garattini, S. (1977). Role of brain monoamines in the anorectic activity of mazindol and d-amphetamine in the rat. *Eur. J. Pharmacol. 43*, 117-124.

Samanin, R., Bendotti, C., Bernasconi, S., and Potaccini, R. (1978). Differential role of brain monoamines in the activity of anorectic drugs. In *Central Mechanisms of Anorectic Drugs*, S. Garattini and R. Samanin (Eds.). Raven, New York, pp. 233-242.

Sanghvi, I., Singer, G., Friedman, E., and Gershon, S. (1975). Anorexigenic effects of d-amphetamine and L-DOPA in the rat. *Pharmacol. Biochem. Behav. 3*, 81-86.

Saper, C. B., Loewy, A. D., Swanson, L. W., and Cowan, W. M. (1976). Direct hypothalamo-autonomic connections. *Brain Res. 117*, 305-312.

Sawchenko, P. E., and Deluca, C. (1978). Subdiaphragmatic vagotomy blocks norepinephrine-induced eating, but does not impede the development of dietary obesity. *Eastern Psychological Association, 49th Annual Meeting*, p. 185.

Schildkraut, J. J. (1970). Tranylcypromine: effects on norepinephrine metabolism in rat brain. *Am. J. Psychiat. 126*, 925-931.

Schmitt, H., Schmitt, H., and Fénard, S. (1973). Action of alpha-adrenergic blocking drugs on the sympathetic centres and their interactions with the central sympatho-inhibitory effect of clonidine. *Arzn. Forsch. 23*, 40-45.

Schneider, B. S., Monahan, J. W., and Hirsch, J. (1979). Brain cholecystokinin and nutritional status in rats and mice. *J. Clin. Invest. 64*, 1348-1356.

Schubert, J., and Sedvall, G. (1972). Accumulation and disappearance of ^{3}H-5-hydroxytryptamine formed in vivo from ^{3}H-tryptophan in various regions of the rat brain. *Eur. J. Pharmacol. 17*, 75-80.

Schwartz, J.-C. (1977). Histaminergic mechanisms in brain. *Annu. Rev. Pharmacol. 17*, 325-339.

Schwartz, W. J. (1978). 6-Hydroxydopamine lesions of rat locus coeruleus alter brain glucose consumption as measured by the 2-deoxy-[^{14}C]glucose tracer technique. *Neurosci. Lett. 7*, 141-150.

Schwartz, W. J., Sharp, F. R., Gunn, R. H., and Evarts, E. V. (1976). Lesions of ascending dopaminergic pathways decrease forebrain glucose uptake. *Nature (London) 261*, 155-157.

Sciorelli, G., Poloni, M., and Rindi, G. (1972). Evidence of cholinergic mediation of ingestive responses elicited by dibutyryl-adenosine-3',5'-monophosphate in rat hypothalamus. *Brain Res. 48*, 427-431.

Sclafani, A. (1971). Neural pathways involved in the ventromedial hypothalamic lesion syndrome in the rat. *J. Comp. Physiol. Psychol. 77*, 70-96.

Sclafani, A., and Berner, C. N. (1977). Hyperphagia and obesity produced by parasagittal and coronal hypothalamic knife cuts: further evidence for a longitudinal feeding inhibitory pathway. *J. Comp. Physiol. Psychol. 91*, 1000-1018.

Setler, P. E. (1975). Noradrenergic and dopaminergic influences on thirst. In *Control Mechanisms of Thirst*, G. Peters and J. T. Fitzsimons (Eds.). Springer-Verlag, New York, pp. 62-68.

Setler, P. E. (1977). The neuroanatomy and neuropharmacology of drinking. In *Handbook of Psychopharmacology*, Vol. 8, *Drugs, Neurotransmitters, and Behavior*, L. L. Iversen, S. D. Iversen, and S. H. Snyder (Eds.). Plenum, New York, pp. 131-158.

Setler, P. E., and Smith, G. P. (1974). Increased food intake elicited by adrenergic stimulation of the diencephalon in rhesus monkeys. *Brain Res. 65*, 459-473.

Severs, W. B., and Daniels-Severs, A. E. (1973). Effects of angiotensin on the central nervous system. *Pharmacol. Rev. 25*, 415-449.

Severs, W. B., Keil, L. C., and Klase, P. A. (1978). Effects of centrally injected vasopressin on urine output and consummatory behavior. *Fed. Proc. 37*, 523.

Sharpe, L. G., and Myers, R. D. (1969). Feeding and drinking following stimulation of the diencephalon of the monkey with amines and other substances. *Exp. Brain Res. 8*, 295-310.

Silverman, H. J., and Zucker, I. (1976). Absence of post-fast food compensation in the golden hamster (*Mesocricetus auratus*). *Physiol. Behav. 17*, 271-285.

Simpson, C. W., DiCara, L. V., and Wolf, G. (1974). Glucocorticoid anorexia in rats. *Pharmacol. Biochem. Behav. 2*, 19-25.

Simpson, J. B., and Routtenberg, A. (1972). The subfornical organ and carbachol-induced drinking. *Brain Res. 45*, 135-157.

Simpson, J. B., and Routtenberg, A. (1974). Subfornical organ: acetylcholine application elicits drinking. *Brain Res. 79*, 157-164.

Simpson, J. B., Epstein, A. N., and Camardo, J. S., Jr. (1978a). Localization of receptors for the dipsogenic action of angiotensin II in the subfornical organ of rat. *J. Comp. Physiol. Psychol. 92*, 581-608.

Simpson, J. B., Mangiapane, M. L., and Dellman, H.-D. (1978b). Central receptor sites for angiotensin-induced drinking: a critical review. *Fed. Proc. 37*, 2676-2682.

Simson, E. L., Gold, R. M., Standish, L. J., and Pellett, P. L. (1977). Axon-sparing brain lesioning technique: the use of monosodium-L-glutamate and other amino acids. *Science 198*, 515-517.

Singer, G., and Kelly, J. (1972). Cholinergic and adrenergic interaction in the hypothalamic control of drinking and eating behaviour. *Physiol. Behav. 8*, 885-890.

Sladek, C., and Robert, J. J. (1978). Cholinergic involvement in the osmotic control of vasopressin release by the organ cultured rat hypothalamo-neurohypophyseal system. *Neurology 28*, 366.

Slangen, J. L., and Miller, N. E. (1969). Pharmacological tests for the function of hypothalamic norepinephrine in eating behavior. *Physiol. Behav. 4*, 543-552.

Smith, G. P., and Cushin, B. J. (1978). Cholecystokinin acts at a vagally innervated abdominal site to elicit satiety. *Neurosci. Abstr. 4*, 180.

Smith, G. P., and Epstein, A. N. (1969). Increased feeding in response to decreased glucose utilization in the rat and monkey. *Am. J. Physiol. 217,* 1083-1087.

Snowdon, C. T. (1969). Motivation, regulation and the control of meal parameters with oral and intragastric feeding. *J. Comp. Physiol. Psychol. 69,* 91-100.

Snowdon, C. T. (1970). Gastrointestinal sensory and motor control of food intake. *J. Comp. Physiol. Psychol. 71,* 68-76.

Snowdon, C. T., and Epstein, A. N. (1970). Oral and intragastric feeding in vagotomized rats. *J. Comp. Physiol. Psychol. 71,* 59-67.

Snyder, J. J., and Levitt, R. A. (1975). Neural activity changes correlated with central anticholinergic blockade of cholinergically-induced drinking. *Pharmacol. Biochem. Behav. 3,* 75-79.

Snyder, S. H. (1975). Amino acid neurotransmitters: Biochemical pharmacology. In *The Nervous System,* Vol. 1, *The Basic Neurosciences,* R. O. Brady (Ed.). Raven, New York, pp. 355-361.

Snyder, S. H. (1978). Peptide neurotransmitter candidates in the brain: focus on enkephalin, angiotensin II, and neurotensin. In *The Hypothalamus,* S. Reichlin, R. J. Baldessarini, and J. B. Martin (Eds.). Raven, New York, pp. 233-243.

Snyder, S. H., Axelrod, J., and Bauter, H. (1964). The fate of C^{14}-histamine in animal tissues. *J. Pharmacol. Exp. Ther. 144,* 373-378.

Sommer, S. R., Novin, D., and LeVine, M. (1967). Food and water intake after intrahypothalamic injections of carbachol in the rabbit. *Science 156,* 983-984.

Soper, W. Y., and Wise, R. A. (1971). Hypothalamically induced eating: Eating from 'non-eaters' with diazepam. *T-I-T J. Life Sci. 1,* 79-84.

Soulairac, A., and Soulairac, M.-L. (1970). Effects of amphetamine-like substances and L-Dopa on thirst, water intake and diuresis. In *Amphetamines and Related Compounds,* E. Costa and S. Garattini (Eds.). Raven, New York, pp. 819-837.

Soulairac, M.-L., and Soulairac, A. (1975). Monoaminergic and cholinergic control of sexual behavior in the male rat. In *Sexual Behavior: Pharmacology and Biochemistry,* M. Sandler and G. L. Gessa (Eds.). Raven, New York.

Sporn, J. R., and Molinoff, P. B. (1976). Beta-adrenergic receptors in rat brain. *J. Cyc. Nucl. Res. 2,* 149-161.

Sprague, J. M., Chambers, W. W., and Stellar, E. (1961). Attentive, affective, and adaptive behavior in the cat. *Science 133,* 165-173.

Stark, P., Turk, J. A., and Totty, C. W. (1971). Reciprocal adrenergic and cholinergic control of hypothalamic elicited eating and satiety. *Am. J. Physiol. 220,* 1516-1521.

Starr, N. J. (1975). The roles of dopaminergic and noradrenergic lateral hypothalamic systems in the modulation of normally-motivated consummatory behavior. Ph.D. thesis, New York University.

Starr, N. J., and Coons, E. E. (1975). The role of dopaminergic, and alpha and beta noradrenergic receptors in the modulation of normally-motivated food intake: Some paradoxical effects. Paper presented at the Eastern Psychological Association, New York.

Steffens, A. B. (1969). Blood glucose and FFA levels in relation to the meal pattern in the normal rat and the ventromedial hypothalamic lesioned rat. *Physiol. Behav. 4,* 215-225.

Stephens, D. N., and Morrissey, S. M. (1975). Hypothalamic stimulation induces acid secretion, hypoglycemia, and hyperinsulinemia. *Am. J. Physiol. 228*, 1206-1209.

Stern, J. J., Caudillo, C. A., and Kruper, J. (1976). Ventromedial hypothalamus and short-term feeding suppression by caerulin in male rats. *J. Comp. Physiol. Psychol. 90*, 484-490.

Stern, J. J., and Zurik, G. (1973). Effects of intraventricular norepinephrine and estradiol benzoate on weight regulatory behavior in female rats. *Behav. Biol. 9*, 605-612.

Stolerman, I. P. (1970). Eating, drinking and spontaneous activity in rats after the administration of chlorpromazine. *Neuropharmacology 9*, 405-417.

Straus, E., and Yalow, R. S. (1979). Cholecystokinin in the brains of obese and nonobese mice. *Science 203*, 68-69.

Stricker, E. M. (1973). Thirst, sodium appetite and complementary physiological contributions to the regulation on intravascular fluid volume. In *The Neuropyschology of Thirst: New findings and advances in Concepts*, A. N. Epstein, H. R. Kissileff, and E. Stellar (Eds.). Winston, Washington, D.C., pp. 73-98.

Stricker, E. M., and Zigmond, M. J. (1976). Brain catecholamines and the lateral hypothalamic syndrome. In *Hunger: Basic Mechanisms and Clinical Implications*, D. Novin, W. Wyrwicka, and G. Bray (Eds.). Raven, New York, pp. 19-32.

Stutinsky, F. (1974). Morphological and physiological reactions of the supraoptic and paraventricular nuclei. In *Neurosecretion: The Final Neuroendocrine Pathway*, F. Knowles and L. Vollrath (Eds.). Springer-Verlag, New York, pp. 15-23.

Subramanian, N., and Mulder, A. H. (1977). Modulation by histamine of the efflux of radiolabeled catecholamines from rat brain slices. *Eur. J. Pharmacol. 43*, 143-152.

Sugrue, M. F., Goodlet, I., and McIndeward, I. (1975). Failure of depletion of rat brain serotinin to alter fenfluramine-induced anorexia. *J. Pharm. Pharmacol. 27*, 950-953.

Swanson, L. W. (1977). Immunohistochemical evidence for a neurophysin-containing autonomic pathway arising in the paraventricular nucleus. *Brain Res. 128*, 346-353.

Swanson, L. W., and Sharpe, L. G. (1973). Centrally induced drinking: comparison of angiotensin II- and carbachol-sensitive sites in rats. *Am. J. Physiol. 225*, 566-573.

Swanson, L. W., and Hartman, B. K. (1975). The central adrenergic system. An immunofluorescence study of the location of cell bodies and their efferent connections in the rat utilizing dopamine-β-hydroxylase as a marker. *J. Comp. Neurol. 163*, 467-506.

Swanson, L. W., Sharpe, L. G., and Griffin, D. (1973). Drinking to intracerebral angiotensin II and carbachol: dose-response relationships and ionic involvement. *Physiol. Behav. 10*, 595-600.

Swanson, L. W., Connelly, M. A., and Hartman, B. K. (1977). Ultrastructural evidence for central monoaminergic innervation of blood vessels in the paraventricular nucleus of the hypothalamus. *Brain Res. 136*, 166-173.

Swanson, L. W., Connelly, M. A., and Hartman, B. K. (1978). Further studies on the fine structure of the adrenergic innervation of the hypothalamus. *Brain Res. 151*, 165-174.

Taylor, K. M. (1975). Brain histamine. In *Handbook of Psychopharmacology*, Vol. 3, *Biochemistry of Biogenic Amines*, L. L. Iversen, S. D. Iversen, and S. H. Snyder (Eds.). Plenum, New York, pp. 327-379.

Teitelbaum, P., and Wolgin, D. L. (1975). Neurotransmitters and the regulation of food intake. In *Progress in Brain Research*, Vol. 42, *Hormones, Homeostasis and the Brain*, W. H. Gispen, Tj. B. Van Wimersma Greidanus, B. Bohus, and D. de Wied (Eds.). Elsevier, Amsterdam, pp. 235-249.

Thompson, D. A., and Campbell, R. G. (1977). Hunger in humans induced by 2-deoxy-D-glucose: Glucoprivic control of taste preference and food intake. *Science 198*, 1065-1068.

Tye, N. C., Nicholas, D. J., and Morgan, M. J. (1976). Chlordiazepoxide and preference for free food in rats. *Pharmacol. Biochem. Behav. 3*, 1149-1151.

Ungerstedt, U. (1971a). Stereotaxic mapping of the monoamine pathways in the rat brain. *Acta Physiol. Scand. Suppl. 367*, 1-48.

Ungerstedt, U. (1971b). Adipsia and aphagia after 6-hydroxydopamine induced degeneration of the nigrostriatal dopamine system. *Acta Physiol. Scand. Suppl. 367*, 95-122.

Ungerstedt, U. (1974). Neuropharmacology of the control of food intake. *Vth International Conference on Physiology of Food and Fluid Intake, Jerusalem*, pp. 33-34.

U'Prichard, D. C., Greenberg, D. A., Sheehan, P., and Snyder, S. H. (1977). Regional distribution of α-noradrenergic receptor binding in calf brain. *Brain Res. 138*, 151-158.

U'Prichard, D. C., Bylund, D. B., and Snyder, S. H. (1978). (±)-[^{3}H] Epinephrine and (−)-[^{3}H] dihydroxyalprenolol binding to β_1- and β_2-noradrenergic receptors in brain, heart, and lung membranes. *J. Biol. Chem. 253*, 5090-5102.

Urano, A., and Kobayashi, H. (1978). Effects of noradrenaline and dopamine injected into the supraoptic nucleus on urine flow rate in hydrated rats. *Exp. Neurol. 60*, 140-150.

Usdin, E., Hamburg, D. A., and Barchas, J. D. (1977). *Neuroregulators and Psychiatric Disorders*. Oxford University Press, New York.

Vandeputte-Van Messom, G., and Peeters, G. (1973). Effects of administration of cholinergic drugs into the 3rd ventricle of goats on water diuresis. *Arch. Int. Pharmacodyn. 206*, 405-406.

van der Gugten, J. (1977). Neurochemical identification of hypothalamic catecholamine neuron systems involved in feeding behavior of the rat. Ph.D. thesis, State University of Utrecht, The Netherlands.

van der Gugten, J., and Slangen, J. L. (1977). Release of endogenous catecholamines from rat hypothalamus in vivo related to feeding and other behaviors. *Pharmacol. Biochem. Behav. 7*, 211-219.

van Houten, M., and Posner, B. I. (1979). Insulin binds to brain blood vessels in vivo. *Nature 282*, 623-625.

Versteeg, D. H. G., van der Gugten, J., de Jong, W., and Palkovits, M. (1976). Regional concentrations of noradrenaline and dopamine in rat brain. *Brain Res. 113*, 563-574.

Vigersky, R. A. (Ed.). (1977). *Anorexia Nervosa*. Raven, New York.

Vijayan, E., and McCann, S. M. (1977). Suppression of feeding and drinking activity in rats following intraventricular injection of thyrotropin releasing hormone (TRH). *Endocrinology 100*, 1727-1730.

von Euler, U. S. (1970). Effect of some metabolic factors and drugs on uptake and release of catecholamines in vitro and in vivo. In *New Aspects of Storage and Release*

Mechanisms of Catecholamines, H. J. Shumann and G. Kroneberg (Eds.). Springer-Verlag, New York, pp. 144-160.

Walsh, L. L., and Grossman, S. P. (1975). Loss of feeding in response to 2-deoxy-D-glucose but not insulin after zona incerta lesions in the rat. *Physiol. Behav. 15,* 481-485.

Walsh, L. L., and Grossman, S. P. (1976). Zona incerta lesions impair osmotic but not hypovolemic thirst. *Physiol. Behav. 16,* 211-215.

Walsh, L. L., Halaris, A. E., Grossman, L., and Grossman, S. P. (1977). Some biochemical effects of zona incerta lesions that interfere with the regulation of water intake. *Pharmacol. Biochem. Behav. 7,* 351-356.

Weinstock, M., Weiss, C., and Gitter, S. (1977). Blockade of 5-hydroxytryptamine receptors in the central nervous system by β-adrenoceptor antagonists. *Neuropharmacology 16,* 273-276.

Weiss, B., Fertel, R., Figlin, R., and Uzunov, P. (1974). Selective alternation of the activity of the multiple forms of adenosine $3',5'$-monophosphate phosphodiesterase of rat cerebrum. *Mol. Pharmacol. 10,* 615-625.

Weissman, A., Koe, B. K., and Tenen, S. S. (1966). Amphetamine effects following inhibition of tyrosine hydroxylase. *J. Pharmacol. Exp. Ther. 151,* 339-352.

Werman, R. (1972). Amino acids as central neurotransmitters. *Res. Publ. ARNMD 50,* 147-180.

Winson, J., and Miller, N. E. (1970). Comparison of drinking elicited by eserine or DFP injected into preoptic area of rat brain. *J. Comp. Physiol. Psychol. 73,* 233-237.

Wise, R. A. (1972). Rebound eating following carbachol-induced drinking in rats. *Physiol. Behav. 9,* 659-661.

Wise, R. A., and Dawson, V. (1974). Diazepam-induced eating and lever pressing for food in sated rats. *J. Comp. Physiol. Psychol. 86,* 930-941.

Wisehart, T. B., and Walls, E. K. (1974). Reduction of stimulus-bound food consumption in the rat following amphetamine administration. *J. Comp. Physiol. Psychol. 87,* 741-745.

Wolf, G., McGovern, J. F., and DiCara, L. V. (1974). Sodium appetite: Some conceptual and methodologic aspects of a model drive system. *Behav. Biol. 10,* 27-42.

Wolfe, B. B., Harden, T. K., and Molinoff, P. B. (1976). β-Adrenergic receptors in rat liver: effects of adrenalectomy. *Proc. Natl. Acad. Sci. U.S.A. 73,* 1343-1347.

Wolny, H. L., Plech, A., and Herman, Z. S. (1974). Diuretic effects of intraventricularly injected noradrenaline and dopamine in rats. *Experientia 30,* 1062-1063.

Woods, S. C., and Porte, D., Jr. (1974). Neural control of the endocrine pancreas. *Physiol. Rev. 54,* 596-619.

Wurtman, J. J., and Wurtman, R. J. (1977). Fenfluramine and fluoxetine spare protein consumption while suppressing calorie intake by rats. *Science 198,* 1178-1180.

Yaksh, T. L., and Myers, R. D. (1972). Hypothalamic "coding" in the unanesthetized monkey of noradrenergic sites mediating feeding and thermoregulation. *Physiol. Behav. 8,* 251-257.

Young, P. M. (1975). The effects of 6-OH-DA upon the ultrastructure and the electrical activity of the paraventricular nucleus of the rat. *J. Anat. 118,* 395.

York, D. H. (1972). Dopamine receptor blockade—a central action of chlorpromazine on striatal neurones. *Brain Res. 37,* 91-99.

Zabik, J. E., Levine, R. M., Spaulding, J. H., and Maickel, R. P. (1977). Interactions of tricyclic antidepressant drugs with deprivation-induced fluid consumption by rats. *Neuropharmacology 16,* 267-271.

Ziance, R. J., Moxley, K., Mullis, M., and Gray, W. (1977). Influence of MAO inhibitors on uptake and release of norepinephrine in rat brain in vitro. *Arch. Int. Pharmacodyn. 228,* 30-38.

Zigmond, M. J., and Stricker, E. M. (1974). Ingestive behavior following damage to central dopamine neurons: implications for homeostasis and recovery of function. In *Neuropsychopharmacology of Monoamines and their Regulatory Enzymes,* E. Usdin (Ed.). Raven, New York, pp. 385-402.

Zis, A. P., and Fibiger, A. P. (1975). Neuroleptic-induced deficits in food and water regulation: similarities to the lateral hypothalamic syndrome. *Psychopharmacologia 43,* 63-68.

Zucker, I. H., Levine, N., and Kaley, G. (1974). Third ventricular injection of hypertonic NaCl: Effect of renal denervation on natriuresis. *Am. J. Physiol. 227,* 35-41.

5
Activity of Hypothalamic and Related Neurons in the Alert Animal

Edmund T. Rolls / University of Oxford, Oxford, England

I. Introduction

Damage to the hypothalamus produces syndromes which provide clues about its function. For example, bilateral lesions of the lateral hypothalamus produce an impairment of eating and drinking so severe that the animal may die, whereas bilateral lesions of the ventromedial hypothalamus result in overeating and obesity (see Grossman, 1967, 1973). Findings of this type led to the dual-center hypothesis of the control of food intake, according to which the lateral hypothalamus contains a feeding control center and the ventromedial hypothalamus a satiety center. The evidence on which this hypothesis was based is now being reexamined. For example, it has been suggested that the aphagia produced by bilateral lateral hypothalamic lesions, which is associated with a complex sensorimotor dysfunction (Marshall et al., 1971), is due to damage to the nigrostriatal dopamine-containing fibers (Ungerstedt, 1971; Marshall et al., 1974; Morgane, 1961c; Gold, 1967) which course past the far-lateral hypothalamus. Damage to the nigrostriatal pathway outside the lateral hypothalamus can lead to aphagia and adipsia (Marshall et al., 1974).

A more direct but complementary way to examine the function of the hypothalamus is to record and analyze the activity of single hypothalamic neurons. This approach is useful in investigating hypothalamic function not only because it provides a direct way to examine neuronal function in the hypothalamus, but also because the activity of single neurons can be analyzed individually despite the presence of intermingled systems with different functions in the hypothalamus. A further advantage of analyzing hypothalamic function by recording the activity of single neurons in the hypothalamus is that it is usually possible to decide on the basis of the wave form and size of the action potential, the presence of an A-B inflection, or the occurrence of injury discharges when the neuron is damaged whether the activity is being recorded from a cell body or from an axon, so that it is possible to include hypothalamic neurons and to exclude fibers of passage from the analysis.

Before the activity of neurons in the hypothalamus of the alert animal is described in Sections II through V, the background of this type of investigation is described in terms of the effects of lesions of the hypothalamus on feeding and the activity of hypothalamic neurons in the anesthetized animal.

A. Lesion Studies

Following the discovery that bilateral lateral hypothalamic lesions lead to aphagia (Anand and Brobeck, 1951a,b), the syndrome of aphagia and adipsia produced by the lesions, and its partial recovery (Teitelbaum and Stellar, 1954; Teitelbaum and Epstein, 1962), has been intensively studied (see Epstein, 1971; Grossman, 1967, 1973, 1975). One important question is whether the effects of the lesions on feeding and drinking are due to the destruction of cells in the hypothalamus or to the interruption of some of the neural pathways which course through and near the lateral hypothalamus. Morgane (1961a,b) showed that far-lateral lateral hypothalamic lesions were particularly effective in producing aphagia and adipsia, that these lesions produce degeneration in pallidofugal pathways, and that bilateral damage to the internal segment of the globus pallidus produces aphagia and adipsia. Ungerstedt (1971) and Marshall et al. (1974) showed that damage to the nigrostriatal dopamine-containing fibers, which are damaged by lateral hypothalamic lesions, leads to aphagia and adipsia. Zeigler and Karten (1974) showed that in the pigeon and rat, damage to ascending trigeminal pathways produced by lateral hypothalamic lesions may contribute to the aphagia and adipsia. Because a number of different pathways are damaged by and contribute to the effects produced by lateral hypothalamic lesions, it is difficult on the basis of the lesion evidence to make any firm conclusion about whether cells in the lateral hypothalamus have functions related to feeding.

B. Exteroceptive Influences on the Activity of Hypothalamic Neurons in the Anesthetized Animal

Exteroceptive stimuli such as flashes of light, clicks, and touch alter the activity of some hypothalamic neurons recorded in the anesthetized animal. For example, in the rabbit and rat anesthetized with urethane, some neurons in a number of hypothalamic areas respond to flashes of light, clicks, touch, pain, or olfactory stimulation (Cross and Green, 1959; Cross and Silver, 1963; Barraclough and Cross, 1963; Cross and Silver, 1966). In cats under light pentobarbital anesthesia, some neurons in the posterior hypothalamus respond to flashes of light, clicks, and electrical stimulation of the sciatic nerve (Dafny et al., 1965; Feldman and Dafny, 1968; Dafny and Feldman, 1970). In the ventromedial nucleus of the hypothalamus of the immobilized cat, light/shadow, back rubbing, tactual stimulation of the buccal and facial regions, as well as gustatory and olfactory stimuli, changed the discharge patterns of different units (Campbell et al., 1969).

Olfactory stimuli affect the activity of some hypothalamic and preoptic neurons. For example, in the preoptic area of the male rat, neurons responded to a range of olfactory stimuli, and the responsiveness of some of these neurons to the odor of female rats' urine was enhanced by testosterone (Pfaff and Pfaffmann, 1969). In a further study, it was found that more units in the preoptic area than the olfactory bulb responded differentially to estrous compared to ovariectomized female rat urine odors. Conversely, in the olfactory bulb, relatively more neurons with differential responses were activated by nonurine odors (Pfaff and Gregory, 1971). In female rats, the responsiveness of hypothalamic neurons to olfactory stimuli is influenced by estrogen and the stage of the estrous cycle (Cross and Silver, 1966; Cross, 1973). These findings suggest that the activity of these preoptic neurons may be related to some of the effects of odors on endocrine function and reproductive behavior (see Findlay, 1972).

Komisaruk and Beyer (1972) showed that in the urethane-anesthetized rat, even the responses to electrical stimulation of the olfactory bulb of hypothalamic neurons were dependent on arousal level, as measured by the electroencephalogram (EEG) and influenced by general stimulation such as tail pinch. This emphasizes that even though hypothalamic neurons are more independent of EEG changes than thalamic, cortical, or midbrain reticular formation neurons (Cross and Silver, 1966; Komisaruk et al., 1967; Komisaruk, 1971; Findlay, 1972), great care must be taken to show that effects of stimuli on hypothalamic neurons are not mediated simply as a result of a change in arousal level produced by the stimuli, or that arousal level has not just altered the effectiveness of stimuli. Gustatory responses have also been recorded from hypothalamic neurons in the rat (Norgren, 1970).

Some hypothalamic neurons also respond to mechanical stimulation of the female genital tract. Distension of the uterus in postpartum cats anesthetized with chloralose causes oxytocin release and an increase in the activity of neurons in the paraventricular region of the hypothalamus (Brooks et al., 1966). In the anesthetized rat, Lincoln (1969) found neurons in the anterior hypothalamus which would respond to cervical stimulation independently of the changes in arousal level produced by the stimulation. In the estrous rat, changes in the activity of hypothalamic as well as midbrain and limbic neurons are produced by vaginal stimulation (Kawakami and Saito, 1967; Kawakami and Kubo, 1971). Responses of hypothalamic neurons to these types of stimulation could participate in the release of oxytocin during labor and copulation, in the effects of such stimuli on ovulation and corpus luteum formation, and in the behavioral responses to such stimulation (Findlay, 1972). In relation to the release of oxytocin and the milk ejection produced by suckling of the mammary glands, it has been found in the lightly anesthetized rat that the oxytocin neurosecretory cells in the paraventricular nucleus, identified by antidromic activation from the neural lobe of the pituitary, show very rapid firing (30-50 spikes/sec) during suckling for a few seconds before milk ejection (Lincoln and Wakerley, 1972; Cross, 1973).

C. Interoceptive Influences on the Activity of Hypothalamic Neurons in the Anesthetized Animal

There are metabolic influences on the activity of hypothalamic neurons. It was reported that the activity of single cells in the ventromedial hypothalamus is increased during increases of glucose utilization and is decreased during hypoglycemia. Single cells in the lateral hypothalamus responded in the opposite way (Anand et al., 1964). The activation of hepatic portal receptors sensitive to glucose and ammonium ion also influences the activity of hypothalamic neurons (Schmitt, 1973). The electro-osmotic application of glucose increased the firing of 16% and decreased the activity of 36% of lateral hypothalamic cells and increased the firing of 38% and decreased the firing of 3% of ventromedial hypothalamic cells (Oomura, 1973; Oomura et al., 1975). The exact functional significance of these interesting findings in relation to endocrine function and feeding behavior remains to be determined. It has also been reported that in the cat the activity of neurons in the ventromedial hypothalamus is increased by stomach distension, with the opposite effect on unit firing in the lateral hypothalamus, and that these effects are abolished by severing the gastric branch of the vagus (Anand and Pillai, 1967).

Influences of hormones on hypothalamic neurons in relation to the control of pituitary function have been reviewed extensively elsewhere (Cross and Silver, 1966; Komisaruk, 1971; Cross, 1973; Dyer, 1974) as well as in these volumes.

These studies on the activity of hypothalamic neurons in the anesthetized animal provide information on the type of signal which can influence their activity. The studies suggest that a wide range of exteroceptive stimuli can influence the activity of hypothalamic neurons. To investigate what stimuli actually influence hypothalamic neurons during feeding and other behavior, and to investigate the activity of hypothalamic neurons in relation to that behavior, recordings must be made from hypothalamic neurons in the alert animal. This type of investigation is now described.

II. The Activity of Neurons in the Hypothalamus and Substantia Innominata of the Alert Animal

A. Background

In the alert monkey (Rolls et al., 1976) and rat (Phillips, 1973; J. T. Koolhaas and E. T. Rolls, in preparation) recordings for periods of several hours can be made from single neurons with movable microelectrodes. It is usually most satisfactory to investigate a relatively rapid response occurring in seconds or at most minutes to a well-defined stimulus, such as the presentation of food, rather than to search for slowly developing changes associated with shifts in homeostatic state. As will be described below, with these techniques it has been possible to analyze the activity of hypothalamic neurons during feeding, drinking, and learning, and to investigate whether the activity of hypothalamic neurons is related to sensory or motor function or to the control of processes such as feeding and autonomic function.

Once the activity of a population of hypothalamic neurons has been found to be associated with the presentation of a stimulus such as food for a hungry animal or a tone which the animal has learned is associated with food, further analysis to clarify the function of the neurons becomes necessary, as shown below. First, a clinical investigation of the factors leading to the neuronal response can be made to determine whether the response is modality-specific, is specific to one class of stimuli, occurs because the animal is making a behavioral response such as a head or limb movement, etc. Second, it is helpful to compare the activity of hypothalamic neurons with the activity of neurons in other structures in the same test situation to determine the way in which hypothalamic neuronal activity and function differs from the activity of, for example, sensory or motor neurons. The recordings described in Sections III and IV from neurons in the visual inferotemporal cortex, and from the globus pallidus and substantia nigra, show how this comparison can be useful. Third, the latency of a hypothalamic neuronal response to an effective stimulus, such as the onset of a visual stimulus indicating food, can be compared with the latency of neuronal responses in sensory and motor structures to determine whether the hypothalamic neurons could produce, or follow, the response of the animal. Fourth, knowledge of the activity of hypothalamic neurons must be considered in relation to other evidence, important among which are the anatomical connections of the neurons. If the neurons are thought to have a particular function, then it must be possible to show how the neurons receive the input necessary for this function and exert their influence on output pathways. In addition to standard anatomical techniques, electrical stimulation of distant brain sites is useful, because it allows the connections of the particular hypothalamic neuron from which a recording is being made to be demonstrated. In some cases it is possible to use electrical stimulation which has particular physiological effects and to determine whether the expression of these effects involves the activation of hypothalamic neurons.

The activity of neurons in the hypothalamus of the rat, cat, and monkey has been studied during ingestive behavior and during learning. Oomura and his colleagues reported that changes in the activity of lateral hypothalamic and ventromedial hypothalamic neurons in the cat occurred before and during eating and were especially pronounced in the hungry cat (Oomura et al., 1969). Hamburg (1971) reported that the firing rates of some units in the lateral hypothalamic area of the hungry rat increased as the animal searched for and approached food and decreased during feeding. Further, in learning experiments in which an auditory stimulus was regularly followed by food, hypothalamic neurons have been found which come to respond to the auditory stimulus (Linseman and Olds, 1973; Olds, 1973). It is difficult in the cat (Ono et al., 1976) and the rat (observations of J. M. Koolhaas and E. T. Rolls, 1976) to analyze the factors which lead to these changes in neuronal activity, partly because many different types of sensory and motor change occur when an animal searches for food in its environment, and partly because, at least in the rat, sniffing behavior itself appears to influence the activity of hypothalamic neurons and to make further analysis difficult. It has been found that in the monkey it is more possible to relate the activity of single neurons to particular processes, and emphasis is placed on investigations in the monkey in the following account of the activity of hypothalamic neurons in the alert animal. In this type of investigation the monkey is sitting in a primate chair and is given or can work for food and can learn and perform visual and auditory discriminations. In addition the factors which lead to a neuronal response can be analyzed in a clinical manner by the experimenter.

B. General Characteristics of the Activity of Hypothalamic Neurons in the Alert Animal

In investigations into the activity of neurons in the lateral hypothalamus and substantia innominata of the alert rhesus and squirrel monkey, in addition to neurons with activity closely related to feeding (Rolls et al., 1976, 1979, 1980; Burton et al., 1976; Mora et al., 1976), neurons with other characteristics were found. First, and in order of frequency of occurrence, the firing rates of some of the neurons increased during states of general arousal and body movement. Second, the activity of some of the neurons was related to particular movements made by the monkey, or to touch. For example, some of the neurons fired when the monkey opened his mouth, or when he protruded his lips to lick, or when he reached, or when he was touched on the face or body. Third, some neurons had activity which was related to visual stimulation by, for example, moving objects, or to eye movement. Fourth, some neurons fired to auditory stimuli, for example, to clicks or to the calls of other monkeys. Fifth, some neurons fired to stimuli which the monkey had learned were aversive, such as the sight of a squeeze bulb from which air was lightly puffed onto the monkey's face or of a syringe from which the monkey was given saline to drink (Rolls et al., 1979). In one sample of 764 neurons in the rhesus and squirrel monkey, 128 (16.8%) were placed in these five categories. Many neurons (69.6% of the sample of 764) in the lateral hypothalamus and substantia innominata of the alert monkey fired regularly and slowly (often in the range 0-30 spikes/sec) and were unaffected by any of the above types of stimulation or movement or by feeding or by visual discrimination learning and performance, or had small responses during these types of behavior which could not be classified into the above categories. The activity of 104 of the neurons (13.6% of the sample) was closely related to feeding, as described below. Although there have been few other studies of the activity of hypothalamic neurons in the alert monkey, Vincent et al. (1972) described some cells with activity related to drinking (see Section II.C) and others with activity related to arousal, and in studies of the globus pallidus, neurons at the ventral border of this structure with the substantia innominata were noted with activity related to feeding (Travis and Sparks, 1968; Travis et al., 1968; De Long, 1971; Sparks and Travis, 1973).

In the alert rat, hypothalamic neurons with activity related to feeding (Hamburg, 1971) and learning (Linseman and Olds, 1973; Olds, 1973) have been described (see Sections II.C and II.F), but at least in the rat lateral hypothalamus, the activity of many neurons alters in relation to sensory or motor changes associated with sniffing behavior, and hence it is difficult in this species to determine what factors are causal in the neuronal responses (observations of J. M. Koolhaas and E. T. Rolls, 1976).

C. Responses of Hypothalamic Neurons Occurring During Eating and Drinking

In the lateral hypothalamus and substantia innominata of the hungry monkey, some neurons alter their activity while food is in the mouth (Rolls et al., 1980; see also Rolls, 1975, 1976; Rolls and Rolls, 1981). Examples of these types of response are shown in Figures 1 and 9. For at least some of these neurons the responses depend on what is being consumed. For example, for the neuron illustrated in Figure 1, even though similar motor responses of drinking were required to ingest the 5% glucose solution and the isotonic sa-

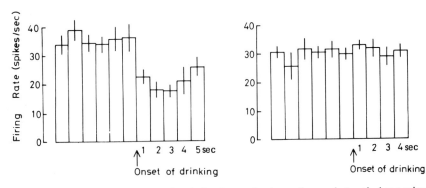

Figure 1 Time course of response of unit in the squirrel monkey substantia innominata to drinking glucose (left histogram). There was no response of the unit before the onset of drinking, but the response was maintained throughout the drinking. The histograms represent the mean of six trials and the vertical lines the standard error (SE) of the response. Right histogram: There was no response to saline (mean and SE of three trials).

Table 1 Numbers of Neurons in the Lateral Hypothalamus and Substantia Innominata of the Rhesus and Squirrel Monkey, with Activity Classified in the Test Situation as Associated with the Sight, Taste, or Sight and Taste of Food, or with Other Factors such as Movement or Touch of the Face or Body

	Number of neurons	
	Decreased rate	Increased rate
Sight of food	26	45
Sight and taste of food	3	16
Taste	9	5
Other responses (see text)	128	
Unclassified or unresponsive	532	
Total	764	

line, the neuron responded (by decreasing its firing rate) only when glucose was being ingested. In this respect the responses of some of these neurons appeared to be associated with taste, and, consistently, depended on the concentration of the glucose (Rolls et al., 1980). Further, when the monkey made mouth movements, this type of neuron did not respond (though others in the hypothalamus did). In the monkey, these neurons were relatively rare, with 14 (1.8%) classified as having responses associated with the taste of food, and an additional 19 (2.5%) classified as having responses associated with the sight as well as the taste of food (see Table 1). Thus in the monkey, in addition to units in the hypothalamus which are active during feeding because of somatosensory or motor changes (see Section II.B), there are other neurons with activity which is more closely related to the ingestion of food, and the responses of these neurons may be associated with the taste of food.

Vincent et al. (1972) have recorded from what may be comparable cells during drinking in the rhesus monkey. In the "lateral and dorsal hypothalamus, in the vicinity of the globus pallidus," they found cells which decreased their firing rates when the thirsty monkeys were drinking water. The firing rates of these cells increased during general arousal and during the intracarotid infusion of hypertonic sodium chloride solution. The responses of these cells did not occur during drinking movements if water was not available; and in this respect their activity appeared to be related to taste rather than to motor or somatosensory changes, although further tests were not described. These cells were a separate population from those in the supraoptic nucleus, which were antidromically activated from the posterior pituitary, and increased their firing rate during water deprivation and decreased it during the drinking of water (see also Arnauld et al., 1975). The activity of these supraoptic neurons was related to the release of antidiuretic hormone, and the immediate decrease of their firing rate produced by drinking water was related to the rapid diuresis produced by the water in this rapid feedback effect (see also Nicolaïdis, 1969).

In the rat, Hamburg (1971) showed that after increasing their firing rate while the animal approached food, the firing rate of some lateral hypothalamic neurons decreased while the animal was eating. The activity of these neurons was not clearly related to taste; for if the rat's food was removed, the activity of the units increased again, even though chewing, and presumably tasting, continued. Also two of three units tested with both sugar-water and water decreased their firing rates during the ingestion of both fluids. Gustatory responses of hypothalamic neurons have been described in the anesthetized rat (Norgren, 1970).

In view of the responses of hypothalamic neurons which occur during feeding and drinking, and some of which appear to be associated with taste, it is of interest that a pathway from the pontine taste area to the lateral hypothalamus, substantia innominata, and central nucleus of the amygdala has recently been described neuroanatomically in the rat (Norgren, 1976).

D. Responses of Hypothalamic Neurons Occurring Before Eating and Drinking

In the monkey, a population of neurons found in the lateral hypothalamus and in a horizontal sheet stretching far laterally in the region of the substantia innominata and basal forebrain nucleus of Meynert alters its activity before the animal ingests food. The response occurs while the hungry monkey is looking at the food, occurs most to highly preferred food (e.g., to the peanut more than to the carrot or grape in Fig. 2), and does not occur because of olfaction or because of mouth, face, or body movements, which occurred when air from the squeeze bulb was lightly puffed on to the monkey's face, or to the sight of an aversive visual stimulus such as the squeeze bulb (Fig. 2; see also Rolls et al., 1976). Care must be taken to determine whether the activity of these neurons is associated with touch to or movements of the mouth, face, or arms; and in doubtful cases, or where the neuron showed apparently a visual and a motor response, the neuron was classified as sensorimotor (see Section II.B). The responses of neurons with activity associated with the sight of food were obtained when the experimenter introduced a food but not a nonfood object into the monkey's visual field and moved it towards and then fed the monkey for the whole period in which the monkey looked at the food (see Fig. 2). The responses also occurred when the food was revealed to the monkey by the opening of a wide-aperture electromagnetic shutter. With the shutter the latencies of the responses to food objects could be measured and were as short as 150-200 msec. In the sample of 764 neurons the number classified as having responses associated with the sight of food was 71 (9.3%) and with the sight and taste of food was 19 (2.5%) (Rolls et al., 1976, 1980; see Table 1).

In the monkey and cat, neurons with activity which occurs immediately before ingestion have also been found by Ono et al. (1976). The factors which led to these responses have not been analyzed, and are difficult to analyze, in the situation used in which the animals were bar-pressing to obtain food. Lateral hypothalamic neurons which alter their activity before ingestion have also been found in the rat (Hamburg, 1971).

E. The Effects of Hunger on the Responses of Hypothalamic Neurons

The hypothalamic neurons in the monkey described in the previous sections respond in association with the sight and/or taste of food if the monkey is hungry and do not respond when the monkey is satiated (Burton et al., 1976; Rolls and Rolls, 1981). For a lateral hypothalamic neuron an example of how the responses associated with the sight of food decrease as satiety progresses is shown in Figure 3, and an example of neuronal firing is

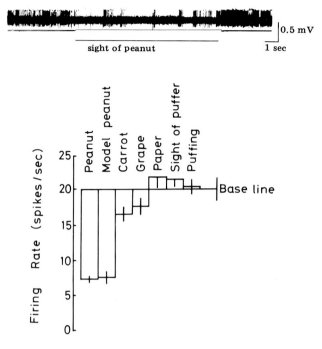

Figure 2 Above: The firing rate of this lateral hypothalamic unit decreased when the squirrel monkey saw a peanut, which was initially at a distance of 1 m from the animal. The peanut was moved gradually toward the animal's mouth, but on this trial the peanut was not given to the monkey. The short bursts of action potentials occurred when the monkey glanced away from the peanut. Below: Firing rate of a squirrel monkey lateral hypothalamic unit during a 10-sec period in which different visual stimuli were shown to the monkey. The histograms show how the firing rate changed (the vertical bars represent the standard error of the firing rate calculated over the 10-sec period in which the stimulus was shown) from the spontaneous base line rate of 20 spikes/sec. The model peanut was made of modeling clay and was used as a control for olfaction. The paper had the same outline and color as the peanut and was used as a control for shape. Air was blown from a squeeze bulb over the monkey's face to control for arousal and general movement (puffing). The squeeze bulb was shown to the monkey to test the effect on the unit of an aversive visual stimulus and of a highly salient visual stimulus which was not food.

shown in Figure 4. A comparable modulation of responsiveness of hypothalamic neurons may occur in the cat, in which changes in unit activity in the lateral and ventromedial hypothalamus associated with searching for food and feeding are more pronounced if the animal is hungry (Oomura et al., 1969). It may also occur in the rat, in that if the anesthetized rat is in a food-deprived state more hypothalamic neurons which respond to the taste of glucose are recorded, and if the rat is in a water-deprived state more hypothalamic neurons which respond to the taste of water are found (Norgren, 1970).

The responsiveness of some neurons in the lateral hypothalamus and substantia innominata is thus increased by hunger. The effects of hunger, thirst, and other changes of state on the average spontaneous firing rates of other hypothalamic neurons to determine

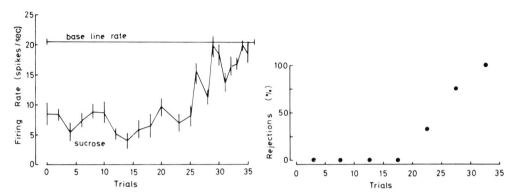

Figure 3 Left: Effect of the sight of a syringe from which the squirrel monkey was fed 2 ml of 20% sucrose solution on the firing rate of a hypothalamic single unit. The rate decreased below the spontaneous baseline rate (the mean and SEM are shown) at the start of the experiment when the monkey was hungry, but not at the end of the experiment when the monkey was satiated. The firing rate and its standard error were measured over a 5-sec period during which the monkey was looking at the syringe. The monkey drank the glucose subsequently. On some trials (left blank) different stimuli were shown, and the monkey was not fed. Right: The time-course of the satiety of the monkey in the same experiment. For each block of trials, the percentage of trials on which the monkey accepted the sucrose solution is shown. Rejection was measured by whether the animal turned its head away from the syringe and failed to reach for the syringe as the syringe was being placed in the mouth after the 5-sec experimental period in which the monkey was looking at the syringe.

whether they could monitor or mediate the effects of the changes of state have been little investigated in the alert animal, although Arnauld et al. (1975) have found that water deprivation increases the mean firing rate of neurons in the supraoptic nucleus of the monkey. In the anesthetized animal, responses of hypothalamic neurons associated with glucose and free fatty acid levels (see Oomura, 1973) and with the activation of hepatic portal receptors sensitive to glucose and ammonium ion (Schmitt, 1973) have been described. In rats and frogs it has been found that gustatory afferent activity recorded in the glossopharyngeal nerve is modulated by the nutritional state of the animal, and it has been suggested that this effect could be mediated by centrifugal pathways from the hypothalamus (Sharma et al., 1975).

F. The Activity of Hypothalamic Neurons During Learning

In the course of the recordings in the monkey hypothalamus, it was observed that the neurons which responded when the animal looked at foods might respond to the sight of oranges, grapes, nuts, bananas, and even the sight of a black syringe from which the animal was fed glucose. Thus, it seemed unlikely that these neurons responded innately to the sight of these different foods and probable that the responses of these neurons could change as a result of learning. To test this, a formal visual discrimination task was set for the monkey while recordings were made from these single units. It was found, for example, that these neurons came to respond to a black syringe from which the animal was fed glucose solution but not to a white syringe from which the animal was fed an aversive (5%) saline solution. Over the same period, the monkey learned to accept the black syringe but to reject the white syringe (see Fig. 5). If the hungry monkey was repeatedly shown food, but was not allowed to eat the food, then the responses of these neurons gradually failed to occur when the food was shown; that is, extinction of the neuronal responses occurred (Mora et al., 1976).

In investigations of neuronal activity during learning, it is important to determine whether any changes of neuronal activity found are due to the altered behavior of the animal or might be involved in mediating the learned alteration in behavior. For example, a hypothalamic neuron with activity related to movement might alter its activity during learning if the animal learned that a stimulus which previously signified food later signified an aversive stimulus, causing the animal to make different movements. To determine whether a neuron might be involved in mediating a learned behavioral change, one criterion is that the neuronal response should be modified before or at the same time as, but not after, the modification of the behavioral response. As shown in Figure 5, the hypothalamic neuronal response changed over the same few trials as the change in acceptance of the food;

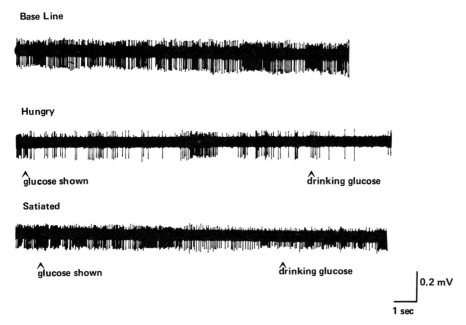

Figure 4 Effect of hunger on the responsiveness to food of a lateral hypothalamic unit in the squirrel monkey. Upper trace: Spontaneous base line firing rate. Middle trace: The firing rate of the unit decreased while the hungry monkey looked at a black glucose-containing syringe (period after "glucose shown") and while the monkey drank the glucose (period after "drinking glucose"). Lower trace: After the monkey had been fed until he was satiated, the firing rate of the unit did not change from the spontaneous firing rate while the monkey looked at the black glucose-containing syringe (period after "glucose shown"), nor did the firing rate change while the monkey drank the glucose (period after "drinking glucose"). (From Burton et al., 1976.)

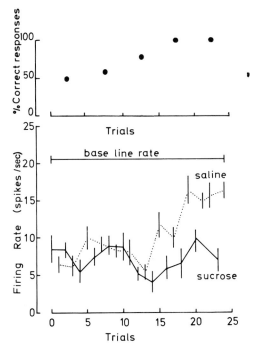

Figure 5 Firing rate of a lateral hypothalamic unit during the learning of a visual discrimination. On alternate trials the monkey saw and was then fed from a black syringe containing 25% sucrose or a white syringe containing 0.9% saline (which was aversive). The firing rate of the unit (± the standard error) over the 10 sec during which the monkey saw a syringe is shown on the ordinate. Between trials 0 and 14 the firing rate of the unit decreased from its base line spontaneous rate (shown below) and the monkey accepted either syringe, so that his performance was approximately 50% correct (shown above). After trial 15 the unit continued to fire at less than its base line rate when the monkey saw the black syringe containing sucrose, but gradually during the next few trials the sight of the white syringe containing saline had less and less effect on the unit. At the same time the monkey learned the visual discrimination; that is, he refused to accept fluid from the white saline-containing syringe but continued to accept the glucose-containing syringe. (From Mora et al., 1976.)

thus this criterion is met by the hypothalamic neurons described. In fact the neuronal and behavioral responses were very closely correlated even on an individual trial if the monkey hesitated during the discrimination. A second criterion is that the latency of the neuronal response to the stimulus to be discriminated must be less than the latency of the behavioral response. This is being investigated for hypothalamic neurons in the rhesus monkey by making recordings while the monkey is performing a visual discrimination to stimuli presented with an electromagnetic shutter (Rolls et al., 1979). On each trial one of two visual stimuli (e.g., either a large red or a small clear syringe) is present behind the shutter when it opens; if the monkey licks a tube in front of him within a short set period, glucose is delivered if the first visual stimulus is present, otherwise saline is delivered. The motor response requires only a licking movement to contact the tube because of the head restraint used for the recording, and the monkey's shortest times for the response in this visual discrimination are typically 400 msec. In this situation the latencies of the responses of the hypothalamic neurons to the food stimulus are approximately 150-200 msec (see Fig. 6). This relatively short latency (compared with latencies of 100-140 msec in the visual inferotemporal cortex in the same test situation—see Section III—and compared with the latency to lick the tube) suggests that these neurons in the hypothalamus and substantia innominata could be involved in the mediation of the responses of the animal to the food. To be sure that the neuronal responses did not simply reflect a motor response already being made by the monkey, the electromyogram (EMG) of the genioglossus muscle, which protrudes the tongue to make the lick response, was measured during these recordings. It was found that on trials on which the monkey licked, activity in the electromyogram started typically 280-300 msec after the shutter opened to reveal the visual stimulus; that is, the EMG in the relevant muscle for the response started 100-120 msec before the lick contact was made. Thus the neuronal response, which occurred 150-200 msec after the visual stimulus was shown, could not have been produced by the motor responses of the monkey. The observation that the responses of these neurons in the lateral hypothalamus and substantia innominata preceded the motor response made to obtain food, and depended on the significance of the visual stimulus, is consistent with the hypothesis that these neurons are involved in the mediation of the responses of the animals to the food. Further information on when activity starts in motor structures is considered in Section IV. Of course, it is also an interesting question to ask where in the nervous system the shortest latency neuronal responses which are modified by learning occur to the stimulus, as this would provide information on which part of the brain mediated this type of learning (see Olds, 1975).

This type of learning is an example of the learning of stimulus-reinforcement associations—that is, of the learning of an association between a stimulus and a reward or a punishment—and is similar to classical or Pavlovian conditioning. In the neurons just described, the learning was of the association between a visual stimulus and food reward. The activity of these neurons became closely associated with stimuli which signified food in that the neurons did not respond when a different visual stimulus, which signified the taste of aversive saline, was presented. The activity of other neurons in the lateral hypothalamus of the monkey is also influenced by this type of learning, but the activity of these neurons became associated with the presentation of stimuli which became aversive because they were followed by punishment. Examples of these neurons noted in Section II.B were the neurons which responded when the monkey was shown the squeeze bulb used to puff air lightly onto its face. Another example of this type was a neuron which responded to the aversive visual stimulus but not to the food-associated visual stimulus during visual discrimination using the shutter. The latency of the response was approximately 150-200 msec (see also Rolls et al., 1979). It was important to investigate whether the activity was associated with any visual stimulus which indicated punishment (the taste of saline in this case), or whether the neuron happened to respond only to the particular visual stimulus paired by chance with saline. To investigate this, it was appropriate to perform a reversal of the significance of the visual stimuli; when this was done, it was found that over a few trials the neuron ceased to respond to the visual stimulus formerly associated with saline and came to respond to the visual stimulus which was presently associated with saline and had formerly been associated with food. Thus the activity of the

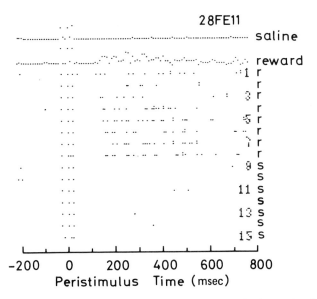

Figure 6 Responses of a neuron in the substantia innominata during a visual discrimination. When the visual stimulus which indicated that fruit juice (r) was available was shown (trials 1-8), the neuron fired (each action potential is indicated by a dot) with a latency of approximately 150 msec on the different trials. The response of contact with a tube by the monkey's tongue made to obtain the fruit juice is shown by a raised dot. When the visual stimulus which indicated the availability of 5% saline (s) was shown (trials 9-15), the neuron did not alter its activity relative to the prestimulus period (-200 to 0 msec), and no lick response was made. The sum of the neuronal responses for the eight reward trials is shown as a peristimulus time histogram beside "reward," and the sum of the neuronal responses for the seven saline trials is shown beside "saline" at the top of the figure. The sum buffers also emphasize that on trials on which the reward stimulus is shown behind the shutter, the neuron responded significantly more than on the saline trials, and that the latency of the neuronal responses was approximately 150 msec after the visual stimulus was shown. The trials have been grouped together into reward and saline only for convenience and were in fact run intermixed. The three columns of dots near 0 msec indicate the pulses used to open the electromagnetic shutter. On trial 2 the animal was not fixating the shutter when it opened, and the response of the unit as well as the response of the animal was delayed by the fixation time.

neuron was related to what was signified by the visual stimulus rather than to the particular physical characteristics of the stimulus used. It was also found that this neuron could be influenced by signals from other modalities, in that it came to respond to a tone when it signified that saline would be delivered and came not to respond to the tone when it signified that food would be delivered.

The activity of neurons in the rat hypothalamus has also been investigated during learning (see Olds, 1975 for review). When a rat learns that one tone signals that food will be delivered into a hopper, the activity of some units in the hypothalamus becomes associated with the tone but not with a different tone which occurs randomly in relation to the delivery of food (Olds et al., 1972; Linseman and Olds, 1973; Olds, 1973). The activity of the units was affected by the learning before the behavioral response of orientation to the food hopper became manifest during learning, and it occurred with a relatively short latency so that it could have been involved in the mediation of the learned responses. At least some of these neurons in the rat are not specifically involved in learning about foods, as some of them show similar changes in activity when a tone is followed by punishment (Linseman, 1974). Multiunit hypothalamic activity associated with an auditory stimulus which indicates that water is available, but not with a stimulus which indicates that water is not available, has been described in the thirsty rat (Sideroff and Bindra, 1976).

G. Activation by Brain-Stimulation Reward of Hypothalamic Neurons

A population of neurons in the lateral hypothalamus and substantia innominata of the monkey is activated by brain-stimulation reward, and it can be shown in the alert monkey that this population includes the neurons with activity associated with the sight and/or taste of food (Rolls et al., 1980). The activation seen is often transsynaptic excitation with a latency of 2-20 msec from a 0.1 or 0.5 msec pulse applied to a self-stimulation site in the lateral hypothalamus, orbitofrontal cortex, or nucleus accumbens. As a compound of excitation and inhibition is also often produced (see Fig. 7), it is difficult to relate whether a neuron decreased or increased its firing rate to food to whether rewarding electrical stimulation produced excitation or inhibition. To investigate whether the neurons activated by food reward and by brain-stimulation reward might be involved in food reward, electrical stimulation was applied through the recording microelectrode while it was in the region of these neurons to determine whether self-stimulation would occur. It was found that self-stimulation occurred while the microelectrode was in the region of these neurons, but it occurred less rapidly and eventually would not occur as the microelectrode was raised above these neurons (see Fig. 8; also Rolls et al., 1980). It was also found that the self-stimulation through the microelectrode was facilitated if the monkey was hungry and was attenuated when the monkey became satiated. These results are consistent with the hypothesis that the electrical stimulation was rewarding because it mimicked the effect of food for a hungry animal, and that the activity of these neurons is related to the rewarding effect which food has for a hungry animal. Evidence consistent with this view is also available in the rat, in which not only has it been shown that hypothalamic neurons are activated from different self-stimulation sites (see Rolls, 1974, 1975; Ito and Olds, 1971; Ito, 1972, 1976) but also that self-stimulation occurs through electrodes at sites from which neurons which respond to taste have been recorded (Norgren, 1970).

H. Analysis of the Function of These Hypothalamic Neurons

Neurons with activity related to feeding and learning have been found in the hypothalamus as described in the previous sections, and it is necessary to ask what the function of these neurons might be. One possibility is that these neurons only fire in relation to the animal's behavior because their activity is determined by responses being made by the animal. Other possibilities are that these neurons are involved in the mediation of the autonomic, endocrine, or feeding responses which occur when food is given to the hungry animal. This type of question is an essential feature of the investigation of the function of neurons

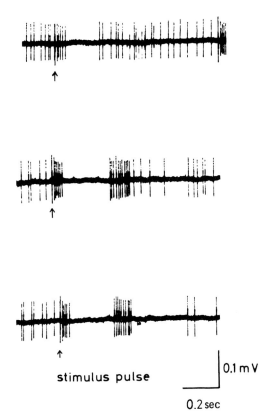

Figure 7 Excitation-inhibition of a lateral hypothalamic neuron produced by stimulation (arrow) at a self-stimulation site in the squirrel monkey. At low currents (just below the current threshold for self-stimulation at the reward site, top trace) excitation lasting approximately 50 msec was followed by inhibition lasting approximately 200 msec. At just higher currents (medium trace, at the threshold current for self-stimulation) this excitation-inhibition was followed by a further phase of excitation-inhibition, and this was more pronounced at higher currents (bottom trace) at current suprathreshold for self-stimulation.

by recording their activity in the alert animal. For hypothalamic neurons this question has been investigated by comparing the activity of hypothalamic neurons with the activity of neurons in sensory and motor structures recorded in the same test situation to determine how closely the function of hypothalamic neurons is related to that in sensory and motor pathways. After the comparison has been made in Sections III and IV, the question of the function of hypothalamic neurons can be more adequately assessed. In addition to a comparison of the activity of hypothalamic neurons with that of neurons in other areas, including comparisons of response latency, other information which is of particular relevance is the connectivity of hypothalamic neurons; and this, together with relevant information using other techniques, will be reviewed in Section V.

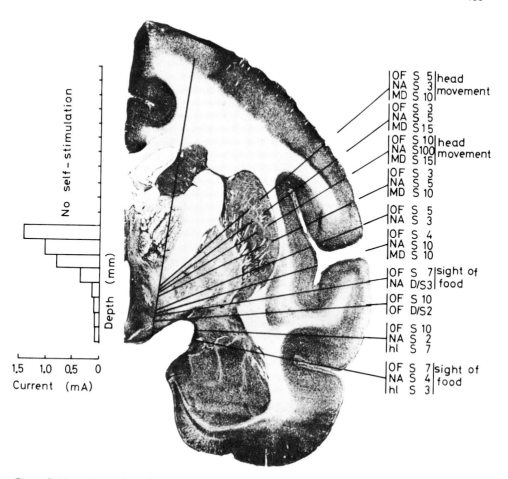

Figure 8 Hypothalamic units activated by both brain-stimulation reward and by natural reward (the sight of food) were recorded at the base of this microelectrode track. Self-stimulation through the recording microelectrode occurred in the region of these units as shown in the histogram of the current threshold for self-stimulation through the recording microelectrode. Both the response of the units to food and the self-stimulation in this region were attenuated if the squirrel monkey was not hungry. Above this region, in the globus pallidus, some activation of units, particularly from reward sites in the orbitofrontal cortex and mediodorsal nucleus of the thalamus, occurred, but the firing of these units occurred in relation to head movements, and self-stimulation of this region was poor (it probably occurred because of current spread to the hypothalamus below). The activation of the units was transsynaptic (S) or direct (D) from self-stimulation sites in the orbitofrontal cortex (OF), nucleus accumbens (NA), mediodorsal nucleus of the thalamus (MD), or lateral hypothalamus (hl), with the latencies shown in msec.

III. Comparison of the Activity of Neurons in the Hypothalamus with the Activity of Neurons in Sensory Input Pathways

To compare with the activity of the hypothalamic neurons which have responses associated with the sight of food, the activity of neurons in visual association cortex has been recorded during feeding in the monkey in the same test situation (Rolls et al., 1977). The part of visual association cortex chosen for analysis was the inferotemporal cortex, because it receives visual inputs after several earlier stages of visual cortical processing (Gross, 1973; Jones and Powell, 1970) and has connections to the lateral amygdala (Jones and Powell, 1970; Herzog and Van Hoesen, 1976) through which visual information could reach the hypothalamus (Jones and Powell, 1970; Nauta, 1961). In relation to feeding, it has been found that bilateral lesions of temporal cortex produce visual discrimination deficits and visual aspects of the Klüver-Bucy syndrome, which include a tendency to select nonfood as well as food objects (Akert et al., 1961; L. Weiskrantz, personal communication). In an anterior part of the inferotemporal cortex, neurons were found with responses which were sustained while the monkeys looked at effective stimuli in the test situation and, in this respect of being sustained, were comparable to the responses of the neurons in the lateral hypothalamus and substantia innominata. But the inferotemporal neurons did not respond only to food objects; and in many cases physical factors such as shape, size, orientation, color, and texture appeared to account for the responses of the neurons. Further it was found that the responses of the neurons were not affected by the significance of the visual stimuli, in that the magnitude of the responses to the effective stimuli was not affected by satiation or by learning that the visual stimulus was associated with food reward (glucose solution) or an aversive taste (5% saline solution). These findings show that responses continue in this part of visual association cortex when they have ceased to occur to nonfood objects in the hypothalamic neurons and indicate that the hypothalamic neurons do not respond as do visual sensory neurons in the same test situation, but rather show responses which are associated with the significance of the visual stimuli. These points are strengthened by the observation that the degree of hunger or satiety did not influence the magnitude of the responses of inferotemporal neurons to their effective visual stimuli, while in the same test situation the responses of hypothalamic neurons are only associated with the sight of food if the monkey is hungry. Another finding was that the shortest latencies of the responses of these anterior inferotemporal neurons were 100-150 msec or more, compared with approximately 150-200 msec for the latencies of hypothalamic neuronal responses in the same test situation with the shutter. This is an indication that the hypothalamic neurons are processing information at a later stage than at least this region of visual association cortex.

If it is correct that visual stimuli might be connected to hypothalamic neurons via the inferotemporal cortex and the amygdala (Jones and Powell, 1970), then it is of interest to determine at which stage subsequent to inferotemporal cortex the significance of the visual stimuli affects the neuronal response. One possible stage is the amygdala, for it has been reported that the activity of amygdaloid neurons is partly, but not uniquely, associated with visual stimuli which signify either food reward or punishment (Fuster and Uyeda, 1971; Ben Ari and Le Gal La Salle, 1972; Sanghera et al., 1979).

Little is known of the activity of neurons in the gustatory or olfactory pathways during feeding in the monkey. In the rat, it appears that in the mitral cell layer of the olfactory bulb the responses of units to food odors (but not to nonfood odors) are enhanced

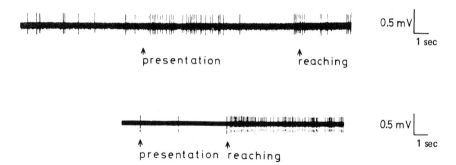

Figure 9 Comparison of the responses of a lateral hypothalamic unit (upper trace) with the response of a unit in the globus pallidus (lower trace). The lateral hypothalamic neuron responded when the animal just saw food and not in association with reaching for food, whereas the pallidal unit did not fire when the monkey first saw food, but only when he reached.

by hunger (Pager et al., 1972). It will be of interest to determine which is the first stage in the gustatory pathways where hunger influences the responses of neurons to gustatory stimuli. It has been reported that in the anesthetized rat, responses to some gustatory stimuli (e.g., sucrose) in the second-order gustatory neurons in the nucleus of the solitary tract are decreased by stomach distension (Glenn and Erickson, 1976).

IV. Comparison of the Activity of Neurons in the Hypothalamus with the Activity of Neurons in Motor Structures

As microelectrodes were lowered at an angle towards the lateral hypothalamus and substantia innominata, it was noted that some neurons in the globus pallidus were active during feeding. The responses of these neurons were usually clearly related to movements such as reaching, mouth movement, or ingestion and swallowing, or to touch to, for example, the monkey's face or tongue (Rolls et al., 1976). (This finding is consistent with that of De Long, 1971, who recorded from pallidal neurons during discrimination behavior.) The responses of a pallidal neuron, with activity related to reaching, are compared in Figure 9 with the responses of a hypothalamic neuron with activity associated with the sight of food. The responses of another pallidal neuron, with activity related to the mouth movement associated with ingestion, is compared in Figure 10 with the responses of a hypothalamic neuron associated with the taste of food. Thus, in contrast to units in the lateral hypothalamus or substantia innominata with responses associated with the sight and/or taste of food, it was usually possible to relate the responses of pallidal units active during feeding to a discrete movement or somatosensory stimulus. That there is a difference between the activity of units in the globus pallidus and the substantia innominata is also emphasized by the frequent observation, noted previously by De Long (1971) and Sparks and Travis (1973), that as the microelectrode is lowered through the region of the ventral border of the globus pallidus with the substantia innominata, there is a change from cells with activity related to movements to cells with activity more closely related to feeding. The latency of the responses of pallidal neurons with activity related to the lick

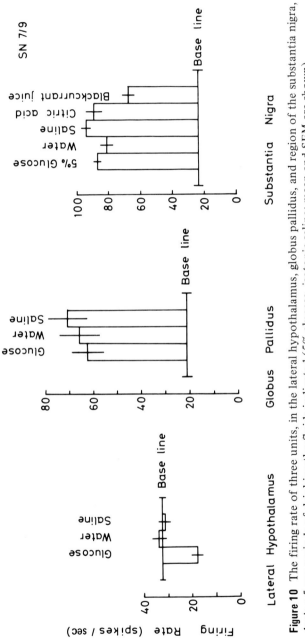

Figure 10 The firing rate of three units, in the lateral hypothalamus, globus pallidus, and region of the substantia nigra, during 5-sec periods of drinking the fluids indicated (5% glucose, isotonic saline; mean and SEM are shown).

response in the visual discrimination described in Section II.F has been measured in a number of pallidal neurons and is approximately 300 msec, with in some cases a single action potential being apparent at 250 msec after the shutter opens to reveal the visual stimulus and the monkey initiates a discriminated lick response (experiments of E. T. Rolls and M. K. Sanghera, 1976). This compares with a latency of 400 msec for the tongue to make contact with the food delivery tube and approximately 150-200 msec for the responses of the hypothalamic neurons in the same test situation. This comparison of latencies also suggests that the hypothalamic units have a function different from that of the pallidal units and is also consistent with the possibility that the hypothalamic neurons have a function which occurs prior to the organization of the motor response. A further observation, consistent with the possibility that the hypothalamic neurons have a function related to motivation and the pallidal neurons with the organization of the motor response, is that the responsiveness of the pallidal units, in contrast to the hypothalamic units, was not affected by transitions from hunger to satiety (Burton et al., 1976).

For another comparison recordings have also been made from single neurons in the region of the substantia nigra during feeding in the monkey (Mora et al., 1977). The substantia nigra is of interest in this context not only for comparison with the lateral hypothalamus, but also because damage to the nigrostriatal bundle produces a syndrome of aphagia, adipsia, and sensorimotor disturbance similar in at least some respects to the lateral hypothalamic syndrome, and because the nigrostriatal bundle is damaged by lateral hypothalamic lesions which produce aphagia and adipsia (Marshall et al., 1974). It was found that some neurons in the region of the substantia nigra had activity associated with feeding and that the activity of these neurons was often associated with discrete movements of the mouth or arms (Mora et al., 1977). The responses of these neurons related to movement were not affected by whether the monkey was hungry or satiated, and similar responses were obtained when fluids with different tastes were ingested (see Fig. 10). Thus, as with pallidal neurons, neurons in the substantia nigra with activity related to feeding usually had responses associated with discrete, identifiable movements. This is in contrast to the responses of hypothalamic neurons classed as having activity associated with the sight and/or taste of food, suggesting that these hypothalamic neurons have a function which is not motor, at least in the form shown by pallidal and nigral neurons (see Rolls et al., 1979). It would also be of interest to determine for the frontal motor cortex the exact timing relationships between these and hypothalamic neurons when the sudden appearance of a visual stimulus leads to the initiation of a discriminated, discrete motor response such as a lick. It would be much more difficult to determine the causal relationships between activity in these two areas if the animal were not in a situation where he was forced to make a discrimination to determine whether food was available and then to initiate rapidly a discrete response to obtain the food, as in the visual discrimination with a lick response described in Section II.F.

V. Synthesis: Nature of the Activity of Neurons Recorded in the Hypothalamus and Substantia Innominata of the Alert Animal

In addition to the neurons described in Section II.B with activity related to somatosensory and motor events, arousal, etc., there are neurons in the lateral hypothalamus and substantia innominata with activity which is closely related to feeding, drinking, and learning.

The function of these neurons will be considered first. The comparisons of the characteristics of the activity and response latency of these neurons with the activity of neurons in visual association cortex and in motor structures such as the globus pallidus and substantia nigra indicate that their function is different from that of sensory and motor structures and precedes that of motor structures in the organization of behavior. The association of the responses of the hypothalamic neurons with the sight of food, the occurrence of the responses only if the animal was hungry, and the modification of the neuronal responses during learning which resulted in responses to stimuli which signified food suggest that these neurons are involved in the organization of the changes produced by food in the hungry animal. These changes include autonomic effects such as salivation, endocrine effects, and food reward so that the hungry animal organizes his behavior to obtain and eat the food. The activity of these neurons could be involved in the mediation of some or all of these effects. In relation to the possibility that the activity of these neurons reflects food reward, it is consistent that these neurons are activated by brain-stimulation reward at a number of sites, that electrical stimulation through a recording electrode in the region of these neurons produces reward, and that this self-stimulation is enhanced by hunger and reduced by satiety (see further Rolls, 1975, 1976, 1980; Rolls and Rolls, 1977; Rolls et al., 1980).

An interesting point is that if these neurons only respond to, for example, the sight or taste of food when the animal is hungry, so that their activity signals food reward, then this signal must be able to influence the animal's behavior to ensure that he will work to obtain the food. This implies that neural pathways travel from these neurons to structures which organize what behavior should be performed next on the basis of the rewards available. It is therefore of interest that efferent connections from the region of the lateral hypothalamus and substantia innominata have recently been described. One set of pathways, from the basal forebrain nuclei of Meynert, projects directly to many cortical areas, including frontal and parietal cortex (Kievit and Kuypers, 1974, 1975; Divac, 1975). The cortex could use information about the rewards available, which would reflect the current motivational and emotional state of the animal, and the effects of reward- and punishment-related learning, to organize the behavior of the animal. Another set of pathways runs from the region of the lateral hypothalamus, lateral preoptic area, substantia innominata, and basal nucleus of Meynert toward the brain stem to structures which include the central gray and substantia nigra (Swanson, 1976). Further analyses of the connectedness of these neurons in the lateral hypothalamus and substantia innominata with activity related to feeding and learning, including latency, electrophysiological mapping, and anatomical evidence, will help to determine what the function of these interesting neurons is by defining what inputs they receive and whether neuronal systems which control autonomic and endocrine function and/or behavior are influenced by their activity.

Two other points should be emphasized about these feeding-related neurons. First, they are not only in the lateral hypothalamus, but are in a sheet which stretches out laterally into the region of the substantia innominata and basal forebrain nuclei of Meynert, in which a number of different types of neuron have been demonstrated anatomically. Second, the mean spontaneous firing rates of these neurons are not related to the nutritional state or hunger of the animal [though the activity of other neurons in this region may be related to such variables (see Oomura et al., 1975; Rolls, 1976)]; so thus they may not signal hunger, but rather the responsiveness of these neurons to food is influenced by hunger.

Less is known about the activity of other neurons in the hypothalamus of the alert animal. The neurons in the supraoptic nucleus recorded in the rhesus monkey by Vincent and his colleagues are an exception in that their activity is closely related to the secretion of antidiuretic hormone. The hypothalamic neurons with activity related to aversive stimuli could function in the autonomic, endocrine, or behavioral responses to such stimuli in a manner analogous to that of the hypothalamic units with activity related to feeding. The many neurons in the hypothalamus which did not alter their activity during feeding, drinking, or learning could have functions related to other processes, including the relatively long-term control of endocrine and autonomic function, the sensing of internal state, and the control of different types of behavior. The electrophysiological activity of these other neurons in the alert, behaving animal is at present almost totally uninvestigated.

Acknowledgments

The author has worked on some of the experiments described here with Drs. M. J. Burton, S. J. Judge, J. M. Koolhaas, G. J. Mogenson, F. Mora, B. J. Rolls, and M. K. Sanghera, and their collaboration is sincerely acknowledged.

References

Akert, K., Gruesen, R. A., Woolsey, C. N., and Meyer, D. R. (1961). Klüver-Bucy syndrome in monkeys with neocortical ablations of temporal lobe. *Brain 84*, 480-498.

Anand, B. K., and Brobeck, J. R. (1951a). Hypothalamic control of food intake in rats and cats. *Yale J. Biol. Med. 24*, 123.

Anand, B. K., and Brobeck, J. R. (1951b). Localization of a feeding center in the hypothalamus of the rat. *Proc. Soc. Exp. Biol. Med. 77*, 323-324.

Anand, B. K., and Pillai, R. V. (1967). Activity of single neurones in the hypothalamic feeding centres: Effect of gastric distension. *J. Physiol.* (London) *192*, 63-77.

Anand, B. K., Chhina, G. S., Sharma, G. S., Dua, S., and Singh, B. (1964). Activity of single neurons in the hypothalamic feeding centers: Effect of glucose. *Amer. J. Physiol. 207*, 1146-1154.

Arnauld, E., Dufy, B., and Vincent, J. D. (1975). Hypothalamic supraoptic nucleus: Rates and patterns of action potential firing during water deprivation in the unanaesthetized monkey. *Brain Res. 100*, 315-325.

Barraclough, C. A., and Cross, B. A. (1963). Unit activity in the hypothalamus of the cyclic female rat: Effect of genital stimuli and progesterone. *J. Endocrinol. 26*, 339-359.

Ben Ari, Y., and Le Gal La Salle, G. (1972). Plasticity at unitary level: II. Modifications during sensory-sensory association procedures. *Electroenceph. Clin. Neurophysiol. 32*, 667-679.

Brooks, C. M., Ishikawa, T., Koizumi, K., and Lu, H. (1966). Activity of neurones in the paraventricular nucleus of the hypothalamus and its control. *J. Physiol.* (London) *182*, 217-231.

Burton, M. J., Rolls, E. T., and Mora, F. (1976). Effects of hunger on the responses of neurons in the lateral hypothalamus to the sight and taste of food. *Exp. Neurol. 51*, 668-677.

Campbell, J. F., Bindra, D., Krebs, H., and Ferenchak, R. P. (1969). Responses of single units of the hypothalamic ventromedial nucleus to environmental stimuli. *Physiol. Behav. 4*, 183-187.

Cross, B. A. (1973). Unit responses in the hypothalamus. In *Frontiers in Neuroendocrinology*, W. F. Ganong and L. Martini (Eds.), Vol. 3. Oxford University Press, New York, pp. 133-171.

Cross, B. A., and Green, J. D. (1959). Activity of single neurones in the hypothalamus: Effect of osmotic and other stimuli. *J. Physiol.* (London) *148*, 554-569.

Cross, B. A., and Silver, I. A. (1963). Unit activity in the hypothalamus and the sympathetic response to hypoxia and hypercapnia. *Expl. Neurol. 7*, 375-393.

Cross, B. A., and Silver, I. A. (1966). Electrophysiological studies on the hypothalamus. *Brit. Med. Bull. 22*, 254-260.

Dafny, N., and Feldman, S. (1970). Unit responses and convergence of sensory stimuli in the hypothalamus. *Brain Res. 17*, 243-257.

Dafny, N., Bental, E., and Feldman, S. (1965). Effect of sensory stimuli on single unit activity in the posterior hypothalamus. *Electroenceph. Clin. Neurophys. 19*, 256-263.

De Long, M. R. (1971). Activity of pallidal neurons during movement. *J. Neurophysiol. 34*, 414-427.

Divac, I. (1975). Magnocellular nuclei of the basal forebrain project to neocortex, brain stem, and olfactory bulb. Review of some functional correlates. *Brain Res. 93*, 385-398.

Dyer, R. G. (1974). The electrophysiology of the hypothalamus and its endocrinological implications. In *Integrative Hypothalamic Activity*, D. F. Swaab and J. P. Schade (Eds.). [*Progress in Brain Research*, Vol. 41.] Elsevier, Amsterdam, pp. 133-147.

Epstein, A. N. (1971). The lateral hypothalamic syndrome: Its implications for the physiological psychology of hunger and thirst. *Progr. Physiol. Psychol. 4*, 263-317.

Feldman, S., and Dafny, N. (1968). Acoustic responses in the hypothalamus. *Electroenceph. Clin. Neurophysiol. 25*, 150-159.

Findlay, A. L. R. (1972). Hypothalamic inputs: Methods, and five examples. In *Topics in Neuroendocrinology*, J. A. Kappers and J. P. Schadé (Eds.). [*Progress in Brain Research*, Vol. 38.] Elsevier, Amsterdam, pp. 163-191.

Fuster, J. M., and Uyeda, A. A. (1971). Reactivity of limbic neurons of the monkey to appetitive and aversive signals. *Electroenceph. Clin. Neurophysiol. 30*, 281-293.

Glenn, J. F., and Erickson, R. P. (1976). Gastric modulation of gustatory afferent activity. *Physiol. Behav. 16*, 561-568.

Gold, R. M. (1967). Aphagia and adipsia following unilateral and bilaterally asymmetrical lesions in rats. *Physiol. Behav. 2*, 211-220.

Gold, R. M. (1973). Hypothalamic obesity: The myth of the ventromedial nucleus. *Science 182*, 488-490.

Gross, C. G. (1973). Visual functions of inferotemporal cortex. In *Handbook of Sensory Physiology*, R. Jung (Ed.), Vol. 7, Pt. 3. Springer, Berlin, pp. 451-482.

Grossman, S. P. (1967). *A Textbook of Physiological Psychology*. Wiley, New York.

Grossman, S. P. (1973). *Essentials of Physiological Psychology*. Wiley, New York.

Grossman, S. P. (1975). Role of the hypothalamus in the regulation of food and water intake. *Psychol. Rev. 82*, 202-224.

Hamburg, M. D. (1971). Hypothalamic unit activity and eating behavior. *Amer. J. Physiol. 220*, 980-985.

Herzog, A. G., and Van Hoesen, G. W. (1976). Temporal neocortical afferent connections to the amygdala in the rhesus monkey. *Brain Res. 115*, 57-69.

Ito, M. (1972). Excitability of medial forebrain neurons during self-stimulating behavior. *J. Neurophysiol. 35,* 652-664.

Ito, M. (1976). Mapping unit responses to rewarding stimulation. In *Brain-Stimulation Reward,* A. Wauquier and E. T. Rolls (Eds.). North-Holland Publ., Amsterdam, Chap. 6, pp. 89-95.

Ito, M., and Olds, J. (1971). Unit activity during self-stimulation behavior. *J. Neurophysiol. 34,* 263-273.

Jones, E. G., and Powell, T. P. S. (1970). An anatomical study of converging sensory pathways within the cerebral cortex of the monkey. *Brain 93,* 793-820.

Kapatos, G., and Gold, R. M. (1973). Evidence for ascending mediation of hypothalamic hyperphagia. *Pharm. Biochem. Behav. 1,* 81-87.

Kawakami, M., and Kubo, K. (1971). Neuro-correlate of limbic-hypothalamo-pituitary-gonadal axis in the rat: Change in limbic-hypothalamic activity induced by vaginal and electrical stimulation. *Neuroendocrinology 7,* 65-89.

Kawakami, M., and Saito, H. (1967). Unit activity in the hypothalamus of the cat: Effect of genital stimuli, luteinizing hormone and oxytocin. *Jap. J. Physiol. 17,* 466-486.

Kievit, J., and Kuypers, H. G. J. M. (1974). Basal forebrain and hypothalamic connections to frontal and parietal cortex in rhesis monkey. *Science 187,* 600-662.

Kievit, J., and Kuypers, H. G. J. M. (1975). Subcortical afferents to the frontal lobe in the rhesus monkey studied by means of retrograde horseradish peroxidase transport. *Brain Res. 85,* 261-266.

Komisaruk, B. R. (1971). Strategies in neuroendocrine neurophysiology. *Amer. Zool. 11,* 741-754.

Komisaruk, B. R., and Beyer, C. (1972). Responses of diencephalic neurons to olfactory bulb stimulation, odor, and arousal. *Brain Res. 36,* 153-170.

Komisaruk, B. R., McDonald, P. G., Whitmoyer, D. I., and Sawyer, C. H. (1967). Effects of progesterone and sensory stimulation on EEG and neuronal activity in the rat. *Exp. Neurol. 19,* 494-507.

Lincoln, D. W. (1969). Responses of hypothalamic units to stimulation of vaginal cervix: Specific versus non-specific effects. *J. Endocrinol. 43,* 638-684.

Lincoln, D. W., and Wakerley, J. B. (1972). Accelerated discharge of paraventricular neurosecretory cells correlated with the reflex release of oxytocin during suckling. *J. Physiol.* (London) *222,* 23P-24P.

Linseman, M. A. (1974). Inhibitory unit activity of the ventral forebrain during both appetitive and aversive Pavlonian conditioning. *Brain Res. 80,* 146-151.

Linseman, M. A., and Olds, J. (1973). Activity changes in rat hypothalamus, preoptic area, and striatum associated with Pavlonian conditioning. *J. Neurophysiol. 36,* 1038-1050.

Marshall, J. F., Turner, B. H., and Teitelbaum, P. (1971). Sensory neglect produced by lateral hypothalamic damage. *Science 174,* 523-525.

Marshall, J. F., Richardson, J. S., and Teitelbaum, P. (1974). Nigrostriatal bundle damage and the lateral hypothalamic syndrome. *J. Comp. Physiol. Psychol. 87,* 808-830.

Mora, F., Rolls, E. T., and Burton, M. J. (1976). Modulation during learning of the responses of neurones in the lateral hypothalamus to the sight of food. *Exp. Neurol. 53,* 508-519.

Mora, F., Mogenson, G. J., and Rolls, E. T. (1977). Activity of neurons in the region of the substantia nigra during feeding. *Brain Res. 133,* 267-276.

Morgane, P. J. (1961a). Electrophysiological studies of feeding and satiety centers in the rat. *Amer. J. Physiol. 201*, 838-844.

Morgane, P. J. (1961b). Evidence of a "hunger motivational" system in the lateral hypothalamus of the rat. *Nature* (London) *191*, 672-674.

Morgane, P. J. (1961c). Medial forebrain bundle and "feeding centers" of the hypothalamus. *J. Comp. Neurol. 117*, 1-26.

Nauta, W. J. H. (1961). Fiber degeneration following lesions of the amygdaloid complex in the monkey. *J. Anat. 95*, 515-531.

Nicolaïdis, S. (1969). Early systemic responses to orogastric stimulation in the regulation of food and water balance: Functional and electrophysiological data. *Ann. N.Y. Acad. Sci. 151*, 1176-1203.

Norgren, R. (1970). Gustatory responses in the hypothalamus. *Brain Res. 21*, 63-77.

Norgren, R. (1976). Taste pathways to hypothalamus and amygdala. *J. Comp. Neurol. 166*, 17-30.

Olds, J. (1975). Unit recordings during Pavlovian conditioning. In *Brain Mechanisms in Mentardation,* UCLA Forum in Medical Sciences, No. 18, N. A. Buchwald and M. A. B. Brazier (Eds.). Academic Press, New York, Chap. 12, pp. 343-371.

Olds, J., Mink, W. D., and Best, P. J. (1969). Single unit patterns during anticipatory behavior. *Electroenceph. Clin. Neurophysiol. 26*, 144-158.

Olds, J., Disterhoft, J. F., Segal, M., Kornblith, C. L., and Hirsh, R. (1972). Learning centres of rat brain mapped by measuring latencies of conditioned unit responses. *J. Neurophysiol. 35*, 202-219.

Olds, M. E. (1973). Short-term changes in the firing pattern of hypothalamic neurons during Pavlovian conditioning. *Brain Res. 58*, 95-116.

Ono, T., Oomura, Y., Sugimori, M., Nakamura, T., Shimizu, N., and Kita, H. (1976). Hypothalamic unit activity related to bar-press food intake in the chronic monkey. In *Hunger: Basic Mechanisms and Clinical Implications*, D. Novin, W. Wyrwicka, and G. A. Bray (Eds.). Raven Press, New York.

Oomura, Y. (1973). Central mechanisms of feeding. *Advan. Biophys. 5*, 65-136.

Oomura, Y., Ooyama, H., Naka, F., Yamamoto, T., Ono, T., and Kobayashi, N. (1969). Some stochastical patterns of single unit discharges in the cat hypothalamus under chronic conditions. *Ann. N.Y. Acad. Sci. 157*, 666-689.

Oomura, Y., Sugimori, M., Nakamura, T., and Yamada, Y. (1975). Contribution of electrophysiological techniques to the understanding of central control systems. In *Neural Integration of Physiological Mechanisms and Behavior*, G. J. Mogenson and F. R. Calaresu (Eds.). University of Toronto Press, Toronto, Canada, pp. 375-395.

Pager, J., Giachetti, I., Holley, A., and Le Magnen, J. (1972). A selective control of olfactory bulb electrical activity in relation to food deprivation and satiety in rats. *Physiol. Behav. 9*, 573-579.

Pfaff, D. W., and Gregory, E. (1971). Olfactory coding in olfactory bulb and medial forebrain bundle of normal and castrated male rats. *J. Neurophysiol. 34*, 208-216.

Pfaff, D. W., and Pfaffmann, C. (1969). Olfactory and hormonal influences on the basal forebrain of the male rat. *Brain Res. 15*, 137-156.

Phillips, M. I. (Ed.) (1973). *Brain Unit Activity During Behavior*. Thomas, Springfield, Ill.

Rolls, E. T. (1974). The neural basis of brain-stimulation reward. *Progr. Neurobiol. 3*, 71-160.

Rolls, E. T. (1975). *The Brain and Reward.* Pergamon, Oxford.

Rolls, E. T. (1976). Neurophysiology of feeding. In *Appetite and Food Intake,* T. Silverstone (Ed.). Dahlem Konferenzen (*Life Sci. Res. Rep. 2,* 21-42), Berlin.

Rolls, E. T. (1980). Processing after the inferior temporal cortex related to feeding, learning, and striatal function. In *Brain Mechanisms of Sensation,* Y. Katsuki, M. Sato, and R. Norgren (Eds.). Academic Press, New York.

Rolls, E. T., and Rolls, B. J. (1977). Activity of neurones in sensory, hypothalamic and motor areas during feeding in the monkey. In *Food Intake and Chemical Senses,* Y. Katsuki, M. Sato, S. Takagi, and Y. Oomura (Eds.). Tokyo University Press, Tokyo, pp. 525-549.

Rolls, E. T., and Rolls, B. J. (1981). Brain mechanisms involved in feeding. In *Human Food Selection,* L. M. Barker (Ed.). AVI, Westport, Connecticut.

Rolls, E. T., Burton, M. J., and Mora, F. (1976). Hypothalamic neuronal responses associated with the sight of food. *Brain Res. 111,* 53-66.

Rolls, E. T., Judge, S. J., and Sanghera, M. (1977). Activity of neurones in the inferotemporal cortex of the alert monkey. *Brain Res. 130,* 229-238.

Rolls, E. T., Thorpe, S. J., Maddison, S., Roper-Hall, A., Puerto, A., and Perrett, D. (1979). Activity of neurones in the neostriatum and related structures in the alert animal. In *The Neostriatum,* I. Divac, and R. G. E. Oberg (Eds.). Pergamon, Oxford, pp. 136-182.

Rolls, E. T., Burton, M. J., and Mora, F. (1980). Neurophysiological analysis of brain-stimulation reward in the monkey. *Brain Res. 199,* in press.

Sanghera, M. K., Rolls, E. T., and Roper-Hall, A. (1979). Visual responses of neurones in the dorsolateral amygdala of the alert monkey. *Exp. Neurol. 63,* 610-626.

Schmitt, M. (1973). Influences of hepatic portal receptors on hypothalamic feeding and satiety centres. *Amer. J. Physiol. 225,* 1089-1095.

Sharma, K. N., Dua-Sharma, S., and Jacobs, H. L. (1975). Electrophysiological monitoring of multilevel signals related to food intake. In *Neural Integration of Physiological Mechanisms and Behavior,* G. J. Mogenson and F. Calaresu (Eds.). University of Toronto Press, Toronto, pp. 194-212.

Sideroff, S., and Bindra, D. (1976). Neural correlates of discriminative conditioning: Separation of associational and motivational processes. *Brain Res. 101,* 378-382.

Sparks, D. L., and Travis, R. P. (1973). Firing patterns of extrapyramidal and reticular neurons of the alert monkey. In *Brain Unit Activity During Behavior,* M. I. Phillips (Ed.). Thomas, Springfield, Ill., pp. 288-300.

Swanson, L. W. (1976). An autoradiographic study of the efferent connections of the preoptic region in the rat. *J. Comp. Neurol. 167,* 227-256.

Tanabe, T., Yarito, H., Iino, M., Ooshima, Y., and Takayi, S. F. (1975). An olfactory projection area in orbitofrontal cortex of the monkey. *J. Neurophysiol. 38,* 1267-1283.

Teitelbaum, P., and Epstein, A. N. (1962). The lateral hypothalamic syndrome: Recovery of feeding and drinking after lateral hypothalamic lesions. *Psychol. Rev. 69,* 74-90.

Teitelbaum, P., and Stellar, E. (1954). Recovery from the failure to eat, produced by hypothalamic lesions. *Science 120,* 894-895.

Travis, R. P., Jr., and Sparks, D. L. (1968). Unitary responses and discrimination learning in the squirrel monkey: The globus pallidus. *Physiol. Behav. 3,* 187-196.

Travis, R. P., Jr., Hooten, T. F., and Sparks, D. L. (1968). Single unit activity related to behavior motivated by food reward. *Physiol. Behav. 3,* 309-318.

Ungerstedt, U. (1971). Adipsia and aphagia after 6-hydroxydopamine induced degeneration of the nigrostriatal dopamine system. *Acta Physiol. Scand. 81* (Suppl. 367), 95-122.

Vincent, J. D., Arnauld, E., and Bioulac, B. (1972). Activity of osmosensitive single cells in the hypothalamus of the behaving monkey during drinking. *Brain Res. 44,* 371-384.

Zeigler, H. P., and Karten, H. J. (1974). Central trigeminal structures and the lateral hypothalamic syndrome. *Science 186,* 636-638.

Author Index

Italic numbers give page on which complete reference is listed.

A

Abboud, F. M., 27, *69*
Abdallah, A. H., 374, *407*
Abdel-Sayed, W. S., 27, *69*
Abele, M., 379, *426*
Abrahams, V. C., 23, *69*
Abrams, R. M., 41, *75*
Adair, E. R., 31, 39, 42, 43, 44, *69, 80*
Adamo, F., 110, *188*
Adolph, E. F., *277*
Aghajanian, G. K., 7, 47, *81*
Ahlquist, R. P., 311, *407*
Ahlskog, J. E., 255, *277*, 318, 331, 332, 333, 335, 363, 365, 378, *407*
Ahn, H. S., 371, *407*
Ahrne, I., 148, 150, *191*
Akert, K., 456, *461*
Akmayev, I. G., 255, *277*
Albert, D. J., 338, *407*
Albert, D. L., 253, *277*
Albert, J. G., 253, *277*, 338, *407*
Alderdice, M. T., 166, *185*
Alexander, G. L., 236, 237, 255, *279*
Alexander, R. W., 312, *407*
Alheid, G. F., 396, 399, *420*
Allen, J. D., 231, *292*
Alpers, B. J., 16, 32, *70*
Alpert, N. R., 85, *206*
Anand, B. K., 22, *81*, 236, 237, 255, 256, 257, 258, 259, 261, 268, 273, *277, 295*, 440, 441, 442, *461*
Andén, N. E., 114, 117, *181, 191*, 313, *407*

Anderson, E., 215, 256, *284*
Anderson, G. H., 327, 341, 344, 376, *407, 408*
Andersson, B., 15, 16, 22, 24, 31, *70*, 327, 340, 342, 387, *407*
Andreas, K., 123, *206*
Anlezark, G. M., 329, *407*
Anlicker, J., 232, *277*
Antelman, S. M., 358, *407*
Antin, J., 402, *407*
Anton, A. H., 22, *78*
Antunes-Rodriques, J., 327, 328, 341, 342, 380, 386, *407, 411, 412, 413, 429*
Appleton, H., 244, *283*
Aprison, M. H., 395, *407*
Aravich, P. F., 338, *408*
Arcomano, A., 315, 316, *423*
Armour, J. A., 85, *188*
Armstrong, S., 106, 168, *181, 188,* 307, *408*
Armstrong, W. F., *417*
Arnauld, E., 444, 445, 448, *461, 465*
Artunkal, A. A., 162, *181*
Ary, M., 175, 174, *181, 187*
Ash, A. S. F., 390, *408*
Ashley, D. V. M., 344, 376, *407, 408*
Ashner, M., *284*
Atkins, A. R., 136, *198*
Atkins, E., 136, 154, *181, 190*
Audillo, C. A., 229, *296*
Auffray, P., 221, 236, 237, *277*
Avanzino, G. L., 314, 357, *410*
Avery, D. D., 106, 107, 121, 122, 167, *181, 182*

Axelrod, J., 102, 180, *191, 202,* 355, 368, 391, 394, *428, 431, 433*
Azmitia, E. C., Jr., 368, *408*

B

Bächtold, H. P., 88, 170, *182, 185*
Baez, L. A., 318, 319, 361, *408*
Baile, C. A., 228, 248, 266, *277, 294,* 306, 370, 382, 402, *408*
Bailey, C. J., 223, 230, 231, *289*
Bailey, P., 236, 247, *277*
Baird, J. A., 106, 123, 130, 158, *182, 191*
Baker, J. C., 174, *190*
Baker, M. A., 7, 32, 40, 47, *70, 73, 75,* 90, 149, *192*
Baker, P. F., 143, *182*
Balagura, S., 217, 265, 270, 271, 274, *278, 281, 284,* 379, *412*
Balch, O. K., 332, *430*
Baldwin, B. A., 42, *70,* 101, *182*
Balinska, H., 221, 225, 232, 236, 261, *278*
Ban, T., 348, *408*
Banerjee, S. P., 319, *408*
Banerjee, U., 88, 174, *182*
Banet, M., 21, *70*
Barbour, H. G., 102, 147, 165, 173, *182*
Barchas, J. D., 96, *182,* 400, *435*
Bard, P., 16, 20, 34, 39, 45, *70, 79,* 151, *182,* 234, 236, 237, *279*
Barker, J. L., 326, 341, 386, *408*
Barletta, M., 346, *417*
Barnett, R. J., 221, 236, 241, 268, 273, *289, 290*
Barney, C. C., 39, *80,* 160, *182*
Barraclough, C. A., 440, *461*
Barrera-Mera, B., 350, *429*
Barris, R. W., *285*
Bartels, M., 214, *278*
Bartholini, G., 314, 357, 361, *424*
Barzaghi, R., 318, 358, *408*
Bates, D., 352, *408*
Bates, M. W., 242, 243, 244, *278*
Battista, A., 90, *182*
Baulu, J., 329, *429*
Baum, H., 163, *195*
Bauter, H., 391, 394, *433*
Bazett, H. C., 16, *70*

Beaton, J. R., 245, *289*
Beatty, W. W., 222, 224, 231, 250, *278, 297*
Beauvallet, M., 110, *182*
Beaven, M. A., 110, *204*
Becker, E. E., 217, 220, 246, *278,* 342, *420*
Beckman, A. L., 49, *70,* 93, 94, 102, 108, 121, 122, 168, *182, 183, 206*
Beleslin, D. B., 84, 97, 98, 99, 101, 130, 139, 171, *183, 200, 201, 208*
Bellinger, L. L., 348, *426*
Bellward, G., 351, *413*
Belt, B. M., 254, *278*
Ben Ari, Y., 456, *461*
Bendotti, C., 318, 364, 373, 374, *431*
Bennett, C. T., 393, 394, *408*
Bennett, I. L., Jr., 148, *183*
Bennett, J. W., 27, 28, *75,* 158, *191*
Benson, M. J., 139, *208*
Bental, E., 440, *462*
Benzinger, T. H., 12, 29, 37, 38, 67, *70,* 83, 136, 175, *183*
Bergquist, E. H., 41, *80*
Bergström, S., 155, *183*
Bernard, B. K., 174, *202*
Bernasconi, S., 318, 364, 373, 374, *431*
Bernardis, L. L., 232, 234, 235, 238-251, 255, 269, 270, *278, 279, 280, 282, 283, 286, 294, 295*
Berner, C. N., 223, 230, 254, *294, 295,* 338, 339, 365, *432*
Bernheim, H. A., 170, 176, *183, 195*
Berthoud, H. R., 238, 239, *279,* 345, 346, 350, *408*
Best, P. J., *464*
Bhargava, A. K., 118, 123, *207*
Bhargava, K. P., 118, 123, *207,* 323, 325, 367, 368, 379, 380, 393, 395, *408*
Bicknell, E. J., 265, 271, *283*
Bierman, E. L., 237, *279*
Bierman, S. M., 22, *78*
Biggart, J. H., 236, 237, 255, *279*
Biggio, G., 362, *408*
Bignall, K. E., 6, 20, 45, 46, 47, 48, 49, 51, 53, 54, 56, 57, 64, *70,* 173, 178, *184*
Bindra, D., 238, 239, 241, 243, 249, 254, *292, 297,* 335, 339, *428,* 440, 453, *461, 465*
Bioulac, B., 444, 445, *465*

Author Index

Birzis, L., 14, 16, 17, 19, *71, 76*
Bisset, G.-W., 323, *409*
Bissette, G., 132, *183,* 402, *427*
Björklund, A., 300, 301, 303, 353, 354, 355, 365, *409, 424*
Black, C. A., 358, *407*
Black, J. W., 390, *409*
Blackmore, W. P., 394, *409*
Blake, D. E., 174, *198*
Blake, W. D., 321, *418*
Blakeborough, L., 371, *426*
Blanc, G., 363, *424*
Blass, E. M., 228, 267, 271, *279, 287, 290*
Bleier, R., 20, *70,* 151, *182*
Bligh, J., 7, 28, 36, 40, *71,* 83, 107, 109, 118, 130, 165, *183, 187, 198*
Block, M. L., 357, 366, 387, *409*
Bloom, F. E., 96, 109, *183,* 300, 313, 335, 351, 353, 357, 398, *409, 421, 426, 430*
Blum, J. C., 221, 237, *277*
Blumberg, J. B., 313, 351, *409*
Blundell, J. E., 325, 339, 343, 362, 365, 370, 372, 373, 374, 375, 376, 377, *409*
Bockaert, J., 313, 351, *410*
Bohuon, C., 168, *191*
Bondareff, W., 308, *430*
Bonnycastle, D. D., 97, 110, *193*
Bonoma, C., 349, *421*
Booth, D. A., 307, 309, 311, 315, 335, 344, 357, *410, 413, 426*
Borbely, A. A., 139, *207*
Borison, H. L., *206*
Boros-Farkas, M., 92, *183*
Borroni, E., 318, 364, 373, 374, *431*
Borsook, P., 94, 95, 101, 109, *184*
Boucher, R., 320, 361, *421*
Boulant, J. A., 6, 9, 13, 15, 29, 30, 36, 37, 46, 47, 48, 49, 51, 52, 53, 54, 55, 56, 57, 58, 59, 60, 61, 63, 64, 65, 66, 67, 68, *71, 78, 80,* 126, 173, 176, 178, *184*
Bouman, P. R., 239, 240, 241, 249, 267, *289*
Bowers, C. Y., 249, *293*
Boyle, P. C., 258, 263, 264, 266, *279, 286*
Bradley, P. B., 314, 357, *410*
Bramwell, B., 214, 215, *279*

Braun, J. J., 260, 263, *279, 287*
Bray, G. A., 211, 215, 221, 222, 237-245, 249, 250, 251, 252, *279, 285, 291, 295, 298, 410*
Breckenridge, B. M., 130, *184,* 351, *410*
Breckenridge, C., 229, *287,* 402, *421*
Breese, G. R., 115, *184, 193,* 318, 327, 344, 345, 373, *410, 418*
Bremer, F., 236, 247, *277*
Brengelmann, G. L., 29, 37, 38, *82*
Bresler, D. E., 344, 345, *413*
Brezenoff, H. E., 134, *184*
Bridges, T. E., 323, 325, 326, 367, 386, *410*
Brimblecombe, R. W., 123, *184*
Brittain, R. T., 92, 113, *184, 195*
Brobeck, J. R., 31, 32, 35, 39, *72, 78, 191,* 214, 217, 220, 221, 222, 223, 226, 232, 234, 235, 236, 237, 238, 242, 243, 244, 247, 256, 257, 258, 259, 261, 268, *277, 279, 282, 283, 290, 296, 297,* 334, 335, *410, 440, 461*
Brodie, B. B., 110, 170, *190, 204*
Brodoff, B. N., 248, *279*
Broekkamp, C. L. E., 344, 373, *410*
Brooks, C. M., 217, 218, 220, 221, 223, 234, 236, 237, 242, 244, 246, 247, 248, 249, 250, *279,* 441, *461*
Brooks, F. P., 241, 242, 246, *293,* 349, *421*
Brophy, P. D., 137, *200*
Brouwer, B., 220, 223, *279*
Browley, W. R., 337-340, *428*
Brown, L. L., 329, 331, 332, 333, 334, 338, 339, 345, 351, 363, 365, 366, *423*
Brownstein, M. J., 102, *202,* 300, 355, 358, 365, 368, *410, 420, 428, 431*
Brück, K., 14, 15, 16, 18, 21, 23, 47, *71, 72, 82,* 101, 107, 122, 173, *184, 203, 207, 210*
Bruinvels, J., 92, 113, 173, *184*
Brut, D. R., 313, 360, *413*
Brutkowski, S., 232, 261, *278*
Brzezinska, Z., 137, 139, *191*
Buckman, J. E., 137, 140, *200*
Buerger, P. B., 384, *410*
Buggy, J., 337, 340, 342, 385, 386, 387, *410, 411*
Bullard, R. W., 37, 38, *79*

Burek, L., 243, *282*
Burkard, W. P., 389, *411*
Burks, T. F., 88, 104, 113, 114, 117, 170, *182, 184, 195, 207*
Burnstock, G., 106, 168, *181, 188*
Burright, R. G., 45, *81*
Burt, B., 329, *417*
Burt, D. R., 360, *411*
Burton, H., 14, *72*
Burton, M. J., 214, 274, *293*, 443, 444, 445, 446, 453, 457, 459, 460, *461, 465*
Butcher, L. L., 378, 380, 381, *411, 428*
Buzzo, H. J., 311, *421*
Byers, S. O., 243, 244, 249, 250, 251, *279, 282*
Bylund, D. B., 312, *411, 435*

C

Cabanac, M., 41, 47, 56, *72*, 153, *184*
Cajal, S. Ramon, 254, *280*
Calaresu, F. R., 23, *72*
Calcutt, C. R., 388, *411*
Caldwell, F. T., 41, *75*
Calvelo, M. G., 27, *69*
Camardo, J. S., Jr., 386, 387, 388, 401, *432*
Camargo, L. A., 327, 341, 380, *411, 412, 429*
Campbell, B. A., 245, *280*
Campbell, C., 49, 56, *78*
Campbell, J. F., 440, *461*
Campbell, R. G., 345, 346, *435*
Campbell, R. J., 352, *408*
Campfield, L. A., 240, 241, *285*
Canal, N., 89, 101, *184*
Cantor, A., 107, *184, 205*
Cardo, B., 363, *424*
Carette, M. B., 108, 160, *185, 203*
Carey, R. J., 333, 363, 375, *411*
Carino, M. A., 132, *185, 193*
Carlisle, H. J., 45, *72*, 122, *182, 196*, 222, 224, 227, 262, 275, *280*, 365, *411*
Carlisle, H. S., 357, *429*
Carlson, L. A., 175, *183*
Carlson, L. D., 101, *185*
Carlsson, A., 314, 315, 361, *411*
Carman, J. S., 403, *414*
Carmel, P. W., 274, *285*

Carmona, A., 349, *411*
Carruba, M. O., 154, 170, *185*
Casby, J. U., 42, *69*
Case, B., 261, *286*
Cassano, G. B., 320, *429*
Caudillo, C. A., 402, *434*
Chaffee, R. R. J., 101, *185*
Chai, C. Y., 25, 26, 31, 34, 35, *72*, 165, 173, *185*
Chambers, W. W., 11, 14, 20, 26, *72*, 151, *185*, 331, *433*
Chan, W. Y., 327, *411*
Chance, W. T., 382, *411*
Chanda, S. K., 319, *408*
Chang, K., 346, *423*
Chang, R. S. L., 389, *411*
Chappius, P., 39, *72*
Chase, T. N., 88, *185*, 315, 396, *427, 430*
Chau, R., 346, *417*
Chawla, N., 113, *185*
Cheh, H. Y., 38, *73*
Chen, H. I., 173, *185*
Cheng, M., 260, *297*
Cherry, G. R., 394, *409*
Chesarek, W., 174, *187*
Cheung, W. Y., 143, 144, *185*
Chhina, G. S., 441, *461*
Chiaraviglio, E., 327, 328, *411*
Chikamori, K., 239, 240, 252, *280*
Childress, A. R., 402, *411*
Chinn, C., 99, 112, 113, 117, *200*
Chiouverakis, C., 235, 238, 241, 248, 269, 270, 271, *280*
Chow, S. L., 240, 249, *284*
Chowers, I., 22, *72*, 166, *185*
Christensen, J. H., 245, 261, *286, 289*
Christophe, J., 242, *280*
Chu, K. C., 235, *284*
Citroni, J., 147, *185*
Clark, G., 147, *204*, 254, 256, *280*
Clark, H. B., 352, *429*
Clark, W. G., 130, 134, 150, 157, 162, 163, 166, 169, 171, 174, *185, 186, 196, 208*, 348, *411*
Clarke, N. P., 23, *72*
Clausen, D. F., 348, *411*
Clay, G. A., 170, *190*
Clement, J. A., 352, *421*
Clinesschmidt, B. V., 318, 373, 384, *412*
Clough, D. P., 35, *72*
Cocchi, D., *198*, 230, *290*, 345, 350, *427*

Cockeram, A. W., 107, 150, 157, *203*
Code, C. F., 211, *280*
Cohn, C., 234, 245, *280*
Colby, J. J., 231, *280*
Coldwell, B. A., 163, 166, *185*
Coleman, D. L., 238, 240, 241, 248, *280*
Colin-Jones, D. G., 274, *280*
Collier, G. H., 247, *280*, 344, *412, 424*
Collier, H. O. J., 165, *186*
Conforti, N., 166, *185*
Connelly, M. A., 352, *434*
Conner, J. D., 20, *73*
Connors, M. H., 262, *280*
Conrad, L. C. A., 126, *186*, 349, *412*
Cook, R. D., 168, *188*
Coons, E. E., 335, 360, *412, 433*
Copper, B. R., 318, 327, 373, *410, 418*
Cooper, K. E., 93, 103, 104, 105, 107, 117, 130, 134, 135, 139, 147, 148, 150, 151, 152, 154, 157, 158, 162, 164, 165, 166, 168, 169, 170, 175, *186, 194, 203, 208*
Coote, J. H., 23, *73*
Coran, N., 338, *429*
Corbit, J. D., 42, 43, *73*, 220, 222, 223, 224, *280*
Corrodi, H., 110, 111, 117, *186*
Cort, R. L., 222, 226, *296*
Coscina, D. V., 252, 274, *278, 280*, 335, 344, 345, 356, 376, 377, 379, *408, 412*
Costa, E., 110, *186*, 313, 374, 375, 398, *412, 415, 416*
Costentin, J., 318, *412*
Cott, A., 267, *297*
Cottle, W. H., 107, 109, *183*
Coury, J. N., 307, 309, 383, 384, *412, 414*
Covian, M. R., 327, 328, 341, 342, 380, *407, 411, 412, 429*
Cow, D., 102, *186*
Cowan, W. M., 349, *431*
Cox, B., 113, 122, 135, 165, 170, 173, 174, 175, 176, *181, 187, 191*
Cox, J. E., 247, *280*
Cox, V. C., 250, 274, *287, 297*, 326, *419*
Coyle, J. T., 398, *417*
Crabai, F., 110, *188*
Crawford, I. L., 20, *73*
Crawford, J. D., 244, *283, 296*
Cranston, W. I., 93, 105, 113, 134, 135, 151, 154, 160, 163, 166, 167, *186, 187,*

[Cranston, W. I.] *204*
Crawshaw, L. I., 41, 42, *73*, 90, 92, 157, 158, 160, 175, *187*
Crayton, J. W., 326, 341, 386, *408*
Creese, I., 313, 360, 371, *411, 413*
Cremer, J. E., 165, *187*
Critchlow, V., 235, *289, 291*
Cronin, M. J., 7, 47, *73*
Crosby, E. C., 7, *73*
Cross, B. A., 274, *280*, 335, *413, 427,* 440, 441, 443, *461, 462*
Crow, T. J., 329, *407*
Cruce, J. A. F., 234, *285,* 317, 344, 403, *413*
Cumby, H. R., 134, 162, 169, *185, 186*
Cunningham, D. J., 6, 47, *73,* 94, 108, *187*
Cushin, B. J., 402, *432*

D

Dahlström, A., 117, *181,* 301, 313, 353, 368, 405, *407, 413*
Dafny, N., 440, *462*
Dale, R., 357, 366, *430*
D'Alecy, L. G., 32, *77,* 176, *195*
Dalglish, F. W., 381, *430*
D'Amour, M. C., 215, 241, *286*
Danforth, E., Jr., 245, *295*
Daniel, J. V., 377, *412*
Daniel-Severs, A. E., 401, *432*
Darling, K. F., 118, *188*
Darling, R. C., 38, *73*
Dascombe, M. J., 131, 163, *188*
Dasler, R., 101, *189*
Davis, D. D., 19, 24, 34, *74*
Davis, H. E., IV, 169, *186*
Davis, J. D., 350, *413*
Davis, J. N., 312, *407, 413*
Davis, J. R., 270, 271, 273, *280,* 307, 309, 350, *413*
Dawe, A. R., 85, *188*
Dawson, V., 375, *436*
Day, T. A., 106, 107, *188*
Dear, W., 249, *288*
Deaux, E., 261, *286*
de Caro, G., 401, *414*
Decastro, J. M., 270, 271, *284*
Defendini, R., 341, 394, 400, *413*
de Groot, J., 218, *297*

de Jong, W., 341, 349, 350, 355, *413*, *435*
Delabaume, S., 318, *412*
De Laey, P., 221, 235, *281*
Delaney, J. P., 28, *82*
de la Torre, J. C., 352, *413*
de la Vergue, P. M., 344, *428*
Dellman, H. D., 386, *432*
De Long, M. R., 444, 457, *462*
Deluca, C., 350, *431*
Demieville, H. N., 9, 49, 52, 54, 55, 56, 63, 64, 65, 67, *71*
Denbow, D. M., 137, *188*
Denny-Brown, D., 20, *73*
Dent, C., 221, 235, *281*
DeRuiter, L., 267, *297*
de Saro, A., 134, *202*
Des Prez, R., 154, *188*
Devenport, L. D., 217, *278*, *281*
De Wied, D., 118, *193*, 325, 326, 341, 386, 393, *413*
Dey, P. K., 169, *188*
DiCara, L. V., 341, *432*, *436*
DiCarlo, V., 102, *193*
Dickenson, A. H., 7, *73*
Dickinson, A., 229, *291*
Di Mascio, A., 400, *424*
Dimick, M. K., 261, *281*
Dimitrijević, M., 139, *183*
Dinarello, C. A., 151, *196*
Disterhoft, J. F., 452, *464*
Divac, I., 460, *462*
Dobrzánski, S., 318, 319, 320, 361, *413*
Doggett, N. S., 173, *188*, 318, 319, 320, 361, *413*
Dogterom, J., 393, *413*
Domino, E. F., 173, *188*
Donovick, P. J., 45, *81*
Dorn, J., 386, *413*
Downey, J. A., 38, *73*
Dragonica, J. A., 307, *425*
Drinsky, R., 229, *287*
Drummond, G. I., 351, *413*
Dua, S., 261, 273, *277*, *295*, 441, *461*
Dua-Sharma, S., 448, *465*
Duce, M., 110, *188*
Ducret, R. P., 373, 375, 377, 378, *418*
Duff, G. W., 163, *187*
Dufour, A. C., 238, 239, 246, *293*
Dufy, B., 445, 448, *461*

Duke, H. N., 323, *413*
Dun, N. J., 380, *420*
Duncan, W. A. M., 390, *409*
Durant, D. J., 390, *409*
Durr, J. C., 250, *290*
Dworkin, S., 20, *73*
Dwyer, P. E., 17, 18, 24, *78*
Dyball, R. E. J., 335, 341, *413*
Dyer, R. G., 335, 341, *413*, 442, *462*

E

Edens, F. W., 137, *188*
Edinger, H. M., 6, 18, 47, *73*
Edwards, C., 351, *413*, *427*
Efendic, S., 300, 353, 368, 395, *418*
Eichler, V. B., 270, *290*
Eichling, J. O., 352, *429*
Eisenman, J. S., 3, 6, 18, 46, 47, 48, 49, 54, 64, *70*, *73*, *79*, 94, 108, 153, 168, *183*, *188*
Ekman, L., 22, *70*
Elde, R., 300, 353, 368, 395, 400, *418*
Elder, J. T., 97, 110, *205*
Elek, S., 249, 250, 251, *282*
Eliel, L. P., 237, 241, *284*
Elizondo, R. S., 39, *80*, 160, *182*
Ellison, G. D., 275, *280*, 344, 345, *413*
Ellman, S. J., 329, *417*
Emery, N., 307, *426*
Eng, R., 252, *294*, 335, *414*
Engle, D. J., 357, 366, *430*
Enns, M., 344, *417*
Epstein, A. N., 212, 218, 229, 230, 257-261, 263-268, 270, 273, *281*, *283*, *287*, *289*, *293*, 295, *297*, 307, 310, 311, 321, 327, 335, 345, 350, 365, 375, 386, 387, 388, 392, 394, 401, 402, *414*, *419*, *430*, *432*, *433*, 440, *462*, *465*
Erb, W. H., 16, 32, *70*
Erdheim, J., 215, *281*
Erickson, R. P., 457, *462*
Erve, P., 139, *209*
Evans, B. K., 106, 168, *181*, *188*
Evans, S. E., *73*
Evarts, E. V., 351, 356, *431*
Evered, M. D., 340, 401, *414*
Ervin, G. N., 318, 373, *418*
Evetts, K. D., 313, 314, *414*

F

Fabry, P., 234, *281*
Falasco, J. D., 228, 229, 275, *288*
Falck, B., 355, *409*
Falk, J. L., 231, *281*
Fallon, J. H., 355, 365, *414*
Farrell, G., 244, *283*
Farrell, R., 329, *417*
Fawcett, A. A., 27, 28, *75*, 158, *191*
Feist, D. D., 97, *188*
Feldberg, W., 84, 86, 88, 89, 90, 92, 97, 101, 102, 103, 117, 137, 149, 151, 155, 157, 158, 159, 162, 169, 172, 174, *182, 188, 189*
Feldman, S., 166, *185*
Feldman, S. E., 261, *281*
Feldman, S. M., 331, *428*, 440, *462*
Fenard, S., 314, 357, *431*
Feng, L. Y., 238, 239, 240, 241, 242, 243, 244, 249, *283, 284*
Ferenchak, R. P., 440, *461*
Ferguson, N. B. L., 223, 260, *281, 288*
Ferguson, T. B., 31, *74*
Fernstrom, J. E., 372, *414*
Fertel, R., 313, *436*
Fessard, M. A., 108, *185*
Fibiger, A. P., 318, 357, 366, *437*
Fibiger, H. C., 245, *280*, 318, 362, 375, *414*
Fibiger, W. C., 340, *430*
Field, F. P., 101, *189*
Figlin, R., 313, *436*
Findlay, A. L. R., 441, *462*
Findlay, J. D., 31, *73*, 86, 118, *188, 189*
Fischer, P., 401, *415*
Fisher, A. E., 321, 357, 366, 367, 383, 384, 385, 387, *409, 410, 414*
Fisher, C., 216, 237, *293*
Fisher, J. C., 250, *290*
Fitzsimons, J. T., 313, 314, 322, 326, 342, 390, 392, 401, 402, *414, 427*
Flannigan, K. P., 271, 273, *294*
Flower, R. J., 155, *189*
Folkow, B., 23, *73*
Fonberg, E., 221, 225, 261, 270, *282, 293, 294*
Fonnum, F., 395, 396, *414*
Fontagne, J., 154, *190*
Ford, D. M., 49, *74*
Forgrave, P., 16, 17, *76*
Forn, J., 313, 351, *420*
Fornaro, P., 320, *429*
Forsling, M. L., 40, *74*
Fossler, D. E., 17, 18, 24, *78*, 131, *196*
Forster, R. E., II., 31, *74*
Forsyth, R. P., 23, *74*
Foster, M. A., 176, *195*
Foster, R. S., 106, 122, 171, 174, *189, 197*
Fox, H., 22, 43, *74*
Fox, K. A., 350, *414*
Fox, R. H., 148, *189*
Francesconi, R. P., 92, 165, *189*
Franklin, K. B. J., 223, *281*, 318, 335, *414, 418*
Frantz, A. G., 274, *285*
Fratta, W., 362, *408*
Freed, W. J., 403, *414*
Freedman, R., 313, *414*
Freeman, W. J., 18, 19, 20, 24, 44, *74, 81*
Fregly, M. J., 22, *78*, 101, *189*, 401, *415*
French, R. G., 221, 236, 241, *289*
Frens, J., 179, *189*
Freund, H., 135, *190*
Frey, H. H., 170, *190*, 318, *415*
Friedman, A. H., 97, *190*
Friedman, E., 307, 311, 317, 319, 360, 361, *415, 416, 431*
Friedman, M., 243, 244, 249, 250, 251, *279, 281, 282*
Friedman, M. I., 250, 266, 267, *281, 296*
Fröhlich, A., 214, *282*
Frohman, L. A., 214, 232, 234, 235, 238, 239, 240, 241, 242, 243, 244, 246, 249, 250, 251, 255, *278, 281, 283, 294*
Frommer, G. P., 224, 227, *288*
Fugazza, J., 110, *182*
Fuhrer, M. J., 38, *74*
Fujiwara, H., 329, 340, *428*
Fukushima, N., 106, *190*
Fuller, C. A., 21, *74*
Fuller, R. W., 174, *190*, 374, *415*
Fulton, J. F., *282*
Fusco, M. M., 2, 16, 24, 31, 36, *74, 75*
Fuster, J. M., 456, *462*
Fuxe, K., 90, 102, 110, 111, 117, *181, 182, 186, 190, 202*, 300, 301, 313, 353, 368, 369, 395, 400, 401, *405, 407, 413, 415, 418*

G

Gál, E. M., 97, *190*
Gale, C. C., 22, 42, 43, *70, 74, 80,* 90, 103, 114, 117, *204, 207*
Gale, K., 398, *415*
Gale, S. K., 230, *295*
Gallagher, T. F., 215, 221, 222, 237-240, 243, 244, *279*
Galli, A., 135, *205*
Gallinek, A., 225, *282*
Galster, W. A., 97, *188*
Ganellin, C. R., 390, *409*
Ganong, W. F., 22, *74,* 346, *420*
Ganten, D., 300, 353, 368, 395, 400, 401, *415, 418*
Ganten, U., 401, *415*
Garattini, S., 97, 174, *190, 205,* 317, 318, 361, 364, 370, 372, 373, 374, 378, 395, 397, *415, 431*
Garay, K. F., 324, 325, 327, 368, *415*
Gardey-Levassort, C., 154, 155, *190, 196, 202*
Gardner, D. R., 41, *74*
Garnett, J. E., 121, *205*
Garver, D. L., 102, *190*
Gaston, M. G., 221, 261, *287*
Gebber, G. L., 23, *74*
Geffen, L. B., 106, 107, *188*
Gentil, C. G., 327, 328, 341, 342, 380, *407, 411, 412*
George, R., 174, *197*
Gerald, M. C., 389, *415*
Gershon, S., 317, 318, 319, 360, 361, *415, 431*
Gessa, G. L., 110, 170, *188, 190,* 362, *408*
Ghezzi, D., 373, *431*
Giachetti, I., 457, *464*
Giarman, N. J., 154, *190*
Gibbs, J., 228, 229, 267, 275, *288, 295,* 350, 402, *407, 415, 420*
Gilman, A. G., 351, *415*
Ginsberg, M., 323, *415*
Girault, J.-M. T., 101, 163, *190, 193, 194*
Gisolfi, C. V., 101, 123, 130, 133, 139, 140, 141, 144, 145, *201, 209*
Gitter, S., 371, *436*
Gladfelter, W. E., 247, *282*
Glavcheva, L., 270, *282*
Glenn, J. F., 457, *462*
Glennon, J. A., 245, *295*
Glick, S. D., 265, *282,* 317, 320, 335, 356, 383, *415, 416*
Glomset, J. A., 237, *279*
Gloor, P., 46, 47, 49, *77*
Glowinski, J., 180, *191,* 313, 314, 315, 320, 351, 360, 361, 363, *410, 416, 424*
Gnegy, M. E., 313, *412, 416*
Godschalk, M., 218, *295*
Godse, D. D., 335, 344, 345, 356, *412*
Gogerty, J. H., 89, 101, *193*
Göing, V. H., 154, *191*
Gokhale, S. D., 313, *416*
Gold, R. M., 218, 220, 230, 232, 234, 235, 252, 253, 254, *282, 285, 294,* 331, 332, 335, 338, 399, *414, 416, 419, 432,* 439, *462, 463*
Goldberg, A. M., 102, 126, 130, *194*
Goldberg, R. S., 250, *287*
Golden, S., 242, 244, *286*
Goldman, H. W., 307, 311, *416*
Goldman, J. K., 232, 234, 235, 239, 240, 241, 242, 243, 244, 246, 249, 250, 251, *278, 282, 283, 286, 294, 295*
Goldman, W., 307, 318, 321, 370, 371, 379, *421*
Goldstein, D. J., 393, *416*
Goldstein, M., 90, *182,* 300, 353, 368, 395, 400, *418*
Gonzalez, R. R., 13, 25, 29, 30, 31, 34, 36, 59, *71, 74,* 77, 93, 105, *191*
Goodall, E. B., 363, 375, *411*
Goodlet, I., 373, *434*
Gordon, R., 110, *191*
Górka, J., 150, 157, 158, *206*
Gorski, R. A., 236, *291*
Gottesman, K. S., 307, *426*
Goudie, A. J., 374, *416*
Goy, R. W., 247, 250, *286*
Grabowska, M., 114, *191*
Grace, J. E., 341, *416*
Graef, I., 236, 237, *283*
Grafe, E., *191*
Graff, H., 222, 224, *283,* 334, *416*
Grandison, L., 395, 399, 402, 403, *416*
Grant, F. C., 256, *297*
Grant, L. D., 327, 379, *410, 412*
Grant, R., 31, *70,* 148, 150, *191*
Gray, W., 316, *436*
Green, J. D., 440, *462*
Green, J. P., 388, 392, *416*
Green, M., 346, *417*
Green, M. D., 134, 135, *191, 197*

Greenberg, D. A., 312, *416, 435*
Greengard, P., 313, 351, 360, 374, 375, *412, 417, 420*
Greenleaf, J. E., 137, 138, 139, *191, 202*
Greenstein, S., 265, *282,* 320, 335, 383, *415, 416*
Greenway, A. P., 329, *407*
Greer, G. L., 41, *74*
Gregory, E., 441, *464*
Griffith, D. R., 250, *285*
Grijalva, C. V., 265, 270, 271, *283, 288*
Grinker, J., 344, *417*
Grodins, F. S., 31, *78*
Gropetti, A., 318, 358, *408*
Gross, C. G., 456, *462*
Grossman, L., 341, *436*
Grossman, S. P., 230, *287,* 306, 307, 321, 335, 338, 339, 340, 341, 345, 356, 379, 381, 383, 396, 399, *412, 417, 420, 427, 435, 436,* 439, 440, *462*
Gruesen, R. A., 456, *461*
Grundman, M. J., 165, *186*
Grzanna, R., 398, *417*
Guerinot, F., 168, *191*
Guerra, F., *191*
Guideri, G., 346, *417*
Guidotti, A., 395, 398, 399, 402, 403, *415, 416*
Guieu, J. D., 15, 34, 46, 47, 49, 56, 64, *74*
Guillemin, R., 249, *288*
Gulati, O. D., 313, *416*
Gumulka, W., 170, 171, *204*
Gunn, R. H., 351, 356, *431*
Gupta, K. P., 158, 169, *188*
Gurman, Y., 389, 390, *417*
Gustafson, J. W., 260, *288*
Gutstein, W. H., 244, *283*

H

Haas, H. L., 351, 392, *417*
Haase, H. J., 357, *417*
Haessler, H. A., 244, *283, 296*
Hain, R. F., 256, 261, *298*
Hainsworth, F. R., 41, *74,* 270, *283*
Halaris, A. E., 341, 358, 369, *427, 436*
Hales, C. N., 238, 239, 240, 243, *283*
Hales, J. R. S., 27, 28, 31, *74, 75,* 106, 158, *182, 191*
Hall, G. D., 383, *427*

Hall, G. H., 86, 87, 123, 124, 125, 130, *191, 192*
Halpern, M., 331, *425*
Halperin, J. A., 393, *416*
Halperin, J. M., 329, *417*
Hamberger, B., 315, *417*
Hamburg, D. A., 400, *435*
Hamburg, M. D., 443, 445, *462*
Hamilton, C. L., 221, 228, 232, 235, 237, 238, 243, *283, 290*
Hamilton, J. W., Jr., *286*
Hamilton, L. W., 246, *294*
Hammel, H. T., 2, 3, 5, 6, 7, 12, 13, 16, 22, 24, 31, 33, 36, 41, 47, 67, *72, 73, 74, 75, 76, 79*
Hammer, N. J., 315, 316, *423*
Hammerschlag, R., 329, 398, *430*
Hammouda, M., 32, *75*
Han, P. W., 232, 234, 235, 238, 239, 240, 241, 242, 243, 244, 246, 247, 249, *283, 284, 285*
Hand, P. J., 387, 388, *426*
Handley, S. L., 113, *184*
Hanegan, J. L., 138, 143, *192*
Hannon, J. P., 85, *206*
Hansen, M. G., 115, *192,* 358, *417*
Harada, Y., 158, *193*
Harden, T. K., 346, *436*
Hardy, J. D., 2, 3, 6, 7, 13, 15, 16, 18, 19, 31, 33, 34, 36, 41, 42, 46, 47, 49, 53, 54, 56, 57, 58, 59, 60, 61, 64, 65, 66, 67, *71-75, 79, 80-82,* 83, 94, 108, 148, 153, 158, 160, 162, 175, 176, 177, *184, 187, 192, 206*
Hare, W. K., 175, *195,* 215, 241, *286*
Harell, L. E., 265, 270, 271, *283, 284*
Harper, A. M., 351, 352, *425*
Harrell, E. H., 217, 244, 249, *284*
Harri, M., 97, 110, 111, *192*
Harris, W. S., 135, *192*
Harrison, F., 31, 32, 39, *78*
Hartman, B. K., 300, 352, *429, 434*
Hartman, R., 338, *429*
Harvey, C. A., 163, 164, *198*
Hasama, B., *75,* 135, *192*
Hashimoto, M., 147, 173, *192*
Hass, W. K., 352, *417*
Hatton, G. I., *417*
Haubrich, R. D., 174, *198*
Hawkins, M., 158, 171, 174, *192, 193, 208*

Hawkins, R. A., 352, *417*
Haymaker, W., 215, 256, 273, *284, 291*
Hayward, J. N., 32, 40, 46, *70, 75,* 90, 149, *192*
Heaney, R. P., 237, 241, *284*
Heater, R. D., 97, *190*
Heath, J. E., 56, *80*
Hefco, V., 275, *284*
Heffner, T. G., 318, 319, 358, 361, *418*
Heinbecker, P., 236, 241, 243, 255, *284,* 335, *418*
Heldenberg, D., *284*
Heller, A., 335, *418*
Heller, E. D., 218, *295*
Heller, H. C., 227, *298*
Hellon, R. F., 6, 7, 14, 46, 47, 49, 56, 57, 64, *75, 76,* 83, 85, 94, 113, 160, 163, 166, 176, *187, 192*
Hellstrøm, B., 12, 13, 24, 31, 36, 37, *76*
Helman, R., 154, *188*
Helms, C. W., 236, 237, *287*
Hemingway, A., 9, 14, 15, 16, 17, 18, 20, 24, 36, 45, *71, 76, 81*
Henatsch, H. D., 15, *77*
Hendler, N. H., 321, *418*
Hendley, E. D., 316, *418*
Hennessy, J. W., 339, *417*
Hensel, H., 7, 21, *70, 76,* 83, 94, 97, 132, *192*
Herberg, L. J., 318, 335, *409, 414, 418*
Herman, Z. S., 325, 367, *436*
Herrera, M. G., 248, *277*
Hertting, G., 315, 379, *426, 429*
Herzog, A. G., 456, *462*
Hetherington, A. W., 215, 216, 222, 232, 234, 235, 246, 247, 249, 253, 254, 256, *284*
Higgins, D., 19, *79,* 101, 123, 141, 142, 143, 145, 151, 177, *201*
Hill, H. F., 170, *193*
Hillhouse, E. W., 323, 367, 386, *410*
Hilton, S. M., 23, *69, 76*
Himsworth, R. L., 274, *280,* 285
Hinman, D. J., 250, *285*
Hirsch, J., 234, *285*
Hirch, R., 452, *464*
Hirsch, E., 227, 247, *280, 295*
Hirsch, J., 402, *431*
Hissa, R., 93, 104, 108, 122, *192, 193, 203*
Hoch, D. B., 87, 123, 180, *200*
Hoebel, B. G., 219, 221, 245, 248, 252,

[Hoebel, B. G.] 255, *277, 285,* 332, 335, 373, 375, 377, 378, 382, *407, 418*
Hoffer, B. J., 313, 398, *409, 414*
Hoffman, P. C., 351, *418*
Hoffman, R. A., *193*
Hoffman, W. E., 401, *418*
Hojnicki, D., 235, 238, 241, 248, *280*
Hökfelt, B., 22, *70*
Hökfelt, T., 90, 102, 110, 111, 117, *186, 190,* 300, 313, 353, 368, 395, 400, *407, 418*
Holaday, J. W., 348, *418*
Holley, A., 457, *464*
Hollister, A. S., 318, 373, *418*
Hollister, L. E., 357, *418*
Holm, H., 232, 234, 235, 241, *285*
Holmes, R. L., 34, *76*
Holmgren, B., 22, *45*
Holt, J., 402, *407*
Holtzman, S. G., 318, 320, *418*
Hongslo, C. F., 241, *285*
Honig, W. M. M., 344, *410*
Honour, A. J., 93, 105, 134, 135, 151, 165, *186*
Hooten, T. F., 444, *465*
Hoover, D. B., 380, *418*
Horeyseck, G., 23, *76*
Hori, T., 47, 49, *76, 79,* 95, 109, 153, 158, *193, 201*
Horita, A., 89, 101, 132, 170, *185, 193*
Horn, A. S., 313, 315, *418*
Horowitz, J. M., 21, *74*
Horton, E. S., 245, *295*
Horton, E. W., 155, *193*
Horwitz, B. A., 21, *74*
Hosko, M. J., 173, *205*
Hösli, L., 314, 357, *410*
Houpt, K. A., 230, 285, 321, *419*
Howard, J. L., 115, *184, 193*
Hoyler, E., 312, *413*
Hrdina, P. D., 173, *202*
Hruska, R. E., 380, *420*
Huang, Y. H., 329, *424, 429*
Hubbard, J. E., 102, *193*
Huckaba, C. E., 38, *73*
Hughes, I. E., 315, *419*
Hulst, S. G. T., 118, *193*
Hummel, K. P., 238, 240, 241, 248, *280*
Humphrey, T., 7, *73*
Humphreys, R. B., 172, *193*

Author Index

Hustvedt, B. E., 232, 234, 235, 239, 240, 241, 242, 243, 244, 251, *285, 286*
Hutchins, M., 235, *289*
Hutchinson, J. S., 401, *415*
Hutchinson, R. R., 321, *419*
Hyatt, R. J., 403, *414*
Hyppa, M., 320, 361, *419*
Hystvedt, B. E., 242, *288*

I

Iino, M., *465*
Illei-Donhoffer, A., 92, *183*
Ingenito, A. J., 97, 110, *193*
Ingram, D. L., 25, 31, 40, 42, 45, *70, 72, 73, 74, 76,* 101, *182*
Ingram, W. R., 216, 236, 237, 256, 261, *285, 293*
Inoue, S., 239, 240, 241, 242, 251, 252, *285*
Iriki, M., 14, 15, 25, 27, 28, 31, *74, 76, 78, 80, 81*
Irwin, D. A., 384, *410*
Isaac, L., 98, 101, *193*
Isaka, K., 239, 240, 252, *280*
Ishikawa, T., 441, *461*
Ito, M., 453, *463*
Itoh, S., 106, *190*
Iversen, L. L., 110, 180, *191, 206,* 315, 320, 351, 360, 361, 395, 396, 398, *416, 419, 429*
Iversen, S. D., 318, *419*
Izumi, H., 239, 240, 252, *280*

J

Jackson, D. C., 2, 6, 7, 46, 47, 48, 49, 64, *73, 75*
Jackson, D. L., 149, 151, *193*
Jackson, H. M., 307, *419*
Jacob, C., 102, *194*
Jacob, J. J., 92, 101, 163, *190, 193, 194*
Jacobs, H. L., 211, 227, *290, 295*
Jacobson, F. H., 13, 17, 24, 31, 32, 36, *77, 80*
Jacobowitz, D. M., 96, 102, 115, 126, 130, *194, 195,* 300, 303, 317, 330, 337, 338, 339, 340, 344, 353, 355, 364, 368, 378, 380, *403, 413, 418, 419, 425, 428, 429*

Jacobj, C., 102, *194*
Jacobs, H. L., 448, *465*
Jahnke, G., 402, *427*
Jalanbo, H., *198*
Jancso, N., 173, *194*
Jancso-Gabor, A., 55, *81,* 173, *194, 207*
Jänig, W., 23, *76*
Jansky, L., 97, *202*
Janssen, P. A., 357, *417*
Jeffcoate, S. L., 300, 353, 368, 395, 400, *418*
Jell, R. M., 46, 47, 49, 54, 56, *77,* 88, 96, 109, 114, 135, 160, 162, *194, 207*
Jellinek, P., 170, *194*
Jeanrenaud, B., 238, 239, 241, 242, 243, 244, 246, 251, *279, 280, 286, 293*
Jenden, D. J., 122, 123, 174, *189, 195, 197*
Jequier, E., 39, *72*
Jeronen, E., 93, *203*
Jessel, T. M., 398, *429*
Jessen, C., 14, 15, 16, 26, 27, 31, 32, 34, 35, 40, *72, 77, 78, 80*
Jewett, R. E., 318, 320, *418*
Jiřička, A., 154, *204*
Jitarin, P., *284*
Jobin, M., 22, 43, *74*
Joel, W., 237, 241, *284*
Johansson, B., 23, *73*
Johansson, O., 300, 353, 368, 395, 400, *418*
Johnson, A. K., 337, 340, 342, 386, 387, *411*
Johnson, C. L., 388, 392, *416*
Johnson, J. M., 29, 37, 38, *82*
Johnson, P. R., 234, *285*
Johri, M. B. L., 113, *185*
Jones, A. P., 254, *282,* 338, *416*
Jones, B. E., 368, 369, *427*
Jones, C. W., 335, 341, *413*
Jones, D. L., 103, 104, 117, 130, 139, *186, 194, 208*
Jones, E. G., 456, *463*
Jones, M. J., 323, 367, 386, *410*
Jonsson, G., 368, 369, *415*
Joo, F., 55, *81,* 173, *207*
Jori, A., 170, 171, *204*
Joseph, D., 234, 245, *280*
Joseph, S. A., 236, 255, *285,* 335, *419*
Jouvet, M., 90, *194*
Joyce, D., 372, *419*
Judah, L. J., 147, *194*

Judge, S. J., 456, *465*
Judy, W. V., 23, *79*

K

Kaciuba-Uscilko, H., 137, 139, *191*
Kadekaro, M., 274, *285*
Kadlecová, O., 154, *198*
Kagan, H., 261, *286*
Kakimoto, Y., 351, *419*
Kakiuchi, S., 130, *194*
Kakolewski, J. W., 261, 274, *286, 297,* 326, *419*
Kaley, G., 327, *437*
Kałuża, Z., 150, *206*
Kamalian, N., 261, *286*
Kanarek, R. B., 346, *419*
Kandasmay, B., 163, *194*
Kanner, M., 350, *419*
Kant, G. J., 329, *420*
Kapatos, G., 220, 235, 253, 254, *282,* 332, 335, 338, 339, 341, *416, 419, 463*
Karakash, C., 238, 239, 241, 242, 243, 244, 246, 251, *286, 293*
Karczmar, A. G., 380, *420*
Karten, H. J., 260, *298,* 440, *465*
Kasatkin, Y. N., 250, *287*
Kasemsri, S., *286*
Katz, J., 242, 244, *286*
Kawa, A., 92, 97, *201*
Kawakami, M., 441, *463*
Kawamura, Y., 9, 15, 16, 18, 20, 24, 45, *81*
Keeler, R., 341, *420*
Keene, W. R., 148, *183*
Keesey, R. E., 216, 223, 232, 247, 250, 254, 258, 259, 261, 263, 264, 265, 266, 267, 268, 271, 273, 274, *278, 281, 286, 289, 292, 296,* 307, 309, 350, *413*
Keil, L. C., 402, *432*
Keller, A. D., 16, 20, 33, 34, 36, *77,* 136, 175, *195,* 215, 236, 241, *286*
Kelley, P. S., 38, *73*
Kelly, J., 321, 381, 383, 396, 399, *420, 432*
Kelly, P. H., 357, 366, *430*
Kelso, S. R., 48, 68, *78*
Kemper, G. C. M., 173, *184*
Kenmitz, J. W., 247, 250, 258, 263, 265, 266, 273, *286*
Kennedy, G. C., 226, 227, 228, 233, 236,
[Kennedy, G. C.] 238-240, 243, 249, *283, 286,* 287
Kennedy, J. I., 149, 151, 168, *196*
Kennedy, M. S., 113, 114, 117, 170, *195, 207*
Kenney, N. J., 402, 403, *420*
Kent, D. L., 368, *420*
Kent, M. A., 231, *287*
Kerwin, R., *187*
Khalaf, F., 221, 261, *287*
Kievit, J., 460, *463*
Killam, K. F., 400, *424*
Kimm, J., 229, *293*
Kimura, H., 398, *420*
King, B. M., 229, 230, 231, *287*
King, J. M., 250, *287*
Kinnard, M. A., 9, *79*
Kipp, S. C., 350, *414*
Kirchner, F., 23, *76*
Kirkpatrick, W. E., 106, 122, 123, 147, 171, *195, 196, 197*
Kirschbaum, W. R., 225, *287*
Kirtland, S. J., 163, *195*
Kissileff, H. R., 212, 217, 220, 246, 261, *278, 281, 287,* 322, 323, 326, 342, *420*
Kita, H., 443, *464*
Kitzinger, C., 12, 37, 38, 67, *70*
Kizer, J. S., 355, 365, 368, *420, 428*
Klase, P. A., 402, *432*
Klee, M. R., 15, *79*
Kluge, L., 220, 229, 230, 231, *294, 295*
Kluger, M. J., 13, 25, 30, 31, 32, 34, 36, *74, 77,* 170, 176, *183, 195*
Klussman, F. W., 15, *77, 79*
Knigge, K. M., 22, *78,* 236, 255, *285,* 335, *419*
Knox, G. V., 49, 56, *78,* 124, *195*
Kobayashi, H., 324, 367, 380, 386, *435*
Kobayashi, R. M., 380, *420*
Koe, B. K., 318, *436*
Koenig, H., *72,* 151, *185*
Koenig, R., 151, *185*
Koh, E. T., 349, *429*
Koizumi, K., 441, *461*
Koltun, W. A., 352, *421*
Komisaruk, B. R., 441, 442, *463*
Komiskey, H. L., 90, 92, 94, *195*
Komorowski, J. M., 238, 239, 241, *287*
Kon, S., 174, *205*
Konijnendijk, W., 239, 240, 241, 249, 267, *289*
Koob, G. F., 329, *420*

Koopmans, H. S., 252, *295*
Kopin, I. J., 315, *427*
Kornblith, C. L., 452, *464*
Kosaka, M., 14, 34, *78*
Kostrzewa, R. M., 115, *195*
Kozawa, E., 25, *76*
Kozlowski, S., 137, 139, *191*
Kraly, F. S., 267, 271, *279, 287,* 350, *420*
Kratskin, I. L., 149, *204*
Krauss, R. M., 228, *287*
Krausz, M., 389, 390, *417*
Krebs, H., 440, *461*
Kreinick, C. J., 260, *288*
Krieger, D. T., 346, *420*
Kriwanek, W., 315, *429*
Kroneberg, G., 154, *195*
Krueger, B. K., 313, 351, 360, *420*
Kruk, Z. L., 92, 113, *195,* 358, 370, 373, *421*
Krukowski, M., 320, 323, *421*
Krunsky, R., 402, *421*
Kruper, J., 229, *296,* 402, *434*
Krupp, P., 165, *210*
Kubikowski, P., 92, 106, *195*
Kubli-Garfias, C., 350, *429*
Kubo, K., 441, *463*
Kuenzel, W. J., 221, 236, 237, 238, *287*
Kühn, E. R., 323, 386, *421*
Kulkaini, A. S., 374, *407*
Kulkosky, P. J., 229, *287,* 402, *421*
Kullmann, R., 16, 25, 27, *77, 78*
Kulshrestha, V. K., 323, 325, 367, 368, 379, 380, 388, *408*
Kung, L. S., 319, *408*
Kuo, P. T., 238, 239, 240, 241, 242, 243, 244, 249, *283, 284*
Kurbjuweit, H. G., 154, *195*
Kuruma, I., 154, *207*
Kuryama, K., 398, *420*
Kuschinsky, W., 352, *421*
Kuypers, H. G. J. M., 460, *463*
Kym, O., 135, *195*

L

Laburn, H. P., 94, 95, 101, 109, 130, 168, *184, 196, 209*
Lahti, H., 122, *203*
Lakey, J. R., 231, *295*
Lambert, E. F., 218, 220, 221, 234, 236, [Lambert, E. F.] 237, 244, 248, 249, 250, *279*
Lanciault, G., 349, *421*
Landis, A. M., 311, *421*
Landis, S. C., 398, *421*
Lang, W. J., 106, 123, 130, *182*
Langelier, P., 320, 361, *421*
Larkin, R. P., 218, *287*
Larsson, K., 117, *181*
Larsson, S., 31, *70,* 262, *292*
Larue, C., 342, *421*
Laszlo, F. A., 325, 326, 386, *413*
Latham, C. J., 343, 362, 375, 376, *409*
Laudenslager, M. L., 122, *196*
Lauer, E. W., 71, *73*
Laverty, R., *196*
Lavyne, M. H., 352, *421*
Lazaris, Y. A., 250, *287*
Leach, L. R., 263, *287*
Leather, S. R., 371, *426*
Le Blanc, J., 101, *182*
Lechat, P., 154, 155, *190, 196, 202*
Leduc, J., 101, *196*
Lee, P. S., 122, *202*
Lee, T. C., 341, *413*
Lee, T. F., 113, 122, *187*
Lefkowitz, R. J., 312, *407, 413, 421*
Le Gal La Salle, G., 456, *461*
Legge, K. F., 25, 31, *76*
Legrand, M., 110, *182*
Lehr, D., 307, 311, 318, 320, 321, 323, 346, 370, 371, 379, *416, 417, 421*
Leibowitz, S. F., 254, *288,* 307, 308, 309, 310, 311, 312, 313, 314, 315, 316, 317, 318, 319, 321, 322, 324, 325, 327, 328, 329, 331, 332, 333, 334, 335, 337, 338, 339, 343, 344, 345, 346, 347, 349, 350, 351, 356, 357, 358, 359, 361, 363, 364, 365, 366, 367, 368, 370, 371, 372, 373, 374, 382, 389, 390, 391, 392, 393, 394, 395, *408, 415, 422, 423*
Le Magnen, J., 322, 326, 342, *414, 424, 457, 464*
Le Marchand, Y., 238, 239, 241, 242, 243, 244, 246, 251, *286, 293*
Le Moal, M., 363, *424*
Lepkovsky, S., 221, 236, 237, 238, *288*
Leschke, E., 147, *185*
Leshem, M. B., 339, 365, 372, 375, *409*
Leshner, A. I., 247, *280,* 344, *412, 424*
Lessin, A. W., 165, *196*
Leung, P. M., 228, *293*

Levi, R., 110, *198*
Leverenz, K., *424*
Levin, R. M., 313, *424*
Le Vine, M., 382, *433*
Levine, M. S., 260, *288*
Levine, N., 327, *437*
Levine, R. J., 163, *208*
Levine, R. M., 320, *436*
Levison, M. J., 224, 227, *288*
Levitsky, S. A., 342, *424*
Levitt, D. R., 274, *288*
Levitt, R. A., 383, 384, *410, 424, 433*
Lewin, R. J., 27, 38, *80*
Lewinska, M. K., 241, *288*
Lewis, M. J., 307, 403, *425*
Lewis, P. J., 175, *204*
Lewis, P. R., 124, 126, 130, *196, 206*, 380, 381, *424*
Li, P., 377, *412*
Liebelt, R. A., 232, 249, 250, *288, 291, 294*
Lim, P. K., 31, *78*
Lin, C. H., 235, *284*
Lin, M. T., 25, 26, 34, 35, 72, 89, 165, *185, 196*
Lincoln, D. W., 335, 341, *413*, 441, *463*
Lindholm, E., 265, 270, 271, *283, 288*
Lindsley, D. B., 19, *78*
Lindvall, O., 300, 301, 303, 353, 354, 355, 365, *409, 424*
Ling, G. M., 173, *202*
Linseman, M. A., 443, 444, 452, 453, *463*
Lints, C. E., 382, *411*
Lipton, J. M., 17, 18, 20, 24, 44, 45, *78*, 135, 147, 149, 150, 157, 158, 168, 171, 172, 174, *192, 193, 196, 208*
Lipton, M. A., 132, *183*, 400, 402, *424, 427*
Lisk, R. D., 130, *184*
Liu, A. C., 234, 235, 250, *283, 284, 298*
Liu, C. M., 228, 266, 275, *288, 298*
Liu, C. N., 11, 14, 20, 26, 72, 151, *185*
Liu, J. C., 11, 14, 20, 26, 72, 88, 151, *185, 196*
Ljungdahl, A., 300, 353, 368, 395, 400, *418*
Lloyd, K. G., 314, 357, 361, *424*
Lockwood, R. A., 217, 218, 220, 223, 256, *279*
Loewy, A. D., 349, *431*
Lomax, P., 49, 56, *78*, 106, 121, 122,

[Lomax, P.] 123, 124, 134, 135, 165, 171, 172, 173, 174, 175, 176, *181, 184, 187, 189, 191, 195, 196, 197*
Long, C. N. H., 151, 169, 170, *204*, 217, 220, 222, 223, 232, 234, 235, 236, 238, 242, 243, 244, 257, *279, 297*
Long, W., 244, *283*
Loosen, P. T., 132, *183*
Lorden, J. F., 255, *288*, 332, *424, 428*
Lorens, S. A., 379, *424*
Lorenz, D., 383, *424*
Lotter, E. C., 239, *288*
Lotti, V. J., 92, 170, 174, *182, 188, 197*, 384, *412*
Louis-Sylvestre, J., 239, 241, *288*
Lovenberg, W., 373, *425*
Lovett, D., 321, *425*
Løvø, A., 232, 234, 235, 239, 240, 241, 242, 243, 244, 246, 251, *285, 286, 288*
Lu, H., 441, *461*
Luduena, F. P., 311, *421*
Ludwig, O., 16, 35, *77*
Luff, R. H., 113, 166, *187, 204*
Luft, R., 300, 353, 368, 395, 400, *418*
Lutherer, L. O., 22, *78*
Lyon, M., 320, 331, 366, *425, 427*
Lytle, L. D., 348, *425*

M

Maas, W. J., 329, *429*
McBurney, P. L., 232, *288*
McCaleb, M. L., 345, 398, *425*
Mc Cann, S. M., 22, 72, 323, 327, 368, 386, 388, 402, *407, 413, 421, 427, 435*
Macht, M. B., 20, 34, 39, *70*
Mc Claskey, E. B., 16, 17, 20, 33, 34, 36, 77, 175, *195*
Maclean, P. D., 9, *79*
Mc Culloch, J., 351, 352, *425*
Mc Geer, E. G., 318, 362, 375, *414*
Mc Govern, J. F., 341, *436*
Mc Guffin, J. C., 318, 373, *412*
Mc Hugh, P. R., 228, 229, 275, *288*
Mc Hugh, P. R., 357, *430*
Mc Indeward, I., 373, *434*
Mc Kay, L. D., 402, 403, *420*
Mac Kenzie, E. T., 351, 352, *425*
McKenzie, R. G., 373, 375, 377, 378, *418*
McLennan, H., 9, *78*

Author Index

Mc Neill, T. H., 326, *425*
Mac Pherson, R. K., 148, *189*
Maddison, S., 402, *425,* 443, 450, 451, 459, *465*
Maeda, T., 102, *197*
Mager, M., 92, 165, *189*
Magoun, H. W., 19, 31, 32, 39, *78,* 147, *204,* 256, *280, 293*
Mah, C. J., 253, *277,* 338, *407*
Maickel, R. P., 135, *197,* 320, 389, *415, 436*
Makman, M. H., 371, *407*
Malkinson, T., 139, *208*
Mallow, J., 320, 323, *421*
Mangiapane, M. L., 384, 386, 387, 388, *425, 432*
Manian, A. A., 313, *428*
Mann, J. F. E., 401, *415*
Mantegazza, P., 154, *185, 198,* 318, 345, 350, 358, *408, 427*
Marcais, H., 318, *412*
Margules, D. L., 255, *288,* 307, 309, 332, 403, *424, 425, 428*
Marks, H. E., 221, 225, 231, 236, 237, *289*
Marley, E., 92, 108, 124, 157, 162, 171, 172, *181, 197, 202*
Marshall, J. F., 246, 258, 260, 261, 274, *289,* 356, 363, *425,* 439, 440, 459, *463*
Marshall, N. B., 245, *289*
Martin, G. E., 99, 100, 174, 180, *197, 209,* 265, 271, *291,* 315, 317, 362, *425*
Martin, J. B., 22, *78*
Martin, J. M., 239, 240, 241, 249, 267, *289*
Martres, M. P., 318, *412*
Mašek, K., 154, *197, 198, 204*
Mashayekhi, M. B., 245, *289*
Maskrey, M., 107, 109, 118, 130, *183, 198*
Massari, W. J., 378, *425*
Mathews, M., 42, *74*
Matthews, J. W., 307, 309, *426*
Masuda, K., 239, 240, 252, *280*
Maudry, M., 318, *412*
Maul, G., 230, 254, *294, 295*
May, K. K., 245, *289*
Mayer, E. Th., 16, 26, 31, 34, *77*
Mayer, J., 221, 228, 229, 232, 236, 242-245, 248, 266, 268, 273, *277, 278, 280, 287, 289, 290, 294, 297,* 346, *419*
Maynert, E. W., 110, *198*

Medon, P. J., 174, *198*
Meeker, R., 398, 399, *426*
Meile, M. J., 228, 229, *291, 293*
Melchior, C. L., 101, 123, 130, 139, 144, 145, *201*
Mendel, V. E., 348, *426*
Menzel, H., 163, 164, *206*
Metcalf, G., 19, *79,* 87, 101, 103, 113, 119, 120, 123, 132, 141, 142, 143, 145, 151, 177, *198, 201*
Meurer, K. A., 14, 15, *78*
Meyer, D. R., 227, *295,* 379, *426,* 456, *461*
Meyer, H. H., 16, *78,* 147, 175, *198*
Meyerhoff, J. L., 329, 348, *418, 420*
Michaud, G., *194*
Middlemis, D. N., 371, *426*
Millard, S. A., 97, *190*
Miller, A. T., Jr., 229, *296*
Miller, C. R., 221, 225, 236, *289*
Miller, J. J., 9, *78*
Miller, N. E., 223, 230, 231, *289,* 307, 308, 311, 313, 314, 315, 322, 357, 383, *423, 426, 432, 436*
Millner, J., 267, *297*
Milloy, S., 317, *416*
Mills, E., 325, *426*
Milton, A. S., 131, 148, 155, 158, 159, 162, 163, 164, 166, 169, *188, 189, 198,* 324, 325, 326, *426*
Mink, W. D., *464*
Mintz, E., 331, *425*
Miselis, R. R., 229, *289,* 387, 388, *426*
Misra, N. K., 7, 14, *76*
Mitchell, D., 136, 160, 163, *187, 198*
Mitchel, J. S., 258, 263, 266, 268, *286, 289*
Mitchell, J. A., 235, *289*
Moberg, G. P., 348, *426*
Mogenson, G. J., 23, *72,* 212, 239, 240, 269, *289, 296,* 340, 345, 346, 350, *408, 414,* 459, *463*
Moisset, B., 403, *425*
Molinoff, P. B., 312, 346, *433, 436*
Molloy, B. B., 374, *415*
Monahan, J. W., 402, *431*
Montemurro, D. G., 235, 244, 250, 257, 258, 270, 271, *290, 296*
Montgomery, R. B., 313, 314, *426*
Mook, D. G., 228, 250, *290*
Mooney, J. J., 151, 154, 169, 170, *190, 204*

Mooney, R. D., 25, 41, *80*
Moore, R. A., 115, *184*
Moore, R. Y., 270, *290,* 300, 353, 355, 365, 368, 369, *414, 426, 427*
Mora, F., 133, 140, 141, *190, 201,* 214, 274, *293,* 443, 444, 445, 446, 449, 450, 453, 459, 460, *461, 463, 465*
Moran, T., 228, 229, 275, *288*
Morgan, L. O., 148, *198*
Morgan, M. J., 375, *435*
Morgane, P. J., 102, *198,* 211, 213, 228, 252, 258, 260, 270, 271, 273, *290, 294,* 368, 370, *427,* 439, 440, *464*
Morris, J. F., 335, 341, *413*
Morris, M., 323, 325, 327, 368, 386, 388, *427*
Morrison, J. E., 174, *197*
Morrison, J. H., 398, *417*
Morrison, S. D., 258, 268, 270, 271, 273, *290*
Morrissey, S. M., 349, *434*
Mosher, R., 139, *209*
Moss, R. L., 325, *427*
Mountford, D., 321, *427*
Mouren-Mathieu, A. M., 237, *292*
Moxley, K., 316, *436*
Mrosovsky, N., 221, 236, *290,* 372, *419*
Mu, J. Y., 31, 35, *72,* 235, *284, 290*
Muelter, P. S., 344, *428*
Mufson, E. J., 263, 265, 271, *290*
Mulder, A. H., 395, *434*
Mullen, Y. S., 241, 252, *285*
Müller, E. E., *198,* 230, *290,* 326, 341, 345, 350, 386, *408, 427*
Mullis, M., 316, *436*
Murakami, N., 6, 47, 49, 53, 57, *73,* 76, 78, *79,* 94, 108, 109, *187, 199*
Murgatroyd, D., 16, 41, 42, *79*
Murphy, D. L., 88, *185*
Muth, E. A., 96, *194,* 380, *418*
Myers, M. G., 175, *204*
Myers, R. D., 9, 18, 19, *79,* 84, 85, 86, 87, 88, 89, 90, 92, 93, 94, 96, 97, 98, 99, 100, 101, 102, 103, 111, 112, 113, 114, 116, 117, 118, 119, 121-127, 130, 132, 133, 134, 136, 137, 138, 139, 140, 141, 142, 143, 144, 145, 147, 148, 149, 150, 151, 153, 155, 157, 160, 161, 163, 164, 165, 166, 168, 169, 171, 172, 174, 175, 177, 178, 179, 180, *183, 189, 190, 191, 197, 198, 199, 200, 201, 205, 206, 208, 209,* 265, 271, *291,* 306, 308, 315,
[Myers, R. D.] 317, 345, 362, 379, 381, 383, 398, *425, 426, 427, 432, 436*
Myhre, K., 41, *75, 79*

N

Nadel, E. R., 29, 37, 38, 39, *79, 80, 81*
Nahorski, S. R., 351, *413, 427*
Naka, F., 443, 446, 447, *464*
Nakajima, T., 351, *419*
Nakamura, K., 115, *201*
Nakamura, T., 255, *291,* 374, *428,* 441, 443, 460, *464*
Nakayama, T., 3, 7, 46, 47, 49, 56, *76, 79,* 95, 109, 153, *193, 201*
Nance, D. M., 230, 236, 250, *291*
Nardi, P., 383, *424*
Nasset, E. S., 228, *291*
Natelson, B. H., 348, *418*
Natterman, R. A., 19, *79,* 101, 123, 141, 142, 143, 145, 151, 177, *201*
Nauss, S. F., 242, 243, 244, *278*
Nauta, W. J. H., 215, 273, *284, 291,* 456, *464*
Nazar, K., 137, 139, *191*
Necker, R., 14, *79*
Needleman, H. L., 344, *427*
Neff, N. H., 110, *186*
Negrin, J., 236, 237, *283*
Negro-Vilar, A., 328, 341, 342, *407*
Neil, D. B., 383, *427*
Neilson, E. B., 320, 366, *427*
Nelson, E. L., Jr., 101, *189*
Nemeroff, C. B., 132, *183,* 402, *427*
Netherton, R. A., 122, *202*
Neville, M., 345, *430*
Newberg, A., 396, 399, *420*
Newberg, D. C., 315, *425*
Newburgh, L. H., 215, *291*
Newman, P. P., 34, *76*
Ng, K. Y., 315, *427*
Nicholas, D. J., 375, *435*
Nickerson, M., 89, 130, *206*
Nicolaïdis, S., 228, 229, *291, 293,* 392, 402, *427, 464*
Nicoll, R. A., 326, 341, 386, *408*
Niederberger, M., 29, 37, 38, *82*
Nielsen, B., 138, 139, *202*
Nielsen, M., 139, *202*
Nielson, H. C., 229, *295*
Nikaido, R. S., 357, 366, *429*

Author Index

Nilsson, G., 300, 353, 368, 395, 400, *418*
Nilton, S. M., 23, *73*
Ninomiya, I., 23, *79*
Nistico, G., 92, 108, 134, 157, 162, 166, 169, 171, 172, *197, 202*
Nobin, A., 355, 365, *409*
Noble, W., 215, *286*
Norelli, C., 373, 375, 377, 378, *418*
Norgren, R., 331, *427,* 441, 446, 447, 453, *464*
Notter, D., 22, 43, *74*
Novin, D., 211, 260, *291, 293,* 382, *433*
Novotná, R., 97, *202*
Nowlis, G., 344, *417*
Nunez, A., 335, *414*
Nutik, S. L., 16, 19, 47, 49, *79,* 124, 128, 178, *202*

O

Oades, R. D., 356, *428*
Oates, J. A., 154, *188*
Öberg, B., 23, *73*
O'Briant, D. A., 250, *278*
O'Donohue, T. L., 337, 338, *428*
Ogawa, M., 90, *182*
Ohga, A., 22, *70*
Oishi, R., 329, 340, *428*
O'Keane, M., 351, *425*
Olds, J., 444, 451, 452, *463, 464*
Olgiati, V. R., 230, *290*
Olive, G., 154, *190, 202*
Oliver, A. P., 335, *409*
Olsen, N. S., 121, *205*
Olson, L., 102, 117, *181, 202*
Olsson, K., 323, 379, 386, *428*
Oltmans, G. A., 255, *288,* 332, *424, 428*
Ono, T., 374, *428,* 443, 446, 447, *464*
Oomura, Y., 211, 212, 214, 255, 274, *291, 297,* 374, *427,* 441, 443, 446, 447, 448, 460, *464*
Ooshima, Y., *465*
Ooyama, H., 443, 446, 447, *464*
Oppenheimer, R. L., 346, *423*
Opsahl, C. A., 240, 242, 250, 251, 252, 270, *291, 292*
Orias, R., 323, 325, 327, 368, 386, 388, *427*
Orloff, J., 235, *296*
Ornesi, A., 89, 101, *184*
Osbahr, A. J., III, 402, *427*

Osumi, Y., 329, 340, *428*
Overstreet, D. H., 122, *202,* 381, *430*

P

Page, I. H., 236, 237, *283*
Pager, J., 457, *464*
Palacios, R., 300, 353, 368, 395, 400, *418*
Palka, Y., *291*
Palkovits, M., 102, 126, *194, 202,* 330, 341, 354, 355, 364, 365, 368, 380, *413, 420, 428, 431, 435*
Palmer, G. C., 313, *428*
Panksepp, J., 212, 213, 228, 229, 230, 245, 255, *291, 292,* 351, 380, 398, 399, *426, 428*
Paolino, R. M., 174, *202*
Papadakos, P. J., 371-375, *423*
Papeschi, R., 320, *429*
Parent, A., 380, *428*
Parikh, H. M., 313, *416*
Parker, R. A., 240, *287*
Parker, S. W., 331, *428*
Parkes, M. W., 165, *196*
Parrott, D. M. V., 249, *287*
Parsons, E. M., 390, *409*
Paterson, A. T., 324, 325, 326, 341, 386, *426*
Pauli, A. I., 344, *410*
Pavan, F., 135, *205*
Pavlides, C., 329, *417*
Paxinos, G., 238, 239, 241, 243, 249, 253, 254, *292, 297,* 335, 339, *428*
Paykel, E. S., 344, *428*
Pearce, W. M., 236, *287*
Pecile, A., 230, *290*
Peck, J. W., 384, 387, *429*
Peeters, G., 323, 386, *435*
Peidaries, R., 92, *194*
Pellett, P. L., 399, *432*
Penaloza-Rojas, 350, *429*
Pengelley, E. T., 85, 97, *206*
Penn, P. E., 107, 122, 167, *182*
Perek, M., 218, 221, 236, 237, *295*
Perez de la Mora, M., 300, 353, 368, 395, 400, *418*
Perlmutter, M. N., 48, 68, *78*
Perlow, M. J., 403, *414*
Perrett, D., 443, 450, 451, 459, *465*
Perry, J. H., 232, 249, *288*

Perry, K. W., 374, *415*
Persson, N., 16, 24, *70*
Pert, A., 393, 394, *408*
Pert, C. B., 403, *425*
Peruzzi, G., *198*
Peters, D. A. V., 173, *202*
Peters, R. H., 220, 231, *287, 292*
Petersdorf, R. G., 148, *183*
Pfaff, D. W., 126, 130, *186,* 255, *292,* 349, *412,* 440, 441, *464*
Pfaffmann, C., 440, *464*
Pflueger, A. B., 373, *412*
Philipp-Dormston, W. K., 158, 163, 164, *202, 203, 206*
Phillips, A. G., 274, *297,* 357, 366, *429*
Phillips, M. I., 140, *190, 209,* 401, *415, 418,* 442, *464*
Phillipson, O. T., 313, *418*
Pickard, J. D., 351, *425*
Pickering, B. T., 335, 341, *413*
Pickford, M., 323, 386, *413, 429*
Pickren, K. S., 352, *421*
Picotti, G. B., 154, *185*
Pierau, Fr. K., 15, *79*
Pilcher, C. W. T., 255, *292*
Pillai, R. V., 255, *277,* 442, *461*
Pillar, A. X., 327, *429*
Pinkston, J. O., 16, *79*
Piras, L., 110, *188*
Pittlet, P., 39, *72*
Pittman, Q. J., 103, 104, 107, 117, 123, 130, 148, 150. 151, 152, 157, 162, 164, 170, *186, 203, 208*
Placheta, P., 315, *429*
Placidi, G. F., 320, *429*
Plech, A., 325, 367, *436*
Pletscher, A., 88, *182*
Plezer, N. L., 345, *430*
Plocher, T. A., 235, 249, *292*
Plum, F., 221, *293*
Poirer, L. J., 237, *292,* 320, 361, *421*
Poitou, P., 168, *191*
Pohorecky, L. A., 173, *203*
Poletti, C. E., 9, *79*
Poloni, M., 382, *431*
Poole, S., 122, 123, *203*
Porceddu, M. L., 362, *408*
Porte, D., Jr., 237, 245, *298,* 349, *436*
Porter, J. H., 231, *292*
Porter, P. B., 229, *295*
Poschel, B.P.H., 274, *292,* 374, 375, *429*

Posner, B. I., 351, *435*
Possani, L., 300, 353, 368, 395, 400, *418*
Potaccini, R., 318, 374, *431*
Poulain, P., 160, *203*
Powley, T. L., 216, 223, 224, 232, 235, 239, 240, 242, 247, 249, 251, 252, 258, 259, 261, 263, 265, 266, 267, 268, 269, 270, 271, 273, 274, 276, *280, 286, 291, 292*
Pradhan, S. N., 383, *429*
Prange, A. J., Jr., 132, *183,* 402, *427*
Pratt, A. W., 12, 37, 38, 67, *70*
Premont, J., 313, 351, 360, *410*
Preskorn, S. H., 352, *429*
Preston, E., 94, 97, 105, 107, 109, 123, 130, 168, 169, *203*
Preziosi, P., 108, *202*
Price, M. T. C., 340, *430*
Price, W. M., 16, 18, 24, 45, *81*
Proppe, D. W., 22, 43, *74, 80*
Protais, P., 318, *412*
Provins, K. A., 7, *76*
Pryol, J. F., 378, 395, 397, *415*
Puerto, A., 443, 450, 451, 459, *465*
Purcell, A. T., 313, 314, *426*
Pyörnilä, A., 93, 104, 122, *192, 203*

Q

Quaade, F., 262, *292*
Quackenbush, P. M., 253, 254, *282*
Quartermain, D., 262, *298,* 335, 344, 346, *410, 412*
Querido, A., 229, 234, *297*
Quick, K. P., 35, *80*
Quock, R. M., 114, 117, *193, 204*

R

Rabin, B. M., 235, 254, *293*
Rabinowitz, J. L., 243, *283*
Rabkina, A. E., 255, *277*
Radsak, K., 163, 164, *206*
Raichle, M. E., 352, *429*
Rall, T. W., 130, *194*
Ralph, J., 265, *278*
Ramsay, D. J., 383, *429*
Randall, P. K., 255, *277*
Randall, W. C., 27, 38, *80*

Author Index

Ransohoff, J., 352, *417*
Ranson, S. W., 16, 18, 31, 32, 39, *78, 80,* 136, *204,* 215, 216, 222, 232, 234, 235, 236, 246, 247, 249, 253, 256, *280, 284, 285, 293*
Rašková, H., 154, *197, 198, 204*
Rasmussen, A. T., 16, 18, 24, 36, *76*
Rasmussen, T., 16, 18, 24, 36, *76*
Rautenberg, W., 93, 108, *193*
Ravona, H., 221, 236, 237, *295*
Rawlins, M. D., 113, 166, 167, *187, 204*
Rawson, R. O., 31, 35, *69, 80*
Reaves, T. A., 56, *80*
Redding, T. W., 249, *293*
Redgrave, P., 19, *79,* 101, 123, 141, 142, 143, 145, 151, 177, *201*
Redgrave, P. C., 120, 121, *198*
Redmond, D. E., Jr., *424*
Redmond, E. D., Jr., 329, *429*
Reeves, A. G., 221, *293*
Rehfeld, J., 300, 353, 368, 395, 400, *418*
Reich, M. J., 231, *292*
Reichl, D., 239, 241, 243, *285*
Reichlin, S., 22, *78*
Reid, I. A., 383, *429*
Reid, J. L., 175, *204*
Reid, W. D., 110, *204*
Reigle, T. G., 93, 121, 171, *204*
Reilly, P., 351, *428*
Reis, D. J., 329, 370, *429, 430*
Remley, N. R., 217, 231, 237, 244, 249, *284, 289*
Renfrew, J. W., 321, *419*
Renold, A. E., 242, *280*
Repin, I. S., 149, *204*
Reubi, J. C., 398, *429*
Rewerski, W. J., 92, 106, 170, 171, *195, 204*
Reynolds, R. W., 218, 229, *293,* 357, *429*
Rezek, M., 260, *293*
Rhodes, B. A., 27, *80*
Ricardo, J. A., 349, *429*
Rice, J. C., 19, *79,* 87, 101, 123, 141, 142, 143, 145, 151, 177, *201*
Richardson, J. S., *289,* 338, *429*
Richer, C. L., 237, *292*
Richter, C. P., 136, *204,* 348, *430*
Ridley, P. T., 228, 241, 242, 246, *291, 293*
Rindi, G., 382, *431*
Rioch, D., Mck., 16, *79*
Ritter, R. C., 307, 310, 311, 332, 335,

[Ritter, R. C.] 343, 345, 375, *430*
Roberge, A. G., 320, 361, *421*
Roberts, D. C. S., 340, *430*
Roberts, E., 329, 396, 397, 398, *430*
Roberts, J. C., 101, *185*
Roberts, J. J., 386, *432*
Roberts, M. F., 29, *81*
Roberts, W. W., 25, 41, *80*
Robertshaw, D., 86, *189*
Robinson, D. W., 307, *419*
Robinson, R. G., 357, *430*
Robinson, T. C. L., 41, *80*
Robinzon, B., 218, 221, 236, 237, *295*
Rodgers, W. L., 259, 262, 267, 273, *293*
Roemer, C., 102, *194*
Rogers, K. J., 351, *413, 427*
Rogers, Q. R., 228, *293*
Rohner, F., 238, 239, 246, *293*
Roizen, M. F., 96, *194*
Rolf, D., 236, 241, 243, 255, *284,* 335, *418*
Rolls, B. J., 357, 366, *430,* 444, 446, 460, *465*
Rolls, E. T., 214, 274, *293,* 357, 366, *430,* 443, 444, 446, 449, 450, 453, 456, 457, 459, 460, *461, 464, 465*
Romaniuk, A., 221, 232, *293*
Rony, H. R., 218, *293*
Roper-Hull, A., 443, 450, 451, 456, 459, *465*
Rosenberg, R. N., 147, *196*
Rosenblum-Blinick, C., 344, 345, 356, *412*
Rosendorff, C., 94, 95, 101, 109, 130, 132, 151, 154, 166, 168, 169, *184, 187, 196, 204, 209*
Rosene, D. L., 352, *421*
Rosenthal, F. E., 135, 169, 170, *204*
Rosenzweig, M. R., 245, *293*
Ross, R. A., 329, *430*
Rossakis, C., 307, 309, 310, 311, 312, 317, 319, 322, 338, 349, 358, 359, 361, 364, 365, 366, 367, *423*
Rotenberg, P., *284*
Roth, S. R., 261, *293*
Roth, T., 383, *429*
Rothschild, R., 380, *420*
Rotiroti, D., 134, 162, 166, *202*
Rotta, J., 154, *197*
Routtenberg, A., 308, 348, 383, 384, 385, 387, *430, 432*

Rowell, L. B., 29, 37, 38, *82*
Rowland, N. E., 228, 258, 261, 271, *293*, 357, 358, 366, *407*, *430*
Rozin, P., 260, *297*
Rozkowska, E., 221, 225, 261, 270, *282*, *293*, *294*
Rozkowska-Ruttimann, E., 47, 49, *70*, 168, *183*
Rudy, T. A., 88, 90, 92, 94, 104, 117, 150, 151, 157, 163, 164, *195*, *201*, *204*, *209*, 383, *427*
Ruf, K. B., 114, *209*, 238, 239, 246, *293*
Ruppel, M., 270, *288*
Rushmer, R. F., 23, *72*
Russek, M., 228, *294*, 350, 382, 429, *430*
Russell, R. W., 381, *430*
Ruwe, W. D., 114, 115, 117, 134, 149, 157, 163, 164, 165, *200*, *205*, *206*
Ruwe, W. E., 116, 117, 165, *200*

S

Saad, W. A., 327, 341, 380, *411*, *412*, *429*
Saavedra, J. M., 102, *202*, 355, 368, *428*, *431*
Sacharoff, G. P., *205*
Said, S., 300, 353, 368, 395, 400, *418*
Salans, L. B., 245, *295*
Salisbury, R., 220, 227, *295*
Saller, C. F., 377, *431*
Salmoiraghi, G. C., 335, *409*
Salt, P. J., *419*
Samanin, R., 174, *205*, 317, 318, 361, 364, 370, 372, 373, 374, 378, 395, 397, *415*, *431*
Samardzić, R., 139, *183*
Samek, D., 97, *206*
Sanders, M. K., 231, *295*
Sanghera, M. K., 456, *465*
Sanghvi, I., 318, 319, 361, *431*
Sanner, J. H., 162, *205*
Sano, I., 351, *419*
Santhakumari, G., 393, 395, *408*
Saper, C. B., 349, *431*
Satinoff, E., 16, *80*, 93, 107, 108, *183*, *184*, *205*
Sauaudeau, C., *194*
Sawchenko, P. E., 252, 254, 282, *294*, 338, 339, 341, 350, *416*, *431*

Saxena, P. N., 103, 104, 113, 137, 155, 157, *185*, *189*, *205*
Schallert, T., 265, 270, 271, 273, 274, *283*, *288*, *294*
Schally, A. V., 249, *293*
Scharrer, E., 228, 266, 267, *294*
Schelling, P., 401, *415*
Schenk, E. A., 228, *291*
Schild, H. O., 390, *408*
Schildkraut, J. J., 316, *431*
Schindler, W. J., 235, 250, *289*, *294*
Schlumpf, M., 398, *421*
Schmeling, W. T., 173, *205*
Schmid, P., 401, *418*
Schmidt, J., 101, *205*
Schmitt, H., 314, 357, *431*
Schmitt, H., 314, 357, *431*
Schmitt, M., 441, 448, *465*
Schnatz, J. D., 235, 239, 240, 241, 242, 243, 244, 249, 250, 251, 269, *278*, *282*, *283*, *286*, *294*
Schneck, D. J., 244, *283*
Schneider, B. S., 402, *431*
Schnell, R. C., 175, *209*
Schoener, E. P., 49, *80*, 152, 153, 162, 166, 169, *205*
Schoenfeld, T. A., 246, *294*
Schönbaum, E., 97, *203*
Schönung, W., 24, 25, 27, *78*, *80*
Schramm, L., 20, 45, *70*
Schreiner, L. H., 19, *78*
Schrier, B. K., 351, *415*
Schubert, J., 372, *431*
Schulz, R., 318, *415*
Schütz, J., 135, *205*
Schwartz, J.-C., 318, 388, 389, 392, *412*, *431*
Schwartz, M., 261, *293*
Schwartz, W. J., 351, 356, *431*
Schwartzbaum, J. S., 260, *288*
Schwennicke, H. P., 23, *71*
Sciorelli, G., 382, *431*
Sclafani, A., 220, 221, 223, 229, 230, 231, 236, 245, 252, 253, 254, *294*, *295*, 335, 338, 339, 351, 365, *408*, *432*
Scott, I. M., 36, *80*
Sedvall, G., 372, *431*
Segal, M., 452, *464*
Seigel, M. S., 11, 14, 20, 26, *72*, 151, *185*
Seller, T. J., 124, *197*
Sensenig, L. D., 231, *292*
Sessions, G. R., 329, *420*

Author Index

Sétaló, G., 240, *295*
Setler, P. E., 306, 313, 314, 321, 366, 383, 384, *414, 432*
Severs, W. B., 401, 402, *432*
Sgaragli, G. P., 135, *205*
Shapiro, R. E., 387, 388, *426*
Sharma, G. S., 441, *461*
Sharma, K. N., *295,* 448, *465*
Sharma, V. K., 319, *408*
Sharp, F., 31, 36, *75,* 351, 356, *431*
Sharp, J. C., 229, *295*
Sharpe, L. G., 94, 121, *200, 205,* 352, 379, 381, 384, 385, *429, 432, 434*
Shaw, G. G., 134, *205*
Shaw, S. G., 357, 366, *430*
Sheehan, P., 312, *435*
Sheikolislam, B. M., 262, *280*
Shellenberger, M. K., 97, 110, *205*
Shemano, I., 89, 130, *206*
Shepherd, J. T., 27, *81*
Sheth, U. K., *206*
Shibuya, H., 403, *425*
Shimizu, N., 102, *197,* 443, *464*
Shirk, T. S., 231, 250, *278*
Shoemaker, W. J., 398, *421*
Shoenberg, K., 261, 273, *277*
Shumway, G. S., 270, *288*
Shute, C. C. D., 124, 126, 130, *196, 206,* 380, 381, *424*
Sideroff, S., 453, *465*
Siders, W. A., 231, 250, *278*
Siegert, R., 158, 163, 164, *203, 206*
Siggens, G. R., 398, *409*
Silva Netto, C. R., 327, 380, *411, 412, 429*
Silver, I. A., 440, 442, *462*
Silverman, H. J., 332, *432*
Simmonds, M. A., 97, 110, 111, 115, *206*
Simon, E., 14, 16, 24, 25, 27, 34, *77, 78, 80, 81*
Simon, H., 363, *424*
Simon, M. L., 134, *191*
Simon-Oppermann, Ch., 40, *77*
Simpson, C. W., 19, *79,* 101, 123, 141, 142, 143, 145, 151, 157, 163, 164, 177, *201, 206, 432*
Simpson, J. B., 348, 383, 384, 385, 386, 387, 388, 401, *425, 432*
Sims, E. A. H., 245, *295*
Simson, E. L., 252, *294,* 399, *432*
Singer, E., 315, *429*
Singer, G., 106, 168, *181, 188,* 307, 313,
[Singer, G.] 314, 318, 319, 321, 345, 361, 381, 383, *408, 425, 426, 431, 432*
Singh, B., *295,* 441, *461*
Singh, D., 227, 231, *295*
Singhal, K. C., 113, *185*
Singhal, R. L., 173, *202*
Sjoerdsma, A., 110, *191*
Skarnes, R. C., 143, *206*
Skelton, F. R., 235, 249, *279*
Sladek, C., 386, *432*
Sladek, J. R., Jr., 102, *190,* 308, 326, 368, *420, 425, 430*
Slangen, J. L., 307, 308, 311, 313, 314, 315, 322, 349, 357, 362, *411, 432, 435*
Slaunwhite, W. R., III, 232, 239, 240, 243, 244, 246, *295*
Smetana, R., 154, *204*
Smiles, K. A., 39, *80*
Smirnova, L. K., 250, *287*
Smith, G. P., 228, 229, 267, 275, *288, 295,* 306, 329, 345, 350, 383, 402, *407, 405, 420, 424, 430, 432*
Smith, J. R., 132, *185*
Smith, M. H., 220, 227, *295*
Smith, N. F., 231, *280*
Smith, P. E., 216, *295*
Smith, R. D., 327, *410*
Smockler, H., 110, *204*
Smutz, E. R., 227, *295*
Smyrl, R., 235, *289*
Snapir, N., 216, 236, 237, *295*
Snellen, J. W., 136, *198*
Snoddy, H. D., 374, *415*
Snowdon, C. T., 261, 263, *296,* 342, 350, *433*
Snyder, D. W., *74*
Snyder, J. J., 384, *433*
Snyder, R. D., 329, *429*
Snyder, S. H., 312, 313, 316, 360, 371, 378, 381, 389, 391, 394, 395, 400, *411, 413, 418, 433, 435*
Sochański, R., 97, *206*
Sommer, S. R., 382, *433*
Soper, W. Y., 374, *433*
Soulairac, A., 320, 322, 370, *433*
Soulairac, M.-L., 320, 322, 370, *433*
South, F. E., 85, *206*
Spafford, D. C., 97, *206*
Sparks, D. L., 444, 457, *465*
Spaulding, J. H., 320, *436*
Spector, H., 139, *209*
Spector, S., 110, *191*

Spence, R. J., 27, *80*
Sperber, M., 221, 236, *294*
Spławiński, J. A., 150, 157, 158, *206*
Sporn, J. R., 312, *433*
Sprague, J. M., 331, *433*
Springer, D., 220, 229, 230, *295*
Spurrier, W. A., 85, *188*
Spyer, K. M., 23, *76*
Squibb, R. L., 344, *412, 424*
Squires, R. D., 13, 17, 24, 31, 32, 36, 77, *80*
Srivastava, Y. P., 323, 325, 367, 368, 379, 380, 388, *408*
Staff, N., 317, 360, 361, *415*
Staib, A. H., 123, *206*
Stancer, H. C., 335, 344, 345, 356, *412*
Standish, L. J., 399, *432*
Stanier, M. W., 40, *74*
Stanley, M. E., 356, *415*
Stanton, T. L., 121, *206*
Stare, F. J., 245, *289*
Stark, P., 382, *433*
Starr, N. J., 360, 367, *433*
Starr, S. E., 244, *296*
Stawarz, R. J., 313, 351, *409*
Steffens, A. B., 239, 240, 245, 267, *296, 297*, 349, *413, 433*
Stein, L., 343, *430*
Steiner, S. S., 329, *417*
Stellar, E., 212, 220, 222, 224, 227, 254, 257, 258, 262, 275, *280, 281, 283, 296, 297*, 331, 334, *416, 433*, 440, *465*
Stenevi, U., 355, *409*
Stephens, D. N., 349, *434*
Stephenson, J. D., 108, 122, 123, 134, 162, *181, 197, 202, 203*
Stern, J. J., 229, 268, *296*, 307, 402, *434*
Stern, W. C., 102, *198*, 273, *290*, 368, 370, *427*
Stern, Y., 329, *417*
Stevens, E. D., 146, *206*
Stevenson, J. A. F., 211, 212, 223, 230, 231, 232, 235, 236, 239, 240, 244, 258, 269, 270, 271, *289, 290, 296*
Stinus, L., 363, *424*
Stitt, J. T., 13, 36, 39, 42, 49, *73, 80, 81,* 148, 157, 158, 160, 162, *187, 207*
Stohmayer, A. J., 267, *295*
Stokes, P. E., 267, *295*
Stolerman, I. P., 307, 309, 357, *426, 434*
Stolwijk, J. A. J., 6, 7, 13, 25, 29, 30, 31, 34, 36, 37, 38, 39, 42, 49, 53, 54, *69,*

[Stolwijk, J. A. J.] 72, 73, 74, 75, *77, 79, 80, 81,* 94, 108, 153, 158, *184, 187, 206*
Storlien, L. H., 252, *277*, 338, *407*
Storm-Mathisen, J., 395, 396, *414*
Stout, H., 237, 241, *284*
Stowe, F. R., Jr., 229, *296*
Straus, E., 402, *434*
Stricker, E. M., 115, *208,* 258, 260, 261, 266, 267, 273, 276, *296,* 318, 319, 356, 357, 358, 361, 363, 377, 391, *418, 431, 434, 436*
Strittmatter, W. J., 312, *413*
Strominger, J. L., 222, 226, *296*
Strømme, S. B., 2, 7, 22, *72, 75*
Strubbe, J. H., 267, *297,* 349, *413*
Strzoda, L., 110, *206*
Stuart, D. G., 9, 14, 15, 16, 18, 20, 24, 45, *76, 81*
Stutinsky, F., 341, *434*
Subramanian, N., 395, *434*
Sugimori, M., 255, *291,* 374, *428,* 441, 443, 460, *464*
Sugrue, M. F., 373, *434*
Sulser, F., 313, 351, *409*
Sumpier, G., 220, *282*
Sun, J. Y., 173, *203*
Sundsten, J. W., 16, 22, 24, *70, 81*
Sundt, T. M., Jr., 352, *408*
Surgeon, J. W., 352, *413*
Sutherland, K., *75*
Suttora, N. L., 96, *194*
Svanes, K., 89, *206*
Švihouec, J., 154, *204*
Swanson, L. W., 300, 349, 352, 384, 385, *431, 434,* 460, *465*
Sweatman, P., 114, 135, 160, 162, *194, 206*
Swies, J., 158, *206*
Szafranowa, H., 154, *190*
Szczepanska-Sadowska, E., 40, *81*
Szelényi, Z., 107, *207*
Szolcsanyi, J., 55, *81,* 173, *194, 207*

T

Tadesada, M., 351, *419*
Takagi, H., 154, *207*
Takaori, S., 329, 340, *428*
Takashima, H., 154, *207*
Taleisnik, W., 327, 328, *411*
Tam, H. S., 38, *73*

Tamir, I., *284*
Tanabe, T., *465*
Tanaka, C., 154, *190*
Tanche, M., 22, *81*
Tangri, K. K., 118, 123, *207*
Tanguy, O., 155, *190*
Tannenbaum, G. A., 238, 239, 241, *297*
Tapia, R., 300, 353, 368, 395, 400, *418*
Tartaglione, R., 267, *297*
Tassin, J. P., 313, 351, 360, 363, *410, 424*
Taylor, G. W., 29, *70*
Taylor, K. M., 388, 392, *434*
Teddy, P. J., 154, *207*
Teitelbaum, P., 219, 221, 222, 223, 224, 227, 230, 231, 245, 248, 259, 260, 261, 262, 266, 267, 268, 273, 274, *285, 288, 289, 293, 297, 298*, 356, *425, 434*, 440, *465*
Tenen, S. S., 318, *436*
Tepperman, J., 217, 220, 222, 223, 232, 234, 235, 236, 238, 242, 243, 244, 257, *279, 297*
Teran, L., 300, 353, 368, 395, 400, *418*
Terenius, L., 300, 353, 368, 395, 400, *418*
Terwelp, D. R., 170, *207*
Tezuka, U., 239, 240, 252, *280*
Tha, S. J., 170, 175, *187*
Thämer, V., 23, *76*
Thauer, R., 14, 15, 20, 34, *78, 81*
Therminarias, A., 22, *81*
Thierry, A. M., 313, 351, 363, *410, 424*
Thoa, N. B., 317, 344, 403, *413*
Thoenen, H., 115, *201*
Thomas, D. W., 220, 228, 229, 267, *294, 297*
Thomas, S., 22, *80*
Thompson, D. A., 345, 346, *436*
Thompson, G. E., 118, *188*
Thorn, N. A., 325, 326, 386, *410*
Thornton, E. W., 374, *416*
Thorpe, S. J., 443, 450, 451, 459, *465*
Timo-Iaria, C., 274, *285*
Tirri, R., 87, 110, 111, *192, 207*
Tizabi, Y., 378, *425*
Tobler, I., 139, *207*
Tofanetti, O., 154, *185*
Toh, C. C., 97, *207*
Toivola, P., 90, 103, *207*
Toh, C. C., 97, *207*
Toner, M. M., 36, *80*
Tószeghi, P., 139, *207*

Totaro, J. A., 373, *412*
Totty, C. W., 382, *433*
Tower, D. B., 396, 397, *430*
Townsend, Y., 163, *187*
Tran, V. T., 389, *411*
Travis, R. P., Jr., 444, 457, *465*
Trulson, M. E., 373, 375, 377, 378, *418*
Trzcinka, G. P., 147, 171, 174, *196, 208*
Tsai, C. T., 227, *298*
Turk, J. A., 382, *433*
Turner, B. H., 258, 274, *289*
Turner, C., 250, *298*
Turner, D. M., 87, *192*
Tye, N. C., 375, *436*
Tyler, P. E., 101, *189*
Tytell, M., 143, *200*

U

Udenfriend, S., 110, *191*, 373, *425*
Udeschini, G., *198*
Ueberacher, H. M., 220, *282*
Ungerstedt, U., 90, 102, 117, *181, 190, 208*, 274, *297*, 331, 353, 355, 356, *435*, 439, 440, *465*
U'Prichard, D. C., 312, *416, 435*
Urano, A., 324, 367, 380, *435*
Urban, I., 335, *427*
Uretsky, N. J., 115, *206*
Usdin, E., 96, *182*, 400, *435*
Uyeda, A. A., 456, *462*
Uzunov, P., 313, *412, 416, 436*

V

Vaernet, K., 262, *292*
Vaidya, A. B., 163, *208*
Valenstein, E. S., 274, *297*, 326, *419*
Valgelli, L., 373, *431*
Valle, L. E. R., 274, *285*
Valzelli, L., 97, *190*
van Bekkum, D. W., 229, 234, *297*
Vance, W. B., 224, 227, *288*
Van den Pol, A. N., 250, *292*
Vandeputte-Van Messom, G., 323, 386, *435*
Van der Gugten, J., 317, 355, 362, *435*
Vander Wal, B., 341, *413*
VanderWeele, D. A., 260, *293*, 350, *414*
Vane, J. R., 155, *189*

Vaněček, J., 154, *204*
Van Hoesen, G. W., 456, *462*
van Houten, M., 351, *435*
Vanhoutte, P. M., 27, *81*
Van Putten, L. M., 229, 234, *297*
Van Rossum, J. M., 344, 373, *410*
Van Wimersma, Tj. B., 393, *413*
Van Zoeren, J. G., 115, *208*
Varagić, V. M., 130, *208*
Vargiu, L., 110, *188*
Vartiainen, A., 88, *189*
Vasquez, G. J., 381, *430*
Vasselli, J. R., 252, *295*
Vaughn, L. K., 166, 176, *195, 208*
Veale, W. L., 9, *79,* 92, 103, 104, 106, 107, 117, 123, 130, 137, 138, 139, 147, 148, 149, 150, 151, 152, 154, 157, 158, 160, 162, 164, 166, 170, *189, 200, 201, 203, 208*
Versteeg, D. H. G., 355, *435*
Vetulani, J., 313, 351, *409*
Vijayan, E., 402, *435*
Vijersky, R. A., *435*
Vilberg, T. R., 231, 250, *278, 297*
Villablanca, J., 148, 151, *208, 209*
Vincent, J. D., 444, 445, 448, *461, 465*
Viswanathan, C. T., 157, 164, *204, 209*
Vitale, J. J., 245, *289*
Vogt, M., *209*
Volicer, L., 110, *204*
von Euler, C., 22, *81,* 100, *209*
von Euler, U. S., 316, *435*

W

Wagner, H., 27, *80*
Wagner, H. N., Jr., 27, *80*
Wagner, J. W., 218, *297*
Wahl, M., 352, *421*
Wakeman, K. A., 45, *81*
Walker, C. A., 97, *190*
Walker, R. R., 352, *413*
Waller, M. B., 85, 87, 89, 90, 92, 99, 100, 101, 103, 124, 126, 127, 128, 130, 153, 160, 163, 164, 165, 168, 169, *200, 209*
Walls, E. K., 361, *436*
Wals, P., 242, 244, *286*
Walsh, L. L., 340, 341, 345, *435, 436*
Walther, O. E., 25, 34, *78, 81*
Wampler, R. S., 231, 261, 263, 265, 271, *290, 296, 297*

Wang, S. C., 35, 49, 54, *72, 80, 81, 82,* 152, 153, 162, 166, 169, *205, 209,* 325, *426*
Warsh, J. J., 377, *412*
Warwick, R., 175, *209*
Waters, D. H., 317, 335, *416*
Watt, J. A., 323, *413*
Wayner, M. J., 211, 267, *297*
Webb-Peploe, M. M., *81*
Weeks, J. R., 155, *183*
Weemaes, A. J. M., 373, *410*
Weick, B. G., 132, *185, 193*
Weil, A., 232, *284*
Weinberg, H., 220, 227, *295*
Weinberger, L. M., 256, *297*
Weinshilboum, R. M., 352, *408*
Weinstein, H., 388, 392, *416*
Weinstock, M., 371, *436*
Weiss, B., 7, 41, 47, *81,* 313, *424, 436*
Weiss, C., 371, *436*
Weissbach, H., 373, *425*
Weissman, A., 318, *436*
Welch, J. P., 150, 157, *196*
Welt, L. G., 235, *296*
Wendlandt, S., 155, 158, 159, 166, 169, *188, 189, 198*
Wenger, C. B., 29, *81*
Werbin, B., *284*
Werdinius, B., 97, *209*
Werman, R., 395, *436*
Werner, A. B., 318, *412*
Wheatley, M. D., 236, 238, *298*
Wheeler, T. J., 374, *416*
Whishaw, I. Q., 358, *417*
White, H. D., 374, *407*
White, H. L., 236, 241, 243, 255, *284,* 335, *418*
White, L. E., 256, 261, *298*
Whitehead, S. A., 114, *209*
Whittow, G. C., 31, *76*
Wiggins, M. L., 217, 218, 220, 223, 256, *279*
Wikoff, H., 16, 18, 24, 36, *76*
Wilcox, R. H., 274, *278*
Wilder, R. M., 215, *298*
Williams, B. A., 138, 143, *192*
Williams, D. R., 227, 266, *298*
Williams, J. W., 157, 169, *205, 209*
Willies, G. H., 94, 95, 101, 109, 130, 132, 168, 169, *184, 196, 209*
Willis, G., 345, *425*
Willis, J. R., 85, *206*

Author Index

Willoughby, J. O., 106, 107, *188*
Wilson, M. F., 23, *79*
Wilson, N. C., 140, *190, 209*
Wilson, W. H., 227, *298*
Windle, W. F., *72,* 151, *185*
Wing, E. S., 102, 173, *182*
Winson, J., 383, *436*
Wise, A., 234, *298*
Wise, C. D., 343, *430*
Wise, R. A., 374, 375, 382, *433, 436*
Wisehart, T. B., 361, *436*
Wishaw, I. Q., 106, 115, 160, 168, *192, 208,* 271, 273, *294*
Wishovsky, T. I., 373, *412*
Wit, A., 49, 54, *81, 82,* 152, *209*
Wojtaszek, B., 157, 158, *206*
Wolf, G., 262, *298,* 341, *432, 436*
Wolf, H. H., 93, 104, 117, 119, 121, 169, 171, *204, 205*
Wolf, P., 351, 392, *417*
Wolfe, B. B., 346, *436*
Wolgin, D. L., 356, *434*
Wolny, H. L., 325, 367, *436*
Wolstencroft, J. H., 34, *76,* 314, 357, *410*
Wood, R. J., 357, 366, *430*
Woods, J. W., 20, *70, 82,* 151, *182*
Woods, S. C., 229, 237, 245, *287, 288, 298,* 349, 402, *420, 421, 436*
Woolf, C., *196*
Woolf, C. J., 130, 132, 168, 169, *196, 209*
Woolf, C. S., 94, 95, 101, 109, *184*
Woolsey, C. N., 456, *461*
Wright, P., 250, *298*
Wrywicka, W., 211, *291*
Wünnenberg, W., 14, 15, 16, 18, 19, 21, 47, 49, *72, 82*
Wurster, R. D., 27, 38, *80*
Wurtman, J. J., 376, *436*
Wurtman, R. J., 352, 372, 376, *414, 421, 436*
Wyss, C. R., 29, 37, 38, *82*

Y

Yaakobi, M., 221, 236, 237, *295*
Yaksh, T. L., 18, *79,* 88, 90, 92, 94, 103, 113, 117, 118, 119, 126, 130, 137, 139, 149, 151, 157, 163, 166, 169, 175, *200,*
[Yaksh, T. L.] *201, 205, 209,* 306, *436*
Yalow, G., 357, *434*
Yamada, Y., 255, *291,* 441, 460, *464*
Yamamoto, T., 443, 446, 447, *464*
Yamamura, H. I., 380, *420*
Yarito, H., *465*
Yasuda, M., 154, *209,* 221, 236, 237, 238, *288*
Yin, T. H., 173, *185,* 227, 228, 235, 266, 275, *288, 290, 298*
York, D. A., 249, 250, *298*
York, D. H., 313, *436*
Young, D. R., 139, *209*
Young, J., 42, *74*
Young, P. M., 326, 341, *436*
Young, R. C., 358, 402, *407, 415*
Young, T. K., 232, 250, *284, 298*
Yu, Y. K., 240, 249, *284*
Yunger, L. M., 379, *424*

Z

Zabik, J. E., 320, *436*
Zaborsky, L., 355, 368, *428*
Zacchei, A. G., 373, *412*
Zacny, E., 150, 157, 158, *206*
Zanick, D. C., 28, *82*
Zbrozyna, A. W., 23, *69, 73*
Zeballos, G., *279*
Zeigler, H. P., 260, 261, *298,* 440, *465*
Zeisberger, E., 14, *72,* 107, 122, *207, 210*
Zemlam, F. P., 373, 375, 377, 378, *418*
Zervas, N. T., 352, *421*
Ziance, R. J., 316, *436*
Ziel, R., 165, *210*
Zighera, C. F., 221, 236, 241, *289*
Zighera, C. Y., 221, *289*
Zigmond, M. J., 258, 260, 261, 266, 267, 273, 276, *296, 298,* 318, 319, 356, 357, 358, 361, 363, *418, 434, 436*
Zimmerman, E. A., 341, 394, 400, *413*
Zis, A. P., 318, 357, 362, 366, 375, *414, 437*
Zomzely, C., *278*
Zucker, I. H., 327, 332, *432, 437*
Zucker, L. M., 234, *285*
Zurhein, G. M., 261, *286*
Zurik, G., 307, *434*

Subject Index

A

Acetaminophen, 155, 158, 165, 169
Acetylcholine
  anatomy, 380
  feeding, 349
  thermoregulation, 18, 86, 101, 117
Acetylcholinesterase, 380
ACTH, 249, 347
Activity, 245, 258, 271, 331
Adenine nucleotides, thermoregulation, 129-132
Adipocyte, 234, 268
Adipose tissue, 22, 232, 242
Adiposogenital syndrome, 214
Adrenals, 249, 270, 272
Adrenergic
  feeding control, 306-320
  pathways, 300
Aldosterone, 341
Alloxan, 250, 350
Amphetamines
  feeding, 229, 314-320, 333, 339, 358, 364, 367, 376
  thermoregulation, 170, 175
  thirst, 382
Amygdala, 456
Angiotensin, 321, 387, 400
Anisomycin, 164
Anterior hypothalamic area, 1, 6, 16, 21, 31, 43-47, 85, 90, 99, 124, 149, 160, 175-177
Antidiuretic response, 39, 324-329, 341, 367, 379, 386, 393, 461
Antipyretics, 149, 160, 165-170, 175-177
Apomorphine, 114, 170, 358, 366
Arousal, 213, 356, 360, 441, 444
Atropine, 123, 139, 382, 386

B

Behavior
  dopaminergic effects, 355
  grooming, 41
  huddling, 41
  hypothalamic single-units, 439-461
  serotonergic effects, 370
  thermoregulatory, 40, 44-46, 64
  VMH rat, 231
Benzodiazepines, 374
Bile, 244
Blood-brain barrier, 89
Blood glucose, 241, 351
Body weight
  defense of, 230, 276
  set-point, 267
Bombesin, 134
Brown adipose tissue, 13, 21
Burimamide, 134
Butaclamol, 114

C

Capsaicin, 55, 173
Carbachol, 118, 321, 381-388
Carcass composition, 232, 239
Carotid rete, 31, 35
Catecholamines
  cerebral circulation, 351
  drinking, 320-323
  feeding, 272, 306-320
  metabolism, 351
  pathways, 300, 328
  salt intake, 326-328
  serotonergic interactions, 375
  thermoregulation, 22, 87, 100-117, 155

[Catecholamines]
  turnover, 317
  water excretion, 323-326
Cats, thermoregulation, 102
Central gray, 331
Central tegmental tract, 305
Chickens, thermoregulation, 107
Chloralose, 171
Chlorcyclazine, 134
Chlorpromazine, 170, 313, 357, 371
Cholecystokinin, 229, 402
Cholesterol, 243
Cholinergic systems, 380
Circadian rhythms, feeding, 220, 261, 307
Circle of Willis, 32
Clonidine, 171, 175, 311
Conditioned taste aversions, 261
Corticosterone, 249
Creatinine, 270
3,5-cyclic AMP, 113, 129-132, 351, 382
Cycloheximide, 164
Cyproheptadine, 89, 374

D

db/db mouse, 248
Decortication, 260
2-Deoxy-D-glucose, 173, 229, 267, 317, 345, 348, 350
Desmethylimipramine, 113, 171
Diabetes, 250
Diaphragm, 236, 242
Diets
  amino-acid imbalanced, 228, 266
  high-fat, 222
  palatability, 222, 259
Dihydroxyphenylalanine, 113
5,6- and 5,7-Dihydroxytryptamine, 99, 173, 332, 377
Dopamine
  anatomy, 353
  feeding, 307, 318, 355-366, 376, 440
  lesion studies, 362
  sensorimotor disturbances, 459
  thermoregulation, 113, 117, 177
Dorsal tegmental tract, 305
Drinking
  catecholaminergic controls, 320-323, 329, 340, 366-368
  cholinergic controls, 383

[Drinking]
  histaminergic controls, 389
  hypothalamic activity, 446
  indoleaminergic controls, 379
  peptidergic controls, 400
Duodenum, 244

E

Electrical stimulation of brain
  reward, 453, 460
  thermoregulation, 9, 15, 19
Endorphins, 134, 403
Energy homeostasis, 211
Epilepsy, 84
Epinephrine, 22, 244, 274
Ergotoxin, 135
Estrus, 250, 441
Evaporative heat loss, 29
Exercise, 8, 35, 139
Exotoxin bacillus, 132

F

Fat, 234, 236, 238, 242, 267
Fatty-acid synthetase, 242
Fenfluramine, 370, 373
Fenoprofen, 160
Fever
  antipyretics, 165
  behavioral controls, 42
  hypothalamic proteins, 164
  hypothalamus, 7, 147-154
  pyrogen-induced, 88, 147-154, 162-164
Food intake
  adrenergic receptor analysis, 311
  caloric dilution, 226
  catacholaminergic controls, 306-320, 334-343
  cholinergic controls, 381
  dietary parameters, 343
  dopaminergic controls, 355-366, 439
  GABA controls, 395-399
  histaminergic controls, 389
  hormonal controls, 346-351
  in humans, 221, 225
  hypothalamic controls, 211, 334, 299-405, 439-461
  lateral hypothalamus, 253, 439
  nutrient preloads, 227

Subject Index

[Food intake]
  patterns, 217-221, 234, 245, 261, 342
  paraventricular and perifornical hypothalamus, 308, 335
  peptidergic controls, 399
  raphe nuclei, 377
  serotonergic controls, 370-376
  single-unit studies, 439-461
  temperature, 231
Fröhlich syndrome

G

GABA, 395
Gastric acid, 241
Gastrointestinal tract, 233, 273, 350
Globus pallidus, 440, 442, 457
Glucagon, 241
Glucocorticoids, 22, 347
Gluconeogenesis, 241
Glucose, 240, 267, 346, 351, 441, 445, 448, 456
Glutamic acid decarboxylase, 396
Glycerol, 244
Glycocides, 173
Glycogen, 236
Gold thioglucose, 248
Growth hormone, 249, 274
Guanethidine, 106
Guinea pigs, thermoregulation, 107

H

Haloperidol, 113, 357, 366
Heart, 236
Hepatic portal receptors, 448
Hibernation, 8, 85, 93, 97, 108, 140
Hierarchical controls
  feeding, 460
  thermoregulation, 43-46
Hippocampus, thermoregulatory effects, 54, 64
Histamine
  anatomy, 388
  autonomic effects, 390
  thermoregulation, 134
Homeostasis
  energy, 256, 274, 344, 348, 442
  glucose, 240, 267, 269
  thermoregulation, 1, 135-137

Humidity, 30
Hunger, 446
Hydrocortisone, 166
6-Hydroxydopamine
  feeding, 265, 313, 332, 344, 355-364, 378
  thermoregulation, 115, 168
  water excretion, 326
5-Hydroxyindoleacetic acid, 98, 155
Hypercholesterolemia, 243
Hyperphagia, 216, 332, 378
Hyperthermia, 94, 99, 118, 162, 170
Hypertriglyceridemia, 243
Hypophysectomy, 249
Hypothalamus
  autonomic controls, 348
  control of blood flow, 30
  food intake, 211-277, 306-320, 334-342, 439-461
  heat-loss responses, 52
  heat-retention responses, 52
  lateral, 17-19, 256
  lesions, thermoregulation, 26
  leukemic infiltration, 237
  posterior, 17-19, 33, 137-139, 141, 177
  single-unit studies, 439-461
  species differences, 93
  thermoregulatory integration, 25, 31, 41-43, 46-55, 83-181
  thermosensitivity, 3-12, 23, 33
  ventromedial, 211-256
Hypothermia, 86, 87, 92, 99, 103, 122, 174
Hypoxia, 25

I

Imipramine, 89
Indomethacin, 162, 169
Inferotemporal cortex, 442, 451, 456
Insulin, 230, 238, 241, 245, 251, 266, 268, 276, 342, 350
Intestinal aborption, 244
Intestinal by-pass, 252
Ions, thermoregulation, 135-147, 169
Iontophoresis
  feeding, 441
  thermoregulation, 88, 95-96, 108, 114, 160
Isoproterenol, 307, 366, 370

K

Kaolin, 226
Kidney, 236
Klüver-Bucy syndrome, 456

L

Lateral hypothalamic syndrome
  cold stress, 265
  dietary dilutions, 266
  energy homeostasis, 256
  etiology of, 270
  finickiness, 262
  glucoprivic challenge, 266, 271
  metabolic disturbances, 271
  motor disturbances, 274
  sensory neglect, 258, 271, 273
Lateral hypothalamus
  feeding, 212, 253, 365, 439
  learning, 451
  water excretion, 327, 367
Learning
  appetitive, 443
  hypothalamic unit activity, 448
  thermoregulatory, 41
Lipogenesis, 242
Liver, 236, 242
Locus ceruleus, 305, 329, 334

M

MAO inhibitors, 316
Mecamylamine, 139
Median eminence, 22
Medulla, 11, 19, 26, 34
Metabolic heat production, 12-13, 16, 20
Metabolism
  glucose, 240, 351
  lipid, 242
Methysergide, 92, 94
Metrazol, 171
Monoamine oxidase, 89
Monoamines
  feeding, 299-405
  thermoregulation, 86
Morphine, 89, 174
Motivation, 230, 443
Movements, 457

Muscarine, 123

N

Neuroglia, 88
Neuronal models, thermoregulation, 2, 10, 48, 59, 62-64, 86
Neurons
  cold-sensitive, 6, 47, 60-62, 94
  heat-loss, 3
  interneurons, 6
  Q_{10}, 3
  temperature-insensitive, 3, 6, 53
  thermoresponse curves, 47
  warm-sensitive, 3, 6, 46-52, 57-60, 94, 108
Neurotensin, 132
Nicotine, 86, 123
Nonshivering thermogenesis, 20, 64, 67
Norepinephrine
  feeding control, 306-320
  pathways, 300
  thermoregulation, 86, 102-116, 133, 177
Nucleus
  accumbens, 107, 114, 453
  anterior hypothalamic, 1, 16, 31, 43-47, 353
  basal forebrain nucleus of Meynert, 446, 460
  interpeduncular, 306, 353
  lateral hypothalamic, 16
  mamillary, 254
  paraventricular, 32, 253, 306, 325, 329, 334, 341, 349, 394
  periventricular, 306, 355
  posterior hypothalamic, 9, 12
  preoptic, 1, 16, 31, 43-47, 441
  raphe, 47, 353, 368, 377
  septal, 9, 12
  supraoptic, 40, 325, 329, 341, 386, 393, 445, 448, 460
  ventromedial, 245, 252, 255, 334

O

Obesity
  body composition, 232
  diencephalic, 225
  hypothalamic, 217, 339

Subject Index

[Obesity]
 ob/ob mouse, 248, 269
Opiate alkaloids, 173
Orbitofrontal cortex, 453
Organum vasculosum of lamina
 terminalis, 385
Ouabain, 175
Oxotremorine, 123
Oxygen consumption, 12, 17, 107, 351
Oxytocin, 441

P

Pair-feeding, 246
Pancreas, 237, 240
Panting, 27-29, 36
Paracetamol, 159
Parachloroamphetamine, 173
Parachlorphenylalanine, 89, 154, 377
Peptides
 feeding, 399
 thermoregulation, 132-134
Phenoxybenzamine, 168, 312
Phentolamine
 feeding, 312, 314, 321, 345
 thermoregulation, 94, 104, 108, 180
Phenylephrine, 104
Picrotoxin, 135, 171
Pimozide, 113, 175, 356, 367, 370
Poikilothermia, 89, 171, 173
Polyphoretin phosphate, 158
Polypnea, 30, 32, 36
Pons, thermoregulation, 19, 26, 33
Pontine taste area, 446
Preoptic area, 1, 6, 9, 21, 43-47, 85, 90,
 97, 149, 180
Primates
 hypothalamic unit activity, 442-460
 thermoregulation, 102
Propranolol
 feeding, 321, 370, 390
 thermoregulation, 104, 108
Prostaglandins, 155, 168, 402
Punishment, 451
Push-pull perfusion, 88, 97, 126, 137-147
Pyrogens, 143, 147-154, 162-165

Q

Q_{10} neurons, 3, 47
Quinine, 346

R

Rabbits, 105
Rats
 feeding, 211-277
 thermoregulation, 106, 121
Receptors
 adrenergic, 312
 temperature, 46-55
Respiration, 133
Respiratory quotient, 242
Reticular formation
 feeding, 331
 thermoregulation, 19, 23
Reward, 453, 460

S

Saccharin, 224
Saliva, 29, 270, 273, 460
Sciatic nerve, 440
Self-stimulation, 453, 460
Septum, 15, 23
Serotonin
 anatomy of, 368
 destruction of, 99
 feeding, 370-375
 heat-production, 89, 95
 lesion studies, 376
 release, 96-99
 pathways, 88
 preoptic area, 97
 pyrogen fever, 154
 thermoregulation, 86, 96-100
Set point
 body-weight, 267
 interneurons, 52
 thermoregulation, 7, 22, 36, 85, 135-140, 178
Shivering, 7, 67
Single-unit studies
 exteroceptive effects on hypothalamic
  cells, 440
 feeding, 439-461
 hypothalamus, 439-461
 learning, 448
 smell, 440
 suckling, 441
 taste, 440
 thermoregulation, 9-10, 19, 48-55, 94
 vaginal stimulation, 441
Skin blood flow, 23, 27

Sleep, 108, 370
Sodium appetite, 326, 341
Sodium pentobarbital, 172
Sodium salicylate, 156, 162, 165, 169
Somatosensory cortex, 7
Species differences
  hypothalamic control of feeding, 221, 224
  hypothalamic neurohumoral coding, 93
  thermoregulation, 102-110
Sphincter of Oddi, 244
Spinal cord
  dorsal root section, 14
  spinal transection, 14, 27, 38
  thermoreceptors, 14
  thermoregulation, 19, 64
  warming of, 35
Spleen, 236
Streptozotocin, 250
Subfornical organ, 384
Substantia innominata, 442, 451, 456, 460
Substantia nigra, 353, 442, 459
Swallowing, 457
Synaptic blocking agents
  adrenergic, 14, 21, 349
  feeding, 311, 358
  ganglionic, 21
Synaptic transmission
  adrenergic, 9, 104, 311
  cholinergic, 9
  dopaminergic, 353-368
  hypothalamus, 85, 311
  serotonergic, 370-375
Sweating
  brain temperature effects, 37
  measurement of, 38
  skin temperature effects, 36

T

Tachypnea, 133
Taste, 445
Tetrabenazine, 313
Theophylline, 169
Thermodes, 12, 43-44
Thermoregulation
  acetylcholine, 18, 86, 117-129
  anterior hypothalamic area, 1, 6, 85, 176
  autonomic controls, 45, 133

[Thermoregulation]
  behavioral controls, 40-45
  blood flow, 11
  brain stem, 24
  calcium ions, 9, 134-146, 169
  central and peripheral interactions, 56
  circulatory effects, 27
  cutaneous vasoconstriction, 3, 18
  decerebration, 25, 34
  dopamine, 113-115
  drug effects on, 165-175
  endocrine controls, 22
  evaporative heat loss, 23
  exercise, 139
  extrahypothalamic, 35, 42
  heat-loss, 9, 117, 175, 392
  heat-production, 9, 100, 117, 126, 163, 175
  hierarchy of control, 11
  hypothalamic control, 41
  lateral and posterior hypothalamus, 24
  lesion studies, 17, 33
  metabolic controls, 22
  neurochemical aspects, 83-210
  neuronal models, 2
  nonshivering, 20, 67
  norepinephrine, 86, 112
  panting, 3, 6, 27, 29, 57
  piloerection, 18, 26
  potassium ions, 85
  preoptic area, 1, 6, 24, 85, 111
  principles of, 11-13
  receptors, 29
  serotonin, 8, 89, 96, 99
  set point, 2, 135-147, 178
  shivering, 3, 7, 13, 16, 26, 67, 90
  skin blood flow, 27, 57
  skin temperature, 27, 85
  sodium ions, 9, 85, 135-147, 169
  spinal cord, 24
  sweating, 29
  sympathetic control, 25
  vasoconstriction, 7, 9, 23
  vasodilation, 23, 25, 41
Thermoreceptors
  abdominal, 35
  chemical coding of, 176-178
  cold-sensitive, 60-62, 153
  hypothalamic, 1-9, 84
  peripheral, 12
  skin, 7
  warm-sensitive, 57-60, 153

Subject Index

Thyroid-stimulating hormone, 22, 132, 250
Thyroxin, 250, 270, 273
Tissue-punch technique, 96
Tract
  central tegmental, 305
  dorsal tegmental, 305, 331
  lateral spinothalamic, 14
  medial forebrain, 16-19, 273, 306
  nigrostriatal, 274, 353, 439
  superior cerebellar peduncles, 305
Tricyclic antidepressants, 315
Tryptamine, 92
Tryptophan, 372

U

Urea, 241
Urine, 324, 329, 386, 440

V

Vagotomy, 251, 272, 350, 442
Vasoactive intestinal peptide, 134
Vasodilation, 23, 25, 41
Vasopressin, 347
Ventromedial hypothalamic syndrome
  body composition, 232, 246
  deprivation response, 229
  dynamic phase, 217

[Ventromedial hypothalamic syndrome]
  etiology of, 245
  finickiness, 222
  gastric acid, 246
  glucose homeostasis, 240
  hyperphagia, 216
  insulin, 238, 246
  meal patterns, 220
  nutrient preloads, 227
  paraventricular nucleus, 335
  pituitary, 215, 253
  sensory reactivity, 246
  static phase, 220
  work, 231
Ventromedial hypothalamus, feeding, 212, 334, 365, 439

W

Water
  body, 233
  conservation, 39
  excretion, 323, 366, 379, 383, 393
  intake, 261, 320, 340, 366, 379, 383, 400

Z

Zona incerta, 340, 353
Zucker "fatty" rat, 248, 344

JUN 2 6 1997